Space-Geodetic Techniques

Space-Geodetic Techniques

Editor

Xiaogong Hu

Basel • Beijing • Wuhan • Barcelona • Belgrade • Novi Sad • Cluj • Manchester

Editor
Xiaogong Hu
Chinese Academy of Sciences
Shanghai, China

Editorial Office
MDPI
St. Alban-Anlage 66
4052 Basel, Switzerland

This is a reprint of articles from the Special Issue published online in the open access journal *Remote Sensing* (ISSN 2072-4292) (available at: https://www.mdpi.com/journal/remotesensing/special_issues/Space_Geodetic_Techniques).

For citation purposes, cite each article independently as indicated on the article page online and as indicated below:

Lastname, A.A.; Lastname, B.B. Article Title. *Journal Name* **Year**, *Volume Number*, Page Range.

ISBN 978-3-0365-9022-6 (Hbk)
ISBN 978-3-0365-9023-3 (PDF)
doi.org/10.3390/books978-3-0365-9023-3

Contents

About the Editor

Xiaogong Hu

Xiaogong Hu received his Ph.D. degree from the Shanghai Astronomical Observatory (SHAO), Chinese Academy of Sciences (CAS), Shanghai, China, in 1999. He is currently a professor at SHAO, CAS. His research interests are focused on spacecraft precise orbit determination and its applications in global navigation satellite systems (GNSS) and deep space exploration.

Article

New Orbit Determination Method for GEO Satellites Based on BeiDou Short-Message Communication Ranging

Xiaojie Li [1,2,3], Rui Guo [1,2,3,*], Jianbing Chen [4], Shuai Liu [1], Zhiqiao Chang [1], Jie Xin [1], Jinglei Guo [1] and Yijun Tian [1]

1   Beijing Satellite Navigation Center, Beijing 100094, China
2   Shanghai Key Laboratory for Space Positioning and Navigation, Shanghai 200030, China
3   Shanghai Astronomical Observatory, Shanghai 200030, China
4   China Top Communication Co., Ltd., Beijing 100088, China
*   Correspondence: shimbarsalon@163.com

Abstract: The radio determination service system (RDSS), a navigation and positioning system independently developed by China, features services such as short-message communication, position reporting, and international search and rescue. The L-band pseudo-range and phase data are the primary data sources in precise orbit determination (POD) for geostationary Earth orbit (GEO) satellite in the BeiDou system, especially in the orbit manoeuvre period. These data are the only data sources in the POD for GEOs. However, when the pseudo-range and phase data is abnormal due to unforeseen reasons, such as satellite hardware failure or monitoring receiver abnormalities, the data abnormality leads to orbit determination abnormalities or even failures for GEOs, then the service performance and availability of the RDSS system are greatly degraded. Therefore, a new POD method for GEOs based on BeiDou short-message communication ranging data has gained research attention to improve the service reliability of the BeiDou navigation satellite system (BDS)-3, realising the deep integration of communication and navigation services of the BDS. This problem has not been addressed so far. Therefore, in this study, a new POD method for GEO satellites is investigated using high-precision satellite laser ranging (SLR) data and RDSS data. The SLR data are used as the benchmark to calibrate the time delay value of RDSS equipment, and RDSS data are only used in the orbit determination process by fixing the corrected RDSS time delay value, and the satellite orbit parameters and dynamic parameters are solved. Experimental analysis is conducted using the measured SLR and RDSS data of the BDS, and the orbit accuracy in this paper is evaluated by the precise ephemeris of the Multi-GNSS pilot project (MGEX) and SLR data. The results show that the orbit accuracy in the orbital arc and the 2-h orbital prediction arc for GEOs are 6.01 m and 6.99 m, respectively, compared with the ephemeris of MGEX, and the short-arc orbit accuracy after 4 h of manoeuvring is 11.11 m. The orbit accuracy in the radial component by SLR data is 0.54 m. The required orbit accuracy for GEO satellites in the RDSS service of the BDS-3 is 15 m. The orbit accuracy achieved in this paper is superior to that of this technical index. This method expands the application field of the RDSS data and greatly enriches the POD method for GEOs. It can be adopted as a backup technology for the POD method for GEOs based on RNSS data, significantly improving the service reliability of the BeiDou RDSS service.

Keywords: BeiDou system (BDS); geostationary earth orbit (GEO) satellite; radio determination satellite service (RDSS); satellite laser ranging (SLR); multi-GNSS pilot project (MGEX)

Citation: Li, X.; Guo, R.; Chen, J.; Liu, S.; Chang, Z.; Xin, J.; Guo, J.; Tian, Y. New Orbit Determination Method for GEO Satellites Based on BeiDou Short-Message Communication Ranging. *Remote Sens.* **2022**, *14*, 4602. https://doi.org/10.3390/rs14184602

Academic Editor: Jianguo Yan

Received: 24 July 2022
Accepted: 8 September 2022
Published: 15 September 2022

Publisher's Note: MDPI stays neutral with regard to jurisdictional claims in published maps and institutional affiliations.

1. Introduction

The BeiDou system (BDS) is a navigation and positioning system developed in China that provides two service modes, the radio navigation satellite service (RNSS) and the radio determination satellite service (RDSS) [1–3]. The benefits of RDSS as a featured part of BDS are to provide regional short message communication, global short message communication, fast location reporting, and international search and rescue (SAR) services. RDSS users can not only know where they are but also tell others where they are. They can also transmit information to other RDSS users and send distress signals to rescue coordination centres (RCC) in distress. Therefore, the RDSS is an important component and a distinctive feature of the BDS. With more than ten years of development, the application scope of the BeiDou RDSS has been continuously expanded, including aviation, maritime, land transportation, and emergency rescue, et al. Therefore, the RDSS is an important component and a distinctive feature of the BDS, as GPS, GLONASS, and Galileo do not feature this component [4–7].

Scholars at home and abroad have conducted extensive research on BeiDou short-message communication ranging data, including the RDSS positioning algorithm, signal design, generalised RDSS design and extension, and combined navigation of the RDSS and strap-down inertial navigation system (SINS). A series of research results have been achieved, and the connotation of the satellite navigation system has been enriched [8]. However, the deep integrated application of BeiDou short-message communication and BeiDou navigation services has not been realised yet. The short-message communication ranging data have not been applied to the field of BeiDou satellite orbit determination, and the orbit accuracy for BeiDou satellites based on RDSS data has not been systematically analysed by the measured short-message communication ranging data.

GEO satellites, as an indispensable part of the BDS constellation composition, bear not only the responsibility of communication transmission but also the task of enhancing the geometric dilution of precision (GDOP) [9,10]. At present, BeiDou GEO satellite orbit determination mainly adopts L-band multi-frequency pseudorange, phase data, and ka-band inter-satellite link data of the RNSS system. The Overlap Orbit Differences (OOD) method is used to evaluate the accuracy of the 3D position of its post-precision products, which is a 2 m level for the BDS-2 GEO satellite based on the L-band multi-frequency pseudorange and phase data [11,12], and about 24 cm for the BDS-3 GEO satellite in the orbital arc based on the L-band data and Ka-band inter-satellite link data [2,13,14]. The laser ranging (SLR) data internationally jointly measured were used to evaluate the orbit accuracy in radial direction, which was 40 cm for the BDS GEO satellites [15–17].

In the case of abnormal RNSS and inter-satellite link data due to uncontrollable factors such as satellite navigation unit failure and inter-satellite link equipment failure, GEO satellite orbit determination accuracy can considerably decrease or orbit determination may fail, which undermines the positioning and timing service performance and availability of the BDS [18]. In case such anomalies occur during the orbit manoeuvre or recovery of GEO satellites, the kinematic orbit determination method is adopted for GEOs. L-band data are the only data resource, if the L-band data are abnormal in this period, the orbit error increases sharply and seriously affects RDSS service performance.

Currently, when the aforementioned problems occur, the out-station beam of this GEO satellite will be closed forcibly, and then the RDSS service of this GEO satellite will be stopped. However, only three GEO satellites of the BDS-3 provide the RDSS services, and the beam coverage areas of the three satellites are different. If the out-station beam of a GEO satellite is closed in the aforementioned situation, the RDSS services cannot be provided to users in the beam coverage area of that satellite. To solve the aforementioned problem, RDSS data can be used for GEO satellite orbit determination, which can be adopted as a backup method during GEO satellite orbit manoeuvre and recovery, improving the reliability of the GEO satellite orbit determination mode.

Dr. Xing adopted the RDSS data to make the orbit determination for the BDS-3 GEO satellite, and the orbit accuracy for GEO is better than 10 m [19]. In the BDS GEO satellite orbit determination method by RDSS data, the RDSS equipment time delay calibration is the first problem to be solved. Dr. Xing adopted the combined orbit determination method using the RNSS data and RDSS data to solve the RDSS equipment time delay. But before the time delay calibration, the time delay value of one RDSS station needs to be accurately calibrated off-line. Then, based on the RNSS data, the equipment time delay value of the other RDSS stations can be solved simultaneously when the orbit and dynamical parameters are solved in the combined orbit determination. Since the RNSS equipment also has a time delay, it is necessary to adopt SLR data or the user equivalent range error (UERE) method to regularly calibrate the RNSS equipment time delay [20], so it needs two equipment delay calibrations to achieve the accurate RDSS equipment time delay calibration.

This study proposes a new precise orbit determination (POD) method for GEO satellites based on BeiDou short-message communication ranging data. The method uses high-precision SLR data as the benchmark for the calibration of RDSS equipment time delay. We need not calibrate the time delay value of a certain RDSS station before the time delay calibration. Based on the precisely calibrated RDSS equipment time delay, RDSS data are used only in the orbit determination process. Thus, the high dependency of the existing GEO satellite orbit determination method on RNSS data can be reduced, the continuity and reliability of the precise orbit products of BeiDou GEO satellites during the orbit manoeuvre and recovery will be improved, and the integrated application of short-message communication and navigation services of BDS will be realised.

In this paper, we first describe the new POD method for the GEO satellites based on RDSS data and the principle of solving time delay parameters of RDSS equipment based on the SLR data. Then, we present the experiments by the measured RNSS, RDSS, and SLR data of the BDS. Finally, we evaluate the accuracy of the estimated RDSS time delay value, the orbit accuracy of GEO satellites based on the observation residuals, overlap orbit differences, and orbit accuracy in radial direction based on the SLR data.

2. Measurement Model of RDSS

BDS adopts an active positioning mode in the RDSS system, which adopts 4-way ranging observations. There are C-band uplink signals and S-band downlink signals in the out-station beam. There are L-band uplink signals and C-band downlink signals in the in-station beam. The out-station C-band uplink signals are transmitted by the master control centre (MCC) to a GEO satellite, the S-band downlink signals are transformed by GEOs to the RDSS user, the L-band uplink signals are transformed by the RDSS user to all observable GEO satellites, and the C-band downlink signals are transformed by GEOs to the MCC. The MCC accomplishes the 4-way ranging observations and estimates the position of the RDSS user within 1 s [5,6]. Figure 1 shows the schematic of the RDSS.

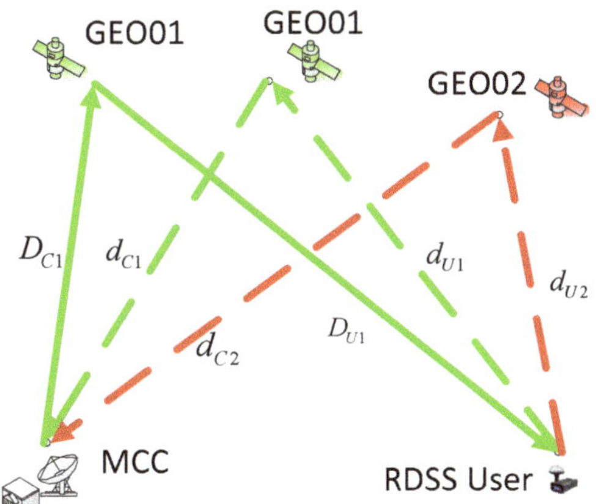

Figure 1. Schematic of the RDSS 4-way ranging observations.

The RDSS observation model is as follows:

$$\begin{aligned}
\rho_{S_1} &= D_{C1} + c \cdot \delta t_{C1O} + c \cdot \delta t_{S1O} + D_{U1} + c \cdot \delta t_U + d_{U1} + c \cdot \delta t_{S1I} + d_{C1} + c \cdot \delta t_{C1I} + \Delta E_{All1} + \varepsilon \\
\rho_{S_2} &= D_{C1} + c \cdot \delta t_{C1O} + c \cdot \delta t_{S1O} + D_{U1} + c \cdot \delta t_U + d_{U2} + c \cdot \delta t_{S2I} + d_{C2} + c \cdot \delta t_{C2I} + \Delta E_{All2} + \varepsilon \\
\tau_{all} &= \delta t_{C1O} + \delta t_{S1O} + \delta t_U + \delta t_{S1I} + \delta t_{C1I}
\end{aligned} \tag{1}$$

$$\begin{aligned}
D_{C1} &= \sqrt{\left(X^{S1}(t_1) - X_C\right)^2 + \left(Y^{S1}(t_1) - Y_C\right)^2 + \left(Z^{S1}(t_1) - Z_C\right)^2} \\
D_{U1} &= \sqrt{\left(X_U(t_2) - X^{S1}(t_1)\right)^2 + \left(Y_U(t_2) - Y^{S1}(t_1)\right)^2 + \left(Z_U(t_2) - Z^{S1}(t_1)\right)^2} \\
d_{U1} &= \sqrt{\left(X^{S1}(t_3) - X_U(t_2)\right)^2 + \left(Y^{S1}(t_3) - Y_U(t_2)\right)^2 + \left(Z^{S1}(t_3) - Z_U(t_2)\right)^2} \\
d_{C1} &= \sqrt{\left(X^{S1}(t_3) - X_C\right)^2 + \left(Y^{S1}(t_3) - Y_C\right)^2 + \left(Z^{S1}(t_3) - Z_C\right)^2} \\
d_{U2} &= \sqrt{\left(X^{S2}(t_4) - X_U(t_2)\right)^2 + \left(Y^{S2}(t_4) - Y_U(t_2)\right)^2 + \left(Z^{S2}(t_4) - Z_U(t_2)\right)^2} \\
d_{C2} &= \sqrt{\left(X^{S2}(t_4) - X_C\right)^2 + \left(Y^{S2}(t_4) - Y_C\right)^2 + \left(Z^{S2}(t_4) - Z_C\right)^2}
\end{aligned} \tag{2}$$

where ρ_{S_1}, ρ_{S_2} are the RDSS observation data for GEO01 and GEO02, respectively. D_{C1} is the geometric range between MCC and GEO01 satellite in the out-station C-band uplink; D_{U1} is the geometric range between GEO01 satellite and RDSS user in the out-station S-band downlink; d_{U1} is the geometric range between GEO01 satellite and RDSS user in the in-station L-band uplink; d_{C1} is the geometric range between MCC and GEO01 satellite in the in-station C-band downlink; d_{U2} is the geometric range between GEO02 satellite and RDSS user in the in-station L-band uplink; d_{C2} is the geometric range between MCC and GEO02 satellite in the in-station C-band downlink. ε is the observation noise. c is the speed of light. δt_{C1O} is the transmission time delay in MCC in out-station downlink; δt_{S1O} is the time delay of the out-station transferring link for GEOs; δt_U is the transmission time delay of the RDSS user; $\delta t_{S1I}, \delta t_{S2I}$ are the time delay of the in-station transferring link for GEO01 and GEO02, respectively; δt_{C1I} and δt_{C2I} are the time delay of the receiver in MCC from GEO01 and GEO02, respectively. τ_{all} is the sum of the time delays of the out-station beam and the in-station beam, including the transmission time delay in MCC and the time delay of the out-station transferring link for GEOs, the transmission time delay of the RDSS user, the time delay of the in-station transferring link for GEOs and the time delay of the receiver in MCC. ΔE_{All1} is the sum of all errors, including the troposphere delays correction, the ionosphere delay correction, the Sagnac effect correction, the general relativistic correction in the out-station C-band uplink, the out-station S-band downlink, the in-station L-band uplink and the in-station C-band downlink and the antenna phase centre correction for GEOs, the RDSS user, and MCC receiver. (X_C, Y_C, Z_C) and (X_U, Y_U, Z_U) are the coordinates of RDSS MCC station and RDSS user, respectively, which are known quantities; (X^{S1}, Y^{S1}, Z^{S1}) and (X^{S2}, Y^{S2}, Z^{S3}) are the orbital positions of satellites GEO01 and GEO02, respectively.

3. Orbit Determination Method for GEO Satellite Based on RDSS Data

3.1. Principles of Orbit Determination

According to the RDSS measurement principle, the orbit accuracy of GEO satellites based on RDSS data considerably depends on the time delay accuracy of RDSS equipment. The time delay for the MCC and satellite is calibrated by a special instrument before being adopted in the BDS. The user equipment time delay is usually calibrated in the special equipment calibration field and the accuracy is better than 10 ns. Owing to equipment aging and hardware replacement, the time delay of RDSS equipment involves the problem of long-term drift [21–23]. The common method is recalibration and retesting; during this process, the equipment of the MCC needs to be off the line and the user equipment needs to be returned to the equipment calibration field for recalibration. Therefore, the time delay recalibration of RDSS equipment is tedious and time-consuming, potentially affecting the RDSS service of BDS, especially since the recalibration of the satellite equipment delay cannot become true [24–27]. For navigation satellites, with centimetre- or millimetre-scale precision, SLR is a high-accuracy measurement method that is independent of radio measurements and is also an important means of evaluating orbit accuracy [10,28,29]. All the BeiDou satellites in orbit are equipped with laser reflectors, laying an important foundation for the accuracy evaluation and time delay calibration of the RDSS data in the BDS.

For the new method of POD for GEO satellites by only using the RDSS data, first the time delay of the RDSS equipment is calculated based on SLR data, and then in the POD process, the corrected RDSS time delay value is fixed. The satellite orbit and dynamic parameters are solved only by using the RDSS data.

3.2. Principles of Calibrating Time Delay for RDSS Equipments Based on SLR Data

For calibrating RDSS equipment time delay based on SLR data, first the SLR data and RDSS data are preprocessed, then a reasonable dynamical model is selected, and, finally, SLR and RDSS data are jointly used for orbit determination. The estimated parameters are the satellite orbit, dynamic parameters, and the RDSS equipment time delay. All GEO satellites are processed together. The normal equations of the SLR data and the RDSS data are formed, respectively. The former only contains the partial derivatives of the satellite orbital parameters and dynamic parameters, while the latter contains the partial derivatives of the satellite orbital parameters, the dynamic parameters, and the partial derivatives of the RDSS time delay parameters. The final normal equations are formed by superpositioning both normal equations, and the joint orbit determination is completed through the least square method. The process flow chart of calibrating RDSS equipment time delay based on SLR data is shown in Figure 2.

For SLR observation r_m^i and RDSS four-way ranging observation ρ_{CU}^{ij}, with the error term removed, the observation equations are as follows:

$$
\begin{aligned}
r_m^i &= 2 * \left| \mathbf{X}_m(t) - \mathbf{X}^i(t) \right| \\
\rho_{CU}^{ij} &= \left| \mathbf{X}_C(t) - \mathbf{X}^i(t) \right| + \left| \mathbf{X}_U(t) - \mathbf{X}^i(t) \right| + \left| \mathbf{X}_U(t) - \mathbf{X}^j(t) \right| + \left| \mathbf{X}_U(t) - \mathbf{X}^j(t) \right| + c \cdot \tau_{all}
\end{aligned}
\tag{3}
$$

where $\mathbf{X}_m(t)$ is the coordinates of the SLR station m, $\mathbf{X}_C(t)$ and $\mathbf{X}_U(t)$ are the coordinates of RDSS MCC station k and user l, respectively, which are known quantities; $\mathbf{X}^i(t)$ and $\mathbf{X}^j(t)$ are the orbital positions of satellites i and j, respectively; τ_{all} is sum of the time delays of the out-station beam and the in-station beam.

The error equation can be concluded by linearising Formula (4).

$$
V(t) = \begin{bmatrix} \dfrac{\partial r_m^i}{\partial \mathbf{X}^i(t_0)} & 0 & 0 \\ \dfrac{\partial \rho_{CU}^{ij}}{\partial \mathbf{X}^i(t_0)} & \dfrac{\partial \rho_{CU}^{ij}}{\partial \mathbf{X}^j(t_0)} & 1 \end{bmatrix} \begin{bmatrix} \mathbf{X}^i(t_0) \\ \mathbf{X}^j(t_0) \\ c \cdot \tau_{all} \end{bmatrix} - L(t)
\tag{4}
$$

where $\mathbf{X}^i(t_0)$ and $\mathbf{X}^j(t_0)$ are the initial orbital positions, velocities, and dynamic model parameters of satellites i and j, respectively. Based on Equation (3), the RDSS equipment time delay is estimated using the least-squares batch method.

Because SLR observation is subject to weather conditions, it is impossible to achieve all-weather observations. Therefore, the POD method by SLR and RDSS data cannot be a routine method for GEOs. While RDSS data can be obtained in real time, this paper adopts the SLR data to calibrate the time delay of the RDSS. RDSS data are used only for GEO satellite orbit determination, for which only the parts corresponding to RDSS data in Formulas (1) and (2) are used.

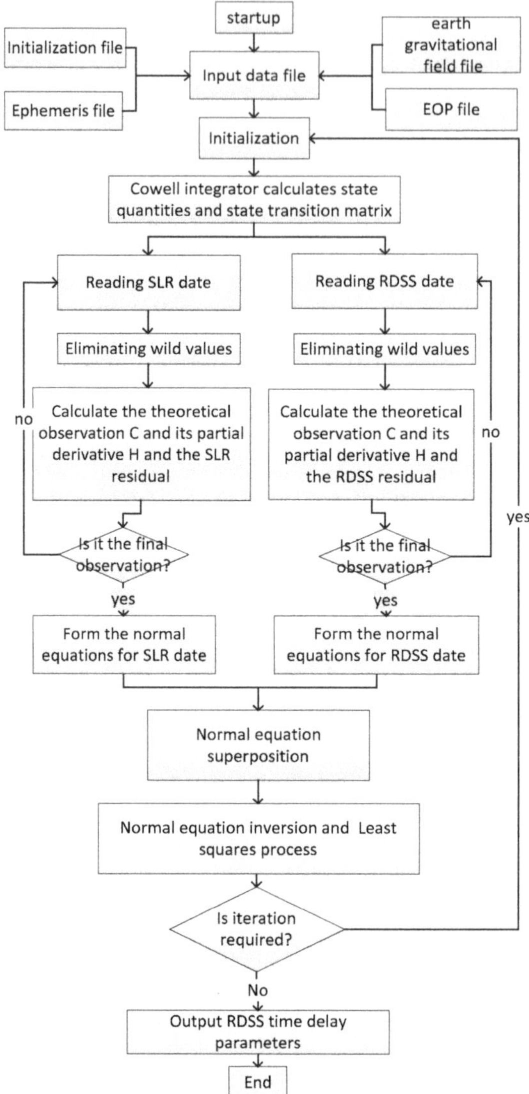

Figure 2. Flow chart of calibrating RDSS equipment time delay based on SLR data.

3.3. Program Design

The RDSS data are obtained from seven RDSS calibration stations deployed in Beijing, Sichuan, Hainan, Southeast China, and Northeast China, and two in Xinjiang. The SLR station was deployed in Beijing. The arc length for orbit determination is three days, and the data sampling rate is 60 s. The estimated parameters include the GEO initial orbit parameters of GEO satellites, solar radiation pressure parameters, and RDSS equipment time delay parameters. The RDSS equipment time delay is the sum of outstation and in-station link equipment time delays τ_{all}. The normal equations of SLR data and the RDSS four-way ranging data are formed. Based on these, the pseudorange/pseudophase data and the RDSS four-way ranging data can be processed independently, thus flexibly processing different types of data.

As for the SLR data preprocessing, first, the phase centres of the antennas of tracking stations are corrected; second, the phase centres of the antennas that transmit satellite navigation signals are attributed to the centre of mass of the satellite; third, the Marini model is used to deduct the tropospheric error; and finally, the errors of general relativity and the tidal effect are corrected. In the RDSS data preprocessing, the phase centres of the satellite and station antenna are corrected; then, the Black model is used to correct the errors of tropospheric delay and the 14-parameter Klobuchar model is used to correct the ionospheric delay; finally, the error terms of the Earth rotation effect and general relativity are corrected.

The orbits for BDS-2 5 GEO satellites, BDS-3 3 GEO satellites are determined. Table 1 shows the strategies for the combined POD methods by RDSS and SLR data.

Table 1. Strategies for the combined POD methods by RDSS and SLR data.

	POD Methods by RDSS and SLR Data
Satellite	8 GEO: C01, C02, C03, C04, C05 in BDS-2 system C59, C60, C61 in BDS-3 system
Stations	Seven RDSS calibration stations deployed in Beijing, Sichuan, Hainan, Southeast China, and Northeast China, and two in Xinjiang One SLR station: Beijing
Arc length	3 days, sampling interval: RDSS data: 60 s SLR data: 1 s
observation	RDSS data in Seven RDSS calibration stations SLR data in Beijing station
Estimated parameter	GEO Initial orbit of the satellite, solar radiation pressure parameters and RDSS equipment time delay parameters
Parameter estimation method	Least squares algorithm
Gravitational field model	EGM 2008 12 × 12
Sun, Moon gravity and the gravitational force of the other planets	Jet Propulsion Laboratory Development Ephemeris 405 (JPL DE405)
Solar radiation pressure (SRP)	An empirical SRP model which is similar to BERNESE ECOM 9 parameter model
Solid tides, ocean tide perturbation	IERS Convention 2003
Precession and nutation	IAU2000R06
EOP parameters	Constraints to the International Earth Rotation and Reference Systems Service (IERS) C04 model

4. Experiment and Analysis

The RDSS data are from 1 to 3 July 2019 and from 1 June to 31 July 2021. The SLR data are on 2 July 2019 and from 1 June to 31 July 2021 in Beijing. These data are jointly used for the time delay calibration of the RDSS equipment, and the RDSS data from 1 to 10 July 2019 and from 29 May to 31 July are used for sliding orbit determination with a 3-day arc length and orbit determination. We evaluate the accuracy of the estimated RDSS time delay value, the orbit accuracy of GEO satellites based on the observation residuals, overlap orbit differences, and orbit accuracy in radial direction based on the SLR data.

4.1. Accuracy Analysis of RDSS Time Delay

Assume that the out-station beam number of an RDSS four-way range is i, the in-station beam number is j, and the channel number of the in-station beam is k. Then, the time delay number of the RDSS data is $240 \times (i - 1) + 24 \times (j - 1) + k$. Since BDS-2 has 10 out-station beams, 10 in-station beams, and 24 channels in one in-station beam, the time delay number ranges from 1 to 2400. For satellite C01, the out-station beam number i is 1 or 2, and the in-station beam number j is 1 or 2. If C01 is used as the out-station and in-station satellite simultaneously, the time delay value is within the following four ranges.

$$
num = \begin{cases}
0 \text{ to } 24 & i = 1, j = 1 \\
24 \text{ to } 48 & i = 1, j = 2 \\
240 \text{ to } 264 & i = 2, j = 1 \\
264 \text{ to } 288 & i = 2, j = 2
\end{cases}
\tag{5}
$$

The out-station beam number i for satellite C02 is 3 or 4, and the in-station beam number j is 3 or 4. Therefore, the time delay number of satellites C01 to C05 can be obtained when they are both out-station and in-station satellites. For C01 to C05 satellites, the time delay number of the RDSS data is as follows:

$$
num = \begin{cases}
0 \text{ to } 48 & and & 240 \text{ to } 288 & for \text{ C01} \\
528 \text{ to } 576 & and & 768 \text{ to } 816 & for \text{ C02} \\
1056 \text{ to } 1104 & and & 1298 \text{ to } 1344 & for \text{ C03} \\
1584 \text{ to } 1632 & and & 1824 \text{ to } 1872 & for \text{ C04} \\
2112 \text{ to } 2160 & and & 2352 \text{ to } 2400 & for \text{ C05}
\end{cases}
\tag{6}
$$

The SLR observations of satellites C01 to C05 were carried out in Beijing. The values and standard deviation (std) of the RDSS combined time delay solved in the POD based on SLR and RDSS data from 1 to 3 July are shown in Figure 3. In the figure, the horizontal axis represents the combined time delay number.

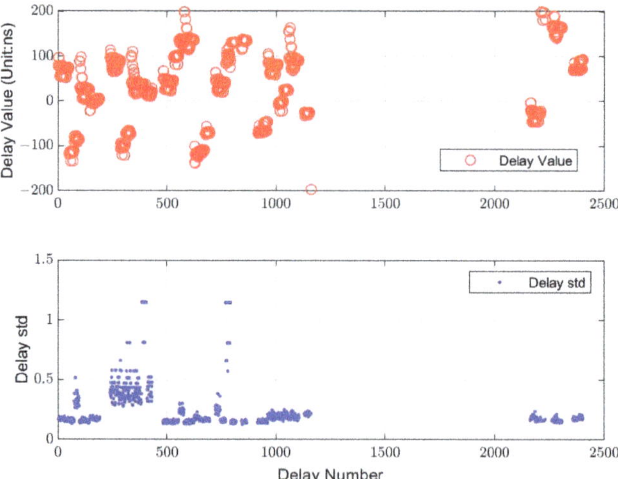

Figure 3. RDSS time delay value solution and post-testing standard deviation.

As the figure indicates, some of the combined time delay parameters cannot be estimated because there are no corresponding RDSS observation data. For example, as satellite C04 does not have in-station and out-station data, the combined time delay parameters of this satellite have no solution. The std of most of the combined delay is less than 1 ns, and the std of some time delay is greater than 1 ns but less than 1.5 ns because of insufficient observation data, which means the internal consistency of the time delay is better than 1 ns. The estimated combined time delay value is within ±200 ns.

4.2. Observation Residuals of RDSS Data

The POD experiments by RDSS data are only made by fixing the RDSS time delay value based on SLR data calibration. Only the satellite orbit and dynamic parameters are solved. The residual sequence of the RDSS data from 1 to 3 July is shown in Figure 4. The root mean square (RMS) value of the RDSS residuals for the GEO satellite is 1.74 m. The residual magnitude is decided by the observation accuracy of the data type. Because the observation accuracy of RDSS data is 1 to 2 m, the residual magnitude is normal.

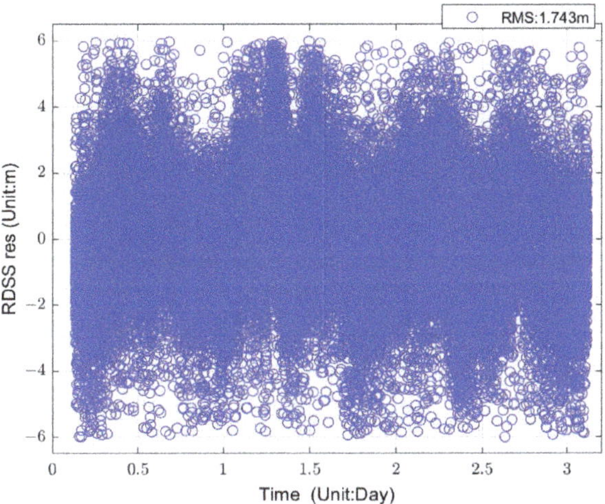

Figure 4. Observation Residuals of RDSS Data.

4.3. Overlap Orbit Differences (OOD)

The International GNSS Service (IGS) and the Multi-GNSS Experiment (MGEX) work together to provide the highest-quality GNSS data and products for free. Among the dozens of analysis centers, the BDS system precision products provided by Wuhan University (WHU, Wuhan, China) are the most complete. The SLR data internationally jointly measured were used to evaluate the orbit accuracy in radial direction, which was 40 cm for the BDS GEO satellites [29]. It is used as the reference orbit for evaluating the orbit accuracy in this paper. The differences between the orbits obtained from RDSS data-based orbit determination and the reference orbit at the same time and the same arc segment show the orbit accuracy by the new POD method based on RDSS data only, including in the orbital arc and the 24 h orbital prediction arc. The signal-in-space range error ($SISRE$) in orbit is determined by the orbit error in the R (radial), T (normal), and N (tangential) directions, which is as follows:

$$SISRE_{orbit} = \sqrt{1.0 \times \Delta R^2 + (0.09 \times \Delta T)^2 + (0.09 \times \Delta N)^2} \tag{7}$$

The OOD Comparisons in the Orbital Arc

If orbit A is determined by the data on days n + 1, n + 2, and n + 3, orbit B is determined by the data on days n + 2, n + 3, and n + 4, and the orbital difference of the overlapping arc segments in the R, T, and N directions on days n + 2 and n + 3 are counted and used as the OOD comparison values for orbit A in the orbital arc. Figures 5 and 6 show the sketch map of the evaluation of the OOD comparisons in the orbital arc and the 2-h orbital prediction arc.

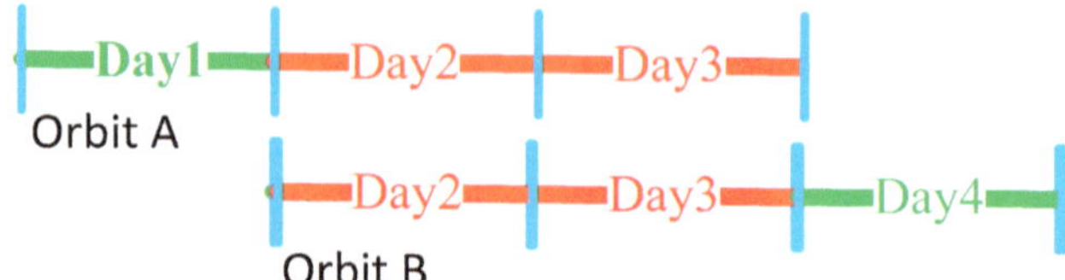

Figure 5. Evaluation of the OOD comparisons in the orbital arc. The red arc segments represent the overlapping arc segments.4.3.2. The OOD Comparisons in the 2-h Orbital Prediction Arc.

Figure 6. Evaluation of the OOD comparisons in the 2-h orbital prediction arc. The red arc segments represent the overlapping arc segments.

Orbit B is used as the reference orbit for evaluating the orbit accuracy in the 2-h orbital prediction arc of orbit A. The differences between the 2-h orbital prediction arc of orbit A and orbit B in the R, T, and N directions are counted and used as the OOD comparison value for orbit A in the 2-h orbital prediction arc.

Tables 2 and 3 give the OOD comparisons for C01, C02, C03, and C05 satellites in the orbital arc and the 2-h orbital prediction arc in the POD of 1 to 3 July 2021, respectively. Figures 7 and 8 give the OOD comparisons for the C02 satellite in the orbital arc and the 2-h orbital prediction arc, respectively. In these tables and figures, R, T, N, Positions and SISRE (orbit) represent the results in the radial direction, along-track direction, cross-track direction, the three-dimensional (3D) position, and SISRE in the orbital arc, respectively.

Table 2. Overlapping orbit differences for GEOs in the orbital arc (unit: m).

Satellites	R	T	N	Positions	SISRE (Orbit)
C01	0.47	6.05	1.81	6.33	0.74
C02	0.41	5.35	1.74	5.64	0.65
C03	0.38	5.37	1.62	5.63	0.63
C05	0.43	6.97	0.63	7.01	0.76
C59	0.48	5.99	1.86	6.29	0.74
C60	0.39	5.32	1.77	5.62	0.64
C61	0.34	5.30	1.59	5.54	0.60
Mean Values	0.41	5.76	1.57	6.01	0.68

Table 3. Overlapping orbit differences for GEOs in the 2-h orbital prediction arc (unit: m).

Satellites	R	T	N	Positions	SISRE (Orbit)
C01	0.56	6.73	2.25	7.12	0.85
C02	0.53	6.49	2.78	7.09	0.82
C03	0.40	5.55	2.54	6.12	0.68
C05	0.58	7.91	2.56	8.34	0.94
C59	0.57	6.77	2.28	7.17	0.86
C60	0.51	6.46	2.74	7.03	0.81
C61	0.36	5.47	2.5	6.03	0.65
Mean Values	0.50	6.48	2.52	6.99	0.80

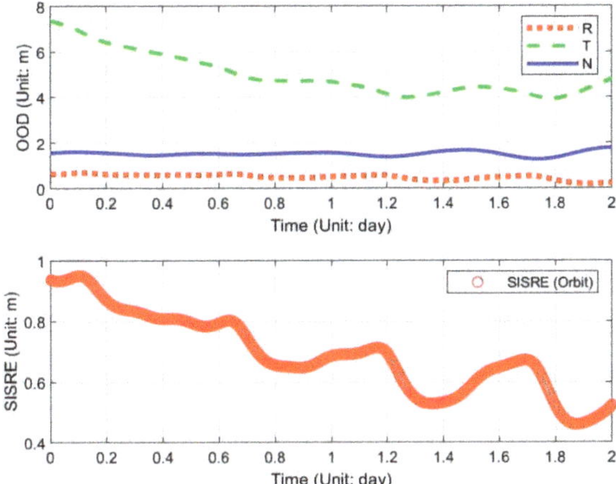

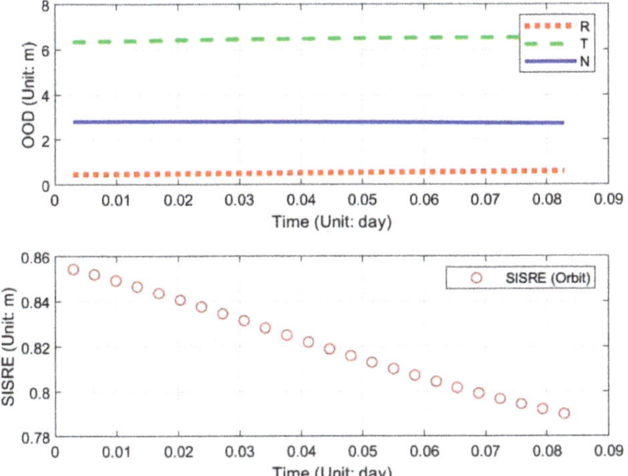

Figure 7. Overlapping orbit differences for C02 in the orbital arc.

Figure 8. Overlapping orbit differences for C02 in the 2-h orbital prediction arc.

The tables and figures show that in the POD for GEO satellites based on only RDSS data, the position error and SISRE in orbit in the orbital arc for GEO satellites are 6.01 m and 0.68 m, respectively; the position error and SISRE in orbit in the 2-h orbital prediction arc are 6.99 m and 0.80 m, respectively. The orbit error in the radial direction is better than 0.6m, but those in the tangential and normal directions are about 6 m and 2 m, respectively. Due to the geostationary characteristics of GEO satellites, only domestic RDSS stations are used for orbit determination, so their geometric configuration is poor. The correlation between the RDSS measurement and the orbit in the radial direction is about 0.98. Those in the normal direction and the tangential direction are very small, especially that in the tangential direction, which is the smallest.

The 3D position error of GEO satellites using RNSS data in the orbital arc and the 2-h orbital prediction arc are 1.46 m and 2.03 m, respectively. The orbit results by the POD based on only RDSS data display slightly low precision due to noise from RDSS data and the ionospheric delay error after the correction by the 14-parameter Klobuchar model with a residual error of 1 to 2 m. The pseudorange data and phase data are used in the POD by RNSS data. The measurement noise of the pseudorange data is about 30 cm and the measurement noise of the phase data is about 2 mm.

The orbit accuracy is in the order of C03, C02, C01, and C05 from high to low. This is related to the observation geometry. Since the domestic stations are approximately distributed from 75°E to 125°E, the C01, C02, C03, C04, and C05 satellites are, respectively, fixed at 140°E, 84°E, 110.5°E, 160°E, and 58.75°E, the station geometry of C03 satellite is the best. C59, C60 and C61 satellites in BDS-3 are fixed at 140°E, 84°E and 110°E, respectively, so the orbit accuracy is equivalent to that of the C01, C02, and C03 satellites.

The combined POD methods by RDSS and SLR data are adopted to solve the RDSS time delay in this paper. The stability of the time delay is important for the orbit accuracy for GEOs. Therefore, the POD experiments over 2 months from 1 June to 31 July 2021 are performed with the time delay estimated on 1 June 2021 to verify the applicability of the time delay and the orbit accuracy over multiple days. Figures 9 and 10 give the OOD results in the R, T, and N, SISRE for C03 and C60 satellites in the orbital arc and the 2-h orbital prediction arc in the POD experiments from 1 to 30 of June, respectively.

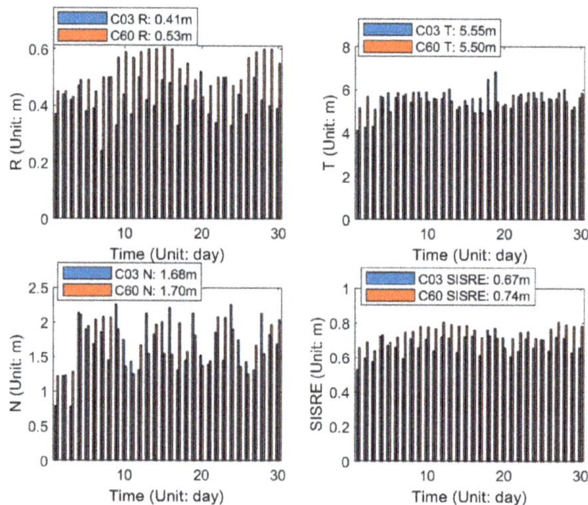

Figure 9. OOD results for C03 and C60 in the orbital arc in the 30 POD experiments.

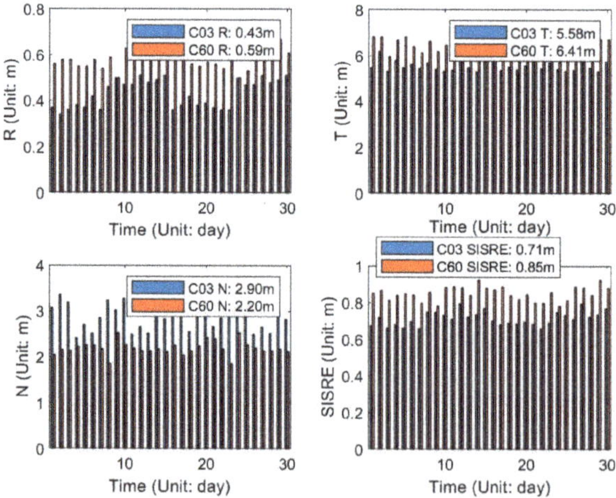

Figure 10. OOD results for C03 and C60 in the 2-h orbital prediction arc in the 30 POD experiments.

The figures show that in the 30 POD experiments, the mean value of the SISRE in orbit in the orbital arc for C03 and C60 satellites are 0.67 m and 0.74 m, respectively; which in the 2-h orbital prediction arc are 0.71 m and 0.85 m, respectively. This analysis shows that the multiday orbit accuracies with the adjusted time delay values are high and stable. The corrected RDSS time delay value is stable in about 30 days. This calibration method is applicable for at least 30 days. We can regularly calibrate the RNSS equipment time delay about every 1 or 2 months. When the orbit accuracy for GEO satellites in BDS is better than 10 m, the horizontal positioning accuracy at different locations averages about 9 m. Orbit accuracy using RDSS data must be suitable for a horizontal positioning precision of less than 20 m of BeiDou-3 RDSS service, with an orbit accuracy higher than 15 m. Therefore, the orbit accuracy achieved in this paper is far superior to the index requirement.

Next, the short-arc dynamic orbit determination experiments are made based on the 4-h RDSS data after the orbit manoeuvre to test the fast orbit recovery capability of GEO satellites after the orbit manoeuvre based on RDSS data. The precise ephemeris of MGEX is used as the benchmark. The experiments are made once in one day from 1 June to 30, 2021, the mean value of these 30 POD experiments is the last results. The orbit accuracies in the short-arc orbit determination for GEO satellites are shown in Table 4.

Table 4. Orbit accuracy in Short-arc orbit determination for GEO satellites based on RDSS data (unit: m).

Satellites	R	T	N	Positions	SISRE (Orbit)
C01	1.95	8.76	5.33	10.44	2.16
C02	2.21	9.78	6.05	11.71	2.44
C03	2.14	9.32	6.14	11.29	2.36
C04	2.80	10.14	7.60	12.98	3.02
C05	2.78	10.06	7.55	12.88	3.00
C59	1.90	8.65	5.13	10.23	2.10
C60	1.78	8.14	4.89	9.66	1.97
C61	1.67	8.13	5.03	9.71	1.88
Mean values	2.15	9.12	5.97	11.11	2.37

As the table shows, in the short-arc dynamic orbit determination based on RDSS data only for GEO satellites, the position accuracy for GEO satellites after 4 h of manoeuvring is 11.11 m. Although this is slightly inferior to the current RNSS data-based orbit determination accuracy during recovery, it can meet the aforesaid accuracy index requirement of 15 m.

4.4. Orbit Accuracy Analysis in Radial Direction Based on the SLR Data

The orbit accuracy in radial direction for satellites C01, C02, C03, and C05 was validated with SLR observation data, and the SLR residual details of satellites C01 and C03 are shown in Figures 11 and 12, respectively. The orbital accuracies in radial direction for satellites C01, C02, C03, and C05 based on RDSS data only are 0.547, 0.569, 0.510, and 0.617 m, respectively, with a mean value of 0.561 m. The accuracies are slightly inferior to the orbit accuracy of BeiDou GEO satellites based on the global network data. The large deviation of SLR residuals comes from the orbital error in the radial direction of GEO satellites. In this paper, the observation stations were deployed in regions in China, and there are only seven stations with a limited observation accuracy of RDSS data of 1 to 2 m. Due to the geostationary characteristics of GEO satellites, the constraints of RDSS data in the radial direction for GEO satellites are limited. Moreover, the constraints in the normal and tangential directions are very small by the satellite to ground measurement data for GEO satellites. Therefore, the orbit errors in the radial, normal, and tangential directions cause a large deviation in the SLR residual error in the radial direction.

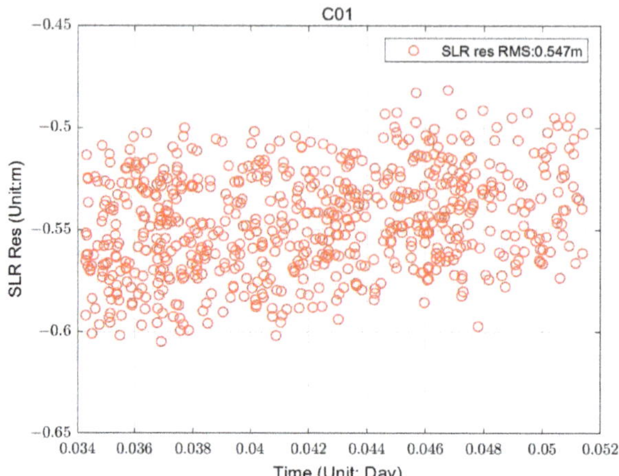

Figure 11. SLR residuals of the orbit for C01 satellite.

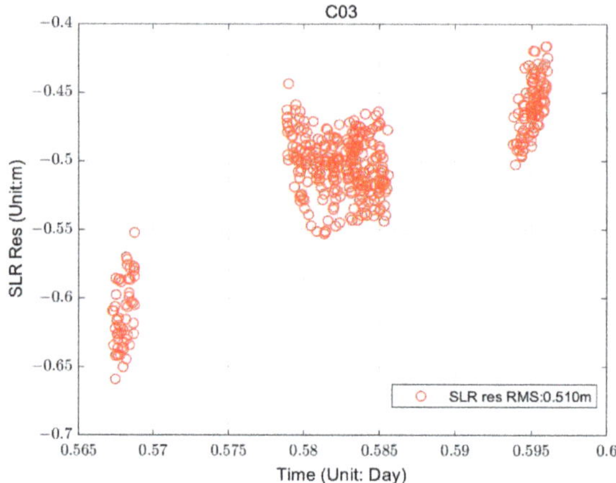

Figure 12. SLR residuals of the orbit for C03 satellite.

The SLR measurements for C01 to C05 satellites are carried out intermittently lasting 2 months from 1 June to 31 July 2021 in Beijing. The experiments are performed to verify the orbit accuracy analysis in a radial direction over multiple days. Table 5 gives the SLR residuals of the orbit for the C01–C05 satellites based on RDSS data. Figure 13 gives the SLR residuals for the C01 and C02 satellites.

Table 5. SLR residuals of the orbit for C01-C05 satellite based on RDSS data (unit: m).

Satellites	2 July 2019	1 June to 31 July 2021
C01	0.55	0.56
C02	0.57	0.54
C03	0.51	0.50
C04	-	0.55
C05	0.62	0.57
Mean values	0.56	0.54

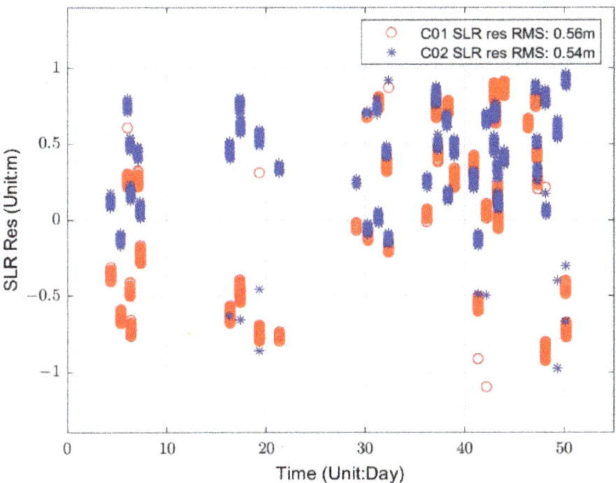

Figure 13. SLR residuals of the orbit for C01, C02 satellite from 1 June to 31 July 2021.

The tables and figures show that the SLR residuals are greater than -1 m but less than 1 m. The RMSs of the SLR residuals for C01 and C02 satellites are 0.56 m and 0.54 m, respectively. The RMS of the SLR residuals for GEO satellites in BDS-2 system is 0.54 m. By contrast, the accuracy of the RNSS phase data of the global network is about 40 cm. There are a large number of observation stations. With the development of the BDS, the improvement of observation accuracy of RDSS data, and the deployment of RDSS calibration stations on a wide scale, the orbit accuracy based on the RDSS data only for GEO satellites is expected to improve by a large margin.

5. Conclusions

In order to solve the problem of unstable orbital results of the BeiDou GEO satellites caused by RNSS data anomalies at different times, this study proposed a new POD for GEO satellites based on RDSS data only, and the time delay of RDSS equipment was calibrated by SLR data. The service performance and reliability of RDSS systems are improved. This study focused on the deep integration and application of short-message communication and navigation services of the BDS. Experiments based on SLR and RDSS data measured in the BDS were performed, and the main conclusions from this study are summarised as follows:

(1) The time delay accuracy of the RDSS data was better than 1 ns, while the observation accuracy of the RDSS was 1 to 2 m. Therefore, the time delay calibration accuracy meets the demand for time delay accuracy in the case of orbit determination for GEO satellites based on RDSS data.

(2) The OOD comparisons in the orbital arc and the 2-h orbital prediction arc for GEO satellites were 6.01 m and 6.99 m, respectively. The orbit accuracy for GEO satellites needs to be higher than 15 m in the RDSS service in BDS. Therefore, the orbit accuracy achieved in this study is considerably higher than this index.

(3) In the short-arc dynamic orbit determination based on RDSS data only for GEO satellites, the position accuracy after 4 h of manoeuvring is 11.11 m. The RDSS ranging data was used to achieve the orbit determination of GEO satellites when RNSS data were unavailable during the recovery period after manoeuvring. Thus, the POD methods of GEO satellites during the recovery period are enriched.

(4) The orbital accuracy in the radial direction based on the SLR data for GEO satellites was 0.54 m. The accuracy depends on the measurement accuracy of RDSS data and the current situation of regional station deployment. In the development of the BDS, we should strive for more and better satellites, orbits, and link resources. In addition, we should continuously improve the technical system of the RDSS and its measurement accuracy, as

well as widely deploy RDSS calibration stations to further improve the orbit accuracy of GEO satellites based on the RDSS data only.

(5) In view of the application of RDSS data in the orbit determination for BeiDou GEO satellites, this study has only conducted preliminary exploration. The orbit accuracy for GEO satellites based on RDSS is lower than that of GEO satellites based on RNSS data, which remains to be studied in depth in the next step. However, it should be noted that the new POD method based on RDSS data only for the GEO satellite proposed in this paper greatly expands the application scope of RDSS data. The functions of positioning, short-message communication, timing service, and orbit determination can be achieved based on RDSS data. The method also enriches the measurement approaches in the orbit determination for GEO satellites, which provides a backup orbit determination technology for BeiDou GEO satellites during the orbit manoeuvre or recovery period.

Author Contributions: Conceptualization, R.G. and J.C.; methodology, J.C. and R.G.; software, S.L.; validation, R.G. and J.C.; formal analysis, J.C; investigation, Z.C., J.X., J.G. and Y.T.; resources, R.G.; data curation, S.L.; writing—original draft preparation, X.L.; writing—review and editing, R.G. and J.C.; visualization, R.G., X.L. and J.C.; supervision, R.G. and J.C.; project administration, R.G.; funding acquisition, R.G. and X.L. All authors have read and agreed to the published version of the manuscript.

Funding: This research was funded by the National Natural Science Foundation of China (Grant Nos.: 41874043, 42004028, 41704037).

Data Availability Statement: Beijing Satellite Navigation Center provides all the test data used in this contribution, including the RDSS, SLR measurements, and the L-band measurements. All data will be made available for scientific research purposes by request to the Beijing Satellite Navigation Centre.

Conflicts of Interest: The authors declare no conflict of interest.

References

1. Yang, Y.; Mao, Y.; Sun, B. Basic performance and future developments of BeiDou global navigation satellite system. *Satell. Navig.* **2020**, *1*, 1–8. [CrossRef]
2. Yang, Y.; Yang, Y.; Hu, X.; Tang, C.; Zhao, L.; Xu, J. Comparison and analysis of two orbit determination methods for BDS-3 satellites. *Acta Geod. Cartogr. Sin.* **2019**, *48*, 831–839.
3. Yang, Y.; Gao, W.; Guo, S.; Mao, Y.; Yang, Y. Introduction to BeiDou-3 navigation satellite system. *Navig. J. Inst. Navig.* **2019**, *66*, 7–18. [CrossRef]
4. Tan, S. Innovative development and forecast of BeiDou system. *Acta Geod. Et Cartogr. Sin.* **2017**, *46*, 1284–1289.
5. Tan, S. *The Engineering of Satellite Navigation and Positioning*; National Defense Industry Press: Arlington, VA, USA, 2011; Volume 1, pp. 12–27.
6. Tan, S. *The Comprehensive RDSS Global Position and Report System*; National Defense Industry Press: Arlington, VA, USA, 2011; Volume 1, pp. 26–32.
7. Tan, S. Theory and application of comprehensive RDSS position and report. *Acta Geod. Cartogr. Sin.* **2009**, *38*, 1–5.
8. Zheng, J. Design of general aviation emergency communication, surveillance and rescue service system based on RDSS. *Mod. Navig.* **2016**, *1*, 1–5.
9. Li, X.; Zhou, J.; Hu, X.; Liu, L.; Guo, R.; Zhou, S. Orbit determination and prediction for Beidou GEO satellites at the time of the spring/autumn equinox. *Sci. China Phys. Mech. Astron.* **2015**, *58*, 089501. [CrossRef]
10. Li, X.; Guo, R.; Hu, X.; Tang, C.; Wu, S.; Chang, Z.; Liu, S. Construction of a BDSPHERE solar radiation pressure model for BeiDou GEOs at vernal and autumn equinox periods. *Adv. Space Res.* **2018**, *62*, 1717–1727. [CrossRef]
11. Zhou, S.; Hu, X.; Wu, B. Orbit Determination and Time Synchronization for a GEO/IGSO Satellite Navigation Constellation with Regional Tracking Network. *Sci. China Phys. Mech. Astron.* **2011**, *54*, 1089–1097. [CrossRef]
12. Tang, C.; Hu, X.; Zhou, S.; Guo, R.; He, F.; Liu, L.; Zhu, L.; Li, X.; Wu, S.; Zhao, G.; et al. Improvement of orbit determination accuracy for Beidou Navigation Satellite System with Two-way Satellite Time Frequency Transfer. *Adv. Space Res.* **2016**, *58*, 1390–1400. [CrossRef]
13. Chen, J.; Hu, X.; Tang, C.; Zhou, S.; Yang, Y.; Pan, J.; Ren, H.; Ma, Y.; Tian, Q.; Wu, B.; et al. SIS accuracy and service performance of the BDS-3 basic system. *Sci. China Phys. Mech. Astron.* **2020**, *63*, 269511. [CrossRef]
14. Yang, Y.; Yang, Y.; Hu, X.; Chen, J.; Guo, R.; Tang, C.; Zhou, S.; Zhao, L.; Xu, J. Inter-Satellite Link Enhanced Orbit Determination for BeiDou-3. *J. Navig.* **2020**, *73*, 115–130. [CrossRef]
15. Li, X.; Zhu, Y.; Zheng, K.; Yuan, Y.; Liu, G.; Xiong, Y. Precise orbit and clock products of Galileo, BDS and QZSS from MGEX since 2018: Comparison and PPP validation. *Remote Sens.* **2020**, *12*, 1415. [CrossRef]

16. Lv, Y.; Geng, T.; Zhao, Q.; Xie, X.; Zhou, R. Initial assessment of BDS-3 preliminary system signal-in-space range error. *GPS Solut.* **2020**, *24*, 16. [CrossRef]
17. Wang, C.; Zhao, Q.; Guo, J.; Liu, J.; Chen, G. The contribution of intersatellite links to BDS-3 orbit determination: Model refinement and comparisons. *Navigation* **2019**, *66*, 71–82. [CrossRef]
18. Li, X.; Zhou, J.; Guo, R. High-precision orbit prediction and error control techniques for COMPASS navigation satellite. *Chin. Sci. Bull.* **2014**, *59*, 2841–2849. [CrossRef]
19. Xing, N.; Tang, C.; Li, X.; Zhang, T.; Ren, H.; Guo, R.; Hu, X. Precision analysis of BDS-3 GEO satellite orbit determination using RDSS. *Sci. China Phys. Mech. Astron.* **2021**, *51*, 019510.
20. Li, X.; Hu, X.; Guo, R.; Tang, C.; Zhou, S.; Liu, S.; Chen, J. Orbit and Positioning Accuracy for the New Generation Beidou Satellites during the Earth Eclipsing Period. *J. Navig.* **2018**, *71*, 1069–1087. [CrossRef]
21. Xing, N.; Su, R.; Zhou, J.; Hu, X.; Gong, X.; Liu, L.; He, F.; Guo, R.; Ren, H.; Hu, G. Analysis of RDSS positioning accuracy based on RNSS wide area differential technique. *Sci. China Phys. Mech. Astron.* **2013**, *56*, 1995–2001. [CrossRef]
22. Yuan, Y.; Huang, J.; Wu, P. Research on the method of satellite selecting in BDS TDOA position reporting. *J. Navig. Position.* **2014**, *2*, 15.
23. Yuan, Y.; Huang, J.; Tao, J. A Beidou RDSS positioning model and its error analysis without elevation. *Comput. Appl. Softw.* **2017**, *34*, 114–118.
24. Li, X.; Hu, X.; Guo, R.; Tang, C.; Liu, S.; Huang, S.; Xin, J.; Pu, J.; Chen, J. A new time delay calibration method to improve the service performance of RDSS in the BDS. *Adv. Space Res.* **2020**, *66*, 2365–2377. [CrossRef]
25. Pang, J.; Zhang, Y.; Zhan, J.; Ou, G. Research on key techniques of the Beidou RDSS receiver test system. *J. Astronaut. Metrol. Meas.* **2016**, *36*, 95–100.
26. Liang, G. Research on EIRP calibration method of RDSS closed-loop test system. *Foreign Electron. Meas. Technol.* **2017**, *36*, 18–20.
27. Montenbruck, O.; Steigenberger, P.; Hauschild, A. Broadcast versus precise ephemerides: A multi-GNSS perspective. *GPS Solut.* **2015**, *19*, 321–333. [CrossRef]
28. Steigenberger, P.; Hugentobler, U.; Hauschild, A.; Montenbruck, O. Orbit and clock analysis of Compass GEO and IGSO satellites. *J. Geod.* **2013**, *87*, 515–525. [CrossRef]
29. Montenbruck, O.; Steigenberger, P.; Prange, L.; Deng, Z.; Zhao, Q.; Perosanz, F.; Romero, I.; Noll, C.; Stürze, A.; Weber, G. The Multi-GNSS Experiment (MGEX) of the International GNSS Service (IGS)–achievements, prospects and challenges. *Adv. Space Res.* **2017**, *59*, 1671–1697. [CrossRef]

remote sensing

Article

Development Status and Service Performance Preliminary Analysis for BDSBAS

Yueling Cao [1,2], Jinping Chen [3,*], Li Liu [3], Xiaogong Hu [1,2], Yuchen Liu [1], Jie Xin [3], Liqian Zhao [4], Qiuning Tian [5], Shanshi Zhou [1,2] and Bin Wu [1,2]

1 Shanghai Astronomical Observatory, Chinese Academy of Sciences, Shanghai 200030, China
2 Shanghai Key Laboratory of Space Navigation and Position Techniques, Shanghai 200030, China
3 Beijing Satellite Navigation Center, Beijing 100094, China
4 Space Star Technology Co., Ltd., Beijing 100095, China
5 Beijing Research Institute of Telemetry, Beijing 100076, China
* Correspondence: paper_2019@sina.com

Abstract: The BeiDou global navigation satellite system (BDS-3) provides positioning, navigation and timing services for global users, moreover, it provides BDS satellite-based augmentation system (BDSBAS) single-frequency (SF) and dual-frequency multi-constellation (DFMC) services for users in China and its surrounding areas. The BDSBAS SF service is in accordance with Radio Technical Commission for Aeronautics (RTCA) standard protocol (RTCA MOPS) and augment GPS constellation, while the BDSBAS DFMC service is in line with SBAS L5 DFMC standard protocol and is aimed at supporting any combination of BDS/GPS/Galileo/GLONASS constellations, including only a single constellation operation. We introduced the development status of the BDSBAS system, including the system architecture and navigation user algorithms. Based on the GPS measurements, the accuracy, integrity and availability of the BDSBAS SF service were evaluated, and with the BDS measurements, the accuracy of the BDSBAS DFMC service was preliminarily analyzed. The integrity and availability of the BDSBAS DFMC service will be discussed in future work as some of the DFMC integrity parameters are still under discussion for optimization. The results show that, for BDSBAS SF service, the horizontal and vertical position accuracy were about 1.0 m and 2.0 m (95%), respectively, which were improved by 39% and 33%, respectively, compared with the GPS SF position accuracy. For BDSBAS DFMC service, the horizontal and vertical position accuracy were about 0.6 m and 1.2 m (95%), respectively, which were improved by about 25% and 20% compared with the BDS dual-frequency position accuracy. No system integrity risk event was detected during the testing period for BDSBAS SF service. The average availability of the BDSBAS SF service was about 98% which was mainly affected by the availability of ionospheric grid delay corrections.

Keywords: BDSBAS SF service; BDSBAS DFMC service; accuracy; integrity; availability

Citation: Cao, Y.; Chen, J.; Liu, L.; Hu, X.; Liu, Y.; Xin, J.; Zhao, L.; Tian, Q.; Zhou, S.; Wu, B. Development Status and Service Performance Preliminary Analysis for BDSBAS. *Remote Sens.* **2022**, *14*, 4314. https://doi.org/10.3390/rs14174314

Academic Editor: Yunbin Yuan

Received: 14 July 2022
Accepted: 25 August 2022
Published: 1 September 2022

1. Introduction

The satellite-based augmentation system (SBAS) can broadcast ephemeris and clock error corrections, ionospheric delay corrections and the corresponding integrity information to users through geostationary orbit satellites (GEO), to improve the accuracy, integrity, continuity and availability of the Global Navigation Satellite System (GNSS) core constellations service. Because of the benefits of the augmentation service, including large service areas and relatively low construction and maintenance costs, many countries and regions have established SBAS systems to meet the navigation performance requirements of high real-time and integrity applications such as aviation users for all phases of flight, from en route through category I approach [1].

Current global SBAS operational systems include the Wide Area Augmentation System (WAAS) of the United States which has been operational since 2003 and can support

CAT I-like approach capability (LPV-200) [2,3]; Japan's Multi-functional Transport Satellite Satellite-based Augmentation System (MSAS) has been operational since 2007 and supported non-precision approach operations [4–6]; the European Geostationary Navigation Overlay Service (EGNOS) system in European Union whose precision approach operations capability began in 2011 [7–9]; both the Indian Global Positioning System (GPS) Aided Geostationary Earth Orit Augmented Navigation (GAGAN) system [10,11] and the Russian System of Differential Correction and Monitoring (SDCM) system [12,13] are in development with plans for horizontal and vertical guidance. In addition to the system already in operation, there are some future SBAS under development: the BeiDou Satellite-Based Augmentation System (BDSBAS) by China, the Korea Augmentation Satellite System (KASS) by South Korea, the SBAS for Africa and Indian Ocean (A-SBAS) and the Australian SBAS (AUSBAS) [14].

BDSBAS is an important part of the BDS system, and provides services for users in China and the surrounding areas. It will be an important supplement to the availability of Global SBAS services. Since civil aviation is the most demanding user for SBAS, the International Civil Aviation Organization (ICAO) established standards and recommended practices (SARPs) providing overarching standards and guidance for global SBAS implementation, and organized the SBAS Interoperability Working Group (IWG) for SBAS service providers to assure common understanding and implementation of the SARPs [15–17]. Incorporation of the ICAO SARPs has become one of the most important jobs of BDSBAS system construction, and BDSBAS provides two kinds of augmentation service: single-frequency (SF) service and dual-frequency multi-constellation (DFMC) service, both in accordance with ICAO standards [18,19].

The BDSBAS SF service augments GPS constellation and meets the RTCA minimum operational performance standards (RTCA MOPS) which defined the GPS L1 C/A signal minimum performance, functions and characteristics. The service provides GPS satellite ephemeris and clock error corrections referenced to the L1 NAV message, ionospheric grid delay corrections, integrity information of corrections, GEO navigation and almanacs, degradation factors and clock–ephemeris covariance matrix, and uses BDSBAS-B1C signal of GEO satellites to broadcast the augmentation messages to GPS/SBAS users, aiming to support APV-I precision approach [19].

The BDSBAS DFMC service is compatible with the SBAS L5 DFMC protocol and the augmentation messages are transmitted by GEO satellites through BDSBAS-B2a signal to augment BDS/GPS/Galileo/GLONASS dual-frequency signals. For BDS, it is recommended to use B1C and B2a dual-frequency signals and the ephemeris corrections are referenced to the B1C CNAV1 navigation message. For GPS, it is recommended to use L1 C/A and L5 dual-frequency signals and the ephemeris corrections are referenced to the L1 NAV navigation message. For Galileo, it is recommended to use E1 and E5a dual-frequency signals and the ephemeris corrections are referenced to the E5a F/NAV navigation message. For GLONASS, it is advised to use L1OC and L3OC dual-frequency signals and the ephemeris corrections reference is to be determined (TBD). BDSBAS DFMC message mainly contains GNSS ephemeris and clock error corrections, integrity messages, clock–ephemeris covariance matrix, degradation parameters, and SBAS satellites ephemeris and almanacs. Compared with SF service, the DFMC service increases the SBAS availability and performance by direct mitigation of ionospheric delay with dual-frequency and inclusion of additional GNSS constellations such as BDS, Galileo and GLONASS, aiming to achieve CAT-I precision approach.

The DFMC service is intended to support any combination of constellations, including only a single constellation operation. At the current test stage, the BDSBAS DFMC service provides BDS and GPS augmentation information, and will gradually increases the dual-frequency augmentation information for the Galileo and GLONASS systems. Table 1 shows the comparison of the BDSBAS SF service and the DFMC service.

Table 1. Comparison of the BDSBAS SF service and the DFMC service.

	SF Service	DFMC Service
Broadcast Signal	BDSBAS-B1C	BDSBAS-B2a
Broadcast Satellites	3GEO	3GEO
Augmentation constellation(s)	GPS	BDS/GPS/Galileo/GLONASS
SBAS Network Time	GPS Time	BDS Time
Clock and Ephemeris slow corrections	Broadcast	Broadcast
Clock fast corrections	Broadcast	Not Broadcast
Clock–Ephemeris correction integrity	Broadcast	Broadcast
Clock–Ephemeris covariance matrix	Broadcast	Broadcast
Ionospheric grid delay corrections and integrity	Broadcast	Not Broadcast
Degradation information	Broadcast	Broadcast
SBAS (GEO) satellites ephemeris and almanacs	Broadcast	Broadcast

The BDSBAS SF service has already begun to provide initial service since 31 July 2020. Reference [20] shows the BDSBAS general design, system time datum, coordinate reference system and signal characteristics, [21] analyzed the pseudorange bias errors of BDSBAS monitoring receivers and their effect on the performance of the BDSBAS service. A general aviation flight test performance of BDSBAS was given by [22]. We focus to introduce the BDSBAS architecture design, user algorithms of broadcast messages and preliminarily evaluated the system performance of accuracy, integrity and availability in accordance with ICAO standards based on monitoring receivers distributed in the China service area with the real observations.

2. BDSBAS Architecture Overview

BDSBAS is operated by the BDS control system and consists of a regional monitoring network in China, two master control stations and three GEO satellites. The two master control stations are backup to each other to enhance the system robustness. To meet the needs of current SBAS users as well as the next generation of SBAS system construction, two parallel threads are dealt with in each master control station processing center for SF service messages and DFMC service messages, respectively.

BDSBAS has chosen a regional network for the SBAS services. To estimate and model the regional ionosphere for SF service, the receiver network is evenly distributed in the service area of China [23]. Each monitoring station is equipped with three independent receivers identified as A/B/C which can receive BDS and GPS pseudorange and carrier phase observation data at present. The observation data from monitoring stations are sent to the master control stations through ground and satellite network links for information processing. Some of monitoring stations equipped with hydrogen atomic clocks are identified as class I stations to ensure the system time reference stability, and the others are identified as class II stations which are widely distributed in the service area. Both the measurements of class I and class II monitoring stations are used to compute differential corrections and integrity information. The workflow of the BDSBAS system is shown in Figure 1.

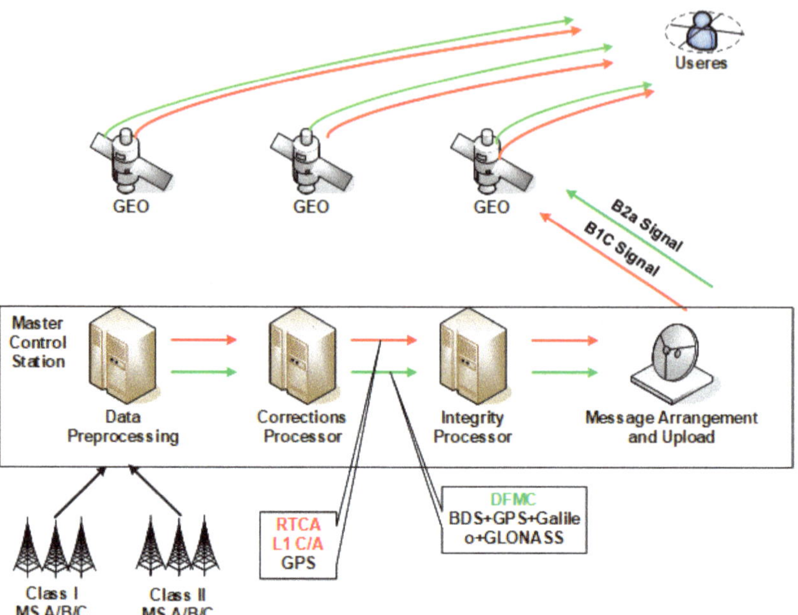

Figure 1. The workflow of the BDSBAS system. The red line indicates the SF service messages processing and the green line indicates the DFMC service messages processing.

The master control station is the data processing center. The received GNSS measurements are first dealt with at the preprocessing facility to identify and eliminate outliers, detect the cycle slip of carrier phase data, correct the code noise and multipath errors in real time, and compute the common errors in the propagation path, such as tropospheric delay, ionospheric delay, receiver antenna phase center corrections. The code noise and multipath errors are corrected in real time with the code noise and multipath correction (CNMC) algorithm which is a kind of carrier-smoothing pseudorange method detail described in [23]. The tropospheric delay is corrected with Black model and the ionospheric delay is corrected with dual-frequency iono-free combination. The preprocessed observations are then sent to the differential correction processing (CP) facility and integrity processing (IP) facility to calculate the differential corrections and evaluate the system output safety.

The CP facility selects preprocessing measurements from A receivers of the monitoring stations, and calculates the differential corrections for SF and DFMC services. For SF service, GPS ephemeris and clock-slow corrections, clock-fast corrections and ionospheric grid delay corrections are determined with respect to L1 C/A signal. For DFMC service, ephemeris and clock corrections are calculated for multiple constellations with dual-frequency. The calculation methods of ephemeris and clock corrections were introduced by [24,25]. GEO Cartesian and Keplerian parameters as well as almanacs are also computed. The corrections are then passed along to the IP facility for evaluation. The IP facility uses preprocessing data screening from B receivers of the monitoring stations to ensure the reliability of SBAS services. The user differential range error (UDRE) monitor determines a confidence bound on the satellite ephemeris and clock corrections with the carrier smoothed iono-free pseudoranges. In parallel, the grid ionospheric vertical error (GIVE) monitor estimates the ionospheric delay confidence bounds for a set of ionospheric grid points (IGPs).

If all the corrections are properly bounded, the corrections and integrity information are then arranged in accordance with RTCA MOPS and SBAS L5 DFMC standards separately and uploaded to the three GEO satellites in a sequence of messages by the message arrangement and upload facility. Both SF and DFMC SBAS messages are broadcast to the

SBAS users with the data rate of 250 bits per second. BDSBAS SF messages are broadcast on BDSBAS-B1C signal and BDSBAS DFMC messages are broadcast on BDSBAS-B2a signal of GEO satellites.

3. Message User Algorithms and Service Performance Evaluation Methods

SBAS can be used in many non-aviation applications and the SARPs proposed by ICAO for the SBAS system can be also extend beyond aviation to other SBAS users. This section describes the user algorithm referred to ICAO SARPs for determination of user position as well as the protection levels with BDSBAS message type (MT) information. The navigation and protection level results of monitoring receivers are used to evaluate the BDSBAS service performance and the evaluation methods are detailed discussed in the following paragraphs.

3.1. Message User Algorithm

The user algorithms include application methods for differential corrections and integrity information. The satellite position and clock-slow corrections will be added to the satellite coordinate vector and clock offset computed from the broadcast ephemeris. The fast corrections will be applied to correct pseudorange measurements. SBAS ionospheric grid delay corrections should be used for single frequency users, to correct the ionospheric delay error of the observation data. For a given satellite without satellite position and clock corrections, it shall not be used in the position equation.

The integrity information is used to indicate an alert condition on one satellite or multiple satellites. Anytime a "don't use" or "not monitored" indication is received, the corresponding satellite should not be used for navigation. Otherwise UDREIs are used for the evaluation of σ^2_{UDRE} indicating the accuracy of combined fast and slow error corrections. The accuracy of ionospheric grid delay corrections indicated in σ^2_{GIVE} is computed form the GIVEIs. These correction accuracies will be applied to determine the weighting matrix for the weighed least squares solution of SBAS user position equation.

Compared with the navigation messages broadcast by the core constellations, the information broadcast by the SBAS system has high frequency update. The fast corrections, slow corrections, and ionospheric corrections are all designed to provide the most recent information to the user. However, there is always the possibility that the user will fail to receive one of these messages. To guarantee integrity even when some messages are not received, SBAS designs the degradation factors of this information to monitor the old data to ensure that they remain valid until they time out. To provide increased availability inside the service volume and increased integrity outside, SBAS broadcast the relative covariance matrix for clock and ephemeris errors to calculate the broadcast σ_{UDRE} degradation values specified as $\delta\mathrm{UDRE}$, as a function of user position. The degradation information is broadcast through different message types and applied to calculate the model variance for the measurements.

The user algorithms of SBAS SF messages are described in detail in RTCA MOPS DO-229D, and the comparable definition of SBAS DFMC operations are given in SBAS L5 DFMC ICD. The calculation processing for SF and DFMC messages are similar. Taking the SF messages as an example, the user computation processing with SBAS messages is introduced as shown in Figure 2.

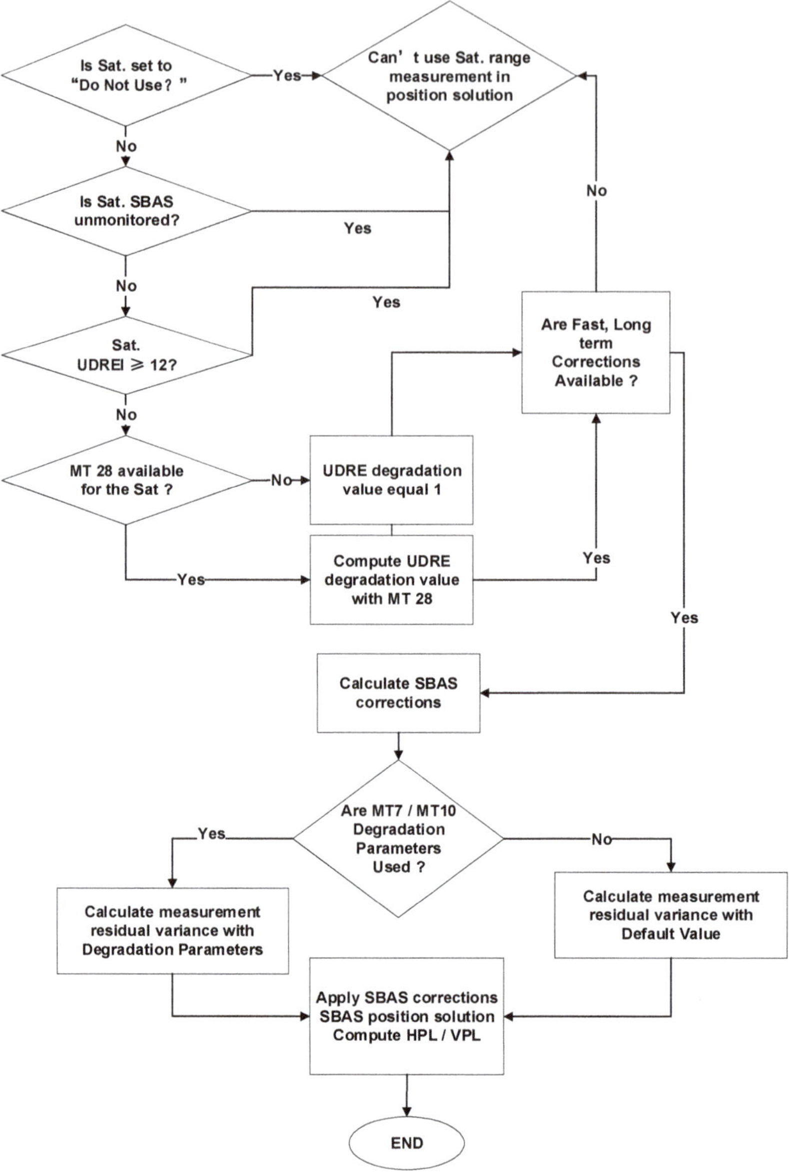

Figure 2. Algorithm flowchart of BDSBAS SF messages [19].

According to Figure 2, after receiving GPS L1 signal and SBAS augmentation messages, the satellites used for navigation are selected according to the satellite vehicle (SV) health status, whether the differential corrections were received and the corresponding integrity information. Once the satellite is unhealthy, or its differential corrections are not received, or the UDREI of the satellite is greater than 12, the measurements of the satellite will not be used. If the MT 28 messages are received normally, the UDRE degradation value δUDRE will be computed with MT 28 as Equation (1), otherwise δUDRE is equal to 1 [19,22].

$$
\begin{aligned}
\delta\text{UDRE} &= \sqrt{l^T \cdot C \cdot l} + \varepsilon_C \\
C &= (E \cdot Fscal)^T \cdot (E \cdot Fscal) \\
Fscal &= 2^{(scale-5)} \\
\varepsilon_C &= C_{covariance} \cdot Fscal \\
E &= \begin{bmatrix} E_{1,1} & E_{1,2} & E_{1,3} & E_{1,4} \\ 0 & E_{2,2} & E_{2,3} & E_{2,4} \\ 0 & 0 & E_{3,3} & E_{3,4} \\ 0 & 0 & 0 & E_{4,4} \end{bmatrix}
\end{aligned}
\tag{1}
$$

where l is the line-of-sight vector from the user to the satellite. ε_C indicates the quantization errors, derived from $C_{covariance}$ which is broadcast in an MT 10 message. If MT 10 data are not available, ε_C is set to zero. C is the relative clock–ephemeris correction covariance matrix reconstructed with the parameters $E_{i,j}$ broadcast in MT 28 message, *scale* is the scale exponent also broadcast in an MT 28 message.

Degradation value δUDRE is applied to model the variance for the measurement residual error σ^2_{fit}. If MT 7 and MT 10 are successfully received, the variance for the measurement residual error σ^2_{fit} will be computed with δUDRE as well as the degradation parameters obtained from MT 7 and MT10. Otherwise, σ^2_{fit} will be computed with δUDRE as well as the default value. The detail computation methods of the measurement residual error variance σ^2_{fit} are shown in [19].

Then, the position results and corresponding protection levels are calculated based on the weighted least squares solutions, as expressed as Equation (2).

$$
X = (G^T W G)^{-1} G^T W L
\tag{2}
$$

where X is the unknown position parameters, G is the observation geometry matrix, L is the observations vector, W is the weight matrix which can be expresses as a diagonal matrix, at least four satellites are visible and useful to ensure that the first term is inversed.

$$
W = diag\left(\frac{1}{\sigma^2_1}, \cdots \frac{1}{\sigma^2_n} \right)
\tag{3}
$$

where σ^2_i is expressed as

$$
\sigma^2_i = \sigma^2_{i,fit} + \sigma^2_{i,UIRE} + \sigma^2_{i,tropo} + \sigma^2_{i,air}
\tag{4}
$$

In the Equation (4), $\sigma^2_{i,fit}$ is the variance for the measurement residual error of satellite I, $\sigma^2_{i,UIRE}$ is the variance of residual of ionospheric model correction, $\sigma^2_{i,tropo}$ is the variance of the residual of tropospheric model correction, and $\sigma^2_{i,air}$ is the variance of the observation noise.

The protection level equations are described in detail by [26].

3.2. Service Integrity Evaluation Method

With the continuous improvement of the service accuracy of GNSS core constellations such as the GPS and BDS, SBAS systems pay more attention to the service integrity augmentation. In the actual navigation processing, users usually do not know their true position error (PE), they can only compute the protection levels (PL), taking all relevant error sources into account associated with the observation geometry and integrity data provided by the SBAS system. The definition of the system integrity performance requirement includes an alert limit (AL) against which the requirement can be assessed. If the PL is too large to be contained within the AL, the SBAS service will be indicated as not available. We used the monitoring receivers with precisely known coordinates evenly distributed in the service area to analyze the BDSBAS service integrity. Once a PE of any monitoring receiver exceeds the AL without receiving an alert in time, while the calculated PL is smaller than the AL, a piece of hazardously misleading information (HMI) exists. The SBAS service integrity orders that the HMI probability (P_{HMI}) should be less than a certain probability. ICAO provides specific requirements of P_{HMI} for different operations. The evaluation method of SBAS service integrity is shown as follows:

(1) During the testing period, the starting and ending epochs are indicated as t_{start} and t_{end}, respectively, and the observation sampling interval is expressed as T.
(2) The position coordinates of the monitoring receivers are calculated with BDSBAS broadcast information. The deviation of the estimate coordinates from the true coordinates of the receivers are computed and divided into horizontal position errors (HPE) and vertical position errors (VPE). The time series of HPE and VPE are taken as the statistical samples.
(3) The associated integrity messages of BDSBAS are used to compute the horizontal protection levels (HPL) and vertical protection levels (VPL) for the monitoring receivers.
(4) At epoch i, comparing the HPE value and HPL value against the horizontal alert limit (HAL) value. If the three values satisfy the relationship as HPL < HAL < HPE, it indicates a horizontal HMI event, $Bool(x)$ is set equal to 1, otherwise, $Bool(x)$ is set equal to 0. Meanwhile, comparing the VPE value, VPL value against the vertical alert limit (VAL) value, if the three values satisfy VPL < VAL < VPE, it indicates a vertical HMI event, and $Bool(y)$ is set equal to 1, otherwise, $Bool(y)$ is set equal to 0.
(5) Using all the experimental samples, the probability of HMI of service integrity is determined as:

$$P_{HMI} = 1 - \frac{\sum\limits_{t=t_{start},inc=Top}^{t_{end}-Top} \left\{ \prod\limits_{i=t,inc=T}^{t+Top} Integrity\_Flag(i) \right\}}{\frac{t_{end}-t_{start}}{Top}}$$

$$Integrity\_Flag(i) = \begin{cases} 0, & (Bool(x)_i + Bool(y)_i) > 0 \\ 1, & other \end{cases} \tag{5}$$

where $Integrity\_Flag(i)$ is the service integrity indicator, where "0" indicates integrity risk occurred and "1" indicates there is no integrity risk; P_{HMI} is the probability of HMI; Top is the service sample interval and usually is set to 150 s for APV I approach; T is observation data sampling interval.

3.3. Service Availability Evaluation Method

Service availability is an indication of the ability of the system to provide usable service within the specified coverage area. It can be expressed as the percentage of time that the position accuracy and integrity of the SBAS service remain within the specified threshold. The availability determined with the GNSS and SBAS observations is described below:

(1) During the testing period, the starting and ending epochs are indicated as t_{start} and t_{end}, respectively, and the observation sampling interval is expressed as T.
(2) The time series of receiver HPE and VPE are calculated with SBAS corrections and the corresponding HPL and VPL are determined with the SBAS integrity information.
(3) At observation epoch i, the receiver HPE and VPE are compared with the service horizontal accuracy threshold ($HPOS_{lim}$) and vertical accuracy threshold ($VPOS_{lim}$), respectively, to determine the accuracy availability indicator at epoch i.

$$Pos\_flag(i) = \begin{cases} 1, & HPE < HPOS_{lim} \& VPE < VPOS_{lim} \\ 0, & other \end{cases} \tag{6}$$

where $Pos\_flag(i)$ is the accuracy availability indicator at epoch i, "1" indicates service accuracy is available, and "0" indicates service accuracy is not available.

(4)   The service integrity indicator is calculated as Equation (5). Considering both the service integrity indicator and accuracy availability indicator, the service availability is given by Equation (7):

$$A_{vail} = \frac{\sum_{t=t_{start},inc=Top}^{t_{end}-Top} \left\{ \prod_{i=t,inc=T}^{t+Top} [Integrity\_Flag(i) \cdot Pos\_flag(i)] \right\}}{\frac{t_{end}-t_{start}}{Top}} \tag{7}$$

where A_{vail} means the service availability, Top is the service sample interval and usually is set to 150 s for APV I approach.

Based on the service integrity and availability evaluation methods, the BDSBAS service performance is preliminarily evaluated with real observation data.

4. BDSBAS Service Performance Evaluation

Currently, the BDSBAS SF service augments GPS L1 C/A signal, and BDSBAS DFMC service augments BDS B1C and B2a dual-frequency signals and GPS L1 C/A and L5 dual-frequency signals. Since only a few of GPS satellites broadcast L5 signal and can be monitored in the Chinese service area, the GPS dual-frequency service performance cannot be reasonably evaluated. In addition, the definitions of some degradation parameters in the BDSBAS DFMC protocol are under discussion for optimization, the integrity and availability of the DFMC service will be discussed in future work. The BDSBAS DFMC service only evaluates the accuracy performance based on BDS constellation.

The performance accuracy, integrity and availability of the BDSBAS SF service as well as the accuracy of the BDSBAS DFMC service are preliminarily evaluated, using the observations of 30 monitoring receivers evenly distributed in China from 1 July to 10 July in 2020. These monitoring receivers are independent from those generating BDSBAS differential corrections.

4.1. BDSBAS SF Service Accuracy

According to the user algorithms of BDSBAS messages, the single point position (SPP) results of the BDSBAS SF service were achieved through 30 monitoring receivers' data. To reduce the observation errors such as multipath error, observation noise, carrier smoothed pseudoranges are used in the SPP estimation. In consideration of calculation efficiency and carrier-smoothed pseudoranges accuracy, the data with 20 s observation sampling interval were adopted. The coordinates of the monitoring receivers are regularly maintained with geodetic survey and GNSS network adjustment calculation, and the 3D coordinate accuracy of these receivers is usually better than 5 cm. As the precise coordinates of the monitoring receivers were already known, the position errors in horizontal and vertical directions were analyzed. The 95th percentile values of absolute value of horizontal and vertical position errors of each day were evaluated, and the time series results during 10 days for 30 receivers were shown in Figure 3.

In Figure 3, the lines in the top subgraph indicate the time series of horizontal position errors of 30 receivers and the lines in the bottom subgraph show the position errors in the vertical direction. Different colors indicate different monitoring receivers. The horizontal axis represents time in days, and the vertical axis represents position errors in meters. It shows that the variance of position errors for the same monitoring receiver is relatively stable in different days. The HPE of different receivers has little difference, varying from 1 m to 1.5 m, while the dispersion of VPE among different receivers is relatively large, varying from 1.5 m to 3 m. The VPE variation of different receivers is mainly caused by the accuracy of corresponding ionospheric grid corrections.

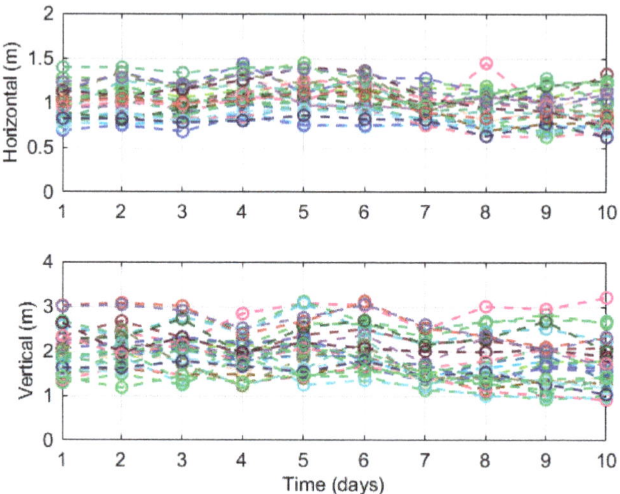

Figure 3. Time series of 95th percentile position errors for 30 monitoring receivers under BDSBAS SF service. One color represents one receiver.

The 24 h time series of position errors of one monitoring receiver are shown as an example, compared between the BDSBAS + GPS SF service and the GPS SF service. The position errors in the north–south, east–west and vertical directions are shown in the top, middle and bottom subgraphs of Figure 4, respectively. In the figure, the red lines indicate the positions errors of the BDSBAS + GPS SF service and the black lines represent the position errors of GPS SF service. The horizontal axis expresses time in hours. As shown in Figure 4, the 95% position errors of the BDSBAS + GPS SF service are 0.638 m, 0.506 m and 1.548 m in the north–south, east–west, and vertical directions, respectively, while the 95% position errors of the GPS SF service are 1.388 m, 0.851 m and 2.427 m in the north–south, east–west, and vertical directions, respectively. The position accuracy of the BDSBAS + GPS SF service is obviously improved compared with the GPS SF service.

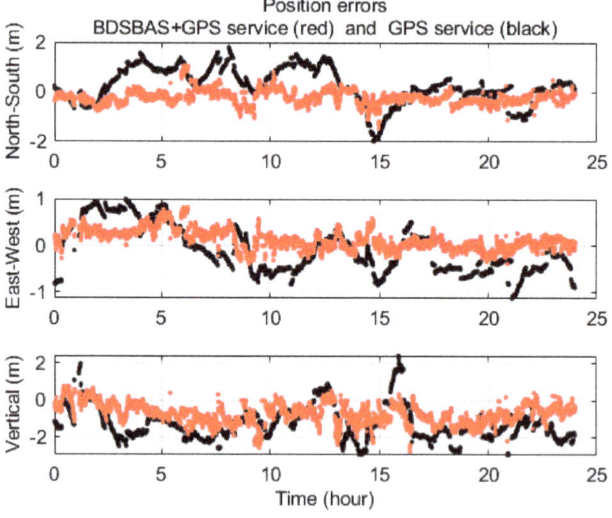

Figure 4. Comparison of position errors between the BDSBAS + GPS SF service (red lines) and the GPS SF service (black lines).

The average value of 95% absolute value of position errors over 10 days is used to evaluate the position accuracy of the monitoring receiver. The position accuracies of the BDSBAS + GPS SF service and the GPS SF service are compared analyzed and the statistics for 30 monitoring receivers are listed in Table 2.

Table 2. The position accuracy statistics of monitoring receivers under the BDSBAS + GPS SF service and the GPS SF service (unit is meters).

Rcv ID	Horizontal (95%)		Vertical (95%)		Rcv ID	Horizontal (95%)		Vertical (95%)	
	BDSBAS + GPS	GPS	BDSBAS + GPS	GPS		BDSBAS + GPS	GPS	BDSBAS + GPS	GPS
1	0.769	1.699	1.837	2.488	16	1.129	1.631	2.662	3.039
2	0.744	1.688	1.739	2.477	17	0.865	1.629	2.503	2.708
3	1.101	1.693	2.444	3.727	18	1.197	1.480	1.910	2.451
4	0.964	1.538	1.398	2.773	19	1.022	1.980	2.358	3.157
5	0.961	1.558	1.338	2.840	20	1.048	1.910	2.072	2.868
6	1.071	1.977	2.736	3.304	21	1.039	1.414	1.617	2.654
7	1.085	2.147	2.270	3.254	22	1.120	1.396	1.686	2.668
8	1.116	1.495	1.829	2.424	23	1.101	1.689	2.107	2.675
9	0.971	1.531	1.823	2.479	24	1.025	1.665	1.931	2.716
10	0.962	1.574	1.451	2.791	25	0.806	1.588	1.559	2.915
11	0.972	1.524	1.393	2.782	26	0.773	1.595	1.526	2.972
12	0.794	1.526	1.331	2.806	27	1.234	1.593	1.682	2.265
13	1.129	1.651	1.600	2.619	28	1.192	1.644	1.393	2.600
14	0.849	1.668	1.782	2.617	29	0.970	1.633	2.656	3.964
15	0.841	1.662	1.868	2.752	30	1.197	1.631	2.656	4.060
Mean	1.00	1.65	1.91	2.86					

The results show that the horizontal and vertical position accuracies of the 30 monitoring receivers are less than 1.5 m and 3 m, respectively, achieved with BDSBAS + GPS SF service. The average accuracies of all 30 receivers in horizontal and vertical directions are about 1 m and 2 m and are improved by about 39% and 33%, respectively, compared with GPS SF service.

4.2. BDSBAS DFMC Service Accuracy

The service accuracy of the BDSBAS DFMC service is similarly analyzed with 30 BDS monitoring receivers. The testing data sampling interval is 20 s. The absolute value of horizontal and vertical position errors with 95% confidence level are calculated every day and the time series of all monitoring receivers 95th percentile position errors are displayed in Figure 5.

In Figure 5, the lines in the top subgraph indicate the time series of horizontal position errors and the lines in the bottom subgraph show the position errors in vertical direction. Different colors indicate different monitoring receivers. The X-axis represents time in days, and the Y-axis represents position errors in meters. It shows that the variance of position errors for the same monitoring receiver is relatively stable in different days, and the position errors of all receivers are relatively consistent. The HPE of different monitoring receivers varies from 0.5 m to 1.0 m, and the VPE of different monitoring receivers varies from 1.0 m to 1.5 m. As most ionospheric delay errors can be eliminated by using dual-frequency observations, the vertical position accuracy of the BDSBAS DFMC service is significantly improved compared with the BDSBAS SF service.

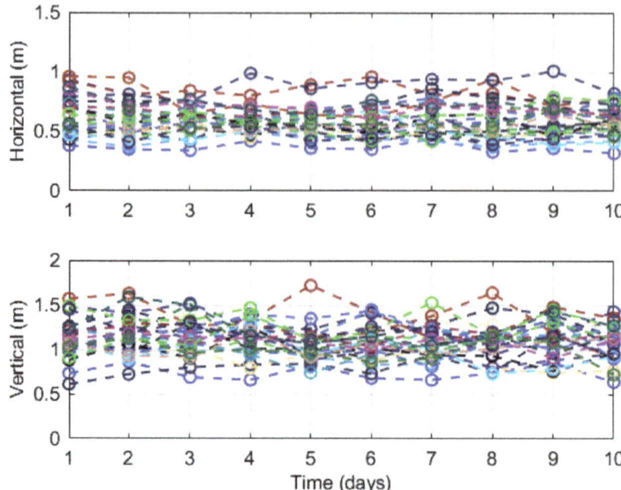

Figure 5. Time series of 95th percentile position errors of 30 monitoring receivers under BDSBAS DFMC service. One color represents one receiver.

The time series of position errors are compared analyzed between the BDSBAS DFMC service and the BDS dual-frequency service and the results of one monitoring receiver are shown in Figure 6 as an example. The position errors in the north–south, east–west and vertical directions are shown in the top, middle and bottom subgraphs, respectively. The red lines indicate the BDSBAS DFMC service and the black lines represent the BDS dual-frequency service. The X-axis represents time in hours. It is noted that the position errors (95%) are 0.51 m, 0.57 m and 1.60 m in north–south, east–west and vertical directions, respectively, under BDS dual-frequency service, and the position errors (95%) are 0.25 m, 0.23 m and 0.77 m in north–south, east–west and vertical directions under BDSBAS DFMC service. The position accuracy of the BDSBAS DFMC service is also clearly improved compared with the BDS dual-frequency service.

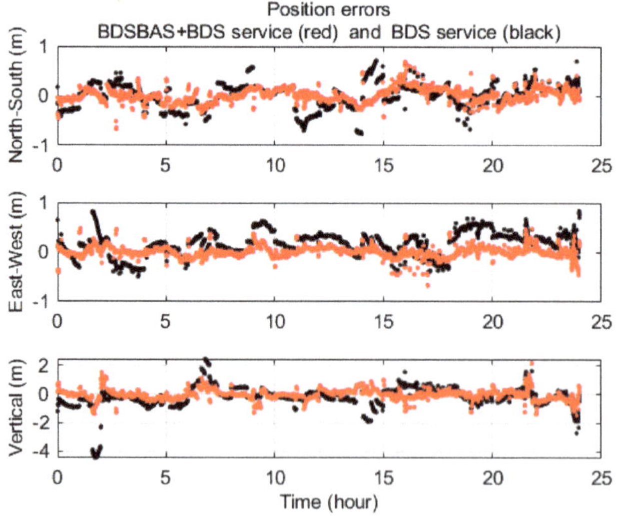

Figure 6. Comparison of position errors between the BDSBAS DFMC service (red lines) and the BDS dual-frequency service (black lines).

The average value of 95% absolute value of position errors of 10 days is calculated to represent the position accuracy of the monitoring receiver. The position accuracies of 30 monitoring receivers under the BDSBAS DFMC service and the BDS dual-frequency service are comparatively analyzed and the statistics results are listed in Table 3.

Table 3. The position accuracy comparison under the BDSBAS DFMC service and the BDS dual-frequency service (unit is meters).

Rcv ID	Horizontal (95%)		Vertical (95%)		Rcv ID	Horizontal (95%)		Vertical (95%)	
	BDSBAS + BDS	BDS	BDSBAS + BDS	BDS		BDSBAS + BDS	BDS	BDSBAS + BDS	BDS
1	0.558	0.765	1.050	1.176	16	0.677	0.931	1.237	1.377
2	0.365	0.642	0.733	1.079	17	0.741	0.928	1.263	1.469
3	0.571	0.758	1.140	1.485	18	0.552	0.844	1.025	1.392
4	0.547	0.748	1.186	1.507	19	0.515	0.737	1.085	1.379
5	0.423	0.649	1.001	1.353	20	0.560	0.706	1.210	1.455
6	0.539	0.784	1.259	1.537	21	0.480	0.653	0.987	1.403
7	0.624	0.892	1.277	1.606	22	0.679	0.845	1.089	1.361
8	0.834	1.023	1.333	1.537	23	0.676	0.835	1.158	1.343
9	0.503	0.730	0.830	1.144	24	0.526	0.701	1.140	1.400
10	0.758	0.904	1.322	1.459	25	0.552	0.756	1.031	1.395
11	0.508	0.708	1.066	1.307	26	0.659	0.855	1.333	1.706
12	0.503	0.754	0.896	1.287	27	0.467	0.716	1.176	1.343
13	0.572	0.715	0.981	1.244	28	0.704	0.859	1.092	1.314
14	0.586	0.767	0.965	1.298	29	0.725	0.915	1.259	1.342
15	0.881	1.133	1.203	1.520	30	0.571	0.789	0.981	1.200
Mean	0.60	0.80	1.11	1.38					

The results show that the horizontal and vertical position accuracies of 30 monitoring receivers are less than 1.0 m and 1.5 m, respectively, under BDSBAS DFMC service. The average values of the 30 monitoring receivers' position accuracies in horizontal and vertical directions are about 0.6 m and 1.2 m, respectively, which are improved by about 25% and 20% compared with BDS dual-frequency service.

4.3. SF Service Integrity and Availability

According to the performance specifications of ICAO Annex 10 for APV-I approach, the HAL and VAL are 40 m and 50 m, respectively, and the horizontal and vertical accuracy thresholds (95%) are required to be above 16 m and 20 m, respectively. The position errors and position protection levels are calculated with the monitoring receiver observations from 1 July to 10 July in 2020. The horizontal integrity and vertical integrity are assessed by the accumulate Stanford diagrams and shown in Figures 7 and 8, respectively.

Stanford diagram outlines the key concepts related to integrity. The X-axis represents the position error and the Y-axis represents the protection level. The misleading information (MI) is defined as when the position error exceeds the protection level but is still less than the required alert limit of the system. If the protection level is less than the alert limit while the position error exceeds the alert limit, it is labelled as hazardously misleading information (HMI). When this occurs, the user is in a potentially dangerous situation as the system is providing dangerous information to the user who is unaware. The integrity risk is the probability that a position error is larger than the alert limit without any alert. If the protection level is less than the system alert limit and the position error is less than the protection level, then the system is available and operating within its normal bounds. If the protection level has exceeded the alert limit, the system service cannot be trusted even though the position error may be acceptable.

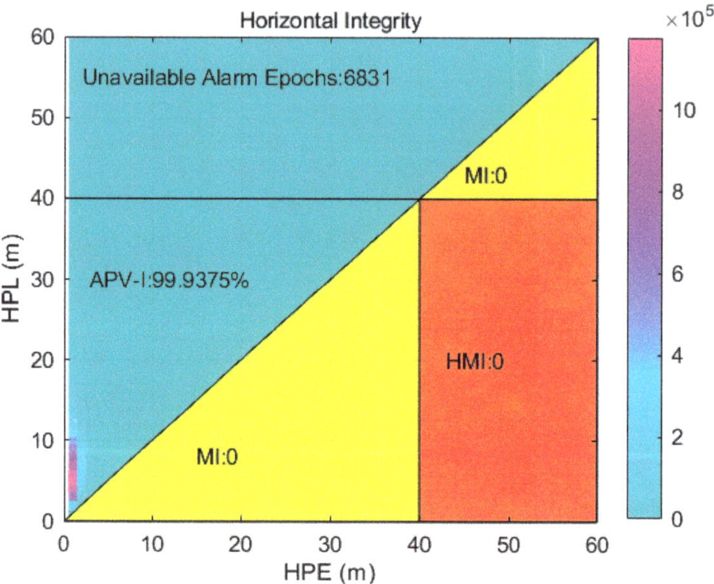

Figure 7. Horizontal integrity assessment with Stanford diagram. No misleading information (MI) and no hazardously misleading information events exist. The 99.9375% testing epochs are operating within their normal bounds in horizontal direction according to APV-I approach requirement.

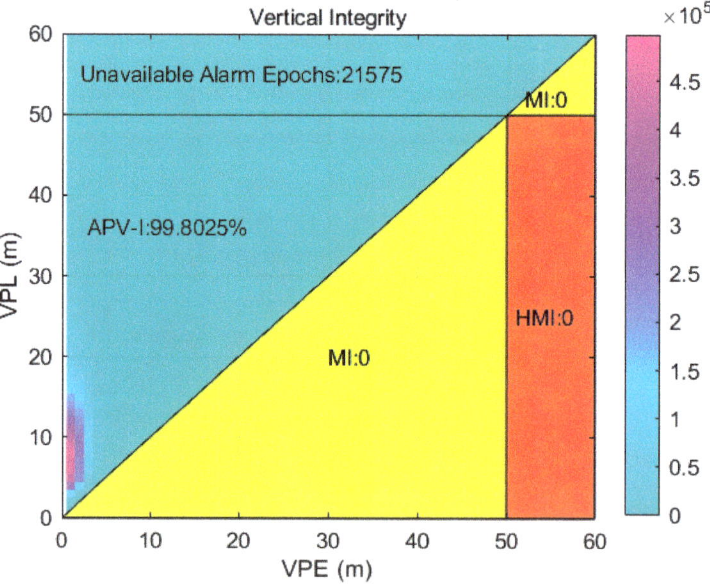

Figure 8. Vertical integrity assessment with Stanford diagram. No misleading information (MI) and no hazardously misleading information events exist. The 99.8025% testing epochs are operating within their normal bounds in vertical direction according to APV-I approach requirement.

It is noted that no integrity risk event occurred during the whole test interval. The availability percentages of the integrity service are about 99.9% and 99.8% in the horizontal and vertical directions, respectively.

To illustrate the detail relation between the protection level and the position error, the 24 h time series of position errors and the corresponding protection levels are shown in Figure 9 at the Beijing station. HPE and HPL are shown in the top subgraph and VPE and VPL are presented in the bottom subgraph. The unit of the X-axis is hours and of the Y-axis is meters. The blue lines indicate position error and the black lines indicate protection level. It shows that the protection level can envelope the position error with 100% confidence, which means that there is no integrity risk event.

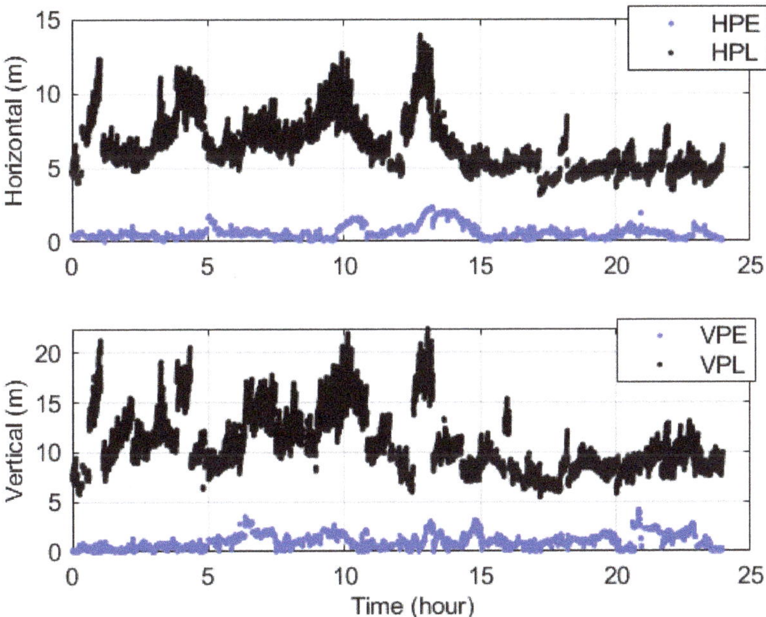

Figure 9. Time series of position errors and protection levels at the Beijing station.

If the position error is less than the system accuracy threshold as well as the integrity service is under normal operation, the system service is considered available. Considering both the accuracy and the integrity performance, the service availabilities of all monitoring receivers are analyzed and shown in Table 4. The results shown that the average value of the service availabilities of the 30 monitoring receivers is about 98%. Some of the receivers located at the service marginal areas have lower service availabilities as their vertical position errors exceed vertical accuracy threshold (20 m), mainly related to the low accuracy of nearby grid corrections as well as the poor dilution of precision (DOP). The accuracy and availability of grid ionospheric corrections at the service marginal areas are low for very few monitoring stations distributed at these areas, which leads to a smaller number of augmented satellites which could be observed in the areas at some time, and leads to the challenge of the service availability.

Table 4. Statistics results of monitoring receiver service availability.

RcvID	Availability	RcvID	Availability
1	0.995	16	0.995
2	0.999	17	0.995
3	0.984	18	0.995
4	0.984	19	0.944
5	0.926	20	0.961
6	0.934	21	0.993
7	0.981	22	0.914
8	0.989	23	0.988
9	0.991	24	0.973
10	0.995	25	0.976
11	0.993	26	0.995
12	0.994	27	0.995
13	0.983	28	0.994
14	0.978	29	0.979
15	0.995	30	0.982
Average value		0.98	

5. Conclusions

We introduce the development status of the BDSBAS system, including the system architecture and the message user algorithms. The accuracies of the BDSBAS SF service and the DFMC service are discussed with system monitoring receiver observations. The preliminary performance of the BDSBAS SF service integrity and availability are evaluated according to ICAO standards. The results show that the position accuracies of the BDSBAS SF service are about 1.0 m and 2.0 m in the horizontal and vertical directions, respectively, which are improved by about 39% and 33% compared with the GPS SF service. The position accuracies of the BDSBAS DFMC service are about 0.6 m and 1.2 m in the horizontal and vertical directions, respectively, which are improved by about 25% and 20% compared with the BDS dual-frequency service. For the BDSBAS SF service, no integrity risk event was detected during the test period, and the average service availability is about 98%. By increasing the effective monitoring areas of grid ionospheric corrections, the availability of the BDSBAS SF service could be further improved, which is an important research field for BDSBAS system upgrading.

Author Contributions: Conceptualization, Y.C., J.C., X.H. and S.Z.; methodology, Y.C., J.X., L.Z. and S.Z.; software, Y.L. and Q.T.; validation, X.H. and B.W.; formal analysis, L.L.; investigation, Y.C. and Q.T.; resources, Y.L.; data curation, Y.L. and J.X.; writing—original draft preparation, Y.C.; writing—review and editing, Y.C., J.C., L.L., X.H. and B.W.; visualization, Y.C.; supervision, J.C. and L.L.; project administration, J.C., L.L., X.H. and B.W.; funding acquisition, Y.C. and S.Z. All authors have read and agreed to the published version of the manuscript.

Funding: This research was funded by the National Key Research Program of China as the "Collaborative Precision Positioning Project", grant number [2016YFB0501900], National Natural Science Foundation of China, grant number [41674041, 12173072 and 11203059] and the Shanghai Key Laboratory of Space Navigation and Position Techniques, grant number [12DZ2273300].

Acknowledgments: The authors are grateful for the comments and remarks of the reviewers, which helped to improve the manuscript.

Conflicts of Interest: The authors declare no conflict of interest.

References

1. FAA. *Global Positioning System Wide Area Augmentation System (WAAS) Performance Standard*, 1st ed.; FAA: Washington, WA, USA, 2008.
2. Bunce, D. Wide Area Augmentation System (WAAS)—Program Update. In Proceedings of the 26th International Technical Meeting of the Satellite Division of The Institute of Navigation (ION GNSS+ 2013), Nashville, TN, USA, 16–20 September 2013; pp. 2299–2326.

3. Walter, T.; Shallberg, K.; Altshuler, E.; Wanner, W.; Harris, C.; Stimmler, R. WAAS at 15. *Navigation* **2018**, *65*, 581–600. [CrossRef]
4. Sakai, T. MSAS Status. In Proceedings of the 26th International Technical Meeting of the Satellite Division of The Institute of Navigation (ION GNSS+ 2013), Nashville, TN, USA, 16–20 September 2013; pp. 2343–2360.
5. Kitamura, M.; Aso, T.; Sakai, T.; Hoshinoo, K. Development of Prototype Dual-frequency Multi-constellation SBAS for MSAS. In Proceedings of the 30th International Technical Meeting of the Satellite Division of The Institute of Navigation (ION GNSS+ 2017), Portland, OR, USA, September 25–29 2017; pp. 997–1007.
6. Kitamura, M.; Sakai, T. DFMC SBAS Prototype System Performance Using Global Monitoring Stations of QZSS. In Proceedings of the ION 2019 Pacific PNT Meeting, Hawaii, HI, USA, 8–11 April 2019; pp. 382–387.
7. Thomas, D. EGNOS V2 Program Update. In Proceedings of the 26th International Technical Meeting of the Satellite Division of The Institute of Navigation (ION GNSS+ 2013), Nashville, TN, USA, 16–20 September 2013; pp. 2327–2342.
8. Fielitz, K.; Meindl, Q.; Breitenacher, A.; Daubrawa, J.; Frankenberger, H.; Braun, R.; Schmitz-Peiffer, A.; Ridings, A.; Marcote, M.; Auz, A.; et al. New Integrity SBAS Test Bed NISTB—DFMC Simulation Results for EGNOS V3 services. In Proceedings of the 31st International Technical Meeting of the Satellite Division of The Institute of Navigation (ION GNSS+ 2018), Miami, FL, USA, 24–28 September 2018; pp. 2104–2118.
9. Bauer, F.; Greze, G.; Haddad, F.; Tourtier, A.; Rols, B.; Urbanska, K. A Study on a New EGNOS V2 Release with Enhanced System Performances. In Proceedings of the 32nd International Technical Meeting of the Satellite Division of The Institute of Navigation (ION GNSS+ 2019), Miami, FL, USA, 16–20 September 2019; pp. 902–919.
10. Rao, K.N.S. GAGAN—The Indian satellite based augmentation system. *Indian J. Radio Space Phys.* **2007**, *36*, 293–302.
11. Tsai, Y.-F.; Low, K.-S. Performance assessment on expanding SBAS service areas of GAGAN and MSAS to Singapore region. In Proceedings of the2014 IEEE/ION Position, Location and Navigation Symposium-PLANS 2014, Monterey, CA, USA, 5–8 May 2014; pp. 686–691.
12. Karutin, S. SDCM grogram status. In Proceedings of the 25th International Technical Meeting of the Satellite Division of the Institute of Navigation (ION GNSS 2012), Nashville, TN, USA, 17–21 September 2012; pp. 1034–1044.
13. Karutin, S. SDCM development strategy. In Proceedings of the 23rd International Technical Meeting of The Satellite Division of the Institute of Navigation (ION GNSS 2010), Nashville, TN, USA, 16–20 September 2013; pp. 2361–2372.
14. Choy, S.; Kuckartz, J.; Dempster, A.G.; Rizos, C.; Higgins, M. GNSS satellite-based augmentation systems for Australia. *GPS Solutions* **2016**, *21*, 835–848. [CrossRef]
15. RTCA. *Minimum Operational Performance Standards for Global Positioning System/Satellite-Based Augmentation System Airborne Equipment*; RTCA: Washington, DC, USA, 2013.
16. IWG. *SBAS L5 DFMC Interface Control Document (SBAS L5 DFMC ICD)*; IWG: Brussels, Belgium, 2016.
17. ICAO. *Annex 10 to the Convention on International Civil Aviation*; ICAO: Montreal, QC, Canada, 2013; Volume 1.
18. China Satellite Navigation Office. *The Application Service Architecture of BeiDou Navigation Satellite System (Version 1.0)*; China Satellite Navigation Office: Beijing, China, 2019.
19. China Satellite Navigation Office. *BeiDou Navigation Satellite System Signal In Space Interface Control Document. Satellite Based Augmentation System Service Signal BDSBAS-B1C (Version 1.0)*; China Satellite Navigation Office: Beijing, China, 2020.
20. Liu, C.; Gao, W.; Shao, B.; Lu, J.; Wang, W.; Chen, Y.; Su, C.; Xiong, S.; Ding, Q. Development of BeiDou Satellite-Based Augmentation System. *Navigation* **2021**, *68*, 405–417.
21. Liu, Y.; Cao, Y.; Tang, C.; Chen, J.; Zhao, L.; Zhou, S.; Hu, X.; Tian, Q.; Yang, Y. Pseudorange Bias Analysis and Preliminary Service Performance Evaluation of BDSBAS. *Remote Sens.* **2021**, *13*, 4815. [CrossRef]
22. Gao, W.; Cao, Y.; Liu, C.; Lu, J.; Shao, B.; Xiong, S.; Su, C. Construction Progress and Aviation Flight Test of BDSBAS. *Remote Sens.* **2022**, *14*, 1218. [CrossRef]
23. Cao, Y.; Hu, X.; Zhou, J.; Wu, B.; Liu, L.; Zhou, S.; Su, R.; Chang, Z.; Wu, X. Kinematic Wide Area Differential Corrections for BeiDou Regional System Basing on Two-Way Time Synchronization. *Lect. Notes Electr. Eng.* **2014**, *305*, 277–288.
24. Liu, J.-L.; Cao, Y.-L.; Hu, X.-G.; Tang, C.-P. Beidou wide-area augmentation system clock error correction and performance verification. *Adv. Space Res.* **2020**, *65*, 2348–2359. [CrossRef]
25. Zhao, L.; Hu, X.; Tang, C.; Cao, Y.; Zhou, S.; Yang, Y.; Liu, L.; Guo, R. Generation of DFMC SBAS corrections for BDS-3 satellites and improved positioning performances. *Adv. Space Res.* **2020**, *66*, 702–714. [CrossRef]
26. Cao, Y.; Chen, J.; Hu, X.; He, F.; Bian, L.; Wang, W.; Wu, B.; Yu, Y.; Wang, J.; Tian, Q. Design of BDS-3 integrity monitoring and preliminary analysis of its performance. *Adv. Space Res.* **2019**, *65*, 1125–1138. [CrossRef]

Article

Impacts of Arc Length and ECOM Solar Radiation Pressure Models on BDS-3 Orbit Prediction

Ran Li [1,2,*], Chunmei Zhao [3], Jiatong Wu [4], Hongyang Ma [5], Yang Zhang [2], Guang Yang [2], Hong Yuan [2] and Hongyu Zhao [6]

1   State Key Laboratory of Geo-Information Engineering and Key Laboratory of Surveying and Mapping Science and Geospatial Information Technology of MNR, Chinese Academy of Surveying and Mapping, Beijing 100830, China
2   Aerospace Information Research Institute (AIR), Chinese Academy of Sciences (CAS), Beijing 100094, China
3   Institute of Geodesy and Geodynamics, Chinese Academy of Surveying and Mapping, Beijing 100830, China
4   Map Supervision Center, Ministry of Natural Resources, Beijing 100830, China
5   School of Geomatics Science and Technology, Nanjing Tech University, Nanjing 210037, China
6   Logistics Department, Space Systems Division, Beijing 100193, China
*   Correspondence: liran@aircas.ac.cn

Abstract: The BeiDou global navigation satellite system (BDS-3) has already provided worldwide navigation and positioning services for which the high-precision BDS-3-predicting orbit is the foundation. The arc length of the observed orbits and the solar radiation pressure (SRP) are two important factors for producing precise orbit predictions. The contribution studies the influences of these factors on BDS-3 orbit prediction. Three-month data from 1 July 2021 to 30 September 2021 are used to analyze optimal arc lengths and different ECOM SRP models for obtaining precise BDS-3 orbit predictions. The results show that the best-fitting arc length for the BDS-3 MEO/IGSO satellite is 42–48 h by comparing the final precise ephemeris and SLR validation. Furthermore, the ECOM9 SRP model shows improved orbit-prediction accuracy than that of the ECOM5 SRP model when the satellites move in and out of the eclipse season. As for the ECOM9 SRP model, the user range error (URE) accuracy of 6 h orbit predictions when satellites are in and outside of the eclipse season is 0.036 m and 0.030 m, respectively. In addition, the orbit prediction accuracy of the BDS-3 satellites does not decrease significantly since BDS-3 satellites apply the continuous yaw-steering (CYS) attitude mode during the eclipse season.

Keywords: BDS-3; precise orbit prediction; satellite laser ranging; eclipse season; ECOM solar radiation pressure model

Citation: Li, R.; Zhao, C.; Wu, J.; Ma, H.; Zhang, Y.; Yang, G.; Yuan, H.; Zhao, H. Impacts of Arc Length and ECOM Solar Radiation Pressure Models on BDS-3 Orbit Prediction. *Remote Sens.* **2022**, *14*, 3990. https://doi.org/10.3390/rs14163990

Academic Editor: Xiaogong Hu

Received: 18 July 2022
Accepted: 12 August 2022
Published: 16 August 2022

Publisher's Note: MDPI stays neutral with regard to jurisdictional claims in published maps and institutional affiliations.

1. Introduction

On 23 June 2020, with the successful launch of the last BeiDou global navigation satellite, the BeiDou global navigation satellite system (BDS-3) was fully completed and includes three geostationary earth orbit (GEO) satellites, three inclined geosynchronous orbit (IGSO) satellites, and twenty-four medium earth orbit (MEO) satellites [1]. BDS-3 can provide global services, such as positioning, velocity, and timing with accuracies of 10 m, 0.2 m/s, and 20 ns, respectively. Since the satellite orbit is the foundation for high-precision services and applications of BDS-3 [2], any global navigation satellite system (GNSS) error will directly affect the accuracy of navigation and positioning solutions, e.g., the emergence of real-time precise point-positioning real-time kinematic (PPP-RTK) technology [3], real-time atmospheric monitoring [4,5], early GNSS-based earthquake warning, and other technologies [6–9]. Therefore, precise real-time GNSS satellite orbits are urgently required for scientific and industrial communities. International GNSS Service (IGS) ultra-rapid (IGU) orbits are always implemented in BDS-3 real-time applications [10], and each IGU orbit product contains the observed orbits of the first 24 h, as well as the predicted orbits of the next 24 h. However, only certain parts of the predicted orbits are useful for real-time users since the IGU orbits are updated every 6 h with a latency of 3 h. Therefore, it is essential to study BDS-3-predicted orbits. The accuracy of the predicted orbits depends on the accuracy of the initial satellite conditions (satellite position and velocity at the initial epoch) and the accuracy of the solar radiation pressure (SRP) models [11].

The initial conditions of the satellite are affected by the fitted arc length of the observed orbit, and the satellite orbits are then predicted based on the initial conditions. Choi et al. (2013) studied Global Positioning System (GPS) ultra-rapid orbit-prediction strategies and found that the optimal arc length of the observed orbits is around 40–45 h [11]. Li et al. (2015) assessed the impact of the arc length on GPS precise point position (PPP) solutions [12], and the results showed that the highest PPP ambiguity fix rates can be achieved when using predicted orbits with an arc length of 42 h. Geng et al. (2018) analyzed the effect of the arc length of the observed orbits on multi-GNSS orbit prediction performances [13], and they found that the optimal arc length is 42–45 h. The optimal arc lengths for predicting orbits of GPS, GLONASS, BDS-2, and GALILEO satellites have been studied for many years; however, few existing manuscripts involve the same research effort for the BDS-3 satellites. How the accuracy of the BDS-3 predicted orbits varies with the arc length of the observed orbit needs to be uncovered. Therefore, it is necessary to study the optimal fitting arc length for BDS-3-predicted orbits.

As one of the major error sources of satellite orbit determination, solar radiation pressure model errors significantly affect the precision of precise satellite orbit determination and prediction [14,15]. Numerous studies have conducted research on the SRP model of GNSS satellites [16–18]. However, the research on the SRP model is limited to precise orbit determination, and few are related to precise orbit prediction. The accuracy of GPS satellites could decrease to 300 m in 3 days because of the existence of solar radiation pressure perturbations [19]. Therefore, investigating different solar radiation pressure models in orbit prediction is necessary. Although the ECOM SRP model is a widely used SRP model in precise GNSS orbit determination [18], the ECOM SRP model is designed for GPS satellites, and its applicability to BDS-3 satellite orbit prediction needs to be further verified. Therefore, this article will concentrate on the impact of the ECOM SRP model on BDS-3 satellite orbit prediction.

The BDS MEO satellites enter the eclipse season twice a year, each time lasting 8–15 days [20]. If this study does not consider the eclipse season and only investigates satellite orbit-prediction accuracies in the non-eclipse season, all orbit prediction starting times are outside the eclipse season. Hence, orbit prediction times are extended to 8–15 days, which would significantly reduce the prediction accuracy of the satellite orbit. Therefore, it is worthy to study the prediction accuracy of BDS-3 satellite orbits during the eclipse season. The China Satellite Navigation Office (CSNO) released BDS-3 satellite metadata in 2019, announcing that the BDS-3 MEO/IGSO satellites adopt continuous yaw steering (CYS) attitude modes during eclipse seasons [18]. The accuracy of current GNSS satellite orbits is significantly degraded during eclipse seasons, particularly for long-arc solutions and orbit predictions [21]. Duan et al. (2019) pointed out that the reason for such degradations is primarily due to the ignorance of thermally imbalanced forces [22]. Xia et al. (2022) also found that orbit degradations result from the unaccounted non-conservative forces, e.g., thermal radiation during Earth's shadow transitions [23]. The above studies are all based on the precise orbit determination of the BDS-3 satellites, and the study of the orbit prediction accuracy of the BDS-3 satellites during the eclipse season is essential as well.

This contribution mainly studies the impact of the arc length and ECOM SRP models on BDS-3 orbit prediction. The article is organized as follows. We first introduce the orbit-prediction method and the experimental data in Section 2. Afterward, the final precise ephemeris and SLR data were used to evaluate BDS-3's orbit prediction accuracy by using different arc lengths in Sections 3.1 and 3.2, and the orbit prediction accuracy of the BDS-3 satellites during and outside of eclipse seasons is provided in Section 3.3. Additionally, we analyze the impact of different ECOM SRP models on the predicted orbit accuracy of the BDS-3 satellites in Section 3.4. Finally, discussions and conclusions are provided in Sections 4 and 5.

2. Orbit Prediction Method and Data Collection

2.1. Orbit Prediction Method

Satellite orbit prediction accuracy relies on the satellite's initial conditions and solar radiation pressure parameters. Hence, in the following paragraph, we will indicate the solution of the satellite's initial conditions and solar radiation pressure parameters over an arc period using the least-squares method.

Let the position, velocity, and dynamic parameters of the satellite at the initial time t_0 be $(r_0, \dot{r}_0, p_0)$. By considering the influence of various perturbation forces, the position and velocity $(r, \dot{r})$ of the satellite at epoch t can be obtained by integrating the satellite dynamics equation. We can predict the satellite orbit at the epoch t as follows:

$$(r, \dot{r}) = F(t, r_0, \dot{r}_0, p_0) \tag{1}$$

where F is the nonlinear integral function. The observation equation can be described by Equation (2):

$$r_t - r = \Phi(t, t_0) \begin{bmatrix} \Delta r_0 & \Delta \dot{r}_0 & \Delta p_0 \end{bmatrix}^T \tag{2}$$

where $\Phi(t, t_0) = \begin{bmatrix} \frac{\partial r}{\partial r_0} & \frac{\partial r}{\partial \dot{r}_0} & \frac{\partial r}{\partial p_0} \end{bmatrix}$ is the state transition matrix; $\Delta r_0, \Delta \dot{r}_0, \Delta p_0$ denote the correction value of the initial position, velocity, and dynamic parameters, respectively.

Assuming that the satellite position at the epoch t_i in the precise ephemeris is r_i during an arc length $[t_i, t_m]$, the m-dimensional observation equation can then be constructed as follows:

$$Y = \theta(\Delta r_0, \Delta \dot{r}_0, \Delta p_0) \tag{3}$$

where the following is the case.

$$Y = \begin{bmatrix} r_{t_1} - r_1 \\ r_{t_2} - r_2 \\ \vdots \\ r_{t_m} - r_m \end{bmatrix}$$

$$\theta(\Delta r_0, \Delta \dot{r}_0, \Delta p_0) = \begin{bmatrix} \theta_1(\Delta r_0, \Delta \dot{r}_0, \Delta p_0) \\ \theta_2(\Delta r_0, \Delta \dot{r}_0, \Delta p_0) \\ \vdots \\ \theta_m(\Delta r_0, \Delta \dot{r}_0, \Delta p_0) \end{bmatrix} = \begin{bmatrix} \Phi(t_1, t_0) \begin{bmatrix} \Delta r_0 & \Delta \dot{r}_0 & \Delta p_0 \end{bmatrix}^T \\ \Phi(t_2, t_0) \begin{bmatrix} \Delta r_0 & \Delta \dot{r}_0 & \Delta p_0 \end{bmatrix}^T \\ \vdots \\ \Phi(t_m, t_0) \begin{bmatrix} \Delta r_0 & \Delta \dot{r}_0 & \Delta p_0 \end{bmatrix}^T \end{bmatrix}$$

The satellite's position, velocity, and dynamic parameters at the initial epoch t_0 can be calculated by applying least-squares estimations in Equation (3). Finally, the predicted orbits are generated through orbit integration using precise satellite initial conditions.

2.2. Data Collection and Dynamic Model

The final orbit products released by the Center for Orbit Determination in Europe (CODE) are used for BDS-3 orbit prediction research studies, and the data's length ranges from 1 July 2021 to 30 September 2021. The final precise ephemeris is not only used as observed orbits but also as reference orbits for evaluating the accuracy of BDS-3-predicted orbits. Table 1 summarizes the precise orbit-prediction strategies in the experiment, including the dynamical models and integration method. It is noteworthy that in order to reduce the influence of Earth orientation parameter (EOP) errors on the predicted orbit, this study uses the final EOP released by the IERS Earth orientation center. Moreover, the same empirical force parameters, integration methods, and integration step sizes are used in this experiment.

Table 1. Dynamical models and integration method.

Models	Reference/Source
Geopotential	EGM 2008, 12×12 degree [24]
N-body gravitation	JPL DE405 ephemeris [25]
EOP	Fixed to IERS EOP 14 C04 dataset
SRP model	ECOM5 model, ECOM7 model, ECOM9 model
Solid earth tides, pole tides	IERS conventions 2010 [26]
Ocean tides	FES2004 [27]
Relativity effect	IERS2010
Integration	Collocation Integration method
Integration step size	900 s

As can be seen on the website of the precise orbit-determination strategies adopted by the major analysis centers of IGS (see https://files.igs.org/pub/center/analysis/, accessed on 17 January 2022), most analysis centers use nine-parameter ECOM or its simplified version to determine precise orbits [28]. Therefore, ECOM and its simplified version were used in our BDS-3 orbit prediction strategy to assess its accuracy in orbit prediction. The ECOM SRP model adopts the form of constant components in addition to periodic terms. Three sets of parameters were used to absorb the influence of solar radiation pressure [29]. The ECOM SRP model formulas are as follows:

$$
\begin{aligned}
\vec{a} &= a_{srp,D} \cdot \vec{e_D} + a_{srp,Y} \cdot \vec{e_Y} + a_{srp,B} \cdot \vec{e_B} \\
a_{srp,D} &= D_0 + D_{Cu} \cos u + D_{Su} \sin u \\
a_{srp,Y} &= Y_0 + Y_{Cu} \cos u + Y_{Su} \sin u \\
a_{srp,B} &= B_0 + B_{Cu} \cos u + B_{Su} \sin u
\end{aligned}
\tag{4}
$$

where $\vec{a}$ represents the acceleration of SRP, $a_{srp,D}$, $a_{srp,Y}$, $a_{srp,B}$ are the SRP accelerations of the D-axis, Y-axis, and B-axis, respectively. u is the argument of the ascending node, and D_0, D_{Cu}, D_{Su}, Y_0, Y_{Cu}, Y_{Su}, B_0, B_{Cu}, and B_{Su} are constant parameters, which are estimated in the orbit's determination process. In the following, when these nine parameters are used for parameter estimations, we call this strategy the ECOM9 SRP model. Springer et al. (1999) introduced a simplified ECOM SRP model containing only 5 parameters (D_0, B_0, Y_0, B_{Cu}, and B_{Su}) [30], which is called the ECOM5 SRP model. The experimental results confirm that when the ECOM SRP model with 7 parameters (D_0, D_{Cu}, D_{Su}, Y_0, B_0, B_{Cu}, and B_{Su}) is used, the orbit's overlap accuracy with resepct to BDS-2's precise orbit determination can be greatly improved [31], which is called the ECOM7 SRP model in the following.

3. Results and Analysis

The accuracies of BDS-3-predicted orbits with different arc lengths are compared to the IGS final precise orbits and SLR data. Three-month data from 1 July 2021 to 30 September 2021 are used in this experiment. The BDS-3 satellite orbit prediction accuracy during and outside of eclipse seasons is provided. Finally, the impacts of different ECOM SRP models on the predicted orbit accuracy of the BDS-3 satellite are discussed. Because of the poor orbital accuracy of the BDS-3 GEO satellites, the BDS-3 GEO satellites will not be considered in this experiment. It should be noted that in Sections 3.1–3.3, we use the ECOM9 SRP model for orbit prediction.

3.1. Impacts of the Arc Length on Orbit Prediction

The user range error (URE) is used to verify the accuracy of the predicted orbit. The satellite URE is associated with the maximum opening angle of the satellite relative to Earth [15]. Without considering the satellite clock error, the MEO satellites' URE is computed as follows [21]:

$$
URE_{MEO} = \sqrt{(0.99\Delta R)^2 + (0.14\Delta T)^2 + (0.14\Delta N)^2}
\tag{5}
$$

where ΔR, ΔT, ΔN are the radial, along-track, and cross-track components of the satellite orbit, respectively. For IGSO and GEO satellites, URE is computed as follows.

$$\mathrm{URE}_{\mathrm{GEO/IGSO}} = \sqrt{(\Delta R)^2 + (0.99\Delta T)^2 + (0.99\Delta N)^2} \tag{6}$$

It can be seen from the Formulas (5) and (6) that the most significant contribution to URE is the radial component of the satellite orbit.

The orbit prediction accuracy of BDS-3 MEO satellites is first calculated in the experiment. The 24–72 h satellite-orbit arc lengths are used as initial observation data. IGS's final precise ephemeris is used as the reference. We use the root mean square (RMS) in the radial (R), along-track (T) and cross-track (N) directions and URE to evaluate the accuracy of BDS-3-predicted orbits. The accuracy of the first 6 h and 24 h of predicted orbits is used to evaluate the impact of the arc length on BDS-3 MEO orbit prediction.

As shown in Figure 1, the smallest orbital URE is in the fitted arc length of 42–48 h for both 6 h and 24 h orbit prediction. It can be observed that the accuracy of the radial, along-track, and cross-track are similar to that of the URE, and the best fitting arc length is 42–48 h. The results also show that when the prediction intervals are 6 h, the URE is the smallest at the arc length of 42 h, and the URE, radial, cross-track, and along-track are 0.029 m, 0.019 m, 0.022 m, and 0.054 m, respectively. When the prediction intervals are 24 h, the URE is also the smallest at the arc length of 42 h, and the URE, radial, cross-track, and along-track are 0.051 m, 0.027 m, 0.034 m, and 0.111 m, respectively.

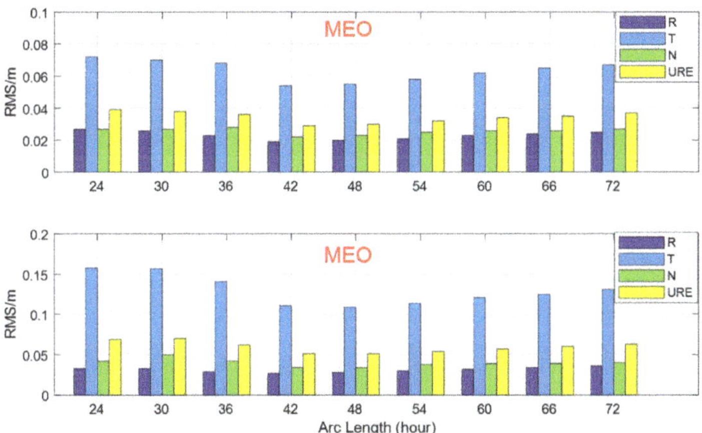

Figure 1. The accuracy of the first 6 h (**top**) and 24 h (**bottom**) of predicted BDS-3 MEO orbits with different arc lengths.

Meanwhile, the orbit prediction accuracy of BDS-3 IGSO satellites is analyzed. BDS-3 has three IGSO satellites (C38, C39, and C40). In this contribution, these BDS-3 IGSO satellites are taken as an example to explore the best-fitting arc length.

As is shown in Figure 2, the results of BDS-3 IGSO are consistent with those of BDS-3 MEO satellites. The smallest orbital URE is in the fitted arc length of 42–48 h for both 6 h or 24 h length orbit prediction. The results show that when the prediction intervals are 6 h, the URE is the smallest at the arc length of 42 h, and the URE, radial, cross-track, and along-track are 0.193 m, 0.125 m, 0.080 m, and 0.125 m, respectively. When the prediction intervals are 24 h, the URE is also the smallest at the arc length of 42 h, and the URE, radial, cross-track, and along-track are 0.231 m, 0.108 m, 0.080 m, and 0.189 m, respectively. It is noteworthy that, compared to BDS-3 MEO satellites, the accuracy of BDS-3 IGSO satellite orbit prediction is significantly reduced when the fitted arc length is less than 36 h. The reason may be that the BDS-3 IGSO satellite has a longer orbital period than the MEO satellite. Therefore, when the fitted arc length is less than 36 h, the accuracy of BDS-3 IGSO satellite orbit predictions will diminish compared to that of the BDS-3 MEO satellites.

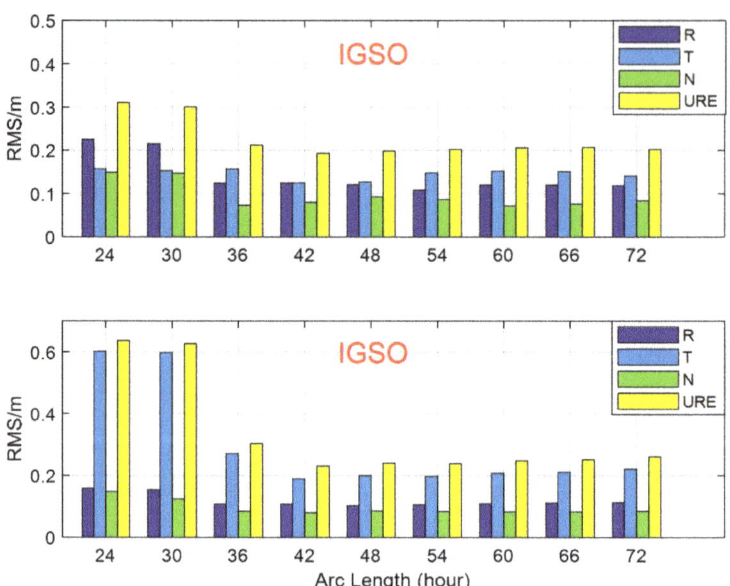

Figure 2. The accuracy of the first 6 h (**top**) and 24 h (**bottom**) of predicted BDS-3 IGSO orbits with different arc lengths.

3.2. Satellite Laser Ranging Validation

Satellite laser ranging is a commonly used method for evaluating orbit accuracy, and its accuracy can reach the millimeter level [32]. Currently, BDS-3 has four MEO satellites (C20, C21, C29, and C30) that are observed by SLR stations coordinated by the International Laser Ranging Service (ILRS). This paper uses the SLR data to verify the BDS-3 satellite's best-fit arc length. The experiment uses three-month SLR data from 1 July 2021 to 30 September 2021. Moreover, we used 24 h predicted orbits for conducting evaluations.

As shown in Figure 3, the best-fitting arc length is 42–48 h for C20, C21, and C30 satellites. The conclusion is consistent with the result of the final IGS precise ephemeris. However, the SLR residual of the C29 satellite is the smallest when the fitting arc is 60 h. It should be noted that the orbit prediction accuracy of the C29 satellite under the fitting arc length of 42–72 h is almost the same as 60 h, with differences within 3 mm.

As is shown in Figure 4, most SLR residuals for BDS-3 satellites are within ±20 cm. There are 361, 287, 238, and 245 normal points for C20, C21, C29, and C30 satellites, respectively. In this paper, the absolute values larger than 50 cm have been removed. After gross error elimination, there are 354, 281, 232, and 227 normal points left; thus, in this experiment, we used more than 96% of the original data for analysis. The accuracy of the predicted orbit is evaluated by SLR, and the RMS values of the SLR residuals for C20, C21, C29, and C30 are 5.36, 5.51, 5.49, and 6.01 cm, respectively. The overall RMS value of SLR residuals for the BDS-3 MEO satellites is 5.59 cm, which shows that the orbital accuracy of the BDS-3 MEO satellites is still maintained at a high level after 24 h of satellite orbit predictions.

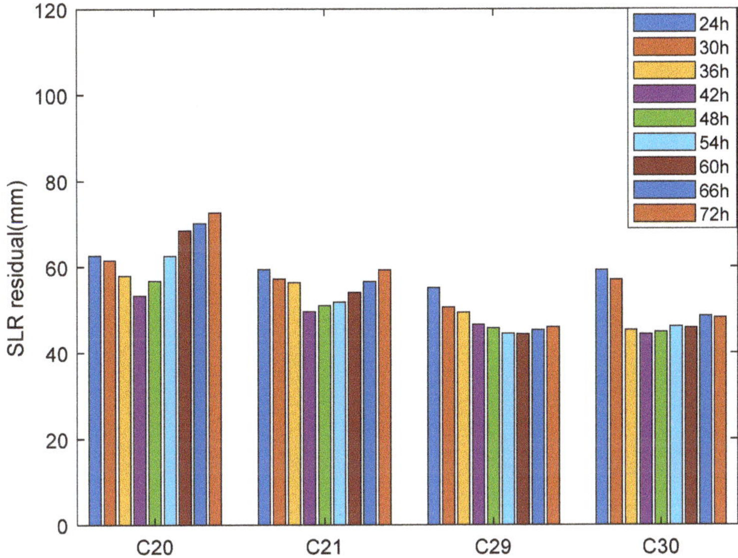

Figure 3. SLR residuals for BDS-3 MEO satellites with different arc lengths.

When the fitting arc length is 42 h, the laser-ranging validation residuals for those four satellites are shown in Figure 4.

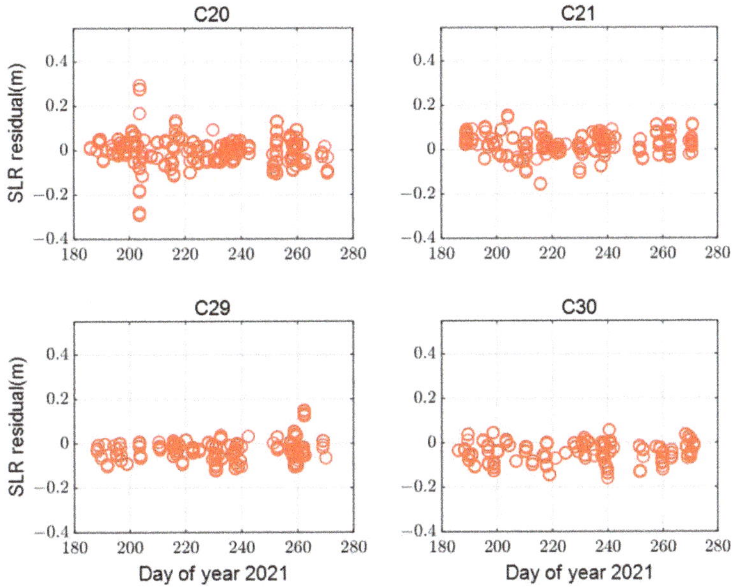

Figure 4. SLR residuals for BDS-3 satellites with prediction interval of 24 h (red circles represent SLR residuals).

3.3. Orbit Prediction Accuracy When Satellites Are during and Outside of Eclipse Season

For the BDS-2 satellite, studies have shown that its orbit determination accuracy will decrease significantly during the eclipse period [20], but there are few studies analyzing the orbit prediction accuracy of the BDS-3 satellite during the eclipse season. Therefore, in the following section, this study explores the orbit prediction accuracy of the BDS-3 satellite during the eclipse season. The BDS-3 IGSO and MEO satellites enter the eclipse season when the $|\beta|$ angle (the elevation angle of the sun above the orbital plane) is less than 8.7° and 12.97°, respectively [33]. Figure 5 shows the eclipse seasons for the BDS-3 satellites. The period of the BDS-3 satellites during the eclipse season is from 1 July 2021 to 30 September 2021.

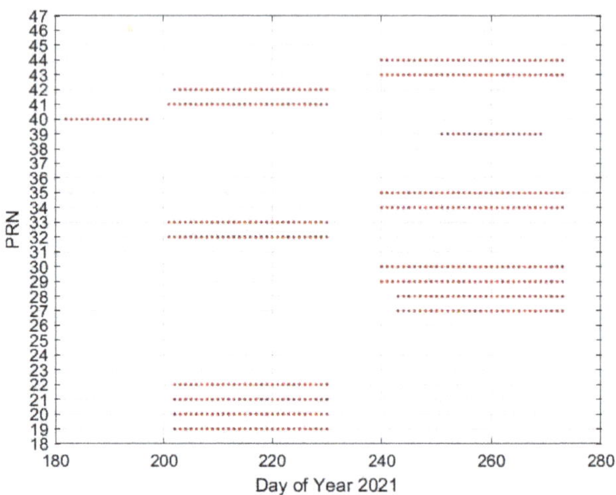

Figure 5. Eclipse seasons for the BDS-3 satellites (1 July 2021 to 30 September 2021, red dots represent eclipse seasons).

As shown in Figure 5, 18 BDS-3 satellites experienced the eclipse seasons during the selected three months. Day of year (DOY) 202 to 230, 2021, was selected, during which eight satellites (C19, C20, C21, C22, C32, C33, C41, and C42) were within the eclipse season. Meanwhile, DOY 243 to 273, 2021 was selected, during which the same eight satellites were outside if the eclipse season. Taking the selected eight BDS-3 satellites as an example, we analyzed the orbit prediction accuracy of BDS-3 satellites using different arc lengths when satellites move in and out of eclipses seasons. It should be pointed out that the selected eight BDS-3 satellites are all MEO satellites. Figure 6 shows the orbit prediction accuracy of different arc lengths when the satellites move in and out of eclipses seasons.

It can be seen from Figure 6 that the arcs of 42–48 h length are the best fitting arc lengths for both satellites moving out of eclipse seasons and those into eclipse seasons. Taking the 48 h fitting arc length as an example, when the prediction intervals are 6 h, the URE increases from 0.0344 m out of eclipse seasons to 0.0419 m during eclipse seasons, which corresponds to an increase of 22%; when the prediction intervals are 24 h, the URE increases from 0.0592 m out of eclipse seasons to 0.0772 m in eclipse seasons, which corresponds to increases of 30%. Compared with the significant decrease in the orbital accuracy of the BDS-2 satellite moving in eclipse seasons, the BDS-3 satellites do not experience a significant decrease in their orbit prediction accuracy. The reason may be due to the advanced attitude control mode of the BDS-3 satellites. In addition to the yaw-steering (YS) mode, the BDS-3 MEO/IGSO satellites adopt the continuous yaw steering (CYS) attitude mode in eclipse seasons [20]. The CYS mode not only meets the requirements of the power supply and thermal control of the BDS-3 MEO/IGSO satellites but also avoids rapid yaw slews near the sun–spacecraft–Earth geometries. The results indicate that the proper attitude mode improved BDS-3's orbit prediction accuracy.

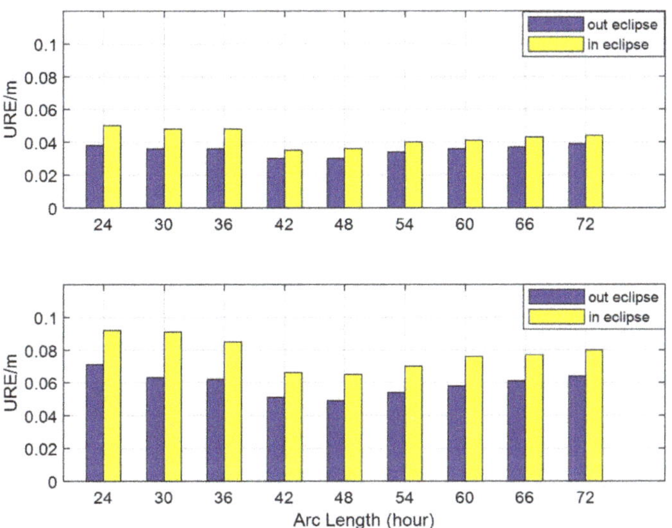

Figure 6. The URE of BDS-3 predicted orbits in and out of eclipse season with prediction intervals of 6 h (**top**) and 24 h (**bottom**).

3.4. Analysis of Different ECOM SRP Models for Orbit Prediction

BDS-3 satellites adopt a different attitude control mode from the BDS-2 satellite. For the BDS-3 MEO/IGSO satellites, the CYS mode is adopted during the eclipse seasons, and the YS mode is adopted out of eclipse seasons.

The BeiDou satellite navigation system has three satellite attitude-control modes: YS mode, orbit-normal (ON) mode, and CYS mode.

The BDS-2 satellite adopts the YS and ON modes. For the YS mode, the satellite uses the solar sensor and the Earth sensor to detect the position of the sun and the Earth; thus, this mode satisfies that requirement in which the satellite antenna points to the center of the earth and the solar panel is always perpendicular to the direction of the sun's illumination. After the BDS-2 IGSO/MEO satellites move into the eclipse season and the β angle (defined by the angle of the sun above the orbit plane) is between $-4°$ and $+4°$, the attitude control mode of the BDS-2 IGSO/MEO satellites changes from the YS mode to the ON mode [34]. As for the ON mode, the yaw angle φ (defined by the angle between the instantaneous velocity and the body-fixed x-axis) is always zero. However, studies have shown that when BDS-2 IGSO/MEO satellites move into the eclipse season, the satellites adopting the ON mode will cause a significant decrease in the accuracy of the satellite's orbit determination [35]. Therefore, the CYS mode is adopted in the BDS-3 MEO/IGSO satellites.

When the BDS-3 MEO/IGSO satellites move in the eclipse season, the satellite adopts the YS and CYS modes. During most of the eclipse season, the BDS-3 MEO/IGSO satellites use the YS mode. The CYS mode includes two periods of midnight-turn maneuver and noon-turn maneuver. Midnight-turn maneuver and noon-turn maneuver are activated when the β angle is between $-3°$ and $+3°$, and the sun's azimuth angle $|\alpha|$ (defined by the angle between Earth-satellite vector and Earth-"noon" vector on the orbit plane) is $\leq 10°$ or $|\alpha \pm 180°| \leq 10°$ [20]. As for the CYS mode, the solar sensor can no longer control the yaw's attitude, and the BDS-3 MEO/IGSO satellites start to yaw with estimated hardware yaw rates in order to make the yaw angle transform $180°$. Yaw attitude departs from the nominal values and it can take up to 30 min to 1 h to correct [36]. Compared with the YS mode, the CYS mode adopts the method of starting the maneuver in advance and ending the maneuver with a delay; thereby, the CYS mode avoids rapid yaw slews near collinear sun–spacecraft–Earth geometries.

The satellite attitude control mode will affect the solar radiation pressure related to the irradiated surface, which will further affect the accuracy of the satellite's precise orbit predictions [37]. Therefore, it is necessary to study the BDS-3 satellite's orbit prediction accuracy for different ECOM SRP models in and out of the eclipse seasons.

Similarly to Section 3.3, eight BDS-3 satellites (C19, C20, C21, C22, C32, C33, C41, and C42) were selected to study the orbital prediction accuracy of the ECOM5, ECOM7, and ECOM9 SRP models when moving in and out of the eclipse seasons.

As shown in Figure 7, the orbit prediction accuracy of the ECOM5, ECOM7, and ECOM9 SRP models is similar in and out of eclipse seasons. The orbital UREs using the ECOM5, ECOM7, and ECOM9 SRP models are 0.035 m, 0.031 m, and 0.030 m outside eclipse seasons, respectively. However, the orbital UREs using the ECOM5, ECOM7, and ECOM9 SRP models are 0.048 m, 0.037 m, and 0.036 m during the eclipse seasons, respectively. The results show that the orbital URE of the ECOM9 model is 19% lower than that of using the ECOM5 SRP model. In addition, the ECOM7 and ECOM9 SRP models show similar orbit prediction accuracy, and both are better than the ECOM5 SRP model. The reason may be that the added parameters absorb more orbital errors.

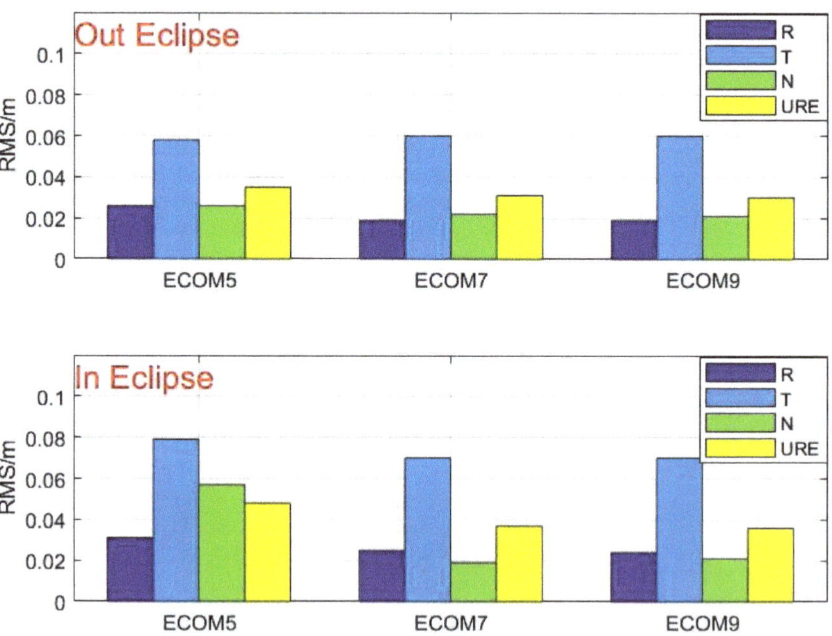

Figure 7. Accuracy of BDS-3 predicted orbits with different ECOM SRP models when satellites move in and out of eclipse seasons (6 h prediction interval).

In addition, we study the orbit prediction accuracy for different ECOM SRP models with different arc lengths. Figure 8 shows the orbit prediction accuracy for different ECOM SRP models out of the eclipse seasons with the prediction interval of 24 h.

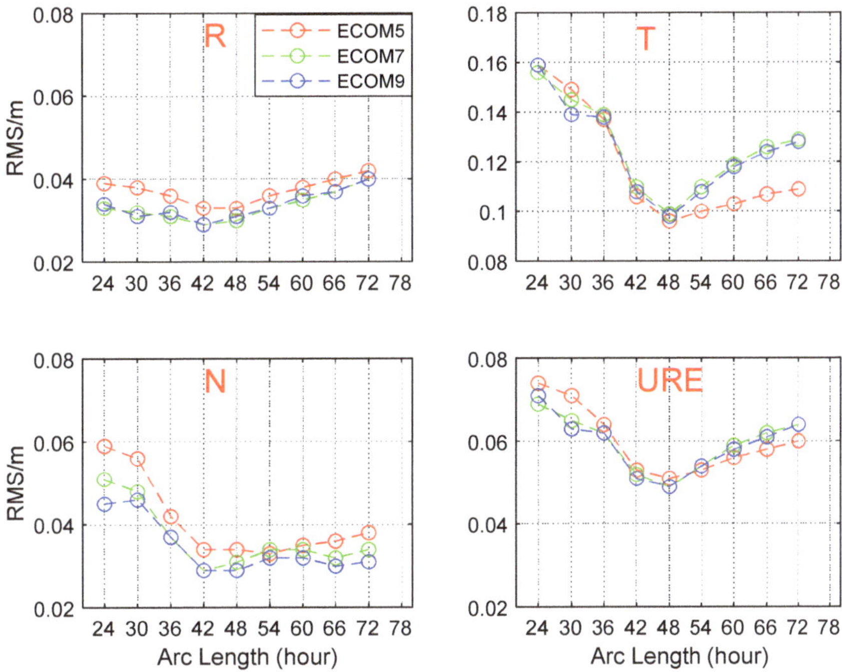

Figure 8. Average RMS and URE of BDS-3 predicted orbits with different ECOM SRP models out of eclipse seasons (24 h prediction interval).

As shown in Figure 8, the optimal arc lengths using the ECOM5, ECOM7, and ECOM9 SRP models are 42–48 h. The difference in orbit prediction accuracy between ECOM7 and ECOM9 SRP models is within 2 mm, and the performances of both ECOM7 and ECOM9 SRP models surpass the ECOM5 SRP model. Moreover, when the fitted arc length is less than 48 h, the orbital accuracy of the ECOM7 and ECOM9 SRP models is better than that of the ECOM5 SRP model. However, when the fitted arc length is longer than 48 h, the ECOM5 SRP model exhibits improved orbit prediction accuracy. The reason is that the ECOM5 SRP model shows improved orbit prediction accuracy in the along-track direction.

Figure 9 shows the orbit prediction accuracy of different ECOM SRP models during the eclipse seasons with the prediction interval of 24 h. The optimal arc lengths using the ECOM5, ECOM7, and ECOM9 SRP models are 42–48 h. The ECOM7 and ECOM9 models have good performances with respect to orbit prediction during the eclipse seasons, for which the orbital UREs are both 0.065 m. However, the orbit prediction accuracy of the ECOM5 SRP model during the eclipse seasons is unsatisfactory, particularly in the cross-track direction. When the arc length is longer than 36 h, the orbit prediction accuracy in the cross-track direction becomes worse, and the cause needs to be further studied. Therefore, using the ECOM9 or ECOM7 SRP model is recommended when performing precise orbit predictions for the BDS-3 MEO satellites during eclipse seasons.

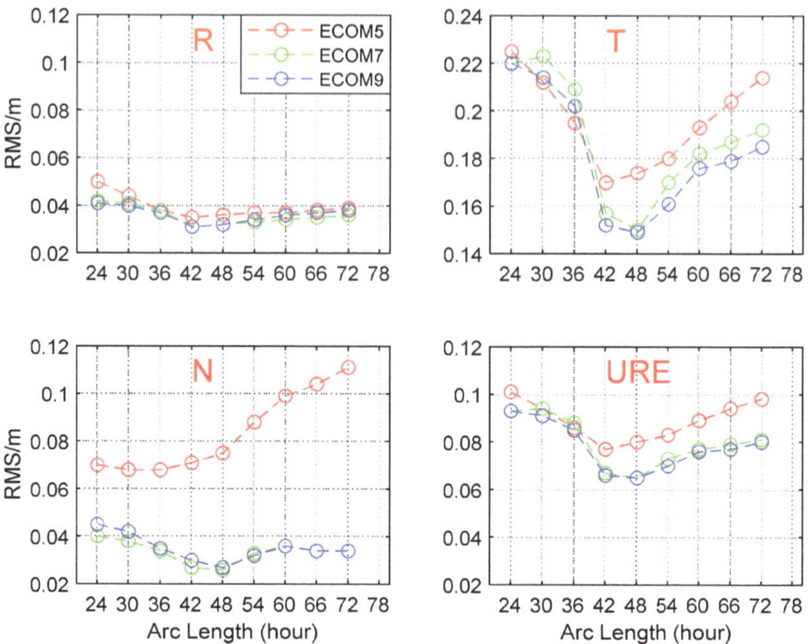

Figure 9. Average RMS and URE of BDS-3 predicted orbits with different ECOM SRP models in eclipse seasons (24 h prediction interval).

4. Discussion

This contribution studies two critical factors (arc length and SRP model) affecting the BDS-3 satellite's orbit-prediction accuracy. We use precise ephemerides and SLR data to verify the BDS-3 satellite's best-fit arc length. Our results show that the best-fitting arc length for the BDS-3 satellites is at 42–48 h. In addition, the BDS-3 satellite's orbit prediction accuracy does not decrease significantly during the eclipse seasons, which should be attributed to the advanced attitude control mode for the BDS-3 satellite. We also find out that the BDS-3 satellite's orbit prediction accuracy of the ECOM5 SRP model during the eclipse seasons is unsatisfactory, especially in the cross-track direction. Therefore, when the BDS-3 satellites are within the eclipse seasons, we recommend using the ECOM9 or ECOM7 SRP model for orbit prediction.

Since the Earth's orientation parameters also have a significant influence on the orbit-prediction accuracy, we will study the use of real-time earth orientation parameters to analyze orbit-prediction accuracies in the future.

5. Conclusions

Three-month data from 1 July 2021 to 30 September 2021 were used to analyze orbit-prediction accuracies under different conditions, including precise ephemerides and SLR data, and to study the effect of arc lengths on orbit-prediction accuracy. We also explore the orbit-prediction accuracy of different ECOM SRP models during and outside of the eclipse seasons. Based on the above results, we can reach the following conclusions.

The final precise ephemerides and SLR validations show that the best-fitting arc length for the BDS-3 satellites is 42–48 h. In contrast to the BDS-3 MEO satellites, the BDS-3 IGSO satellites have diminished orbit-prediction accuracies when the fitting arc length is shorter than 36 h. The result shows that the orbital accuracy of the BDS-3 MEO satellites maintains a high level after 24 h of satellite orbit predictions. The UREs of BDS-3 MEO- and IGSO-predicted orbits are 0.051 m and 0.231 m, respectively. SLR validation shows that the BDS-3 satellites have the best orbit prediction accuracy for the arc length of 42 h; thus, the average SLR residual is approximately 5.59 cm with a prediction interval of 24 h.

Regardless of whether the satellites are within or outside of the eclipse seasons, the best fitting arc length is 42–48 h. Compared with the BDS-2 satellite's attitude control mode, the BDS-3 satellites adopt the CYS mode, which does not significantly reduce orbit-prediction accuracies during the eclipse seasons. When the arc length is 42 h and the orbit is predicted for 24 h, the UREs of the BDS-3 MEO satellite are 0.077 m and 0.059 m during and outside of the eclipse seasons, respectively.

The experimental results show that the ECOM9 SRP model has the best orbit prediction accuracy during and outside of eclipse seasons. The UREs of the 6 h predicted orbit are 0.036 m and 0.030 m, respectively. Compared with ECOM7 and ECOM9 SRP models, the ECOM5 SRP model has worse orbit prediction accuracies during the eclipse seasons.

Author Contributions: Methodology, R.L. and C.Z.; software, H.M. and J.W.; validation, Y.Z. and G.Y.; investigation, J.W.; writing—original draft preparation, R.L.; writing—review and editing, J.W. and H.Y.; visualization, H.M.; supervision, J.W., Y.Z., G.Y. and H.Z. All authors have read and agreed to the published version of the manuscript.

Funding: This work was supported by State Key Laboratory of Geo-Information Engineering and Key Laboratory of Surveying and Mapping Science and Geospatial Information Technology of MNR, CASM (No. 2021-01-07, 2022-01-09), the National Natural Science Foundation of China (No. 42122026, 42174033, 42074043, 42174038), Youth Innovation Promotion Association of Chinese Academy of Sciences (2022126), and Major Special Project of China's Second Generation Satellite Navigation System (JZX2B202012GG0110).

Data Availability Statement: Precise satellite orbit products can be found in ftp://cddis.gsfc.nasa. gov/pub/gps/products/. SLR observation data can be found in ftp://cddis.gsfc.nasa.gov/pub/slr/ data/npt_crd/, all accessed on 17 January 2022.

Acknowledgments: The authors acknowledge the IGS MGEX and ILRS for providing the multi-GNSS and SLR tracking data.

References

1. Yang, Y.; Mao, Y.; Sun, B. Basic performance and future developments of BeiDou global navigation satellite system. *Satell. Navig.* **2020**, *1*, 1–8. [CrossRef]
2. Li, R.; Wang, N.; Li, Z.; Zhang, Y.; Wang, Z.; Ma, H. Precise orbit determination of BDS-3 satellites using B1C and B2a dual-frequency measurements. *GPS Solut.* **2021**, *25*, 1–14. [CrossRef]
3. Ma, H.; Zhao, Q.; Verhagen, S.; Psychas, D.; Liu, X. Assessing the performance of multi-GNSS PPP-RTK in the local area. *Remote Sens.* **2020**, *12*, 3343. [CrossRef]
4. Li, Z.; Wang, N.; Hernández-Pajares, M.; Yuan, Y.; Krankowski, A.; Liu, A.; Zha, J.; García-Rigo, A.; Roma-Dollase, D.; Yang, H. IGS real-time service for global ionospheric total electron content modeling. *J. Geod.* **2020**, *94*, 1–16. [CrossRef]
5. Ma, H.; Verhagen, S. Precise point positioning on the reliable detection of tropospheric model errors. *Sensors* **2020**, *20*, 1634. [CrossRef] [PubMed]
6. Li, K.; Zhou, X.; Guo, N.; Zhou, S. Effect of PCV and attitude on the precise orbit determination of Jason-3 satellite. *J. Appl. Geod.* **2022**, *16*, 143–150. [CrossRef]
7. Li, R.; Li, Z.; Wang, N.; Tang, C.; Ma, H.; Zhang, Y.; Wang, Z.; Wu, J. Considering inter-receiver pseudorange biases for BDS-2 precise orbit determination. *Measurement* **2021**, *177*, 109251. [CrossRef]
8. Wang, Z.; Li, Z.; Wang, L.; Wang, N.; Yang, Y.; Li, R.; Zhang, Y.; Liu, A.; Yuan, H.; Hoque, M. Comparison of the real-time precise orbit determination for leo between kinematic and reduced-dynamic modes. *Measurement* **2022**, *187*, 110224. [CrossRef]
9. Zhang, Y.; Li, Z.; Li, R.; Wang, Z.; Yuan, H.; Song, J. Orbital design of LEO navigation constellations and assessment of their augmentation to BDS. *Adv. Space Res.* **2020**, *66*, 1911–1923. [CrossRef]
10. Chen, Q.; Song, S.; Zhou, W. Accuracy Analysis of GNSS Hourly Ultra-Rapid Orbit and Clock Products from SHAO AC of iGMAS. *Remote Sens.* **2021**, *13*, 1022. [CrossRef]
11. Choi, K.K.; Ray, J.; Griffiths, J.; Bae, T.-S. Evaluation of GPS orbit prediction strategies for the IGS Ultra-rapid products. *GPS Solut.* **2013**, *17*, 403–412. [CrossRef]
12. Li, Y.; Gao, Y.; Li, B. An impact analysis of arc length on orbit prediction and clock estimation for PPP ambiguity resolution. *GPS Solut.* **2015**, *19*, 201–213. [CrossRef]

13. Geng, T.; Zhang, P.; Wang, W.; Xie, X. Comparison of ultra-rapid orbit prediction strategies for GPS, GLONASS, Galileo and BeiDou. *Sensors* **2018**, *18*, 477. [CrossRef] [PubMed]
14. Tang, C.; Hu, X.; Zhou, S.; Guo, R.; He, F.; Liu, L.; Zhu, L.; Li, X.; Wu, S.; Zhao, G. Improvement of orbit determination accuracy for Beidou navigation satellite system with two-way satellite time frequency transfer. *Adv. Space Res.* **2016**, *58*, 1390–1400. [CrossRef]
15. Zhou, S.; Hu, X.; Zhou, J.; Chen, J.; Gong, X.; Tang, C.; Wu, B.; Liu, L.; Guo, R.; He, F. Accuracy analyses of precise orbit determination and timing for COMPASS/Beidou-2 4GEO/5IGSO/4MEO constellation. In *China Satellite Navigation Conference (CSNC) 2013 Proceedings*; Springer: Berlin/Heidelberg, Germany, 2013.
16. Yan, X.; Liu, C.; Huang, G.; Zhang, Q.; Wang, L.; Qin, Z.; Xie, S. A Priori Solar Radiation Pressure Model for BeiDou-3 MEO Satellites. *Remote Sens.* **2019**, *11*, 1605. [CrossRef]
17. Wang, C.; Guo, J.; Zhao, Q.; Ge, M. Improving the Orbits of the BDS-2 IGSO and MEO Satellites with Compensating Thermal Radiation Pressure Parameters. *Remote Sens.* **2022**, *14*, 641. [CrossRef]
18. Zhao, Q.; Guo, J.; Wang, C.; Lyu, Y.; Xu, X.; Yang, C.; Li, J. Precise orbit determination for BDS satellites. *Satell. Navig.* **2022**, *3*, 1–24. [CrossRef]
19. Kaplan, E.D.; Hegarty, C. *Understanding GPS/GNSS: Principles and Applications*; Artech House: Norwood, MA, USA, 2017.
20. Li, X.; Hu, X.; Guo, R.; Tang, C.; Zhou, S.; Liu, S.; Chen, J. Orbit and positioning accuracy for new generation BeiDou satellites during the Earth eclipsing period. *J. Navig.* **2018**, *71*, 1069–1087. [CrossRef]
21. Rodriguez-Solano, C.; Hugentobler, U.; Steigenberger, P.; Allende-Alba, G. Improving the orbits of GPS block IIA satellites during eclipse seasons. *Adv. Space Res.* **2013**, *52*, 1511–1529. [CrossRef]
22. Duan, B.; Hugentobler, U.; Chen, J.; Selmke, I.; Wang, J. Prediction versus real-time orbit determination for GNSS satellites. *GPS Solut.* **2019**, *23*, 1–10. [CrossRef]
23. Xia, F.; Ye, S.; Chen, D.; Tang, L.; Wang, C.; Ge, M.; Neitzel, F. Advancing the Solar Radiation Pressure Model for BeiDou-3 IGSO Satellites. *Remote Sens.* **2022**, *14*, 1460. [CrossRef]
24. Pavlis, N.K.; Holmes, S.A.; Kenyon, S.C.; Factor, J.K. The development and evaluation of the Earth Gravitational Model 2008 (EGM2008). *J. Geophys. Res. Solid Earth* **2012**, *117*, B4. [CrossRef]
25. Standish, E. JPL Planetary and Lunar Ephemerides, DE405/LE405. JPL IOM 312. F-98_048. 1998; pp. 42–196. Available online: https://ssd.jpl.nasa.gov/ftp/eph/planets/bsp/ (accessed on 17 January 2022).
26. Petit, G.; Luzum, B. IERS Conventions. 2010, Bureau International des Poids et Mesures Sevres (France). Available online: https://www.iers.org/IERS/EN/Publications/TechnicalNotes/tn36.html (accessed on 17 January 2022).
27. Lyard, F.; Lefevre, F.; Letellier, T.; Francis, O. Modelling the global ocean tides: Modern insights from FES2004. *Ocean. Dyn.* **2006**, *56*, 394–415. [CrossRef]
28. Guo, F.; Li, X.; Zhang, X.; Wang, J. Assessment of precise orbit and clock products for Galileo, BeiDou, and QZSS from IGS Multi-GNSS Experiment (MGEX). *GPS Solut.* **2017**, *21*, 279–290. [CrossRef]
29. Beutler, G.; Brockmann, E.; Gurtner, W.; Hugentobler, U.; Mervart, L.; Rothacher, M.; Verdun, A. Extended orbit modeling techniques at the CODE processing center of the international GPS service for geodynamics (IGS): Theory and initial results. *Manuscr. Geod.* **1994**, *19*, 367–386.
30. Springer, T.; Beutler, G.; Rothacher, M. A new solar radiation pressure model for GPS satellites. *GPS Solut.* **1999**, *2*, 50–62. [CrossRef]
31. Yue, M.; Song, X.; Jia, X.; Ruan, R. Analysis about parameters selection strategy of ECOM solar radiation pressure model for BeiDou satellites. *Acta Geod. Cartogr. Sin.* **2017**, *46*, 1812.
32. Montenbruck, O.; Steigenberger, P.; Kirchner, G. GNSS satellite orbit validation using satellite laser ranging. In Proceedings of the ILRS Workshop Proceedings, Fujiyoshida, Japan, 11–15 November 2013.
33. Wang, C.; Guo, J.; Zhao, Q.; Liu, J. Yaw attitude modeling for BeiDou I06 and BeiDou-3 satellites. *GPS Solut.* **2018**, *22*, 1–10. [CrossRef]
34. Zhao, Q.; Wang, C.; Guo, J.; Wang, B.; Liu, J. Precise orbit and clock determination for BeiDou-3 experimental satellites with yaw attitude analysis. *GPS Solut.* **2018**, *22*, 1–13. [CrossRef]
35. Wang, W.; Chen, G.; Guo, S.; Song, X.; Zhao, Q. A Study on the Beidou IGSO/MEO Satellite Orbit Determination and Prediction of the Different Yaw Control Mode. In *China Satellite Navigation Conference (CSNC) 2013 Proceedings: Volume III. Lecture Notes in Electrical Engineering*; Springer: Berlin/Heidelberg, Germany, 2013; pp. 31–40.
36. Li, X.; Guo, R.; Wu, S.; Chang, Z.; Liu, S.; Chen, J. Orbit Determining Strategy Analysis for BeiDou Satellite in Different Attitude Control Modes. *Geomat. Inf. Sci. Wuhan Univ.* **2019**, *44*, 1465–1471.
37. Dai, X.; Ge, M.; Lou, Y.; Shi, C.; Wickert, J.; Schuh, H. Estimating the yaw-attitude of BDS IGSO and MEO satellites. *J. Geod.* **2015**, *89*, 1005–1018. [CrossRef]

Article

Analysis and Performance Evaluation of BDS-3 Code Ranging Accuracy Based on Raw IF Data from a Zero-Baseline Experiment

Yu Liu [1], Fan Gao [1,2,*], Junxiang Li [1], Yunqiao He [1], Baojiao Ning [1], Yang Liu [1], Sijia Chen [1] and Yanqing Qiu [1]

[1] School of Space Science and Physics, Shandong University, Weihai 264209, China; 201900810123@mail.sdu.edu.cn (Y.L.); 201900810112@mail.sdu.edu.cn (J.L.); heyunqiao@mail.sdu.edu.cn (Y.H.); 202017724@mail.sdu.edu.cn (B.N.); 202000810242@mail.sdu.edu.cn (Y.L.); 201900800335@mail.sdu.edu.cn (S.C.); 201900830050@mail.sdu.edu.cn (Y.Q.)

[2] Institute of Space Science, Shandong University, Weihai 264209, China

* Correspondence: gaofan@sdu.edu.cn

Abstract: China's BDS-3 global navigation satellite system has been built and is providing official open Positioning, Navigation, and Timing (PNT) service with full operational capability (FOC) since July 2020. The main new civil B1C and B2a ranging code signals are broadcasted on the two carriers with central frequencies of 1575.42 MHz and 1176.45 MHz, which were shared by other GNSSs. Compared with traditional signals, such as GPS L1 C/A and BDS B1I, the new civil signals have better modulation and wider bandwidth to be expected to achieve a better range performance. In order to evaluate code ranging accuracies directly, a zero-baseline experiment using a geodetic GNSS antenna and a four-channel intermediate frequency (IF) signal recorder was conducted. Two channels were used to receive the signals with a central frequency of 1575.42 MHz at a 62 MHz sampling rate, and the other two channels are for 1176.45 MHz. The raw IF data were post-processed using a software-defined receiver (SDR) to compute the code signal path differences between two channels with the same frequencies. Compared with the traditional hardware receiver, SDR has the characteristics of flexible use and good operability, but its running speed is slow. The root-mean-square (RMS) and bias values of the path differences from BDS B1C, BDS B2a, and GPS L5C were used to evaluate their accuracies. The results show that there is a weak negative correlation between the satellite elevation and the ranging accuracy when the satellite elevation ranges from 30° to 90°. The ranging accuracy of the B1C signal is lower than that of B2a, which may be caused by different code rates, bandwidth, and signal structure. The GPS L5C is used for precision analysis as a comparison. It shows that the code signal path differences accuracy of L5C is close to the B2a.

Keywords: BDS-3; ranging code; zero-baseline; raw IF data

Citation: Liu, Y.; Gao, F.; Li, J.; He, Y.; Ning, B.; Liu, Y.; Chen, S.; Qiu, Y. Analysis and Performance Evaluation of BDS-3 Code Ranging Accuracy Based on Raw IF Data from a Zero-Baseline Experiment. *Remote Sens.* **2022**, *14*, 3698. https://doi.org/10.3390/rs14153698

Academic Editor: Xiaogong Hu

Received: 5 July 2022
Accepted: 28 July 2022
Published: 2 August 2022

Publisher's Note: MDPI stays neutral with regard to jurisdictional claims in published maps and institutional affiliations.

1. Introduction

Since the 1980s, China has been committed to the development of its own global satellite navigation system with independent intellectual property rights, which was named the BeiDou Navigation Satellite System (BDS). The construction of the BDS is divided into three stages: the BeiDou double satellites positioning system (BDS-1), the BeiDou regional navigation system (BDS-2), and the global system (BDS-3) [1]. BDS-1 was announced in mid-2003 to operate two geostationary satellites and a backup satellite [2], which has now been deactivated. BDS-2 has officially started to provide Positioning, Navigation, and Timing (PNT) services to the Asia Pacific region on 27 December 2012 [3]. The BDS-3 was put into use after the successful launch of the last networking satellite in June 2020. After that, the number of BDS satellites that can provide navigation signals in space reached 55, including 30 networking satellites of BDS-3 [4]. Among the 30 networking satellites of BDS-3, there are 24 medium earth orbit satellites (MEO), three inclined geosynchronous orbit satellites (IGSO), and three geostationary orbit satellites (GEO) [1–3]. It can provide global users with PNT services, short message communication, and other services, transmitting five public service signals centered at B1I (1561.098 MHz), B3I (1268.52 MHz), B1C (1575.42 MHz), B2a (1176.45 MHz), and B2b (1207.14 MHz) [5–9]. Compared with BDS-2, BDS-3 has significantly improved system coverage, spatial signal accuracy, spatial signal continuity and availability [10–12]. The continuity and availability of BDS-3 space signal are about 99.99% and 99.78%, respectively [13]. The signal transmitted at each frequency point is composed of ranging code and navigation data modulated on the carrier signal. Users can compute the propagation time of the signal at each time to calculate their position. According to the public system documents, the positioning accuracy of the Beidou satellite navigation system in the world will reach 10 m horizontally and 10 m vertically (95%). In the Asia-Pacific region, the positioning accuracies are 5 m horizontally and 5 m vertically (95%) [3].

In addition to the BeiDou navigation system, the GNSS has another three major global navigation systems. The Global Positioning System (GPS) was launched by the US Department of Defense in 1973. Its space segment consists of 24 satellites and 3 standby satellites. [14]. GPS III, a new generation of modern GPS satellites, has higher navigation accuracy, transmission power, and signal integrity with the new L1C civil signal [15]. The Galileo system consists of 24 operational satellites and 6 standby satellites [14]. The Galileo system plans to carry out extensive infrastructure development and deployment activities to achieve full operational capability (FOC) [16]. The GLONASS system launched its first networking satellite in 1982. Its fully operational constellation consists of 24 satellites distributed in three orbital planes. It will deploy GLONASS-K2 satellite and improve the tracking network to meet the needs of modernization [17].

With the advent and wide application of BDS-3, people have carried out all-round research on BDS-3, including signal characteristics, positioning accuracy, precision point positioning (PPP), etc. The comparison of BDS-3 features with other satellite navigation systems has also been widely promoted. Xie et al. [18] points out that the signal strength of the same type of satellite BDS-3 is greater than that of BDS-2. The MP combination is used to study the deviation between the pseudo-range and carrier phase measurement of BDS-3 and BDS-2. The authors point out that the BDS-2 coding is affected by the satellite-induced deviation of 1 m from the horizon to the zenith. Liu et al. [19] calculated the signal-in-space range error (SISRE) of different types of navigation satellites. The average SISREs of BDS-3 MEO, IGSO, and GEO satellites are 0.52 m, 0.90 m, and 1.15 m. The SISRE of BDS-3 MEO satellite is slightly lower than that of Galileo by 0.4 m, slightly higher than that of GPS by 0.59 m, and significantly better than that of GLONASS by 2.33 m. BDS-3 can achieve significant positioning accuracy. Fan [20] uses the user equivalent range error (UERE) to describe the signal quality of BDS-3 satellite in polar and global conditions. The UERE of BDS-3 is in the range of 0.5–1.0 m. To eliminate the ionosphere delay, orbit error, multipath error, troposphere delay, and satellite clock error, the zero-baseline experiment can be used for analysis. It can also evaluate the ranging accuracy of the signal. Yang et al. [21], using zero-baseline inter-station single-difference (SD) and ultra-short-baseline SD methods determines that the measurement accuracy of the BDS-3 system code and carrier phase are 33 cm and 2 mm, respectively. Deng et al. [22] used three zero-baseline experiments to find that the noise of code phase measurement and carrier measurement of BDS-3 are improved compared with BDS-2, and millimeter-level noise is measured on the B3I signal. Roberts et al. [23] compared and fused the observation accuracy of BeiDou and GPS by using the zero-baseline experimental data. The relative positioning accuracy of BeiDou can reach millimeter level, but it is lower than that of GPS. The integrated use of BeiDou and GPS can improve positioning accuracy. In addition, the initial positioning performance of the BDS-3 under different modes such as single-point positioning and RTK is also studied [24–26]. The above research on ranging performance and accuracy often requires a long time of complex observation and tedious post-processing of data operation, so it is impossible to directly obtain an index to quantitatively measure the signal ranging performance.

In this contribution, the ranging code accuracy and performance of two new BDS-3 civil signals B2a and B1C are preliminarily analyzed and compared with GPS L5C. The four-channel IF signal recorder is used by us to collect the raw IF signals of two BDS frequency points and conduct the zero-baseline experiment. The code ranging accuracy is analyzed by calculating the code signal path difference based on the waveform's extreme value difference, which can directly reflect the ranging performance, avoiding the complicated observation and data processing. This ranging accuracy evaluation method is first applied to the ranging accuracy analysis of BDS-3 B1C and B2a.

A software-defined receiver (SDR) was used to process the raw intermediate frequency (IF) signal to calculate the code-level path differences. SDR, which is a general framework for highly flexible multiple system solutions, is also applicable to general GNSS equipment [27]. It realizes all signal processing processes through programmable software or microprocessors Its advantage is particularly important in the case of GNSS because of the series of signals used in PNT services. If traditional hardware receivers are used, new hardware components must be developed or purchased, while SDR can include the processing function of new signals only by updating the software [28]. Our research covers multiple signal frequency bands, so SDR is selected for signal processing. However, although SDR has greater flexibility, it has an obvious disadvantage that the signal processing speed is extremely slow. The amount of raw IF signal data used in this study is large.

The remaining chapters of this paper are arranged as follows. In Section 2, the two civil signals of BDS-3 and other relevant details with the rationale of the zero-baseline experiment are described. In Section 3, the specific steps of our zero-baseline experiment will be described. In Section 4, we discuss the sampling results and the accuracy comparison among the three signal frequency points in detail. Finally, the conclusions and summary of the study, as well as the unsolved problems of this research, are discussed in Section 5.

2. Materials and Methods

2.1. Signals

BDS-3 has been provided official open PNT service since 31 July 2020. The BDS-2 regional satellite navigation system was built earlier than BDS-3, which has opened three civil signals for B1I (1561.098 MHz), B2I (1207.14 MHz), and B3I (1268.52 MHz). The open civil signals of BDS-2 are mainly modulated by Binary Phase Shift Keying (BPSK). After the completion of BDS-3, the BDS-3 inherits the B1I and B3I signals of BDS-2, and the signal modulation methods of B1I and B3I are strictly consistent with BDS-2 to ensure the interoperability of system signals. BDS-3 began to adopt new civil signal frequency points such as B1C (1575.42 MHz), B2a (1176.45 MHz), and B2b (1207.14 MHz) [5–9], and the observational accuracy of the BDS-3 signals is comparable to GPS L1/L2/L5 and Galileo E1/E5a/E5b, according to the report [29]. Frequency multiplexing exists among GNSSs. Figure 1 shows the frequency distribution of the current four global satellite navigation systems. If the frequency band numbers in Figure 1 have the same position on the frequency coordinate axis, they have the same frequency center, and there is a potential possibility of interoperability between systems. Frequency overlap puts forward higher requirements for compatibility and interoperability among systems.

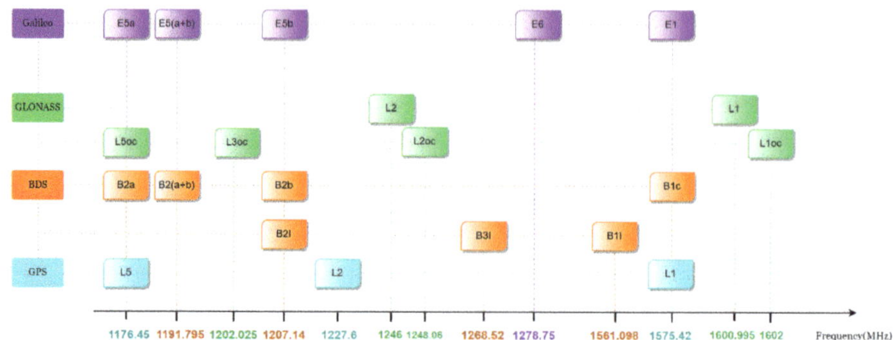

Figure 1. Frequency distribution of GNSSs.

B1C and B2a are the service signals that BDS-3 mainly opens to users. B1C and B2a can broadcast signals through 24 medium earth orbit satellites (MEO) and 3 inclined geosynchronous orbit satellites (IGSO). The signals broadcasted by the satellite are distinguished by capturing different pseudo-random noise codes (PRN). PRN numbers ranging from 19 to 46 belong to BDS-3 satellites, of which 38, 39, and 40 are IGSO satellites. The code length of the two signals is 10230, uniformly. However, the difference is that the coding rate of B2a is 10.23 Mcps, while that of B1C is 1.023 Mcps [7,8].

By referring to the official spatial signal interface document (ICD) of the BeiDou system, the B1C signal is composed of data component and pilot component [8]. The carrier center frequency of both is 1575.42 MHz, where the GPS L1 and Galileo E1 are also broadcasted on it. The structural characteristics of B1C can be listed in Table 1.

Table 1. The characteristics of the B1C signal.

Signal Component	Center Frequency/(MHz)	Modulation	Rate Speed/(sps)
B1C_data	1575.42	BOC (1,1)	100
B1C_pilot	1575.42	QMBOC (6,1,4/33)	0

The pilot component adopts quadrature multiplexed binary offset carrier (QMBOC) modulation, which allows the signal to be composed of a narrow-band component with larger power distribution and a broadband component with smaller power distribution [30]. This shows that the B1C signal can be compatible with high-performance receiving mode and low-complexity receiving mode, which is a relatively novel feature of the BDS-3 B1C signal.

The complex envelope form of B1C signal [31] can be written as:

$$S_{(B1C)} = s_{(B1C_D)}(t) + js_{(B1C_P)}(t) \tag{1}$$

where $s_{(B1C_D)}(t)$ is the data component of the signal, which is generated through the modulation of navigation message $D_{(B1C_D)}(t)$, ranging code $C_{(B1C_D)}(t)$, and subcarrier signal. $s_{(B1C_P)}(t)$ is the pilot component of the signal, which is modulated by subcarrier and ranging code $C_{(B1C_P)}(t)$, the navigation message is not modulated in it. $s_{(B1C_D)}(t)$ and $s_{(B1C_P)}(t)$ are distributed in the power ratio of 1:3. The modulated B1C signal can finally be written as the following expression in bandpass form [31]:

$$S^{(i)}_{(B1C)}(t) = \sqrt{2P_{(B1C)}} \left[\frac{1}{2} D^{(i)}_{(B1C)}(t) C^{(i)}_{(B1C_D)}(t) \cos(2\pi ft) + \frac{\sqrt{3}}{2} C^{(i)}_{(B1C_P)}(t) \sin(2\pi ft) \right] \tag{2}$$

In the formula, $S^{(i)}_{(B1C)}(t)$ represents the signal from the satellite labeled i, $D^{(i)}_{(B1C)}(t)$ represents the modulated navigation message, $P_{(B1C)}$ represents the signal power of B1C,

$C_{(B1C_D)}^{(i)}(t)$ is the ranging code of the data component, and $C_{(B1C_P)}^{(i)}(t)$ is the ranging code of the pilot component. The design of this data component and pilot component will greatly improve the ranging ability of the signal.

As for the ranging code, it is a layered code structure of B1C that the sub-code chip is strictly aligned with the time of the first code chip of the main code. The code rate of the B1C ranging code is 1.023 Mbps. It is obtained by truncating a Weil code with a length of 10243 chips. The Weil code sequence with length n is defined as follows:

$$W(t, w) = L(t) \oplus L(t + w) \tag{3}$$

$L(t)$ is a Legendre sequence of fixed length N, w is the phase difference between two Legendre sequences, and the value is 1 to 5121. The Legendre sequences of length N are defined as follows:

$$L(t) = \begin{cases} 1; & t \neq 0, \text{intx}; \ t = x^2 \text{mod} N \\ 0; & t = 0 \\ 0; & \text{else} \end{cases} \tag{4}$$

The B1C ranging code can be obtained by cyclic interception of the Weil code.

B2a is another public navigation service signal broadcast by BDS-3. The BPSK(10) method is selected for its signal modulation in the data component and pilot component. Similarly, according to the spatial interface document of the BDS, we can express the signal in the form of a complex envelope [32]:

$$S_{(B2a)} = s_{(B2a_D)}(t) + js_{(B2a_P)}(t) \tag{5}$$

where $s_{(B2a_D)}(t)$ represents the data component of the signal, and $s_{(B2a_P)}(t)$ represents the pilot component of the data. We can also write the signal in bandpass form [32]:

$$S_{(B2a)}^{(i)}(t) = \sqrt{2P_{(B2a)}} \left[D_{(B2a)}^{(i)}(t) C_{(B2a_D)}^{(i)}(t) \cos(2\pi ft) - C_{(B2a_P)}^{(i)}(t) \sin(2\pi ft) \right] \tag{6}$$

In this formula, $D_{(B2a)}^{(i)}(t)$ represents the modulated navigation message data, $C_{(B2a_D)}^{(i)}(t)$ is the data code of the data component, and $C_{(B2a_P)}^{(i)}(t)$ is the data code of the pilot component. $P_{(B2a)}$ gives the power of the B2a signal. The code rate of B2a ranging code is 10.23 Mcps and the code length is 10,230. Two thirteen-level linear shift registers $g_{1(x)}$ and $g_{2(x)}$ are used to generate two extended gold codes of 10,230 chips, and both of them obtain B2a ranging code through modulo-2 addition. The two polynomials are:

$$g_1(x) = 1 + x + x^5 + x^{11} + x^{13} \tag{7}$$

$$g_2(x) = 1 + x^3 + x^5 + + x^9 + x^{11} + x^{12} + x^{13} \tag{8}$$

for the data component and are:

$$g_1(x) = 1 + x^3 + x^6 + x^7 + x^{13} \tag{9}$$

$$g_2(x) = 1 + x + x^5 + + x^7 + x^8 + x^{12} + x^{13} \tag{10}$$

for the pilot component.

For the B2 band of BDS-3 new signal, it also broadcasts B2b signals. Both signals are composed of two orthogonal components. Therefore, the asymmetric constant envelope binary offset carrier technology (ACE-BOC) can be used to synthesize the two signals to form a constant envelope signal, which can reduce the payload and multiplexing loss. The use of ACE-BOC makes the signal receiving of the B2 signal highly flexible: the above two signals in the B2 band can be received as broadband signals or two separate quadrature phase-shift keying (QPSK) signals.

2.2. Principles in Zero-Baseline Experiments

Zero-baseline means that the antenna phase centers corresponding to multiple signal channels are consistent, i.e., the same antenna is used to receive signals. In the study, the GNSS signal recorder used in the zero-baseline experiment has four signal input channels corresponding to different frequencies and signal branches. As shown in Figure 2, after the signal is led out from the antenna, the power distributor is used to connect the signal to the four signal ports through the same transmission conditions, and the zero-baseline condition is formed between the four signal receiving terminals. For the actual recorder, the signal channel 1 and signal channel 2 will operate on the same frequency point, and signal channel 3 and signal channel 4 will operate on another frequency point.

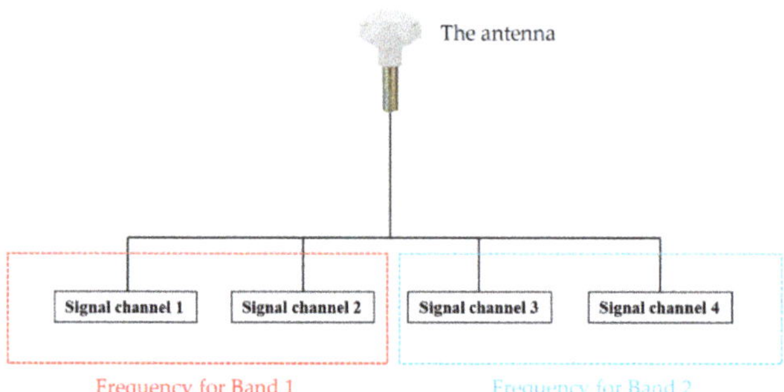

Figure 2. Diagram of the zero-baseline experiment in this study.

Since the four signal channels were connected with the same antenna, the error influence is consistent among the four signal channels in the process of the GNSS signal transmission through satellite, ionosphere, and atmosphere propagation and the signal line before obtaining access to the recorder. Therefore, theoretically, the data collected by signals of the same signal frequency between different signal channels should be consistent. However, in practice, owing to the influence of the structure of the ranging code itself, the performance of the ranging code cannot be fully idealized. It leads to certain differences in the ranging code data of different signal channels at the same epoch and frequency, which can be used to reflect the ranging performance of the signal frequency ranging code directly.

The code signal path differences can quantitatively reflect the difference in the ranging code data. In order to obtain the values of code signal path differences, the code-level waveform diagram of the two signal channels were obtained by phase correlation between local codes and two-channel signals. Figure 3 shows the code-level waveform diagram obtained after two signal channels pass through the same epoch. Generally, the waveform figure at a single epoch has a power maximum near the central position, where the path delay is 0 and the power on both sides decreases with the increase in the absolute value of the path. Delay difference between the extreme positions of the waveform between two channels code signal path differences of two signal channels. In the observation period, the code signal path differences of each epoch can be obtained through the same data process under different epochs that meet the observation requirements, and the changing image of code signal path differences with time can be obtained.

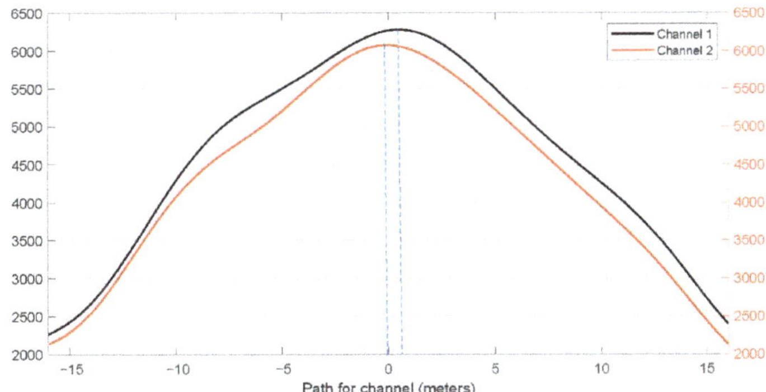

Figure 3. The calculation of code signal path differences. Code signal path differences are the path difference between the two blue lines in the figure.

Ideally, the code signal path difference of the waveform signals in the same epoch should be strictly 0 due to the zero-baseline environment. Therefore, the code signal path differences can be used to evaluate the performance and ranging accuracy of the ranging code directly. The smaller the code signal path differences, the better the performance and ranging accuracy of its corresponding ranging code. In order to obtain the code-level signal path differences for each signal channel, we use a 4-channel IF signal recorder corresponding to Figure 2 that can receive B2 and B1 band signals simultaneously to collect the raw IF signal. There is two signal receiving channels under each signal frequency point. The collected raw IF signal is processed by a self-developed SDR based on MATLAB to obtain the code signal path differences of two signal input channels at the same frequency with a zero-baseline. The raw IF data processing of GNSS signal recorder under a single channel can be shown in Figure 4.

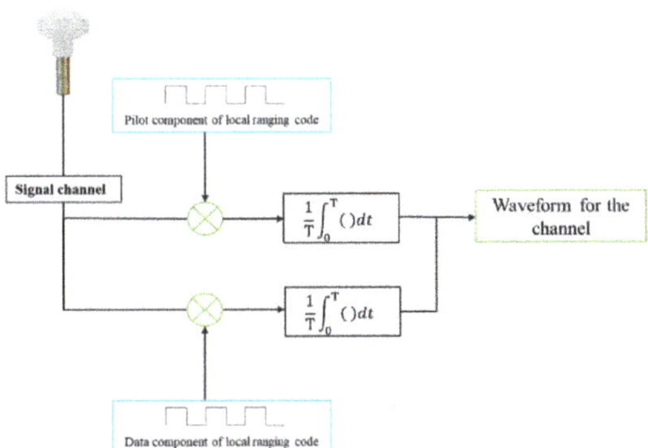

Figure 4. The IF data acquisition and process for one channel.

After the signal is transmitted to an input port through a four-power divider, it is first divided into two sub-channels inside the receiver channel for further processing. SDR will process the data epoch by epoch. One sub-channel of the signal is cross-correlated with the pilot component signal of the local code, and the other sub-channel of the signal is cross-correlated with the data component signal of the local code. A set of waveform data are obtained from the coherent integration of two signals for a certain epoch. According to the ICD files of B1C and B2a, the waveform data of the pilot component and the data component are superimposed coherently according to the power ratio of the photographic frequency point to obtain the waveform data under this channel.

3. The Zero-Baseline Experiment

Our experiment was conducted on the roof of a high-rise building on the campus of Shandong University (Weihai), Weihai, Shandong Province, China. The surrounding environment of the experiment is open, which is convenient for conducting zero-baseline experiments. Our experiment was conducted on 7 March 2022. The day was sunny, and the weather conditions remained stable throughout the day, which was conducive to the operation of the zero-baseline experiment. The data receiving and acquisition system used in the zero-baseline experiment is mainly composed of a coking-ring GNSS antenna and a four-channel IF signal recorder. Figure 5a,b record the field environment of the zero-baseline experiment.

(a)

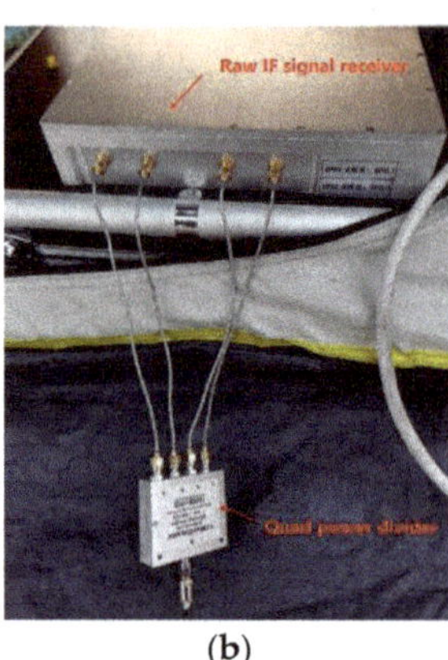

(b)

Figure 5. The field of the zero-baseline experiment: (**a**) Horizontally placed geodetic GNSS antenna and its tripod; (**b**) Quad-power divider and raw IF signal recorder.

In Figure 5a, the geodetic GNSS antenna is erected and fixed by a tripod to keep it horizontal. The signal is led out from the antenna by one shielded cable. It can be noted that there is a fence on the top of the building, the horizontal plane of the erected antenna is higher than the height of the fence, and the other positions remain open. Therefore, it can be considered that the antenna signal is not affected by the surrounding obstacles. Figure 5b shows the wiring of the signal receiving and processing terminal. The shielded cable is led out from the antenna and connected to the four power dividers in Figure 5b. Before entering the recorder, the signal is divided into the same four branches by the four power dividers and connected to the four signal receiving ports of the recorder, respectively. It should be noted that the four data lines connecting the power divider and the recorder should be completely consistent to eliminate the difference in the line's thermal noise. The bandwidth of the recorder is 20.46 MHz, of which two signal channels will receive the B1 band with a frequency center of 1529 MHz, which can collect B1C signals with the center frequency of 1575.42 MHz. The other two channels will receive the B2 band with a frequency center of 1130 MHz, which can collect B2a signals with the center frequency of 1176.45 MHz. In Figure 5, the two input interfaces on the right correspond to the B1 band of the signal. It can be considered as the signal channel 1 and signal channel 2 of Figure 2. The two input interfaces on the left correspond to the B2 band of the signal. They can be considered as signal channel 3 and signal channel 4 of Figure 2. The recorder works at 62 MHz continuous sampling, and the collected IF data are quantized by 2 bits. After the signal is processed by the original IF signal receiver, the original IF signal data are stored in the data hard disk connected to the receiver at an interval of two minutes for subsequent data processing. In addition, in the later data processing, in order to minimize the impact of the multipath, the data with satellite elevation above 30 degrees were selected for processing. Thanks to the open surrounding environment during the experiment, the data can be arbitrarily selected from the satellite azimuth of 0 to 360 degrees.

4. Results

4.1. Satellite and the Acquisition of Data

The time range of the experimental data collection is 11:02 to 18:00 on 7 March 2022 (Beijing time). Data processing and analysis shall be carried out within the time range when the satellite elevation meets the requirements. Our selected satellite elevation mainly ranges from 30° to 70°, and the satellite elevation can be calculated from the satellite's broadcast ephemeris. The duration of raw IF data used for the data processing shall be greater than 1 h to ensure the effectiveness of the analysis results. When selecting the period, we try to ensure that the elevation of the selected satellite is within a certain similar range, so as to reduce the impact of satellite elevation when analyzing the ranging accuracy alone under a specific PRN. At the same time, considering the slow processing speed of SDR and the huge storage space required for data results, we basically choose an interval of 1 to 3 h for further processing, which not only ensures that the satellite elevation conditions are met, but also ensure the processing process is not too long. As the structure of the ranging code broadcasted from a satellite system is similar under a frequency point, the ranging performance of the system can be reflected by processing at least one set of satellite data, and the conclusion is not affected by the number of satellites and satellite orbit type. Therefore, only limited satellite data have been processed and analyzed in our research.

The signals from satellites with PRN 37 (MEO) and PRN 39 (IGSO) were used for the ranging accuracy analysis of B2a. The satellite with PRN 37 is an MEO satellite. Its satellite elevation has a large variation range, so we can also collect and process its data for a long time to study the relationship between ranging accuracy and the satellite elevation. When processing B1C frequency points, we use satellite PRN 39 for data processing, which is consistent with the selection of B2a satellites. This is an IGSO satellite, which can ensure a high satellite elevation and a long time in the observation period. At the same time, thanks to the good interoperability and compatibility of BDS-3, the B2a is consistent with the center frequency of GPS L5C. Therefore, we use the same raw IF signal to analyze the ranging accuracy of L5C PRN 4 for comparison with the BDS-3. GPS PRN 4 satellite is also an MEO satellite, which has a large range of elevation changes. We can use this satellite to compare the ranging performance of the GPS L5 with that of the BDS-3 frequency point of B2a and B1C.

In the experiment, the zero-baseline experimental environment can only ensure that all conditions are consistent before the signal is connected to each channel of the IF signal recorder. There are four channels and two crystal oscillators in our recorder. Two of the channels share one crystal oscillator and a frequency point, while four channels share a clock. So, the bias between the channels at the same frequency point is small. In our work, the path difference between two channels at the same frequency point are calculated to investigate the accuracy of the ranging code. Therefore, in the data processing, we ignore the system error of the receiver and consider the hardware delay equal in each signal channel. We calculate the code signal path differences directly after mapping the waveform figure as Figure 3. By analyzing the solutions in Table 2, we find that there is no bias for the two channels at the frequency of 1176.45 MHz, while there is a 26 cm bias for the two channels at the frequency of 1575.42 MHz. When signals of all channels are processed by the GNSS-SDR according to Figure 4, the coherence integration time is set as 10 ms to enhance the signal-to-noise ratio. After processing 10ms data, the complete waveform data can be obtained, and the code signal path differences can be calculated by the differences between the extreme values of the waveform. As the calculation efficiency of GNSS-SDR is relatively low, we conduct data calculation at an interval of 50 ms, and 19 waveform data will be obtained in a single channel within 1 s. After the code signal path differences calculation of each of the 19 waveform data, the median value is taken as the code signal path differences value in this second. The standard deviation of 19 groups of data within 1 s can also be calculated accordingly, which reflects the dispersion degree of internal code signal path differences in 1 s, and its satellite elevation-varying figures can also be given. By arranging the results of each second in the processing period, we can obtain the corresponding time sequence change diagram.

Table 2. Statistical characteristics of code delay.

PRN	GPS L5C	BDS B2a		BDS B1C
	4	37	39	39
RMS(m)	0.1603	0.1526	0.1527	0.2674
STD(m)	0.1585	0.1309	0.1485	0.1779
Mean(m)	-0.0234	-0.0785	-0.0358	0.1996

4.2. Influence of the Change of Satellite Elevation

Firstly, we analyze the influence of the satellite's high elevation on the ranging accuracy of a single frequency point. As mentioned above, the BDS-3 satellite of PRN 37 is an MEO satellite, which satellite elevation changes greatly. We use the SDR of the B2a frequency point to process the raw IF data of PRN 37. The selected processing period is 13:09–17:23 (Beijing time). During this period, the satellite elevation gradually increased from 30° to the maximum, and then gradually decreased back to 30 degrees. According to Section 4.1, the code signal path differences of B2a and the satellite elevation time series change diagram can be seen in Figure 6.

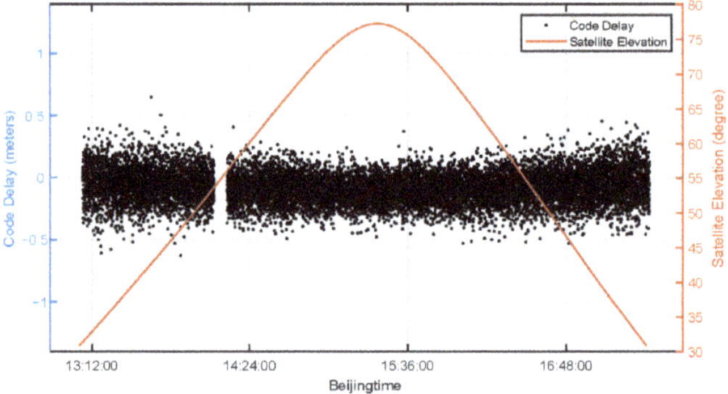

Figure 6. Time distribution of B2a PRN 37 satellite code signal path differences and satellite elevation.

It should be noted that the time starting point and length of the data processing period selected at each frequency point is different. As described in Section 4.1, this is to meet the requirements of satellite elevation similarity. In addition, the processing speed of SDR is slow. Choosing 1–3 h instead of the whole range of data for processing can improve the processing efficiency and ensure the accuracy of the results.

It should be noted that there are certain data interruptions in Figure 6, which are empty data records caused by loose data connection ports. It can be preliminarily seen from the Figure 6 that the absolute value of the code signal path differences remains within 0.5 m during the observation period. The code signal path differences converge towards 0 m in the state of high satellite elevation, and they are relatively discrete in the state of low satellite elevation. Standard deviation (STD) can be used to quantitatively describe the degree of data dispersion, which is an indicator of ranging accuracy. In the process of processing, since 19 code signal path differences will be recorded in one second, the STD of these 19 data in one second can reflect the dispersion of the code signal path differences in that second. The same processing is performed on the data in the selected period to obtain all the STD distribution data at the 1 s interval. STD data with satellite elevation are plotted as shown in Figure 7.

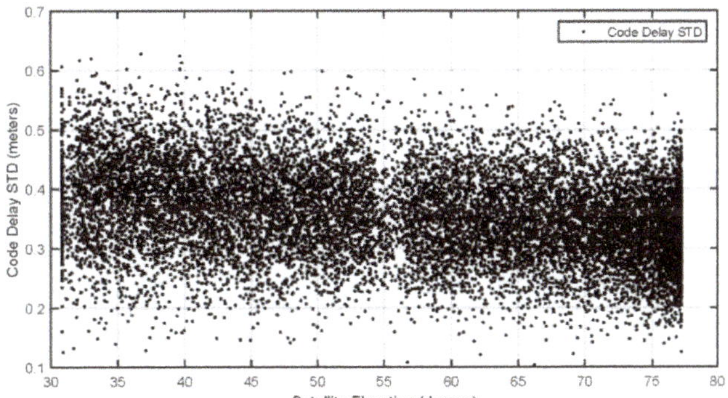

Figure 7. Satellite elevation distribution of B2a PRN 37 satellite code signal path differences STD.

In Figure 7, with the increase in satellite elevation, it can be seen that the STD value has a very obvious downward trend. It is noted that there are many data points recorded at high satellite elevation, which is caused by the slow change in satellite elevation, and a large number of data points are recorded when the satellite operates at high elevation. The relationship between STD and satellite elevation will be described quantitatively. According to the data in Figure 7, the data correlation analysis can be carried out. The Pearson correlation coefficient matrix is used to describe this quantitative correlation. The STD value and satellite elevation can be regarded as two variables, and the relationship between them can be described by a two-dimensional matrix P:

$$P = \begin{bmatrix} 1 & -0.308 \\ -0.308 & 1 \end{bmatrix} \tag{11}$$

In this matrix P, the main diagonal element represents the correlation coefficient of the two variables; that is, 1 represents a complete correlation. The sub-diagonal represents the Pearson correlation coefficient between STD value and satellite elevation. The closer the absolute value of the Pearson correlation coefficient is to 1, the stronger the correlation between the two variables they have. The Pearson correlation coefficient between STD and satellite elevation is -0.308. Therefore, we can conclude that there is a weak negative correlation between the satellite elevation and the STD value; this conclusion is only satisfied in this paper when the satellite elevation is between $30°$ and $90°$. The period with high satellite elevation will produce better ranging accuracy, but the accuracy changes slightly with the satellite elevation.

The higher the satellite elevation, the greater the signal power and the lesser the multi-path effect. So, the high satellite altitude angle has a smaller code signal path difference. Since the satellite altitude angle only affects the code signal path differences of a single satellite within a period and has nothing to do with the signal structure differences between different frequency points. We can draw the above conclusions by processing only one satellite for each frequency point.

4.3. Ranging Accuracy Analysis

In this subsection, the ranging accuracies of BDS-3 B2a and B1C according to code signal path differences are analyzed. We processed the measurements from PRN 39 for B2a from 11:05 to 13:40 and PRN 39 for B1C from 12:00 to 13:15, while plotting the code signal path differences and the time distribution of satellite elevation. In addition, the center frequency of the GPS L5C frequency point is the same as that of BDS B2a. We processed its data from 11:05 to 13:09 for comparisons. The corresponding calculation results are shown in Figures 8–10. Figure 8 shows the calculation results of B2a PRN39 satellite; Figure 9 shows the calculation results of B1C PRN39 satellite; Figure 10 shows the calculation results of L5C PRN4 satellite.

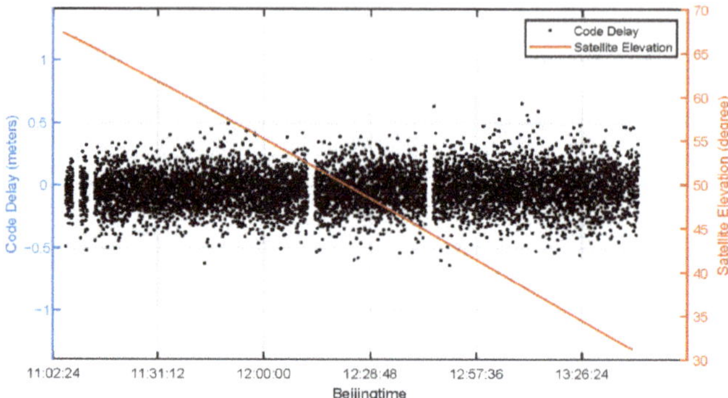

Figure 8. Time distribution of B2a PRN 39 satellite code signal path differences and satellite elevation.

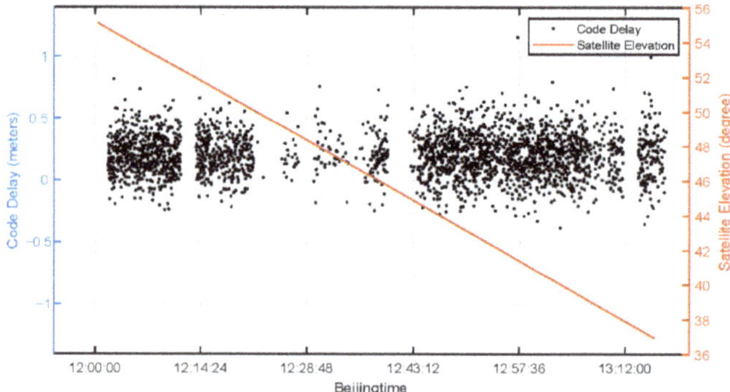

Figure 9. Time distribution of B1C PRN 39 satellite code signal path differences and satellite elevation.

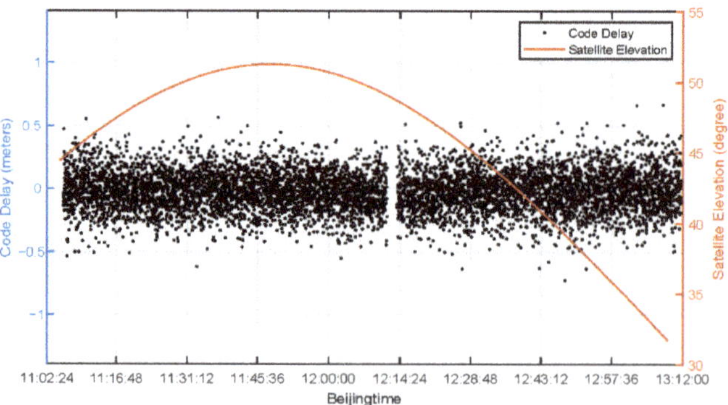

Figure 10. Time distribution of L5C PRN 4 satellite code signal path differences and satellite elevation.

It is noted that when processing the data of B1C, the figure is relatively loose, which is mainly caused by two reasons below. One is that the loose data interface makes certain empty data records exist during data sampling. On the other hand, the data with code signal path differences greater than 0.8 m or greater than three times RMS is considered as gross error and removed. The distribution of these data is messy and not the main component of the data composition, so they should be regarded as gross errors and discarded. In addition, the data observation duration of B1C is about 1 h shorter than that of B2a. This is because the data point processing results in the selected period of B1C are denser than those in other periods. The variation range of the satellite elevation data is similar to that of B2a, which can more intuitively reflect the distribution of code signal path differences. The observation duration meets the minimum 1 h requirement described in Section 4.1. In this way, the overall characteristics of B1C signal code signal path differences can be seen during this period of time, and the decline in the amount of data will not affect the main conclusions. It is worth noting that there is an obvious overall data deviation in the calculation results of B1C. Except for the B1C signal, the code signal path difference of other signals is between -0.5 m and 0.5 m, which means that their ranging accuracy remains basically unchanged. However, compared with other analyzed frequency points, the calculation result of B1C has a very significant deviation from the theoretical value 0 m, and the whole calculation result shows an upward trend. To quantitatively describe this change, the overall code signal path differences RMS, mean value, and STD of each frequency point in the observation processing time can also be calculated. The corresponding processed figures and data results are given in Table 2.

The RMS, STD, and mean value of satellite data in the whole period are calculated. It has been pointed out in Section 4.2 that there is only a weak negative correlation between satellite elevation and ranging accuracy when the satellite elevation is between $30°$ and $90°$ in our study. Therefore, the influence of satellite elevation can be ignored if the results of long-term observation are used. Moreover, the conclusion of ranging accuracy will not be affected by the number of satellites and processing time. It can be considered that the conclusions drawn from the satellites selected in this experiment are universal. Under the BDS B2a frequency point, the RMS values are 0.1526 m for PRN 37 and 0.1527 m for PRN 39, respectively. We note that BDS B2a covers two satellite orbit types, MEO and IGSO, and their data results are basically consistent. Therefore, the difference of satellite orbit types can be ignored, and it can be considered that the result of ranging accuracy is only related to the signal structure of different frequency point. Under the BDS B1C frequency point, our RMS value is 0.2674. We can note that the RMS value of B1C is lower than the RMS value of B2a. To enhance the comparison of data, we use the same raw IF data and processing method to calculate the code signal path differences of the GPS L5C band. The results show that the ranging accuracy of BDS-3 B2a and GPS L5C are roughly the same, and both of them are significantly better than the ranging accuracy of B1C. For the average value of code signal path differences, the calculation results of GPS L5C and BDS B2a are on the order of 10^{-2} m, while the average value of the B1C signal is 0.1996 m, which is far from the theoretical value of zero. The experimental data also show that B2a and GPS L5C have similar ranging performances. The code rate and code length of the GPS L5C signal are consistent with those of B2a, so it has a similar ranging performance, which is mutually confirmed with the experimental results. Therefore, this zero-baseline experiment has certain reliability. In addition, the standard deviation of B1C code signal path differences are also significantly greater than that of other frequency points. These results show that B1C has poor ranging accuracy in the zero-baseline experiment. We try to explain this from the difference in signal structure.

The particularity of the B1C signal structure may cause some deviation in the ranging results. In Section 2.1, we know that the pilot component of the B1C signal is modulated by QMBOC. In fact, the QMBOC subcarrier is composed of two bipolar subcarriers, so the B1C signal can be regarded as a combination of three bipolar components [31]:

$$\begin{aligned}
S^{(i)}_{(B1C)}(t) = {}& \tfrac{1}{2}D^{(i)}_{(B1C)}(t)C^{(i)}_{(B1C_D)}(t)\cdot\mathrm{sign}(\sin(2\pi f_a t)) \\
& + \sqrt{\tfrac{1}{11}}C^{(i)}_{(B1C_P)}(t)\cdot\mathrm{sign}(\sin(2\pi f_b t)) \\
& + j\sqrt{\tfrac{29}{44}}C^{(i)}_{(B1C_P)}(t)\cdot\mathrm{sign}(\sin(2\pi f_a t))
\end{aligned} \tag{12}$$

where $f_a = 1.023$ MHz and $f_b = 6.138$ MHz. In some research, the effect of the subcarrier is interpreted as a selective communication channel that distorts the useful signal [33]. The existence of subcarriers may affect the accuracy of ranging code at SDR reception. For ranging codes, B1C and B2a have the same code length, but the code rate of B2a is ten times that of B1C. A lower code rate may affect the accuracy of signal ranging. In this study, the coherence time of B2a and B1C is 10 ms. B1C signal has wider bandwidth. Although the wide bandwidth of B1C is conducive to combating multipath effects, the existence of subcarriers and lower code rate causes the positioning result to shift. Affected by the above factors, B1C has lower ranging accuracy in this zero-baseline experiment. Although some studies indicate that the ranging accuracy of B2a is lower [34], its testing environment should be affected by adverse environmental factors, which needs further discussion. We will conduct a more detailed study on this issue in future work.

5. Discussion and Conclusions

The zero-baseline experiments were used to evaluate the GNSS positioning performance. This paper presents a method to figure out the accuracies of BDS-3 new civil ranging codes directly using the data collected from a zero-baseline experiment. The results show that the B2a code ranging accuracy is better than that of B1C, and similarly with GPS L5C code.

About 7 h of raw IF data were collected on two frequencies. For a certain frequency point, at least one hour of data were selected for processing. The final calculation results show that the precision of the new civil signal B1C is in correspondence with B2a, while B1C has a larger STD value. The difference between B2a and B1C signal structures has a great impact on the ranging accuracy. B2a has a higher code rate, so it has higher ranging accuracy. Although B1C has a wide bandwidth, its accuracy is reduced due to its slow code rate. At the same time, compared with GPS signals, it can be found that B2a and L5C have similar ranging accuracy, owing to the same code rate and code length. In addition, the dispersion of the ranging results has weak negative correlation between satellite elevations when the satellite elevation meets $30°$ to $90°$, for which the Pearson correlation coefficient is -0.308. Under the condition of high satellite elevation, the data onto one second have a relatively small STD value, that is, higher ranging accuracy.

In this work, the amount of the raw IF data storage is too large, and the speed of processing the data using the GNSS-SDR software receiver is too slow. We only processed and analyzed parts of the data in the environment with good weather and an open surrounding area. The analysis of this work only involves the B2a and B1C signals of BDS-3 with the L5 signal of GPS. Therefore, we also plan to compare the ranging performance of each frequency band signal of other GNSSs in future work to expand the universality of the conclusion.

Author Contributions: Conceptualization, Y.L. (Yu Liu) and F.G.; data curation, Y.L. (Yu Liu), J.L., Y.L. (Yang Liu) and Y.Q.; formal analysis, Y.L. (Yu Liu); funding acquisition, F.G.; investigation, Y.L. (Yu Liu), J.L., B.N., Y.L. (Yang Liu) and Y.Q.; methodology, Y.L. (Yu Liu); project administration, F.G.; resources, Y.L. (Yu Liu); software, Y.L. (Yu Liu), J.L., B.N. and S.C.; supervision, F.G.; validation, Y.L. (Yu Liu), J.L., Y.L. (Yang Liu) and Y.Q.; visualization, J.L.; writing—original draft, Y.L. (Yu Liu); writing—review and editing, F.G., J.L. and Y.H. All authors have read and agreed to the published version of the manuscript.

Funding: This research was funded by Wenhai Program of the S&T Fund of Shandong Province for Pilot National Laboratory for Marine Science and Technology (Qingdao) (NO.2021WHZZB1004) and the Program of the National Natural Science Foundation of China, (grant number 41604003, 41704017).

Data Availability Statement: The data used in the experiment are managed by the Institute of Space Science of Shandong University, and the author can be contacted to obtain the experimental data.

Acknowledgments: The author thanks the satellite navigation and remote sensing research group of the Institute of Space Science of Shandong University for its site and technical support.

References

1. The State Council Information Office. China's BeiDou Navigation Satellite System. Available online: http://www.BeiDou.gov. cn/xt/gfxz/201712/P020210810469870523027.pdf (accessed on 25 May 2022).
2. Yang, Y. Progress, Contribution and Challenges of Compass/BeiDou Satellite Navigation System. *Acta Geod. Cartogr. Sin.* **2010**, *39*, 1–6.
3. The State Council Information Office. Development of the BeiDou Navigation Satellite System (Version 4.0). Available online: http://www.BeiDou.gov.cn/xt/gfxz/201912/P020191227430565455478.pdf (accessed on 25 May 2022).
4. Xiao, W.; Liu, W.; Sun, G. Modernization milestone: BeiDou M2-S initial signal analysis. *GPS Solut.* **2016**, *20*, 125–133. [CrossRef]
5. The State Council Information Office. BeiDou Navigation Satellite System Signal in Space Interface Control Document Open Service Signal B1I (Version 3.0). Available online: http://www.BeiDou.gov.cn/xt/gfxz/201902/P020190227593621142475.pdf (accessed on 25 May 2022).
6. The State Council Information Office. BeiDou Navigation Satellite System Signal in Space Interface Control Document Open Service Signal B3I (Version 1.0). Available online: http://www.BeiDou.gov.cn/xt/gfxz/201802/P020180209623601401189.pdf (accessed on 25 May 2022).
7. The State Council Information Office. BeiDou Navigation Satellite System Signal in Space Interface Control Document Open Service Signal B2a (Version 1.0). Available online: http://www.BeiDou.gov.cn/xt/gfxz/201712/P020171226742357364174.pdf (accessed on 25 May 2022).
8. The State Council Information Office. BeiDou Navigation Satellite System Signal in Space Interface Control Document Open Service Signal B1C (Version 1.0). Available online: http://www.BeiDou.gov.cn/xt/gfxz/201712/P020171226741342013031.pdf (accessed on 25 May 2022).
9. The State Council Information Office. BeiDou Navigation Satellite System Signal in Space Interface Control Document Open Service Signal B2b (Version 1.0). Available online: http://www.BeiDou.gov.cn/xt/gfxz/202008/P020200803362059116442.pdf (accessed on 25 May 2022).
10. Yang, Y.F.; Yang, Y.X.; Hu, X.; Chen, J.; Guo, R.; Tang, C. Inter-satellite link enhanced orbit determination for BeiDou-3. *Navigation* **2019**, *66*, 115–130. [CrossRef]
11. Yang, Y.X.; Gao, W.; Guo, S.; Mao, Y.; Yang, Y. Introduction to BeiDou-3 navigation satellite system. *Navigation* **2019**, *66*, 7–18. [CrossRef]
12. Yang, Y.X.; Yang, Y.; Hu, X.; Tang, C.; Zhao, L.; Xu, J. Comparison and analysis of two orbit determination methods for BDS-3 satellites. *Acta Geod. Cartogr. Sin.* **2019**, *48*, 831–839. [CrossRef]
13. Yang, Y.; Mao, Y.; Sun, B. Basic performance and future developments of BeiDou global navigation satellite system. *Satell. Navig.* **2020**, *1*, 1. [CrossRef]
14. Mladen, Z.; Đuro, B.; Kristina, M. Development and Modernization of GNSS. *Geod. List* **2019**, *73*, 45–65. Available online: https://hrcak.srce.hr/218855 (accessed on 4 July 2022).
15. Hein, G.W. Status, perspectives and trends of satellite navigation. *Satell. Navig.* **2020**, *1*, 22. [CrossRef]
16. Benedicto, J. Directions 2020: Galileo Moves Ahead. GPS World. 2019. Available online: https://www.gpsworld.com/directions-2020-galileo-moves-ahead/ (accessed on 26 July 2022).
17. Langley, R.B. Innovation: GLONASS—Past, Present and Future. GPS World. 2017. Available online: https://www.gpsworld.com/innovation-glonass-past-present-and-future/ (accessed on 26 July 2022).
18. Xie, X.; Geng, T.; Zhao, Q.; Liu, J.; Wang, B. Performance of BDS-3: Measurement Quality Analysis, Precise Orbit and Clock Determination. *Sensors* **2017**, *17*, 1233. [CrossRef]
19. Liu, W.; Jiao, B.; Hao, J.; Lv, Z.; Xie, J.; Liu, J. Signal-in-space range error and positioning accuracy of BDS-3. *Sci. Rep.* **2022**, *12*, 8181. [CrossRef]
20. Fan, S. Analysis of the BDS-3 Complete System on Positioning Performance in Polar Region. In Proceedings of the China Satellite Navigation Conference (CSNC 2021), Nanchang, China, 22–25 May 2021; pp. 174–186. [CrossRef]
21. Yang, Y.; Li, J.; Wang, A.; Xu, J.; He, H.; Guo, H.; Shen, J.; Dai, X. Preliminary assessment of the navigation and positioning performance of BeiDou regional navigation satellite system. *Sci. China Earth Sci.* **2014**, *57*, 144–152. [CrossRef]
22. Deng, C.; Qi, S.; Li, Y. A comparative analysis of navigation signals in BDS-2 and BDS-3 using zero-baseline experiments. *GPS Solut.* **2021**, *25*, 143. [CrossRef]
23. Roberts, G.W.; Tang, X.; He, X. Accuracy analysis of GPS/BDS relative positioning using zero-baseline measurements. *J. Glob. Position. Syst.* **2018**, *16*, 7. [CrossRef]
24. Lv, Y.; Geng, T.; Zhao, Q.; Xie, X.; Zhou, R. Initial assessment of BDS-3 preliminary system signal-in-space range error. *GPS Solut.* **2020**, *24*, 16. [CrossRef]
25. Zhang, X.; Li, X.; Lu, C.; Wu, M.; Pan, L. A comprehensive analysis of satellite-induced code bias for BDS-3 satellites and signals. *Adv. Space Res.* **2019**, *63*, 2822–2835. [CrossRef]

26. Shi, J.; Ouyang, C.; Huang, Y.; Peng, W. Assessment of BDS-3 global positioning service: Ephemeris, SPP, PPP, RTK, and new signal. *GPS Solut.* **2020**, *24*, 81. [CrossRef]
27. Soltanian, B.; hagh ghadam, A.S.; Renfors, M. Utilization of multi-rate signal processing for GNSS-SDR receivers. *EURASIP J. Adv. Signal Process.* **2014**, *2014*, 42. [CrossRef]
28. Mitola, J. The software radio architecture. *IEEE Commun. Mag.* **1995**, *33*, 26–38. [CrossRef]
29. Zhang, X.; Wu, M.; Liu, W. Initial assessment of the COMPASS/BeiDou-3: New-generation navigation signals. *J. Geod.* **2017**, *91*, 1225–1240. [CrossRef]
30. Lu, M.; Li, W.; Yao, Z.; Cui, X. Overview of BDS III new signals. *Navigation* **2019**, *66*, 19–35. [CrossRef]
31. Lvyang, Y. Analog Generation and Performance Analysis of BeiDou-3 B1C Signal. Master's Thesis, University of Chinese Academy of Sciences (National Time Service Center of Chinese Academy of Sciences), Beijing, China, 4 June 2019. Available online: https://kns.cnki.net/KCMS/detail/detail.aspx?dbname=CMFD201902&filename=1019611492.nh (accessed on 26 July 2022).
32. Yumeng, T. Design of BeiDou Third Generation B2a Frequency Point Software Receiver. Master's Thesis, Xi'an University of Technology, Xi'an, China, 12 April 2019. Available online: https://kns.cnki.net/KCMS/detail/detail.aspx?dbname=CMFD201902&filename=1019858483.nh (accessed on 26 July 2022).
33. Anantharamu, P.B.; Borio, D.; Lachapelle, G. Sub-carrier shaping for BOC modulated GNSS signals. *EURASIP J. Adv. Signal Process.* **2011**, *2011*, 133. [CrossRef]
34. Yuan, Y.; Mi, X.; Zhang, B. Initial assessment of single and dual-frequency BDS-3 RTK positioning. *Satell. Navig.* **2020**, *1*, 31. [CrossRef]

Article

Spatial–Temporal Variability of Global GNSS-Derived Precipitable Water Vapor (1994–2020) and Climate Implications

Junsheng Ding [1,2], **Junping Chen** [1,2,3,*], **Wenjie Tang** [1,2] **and Ziyuan Song** [1,2]

1   Shanghai Astronomical Observatory, Chinese Academy of Sciences, Shanghai 200030, China; dingjunsheng@shao.ac.cn (J.D.); tangwj@shao.ac.cn (W.T.); songziyuan@shao.ac.cn (Z.S.)
2   School of Astronomy and Space Science, University of Chinese Academy of Sciences, Beijing 100049, China
3   Shanghai Key Laboratory of Space Navigation and Positioning Techniques, Shanghai 200030, China
*   Correspondence: junping@shao.ac.cn

Abstract: Precipitable water vapor (PWV) is an important component in the climate system and plays a pivotal role in the global water and energy cycles. Over the years, many approaches have been devised to accurately estimate the PWV. Among them, global navigation satellite systems (GNSS) have become one of the most promising and fastest-growing PWV acquisition methods because of its high accuracy, high temporal and spatial resolution, and ability to acquire PWV in all weather and in near real time. We compared GNSS-derived PWV with a 5 min resolution globally distributed over 14,000 stations from the Nevada Geodetic Laboratory (NGL) from 1994 to 2020 with global radiosonde (RS) data, temperature anomalies, and sea height variations. Then, we examined the temporal and spatial variability of the global PWV and analyzed its climate implications. On a global scale, the average bias and root mean square error (RMSE) between GNSS PWV and RS PWV were ~0.72 ± 1.29 mm and ~2.56 ± 1.13 mm, respectively. PWV decreased with increasing latitude, and the rate of this decrease slowed down at latitudes greater than 35°, with standard deviation (STD) values reaching a maximum at latitudes less than 35°. The global average linear trend was ~0.64 ± 0.81 mm/decade and strongly correlated with temperature and sea height variations. For each 1 °C and 1 mm change, PWV increased by ~2.075 ± 0.765 mm and ~0.015 ± 0.005 mm, respectively. For the time scale, the PWV content peaked ~40 days after the maximum solar radiation of the year (the summer solstice), and the delay was ~40 days relative to the summer solstice.

Keywords: precipitable water vapor (PWV); global navigation satellite system (GNSS); temporal and spatial variability; radiosonde (RS)

Citation: Ding, J.; Chen, J.; Tang, W.; Song, Z. Spatial–Temporal Variability of Global GNSS-Derived Precipitable Water Vapor (1994–2020) and Climate Implications. *Remote Sens.* **2022**, *14*, 3493. https://doi.org/10.3390/rs14143493

Academic Editor: Chung-yen Kuo

Received: 22 June 2022
Accepted: 19 July 2022
Published: 21 July 2022

Publisher's Note: MDPI stays neutral with regard to jurisdictional claims in published maps and institutional affiliations.

1. Introduction

Precipitable water vapor (PWV), typically expressed in terms of height in millimeters, is the amount of water found in a vertical column of unit cross-sectional area, extending all the way from the Earth's surface to the upper edge of the atmosphere, that is potentially available for precipitation [1–3]. Although PWV accounts for less than 5% of the volume of the whole atmosphere, it is the most variable (by more than three orders of magnitude) of the major constituents of the atmosphere. In addition, PWV contributes more than any other component of the atmosphere to the greenhouse effect, playing a crucial role in the global water and energy cycles [4–7]. Therefore, the acquisition of PWV content in the air and the study of its variation over time and space are indispensable for understanding climate change and weather processes.

Traditional PWV measurement methods include radiosonde (RS), ground-based microwave radiometers, weather radar, etc. [8–13]. However, because of their high cost, most of these methods have not been widely rolled out and used on a global scale except for RS. Radiosondes are mounted on sounding balloons to record data from different pressure layers and transmit them to the ground by radio. As the most classic means of water vapor acquisition, RS has the longest record and has been considered to have the widest and most comprehensive global coverage [14–16], so RS PWV data are among the most suitable data for global analysis. In addition, because of the accuracy and reliability of RS data, they are often used as a standard to verify the observations of other instruments [17]. However, because of the high maintenance cost, the spatial and temporal resolution of sounding instrument observations is still low. There are only more than a thousand sounding stations worldwide, and usually, only two observations are made per day, mostly at 0:00 and 12:00 UTC [18,19]. Furthermore, there is a significant lack of RS data in less developed regions and in regions where extreme weather conditions exist year-round. These factors limit the application of RS for high-spatial- and -temporal-resolution water vapor monitoring and analysis. Note that numerical weather models (NWMs) are also among main sources of PWV data, and most existing PWV studies have also been based on the evaluation and analysis of NWMs. However, NWMs do not represent direct measurement, as RS and GNSS do, and may be less reliable for areas where no or limited data assimilation observations are available, so they are not discussed in this research.

As early as 1992, Bevis et al. [4] proposed the concept of global positioning system (GPS) meteorology and experimentally verified the possibility of inversion of water vapor by GPS. Ground-based GPS can operate in all weather, is virtually impervious to weather extremes, and can operate consistently over long periods of time at very low maintenance costs, which gives the PWV data obtained by this method high degrees of continuity and integrity. With the development and improvement of global and regional satellite navigation systems such as GLONASS, the BeiDou Navigation Satellite System (BDS), and Galileo, an increasing number of scholars in related fields have conducted experiments to evaluate the reliability of this technology and to verify the accuracy of GNSS (global navigation satellite system) PWV. Rocken et al. [20] assessed PWV measured with six GPS receivers for 1 month at sites in Colorado, Kansas, and Oklahoma and concluded that GPS and water vapor radiometer (WVR) estimates agreed to 1–2 mm root mean square error (RMSE). Using one and a half years of data from the western Mediterranean region, Hasse et al. [21] obtained a standard deviation (STD) difference of 1.2 cm between the RS and GNSS zenith total delay (ZTD), which was equivalent to a PWV difference of 2 mm [22]. Zhang et al. [23] obtained an STD of 3.27 mm between GPS PWV and RS PWV from 2011–2013 in Tahiti, French Polynesia, located in the tropical southern–central Pacific Ocean. The growing number of GNSS PWV studies has opened up the possibility of using GNSS to analyze high-spatial- and -temporal-resolution PWV. However, limited by experimental materials, these studies have often been conducted only regionally at only a few to a few dozen stations [24–31], and the results obtained in this way may not be applicable on a global scale. These results have validated GNSS as a good technique for acquiring PWV but have not reflected the strengths and potential of GNSS in global-scale PWV monitoring and analysis. Fortunately, the publication of the Nevada Geodetic Laboratory (NGL) troposphere products (which is probably the most comprehensive GNSS database currently available) solved this dilemma.

In this study, we examined the RS dataset from Integrated Global Radiosonde Archive Version 2 (IGRA2) and the GNSS products from the NGL. Then, the methods of PWV retrieval from these two observation techniques were introduced. To evaluate the reliability of the NGL products, the IGRA2 PWV for the 1994–2020 period was used as a reference, and the mean deviation (bias) and RMSE of the NGL PWV were calculated. Finally, the PWV data derived from more than 14,000 NGL GNSS stations over 20 years were analyzed in terms of temporal and spatial variation characteristics, anomalies in PWV were analyzed for correlations with global temperature and sea level data, and some valuable conclusions were drawn that are discussed and summarized herein.

2. PWV Data and Analysis Methods

2.1. PWV Retrieval

As PWV is not a raw observation but needs to be derived by certain methods and models, which are not exactly the same for different databases, it is necessary to give a brief introduction to the method of PWV retrieval.

2.1.1. Radiosonde PWV Retrieval

The RS provides pressure level data from the ground to the top of the atmosphere, including temperature, pressure, relative humidity, geopotential height, dewpoint depression, wind direction and wind speed. RS-derived PWV is described as follows [32,33]:

$$e = 6.112 \exp\left(\frac{17.6T_d}{T_d + 243.15}\right) \tag{1}$$

$$PWV = \frac{1}{g} \int_{p1}^{p2} \frac{0.622e}{p - e} dp \tag{2}$$

where T_d is the dewpoint temperature ($^\circ$C); e is the vapor pressure (hPa); p is the atmospheric pressure (hPa); p_1 and p_2 are the atmospheric pressures at the lower and upper layers (hPa), respectively; and g is the acceleration of gravity (m/s^2).

2.1.2. GNSS PWV Retrieval

ZTD can be divided into two components, known as zenith hydrostatic delay (ZHD) and zenith wet delay (ZWD). PWV can be derived from ZWD by the following formula [34]:

$$PWV = \Pi \times ZWD \tag{3}$$

$$\Pi = \frac{10^6}{\rho R_v[(k_3 / T_m) + k_2']} \tag{4}$$

where ρ is the density of liquid water, R_v is the gas constant of water vapor, T_m denotes the weighted mean temperature of the atmosphere, and k_2' and k_3 are constants determined experimentally. NGL products obtained T_m from VMF1 (Vienna Mapping Function 1) gridded NWM data [35].

It should be noted that since the vast majority of GNSS stations are not equipped with meteorological sensors, ZHD is usually obtained from a model in GNSS data processing. However, this may result in model errors being absorbed into the ZWD and then into the PWV. The NGL solution used the VMF1 grid products as the ZHD input. The VMF1 was computed from the NWM by ray-tracing methods. When obtaining the site ZHD from the VMF1 grid products, the grid point ZHD was first converted to the corresponding height pressure, the pressure was lifted from grid height to site height, and finally, the lifted pressure was converted to ZHD again [36].

Another way to obtain high-accuracy ZHD is with reanalysis data. Reanalysis data give the best estimates where no data are recorded, in between stations and observation times. It is an "optimal" reflection of the atmospheric conditions, but not a true reflection. According to experiments by Fernandes et al. [37] and Urquhart et al. [38], the bias of the ZHD obtained from the VMF1 ZHD and the in-situ measurement was approximately 3 mm, which translated into a PWV of less than 0.5 mm. Furthermore, according to the experiments of Urquhart et al. [38] and Wang et al. [39], the bias of the ZHD difference between VMF1 and reanalysis data into PWV was around 0.3–0.4 mm. The results of these studies showed that there was no significant difference between using VMF1 and reanalysis data to obtain ZHD in regard to the accuracy of the derived PWV.

2.2. RS and GNSS PWV

In this section, the RS and GNSS data used and the data preprocessing methods used in this study are introduced.

2.2.1. IGRA2 RS PWV

The Integrated Global Radiosonde Archive (IGRA) is an RS dataset maintained, archived, and distributed by the National Oceanic and Atmospheric Administration (NOAA) National Centers for Environmental Information (NCEI), formerly National Climatic Data Center (NCDC) [40]. IGRA2 is the second version of IGRA and consists of quality-controlled RS observations at stations across all continents. Its data are drawn from more than 30 different sources. The earliest year of data is 1905, and the data are updated on a daily basis. In this study, the IGRA2 RS PWV data from 1994 to 2020 were selected, covering the same period as the GNSS PWV data. To ensure the collocation of the sounding stations used for reference, the following two constraints were applied: (i) the distance between the two types of stations was less than 40 km, and the height difference was less than 100 m (existing research on this restriction has proven it to be sufficiently stringent, feasible, and effective [12]); (ii) the common available data covered at least one year. After the selection, a total of 1534 IGRA2 stations were available.

2.2.2. NGL GNSS PWV

At the end of 2017, the Nevada Geodetic Lab announced the new public availability of over 34 million station-days of their tropospheric products (total zenith delay, north gradient, and east gradient every 5 min since 1996) from over 16,000 stations. The data processing strategy was precise point positioning (PPP). In March 2020, the products were updated with an earlier starting year of 1994, more than 19,000 stations, over 43 million station-days, and the addition of ZWD, T_m, and PWV data [35,41]. Table 1 summarizes the properties of NGL products, including the adopted models, strategies, and products used. In addition, more detailed information can be obtained from NGL's solution strategy file: http://geodesy.unr.edu/gps/ngl.acn.txt (accessed on 8 August 2021). The NGL products provided the corresponding STD of the elements listed in the table, with the exception of T_m, as this was provided by the VMF1 gridded numerical weather model. In addition, although the database is updated weekly, the data delay is approximately three weeks. Although it has been open for only a short time, NGL's GNSS products surpass common databases, such as the International GNSS Service (IGS) and Crustal Movement Observation Network of China (CMONOC), in terms of data volume and timespan. In particular, for the number of stations, NGL surpasses the first two by two orders of magnitude. In addition, the coverage and density of GNSS are wider and denser than those of RS in a large number of less developed and bipolar regions, which makes the NGL tropospheric products possibly the most comprehensive and extensive PWV database currently available.

Table 1. Adopted models and strategies of Nevada Geodetic Laboratory products.

Items	Properties
Service time	5 November 2017—present
Timespan	1 January 1994—present
Sampling interval	5 min
Elevation cutoff angle	7 degrees
Trop mapping function	VMF1 [42]
Elements	total zenith delay, north gradient, east gradient, water vapor, and weighted mean temperature
Orbit	JPL's Repro 3.0 orbits
Clock	JPL's Repro 3.0 clocks
Software	JPL's GipsyX 1.0 [43]
System	GPS-only
Data amount	>46 million days
Number of stations	>19,000
Update frequency	1 week (with new incoming data, as well as newly discovered stations)
Station growth rate	About 1000 per year

The number of valid NGL stations, the number of total stations, and the total number of products in the database were counted for each day from 1994 to 2020 and are displayed in Figure 1. Figure 1a shows the variation in these quantities with time, and Figure 1b shows the effective time for each station (blue is the background color, each yellow horizontal line represents a station, and stations are in alphabetical order from top to bottom). The following information is relevant to Figure 1: (i) on 4 December 2004, the number of valid stations dropped abruptly from approximately 2000 to 102 and recovered immediately afterwards, probably because of a network outage; (ii) the sudden increase in the total number of stations on 1 January 2009 was due to the addition of a large number of stations from Japan to the database, with numbers ranging from J001 to J999. In addition, starting in 2009, the number of stations entered a period of rapid growth, reaching approximately 1000 new stations added per year; (iii) stations J001–J999, located in Japan, and P001–P821, on the west coast of the United States, had good observation quality, with fewer missing data and interruptions than other stations.

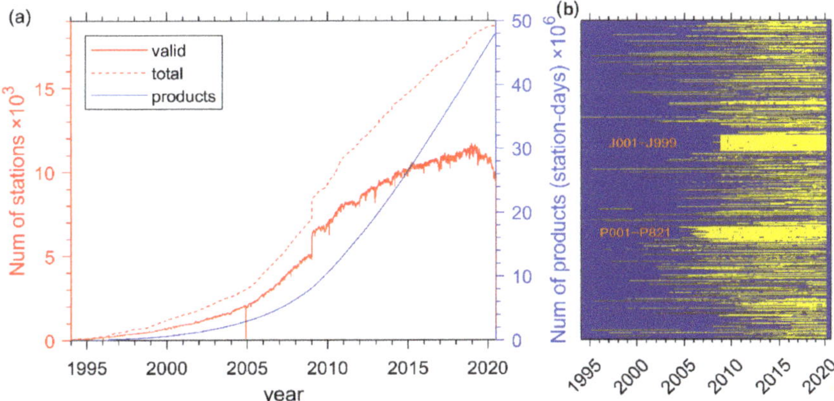

Figure 1. Variability in the numbers of Nevada Geodetic Laboratory (NGL) stations and products. (**a**) The number of valid NGL stations (red solid line), the total number of stations (red dotted line), and the total number of products in the database data (blue solid line); (**b**) the effective time for each station (each yellow horizontal line represents a station, and station names are in alphabetical order from top to bottom).

Information on the distribution of the stations is shown in Figure 2. The colors of the dots indicate the ellipsoid heights of the corresponding stations. Per the map, the majority of stations were below 1 km in ellipsoid height, and the stations with higher ellipsoid heights were mainly concentrated on the west coast of the American continent, which is where the Cordillera de los Andes is located. In addition, there were a number of stations in the Antarctic and Greenland, which are valuable for studying PWV changes in both polar regions. It should be noted that because of the large number of stations, there are inevitably some overlaps in the display. For better visualization, important information is displayed first as needed.

2.3. Product Validation and Analysis Methods

In this section, a comparison between NGL GNSS PWV and IGRA2 RS PWV is presented, and the extraction of seasonal and interannual variation signals of PWV is introduced.

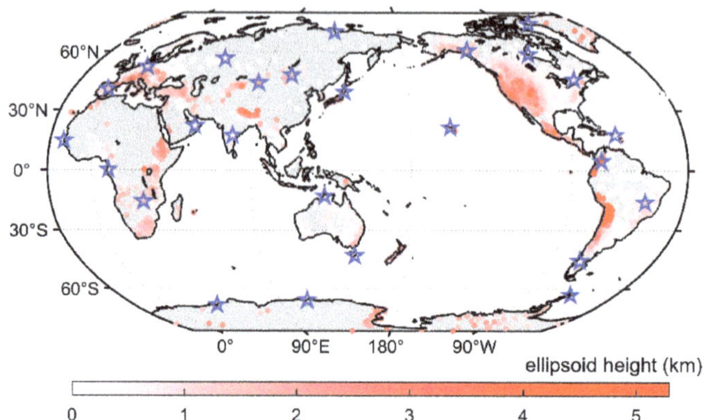

Figure 2. Distribution of global NGL stations. Colors represent the ellipsoids of the corresponding sites. The blue pentagrams represent the stations used for power spectral density (PSD) analysis.

2.3.1. Comparison with RS PWV

A total of 1534 pairs of NGL and IGRA2 stations were screened, each pair within 40 km on the ellipsoid surface and within 100 m in height, with a common data validity of more than one year (pairs of stations that did not have a full year of common observations because of missing data were excluded). The bias values, relative bias, RMSE values, and normalized RMSE values of these stations are presented in Figure 3. The biases of most of the stations were positive, and stations with negative bias values (GNSS PWV less than RS PWV) are mainly concentrated on the west coast of America and inland on the Eurasian continent. The RMSE value showed a negative correlation that decreased with increasing latitude, and for relative bias and normalized RMSE, many large values appeared in the polar regions, which indicated that the difference between the two PWV acquisition methods was larger in the high-latitude regions with low PWV content than in other regions.

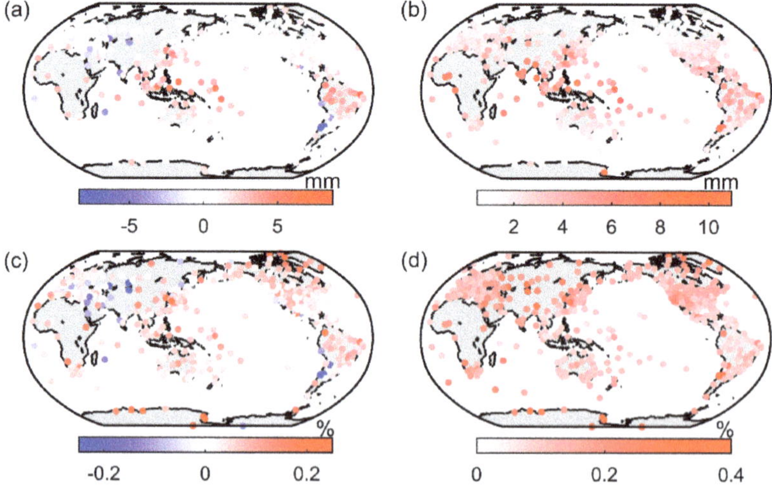

Figure 3. Bias (**a**), relative bias (**c**) (in % relative to mean PWV), root mean square error (RMSE) (**b**), and normalized RMSE (**d**) of 1534 pairs of NGL global navigation satellite system (GNSS) and Integrated Global Radiosonde Archive Version 2 (IGRA2) RS stations.

The statistical results of the data are shown in Figure 4. By fitting a normal distribution to the frequency histogram shown in Figure 4, we revealed that for NGL GNSS PWV, the global average bias was ~0.72 ± 1.29 mm, the mean RMSE is ~2.56 ± 1.13 mm, the global average relative bias was ~0.03%, and the normalized mean RMSE was ~0.14%. Based on these results, we screened NGL stations and eliminated stations with RMSEs lower than 0.14%. It should be noted that the comparison here was only to analyze how the NGL products differed from the mainstream PWV products (there have been a large number of studies based on the RS products) and not to consider whether the RS products had higher accuracy. In fact, although NCEI scientists applied a comprehensive set of quality control procedures, IGRA data still suffered from jumps due to changes in instrumentation, etc., and the unhomogenized RS data even affected most of the reanalysis data [12]. In the GNSS solution, however, the tropospheric delay and the station coordinates were estimated together, and the errors caused by these issues were mainly absorbed by the station coordinates and hardly affected the tropospheric delay. In addition, since GNSS satellites are more than 20,000 km from the ground, they cover the full length of the water vapor path and do not need to be imputed to the top atmosphere in the same way as RS data.

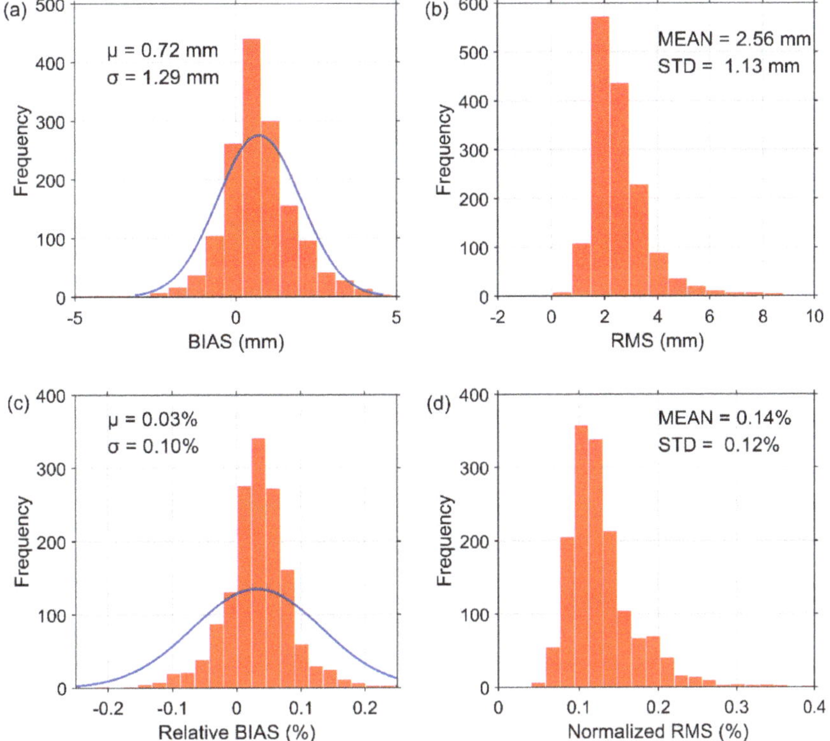

Figure 4. Frequency histograms and normal fittings of bias (**a**), RMSE (**b**), relative bias (**c**), and normalized RMSE (**d**) of 1534 stations during 1994–2020.

2.3.2. Acquisition of PWV Spatial–Temporal Distribution Features

Significant annual periodic terms can be found in the PWV time series. To obtain all the periodic terms in the PWV series, 26 globally distributed stations (the blue pentagrams in Figure 2) were selected, which were well distributed and from which the available data were more than 20 years old. The power spectral densities (PSDs) of the PWV time series from these stations were obtained and are plotted in Figure 5. In Figure 5, the gray line is the PSD of these 26 stations, and the red line is the average PSD of the 26 stations. This figure shows that in addition to the annual cycle term, there was a significant semiannual cycle term in the PWV series. Theoretically, high-temporal-resolution GNSS data would allow smaller-period features to be extracted. However, some scholars have claimed that smaller-period temporal characteristics, such as diurnal or semidiurnal cycle characteristics, can be extracted from the reanalysis data [44]. Such features were not obtained from GNSS PWV, probably because these period amplitudes were very small (less than PWV accuracy) and were masked by observation noise.

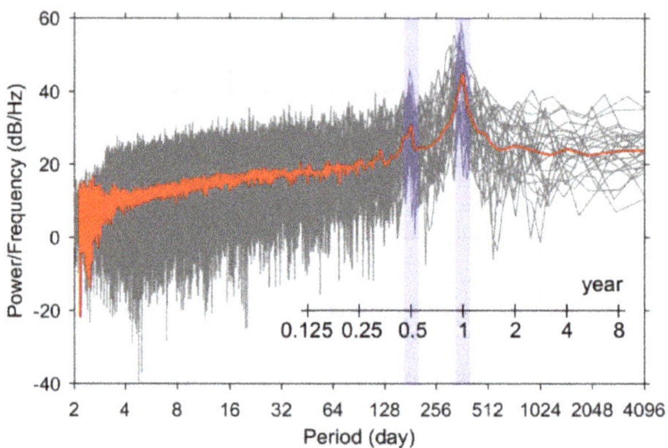

Figure 5. Time series of the power spectral density (PSD) of PWV. Each gray line represents the PSD of a station, and the red line represents the average PSD value.

Trend fitting and seasonal cycle fitting were performed using Equations (5) and (6), respectively:

$$PWV(t) = a + b \cdot t \tag{5}$$

where a is the intercept, b is the growth rate of the PWV, and t is the decimal year;

$$PWV(doy) = A_1 \cos(2\pi\frac{doy + P_1}{365.25}) + A_2 \cos(4\pi\frac{doy + P_2}{365.25}) + C \tag{6}$$

where A_1 and P_1 are the annual amplitude and phase of the PWV, respectively; A_2 and P_2 are the semiannual amplitude and phase, respectively; C is a constant term that represents the annual average value of PWV; and doy is the day of the year.

3. Results

3.1. Spatial Analysis

PWV data were extracted from the ZWD while retaining most of the spatial characteristics of the ZWD. We counted all PWV data according to the latitudes and ellipsoidal heights of the stations from which they were collected. Since the temporal resolution of GNSS PWV was 5 min, this led to an overwhelming amount of data that was hard to process, so we calculated first the daily mean value and STD of the PWV and then the average value of each station's means and STDs for the 1994–2020 period. Subsequently, the PWV means and STDs were calculated by group according to latitude and ellipsoid height, and the results of the means and STDs for each group are shown in Figure 6.

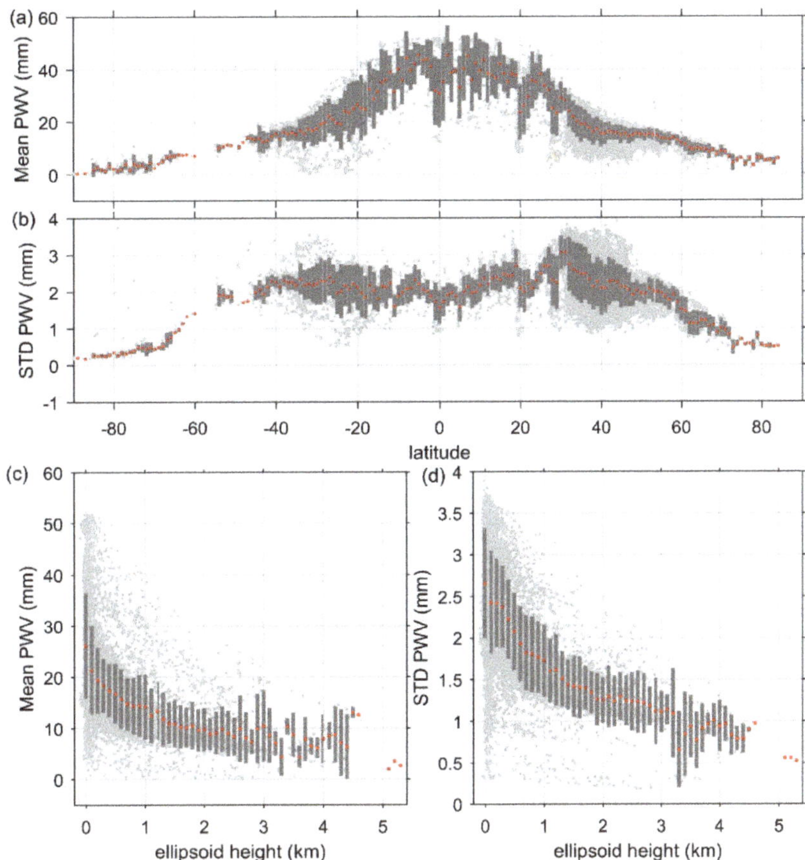

Figure 6. Distribution of mean PWV and PWV standard deviation (STD) with latitude and ellipsoid height. (**a**) PWV mean with latitude, (**b**) PWV STD with latitude, (**c**) PWV mean with ellipsoid height, (**d**) PWV STD with ellipsoid height. Light gray dots indicate each station, red dots indicate group means, and dark gray bars represent group STD. Note: the groups for latitude and ellipsoid height are in intervals of 1° and 100 m, respectively.

Figure 6a shows that: (i) PWV mean values were symmetrically distributed in the Northern and Southern Hemispheres and negatively correlated with latitude, and their relationship with latitude could be fitted by a segmented linear function in which the absolute slope value for 0–35 degrees latitude was greater than that for 35–90 degrees; (ii) STD in the latitude groups decreased with increasing latitude, which indicated that the fluctuation of PWV mean values in different regions at the same latitude increased with decreasing latitude. Figure 6b shows that: (i) the mean value of the PWV daily STD decreased with increasing latitude in the regions with latitudes greater than 35°, but there was no significant correlation in the regions with latitudes less than 35° which indicated that the PWV fluctuations reached the limit in low- and mid-latitude regions; (ii) in the group where the maximum value was, i.e., at approximately 35°, the degree of variation in fluctuations was the greatest, which also indicated that the variation in PWV was the greatest among the geographical areas around this group.

From Figure 6c,d: (i) the mean PWV and STD were exponentially negatively correlated with the ellipsoid height of the stations, the PWV values varied greatly between stations in the low ellipsoid height region, and the difference decreased as the height increased;

(ii) the rate of decrease in PWV with increasing ellipsoid height decreased with increasing ellipsoid height, because water vapor is mainly concentrated on the Earth's surface.

3.2. Seasonal Cycle

The seasonal cycle was the most intuitive and significant temporal feature exhibited by the PWV time series. We fit the PWV using the annual periodicity + semiannual periodicity function of (11), and the fitting residuals and fitting parameters are shown in Figure 7 (Figures A1–A4 in Appendix A show the PWV fit of four stations at different latitudes). It should be noted that, considering that accurate periodicity can be extracted only from data with a sufficient timespan, we excluded stations with less than two years of data, and the fitting results for the remaining >14,000 stations are shown in Figure 7.

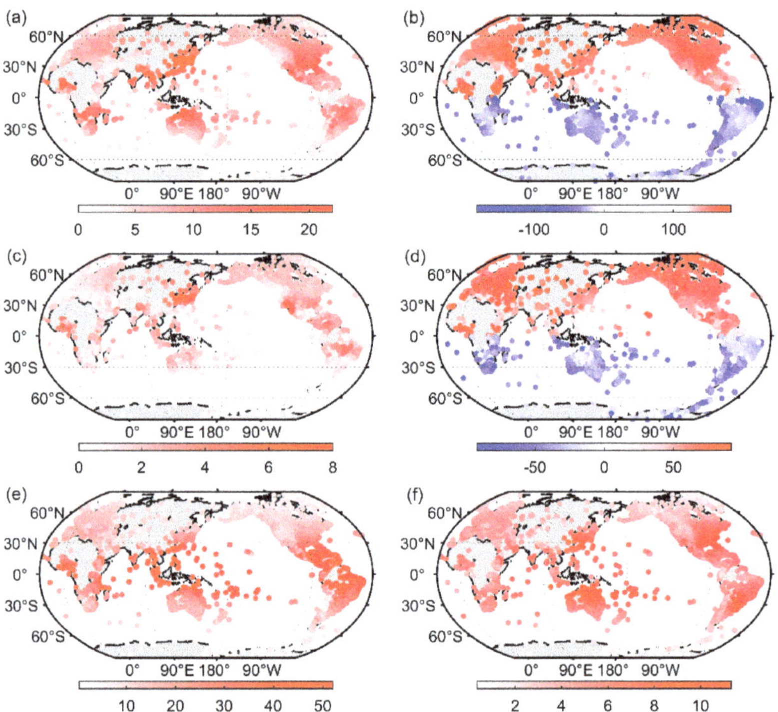

Figure 7. Global distribution of PWV fitting parameters and residuals. (**a**) Annual amplitude A_1; (**b**) annual phase P_1; (**c**) semiannual amplitude A_2; (**d**) semiannual phase P_2; (**e**) constant term C; (**f**) RMSE of the fitting residuals.

Figures 7a and 7b show the annual amplitude A1 and phase P1, respectively, and Figures 7c and 7d display the semiannual amplitude A2 and phase P2, respectively. The annual amplitude was between 0 and 22 mm and showed an overall pattern of increasing from high latitudes to low latitudes. However, the maximum value appeared not in the equatorial region but near the Tropic of Cancer, especially in the coastal areas of the region. This was because these regions have sufficient water vapor and a large temperature difference during the year, while the equatorial region has a small temperature difference throughout the year, and the fluctuation of water vapor content is relatively small.

The annual phase can reflect the time of the year when the PWV content is highest and the time when it is lowest. As shown in Figure 7b, the phase sign was opposite in the Northern and Southern Hemispheres. The phase value was concentrated in the periods between late May and early June in the Northern Hemisphere and in the periods between late November and early December in the Southern Hemisphere, and the semiannual cycle phase was similar to the annual phase. To obtain the distribution characteristics of the phases, statistical plots of the frequency distribution of the phase values of all the stations are shown in Figure 8. Figure 8 shows that both the annual and semiannual phases showed the phenomenon of clustering towards different values in the Northern and Southern Hemispheres. The mean value of the annual phase in the Northern Hemisphere stations was ~152.1, and the mean value of the semiannual phase was ~61.6, while the mean value of the annual phase in the Southern Hemisphere was ~-34, and the mean value of the semiannual phase was ~-40. The conversion led to the conclusion that PWV levels in the Northern Hemisphere peaked around August 1 (the 213th day of the year) and reached a trough around January 31 each year, while PWV levels in the Southern Hemisphere peaked around February 3 (the 34th day of the year) and reached a trough around August 5 each year. This indicated that the water vapor content in the atmosphere peaked approximately 40 days after the solar radiation reached its maximum (on the summer solstice) during the year. These results may be more informative for global water and energy cycle studies than some studies of global temperature through surface sensor data [45–47].

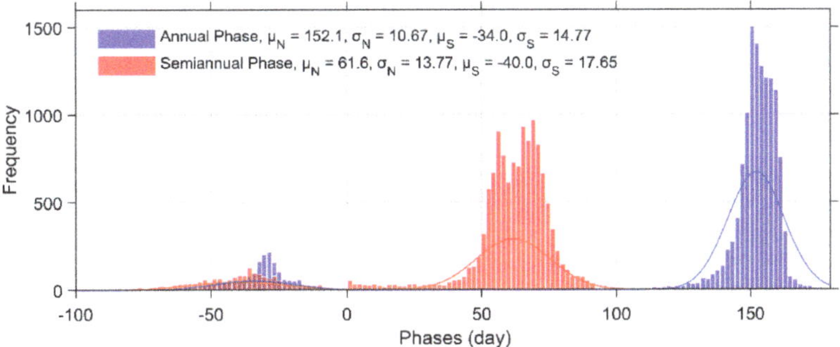

Figure 8. Statistical histogram and normal fitting of the annual phase (blue) and semiannual phase (red) in the Northern and Southern Hemispheres. The phase in the Southern Hemisphere was less than zero, the phase in the Northern Hemisphere was greater than zero, and the subscripts N and S represent the Northern and Southern Hemispheres, respectively.

The semiannual amplitude ranged from 0 to 8 mm, which was much smaller than the annual amplitude, and the decreasing pattern with increasing latitude was not as significant as that of the annual amplitude. The semiannual amplitude was significantly larger on the southeast coast of China–Japan and in the California axis region of the west coast of the United States than in the other regions.

The constant term C (Figure 7e) represents the annual average value of PWV, from which it was found that the annual average value of PWV ranged from 0 to 50 mm and that the values decreased with increasing latitude, from more than 40 mm in the tropical regions to less than 10 mm in the polar regions. Figure 7f shows the RMSE value of the fitted residuals, from which it was found that the polar regions had small RMSE values, larger values of RMSE appeared on the west coasts of the Pacific and Atlantic Oceans, and the global average RMSE was ~5.72 $\pm$ 1.89 mm. This indicated that these regions had greater daily variation around the seasonal means. These were also the areas where severe weather, such as typhoons and rainstorms, often occurs.

3.3. Trend Analysis

The trend signal reflects changes in the time series over long periods of time, and we used (5) to obtain a linear trend of PWV time series for over 5000 NGL stations between 2009 and 2020. All of the selected stations had more than ten years of available data (stations with slight data missing were complemented by the fitting value of (11), while stations with serious missing data were eliminated). After calculating the interannual variation in the annual mean value of each station, we found that the PWV trend had no regional distribution characteristics. Considering the large variation in PWV values between seasons, we grouped the statistics according to four seasons: spring (March–April–May for the Northern Hemisphere (NH) and September–October–November for the Southern Hemisphere (SH)), summer (June–July–August for the NH and December–January–February for the SH), autumn (September–October–November for the NH and March–April–May for the SH), and winter (December–January–February for the NH and June–July–August for the SH). The results are presented in Figure 9. In addition, stations with positive values in all four seasons and stations with negative values in all four seasons are shown in Figures 9e and 9f, respectively.

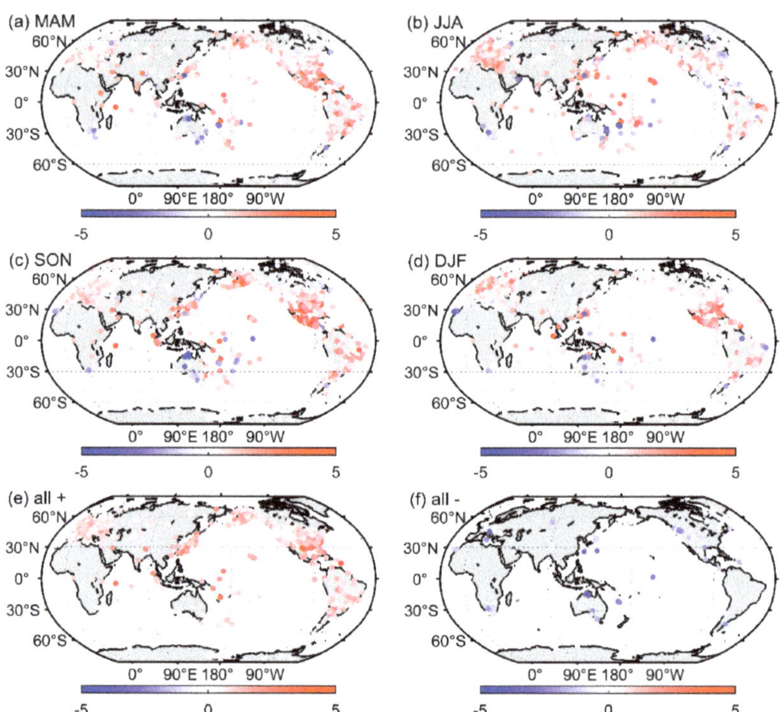

Figure 9. Spatial pattern of seasonal PWV trends during 1994–2020. (**a**) March–April–May (MAM), (**b**) June–July–August (JJA), (**c**) September–October–November (SON), (**d**) December–January–February (DJF), (**e**) stations with positive values in all four seasons, (**f**) stations with negative values in all four seasons.

From Figure 9a,d, we found that more stations had positive than negative values in all four seasons. However, in regard to the stations with positive (e) and negative (f) values in each of the four seasons, we found that the number of stations with positive PWV growth rates was much larger than the number with negative growth rates. In addition, the following patterns were found: (i) the PWV growth rates in Antarctica, the tropical region of South America, and Alaska were almost all positive, indicating that the PWV values in these regions were increasing year by year; (ii) a large number of stations in the Canada and South Africa regions had negative PWV growth rates for all four seasons, while very few stations had positive values, indicating that the PWV values in these regions were decreasing year by year.

Figure 10 shows the frequency distribution of the PWV growth rate of all stations in the four seasons. From the figure, it was found that all four seasons conformed to the normal distribution, and by fitting the normal distribution, we obtained mathematical expectations of the PWV growth rate of ~0.68 ± 0.92 mm/decade, ~0.43 ± 1.14 mm/decade, ~0.80 ± 1.36 mm/decade, and ~0.64 ± 1.02 mm/decade in spring, summer, autumn, and winter, respectively. This result was numerically consistent with studies using data obtained from other data sources, including reanalysis data [48], RS data [49], MODIS (moderate resolution imaging spectroradiometer) data [50], etc. This indicated that: (i) the global PWV content showed a yearly increasing trend in all four seasons, which was most significant in autumn and least significant in summer, and (ii) the dispersion (standard deviation) of the PWV growth rate among stations increased gradually in spring, summer, and autumn but was lowest in winter.

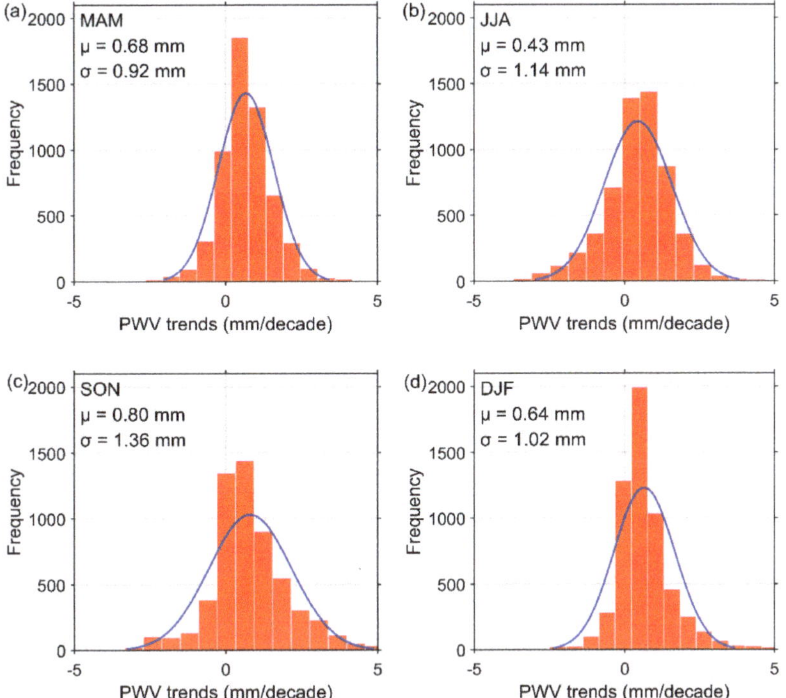

Figure 10. Seasonal frequency histograms of PWV trends during 1994–2020 for (**a**) MAM, (**b**) JJA, (**c**) SON, and (**d**) DJF.

In addition, it was found from the results that the PWV growth rate had large uncertainty and that its spatial distribution was not continuous in some regions. This phenomenon was also found in some studies of long-term trends in PWV obtained using other methods [50,51]. Combining these studies, the following possible reasons for this phenomenon are listed: (i) PWV is highly correlated with land cover type [52], and the density of discrete GNSS stations is still low; (ii) differences in analysis periods and spatial coverage could be responsible for these discrepancies [49].

Table 2 shows the percentages of stations with positive and negative PWV growth rates in each of the four seasons to the total number of stations. From the table, it was found that in each season, the proportion of positive values was approximately triple to quadruple that of negative values, and the proportion of stations with positive values in all seasons was more than 20 times the proportion of stations with negative values in all seasons. This indicated that PWV was increasing year by year on a global scale.

Table 2. Percentages of positive and negative values for the four seasons and all seasons.

	Spring	Summer	Autumn	Winter	All Seasons
positive	82.72%	70.72%	78.22%	82.77%	45.73%
negative	17.28%	29.28%	21.78%	17.23%	2.15%

The relationships between the PWV growth rate and both latitude and ellipsoid height were likewise analyzed, with the latitude and ellipsoid height grouped in the same way as above, and the results are presented in Figure 11. It should be noted that the increase in PWV did not mean that the evaporation in this area had also increased, but it meant that increased water vapor was concentrated in these areas. The results in Figure 11 showed that the PWV growth rate was not simply positively or negatively correlated with latitude and ellipsoid height. It was found that the increased water vapor was more concentrated near the equator, the Tropic of Cancer, and 60° latitude, while from the ellipsoid height, the increased water vapor was more concentrated in the area within 1 km from the ground (these areas all had growth rates higher than 0.5 mm/decade). This suggested that these regions are more likely to suffer drastic weather changes due to uneven increases in water vapor.

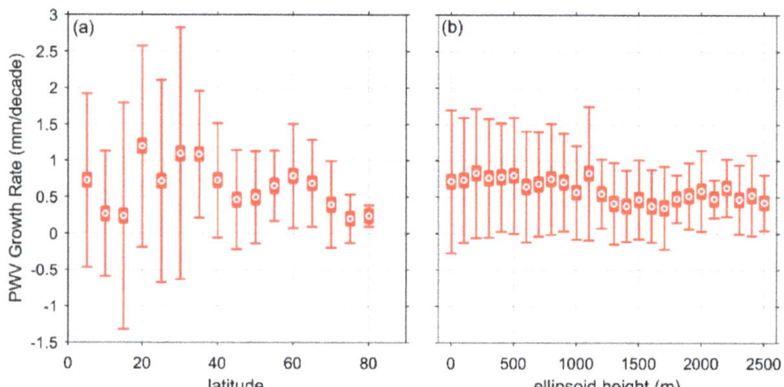

Figure 11. Distribution of PWV growth rates with latitude and ellipsoid height. (**a**) PWV growth rate with latitude; (**b**) PWV growth rate with ellipsoid height. The error bars are STDs of PWVs of stations in the same latitude/height grouping area. Note: the groups for latitude and ellipsoid height are in intervals of 5° and 100 m, respectively.

Under a warming climate, increasing temperatures exacerbate the evaporation of water, a process that in turn is enhanced by increasing levels of water vapor, the world's most dominant greenhouse gas. To establish the link between the growth rates of PWV and temperature, etc. and to analyze the correlation between them, we obtained global temperature anomalies and sea height variations from Berkeley Earth [53] and NASA's Goddard Institute for Space Studies (GISS) [54] for the last two decades and compared PWV anomalies with them. After data processing, the results of the comparison are shown in Figure 12, where the red points represent the global annual values of anomalies (variations), the red error bars are the yearly fluctuations in temperature anomalies and sea height variations, and the blue error bars are the yearly fluctuation of PWV anomalies. The calculation of the PWV anomalies followed that of temperature anomalies and sea height variations, with all being calculated as the differences relative to the first value. Note that the data used here were annual averages of the global data, which were calculated from monthly averages, and the yearly fluctuation is the STD of these data. The results shown in Figure 12 showed strong positive correlations between PWV anomalies and both temperature anomalies and sea level variations, with correlation coefficients of 0.81. This result was similar to that obtained by Kwon et al. [50] with a correlation coefficient of 0.8 using MODIS PWV compared with temperature anomalies. In addition, the results of the linear fit indicated that a 1 °C temperature increase corresponded to a PWV increase of ~2.075 ± 0.765 mm, and a 1 mm sea level height increase corresponded to a PWV increase of ~0.015 ± 0.005 mm.

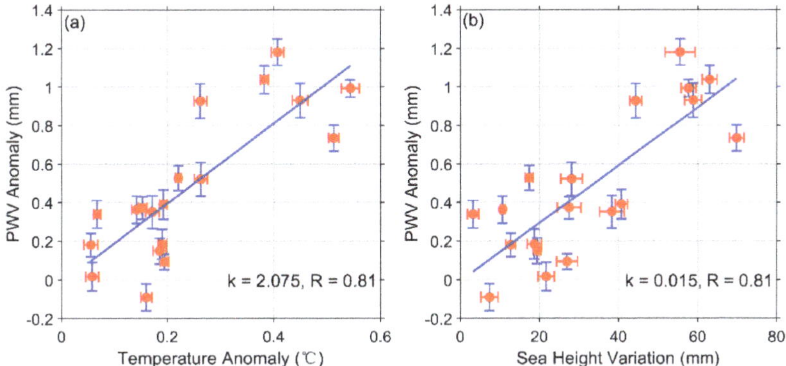

Figure 12. Comparison of PWV anomalies with temperature anomalies (**a**) and sea height variations (**b**). The red dots are the annual averages, the red error bars are the yearly fluctuations in temperature anomalies and sea height variations, and the blue error bars are the yearly fluctuations in PWV anomalies. k represents the slope of the linear fit, and R represents the correlation coefficient.

4. Discussion

In this contribution, the spatial and temporal distribution characteristics of GNSS-derived PWV for the last two decades from more than 14,000 stations distributed globally were extracted and analyzed. In the spatial analysis, we found that the annual STDs of PWV tended to be flat in the middle- and low-latitude regions, especially at low latitudes, which may be due to the saturation of water vapor content in this region. In the seasonal analysis, the amplitude and phase of each station was extracted, which could be used as model values for PWV recovery, as were the spatial distribution characteristics of each station. In the trend analysis, the interannual variability of each station was extracted, and correlations with the interannual variability of temperature and sea level were obtained, which can be used as additional supporting evidence for global warming.

Compared with existing studies on PWV (which have mostly been based on reanalysis data or RS data), this study did not have an advantage in terms of the timespan of the data because of the short development time of GNSS. Furthermore, it is difficult to obtain long-term GNSS PWV signatures for marine areas, since ground-based GNSS stations are usually fixed to bedrock.

Finally, although there are already many stations in the NGL database, they are still not evenly distributed spatially and are mainly concentrated in developed regions. Therefore, greater efforts need to be made to build stations, especially in northern–central Africa, the Asian continent, and the polar regions. With the development of low-cost, small GNSS receivers, the acquisition of GNSS-based PWV will become more and more convenient. The data records of GNSS PWV will become longer, and the coverage will become wider. GNSS will become one of the most mainstream ways to acquire PWV.

5. Conclusions

Reliable temporal and spatial variability information on water vapor is crucial to understanding climate processes and the global water and energy cycles. This study used PWV data derived from IGRA2 RS to evaluate NGL GNSS PWV products in 1994–2020. A total of 1534 RS sites were selected as the reference. The comparison showed a very high level of agreement between these two types of PWV data. Then, the temporal and spatial distribution characteristics of PWV from 1994–2020 for over 14,000 NGL stations worldwide were analyzed and discussed. In addition, long-term PWV anomalies were analyzed in comparison with temperature anomalies and sea surface height variability data. The following conclusions were obtained:

1. Compared with the RS data, the global average bias of NGL GNSS PWV was ~0.72 $\pm$ 1.29 mm, and the global average RMSE was ~2.56 $\pm$ 1.13 mm. This result showed that GNSS PWV had a very high agreement with RS PWV, and together with the advantages of GNSS, such as high temporal resolution and all-weather operation, GNSS is becoming one of the most important methods of PWV acquisition, which should be valuable for water vapor synthesis and reanalysis.
2. On a global scale, the PWV value tended to increase year by year, and the global average growth rate was ~0.64 $\pm$ 0.81 mm·decade^{-1}. PWV growth showed strong correlations with temperature anomalies and sea height variation. For each 1 °C and 1 mm change, PWV responded with ~2.075 $\pm$ 0.765 mm and ~0.015 $\pm$ 0.005 mm, respectively.
3. The PWV mean value tended to decrease in segments with increasing latitude and decreased more rapidly below 35° latitude than above 35° latitude. The PWV STD value reached a maximum at low latitudes and did not show the same complete negative correlation as PWV.
4. The mean value of GNSS PWV was between 0 and 50 mm and was negatively correlated with latitude. The annual cycle amplitude of PWV was also negatively correlated with latitude but reached its maximum near the Tropic of Cancer, especially in the coastal areas of the region, which can be explained by the eddy transport being more efficient there and frontal systems mixing polar air with tropical air.
5. PWV levels in the Northern and Southern Hemispheres reached their peak around August 1 and January 3 each year, respectively, and reached troughs around January 31 and August 5 each year, respectively. There was a delay of approximately 40 days relative to the time of year when the solar radiation reached its maximum (the summer solstice).

Author Contributions: Conceptualization, J.D.; writing—original draft preparation, J.D.; writing—review and editing, J.C., J.D., W.T. and Z.S.; visualization, J.D.; validation, J.D. and J.C.; supervision, J.C.; funding acquisition, J.C. All authors have read and agreed to the published version of the manuscript.

Funding: This research was funded by the Program of Shanghai Academic/Technology Research Leader (No.20XD1404500); the National Natural Science Foundation of China (No.11673050); the Key Program of Special Development funds of Zhangjiang National Innovation Demonstration Zone (Grant No. ZJ2018-ZD−009); the National Key R&D Program of China (No. 2018 YFB0504300); and the Key R&D Program of Guangdong province (No. 2018 B030325001).

Institutional Review Board Statement: Not applicable.

Informed Consent Statement: Not applicable.

Data Availability Statement: The radiosonde data were provided by the National Oceanic and Atmospheric Administration's (NOAA's) NCEI (National Centers for Environmental Information) and are available from ftp://ftp.ncdc.noaa.gov/pub/data/igra/ (accessed on 8 August 2021). The GNSS PWV data were provided by the Nevada Geodetic Lab at http://geodesy.unr.edu/gps_timeseries/trop/ (accessed on 17 August 2021). The global temperature and sea level data were provided by Berkeley Earth and NASA's Goddard Institute for Space Studies (GISS) at http://berkeleyearth.org/data/ (accessed on 9 November 2021) and https://climate.nasa.gov/vital-signs/ (accessed on 9 November 2021), respectively.

Acknowledgments: The authors gratefully acknowledge the NOAA's NCEI, the Nevada Geodetic Lab, Berkeley Earth, and NASA's Goddard Institute for Space Studies for providing the datasets.

Conflicts of Interest: The authors declare no conflict of interest.

Appendix A

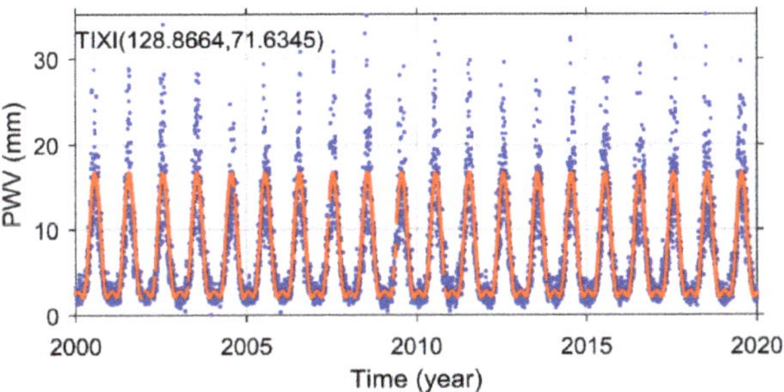

Figure A1. PWV time series (blue dots) and its fitting (red line) from the GNSS station TIXI.

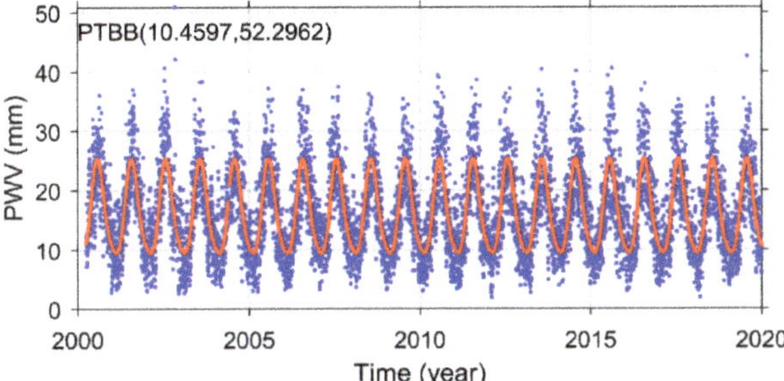

Figure A2. PWV time series (blue dots) and its fitting (red line) from the GNSS station PTBB.

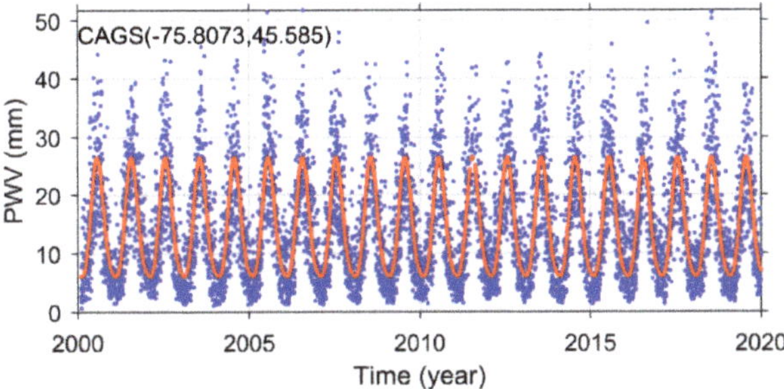

Figure A3. PWV time series (blue dots) and its fitting (red line) from the GNSS station CAGS.

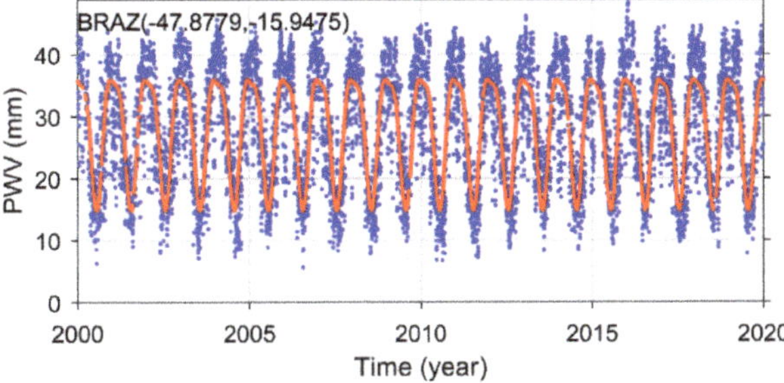

Figure A4. PWV time series (blue dots) and its fitting (red line) from the GNSS station BRAZ.

References

1. Cassel, D.; Thapa, B. Water Cycle. *Encyc. Soil. Environ.* **2005**, 258–264. [CrossRef]
2. Li, X.; Long, D. An improvement in accuracy and spatiotemporal continuity of the MODIS precipitable water vapor product based on a data fusion approach. *Remote Sens. Environ.* **2020**, *248*, 111966. [CrossRef]
3. Zhang, B.; Yao, Y. Precipitable water vapor fusion based on a generalized regression neural network. *J. Geod.* **2020**, *95*, 1–14. [CrossRef]
4. Bevis, M.; Businger, S.; Herring, T.; Rocken, C.; Anthes, R.A.; Ware, R.H. GPS meteorology: Remote sensing of atmospheric water vapor using the global positioning system. *J. Geophys. Res.-Atmos.* **1992**, *97*, 15787–15801. [CrossRef]
5. Wallace, J.; Hobbs, P. *Atmospheric Science: An Introductory Survey*, 2nd ed.; Academic Press: New York, NY, USA, 2006; pp. 15–36.
6. Vey, S.; Dietrich, R.; Rulke, A.; Fritsche, M.; Steigenberger, P.; Rothacher, M. Validation of Precipitable Water Vapor within the NCEP/DOE Reanalysis Using Global GPS Observations from One Decade. *J. Clim.* **2010**, *23*, 1675–1695. [CrossRef]
7. Smith, T.M.; Arkin, P.A. Improved Historical Analysis of Oceanic Total Precipitable Water. *J. Clim.* **2015**, *34*, 4693–4710. [CrossRef]
8. Pacione, R.; Fionda, E.; Lanotte, R.; Sciarretta, C.; Vespe, F. Comparison of atmospheric parameters derived from GPS, VLBI and a ground-based microwave radiometer in Italy. *Phys. Chem. Earth Part A/B/C* **2002**, *27*, 309–316. [CrossRef]
9. Kern, A.; Bartholy, J.; Borbás, E.; Barcza, Z.; Pongrácz, R.; Ferencz, C. Estimation of vertically integrated water vapor in Hungary using MODIS imagery. *Adv. Space Res.* **2008**, *41*, 1933–1945. [CrossRef]
10. Renju, R.; Raju, C.; Mathew, N.; Antony, T.; Moorthy, K. Microwave radiometer observations of interannual water vapor variability and vertical structure over a tropical station. *J. Geophys. Res.-Atmos.* **2015**, *120*, 4585–4599. [CrossRef]
11. Gui, K.; Che, H.; Chen, Q.; Zeng, Z.; Liu, H.; Wang, Y.; Zheng, Y.; Sun, T.; Liao, T.; Wang, H.; et al. Evaluation of radiosonde, MODIS-NIR-Clear, and AERONET precipitable water vapor using IGS ground-based GPS measurements over China. *Atmos. Res.* **2017**, *197*, 461–473. [CrossRef]
12. Zhang, W.; Lou, Y.; Huang, J.; Zheng, F.; Cao, Y.; Liang, H.; Shi, C.; Liu, J. Multiscale Variations of Precipitable Water over China Based on 1999–2015 Ground-Based GPS Observations and Evaluations of Reanalysis Products. *J. Clim.* **2018**, *31*, 945–962. [CrossRef]
13. Zhao, Q.Z.; Ma, X.W.; Yao, W.Q.; Liu, Y.; Yao, Y.B. A Drought Monitoring Method Based on Precipitable Water Vapor and Precipitation. *J. Clim.* **2020**, *33*, 10727–10741. [CrossRef]
14. Zhao, T.B.; Dai, A.G.; Wang, J.H. Trends in Tropospheric Humidity from 1970 to 2008 over China from a Homogenized Radiosonde Dataset. *J. Clim.* **2012**, *25*, 4549–4567. [CrossRef]
15. Zhang, Y.; Xu, J.; Na, Y.; Lan, P. Variability and Trends in Global Precipitable Water Vapor Retrieved from COSMIC Radio Occultation and Radiosonde Observations. *Atmosphere* **2018**, *9*, 174. [CrossRef]
16. Zhou, C.; Wang, J.H.; Dai, A.G.; Thorne, P.W. A New Approach to Homogenize Global Subdaily Radiosonde Temperature Data from 1958 to 2018. *J. Clim.* **2021**, *34*, 1163–1183. [CrossRef]
17. Wang, M. The Assessment and Meteorological Applications of High Spatiotemporal Resolution GPS ZTD/PW Derived by Precise Point Positioning. Ph.D. Thesis, Tongji University, Shanghai, China, 2019; p. 4.
18. Chrysoulakis, N.; Proedrou, M.; Cartalis, C. Variations and trends in annual and seasonal means of precipitable water in Greece as deduced from radiosonde measurements. *Toxicol Environ. Chem.* **2001**, *84*, 1–6. [CrossRef]
19. Pralungo, R.L.; Haimberger, L.; Stickler, A.; Brönnimann, S. A global radiosonde and tracked balloon archive on 16 pressure levels (GRASP) back to 1905-Part 1: Merging and interpolation to 00:00 and 12:00 GMT. *Earth Syst. Sci. Data.* **2014**, *6*, 185–200. [CrossRef]
20. Rocken, C.; Hove, T.; Johnson, J.; Solheim, F.; Ware, R.; Bevis, M.; Chiswell, S.; Businger, S. GPS/STORM—GPS sensing of atmospheric water vapor for meteorology. *J. Atmos. Ocean. Technol.* **1995**, *12*, 468–478. [CrossRef]
21. Haase, J.; Vedel, H.; Ge, M.; Calais, E. GPS zenith tropospheric delay (ZTD) variability in the Mediterranean. *Phys. Chem. Earth Pt. A-Solid Earth Geod.* **2001**, *26*, 439–443. [CrossRef]
22. Haase, J.; Ge, M.; Vedel, H.; Calais, E. Accuracy and variability of GPS tropospheric delay measurements of water vapor in the western Mediterranean. *J. Appl. Meteorol.* **2003**, *42*, 1547–1568. [CrossRef]
23. Zhang, F.; Barriot, J.; Xu, G.; The, T. Metrology Assessment of the Accuracy of Precipitable Water Vapor Estimates from GPS Data Acquisition in Tropical Areas: The Tahiti Case. *Remote Sens.* **2018**, *10*, 758. [CrossRef]
24. Wang, X.; Zhu, W.; Yan, H.; Cheng, Z.; Ding, J. Preliminary Results of Precipitable Water Vaopr Monitored by Ground-Based GPS. *Chinese J. Atmos. Sci.* **1999**, *23*, 605–612. [CrossRef]
25. Wang, J.H.; Zhang, L.Y. Systematic Errors in Global Radiosonde Precipitable Water Data from Comparisons with Ground-Based GPS Measurements. *J. Clim.* **2008**, *21*, 2218–2238. [CrossRef]
26. Roman, J.A.; Knuteson, R.O.; Ackerman, S.A.; Tobin, D.C.; Revercomb, H.E. Assessment of Regional Global Climate Model Water Vapor Bias and Trends Using Precipitable Water Vapor (PWV) Observations from a Network of Global Positioning Satellite (GPS) Receivers in the U.S. Great Plains and Midwest. *J. Clim.* **2012**, *25*, 5471–5493. [CrossRef]
27. Isioye, O.; Combrinck, L.; Botai, J. Retrieval and analysis of precipitable water vapour based on GNSS, AIRS, and reanalysis models over Nigeria. *Int. J. Remote Sens.* **2017**, *38*, 5710–5735. [CrossRef]
28. Liu, Z.; Li, Y.; Fei, L.; Guo, J. *Estimation and Evaluation of the Precipitable Water Vapor from GNSS PPP in Asia Region*; China Satellite Navigation Office: Beijing, China, 2017; pp. 85–95.

29. Wang, R.; Fu, Y.F.; Xian, T.; Chen, F.J.; Yuan, R.M.; Li, R.; Liu, G.S. Evaluation of Atmospheric Precipitable Water Characteristics and Trends in Mainland China from 1995 to 2012. *J. Clim.* **2017**, *30*, 8673–8688. [CrossRef]
30. Chen, B.; Dai, W.; Liu, Z.; Wu, L.; Xia, P. Assessments of GMI-Derived Precipitable Water Vapor Products over the South and East China Seas Using Radiosonde and GNSS. *Adv. Meteorol.* **2018**, *2018*, 1–12. [CrossRef]
31. Wang, X.; Zhang, K.; Wu, S.; Li, Z.; Chen, Y.; Li, L.; Hong, Y. The correlation between GNSS-derived precipitable water vapor and sea surface temperature and its responses to El Nio–Southern Oscillation. *Remote Sens. Environ.* **2018**, *216*, 1–12. [CrossRef]
32. Bolton, D. The computation of equivalent potential temperature. *Mon. Weather Rev.* **1980**, *108*, 1046–1053. [CrossRef]
33. Wong, M.; Jin, X.; Liu, Z.; Nichol, J.; Chan, P. Multi-sensors study of precipitable water vapour over mainland China. *Int. J. Climatol.* **2015**, *35*, 3146–3159. [CrossRef]
34. Bevis, M.; Businger, S.; Chiswell, S.; Herring, T.A.; Anthes, R.A.; Rocken, C.; Ware, R.H. GPS meteorology: Mapping zenith wet delays onto precipitable water. *J. Appl. Meteorol.* **1994**, *33*, 379–386. [CrossRef]
35. Blewitt, G.; Hammond, W.C.; Kreemer, C. Harnessing the GPS data explosion for interdisciplinary science. *Eos* **2018**, 99. [CrossRef]
36. Kouba, J. Implementation and testing of the gridded Vienna Mapping Function 1 (VMF1). *J. Geod.* **2008**, *82*, 193–205. [CrossRef]
37. Fernandes, M.J.; Pires, N.; Lázaro, C.; Nunes, A.L. Tropospheric delays from GNSS for application in coastal altimetry. *Adv. Space Res.* **2013**, *51*, 1352–1368. [CrossRef]
38. Urquhart, L.; Santos, M.C.; Nievinski, F.G.; Böhm, J. Generation and Assessment of VMF1-Type Grids Using North-American Numerical Weather Models. In *Earth on the Edge: Science for a Sustainable Planet*; Rizos, C., Willis, P., Eds.; International Association of Geodesy Symposia; Springer: Berlin, Heidelberg, 2014; Volume 139, pp. 3–9. [CrossRef]
39. Wang, X.; Zhang, K.; Wu, S.; He, C.; Cheng, Y.; Li, X. Determination of zenith hydrostatic delay and its impact on GNSS-derived integrated water vapor. *Atmospheric Meas. Tech.* **2017**, *10*, 2807–2820. [CrossRef]
40. Durre, I.; Vose, R.; Wuertz, D. Overview of the integrated global radiosonde archive. *J. Clim.* **2006**, *19*, 53–68. [CrossRef]
41. Ding, J.; Chen, J. Assessment of Empirical Troposphere Model GPT3 Based on NGL's Global Troposphere Products. *Sensors* **2020**, *20*, 3631. [CrossRef]
42. Boehm, J.; Werl, B.; Schuh, H. Troposphere mapping functions for GPS and very long baseline interferometry from European Centre for Medium-Range Weather Forecasts operational analysis data. *J. Geophys. Res.-Solid Earth.* **2006**, *111*, 2006. [CrossRef]
43. Bertiger, W.; Bar-Sever, Y.; Dorsey, A.; Haines, B.; Harvey, N.; Hemberger, D.; Heflin, M.; Lu, W.; Miller, M.; Angelyn, W.; et al. GipsyX/RTGx, a new tool set for space geodetic operations and research. *Adv. Space Res.* **2020**, *66*, 469–489. [CrossRef]
44. Li, T.; Wang, L.; Chen, R.; Fu, W.; Xu, B.; Jiang, P.; Liu, J.; Zhou, H.; Han, Y. Refining the empirical global pressure and temperature model with the ERA5 reanalysis and radiosonde data. *J. Geod.* **2021**, *95*, 31. [CrossRef]
45. Benestad, R.E.; Erlandsen, H.B.; Mezghani, A.; Parding, K.M. Geographical Distribution of Thermometers Gives the Appearance of Lower Historical Global Warming. *Geophys. Res. Lett.* **2019**, *46*, 7654–7662. [CrossRef]
46. Blesić, S.; Zanchettin, D.; Rubino, A. Heterogeneity of Scaling of the Observed Global Temperature Data. *J. Clim.* **2019**, *32*, 349–367. [CrossRef]
47. Huang, B.; Thorne, P.W.; Banzon, V.F.; Boyer, T.; Zhang, H.M. Extended Reconstructed Sea Surface Temperature, Version 5 (ERSSTv5): Upgrades, Validations, and Intercomparisons. *J. Clim.* **2017**, *28*, 911–930. [CrossRef]
48. Rinke, A.; Segger, B.; Crewell, S.; Maturilli, M.; Naakka, T.; Nygard, T.; Vihma, T.; Alshawaf, F.; Dick, G.; Wickert, J.; et al. Trends of Vertically Integrated Water Vapor over the Arctic during 1979–2016: Consistent Moistening All Over? *J. Clim.* **2019**, *32*, 6097–6116. [CrossRef]
49. Imke, D.; Williams, C.N., Jr.; Yin, X.; Russell, S.V. Radiosonde-based trends in precipitable water over the Northern Hemisphere: An update. *J. Geophys. Res. Atmos.* **2009**, *115*, 162009. [CrossRef]
50. Kwon, C.; Lee, D.; Lee, K.-S.; Seo, M.; Seong, N.-H.; Choi, S.; Jin, D.; Kim, H.; Han, K.-S. Long-term variability of Total Precipitable Water using a MODIS over Korea. *Korean J. Remote Sens.* **2016**, *32*, 195–200. [CrossRef]
51. Ziv, S.Z.; Alpert, P.; Reuveni, Y. Long-term variability and trends of precipitable water vapour derived from GPS tropospheric path delays over the Eastern Mediterranean. *Int. J. Climatol.* **2021**, *41*, 6433–6454. [CrossRef]
52. Ma, X.; Yao, Y.; Zhang, B.; He, C. Retrieval of high spatial resolution precipitable water vapor maps using heterogeneous earth observation data. *Remote Sens. Environ.* **2022**, *278*, 113100. [CrossRef]
53. Rohde, R.A.; Hausfather, Z. The Berkeley Earth Land/Ocean Temperature Record. *Earth Syst. Sci. Data* **2020**, *12*, 3469–3479. [CrossRef]
54. Beckley, B.; Yang, X.; Zelensky, N.P.; Holmes, S.A.; Lemoine, F.G.; Ray, R.D.; Mitchum, G.T.; Desai, S.; Brown, S.T. GSFC 2020, Global Mean Sea Level Trend from Integrated Multi-Mission Ocean Altimeters TOPEX/Poseidon, Jason-1, OSTM/Jason-2, and Jason-3 Version 5.0 Ver. 5.0. PO.DAAC: Pasadena, CA, USA. [CrossRef]

Article

Unified Land–Ocean Quasi-Geoid Computation from Heterogeneous Data Sets Based on Radial Basis Functions

Yusheng Liu and Lizhi Lou *

College of Surveying and Geo-Informatics, Tongji University, 1239 Siping Road, Shanghai 200092, China; lyssyl@tongji.edu.cn
* Correspondence: llz@tongji.edu.cn

Abstract: The determination of the land geoid and the marine geoid involves different data sets and calculation strategies. It is a hot issue at present to construct the unified land–ocean quasi-geoid by fusing multi-source data in coastal areas, which is of great significance to the construction of land–ocean integration. Classical geoid integral algorithms such as the Stokes theory find it difficult to deal with heterogeneous gravity signals, so scholars have gradually begun using radial basis functions (RBFs) to fuse multi-source data. This article designs a multi-layer RBF network to construct the unified land–ocean quasi-geoid fusing measured terrestrial, shipborne, satellite altimetry and airborne gravity data based on the Remove–Compute–Restore (RCR) technique. EIGEN-6C4 of degree 2190 is used as a reference gravity field. Several core problems in the process of RBF modeling are studied in depth: (1) the behavior of RBFs in the spatial domain; (2) the locations of RBFs; (3) ill-conditioned problems of the design matrix; (4) the effect of terrain masses. The local quasi-geoid with a 1′ resolution is calculated, respectively, on the flat east coast and the rugged west coast of the United States. The results show that the accuracy of the quasi-geoid computed by fusing four types of gravity data in the east coast experimental area is 1.9 cm inland and 1.3 cm on coast after internal verification (the standard deviation of the quasi-geoid w.r.t GPS/leveling data). The accuracy of the quasi-geoid calculated in the west coast experimental area is 2.2 cm inland and 2.1 cm on coast. The results indicate that using RBFs to calculate the unified land–ocean quasi-geoid from heterogeneous data sets has important application value.

Keywords: fusion of heterogeneous data; unified land–ocean quasi-geoid; radial basis functions; Remove–Compute–Restore; EIGEN-6C4

Citation: Liu, Y.; Lou, L. Unified Land–Ocean Quasi-Geoid Computation from Heterogeneous Data Sets Based on Radial Basis Functions. *Remote Sens.* **2022**, *14*, 3015. https://doi.org/10.3390/rs14133015

Academic Editor: Xiaogong Hu

Received: 16 May 2022
Accepted: 21 June 2022
Published: 23 June 2022

Publisher's Note: MDPI stays neutral with regard to jurisdictional claims in published maps and institutional affiliations.

1. Introduction

The geoid is a fundamental element in determining the shape of the Earth [1]. The calculation of a high-quality geoid requires firstly constructing the Earth's gravity field with high precision and resolution. With the enrichment of measurement means, the breadth and depth of geospatial data are being continuously improved by terrestrial, shipborne, airborne and satellite gravity surveys, etc. [2–7]. The unified land–ocean quasi-geoid can be constructed by fusing heterogeneous data sets in the boundary areas between land and ocean, which is beneficial to advance the construction of land–ocean integration. At present, calculation methods of the gravimetric geoid are mainly divided into analytical methods and statistical methods. Analytical methods include the classical Stokes, Hotine, Molodensky and Helmert integral algorithms [8–12]. The defects of analytical methods lie in the strict requirement of the boundary surface and the difficulty fusing multi-source gravity signals. Statistical methods represented by the least-squares collocation (LSC) have significant advantages in fusing heterogeneous data sets. LSC is first introduced into the study of local gravity field approximation by Krarup and Moritz. Tscherning, Rapp, Hwang and other scholars have performed a lot of later research [13–17]. The defect of LSC is that it is difficult to construct an appropriate and accurate local covariance model.

However, as a statistical algorithm, LSC is still a good choice. This article mainly focuses on analytical methods.

Radial basis functions have been widely used in local gravity field modeling in recent years due to their simple function forms and the ability to fuse multi-source data. RBFs are first proposed as a function approximation theory in mathematics [18]. The Multiquadric (MQ) kernel function proposed by Hardy and the point mass function are first introduced into Earth's gravity field approximation [19]. Weightman is one of the first scholars to apply the point mass function in physical geodesy. This new method can replace the classical spherical harmonic function to express the Earth's gravity field. Reilly and Herbrechtsmeier apply the point mass model to the fusion of multi-source data earlier. They use simulated altimetry data to invert marine gravity anomalies and combine the measured gravity anomalies on land to construct a unified local gravity field, whose accuracy reaches 20 mGal after being verified [20]. Barthelmes designs a free-positioned point mass optimization algorithm based on the least-squares adjustment by setting four free parameters on each mass point, which effectively reduces the number of mass points and significantly improves the calculation efficiency [21]. Lehmann further studies the free-positioned algorithm, which can be used more flexibly to determine the local geoid with measured gravity data [22]. These studies lay the foundation of radial basis function modeling. Then, a series of high-order RBFs such as radial multipoles and Poisson wavelets are derived from the point mass function [23,24]. In addition, some scholars introduce the classical Blackman kernel, the Shannon kernel and the spherical spline function into the RBF model [25,26]. After years of development, scholars have performed a lot of research on the application of various RBFs in local gravity field modeling. The research contents mainly focus on the selection of RBF types, the treatment of ill-conditioned problems and the determination of the spatial position of RBFs.

Tenzer and Klees carry out experiments in the plain area to compare the performance of the point mass kernel, the Poisson kernel, radial multipoles and Poisson wavelets. According to the results, they conclude that these RBFs can obtain almost the same accuracy of gravity field modeling when the depth of RBFs is chosen properly [27]. Bentel et al. use simulated data to model the local gravity field based on the Shannon low-pass kernel, the Shannon high-pass kernel, the Blackman low-pass kernel, the cubic polynomial kernel, the Poisson multipoles kernel, the Abel–Poisson kernel truncated and the Abel–Poisson kernel. The experimental results show that the Blackman low-pass kernel, the cubic polynomial kernel and the Abel–Poisson kernel truncated perform best [28]. In the process of RBF modeling, the design matrix may be ill conditioned due to the uneven distribution of observations and excessive number of RBFs. Tikhonov regularization is currently the mainstream method to deal with ill-conditioned problems. Wu et al. adopt zero-order and first-order Tikhonov regularization to solve ill-posed equations and prove that first-order regularization has better performance [29]. Based on the Tikhonov regularization, Liu et al. analyze the defects of the L-curve method and variance component estimation (VCE) in determining regularization parameters and then propose two combined methods, VCE-Lc and Lc-VCE, which are proven to be superior to traditional methods [30,31]. The spatial positions of RBFs have a great influence on the modeling result. Eicker uses various spherical grids to place RBFs such as the Reuter grid, the Triangle-Center grid and the Triangle-Vertex grid, which are more evenly distributed than the geographical grid [32]. Klees and Witter propose an adaptive selection of RBFs before parameter estimation according to the distribution, signal variation and noise of observations [33]. Tenzer and Klees, respectively, use the GCV and RMS minimization methods to determine the optimal depth of RBFs and establish a linear functional relationship between the depth and the correlation length of RBFs. Experiment results prove that the optimal depth is related to the type of RBFs [27]. Tenzer et al. subsequently find in the study of mountainous areas that the RBF model solution is very sensitive to change in depth of even several hundred meters when regional shape fluctuation and gravity signals vary greatly [34]. To sum up the current research by scholars, the main problems are as follows: (1) the selection of RBF types and the determination of spatial positions have not been standardized, which needs further research; (2) the studies carried out using real gravity data and the experiments carried out in the land–ocean boundary areas are still too few.

Considering the above problems, we propose a multi-layer RBF model to fuse heterogeneous data sets for the unified land–ocean quasi-geoid. In Section 2, two coastal areas and relevant experiment data are detailed. The modeling workflow is given, and the modeling methodologies are introduced in detail, including (1) the characteristics of RBFs in the spatial domain; (2) the spatial position of RBFs, i.e., the resolution and depth of spherical grids; (3) Tikhonov regularization to deal with ill-conditioned problems of the model design matrix; (4) residual terrain model (RTM) to represent the impact of terrain mass. The local quasi-geoid with a 1′ resolution is calculated by setting up multi-layer RBF networks, respectively, on the flat east coast and the rugged west coast of the United States in Section 3. In Section 4, we discuss the results and shortcomings of the above research and propose an outlook for future work. Section 5 summarizes the main research content and conclusions of this article.

2. Data and Method

This article studies the RBF model to fuse multi-source gravity data. Numerical experiments are carried out in the land–ocean junction areas. The experiments are conducted in two topographically different areas, one on the east and the other on the west coast of the United States. The east coast of the United States is relatively flat, expending inland from the coastline into a broad plain. On the west coast, under the influence of the Rocky Mountains, the land elevation begins to rise rapidly not far from the coastline. Details of the relevant experiment data are described in this section. In addition, the general modeling process is summarized, and the core modeling methodologies are studied in depth.

2.1. Data Preparation

The east coast experiment area is located in North Carolina, USA, with an average elevation of approximately 8 m. The target area range is between $34°$ and $37°$ latitude and between $-78°$ and $-74°$ longitude. The west coast experiment area is in the border area of Oregon and Washington State in the northwest of the United States. Its land part is rugged with an average elevation of approximately 334 m. The target area range is between $44°$ and $47°$ latitude and between $-126°$ and $-122°$ longitude. The regional topography is shown in Figure 1.

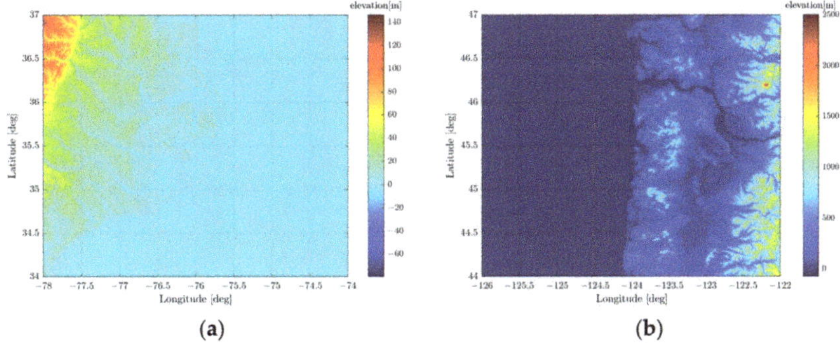

Figure 1. Regional terrain in two coastal areas of the United States: (**a**) east coast; (**b**) west coast.

The gravity data used in the experiments include measured terrestrial, shipborne, airborne gravities and gravity anomalies derived from satellite altimetry. Terrestrial and shipborne gravity data come from the NGS99 gravity data set provided by the National Geodetic Survey (NGS). NGS99 is a compilation of the measured terrestrial and shipborne gravity data across the USA (data source: https://www.ngdc.noaa.gov/mgg/gravity/19

99/data/regional/ngs99, accessed on 1 March 2021). The product provides free-air gravity anomalies (FAA), Bouguer anomalies, etc., after a series of preprocessing. In the east coast experiment area, we use 2860 terrestrial gravity points and 1678 shipborne gravity points, which are shown in Figure 2a. The distribution of shipborne gravity points is uneven, which is reflected in the dense distribution of data on the shipping route but many gaps outside the route. The shipborne gravity signals vary greatly, while the terrestrial gravity signals vary more gently and are densely distributed in most areas. In the west coast experiment area, there are 2814 terrestrial gravity points and 2288 shipborne gravity points, as shown in Figure 2b. Different from the east coast experiment area, the distribution of shipborne gravity points in the west coast area is more even.

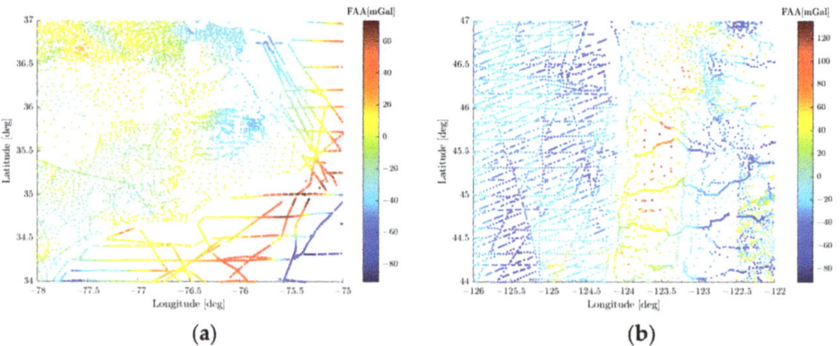

Figure 2. Distribution of terrestrial and shipborne gravity points: (**a**) east coast; (**b**) west coast.

Due to the data gaps between ship lines, satellite altimetry and airborne gravities with more uniform distribution are needed as supplements. Satellite altimetry used in this article are DTU15 gravity anomalies with a 2′ resolution (data source: https://ftp.space.dtu.dk/pub/DTU15, accessed on 1 March 2021). Under the influence of the complex terrain on the ground, the accuracy of gravity anomalies inversed from satellite altimetry in inshore areas is not so good, but their high resolution and uniform distribution can make up for the deficiency of shipborne gravity to a certain extent. The distribution of DTU15 on sea is shown in Figure 3.

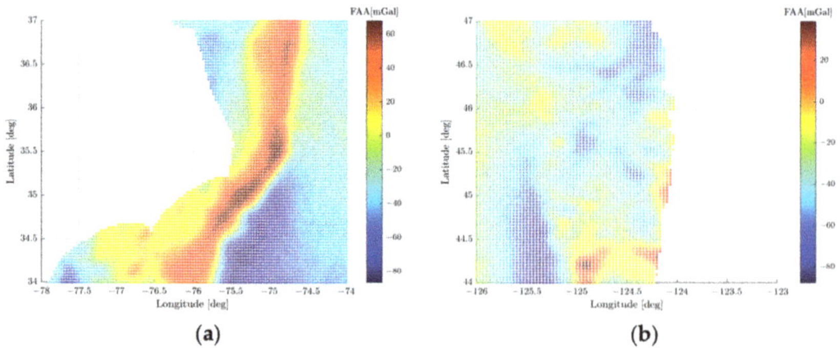

Figure 3. Distribution of DTU15 gravity points: (**a**) east coast; (**b**) west coast.

Airborne gravity data come from the GRAV-D project, which is developed by the NGS to redefine the vertical datum of the USA. GRAV-D currently covers most areas of the USA, providing full-field gravity data, which can be turned into free-air gravity disturbances (FAD) or free-air gravity anomalies (data source: https://www.ngs.noaa.gov/GRAV-D, accessed on 1 March 2021). In the east coast experiment area, 5367 airborne gravity points are selected, and their average flight altitude is approximately 5460 m. In the west coast experiment area, the number of airborne gravity points is 7458, and the average flight altitude is approximately 6938 m. The distribution of airborne gravity data is shown in Figure 4. Airborne gravity data have high resolution and uniform distribution, so they are not limited by ground terrain conditions. The shortcoming of airborne data is that the high flight altitude leads to low data accuracy, thus it is difficult to simulate the full-band gravity signal on ground points.

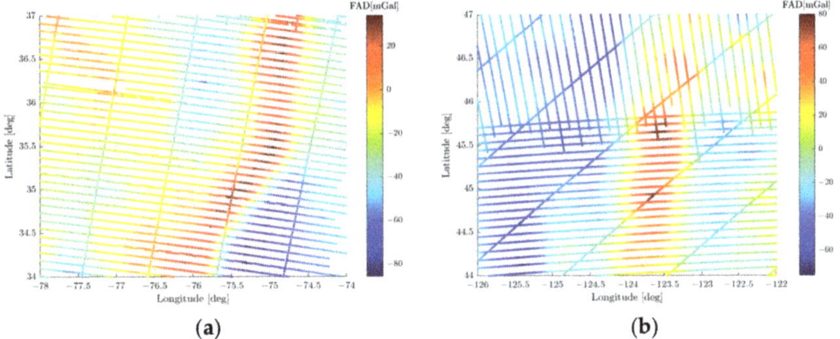

Figure 4. Distribution of airborne gravity points: (**a**) east coast; (**b**) west coast.

GPS/leveling data are used to verify the accuracy of the unified land–ocean quasi-geoid (data source: https://geodesy.noaa.gov/GPSonBM, accessed on 1 March 2021). In the east coast experiment area, due to the flat land terrain, there are many measured GPS/leveling points with a total of 807, which is 177 in the west coast experiment area, as shown in Figure 5. GPS and leveling surveys are impossible to be carried out offshore. So, this article uses GPS/leveling points along the coast to check the accuracy level of the unified land–ocean quasi-geoid on sea.

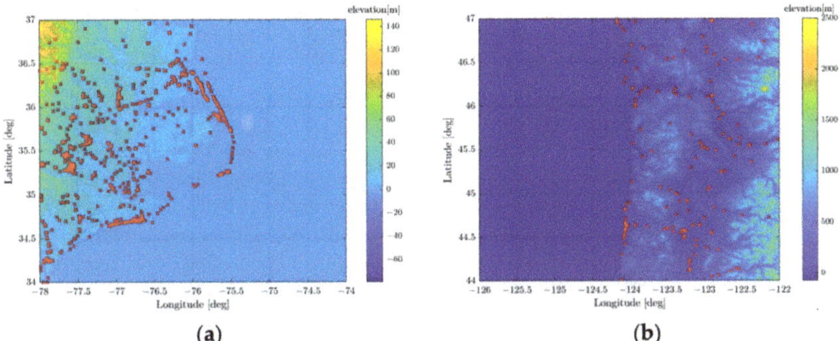

Figure 5. Distribution of GPS/leveling points: (**a**) east coast; (**b**) west coast.

2.2. RBF Modeling Strategies

In this article, the local quasi-geoid is calculated based on the RBF model following the basic framework of the RCR technique [35]. After removing GGM and terrain effects from gravity observations, the residual gravity is included in the RBF model as input data. Helmert VCE is used to evaluate various observations [36]. Tikhonov regularization is used to solve ill-conditioned problems that may occur in the design matrix. A small number of points are selected from GPS/leveling data as control points, and the remaining data are as internal checkpoints. Under the constraint of control points, the optimal RBF network is determined to construct the model. The quasi-geoid calculated is compared with internal checkpoints, and the standard deviation (STD) of the difference between them is taken as the accuracy indicator. The flowchart is shown in Figure 6.

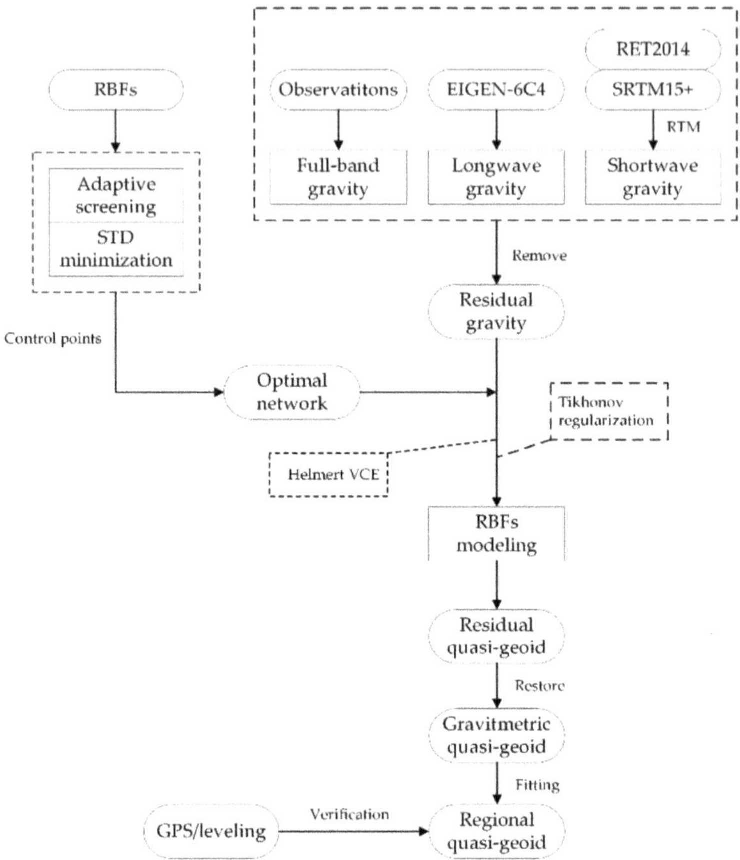

Figure 6. Flowchart of RBF modeling.

When the spherical grid is used to place the RBFs, not all grid points should be included in the model, which is because the RBFs have strong localization characteristics, and their energies are mainly concentrated around the center. So, the observations near RBFs are the main contribution to the simulated gravity signal. If the RBFs without enough observations around are included in the model, the local gravity field will be over-parameterized, and the reliability of the model solution will be reduced. In order to locate and eliminate the redundant RBFs, it is usually necessary to introduce a filter radius R_I:

$$R_I = \chi \cdot v_{0.5} \tag{1}$$

where I is the number of RBFs; $v_{0.5}$ denotes the correlation length of RBFs, which specifically refers to the straight-line distance between the RBFs and observations when the absolute value of RBFs decays to half of the maximum value. χ represents correlation length parameter. With the center of RBFs as the spherical center, if the number of all observations within the radius R_I is greater than q, the RBFs will remain; otherwise, they will be excluded. χ and q need to be determined based on the actual situation. In this article, we specify $q = 0$. R_I ranges between 50 and 90 km. We determine the suitable value of χ through some trial calculations, ensuring R_I meets the modeling requirements. The general adaptive screening process is shown in Figure 7.

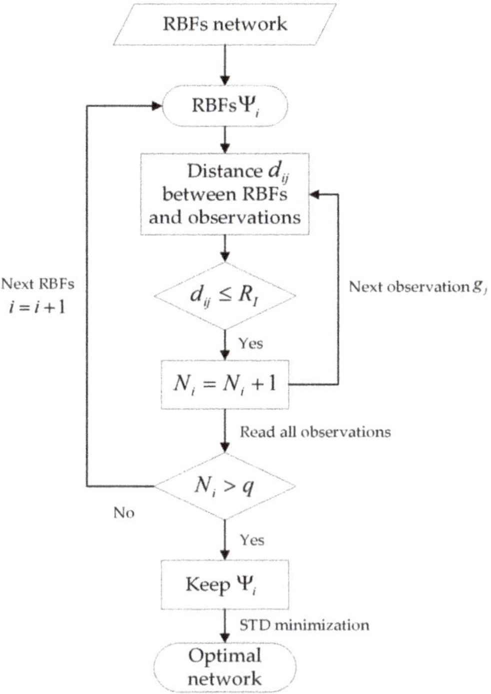

Figure 7. Processing flow of adaptive screening technique.

Considering the altitude of airborne gravity data is much higher than the terrestrial, shipborne and satellite altimetry data, the one-layer RBF network cannot simulate the full gravity signals well. Therefore, we design a multi-layer RBF network to fuse these four types of data, which are shown in Figure 8. We are currently only using two layers, and more layers are of course encouraged to be used.

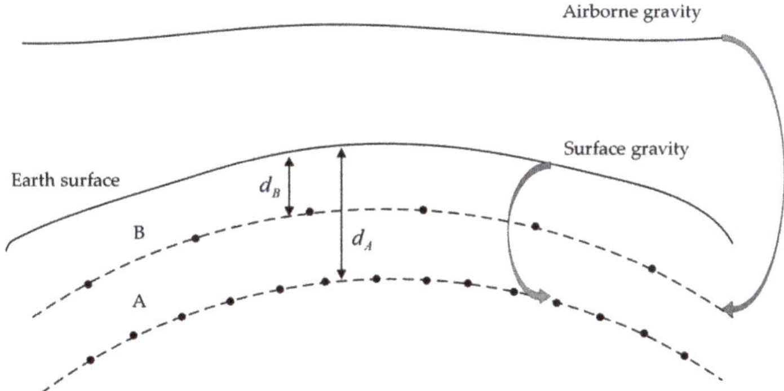

Figure 8. Multi-layer RBF networks.

In Figure 9, surface gravity denotes terrestrial, shipborne and satellite altimetry data; d_A and d_B denote the depth of grid A and B. The specific modeling process is as follows:

1. GGM and RTM are removed from the terrestrial, shipborne and satellite altimetry observations to obtain residual gravity anomalies, $\Delta g_A = \Delta g - \Delta g_{GGM} - \Delta g_{RTM}$. The RBF network A is determined by the STD minimization. After the three kinds of gravity data are fused to calculate the RBF model parameters, the airborne gravity

points are taken as prediction points and the corresponding model gravity disturbance δg_A will be calculated.

2. Remove GGM and δg_A from the airborne gravity to obtain residual gravity disturbances, $\delta g_B = \delta g - \delta g_{GGM} - \delta g_A$. The RBF network B is determined by STD minimization and then the corresponding RBF model parameters will be calculated.

3. Based on the RBF networks A and B, the height anomalies on unknown points are computed, respectively, and added together. Then GGM and RTM signals are restored then to obtain the final quasi-geoid.

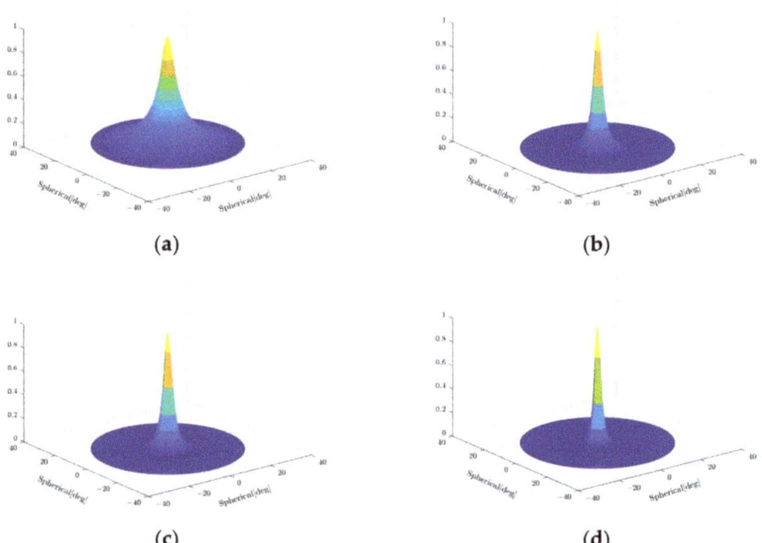

Figure 9. Behavior of RBFs in the spatial domain: (**a**) the IMQ kernel; (**b**) the Poisson kernel; (**c**) radial multipoles of order 1; (**d**) Poisson wavelets of order 1.

2.3. RBF Modeling Methodology

RBF is a nonlinear function with the local characteristic and radial symmetry. In essence, the RBF model only depends on the relative position relationship between the computed points and the center of RBFs. Represent the spatial position of RBFs as $y(y_1, y_2, y_3)$, which is usually placed on a sphere inside the Earth such as the Bjerhammar sphere. The point $x(x_1, x_2, x_3)$ outside the sphere is used to represent the spatial position of the observations. RBFs can be expressed as:

$$\Psi(x,y) = \sum_{m=0}^{\infty} \psi_m(2m+1)\left(\frac{R_B}{|x|}\right)^{m+1} P_m\left(\hat{x}^T \hat{y}\right), |y| \langle R_B, |x|\rangle R_B \tag{2}$$

where R_B denotes the radius of the Bjerhammar sphere; P_m denotes the Legendre polynomial of degree m; ψ_m is the Legendre coefficient, which is the shape factor of RBFs, determining their properties in the spatial and frequency domain; $\hat{x}$ and $\hat{y}$ are the unit vector of x and y. Define the depth of RBFs as: $d_s = R_B - |y|$.

According to the Runge–Krarup principle, the external disturbing potential T can be approximated by a harmonic function completely embedded inside the Earth. So, the linear combination of RBFs can be used to approximate T:

$$T(x) = \frac{GM}{R_B} \sum_{i=1}^{K} \beta_i \Psi(x, y_i) \tag{3}$$

where K denotes the number of RBFs; β_i denotes the unknown coefficient of the RBF model. Based on the RCR technique, T can be decomposed into three components:

$$T = T_{GGM} + T_{Terrian} + T_{res} \tag{4}$$

where T_{GGM} denotes long-wave signals represented by GGM; $T_{Terrian}$ denotes high-frequency gravity information implied by terrain masses, used to represent short-wave signals; T_{res} denotes residual disturbing potential. According to the functional relationship between the observations and T, the linear combination of RBFs can be further used to simulate the observations:

$$\Delta g_{res}(x) = \sum_{i=1}^{K} \mu_i \left(-\frac{\partial}{\partial |x|} \Psi(x, y_i) - \frac{2}{|x|} \Psi(x, y_i) \right)$$
$$\delta g_{res}(x) = -\sum_{i=1}^{K} \mu_i \frac{\partial}{\partial |x|} \Psi(x, y_i) \tag{5}$$
$$\zeta_{res}(x) = \sum_{i=1}^{K} \mu_i \frac{\Psi(x, y_i)}{\gamma}$$

where $\mu_i = \frac{GM}{R_B} \beta_i$; $\Delta g_{res}(x)$ denotes residual gravity anomaly; $\delta g_{res}(x)$ denotes residual gravity disturbance; $\zeta_{res}(x)$ denotes residual height anomaly. In this article, all geoid undulations have been converted to height anomalies (GPS/leveling observations), ensuring the unity of elevation datum:

$$\zeta \approx N - \frac{\Delta g_B}{\bar{\gamma}} H \tag{6}$$

where Δg_B denotes bouguer gravity anomaly; $\bar{\gamma}$ denotes mean normal gravity; H denotes the topographic height. It is worth noting that the above formula is an approximate formula.

Equation (5) can be abstracted as a linear observation model:

$$y_p = A_p x + \varepsilon_p \tag{7}$$

where y_p denotes the observation vector of class p; x is the vector of model coefficients; ε_p is the error vector. The equation can be solved using the least-squares estimation. Due to the lack of the prior information related to the error of observations, the weight of all kinds of observations can be determined by the VCE technique.

2.3.1. Characteristics of RBFs in the Spatial Domain

Several types of RBFs widely used in local gravity field modeling are as follows: the IMQ kernel, the Poisson kernel, the radial multipoles kernel and the Poisson wavelets kernel, represented by Ψ_{IMQ}, Ψ_{pk}, Ψ_{rm} and Ψ_{pw} respectively. We use these RBFs' analytical expressions to construct the model. See the related research for their specific formulas [37]. Convert the analytical formula of RBFs to the unit sphere, and then, respectively, draw their pictures in the spatial domain when $d_s = 300$ km, as shown in Figure 9. It can be seen from the figure that RBFs have obvious localization characteristics. In the position far from the center of RBFs, the signals' energies rapidly decay, which is conductive to the concentration of local gravity signals.

2.3.2. RBF Networks

The design of the RBF network is divided into the determination of the sphere position and the buried depth. There are mainly two strategies to determine the sphere position of RBFs. One is to directly place RBFs under observations, which is extremely dependent on the spatial distribution of observations. The other is to place RBFs on the regular grid, which can make RBFs evenly distributed in the computation areas. At present, the second strategy is more commonly used, which is also adopted in this article. The simplest regular spherical grid is a geographical grid, which is one kind of isogonal grids. Geographical grid nodes space along the longitude circle and latitude circle with equal-angle intervals. Figure 10 shows its distribution globally and in the Arctic.

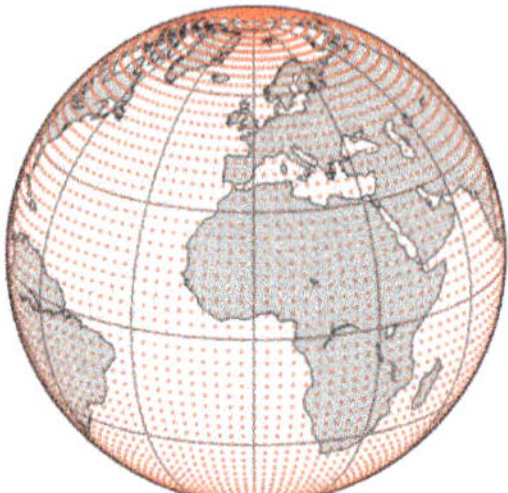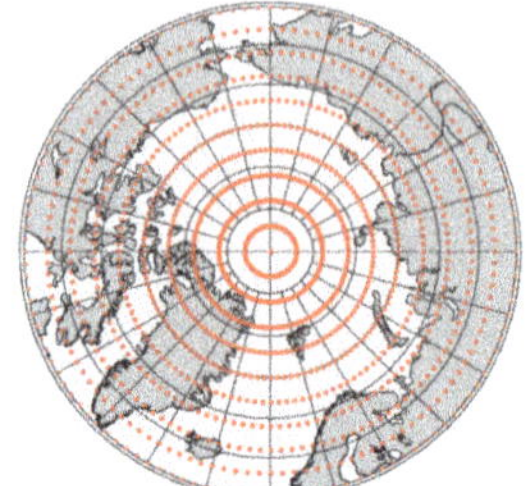

Figure 10. Distribution of the geographical grid.

It can be clearly seen from Figure 11 that the geographical grid is very unevenly distributed in high-latitude regions. The higher the latitude is, the denser the grid dots are. In a slightly large area, such a defect will become more obvious. In view of this phenomenon, scholars propose a kind of spherical isometric grid, the Reuter grid, which ensures that the spherical distances between grid dots in meridian and latitude direction are equal, so that RBFs can be distributed evenly in high-latitude regions, as shown in Figure 11. The specific formulas can be found in the relevant articles [32].

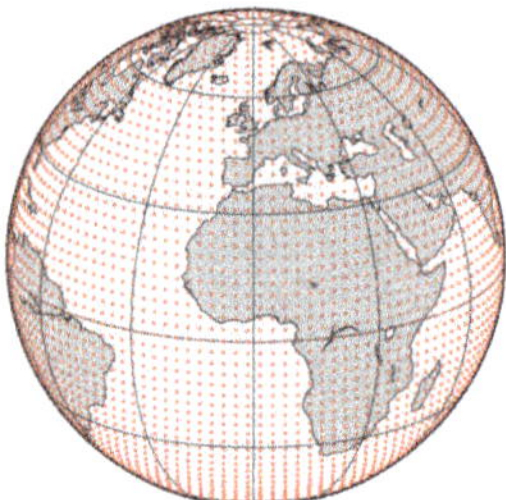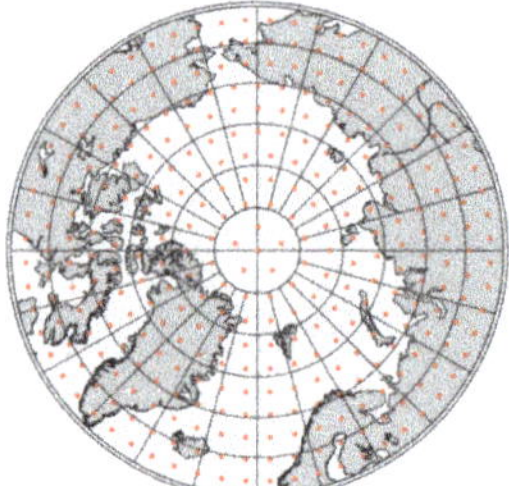

Figure 11. Distribution of the Reuter grid.

The determination of the optimal RBF network is always the focus and difficulty in RBF modeling. When RBFs are buried deep, the gravity signals simulated by the model are mainly concentrated at low frequencies. This kind of model has strong stability but leads to the simulated gravity signals being coarse due to omission errors. The shallower the buried depth of RBFs is, the more sensitive they are to the high-frequency signals such as terrain masses. The stability of the model will be reduced, and ill-conditioned problems may occur. When the number of RBFs is too large caused by high grid resolution, the overlap between RBFs will increase, which leads to the RBFs model being over-parameterized. On the contrary, too few RBFs are insufficient to simulate full gravity signals.

When the RBF model is constructed using discrete gravity points, it is difficult to uniquely determine the appropriate position of RBFs by theories before modeling. Scholars generally use posterior statistical methods such as GCV, RMS or STD minimization technique to screen out the optimal RBF network [27]. In this article, the STD minimization technique is adopted to design the RBF network. This method is simple and efficient, but it lacks strict a theoretical basis. The STD minimization technique divides a certain range of grid resolutions and depths into different combinations according to a certain step. These combinations are, respectively, used for modeling and compared with the control points. The combination achieving the smallest standard deviation is determined as the optimal RBF network.

2.3.3. Tikhonov Regularization Technique

When directly using discrete data for modeling, the ill condition of the design matrix may occur due to the uneven distribution of observations and the excessive number of RBFs. Considering the lack of the sufficient prior information of observations, Tikhonov regularization is introduced to deal with the ill-conditioned problems [38]. Tikhonov regularization meets the following estimation criteria:

$$\min_{x} : \Phi(x) = \|y - Ax\|_P^2 + \alpha\|x\|_Q^2 \tag{8}$$

The general estimation formula is written as:

$$\hat{x}_\alpha = \left(A^T PA + \alpha Q\right)^{-1} A^T Py \tag{9}$$

where Q is the regularization matrix; α is the regularization parameter. In this article, Q is assumed to be the identity matrix, i.e., $Q = I$. After singular value decomposition (SVD), the regularization solution can be obtained as:

$$\widetilde{x}_\alpha = \sum_{i=1}^{n} \frac{\sigma_i^2}{\sigma_i^2 + \alpha} \frac{u_i^T \widetilde{y}}{\sigma_i} v_i \tag{10}$$

The regularization parameter is determined by the following two methods: the MSE and L-curve techniques. The basic principle of the MSE technique is as follows:

$$\min_{\alpha} : Tr(MSE) = \sum_{i=1}^{n} \frac{\sigma_0^2 \sigma_i^2 + \alpha^2 z_i^2}{\left(\sigma_i^2 + \alpha\right)^2} \tag{11}$$

where $z_i = v_i^T \overline{x}$. $\overline{x}$ denotes the truth value of unknown parameter x. We usually use the estimated result $\hat{x}$ to replace $\overline{x}$, which will affect the optimality of regularization parameter α to some extent.

The L-curve technique is used to find the best one from different regularization parameters to achieve the optimal balance between the residual sum of squares $V^T PV$ and the smoothness of regularization function $\hat{x}^T Q\hat{x}$. Construct a curve with $\mu(\alpha)$ as abscissa and $\lambda(\alpha)$ as ordinate. The point with maximum curvature (inflection point) corresponds to the optimal regularization parameter.

$$\mu(\alpha) = \log(\|y - Ax\|_P), \ \lambda(\alpha) = \log(\|\hat{x}\|_Q) \tag{12}$$

2.3.4. Residual Terrain Model

The high-frequency gravity signals implied by the terrain mass have an important influence on local gravity field modeling, especially in mountainous areas [39]. RTM can be used to represent terrain effects [40–42]. The RTM technique is essentially the differences between the real terrain and the reference terrain, as shown in Figure 12. The real terrain surface usually denotes digital terrain model (DEM) with a high resolution. The reference terrain is relatively smooth with a lower resolution, which can be computed from the spherical harmonic terrain model such as RET2014 [43]:

$$z^{RET2014}(\varphi, \lambda) = \sum_{n=0}^{n_{max}} \sum_{m=0}^{n} \left(\overline{HC}_{nm} \cos m\lambda + \overline{HS}_{nm} \sin m\lambda\right) \overline{P}_{nm}(\sin \varphi) \tag{13}$$

where $n_{max} = 2160$; $\left(\overline{HC}_{nm}, \overline{HS}_{nm}\right)$ denotes full normalized height coefficients. Based on Forsberg's classic TC program, RTM results can be well calculated.

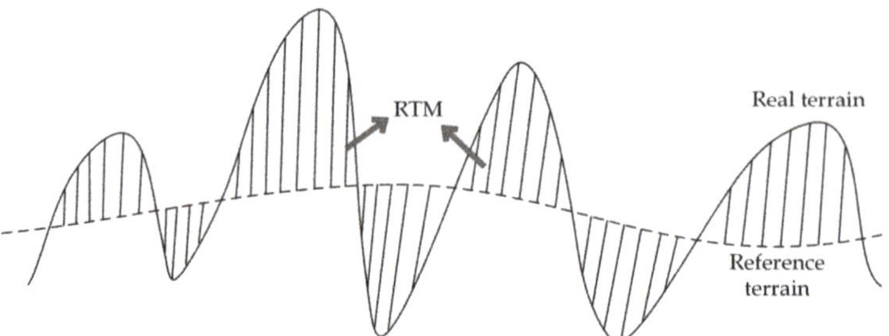

Figure 12. Residual terrain model.

3. Results

This section describes in detail the experiment process and the results of calculating the unified land–ocean quasi-geoid from heterogeneous data sets based on RBFs. Compared with the results of the Stokes integral, the advantages of RBFs in multi-source data fusion are proved.

3.1. The East Coast Experiment Area of the USA

Based on the RCR technique, EIGEN-6C4 (degree 2190) gravity anomalies are removed from the measured terrestrial, shipborne and DTU15 data, as shown in Figure 13. Detailed statistics of the residual gravity anomalies are shown in Table 1, which shows that the residual terrestrial gravity signals are the smoothest. The amplitude of shipborne residual gravity anomalies is the largest, whose STD reaches 5.261 mGal. The terrain of the experiment area is quite flat, so the impact of terrain masses can be ignored, seen from the value of $\Delta g_{terrestrial} - \Delta g_{EIGEN-6C4}$ and $\Delta g_{terrestrial} - \Delta g_{EIGEN-6C4} - \Delta g_{RTM}$ in Table 1.

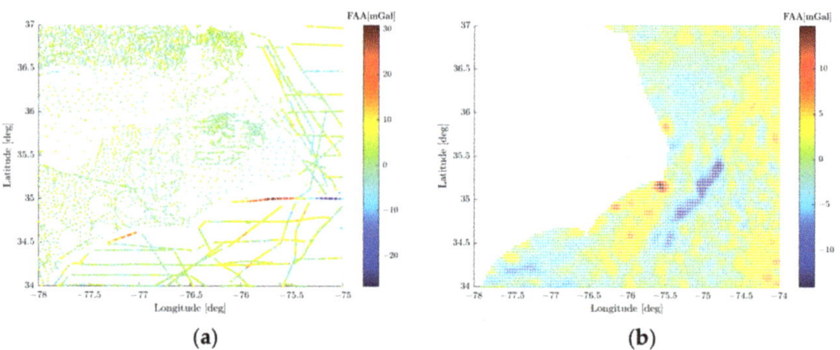

Figure 13. Distribution of residual gravity anomalies in the east coast experiment area: (**a**) terrestrial and shipborne; (**b**) DTU15.

Table 1. Residual gravity anomalies in the east coast experiment area (mGal).

	Mean	Min	Max	Std	Rms
$\Delta g_{terrestrial} - \Delta g_{EIGEN-6C4}$	0.579	−7.563	9.946	1.982	0.579
$\Delta g_{terrestrial} - \Delta g_{EIGEN-6C4} - \Delta g_{RTM}$	3.697	−5.449	12.058	2.109	4.257
$\Delta g_{shipborne} - \Delta g_{EIGEN-6C4}$	2.385	−26.723	30.875	5.261	2.385
$\Delta g_{DTU15} - \Delta g_{EIGEN-6C4}$	−0.314	−14.073	14.717	2.724	−0.314

Firstly, we use terrestrial and DTU15 gravity anomalies for modeling, considering the shipborne data in this zone are few and unevenly distributed. Set the grid resolution range to 0.1°~0.9° and the step size be 0.1°; set the depth range to 10~50 km and the step size be

1 km. The (part of) condition number of the design matrix is counted as shown in Figure 14. It can be seen from the figure that the lower the grid resolutions and the deeper the depths are, the smaller the condition number is. Otherwise, the condition number will be larger, making the ill condition more serious.

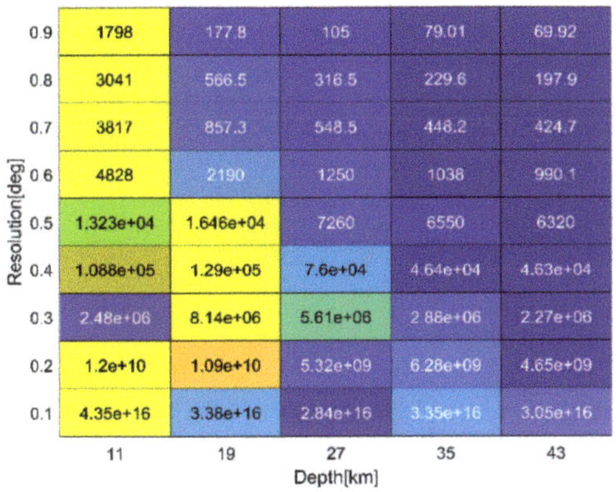

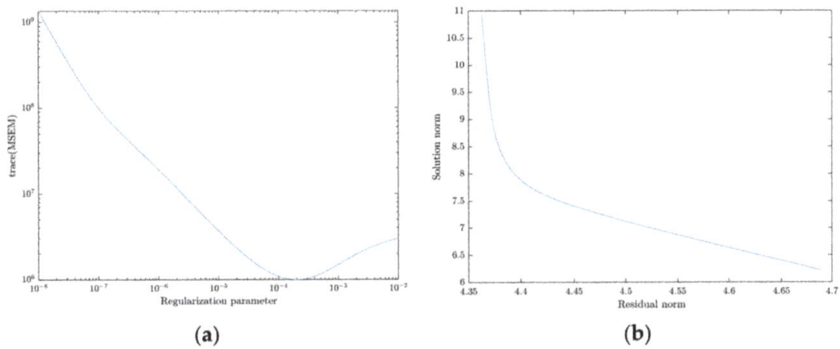

Figure 14. Condition number of the design matrix.

Taking the combination with resolution of 0.1° and depth of 11 km as an example, the number of RBFs is 2324 and the condition number of the design matrix is 4.35×10^6. The accuracy of the ill-conditioned model is 2.454×10^4 m checked by control points. Use the MSE method and the L-curve method, respectively, to determine the regularization parameter, as shown in Figure 15. The results show that the optimal regularization parameters determined by the MSE and L-curve methods are all 1.995×10^{-4}, and the accuracy of the geoid calculated using this parameter is 6.9 cm, which is a great improvement compared with the original modeling result.

Figure 15. Determination of the optimal regularization parameter: (**a**) MSE; (**b**) L-curve.

The calculation results of all combinations of resolutions and depths are shown in Figure 16. Figure 16a shows the accuracy of original modeling without regularization; the accuracy of RBF modeling using Tikhonov regularization is shown in Figure 16b. As can be seen in Figure 16, when the condition number is greater than 1000, the serious ill-conditioned problems lead to the gradual deviation of the modeling results from the normal values. In particular when the grid resolution is over 0.3°, the accuracy of the model geoid is lower than the meter level. Using the Tikhonov regularization technique, the accuracy of the ill-conditioned model is greatly improved to the centimeter level, which improves that Tikhonov regularization can effectively solve ill-conditioned problems in the RBF model.

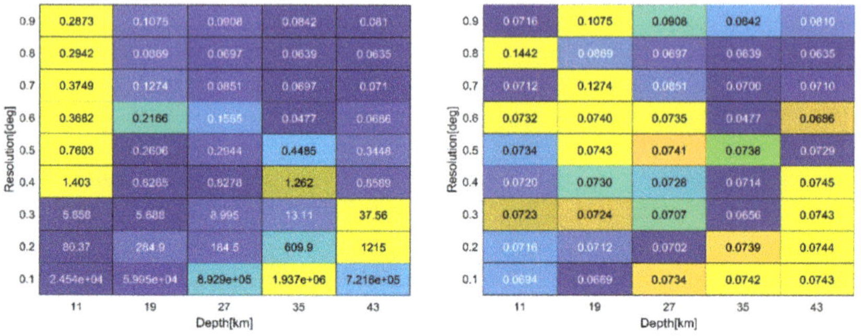

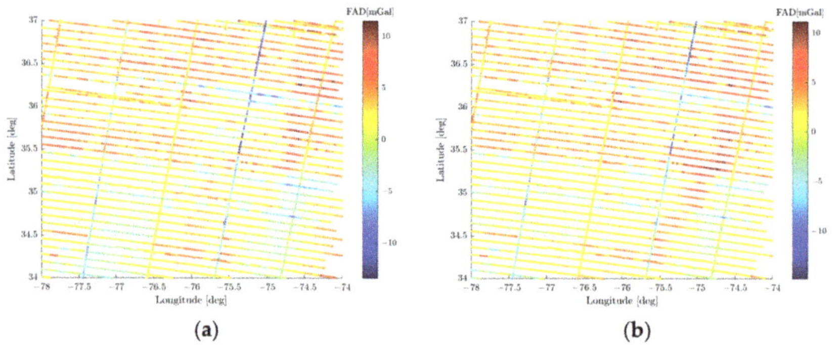

Figure 16. Accuracy of RBF modeling: (**a**) without regularization; (**b**) with regularization.

Then shipborne gravities are added to terrestrial and DTU15 data sets. The weight ratio of terrestrial, DTU15 and shipborne data is 1:0.9410:0.0989 determined by the Helmert VCE technique. Fuse the three data sets to construct the RBF model with a grid resolution of 0.4° and a depth of 45 km, calculating regional gravity field at airborne points.

The distribution of airborne gravity disturbances after removing EIGEN-6C4 values is shown in Figure 17a. The fluctuation of airborne gravity signals is relatively stable. Gravity disturbances δg_A at airborne points are calculated based on the network A determined by terrestrial, DTU15 and shipborne data. Then δg_A are removed from the residual gravity in Figure 17a and as input data for the RBF modeling, as shown in Figure 17b. The statistical information of the two residual gravity disturbances is shown in Table 2.

Figure 17. Distribution of residual airborne gravity disturbances in the east coast experiment area: (**a**) $\delta g_{airborne} - \delta g_{EIGEN-6C4}$; (**b**) $\delta g_{airborne} - \delta g_{EIGEN-6C4} - \delta g_A$.

Table 2. Residual airborne gravity disturbances in the east coast experiment area (mGal).

	Mean	Min	Max	Std	Rms
$\delta g_{airborne} - \delta g_{EIGEN-6C4}$	2.040	−13.427	11.490	3.184	3.781
$\delta g_{airborne} - \delta g_{EIGEN-6C4} - \delta g_A$	1.679	−14.825	11.158	3.022	3.457

Based on the STD minimization technique, the network B is determined with a resolution of 0.8° and a depth of 16 km using $\delta g_{airborne} - \delta g_{EIGEN-6C4} - \delta g_A$ as input data. The spatial distribution of networks A and B is shown in Figure 18, where the bottom blue dots refer to the network A; the red dots in the middle refer to the network B; the top yellow dots refer to the geoid grid points computed by the RBF model.

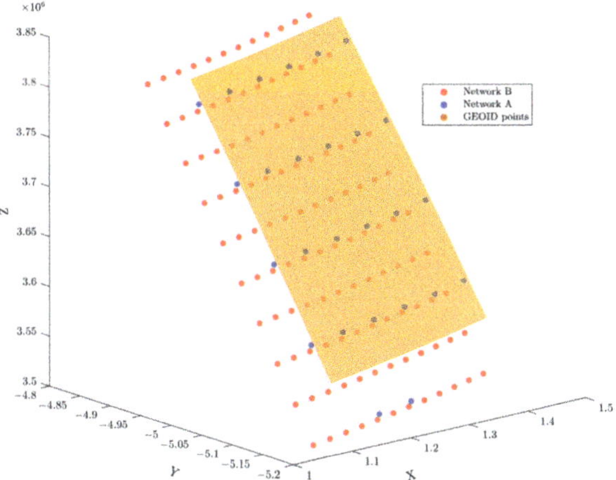

Figure 18. Distribution of networks A and B.

The RBF model is then constructed based on the network B, which is added together with the modeling result calculated by terrestrial, DTU15 and shipborne data based on the network A to obtain the final quasi-geoid, as shown in Figure 19a.

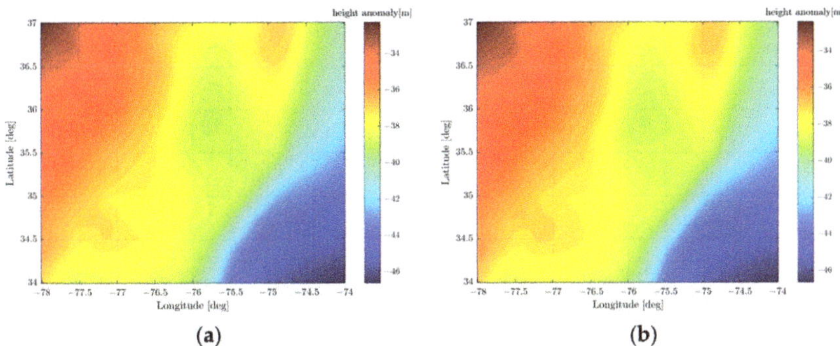

(a)

(b)

Figure 19. The unified land–ocean quasi-geoid from heterogeneous data sets in the east coast experiment area: (**a**) terrestrial + shipborne + DTU15 + airborne gravity; (**b**) terrestrial + shipborne + airborne gravity.

The quasi-geoid calculated by the RBF model is essentially a gravimetric geoid. It has a different datum than the GPS/leveling geoid, resulting in significant systematic errors. We use simple polynomial fitting to convert the datum of the gravimetric geoid to GPS/leveling geoid. The formula is as follows:

$$\Delta N = \alpha_0 + \alpha_1(\varphi - \varphi_m) + \alpha_2(\lambda - \lambda_m) + \alpha_3(\varphi - \varphi_m)^2 + \alpha_4(\varphi - \varphi_m)(\lambda - \lambda_m) + \alpha_5(\lambda - \lambda_m)^2 + \cdots \tag{14}$$

where $\alpha_0, \alpha_1, \alpha_2 \cdots$ denote fitting coefficients, i.e., bias parameter and tilt parameter; ΔN denotes the differences between the gravimetric geoid and the GPS/levelling geoid. (φ_m, λ_m) denotes the center longitude and latitude. The order of the formula can be taken very high, but generally two orders is enough.

Considering the DTU15 data have already been introduced by the EIGEN-6C4 model, we also calculate the results without DTU15 gravity data, shown in Figure 19b. The calculation methods of the two geoids are the same.

The accuracy of the two quasi-geoids shown in Figure 19 is checked by the GPS/leveling data, respectively, on inland and coast, as shown in Table 3. On inland, the accuracy of the quasi-geoid fusing terrestrial, shipborne, DTU15 and airborne data is 1.9 cm, which is 1.3 cm on sea. After the DTU15 gravities are removed, the modeling accuracy is 1.9 cm inland and 1.2 cm on coast, not much different from the previous result.

Table 3. Accuracy of the unified land–ocean quasi-geoid in the east coast experiment area (cm).

		Mean	Min	Max	Std	Rms
inland	Terrestrial + DTU15 + shipborne + airborne	−0.2	−6.3	6.6	1.9	1.9
	Terrestrial + shipborne + airborne	−0.3	−7.2	6.4	1.9	2.0
coast	Terrestrial + DTU15 + shipborne + airborne	−0.3	−3.8	3.5	1.3	1.4
	Terrestrial + shipborne + airborne	−0.2	−4.1	3.1	1.2	1.2

3.2. The West Coast Experiment Area of the USA

The west coast of the USA has quite different topographic features from the east coast. Affected by the overall terrain high in the west and low in the east, the mountains in the west extend to the coastal zone, making the land–ocean boundary areas rugged, which brings many difficulties to the geoid calculation. Based on the RCR technique, EIGEN-6C4 gravity anomalies are removed from the measured terrestrial and shipborne gravity, as shown in Figure 20a. It can be seen from the figure that after the EIGEN-6C4 value is removed, shipborne gravity signals become relatively flat. There are serious omission errors in EIGEN-6C4 on terrestrial gravity points affected by the topographic relief. This article calculates RTM based on the prism integral to simulate high-frequency gravity signals and compensate for the omission errors existing in GGM.

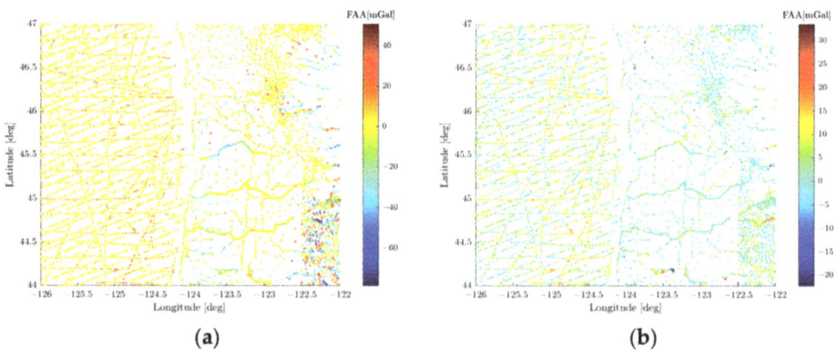

(a)   **(b)**

Figure 20. Distribution of residual terrestrial and shipborne gravity anomalies in the west coast experiment area: (**a**) without RTM; (**b**) with RTM.

In the calculation of RTM, the DEMs are usually expanded by approximately 2° from the edge of the gravity data area to avoid edge effects, as shown in Figure 21. The calculation efficiency can be improved by setting inner and outer computing regions. For example, the inner region within a 100 km radius uses high-resolution DEMs and the outer region within a 200 km radius uses low-resolution DEMs. The impact of terrain mass outside the outer region can be ignored due to the oscillating nature of RTM elevations [42].

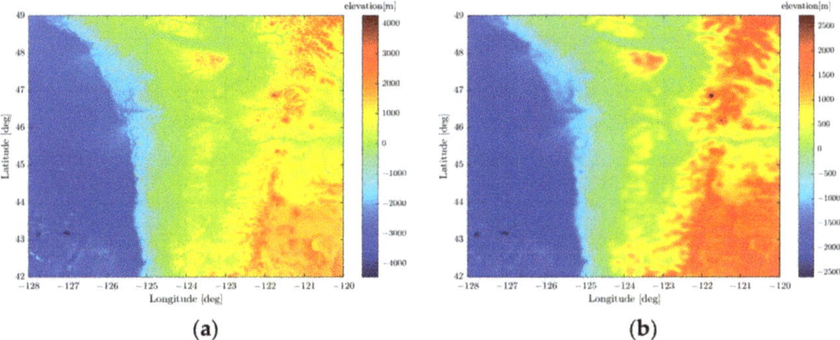

Figure 21. DEMs for the RTM technique: (**a**) SRTM15+; (**b**) RET2014.

The residual gravity anomalies obtained by removing RTM from the terrestrial observations are shown in Figure 20b, and their STD is reduced from 15.983 mGal to 4.134 mGal, indicating that the compensation for EIGEN-6C4 omission errors by RTM reaches approximately 74%. The residual DTU15 gravity anomalies are shown in Figure 22. The statistical information of various residual gravity data is shown in Table 4, which indicates that the variation range of residual DTU15 gravity anomalies is the smallest. The variation amplitude of residual terrestrial gravity anomalies is close to shipborne data.

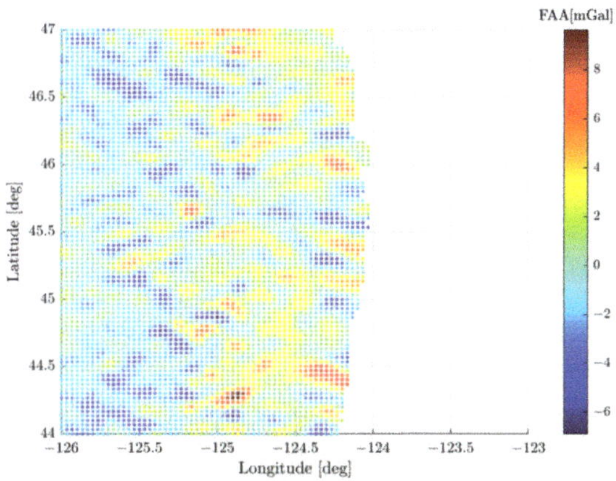

Figure 22. Distribution of residual DTU15 gravity anomalies in the west coast experiment area.

Table 4. Residual gravity anomalies in the west coast experiment area (mGal).

	Mean	Min	Max	Std	Rms
$\Delta g_{terrestrial} - \Delta g_{EIGEN-6C4}$	−7.663	−78.808	50.652	15.983	−7.663
$\Delta g_{terrestrial} - \Delta g_{EIGEN-6C4} - \Delta g_{RTM}$	1.585	−22.311	33.455	4.134	1.585
$\Delta g_{shipborne} - \Delta g_{EIGEN-6C4}$	4.591	−8.221	23.001	4.009	4.591
$\Delta g_{DTU15} - \Delta g_{EIGEN-6C4}$	0.049	−6.878	9.645	2.354	0.049

Firstly, terrestrial and shipborne gravity data are used for modeling. The weight ratio of the two observations is 1:1.004 computed by the Helmert VCE technique, which shows that their accuracy level is close. Based on the STD minimization technique, the optimal RBF grid resolution is determined as 0.6° and the optimal depth is 22 km. Then we add residual DTU15 gravity anomalies to the terrestrial and shipborne gravity to construct the RBF model with an optimal grid resolution of 0.8° and an optimal depth of 26 km.

The distribution of airborne gravity disturbances after removing the EIGEN-6C4 value is shown in Figure 23a. The variation amplitude of residual airborne gravities is relatively small. Based on the RBF network A determined by terrestrial, shipborne and DTU15 data, the model gravity disturbances are calculated on airborne points. Then δg_A are removed from the residual gravities in Figure 23a and as input data for the RBF modeling, as shown in Figure 23b. The statistical information of the two kinds of residual gravity disturbances is shown in Table 5.

Table 5. Residual airborne gravity disturbances in the west coast experiment area (mGal).

	Mean	Min	Max	Std	Rms
$\delta g_{airborne} - \delta g_{EIGEN-6C4}$	2.126	-5.884	14.753	2.321	3.147
$\delta g_{airborne} - \delta g_{EIGEN-6C4} - \delta g_A$	-0.190	-7.867	11.710	2.394	2.402

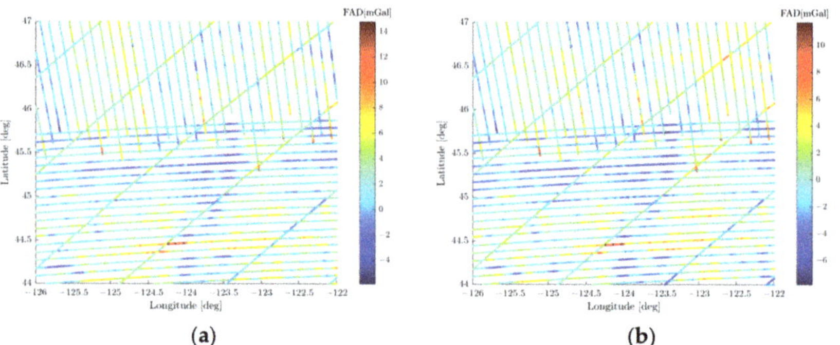

Figure 23. Distribution of residual airborne gravity disturbances in the west coast experiment area: (**a**) $\delta g_{airborne} - \delta g_{EIGEN-6C4}$; (**b**) $\delta g_{airborne} - \delta g_{EIGEN-6C4} - \delta g_A$.

Based on the STD minimization technique, network B is determined with a resolution of $0.8°$ and a depth of 16 km. The RBF model is then constructed based on the network B, which is added with the modeling result based on the network A to obtain the final quasi-geoid, as shown in Figure 24a. We also calculate the results without DTU15 gravity data, shown in Figure 24b.

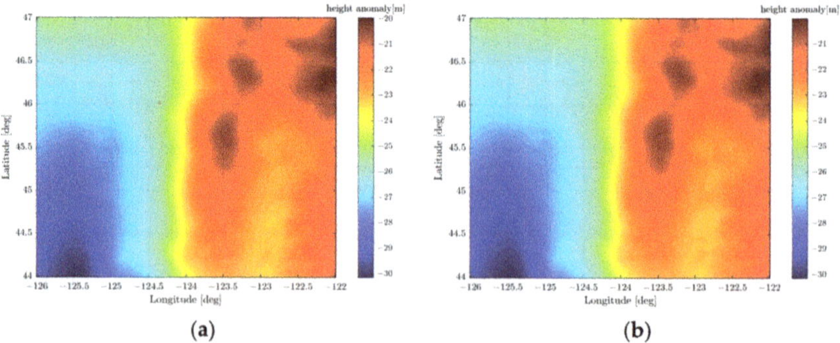

Figure 24. The unified land–ocean quasi-geoid from heterogeneous data sets in the west coast experiment area: (**a**) terrestrial + shipborne + DTU15 + airborne gravity; (**b**) terrestrial + shipborne + airborne gravity.

Check the accuracy of the three quasi-geoids shown in Figure 24 by 64 GPS/leveling points inland and 36 GPS/leveling points on coast, as shown in Table 6. The accuracy of the RBF model quasi-geoid is 2.2 cm inland, which is 2.1 cm on coast. After the DTU15 data are removed, the modeling accuracy is 2.0 cm inland and 2.1 cm on coast.

Table 6. Accuracy of the unified land–ocean quasi-geoid in the west coast experiment area (cm).

		Mean	Min	Max	Std	Rms
inland	Terrestrial + DTU15 + shipborne + airborne	0.0	−6.6	5.1	2.2	2.2
	Terrestrial + shipborne + airborne	−0.2	−6.9	4.0	2.0	2.0
coast	Terrestrial + DTU15 + shipborne + airborne	−0.3	−4.2	5.8	2.1	2.0
	Terrestrial + shipborne + airborne	−0.5	−4.6	5.6	2.1	2.1

According to the results of the two experiments on the east and west coast, we consider that the accuracy of the unified land–ocean quasi-geoid can be improved by fusing heterogeneous data sets. Fusing terrestrial, shipborne, DTU15 and airborne gravities based on a multi-layer RBF network achieves great modeling accuracy both inland and on coast.

4. Discussion

Due to the complex topography of land–ocean junction areas, terrestrial and shipborne gravity measurements cannot be fully carried out. As it is also affected by the terrain, the accuracy of satellite altimetry data inshore is reduced. The airborne gravity is not limited by the terrain, but its accuracy will be lost with downward continuation. Therefore, the important premise of constructing a high-precision unified land–ocean quasi-geoid is to effectively fuse heterogeneous data sets. Considering the shortcomings of the commonly used integral algorithm, we use the RBFs to calculate the unified land–ocean quasi-geoid, the accuracy of which is improved by fusing more types of data sets. However, there are still some problems to be discussed and further studied, as follows:

(1) The RBF used to calculate the land–ocean quasi-geoid in this article is the IMQ kernel, which has simpler function forms than the Poisson kernel, the radial multipoles kernel and the Poisson wavelets kernel. It can be seen from Figure 9 that the first-order Poisson wavelets kernel has the strongest localization characteristics when the depth is 300 km. However, the buried depth of RBFs will not be so deep generally when dealing with the measured data. When the buried depth is less than 100 km, the localization characteristics of various RBFs are all strong and their differences are quite small. By further analyzing the spectrum characteristics of RBFs, as shown in Figure 25, it can be seen that the Poisson kernel, the radial multipoles kernel and the Poisson wavelets kernel all have band-pass characteristics, while the IMQ kernel presents the characteristics of low-pass filtering. We have carried out some experiments to compare the modeling accuracy of the four RBFs. We preliminarily find that the low-pass characteristics of the IMQ kernel can help it filter out more high-frequency noise when dealing with the terrestrial gravity, making its modeling results slightly better than the other RBFs. The specific experimental results will not be presented in this article. Theoretically, the band-pass characteristics can help the RBFs simulate gravity signals more accurately. In this paper, the IMQ kernel is used to calculate the quasi-geoid for the time being, and the modeling differences between various RBFs will be more comprehensively analyzed in the future.

(2) The STD minimization technique used to determine the optimal RBF network lacks a strictly theoretical basis, which is a compromise method in view of the lack of other more effective strategies. The reason why it is difficult to determine the RBF networks is that it is difficult to directly determine the appropriate positions of RBFs based on the prior information of discrete observations. If the gravity signals are gridded before modeling, we may directly determine the spatial position of the RBF points according to the resolution and height of the gravity grid. However, our goal is to model using discrete observations. The STD minimization technique can obtain

good modeling results by screening lots of RBF networks, but it increases too much redundant calculations, which hinders the solution efficiency of the RBF model to a great extent. So, it is very important to develop a rigorous and logical method to determine the RBF networks.

(3)   The accuracy of the quasi-geoid fusing terrestrial, shipborne and DTU15 data is quite high, but the improvement is not obvious after adding airborne gravity. The main reason is that the gravity signals on airborne points simulated by network A is insufficient, resulting in little change in residual gravity disturbances after removing δg_A, as shown in Tables 2 and 5. Theoretically, if the result of $\delta g_B = \delta g - \delta g_{GGM} - \delta g_A$ is significantly reduced, the function of the network B will be more obvious. In the future, we can further improve the multi-layer RBF network and try to set more layers of RBF grids, simulating the gravity signals more accurately.

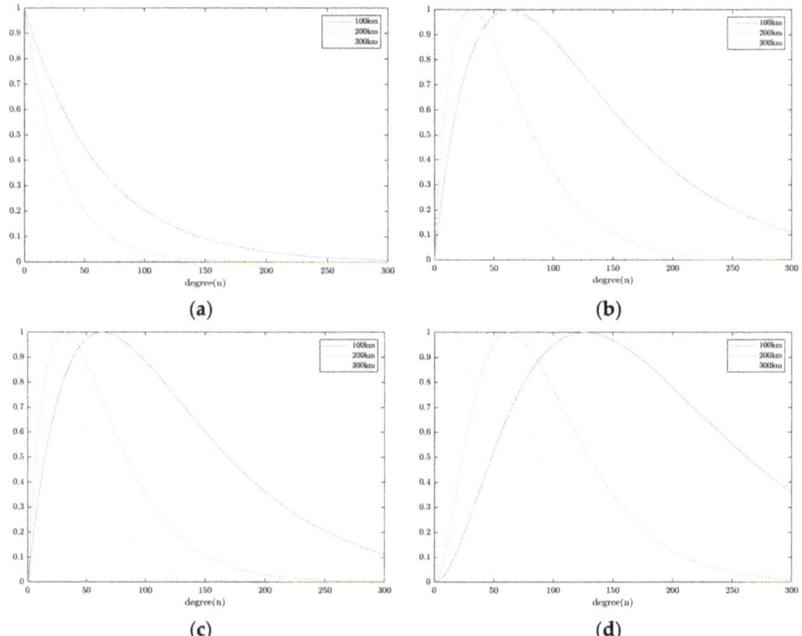

Figure 25. Behavior of RBFs in the spectral domain: (**a**) the IMQ kernel; (**b**) the Poisson kernel; (**c**) radial multipoles of order 1; (**d**) Poisson wavelets of order 1.

5. Conclusions

This article designs a multi-layer RBF model to construct the unified land–ocean quasi-geoid fusing the measured terrestrial, shipborne, satellite altimetry and airborne gravity data in coastal areas. Several core problems in the process of RBF modeling are studied in depth. The local quasi-geoid with a $1'$ resolution is calculated, respectively, on the flat east coast and the rugged west coast of the United States. Several conclusions are summarized as follows:

(1)   The behavior of four types of RBFs—the IMQ kernel, the Poisson kernel, radial multipoles and Poisson wavelets—is analyzed in the spatial domain. The figures show that RBFs have significant localization characteristics in the spatial domain, which is helpful to concentrate more gravity signals in local gravity field approximation. Placing RBFs on the geographic or the Reuter grid, the optimal RBF network, i.e., the optimal grid resolution and depth can be effectively determined based on the STD minimization technique.

(2)   The ill condition of the design matrix may occur due to the uneven distribution of observations and the excessive number of RBFs. Using the Tikhonov regularization technique, the accuracy of the ill-conditioned model is greatly improved to the centimeter level. The regularization parameters determined by the MSE and L-curve methods are basically the same. RTM calculated based on the prism integral algorithm can effectively simulate the high-frequency gravity signals implied by terrain masses and compensate for the omission errors existing in EIGEN-6C4. In the west coast experiment area, the compensation effect can reach approximately 74%. Therefore, in areas with large topographic relief, it is necessary to consider the influence of terrain masses on geoid calculations.

(3)   The local gravity quasi-geoids with a $1'$ resolution are calculated by setting up multilayer RBF networks based on the IMQ kernel, respectively, on the east and west coast of the United States. The results show that the accuracy of the quasi-geoid computed by fusing the terrestrial, shipborne, satellite altimetry and airborne gravity data in the east coast experimental area is 1.9 cm inland and 1.3 cm on coast after internal verification. The accuracy of the quasi-geoid calculated in the west coast experimental area is 2.2 cm inland and 2.1 cm on coast.

Through the above research results, we can roughly summarize the advantages of RBFs compared with other geoid calculation methods: (a) RBFs are simple and easy for multi-source data fusion; (b) RBFs can directly deal with discrete observations without gridding or downward continuation; (c) RBFs have a strong adaptive ability. In general, the RBF model has an important application value to calculate the unified land–ocean quasi-geoid from heterogeneous data sets.

Author Contributions: Conceptualization, Y.L. and L.L.; methodology, Y.L.; funding acquisition and project administration, L.L.; investigation and software, Y.L.; supervision and validation, L.L.; writing—original draft, Y.L.; writing—review and editing, L.L. All authors have read and agreed to the published version of the manuscript.

Funding: This research was funded by the National Natural Science Foundation of China, grant number 41874004.

Data Availability Statement: All data for this paper are properly cited and referred to in the reference list.

Acknowledgments: The authors are very grateful to the NGS, the DTU, the NOAA and the ICGEM for providing gravity, GPS/leveling, geoid and GGM data. The authors express thanks to Hirt and Tozer for providing DEM data. The software Gravsoft shared by Forsberg also helped a lot.

Conflicts of Interest: The authors declare no conflict of interest.

References

1. Forsberg, R. Modelling the Fine-Structure of the Geoid: Methods, Data Requirements and Some Results. *Surv. Geophys.* **1993**, *14*, 403–418. [CrossRef]
2. Alberts, B.; Klees, R. A Comparison of Methods for the Inversion of Airborne Gravity Data. *J. Geodesy* **2004**, *78*, 55–56. [CrossRef]
3. Sabri, L.M.; Sudarsono, B.; Pahlevi, A. Geoid of South East Sulawesi from Airborne Gravity Using Hotine Approach. *IOP Conf. Ser. Earth Environ. Sci.* **2021**, *731*, 012014. [CrossRef]
4. Strykowski, G.; Forsberg, R. Operational Merging of Satellite, Airborne and Surface Gravity Data by Draping Techniques. In *Geodesy on the Move*; Forsberg, R., Feissel, M., Dietrich, R., Eds.; International Association of Geodesy Symposia; Springer: Berlin/Heidelberg, Germany, 1998; Volume 119, pp. 243–248. [CrossRef]
5. Knudsen, P.; Andersen, O.B. Improved Recovery of the Global Marine Gravity Field from the GEOSAT and the ERS-1 Geodetic Mission Altimetry. In *Gravity, Geoid and Marine Geodesy*; Segawa, J., Fujimoto, H., Okubo, S., Eds.; International Association of Geodesy Symposia; Springer: Berlin/Heidelberg, Germany, 1997; Volume 117, pp. 429–436. [CrossRef]
6. Eppelbaum, L.; Katz, Y. A New Regard on the Tectonic Map of the Arabian–African Region Inferred from the Satellite Gravity Analysis. *Acta Geophys.* **2017**, *65*, 607–626. [CrossRef]
7. Braitenberg, C.; Ebbing, J. New Insights into the Basement Structure of the West Siberian Basin from Forward and Inverse Modeling of GRACE Satellite Gravity Data. *J. Geophys. Res. Earth Surf.* **2009**, *114*, 1–15. [CrossRef]

8. Neyman, Y.M.; Li, J. Modification of Stokes and Vening-Meinesz Formulas for the Inner Zone of Arbitrary Shape by Minimization of Upper Bound Truncation Errors. *J. Geod.* **1996**, *70*, 410–418. [CrossRef]

9. Abbak, R.A.; Sjöberg, L.E.; Ellmann, A.; Ustun, A. A Precise Gravimetric Geoid Model in a Mountainous Area with Scarce Gravity Data: A Case Study in Central Turkey. *Stud. Geophys. Geod.* **2012**, *56*, 909–927. [CrossRef]

10. Featherstone, W.E. Deterministic, Stochastic, Hybrid and Band-Limited Modifications of Hotine's Integral. *J. Geodesy* **2013**, *87*, 487–500. [CrossRef]

11. Wichiengaroen, C. A Comparison of Gravimetric Undulations Computed by the Modified Molodensky Truncation Method and the Method of Least Squares Spectral Combination by Optimal Integral Kernels. *J. Geodesique* **1984**, *58*, 494–509. [CrossRef]

12. Nahavandchi, H.; Sjöberg, L.E. Precise Geoid Determination over Sweden Using the Stokes-Helmert Method and Improved Topographic Corrections. *J. Geodesy* **2001**, *75*, 74–88. [CrossRef]

13. Tscherning, C.C.; Rapp, R.H. *Closed Covariance Expressions for Gravity Anomalies, Geoid Undulations, and Deflections of the Vertical Implied by Anomaly Degree Variance Models*; Scientific Interim Report Ohio State University: Columbus, GA, USA, 1974.

14. Rapp, R.H. Gravity Anomalies and Sea Surface Heights Derived from a Combined GEOS$_3$/Seasat Altimeter Data Set. *J. Geophys. Res.* **1986**, *91*, 4867. [CrossRef]

15. Hwang, C. Analysis of Some Systematic Errors Affecting Altimeter-Derived Sea Surface Gradient with Application to Geoid Determination over Taiwan. *J. Geod.* **1997**, *71*, 113–130. [CrossRef]

16. Hwang, C.; Guo, J.; Deng, X.; Hsu, H.-Y.; Liu, Y. Coastal Gravity Anomalies from Retracked Geosat/GM Altimetry: Improvement, Limitation and the Role of Airborne Gravity Data. *J. Geodesy* **2006**, *80*, 204–216. [CrossRef]

17. Olesen, A.V.; Andersen, O.B.; Tscherning, C.C. Merging of Airborne Gravity and Gravity Derived from Satellite Altimetry: Test Cases Along the Coast of Greenland. *Stud. Geophys. Geod.* **2002**, *46*, 387–394. [CrossRef]

18. Stein, E.M.; Weiss, G. *Introduction to Fourier Analysis on Euclidean Spaces (PMS-32)*; Princeton University Press: Princeton, NJ, USA, 2016; Volume 32. [CrossRef]

19. Hardy, R.L. Multiquadric Equations of Topography and Other Irregular Surfaces. *J. Geophys. Res.* **1971**, *76*, 1905–1915. [CrossRef]

20. Reilly, J.P.; Herbrechtsmeier, E.H. A Systematic Approach to Modeling the Geopotential with Point Mass Anomalies. *J. Geophys. Res.* **1978**, *83*, 841. [CrossRef]

21. Barthelmes, F. Local Gravity Field Approximation by Point Masses with Optimized Positions. In Proceedings of the 6th International Symposium "Geodesy and Physics of the Earth", Potsdam, Germany, 22–27 August 1988. [CrossRef]

22. Lehmann, R. The Method of Free-Positioned Point Masses—Geoid Studies on the Gulf of Bothnia. *Bull. Géodésique* **1993**, *67*, 31–40. [CrossRef]

23. Marchenko, A.N.; Barthelmes, F.; Meyer, U.; Schwintzer, P. Regional Geoid Determination: An Application to Airborne Gravity Data in the Skagerrak. 2001, pp. 1–48. Available online: https://gfzpublic.gfz-potsdam.de/rest/items/item_8522_3/component/file_8521/content (accessed on 1 March 2021).

24. Holschneider, M.; Iglewska-Nowak, I. Poisson Wavelets on the Sphere. *J. Fourier Anal. Appl.* **2007**, *13*, 405–419. [CrossRef]

25. Schmidt, M.; Fengler, M.; Mayer-Gürr, T.; Eicker, A.; Kusche, J.; Sánchez, L.; Han, S.-C. Regional Gravity Modeling in Terms of Spherical Base Functions. *J. Geod.* **2006**, *81*, 17–38. [CrossRef]

26. Freeden, W. Spherical Spline Interpolation—Basic Theory and Computational Aspects. *J. Comput. Appl. Math.* **1984**, *11*, 367–375. [CrossRef]

27. Tenzer, R.; Klees, R. The Choice of the Spherical Radial Basis Functions in Local Gravity Field Modeling. *Stud. Geophys. Geod.* **2008**, *52*, 287–304. [CrossRef]

28. Bentel, K.; Schmidt, M.; Gerlach, C. Different Radial Basis Functions and Their Applicability for Regional Gravity Field Representation on the Sphere. *Int. J. Geomath.* **2013**, *4*, 67–96. [CrossRef]

29. Wu, Y.; Zhong, B.; Luo, Z. Investigation of the Tikhonov Regularization Method in Regional Gravity Field Modeling by Poisson Wavelets Radial Basis Functions. *J. Earth Sci.* **2018**, *29*, 1349–1358. [CrossRef]

30. Liu, Q.; Schmidt, M.; Sánchez, L.; Willberg, M. Regional Gravity Field Refinement for (Quasi-) Geoid Determination Based on Spherical Radial Basis Functions in Colorado. *J. Geod.* **2020**, *94*, 99. [CrossRef]

31. Liu, Q.; Schmidt, M.; Pail, R.; Willberg, M. Determination of the Regularization Parameter to Combine Heterogeneous Observations in Regional Gravity Field Modeling. *Remote Sens.* **2020**, *12*, 1617. [CrossRef]

32. Eicker, A. *Gravity Field Refinement by Radial Basis Functions from In-Situ Satellite Data*; Bonn University: Bonn, Germany, 2008.

33. Klees, R.; Wittwer, T. A Data-Adaptive Design of a Spherical Basis Function Network for Gravity Field Modelling. In *Dynamic Planet*; Tregoning, P., Rizos, C., Eds.; International Association of Geodesy Symposia; Springer: Berlin/Heidelberg, Germany, 2007; Volume 130, pp. 322–328. [CrossRef]

34. Tenzer, R.; Klees, R.; Wittwer, T. Local Gravity Field Modelling in Rugged Terrain Using Spherical Radial Basis Functions: Case Study for the Canadian Rocky Mountains. In *Geodesy for Planet Earth*; Kenyon, S., Pacino, M.C., Marti, U., Eds.; International Association of Geodesy Symposia; Springer: Berlin/Heidelberg, Germany, 2012; Volume 136, pp. 401–409. [CrossRef]

35. Sjöberg, L.E. A Discussion on the Approximations Made in the Practical Implementation of the Remove–Compute–Restore Technique in Regional Geoid Modelling. *J. Geodesy* **2005**, *78*, 645–653. [CrossRef]

36. Li, B.; Shen, Y.; Lou, L. Efficient Estimation of Variance and Covariance Components: A Case Study for GPS Stochastic Model Evaluation. *IEEE Trans. Geosci. Remote Sens.* **2011**, *49*, 203–210. [CrossRef]

37. Wittwer, T. *Regional Gravity Field Modelling with Radial Basis Functions*; Netherlands Geodetic Commission: Delft, The Netherlands, 2009.

38. Kusche, J.; Klees, R. Regularization of Gravity Field Estimation from Satellite Gravity Gradients. *J. Geodesy* **2002**, *76*, 359–368. [CrossRef]
39. Hirt, C.; Kuhn, M. Band-limited Topographic Mass Distribution Generates Full-spectrum Gravity Field: Gravity Forward Modeling in the Spectral and Spatial Domains Revisited. *J. Geophys. Res. Solid Earth.* **2014**, *119*, 3646–3661. [CrossRef]
40. Hirt, C. RTM Gravity Forward-Modeling Using Topography/Bathymetry Data to Improve High-Degree Global Geopotential Models in the Coastal Zone. *Mar. Geodesy* **2013**, *36*, 183–202. [CrossRef]
41. Hirt, C.; Kuhn, M.; Claessens, S.; Pail, R.; Seitz, K.; Gruber, T. Study of the Earth's Short-Scale Gravity Field Using the ERTM2160 Gravity Model. *Comput. Geosci.* **2014**, *73*, 71–80. [CrossRef]
42. Hirt, C.; Yang, M.; Kuhn, M.; Bucha, B.; Kurzmann, A.; Pail, R. SRTM2gravity: An Ultrahigh Resolution Global Model of Gravimetric Terrain Corrections. *Geophys. Res. Lett.* **2019**, *46*, 4618–4627. [CrossRef]
43. Hirt, C.; Rexer, M. Earth2014: 1 Arc-Min Shape, Topography, Bedrock and Ice-Sheet Models—Available as Gridded Data and Degree-10,800 Spherical Harmonics. *Int. J. Appl. Earth Obs. Geoinf.* **2015**, *39*, 103–112. [CrossRef]

Article

Orbit Determination for All-Electric GEO Satellites Based on Space-Borne GNSS Measurements

Wenqiang Lu [1,2,3], Haoguang Wang [4], Guoqiang Wu [4] and Yong Huang [1,3,*]

1   Shanghai Astronomical Observatory, Chinese Academy of Sciences, Shanghai 200030, China; luwq@cma.gov.cn
2   National Satellite Meteorological Center, Beijing 100081, China
3   University of Chinese Academy of Sciences, Beijing 100049, China
4   Innovation Academy for Microsatellites Chinese Academy of Sciences, Shanghai 200135, China; wanghg@microsate.com (H.W.); wugq@microsate.com (G.W.)
*   Correspondence: yongh@shao.ac.cn

Abstract: Orbit accuracy of the transfer orbit and the mission orbit is the basis for the orbit control of all-electric-propulsion Geostationary Orbit (GEO) satellites. Global Navigation Satellite System (GNSS) simulation data are used to analyze the main factors affecting GEO satellite orbit prediction accuracy under the no-thrust condition, and an electric propulsion calibration algorithm is designed to analyze the orbit determination and prediction accuracy under the thrust condition. The calculation results show that the orbit determination accuracy of mission orbit and transfer orbit without thrust is better than 10 m using onboard GNSS technology. The calibration accuracy of electric thrust is about 10^{-9} m/s^2 and 10^{-7} m/s^2 with 40 h and 16 h arc length, respectively, using the satellite self-positioning data of 100 m accuracy to calibrate the electric thrust. If satellite self-positioning data accuracy is at the 10 m level, the electric thrust calibration accuracy can be improved by about one order of magnitude, and the 14-day prediction accuracy of the transfer orbit with thrust is better than 1 km.

Keywords: all-electric propulsion GEO satellite; transfer orbit; onboard GNSS; electric propulsion calibration; precise orbit determination (POD)

Citation: Lu, W.; Wang, H.; Wu, G.; Huang, Y. Orbit Determination for All-Electric GEO Satellites Based on Space-Borne GNSS Measurements. *Remote Sens.* **2022**, *14*, 2627. https://doi.org/10.3390/rs14112627

Academic Editors: Jianguo Yan and Qile Zhao

Received: 7 February 2022
Accepted: 28 May 2022
Published: 31 May 2022

Publisher's Note: MDPI stays neutral with regard to jurisdictional claims in published maps and institutional affiliations.

1. Introduction

GEO all-electric propulsion satellites completely rely on the electric propulsion system to change their orbits into the mission orbit after separation of the satellite and the rocket; they also use their electric propulsion system to maintain position after entering mission orbit [1,2]. Compared with traditional chemical propulsion control, electric propulsion technology has the advantages of long life, high specific impulse, widely adjustable range, simple structure, and high reliability, which can significantly reduce the mass of propellant carried by spacecraft, providing more capacity for payload and significantly enhancing economic benefits [3–6]. Early electric propulsion technology research and product development mainly targeted Low Earth Orbit (LEO) satellites and the north–south station-keeping of GEO satellites [7–14]. Due to the significant advantages of electric propulsion systems, research and practice of transfer orbit of GEO satellites and deep space exploration and other missions have been carried out successively [15–24]. A U.S. space exploration company realized the world's first application of all-electric-propulsion GEO satellite on 1 March 2015; it carried four XIPS-25 ion electric thrusters on board and delivered the satellite into geostationary orbit after about 8 months of orbital transfer [25,26]. Due to the use of an electric propulsion system for orbital transfer, the weight of the satellite was reduced from 4 tons to 2 tons, and the launch cost was directly reduced by USD 50 million. Subsequently, many countries have successively achieved the practical application of electric propulsion platforms, and electric propulsion satellites have become an important direction for the development of geostationary satellites [27]. The SJ-9A satellite launched by China in 2012 is equipped with an LIPS-200 ion thruster and HET-40 hall thruster for on-orbit experiments to undertake the mission of maintaining north–south station keeping during the full life cycle of the satellite, which is the first space application of Chinese electric propulsion products [28–30]. In 2020, the LIPS-200 ion thruster electric propulsion system of the Asia–Pacific 6D satellite was used for north–south station keeping, which is the first commercial application of electric propulsion technology in China [31]. Chinese all-electric satellites will also be launched soon [32].

With the characteristics of real-time, high accuracy, and low cost, onboard GNSS orbit determination technology is currently the main technique for spacecraft autonomous orbit determination internationally [33–35]. In the past decades, GNSS-based orbit determination technology has been commonly used in LEO spacecraft, such as TOPEX, GRACE, NOAA, METOP, FY, HY, ZY, and GF, to provide real-time position or final precision orbit with an accuracy within 10 cm [36–41],which is largely attributed to the fact that LEO satellites provide continuous tracking and multi-directional observation geometry. In the field of high orbit spacecraft, GNSS receiving antennas pointing to the center of the Earth receive primary and secondary signals from navigation satellites on the other side of the Earth for orbit determination. Europe and the United States started earlier in this field and achieved operational applications on GOES-R meteorological satellites. In recent years, China has used CE-5T1 and GEO satellites such as SJ-17 and TJS-2 to validate GNSS orbit determination technology, and improved the accuracy of real-time orbit determination by 30 m [42–52].

Since satellites usually spend several months in transfer orbit before entering geostationary orbit by electric propulsion, during which frequent maneuvers are required according to the orbit design [53–57], this poses a new challenge for satellite precision orbit determination. However, previous studies on precision orbit determination have generally focused only on LEO and GEO satellites, and few papers have studied GNSS precision orbit determination for transfer orbits. In this paper, we study the ground orbit determination algorithm for an all-electric propulsion GEO satellite in transfer orbit and mission orbit by analyzing different conditions, and design an electric propulsion calibration algorithm to analyze its orbit prediction accuracy, which provides important support for the orbit control of all-electric-propulsion GEO satellite in transfer orbit and mission orbit.

2. Materials and Methods

2.1. Satellite Electric Propulsion and Orbit Measurement System

Satellite use the electric propulsion system to realize orbit transfer and finally enter mission orbit. During the satellite orbit transfer process, the satellite goes through the initial orbit, the first stage of transfer orbit, the second stage of transfer orbit, and the final mission orbit. The changes in satellite height are shown in Figure 1. Firstly, the perigee and apogee of the satellite are all raised, and the apogee can reach up to 70,000 km in the first stage of the transfer orbit, while the optimal control strategy is adopted to finally transfer to the mission orbit in the second stage of the transfer orbit.

Four HET300 Hall electric thrusters are used in the electric propulsion system to realize the orbit control of the satellite. These are installed on the $-Z$ side of the satellite, and the thrust direction of each thruster is through the center of mass of the satellite. The installation layout is shown in Figure 2.

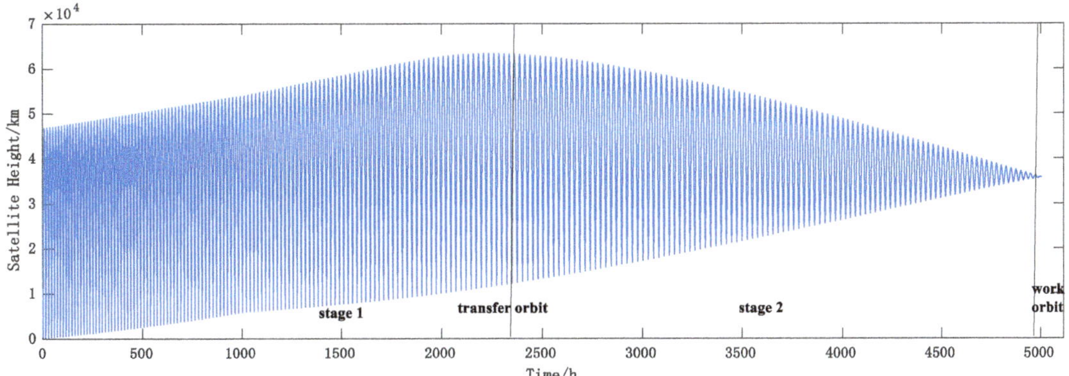

Figure 1. Time-dependent graph of satellite height.

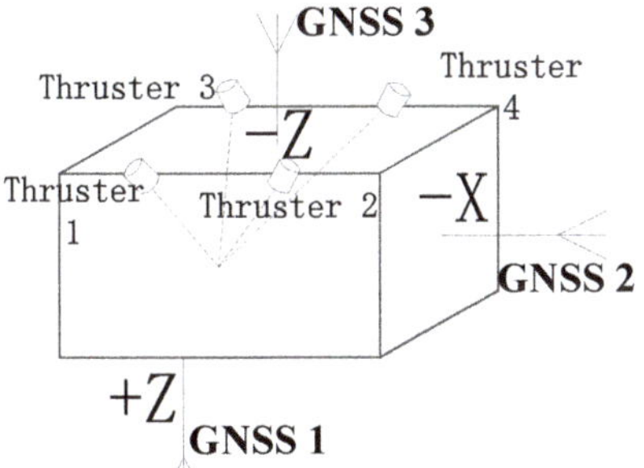

Figure 2. Installation diagram of thruster and GNSS antennas.

The satellite adopts an all-electric-thrust orbit control strategy with a control period of 14 days. No thrust control is applied to the satellite during the satellite orbit determination period (the first day), and then satellite orbit control takes place over the next 13 days.

- In the transfer orbit, the satellite adopts the orbit transfer control strategy, the perigee altitude is raised as soon as possible to cross the core region of the inner radiation belt in the first stage; the satellite is transferred to the mission orbit in the optimal transfer time in the second stage. The satellite thrusters are in mission state except for the ground shadow region during the satellite orbit control period.
- In the mission orbit, the satellite adopts an orbit-holding control strategy. The satellite orbit is maintained at the target geographic longitude by applying thrust control at a specific orbital phase. The satellite thrusters are operated for no more than two hours per day during the satellite orbit control period.

The satellite carries only three GNSS antennas for orbit determination, and the installation layout is shown in Figure 2. Antenna 1 and antenna 2 are both high-gain antennas located in the +Z and −X direction of the satellite body-fixed system, receiving GNSS signals on the opposite side of the Earth. Antenna 3 is a low-gain antenna located in the −Z direction of the satellite body-fixed system, receiving GNSS signals overhead, which ensures that the satellite receives GNSS signals all the time in the transfer orbit and the mission orbit. GNSS receiver technical indicators are shown in Table 1.

Table 1. GNSS receiver indicators.

Type	Indicators
Operating frequency	GPS L1/GLONASS L1/BDS B1
Time accuracy	<5 us
Pseudo-Range Measurement Accuracy (RMS)	<1 m (height < 1000 km), <10 m (height < 36,000 km)
Carrier Phase Accuracy (RMS)	<5 mm (height < 1000 km), <3 cm (height < 36,000 km)
Measurement extraction interval	5 min

2.2. Orbit Determination Principle

According to the theory of artificial satellite orbit [58], the motion equation of the satellite in the inertial coordinate system is as follows:

$$\ddot{\vec{r}} = \vec{f}_0 + \vec{f}_\varepsilon + \vec{f}_{Thrust} \tag{1}$$

where $\vec{r}$ is the position vector of the satellite in the inertial coordinate system, and the right end of the equation is the force acting on the satellite per unit mass; $\vec{f}_0$ is the two-body force; $\vec{f}_\varepsilon$ is the sum of the natural regenerative forces on the satellite, including the N-body regenerative force, the Earth's non-spherical gravity, atmospheric drag, solar radiation pressure, etc.; $\vec{f}_{Thrust}$ is the electric thrust force on the satellite.

The motion states $\vec{r}$ and $\dot{\vec{r}}$ of the satellite at any moment $t \geq t_0$ can be obtained from the motion states $\vec{r}_0$ and $\dot{\vec{r}}_0$ of the satellite at initial moment t_0. The solution of the motion equation is generally obtained numerically, and its initial conditions are:

$$\vec{r}(t_0) = \vec{r}_0 \tag{2}$$

$$\dot{\vec{r}}(t_0) = \dot{\vec{r}}_0 \tag{3}$$

Assuming that the electric thrust $\vec{f}_{Thrust}$ has three directional components in the satellite body coordinate system: f_x, f_y, and f_z. A thrust model is built in the satellite body coordinate system since the satellite thrust has a fixed direction (+Z), and the electric thrust is converted to the inertial system by attitude information and other force models (solar radiation pressure, gravity field, N-body). The M matrix is the conversion matrix from the satellite body coordinate system to the inertial coordinate system. Then

$$\begin{bmatrix} f_{0x} \\ f_{0y} \\ f_{0z} \end{bmatrix} = M \cdot \begin{bmatrix} f_x \\ f_y \\ f_z \end{bmatrix} \tag{4}$$

Let the thrust parameter to be solved $\vec{p} = [f_{0x}, f_{0y}, f_{0z}]$. Because $\vec{p}$ is a constant, $\vec{p}' = 0$. Let

$$X = \begin{bmatrix} \vec{r} \\ \dot{\vec{r}} \\ \vec{p} \end{bmatrix} \quad F = \begin{bmatrix} \dot{\vec{r}} \\ \ddot{\vec{r}} \\ 0 \end{bmatrix} \tag{5}$$

Then the kinetic equation can be written as

$$\begin{cases} \dot{X} = F(X, t) \\ X(t_0) = X_0 \end{cases} \tag{6}$$

where X is the state vector of the satellite at time t, and the partial derivative $\dot{X}$ of the satellite state quantity with respect to time is a function of the satellite state X at time t, called the state function, denoted as $F\,(X,t)$. The state quantity $X\,(t_0)$ of the satellite at time t_0 is recorded as X_0. Linearize it:

$$\begin{cases} \dot{x}\,(t) = \left(\frac{\partial F}{\partial X}\right)^{*} x\,(t) \\ \quad x\,(t_0) = x_0 \end{cases} \tag{7}$$

where $x\,(t) = X\,(t) - X^{*}(t)$.

The observation equation can be written as:

$$Y = G\,(X,t) + \varepsilon \tag{8}$$

where $G\,(X,t)$ is the true value corresponding to the observed data Y, and ε is the random noise of Y. After linearizing the above equation, it can be written as:

$$y = Hx + \varepsilon \tag{9}$$

where $H = \left(\frac{\partial G}{\partial X}\right)^{*} \Phi\,(t,t_0)$, $\Phi\,(t,t_0)$ is the solution of the following matrix differential equation:

$$\begin{cases} \dot{\Phi}\,(t,t_0) = \left(\frac{\partial F}{\partial X}\right)^{*} \Phi\,(t,t_0) \\ \quad \Phi\,(t,t_0) = I \end{cases} \tag{10}$$

The observed data are used to solve the optimal valuation of parameter $\hat{X}_0$ to solve the thrust parameter according to the motion equation. Assuming that a series of unequal precision observations $(y_1, y_2, \cdots, y_N)$ are obtained from time t_1 to t_N with a covariance matrix R,

$$R = \begin{bmatrix} R_1 & & & \\ & R_2 & & \\ & & \cdot & \\ & & & \cdot \\ & & & & R_N \end{bmatrix} \tag{11}$$

It is known that the prior value of the state deviation x_0 at the initial moment is $\overline{x}_0$, and the prior covariance matrix is $\overline{P}_0$. Then

$$\begin{cases} y = Hx_0 + \varepsilon \\ \overline{x}_0 = x_0 + \eta_0 \end{cases} \tag{12}$$

The following statistical characteristics are met:

$$\begin{cases} E\,(\varepsilon_i) = E(\eta_0) = 0 \\ E\,(\varepsilon_i\varepsilon_j{}^{T}) = R_i\delta_{ij}\,, \quad \begin{matrix} \delta_{ij} = 1, i = j \\ \delta_{ij} = 0, i \neq j \end{matrix} \\ E\,(\eta_0\eta_0{}^{T}) = \overline{P}_0 \\ E\,(\eta_0\varepsilon_i{}^{T}) = 0 \end{cases} \tag{13}$$

Under the above conditions, the linear unbiased minimum variance is estimated as

$$\hat{x}_0 = \left(H^{T}R^{-1}H + \overline{P}_0{}^{-1}\right)^{-1}\left(H^{T}R^{-1}y + \overline{P}_0{}^{-1}\overline{x}_0\right) \tag{14}$$

3. Results

The orbit prediction accuracy of an all-electric GEO satellite in the mission and transfer orbit are analyzed using simulation data. Firstly, the "real orbit" is obtained based on the integration of the dynamics model, and then the simulated GNSS pseudorange and phase observation data with certain noise are obtained based on the GNSS navigation satellite precision ephemeris and clock difference products (from the Wu Han University (WHU) products of Multi-GNSS Experiment (MGEX), including GPS, BDS, and GLONASS, as shown in Table 2). The orbit of an all-electric GEO satellite determined from the simulated GNSS observation is compared with the "real orbit" to obtain the determination and forecasting accuracy.

Table 2. GNSS satellites used in the simulation.

Type	PRN	NUM
GPS	G01~G32 (except G04/G19)	30
GLONASS	R01~R24 (except R06/R12)	22
BDS	C01~C37 (except C15/C16/C17/C18/C20/C28/C30/C31)	29

GNSS signal visibility means the line of sight between the GNSS satellite and the receiver is not blocked by the Earth and the signal's power at the receiver satisfies the signal capture tracking threshold when generating simulated observations. Thus, visibility analysis mainly considers two aspects: the geometric relationship among GNSS satellites, GNSS receiver antennas of the user satellite and the Earth, and whether the received signal power level meets the threshold. The angle of the main beam is different for the frequency and the generation (e.g., block III) of the satellite. For example, L-band antennas of GPS satellites can transmit L1, L2, and L5 with a three-frequency carrier antenna array, and the center of the antenna direction is aligned with the center of the earth, with main beam half-angle of 21.3° and half-cone angle of the earth blocking the GPS signal of 13.9°. Therefore, receivers with orbit height higher than the GNSS constellation can only receive the signal within a circular cone of about 8° at the edge of the main beam of the transmitting antenna, or use the bypass beam signal, as shown in Figure 3.

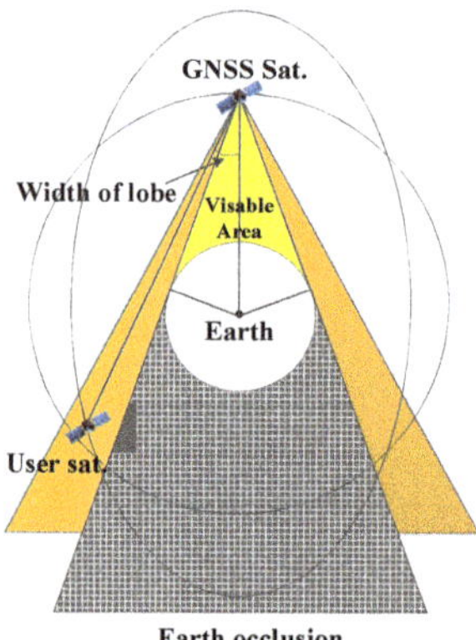

Figure 3. GNSS satellite geometry visible schematic.

3.1. Analysis of Orbit Determination Accuracy of Mission Orbit

The simulation time is from 15 June 2019 00:00:00 to 16 June 2019 00:00:00. the noise level of simulated onboard GNSS pseudorange and phase data is given according to the receiver design indicators in Table 1, and the simulation strategy is shown in Table 3.

Table 3. GNSS observation data simulation strategy.

Type	Value
Arc length	24 h
Measurement extraction interval	5 min
Pseudorange noise (RMS)	10 m
Phase noise (RMS)	0.03 m
GNSS satellite ephemeris	MGEX-WHU
GNSS satellite clock difference	MGEX-WHU
Earth gravity field	EIGEN_GL04C (100×100)
N body	DE421
Solid tide	IERS 2003
Sea tide	FES2004
Solar radiation pressure	Fixed surface-to-mass ratio model

Only antenna 1 located in +Z direction of the satellite's body system can receive GNSS signals when the satellite is in GEO mission orbit, and at least 24 GNSS satellites can be observed in the full arc, among which there are about 8–12 GPS, and Geometric Dilution of Precision (GDOP) is around 13, as shown in Figure 4.

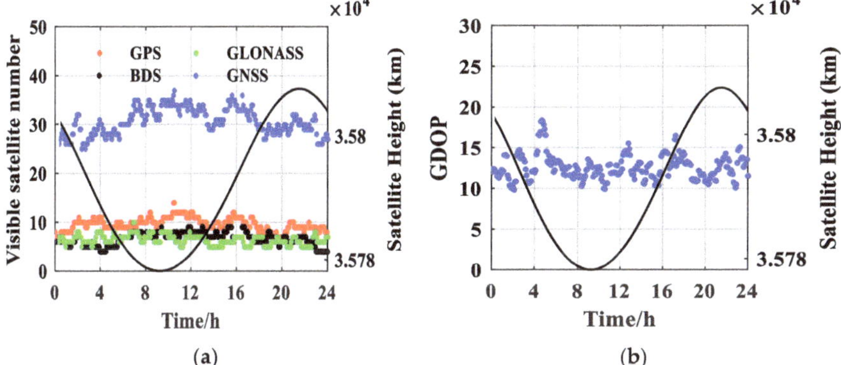

Figure 4. Visible GNSS number and GDOP of antenna 1. (**a**) Visible GNSS number of antenna 1. (**b**) GDOP of antenna 1.

The weighted least squares method is adapted using the pseudorange and carrier phase, and the precise orbit determination processing strategy is shown in Table 4. Using all GNSS satellites, the position accuracy is about 0.23 m with only measurement noise and 0.1 m GNSS satellite ephemeris error.

Observation accuracy and dynamics model error are important factors affecting POD accuracy. POD accuracy is anlyzed by selecting different observation noise levels, observation arc lengths, ephemeris errors, and dynamic model errors, separately. The results are shown in Table 5.

The analysis results show that the arc length and noise of the observation data have a great influence on the orbit determination accuracy. The orbit determination accuracy of using GPS satellites alone is 0.27 m, which is close to the accuracy when using multiple navigation satellites. The orbit determination accuracy drops to 1.34 m when precision ephemeris error increases to 1 m. For gravity field model, degree, and order, five is enough for POD accuracy. For solar radiation pressure, a 5% error will impact the accuracy of orbit determination.

Table 4. Processing strategy for POD.

Model	Describe
Arc length	24 h
Measurement extraction interval	5 min
Pseudorange noise (RMS)	10 m
Phase noise (RMS)	0.03 m
Earth gravity field	EIGEN_GL04C (100 × 100)
N body	DE421
Solid tide	IERS 2003
Sea tide	FES2004
Solar radiation pressure	Fixed surface-to-mass ratio model
Estimated parameters	Satellite initial state (position, velocity) receiver clock error, phase ambiguity
Observation data weights	The weight ratio of code observation to phase observation is 1/100
GNSS navigation satellite ephemeris	Precision ephemeris (add 0.1 m position error)

Table 5. POD accuracy with different model errors (RMS).

	Model Errors		POD Accuracy (m)
Observation data	Pseudorange noise (6 m)		0.15
	5 h arc length		1.19
	Pseudorange only		1.2
GNSS navigation satellite ephemeris	GPS only		0.27
	Precision ephemeris (add 1 m position error)		1.34
Dynamics model	Earth gravity field	5 × 5	0.24
		3 × 3	0.44
		2 × 2	5.66
	Solar radiation pressure	5% error	0.66
		10% error	1.32

Since the real-time POD uses the broadcast ephemeris, which has a position error of several meters, the orbit is determined based on the GPS broadcast ephemeris. Figure 5 shows that the accuracy of the position and velocity is 3.96 m and 0.03 mm/s, respectively, in the RTN coordinate system, which is similar to the results of Wu and Fan [59,60]. They analyzed the orbit determination accuracy of the XY-2 satellite and the SJ-17 satellite using onboard GNSS data.

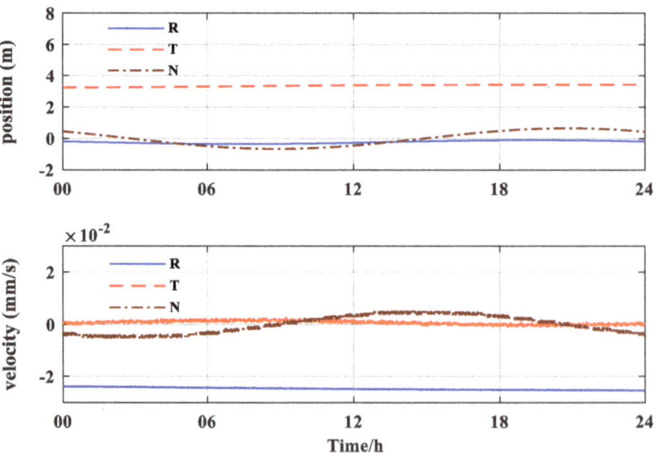

Figure 5. Orbit overlap difference using GPS broadcast ephemeris.

3.2. Accuracy Analysis of Non-Thrust Orbit Determination for Transfer Orbit

The all-electric-propulsion GEO satellite will fly for several months in the transfer orbit due to the small thrust before entering the mission orbit, as shown in Figure 1. The typical arc of the all-electric-propulsion GEO satellite transfer orbit is selected from 00:00:00 on 1 July to 00:00:00 on 2 July 2019. The simulation strategy is the same as the mission orbit, and the POD accuracy and the orbit prediction accuracy for 14 days are analyzed with non-thrust mode.

As shown in Figure 6, satellite height varied from 2030 km to 49,422 km, with an orbit period of about 16 h. In arcs above 8000 km, antennas 1 and 2 can receive GNSS signals opposite the earth, the visible GNSS number of antenna 1 is stable at 16~34, and the GDOP of antenna 1 is around 20. In the arc below 8000 km, only antenna 3 can receive the overhead GNSS signal, the visible GNSS number of antenna 3 is stable at 12–18, and the GDOP of antenna 3 is around 1.5.

The simulation data of the transfer orbit are used to analyze the influence of the observation data arc length, dynamic model errors, and other factors on the orbit determination accuracy. Solar-pressure model error is fixed at 10%, and the pseudorange noise, phase noise, and measurement extraction interval are 10 m, 0.03 m, and 5 min, respectively. The orbit determination accuracy and the 14-day prediction accuracy are analyzed. The calculation results are shown in Table 6. The factors considered include:

(1)   GNSS (GPS + BDS + GLONASS), single GPS, single BDS;
(2)   Final precision ephemeris (0.1 m), broadcast ephemeris;
(3)   Arc length 24 h/10 h/5 h;
(4)   For the arc length of 24 h, consideration that GNSS data are not continuous

It can be seen from Table 6 that whether the observation data are continuous or not has little effect on the accuracy of orbit determination prediction, but the arc length of the observation data has a greater impact on the accuracy of orbit determination. In terms of ephemeris products, the accuracy of orbit determination and prediction of using the GPS alone is comparable to that of multi-navigation satellite ephemeris products, while the use of broadcast ephemeris has a greater impact on the accuracy of orbit determination prediction. In order to meet the prediction accuracy of 1 km for 14 days, the orbit determination arc length needs at least 5 h (if using the precise ephemeris).

Table 6. Orbit determination and prediction of different conditions.

Navigation System	Ephemeris	Arc Length	Orbit Determination Accuracy (m)	14-Day Prediction Accuracy (m)
GNSS (GPS + BDS + GLONASS)	Final precision ephemeris (0.1 m error)	24 h continuous	1.56	272.29
		24 h discontinuity	1.94	628.6
		10 h	0.97	552.75
		5 h	1.4	858.84
	Broadcast ephemeris	24 h continuous	5.03	396.87
		24 h discontinuity	6.19	928.49
		10 h	5.02	1213.25
		5 h	5.43	3147.11
GPS	Final precision ephemeris (0.1 m error)	24 h continuous	1.89	411.81
		24 h discontinuity	1	561.61
		10 h	0.59	667.52
		5 h	0.85	743.44
	Broadcast ephemeris	24 h continuous	4.86	477.74
		24 h discontinuity	4.61	685.85
		10 h	4.35	1369.57
		5 h	5.23	2061.98

Table 6. *Cont.*

Navigation System	Ephemeris	Arc Length	Orbit Determination Accuracy (m)	14-Day Prediction Accuracy (m)
BDS	Final precision ephemeris (0.1 m error)	24 h continuous	1.94	253.21
		24 h discontinuity	2.96	469.1
		10 h	1.37	834.45
		5 h	1.96	997.81
	Broadcast ephemeris	24 h continuous	5.3	303.39
		24 h discontinuity	5.92	747.34
		10 h	7.07	692.47
		5 h	5.23	2521.03

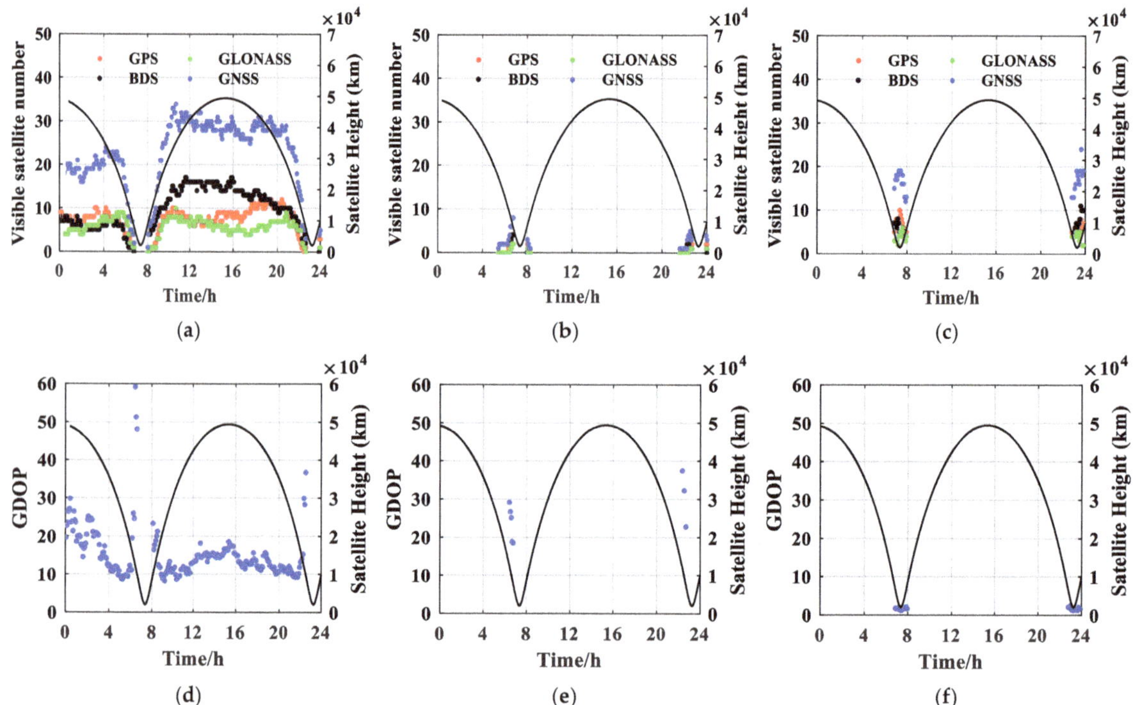

Figure 6. Visible GNSS number and GDOPs of antennas 1, 2, and 3. (**a**) Visible GNSS number of antenna 1. (**b**) Visible GNSS number of antenna 2. (**c**) Visible GNSS number of antenna 3. (**d**) GDOP of antenna 1. (**e**) GDOP of antenna 2. (**f**) GDOP of antenna 3.

3.3. Accuracy Analysis of Thrust Orbit Determination for Transfer Orbit

The typical arcs with thrust in the transfer orbit of the all-electric-propulsion GEO satellite from 00:00:00 on 25 July 2019 to 00:00:00 on 26 July 2019 are selected, and the real orbit is determined by integration based on the initial orbit + electric thrust + other dynamics. Then, the GNSS pseudorange and phase data are generated by simulation using the noise level according to the indications of the GNSS receiver in Table 2 and the simulation strategy shown in Table 3. The orbit determination accuracy under thrust in the transfer section is analyzed.

Analysis of the observation data generated by the simulation shows that the highest satellite altitude is 53,721 km and the lowest is 5649 km. The GNSS data of the three integrated antennas are obtained, and the arcs with at least four GPS satellites are 02:16–06:17

and 20:45–24:00. The arcs with at least four BDS satellites are 02:13–12:11 and 21:43–24:00. The arcs with at least four GLONASS satellites are 02:02–04:46 and 21:17–23:44. The arcs with at least four GNSS satellites are 00:30–12:30, 14:30–14:47, and 20:23–24:00.

As shown in Figure 7, in the arcs below 8000 km, the visible GNSS number of antenna 1 is 0. From 8000 km to 20,000 km, the visible GNSS number of antenna 1 increases rapidly from 4 to 14, and the GDOP of antenna 1 gradually increases from 2 to 50 as the altitude increases. In arcs above 20,000 km, the visible GNSS number of antenna 1 rapidly decreases to 4 with the increase in altitude. In arcs within 8000 km, the visible GNSS number of antenna 2 is 0. From 8000 km to 20,000 km, the visible GNSS number of antenna 2 also increases rapidly from 4 to 50, and the GDOP of antenna 2 gradually increases from 1 to 8 as the altitude increases. In the 20,000 km–30,000 km arcs, the visible GNSS number of antenna 2 rapidly decreases to less than 10, and the GDOP decreases to 25 with the increase of altitude. In arcs above 30,000 km, the visible GNSS number of antenna 2 is 0. In arcs within 8000 km, the visible GNSS number of antenna 3 is stable at 14–34, and the GDOP is stable at around 2. In arcs above 8000 km, the visible GNSS number of antenna 3 is 0.

Simulation data of transfer orbit with thrust are used to analyze the orbit determination accuracy. The initial value is taken from the ephemeris on 25 July 2019, and the error of the solar pressure model is fixed at 10%, pseudorange noise is 10 m, phase noise is 0.03 m, and the measurement extraction interval is 5 min. The orbit determination and forecast accuracy are analyzed for 10 h by solving for and fixing the empirical force [61,62].

The observation data from 00:30–12:30 is used to determine the orbit and to solve for empirical force, the orbit determination accuracy is 712 m, with a maximum error of 300 km for the 10 h forecast accuracy. A set of empirical forces to be solved are fixed as a priori values, and the orbit determination is carried out using the observation data from 00:30–12:30. Orbit determination accuracy is 709 m, with a maximum error of 20 km for the 10 h forecast accuracy, as shown in Table 7.

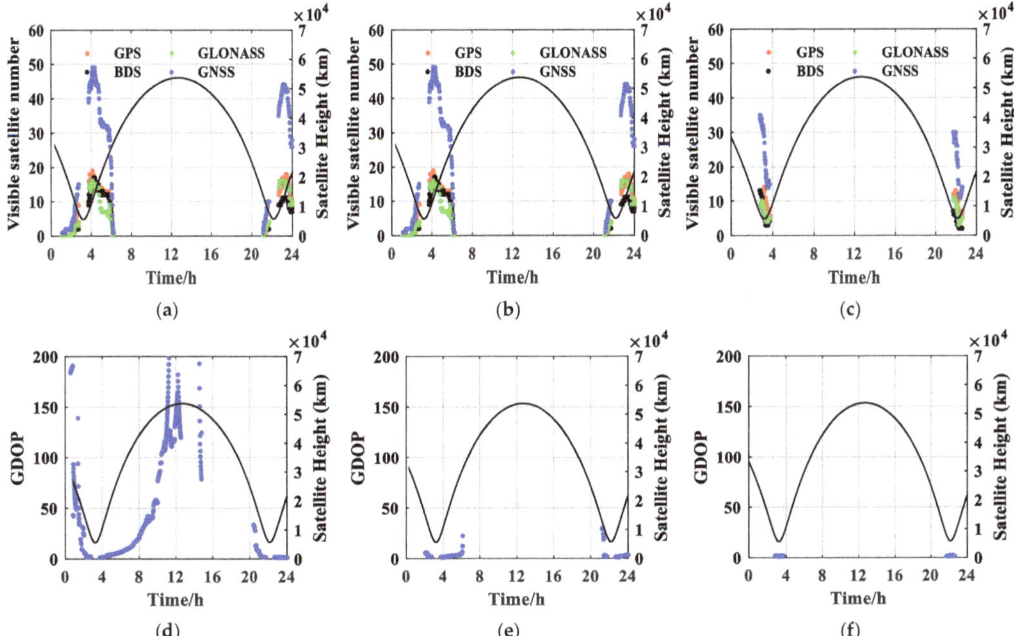

Figure 7. Visible GNSS number and GDOP of antennas 1, 2, and 3 (**a**) Visible GNSS number of antenna 1. (**b**) Visible GNSS number of antenna 2. (**c**) Visible GNSS number of antenna 3. (**d**) GDOP of antenna 1. (**e**) GDOP of antenna 2. (**f**) GDOP of antenna 3.

Table 7. Analysis of orbit determination and prediction accuracy.

Empirical Force Processing Strategy	Orbit Determination Accuracy (m)	10 h Prediction Accuracy (km)
Solve	712	300
Fix after solving	709	20

Compared with the direct calculation of the empirical force during orbit determination, the fixed use of the solved empirical force for orbit determination has little effect on the orbit determination accuracy but can significantly improve the prediction accuracy.

4. Discussion

4.1. Influence Analysis of Electric Thrust

Electric thrust, like other perturbation forces (such as solar radiation pressure, N-body perturbation, and nonspherical gravitational perturbation), will impact the satellite's orbit, and the cumulative impact increases over time. Thus, it needs to be considered when determining the orbit in order to avoid it affecting the orbit determination accuracy. If the electric thrust can be accurately modeled, it will not affect the POD accuracy. However, due to the influence of various errors, such as control and installation, there is a big difference between the actual force and the theoretical model, resulting in a systematic deviation of the thrust model to affect orbit accuracy.

In order to improve the prediction accuracy of satellites that use electric thrust, it is necessary to use on-orbit data to calibrate the electric thrust model and to use the calibrated parameters in the subsequent orbit determination to improve the orbit determination accuracy. The calibration method can be determined by adjusting the size of factors on orbital experiments to determine the optimal factors, or some parameters can be solved and compared with the prior parameter values. Calibration of the electric thrust model is coupled with orbit determination, which uses external data to fit the relevant parameters in the thrust model.

Two typical arcs of the transfer orbit of the all-electric GEO satellite are selected: the initial orbit is 00:00:00 on 15 December 2019 and 00:00:00 on 24 February 2020. First, the reference orbit is obtained by numerical integration, and then the electric thrust is added for integration. As shown in Figure 8, the influence of the electric thrust on orbit determination is as big as about 60,000 km, far exceeding other perturbations.

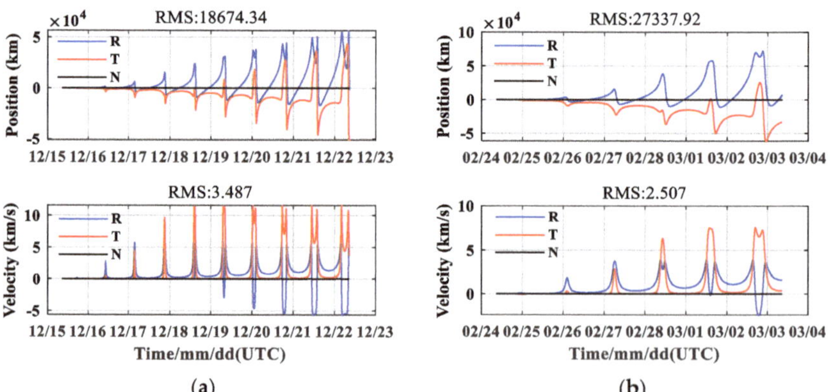

Figure 8. Electrical thrust effect. (**a**) Electrical thrust effect of phase 1. (**b**) Electrical thrust effect of phase 2.

4.2. Simulation Analysis of Electric Thrust Calibration

Satellite self-positioning data are obtained by using pseudorange data with single point position (SPP), which is not affected by satellite dynamics model error, so it can be used for electric thrust calibration. Simulation data are used to analyze the calibration accuracy of the electric thrust model; the simulation process is as follows:

(1) According to the initial orbit, electric thrust, and other dynamic models, the real orbit of many days is obtained, and the real acceleration time series of electric thrust is output at the same time.

(2) Simulate the self-positioning data of the spaceborne GNSS receiver, add a certain random error (10/100 m) in each direction (XYZ in the Earth-fixed coordinate system), and set the data measurement extraction interval as 5 min;

(3) Use the self-positioning data of a certain arc to determine the orbit and the electric thrust model parameters. The calculated electric thrust model parameters are compared with the real thrust values to evaluate the calibration accuracy of the electric thrust.

(4) The attitude error in the evaluation is considered to be 0.1 degrees.

Selecting the typical arcs used in the previous section, the simulation accuracy of electric thrust calibration is analyzed using the self-positioning data of different accuracy and arc length, and the results are shown in Table 8.

Table 8. Analysis of 14-day prediction and calibration accuracy.

Self-Positioning Data Error (m)	Arc Length (hours)	14-Day Prediction Accuracy (km)	Calibration Accuracy (m/s^2)
10	16	20	1.2 d$-$08
10	40	1	1.9 d$-$10
100	16	200	1.2 d$-$07
100	40	4	1.9 d$-$9

The calibration results show that the satellite self-positioning data accuracy and data arc length are the main factors affecting the calibration accuracy. Under the condition of 100 m accuracy of satellite self-positioning data and the arc length of 40 h orbit determination, the calibration accuracy of the electric thrust is about 10^{-9} m/s^2, and for the arc length of 16 h orbit determination, the calibration accuracy of the electric thrust is about 10^{-7} m/s^2. If the accuracy of satellite self-positioning data can reach 10 m, the accuracy of electric thrust calibration can also be improved by about one order of magnitude, and the 40 h orbit prediction can reach 1 km in 14 days.

5. Conclusions

The prediction accuracy of the transfer orbit and mission orbit of the all-electric propulsion GEO satellite based on onboard GNSS is studied by simulating the GNSS information of an all-electric propulsion GEO satellite. The main factors affecting the prediction accuracy of GEO satellite orbit determination under the condition of no thrust are also analyzed. An electric propulsion calibration algorithm is designed to analyze the prediction accuracy of orbit determination under thrust. The calculation results show that, using onboard GNSS technology, the orbit determination accuracy of all-electric GEO satellites is better than 10 m, the orbit determination accuracy of transfer orbit is better than 10 m, and the 14-day forecast accuracy is better than 1 km, which can provide support for transfer and mission orbit controls. The satellite self-positioning data are used to calibrate the electric thrust during satellite self-positioning data to 10 m accuracy. Using the 40 h orbit determination arc length data, the absolute calibration accuracy of the electric thrust is about 10^{-10} m/s^2 magnitude, and the 14-day orbit forecast predicted accuracy can reach 1 km.

Author Contributions: Conceptualization, H.W. and G.W.; methodology, Y.H.; software, Y.H.; validation, W.L.; formal analysis, W.L.; resources, H.W.; data curation, W.L.; writing—original draft preparation, W.L. and Y.H.; writing—review and editing, G.W.; funding acquisition, Y.H. All authors have read and agreed to the published version of the manuscript.

Funding: This research is supported by the National Natural Science Foundation of China (grant no. U1931119) and the Preresearch Project on Civil Aerospace Technologies funded by the China National Space Administration (grant no. D010105).

Data Availability Statement: The data sources are supported by Shanghai Astronomical Observatory, Chinese Academy of Sciences.

Acknowledgments: The authors thank the Innovation Academy for Microsatellites Chinese Academy of Sciences for the satellite data, and WHU for MGEX ephemeris product support. The authors also thank Jianfeng Cao from Beijing Aerospace Control and Command Center and Xinglong Zhao from the China Academy of Space Technology for orbit determination software support.

Conflicts of Interest: The authors declare no conflict of interest.

References

1. Martinez-Sanchez, M.; Pollard, J. Spacecraft electric propulsion-an overview. *J. Propuls. Power* **1998**, *14*, 688–699. [CrossRef]
2. Jahn, R.G. *Physics of Electric Propulsion*; Courier Corporation: Chelmsford, MA, USA, 2006.
3. Goebel, D.; Katz, I. *Fundamentals of Electric Propulsion: Ion and Hall Thrusters*; John Wiley & Sons: Hoboken, NJ, USA, 2008.
4. O'Reilly, D.; Herdrich, G.; Kavanagh, D. Electric propulsion methods for small satellites: A review. *Aerospace* **2021**, *8*, 22. [CrossRef]
5. Duan, C.; Chen, L. Research and Inspiration of All-electric Propulsion Technology for GEO Satellites. *Spacecr. Eng.* **2013**, *22*, 99–104.
6. Duchemin, O.; Caratge, A.; Cornu, N.; Sannino, J.M.; Lorand, A. Ariane 5-ME and electric propulsion: GEO insertion options. In Proceedings of the 47th AIAA/ASME/SAE/ASEE Joint Propulsion Conference & Exhibit, San Diego, CA, USA, 31 July–3 August 2011; p. 6084.
7. Yu, D.; Qiao, L.; Jiang, W.; Liu, H. Development and prospect of electric propulsion technology in China. *J. Propuls. Technol.* **2020**, *41*, 1–11.
8. Tang, Z.; Zhou, C.; Han, D.; Ma, X.; Chen, T. Study on high power all-electric propulsion system for spacecraft orbit transfer. *J. Deep. Space Explor.* **2018**, *5*, 367–373.
9. Hu, Y.; Mao, W.; Li, D.; Wei, L.; Wei, Y. The hall propulsion technology oriented all-electric-propulsion satellites. *Aerosp. Control Appl.* **2017**, *43*, 73–78.
10. Wei, B.; Sun, X.; Wang, X. The Review of All-Electric Propulsion Platform on Satellite. *Vac. Cryog.* **2016**, *22*, 301–305+310.
11. Yang, D. *Research on Key Technologies of Orbit Design and Control for All Electric Satellite*; Nanjing University of Aeronautics and Astronautics: Nanjing, China, 2016.
12. Hu, Z.; Wang, M.; Yuan, J. A review of the development of all-electric propulsion platform in the world. *Spacecr. Environ. Eng.* **2015**, *32*, 566–570.
13. Zuo, K.; Wang, M.; Li, M.; Tang, H. Research overview of commercial satellite platform with all-electric propulsion system. *J. Rocket. Propuls.* **2015**, *41*, 13–20.
14. Sun, X.; Zhang, T.; Wang, X.; Wang, L.; Liu, M. The Research of Electric Propulsion's Application of Electric Propulsion Satellite Platform. *Vac. Cryog.* **2015**, *21*, 6–10.
15. Nishi, K.; Ozawa, S.; Matunaga, S. Design and guidance for robust orbit raising trajectory of all-electric propulsion geostationary satellites. *Trans. Jpn. Soc. Aeronaut. Space Sci. Aerosp. Technol. Jpn.* **2021**, *19*, 553–561. [CrossRef]
16. Jin, G.; Kang, L. Design of Hall Electric Propulsion System in "Tianhe" Core Module. *Aerosp. China* **2021**, *8*, 22–27.
17. Chen, M.; Liu, X.; Zhou, H.; Chen, C.; Jia, H. Research and development of micro electric propulsion technology for micro/nano satellites. *J. Solid Rocket. Technol.* **2021**, *44*, 188–206.
18. Yang, F.; Wang, C.; Hu, J.; Zhang, H.; Wu, C.; Zhang, X.; Geng, H.; Fu, D. Technical project of ion propulsion for satellites in super low Earth orbit. *Chin. Space Sci. Technol.* **2021**, *41*, 52–59.
19. Wang, Z.; Zhang, T.; Peng, K. Design of Electric Propulsion System for Ground-month Single-ship Cargo Ship. *Vac. Cryog.* **2020**, *26*, 317–322.
20. Wang, S.; Xiong, S. Research on transfer orbit based on electric propulsion satellite to halo orbit. *Chin. J. Space Sci.* **2019**, *39*, 489–493.
21. Ren, Y.; Wang, X. Development of High-performances Electric Propulsion System and the Application on GEO Satellite Platforms. *Vac. Cryog.* **2018**, *24*, 60–65.
22. Kuai, Z. *Research on Station Keeping and Orbital Transfer of Geostationary Satellites by Impulse and Electric Propulsion*; University of Science and Technology of China: Hefei, China, 2017.
23. Tian, B.; Xue, D.; Huang, M. Orbit Transfer Strategies for GEO Satellites Using All-electric Propulsion. *Spacecr. Eng.* **2015**, *24*, 7–13.
24. Li, S. *The Research on the Application of Electrical Propulsion Technology in GEO Satellite Orbit Control*; National University of Defense Technology: Changsha, China, 2015.
25. Feuerborn, S.A.; Perkins, J.; Neary, D.A. Finding a way: Boeing's all electric propulsion satellite. In Proceedings of the 49th AIAA/ASME/SAE/ASEE Joint Propulsion Conference, San Jose, CA, USA, 14–17 July 2013; p. 4126.

26. Casaregola1, C. Electric Propulsion for Station Keeping and Electric Orbit Raising on Eutelsat Platforms. In Proceedings of the Japan International Electric Propulsion Conference, Kobe, Japan, 4–10 July 2015.
27. Zhou, Z.; Gao, J. Development Approach to All-Electric Propulsion GEO Satellite Platform. *Spacecr. Eng.* **2015**, *24*, 1–6.
28. Zhang, T. The LIPS-200 Ion Electric Propulsion System Development for the DFH-3B Satellite Platform. In Proceedings of the 64th International Astronautical Congress, Beijing, China, 23–27 September 2013.
29. Zhang, T. Initial Flight Test Results of the LIPS-200 Electric Propulsion System on SJ-9A Satellite. In Proceedings of the International Electric Propulsion Conference, Washington, DC, USA, 6–10 October 2013.
30. Zhang, T. New progress of electric propulsion technology in LIP. *J. Rocket. Propuls.* **2015**, *41*, 7–12.
31. Wen, Z.; Shi, M.; Geng, H.; Zhang, W.; Li, Z. Application and Fight Testing of Ion Electric Propulsion in High Throughout Satellite. *Vac. Cryog.* **2022**, *28*, 79–86.
32. Wang, H.; Li, G.; Shi, B.; Zhang, J.; Jiang, G.; Shen, Y.; Wu, G. Orbit Transfer Method of All-electric Propulsion SmallGEO Satellite Based on GNSS. *Radio Commun. Technol.* **2020**, *46*, 527–533.
33. Hofmann-Wellenhof, B.; Lichtenegger, H.; Wasle, E. *GNSS–Global Navigation Satellite Systems: GPS, GLONASS, Galileo, and More*; Springer: Berlin, Germany, 2007.
34. Groves, P. Principles of GNSS, inertial, and multisensor integrated navigation systems. *IEEE Aerosp. Electron. Syst. Mag.* **2015**, *30*, 26–27. [CrossRef]
35. Li, X.; Ge, M.; Dai, X.; Ren, X.; Fritsche, M.; Wickert, J.; Schuh, H. Accuracy and reliability of multi-GNSS real-time precise positioning: GPS, GLONASS, BeiDou, and Galileo. *J. Geod.* **2015**, *89*, 607–635. [CrossRef]
36. Zhou, X.; Wang, X.; Zhao, G.; Peng, H.; Wu, B. The precise orbit determination for hy2a satellite using GPS, DORIS and SLR data. *Geomat. Inf. Sci. Wuhan Univ.* **2015**, *40*, 1000.
37. Gong, X.; Wang, F. Autonomous orbit determination of hy2a and zy3 missions using space-borne GPS measurements. *Geomat. Inf. Sci. Wuhan Univ.* **2017**, *42*, 309–313.
38. Yuan, J.; Zhao, C.; Wu, Q. Phase center offset and phase center variation estimation in-flight for zy-3 01and zy-3 02spaceborne GPS antennas and the influence on precision orbit determination. *Acta Geod. et Cartogr. Sin.* **2018**, *47*, 672–682.
39. Li, M.; Li, W.; Shi, C.; Jiang, K.; Guo, X.; Dai, X.; Meng, X.; Yang, Z.; Yang, G.; Liao, M. Precise orbit determination of the Fengyun-3C satellite using onboard GPS and BDS observations. *J. Geod.* **2017**, *91*, 1313–1327. [CrossRef]
40. Li, W.; Li, M.; Zhao, Q.; Shi, C.; Guo, X.; Meng, X.; Yang, Z. FY3C Satellite Onboard BDS and GPS Data Quality Evaluation and Precise Orbit Determination. *Acta Geod. et Cartogr. Sin.* **2018**, *47*, 9–17.
41. Zhao, X.; Zhou, S.; Ci, Y.; Hu, X.; Cao, J.; Chang, Z.; Tang, C.; Guo, D.; Guo, K.; Liao, M. High-precision orbit determination for a LEO nanosatellite using BDS-3. *GPS Solut.* **2020**, *24*, 102. [CrossRef]
42. Powell, T.; Martzen, P.; Sedlacek, S.; Chao, C.-C.; Silva, R. GPS Signals in a Geosynchronous Transfer Orbit:"Falcon Gold" Data Processing. In Proceedings of the 1999 National Technical Meeting of the Institute of Navigation, San Diego, CA, USA, 25–27 January 1999; pp. 575–585.
43. Bauer, F.; Moreau, M.; Davis, E.; Carpenter, J.; Kelbel, D.; Davis, G.; Axelrad, P. Results from the GPS flight experiment on the high earth orbit amsat oscar-40 spacecraft. In Proceedings of the 15th International Technical Meeting of the Satellite Division of the Institute of Navigation, Portland, OR, USA, 24–27 September 2002.
44. Unwin, M.; Blunt, P.; de Vos van Steenwijk, R. Navigating above the GPS constellation—Preliminary results from the SGR-GEO on GIOVE-A. In Proceedings of the 26th International Technical Meeting of the Satellite Division of the Institute of Navigation, Nashville, TN, USA, 16–20 September 2013.
45. Barker, L.; Frey, C. GPS at GEO: A first look at GPS from SBIRS GEO. *Adv. Astronaut. Sci.* **2012**, *144*, 199–212.
46. Fan, M.; Hu, X.; Dong, G.; Huang, Y.; Cao, J.; Tang, C.; Li, P.; Shengqi, C.; Yu, Y. Orbit improvement for Chang'e-5T lunar returning probe with GNSS technique. *Adv. Space Res.* **2015**, *56*, 2473–2482. [CrossRef]
47. Wang, M.; Shan, T.; Ma, L.; Tao, R.; Chen, C. Performance of GPS and GPS/SINS navigation in the CE-5T1 skip re-entry mission. *GPS Solut.* **2018**, *22*, 1–12. [CrossRef]
48. Wang, M.; Shan, T.; Wang, D. Development of GNSS technology for high earth orbit spacecraft. *Acta Geod. et Cartogr. Sin.* **2020**, *49*, 1158–1167.
49. Li, B.; Liu, L.; Wang, M. Performance demonstration and analysis of GNSS navigation in geo satellites. *Aerosp. Shanghai* **2017**, *34*, 133–143.
50. Jiang, K.; Min, L.; Wang, M.; Zhao, Q.; Li, W. Tjs-2 geostationary satellite orbit determination using onboard GPS measurements. *GPS Solut.* **2018**, *22*, 1–14. [CrossRef]
51. Zhu, J.; Wang, C.; He, Y.; Yang, Y. High Accuracy Navigation for Geostationary Satellite TTS-II via Space-borne GPS. In Proceedings of the 39th Chinese Control Conference, Shenyang, China, 27–30 July 2020; pp. 3362–3367.
52. Winternitz, L.; Bamford, W.; Heckler, G. A GPS receiver for high-altitude satellite navigation. Selected Topics in Signal Processing. *IEEE J.* **2009**, *3*, 541–556.
53. Han, M.; Wang, Y. Optimization method for orbit transfer of all-electric propulsion satellite based on reinforcement learning. *Syst. Eng. Electron.* **2022**, *44*, 1652–1661.
54. Duan, C.; Ren, L.; Chang, Y.; Bai, Q.; An, R.; Huang, Y. All-electric propulsion satellite trajectory optimization by homotopic approach. *Chin. Space Sci. Technol.* **2020**, *40*, 42–48.

55. Wang, M.; Li, Q.; Liang, X.; An, R. Low-Thrust Orbit Transfer Strategy On-Board Computation for All Electric Propulsion Satellite. *J. Propuls. Technol.* **2020**, *41*, 180–186.
56. Cui, Z.; Zhou, L.; Hu, S.; Gong, J. An electric thrust vector calibrating algorithm using angular momentum. *Chin. Space Sci. Technol.* **2019**, *39*, 1–8.
57. Li, H.; Topputo, F.; Baoyin, H. Autonomous time-optimal many-revolution orbit raising for electric propulsion geo satellites via neural networks. *arXiv* **2019**, arXiv:1909.08768.
58. Montenbruck, O.; Gill, E. *Satellite Orbits*; Springer: Berlin/Heidelberg, Germany, 2000.
59. Wu, Z. *Researches on Precise Orbit Determination Based GNSS and Its Application*; University of Chinese Academy Sciences: Beijing, China, 2018.
60. Fan, M. *Researches of GNSS-Based Navigation for Lunar Missions*; University of Chinese Academy Sciences: Beijing, China, 2017.
61. Zhu, J.; Wang, J.; Chen, J.; He, Y. Centimeter precise orbit determination for HY-2 via DORIS. *J. Astronaut.* **2013**, *34*, 163–169.
62. Montenbruck, O.; Helleputte, T.; Kroes, R.; Gill, E. Reduced Dynamic Orbit Determination using GPS Code and Carrier Measurements. *Aerosp. Sci. Technol.* **2005**, *9*, 261–271. [CrossRef]

Communication

Research on the Impact of BDS-2/3 Receiver ISB on LEO Satellite POD

Xinglong Zhao [1,2], Shanshi Zhou [2,*], Jianfeng Cao [3], Junjun Yuan [2,4], Ziqian Wu [5] and Xiaogong Hu [2]

[1] Institute of Telecommunication and Navigation Satellites, China Academy of Space Technology, Beijing 100094, China; zhao_xing_long@yeah.net or zhaoxinglong15@mails.ucas.ac.cn

[2] Shanghai Astronomical Observatory, Chinese Academy of Sciences, Shanghai 200030, China; yuanjunjun@shao.ac.cn (J.Y.); hxg@shao.ac.cn (X.H.)

[3] Beijing Aerospace Control and Command Center, Beijing 100094, China; jfcao@foxmail.com

[4] University of Chinese Academy of Sciences, Beijing 100049, China

[5] The 54th Research Insitute of CETC, Shijiazhuang 050081, China; ziqian2009@163.com

* Correspondence: sszhou@shao.ac.cn

Abstract: In recent years, the multi-GNSS positioning application is becoming more and more popular, same to the low Earth orbit (LEO) satellite precise orbit determination (POD) based on the onboard multi-GNSS measurements. The third-generation Beidou navigation satellite system (BDS-3) provides a new option to obtain the LEO satellite orbit solutions. However, the receiver intersystem bias (ISB) of different GNSS is unavoidable in multi-GNSS data processing. This paper's main goal is absorption of the impact of the ISB between BDS-3 and BDS-2 on the LEO satellite POD. Taking GPS-based POD solutions for the reference orbit, this paper evaluates the orbit accuracy of BDS-2-based POD, BDS-3-based POD, BDS-2 and BDS-3 combined PODs with/without ISB. The BDS-3-based POD accuracy is 6.57 cm in the 3D direction, a 56% improvement over BDS-2-based POD. When the ISB between BDS-3 and BDS-2 is estimated, the BDS-2/3 combined POD accuracy of 5.37 cm in the 3D direction is better than that without ISB, which is a 64% improvement over BDS-2-based POD and 18% improvement over BDS-3-based POD. For GPS and BDS-2/3 combined POD, the GPS and BDS-3 joint POD solutions have the smallest RMS differences in overlapping consistency and smallest RMS differences compared to GPS-based POD. This study indicates that estimating the BDS-2/3 receiver ISB in BDS-2/3 joint POD could improve the orbit accuracy, and the GPS and BDS-3 joint POD solution is better than another combined POD. This paper will provide meaningful references for the LEO satellite multi-GNSS-based POD.

Keywords: low Earth orbit; precise orbit determination; intersystem bias; BDS-2/3-based POD; GPS and BDS joint POD

Citation: Zhao, X.; Zhou, S.; Cao, J.; Yuan, J.; Wu, Z.; Hu, X. Research on the Impact of BDS-2/3 Receiver ISB on LEO Satellite POD. *Remote Sens.* **2022**, *14*, 2514. https://doi.org/10.3390/rs14112514

Academic Editor: Yunbin Yuan

Received: 20 February 2022
Accepted: 22 May 2022
Published: 24 May 2022

Publisher's Note: MDPI stays neutral with regard to jurisdictional claims in published maps and institutional affiliations.

1. Introduction

At present, it is mainstream to use the onboard Global Positioning System (GPS) measurements to obtain precise orbit solutions for low Earth orbit (LEO) satellites and other spacecrafts [1–8]. In 2012, the Beidou regional navigation satellite system (BDS-2) was built to provide positioning, navigation and timing (PNT) services to Asia-Pacific users, which accelerated the application and investigation of LEO satellite BDS-based orbit determination. Fengyun-3 employed a Global navigation satellite system occultation sounder (GNOS) to collect BDS-2/GPS measurements and opened the door to LEO orbit determination based on BDS [9–13]. Li Min accomplished an 8.4-cm precision orbit solution using the onboard BDS-2 measurements for the Fengyun-3C satellite [14]. The LING QIAO satellite carried a space-borne BDS receiver, an experimental payload, to complete in-orbit positioning experiment [15]. The Luojia-1A satellite is a navigation signal augmentation experimental satellite obtained centimeter-level orbit overlap consistence [16] with onboard

BDS observations. Some researchers studied the GPS/BDS-2 differential code bias estimation based on Fengyun-3C/3D onboard measurements [17–19]. Lu Cuixian estimated the phase center variation (PCV) of BDS-2/GPS satellites and used them to improve the orbit precision of the Fengyun-3C satellite [20].

In August 2020, the third-generation Beidou navigation satellite system (BDS-3) was accomplished to work for users worldwide, which promotes the investigation and application of a Global navigation satellite system (GNSS). Yang Yuanxi explained in detail BDS-3, including the architecture, time system, coordinate reference frame, signals and the planned services [21]. Li Liu introduced the BDS-3 and presented the innovative designs and usage of BDS-3 navigation messages broadcast [22]. Some researchers have assessed the SIS (signals in space) accuracy and service performance of BDS-3 [23–25]. Zhao Xinglong first assessed the quality of onboard B1I/B3I measurements from the Tianping-1B nanosatellite. The study results showed that a 4.57-cm precision orbit was solved only using the onboard BDS-3 B1I/B3I ionosphere-free combined measurements. It is noteworthy that the BDS-3 precise products of orbit and clock offset used in LEO POD (precise orbit determination) are from the separate dedicated BDS POD result enhanced by BDS-3 intersatellite link measurements [26].

Lu Mingquan elaborated the modulation techniques and the detailed signal structures of BDS-3. In order to achieve compatibility and interoperability with other GNSSs, the BDS-3 broadcast the B1C signal was applied to a new quadrature multiplexed binary offset carrier (QMBOC) (6, 1, 4/33) modulation in B1 band at the 1575.42 MHz carrier frequency [27]. Simultaneously, for the BDS-3 satellite, the B1I signal uploaded at 1561.098 MHz carrier frequency was kept to combine the BDS-3 B1C signal into a multicarrier constant envelope composite navigation signal in the B1 band, sharing one common transmitter chain. The BDS-3 B1I signal strongly coherent with the B1C signal should be specially processed to user receivers [28,29], not the same as the BDS-2 B1I signal.

More researchers have studied the ISBs (intersystem biases) of multi-GNSS [30–34]. Many researchers have indicated that the ISB is a receiver-type dependent [35,36]. Li Xingxing studied the four-system positioning model, pointing out that the code hardware delay is set to zero to eliminate the singularity between the receiver clock and code hardware delay; the other GNSS would estimate the related code hardware delay bias with respect to the GPS [37], and the corresponding phase ambiguity parameters would absorb the phase hardware delay bias for the phase measurement. Wang Ningbo assessed the quality of the GPS, Galileo and BDS-2/3 satellite broadcast group delay and found a systematic B1I-B3I TGD (time group delay) offset between BDS-2 and BDS-3 [38]. Zhang Yize and Jiao Guoqiang found a clock offset and TGD bias between BDS-2 and -3 [39–41]. The clock bias was similar to each receiver, while the B1I-B3I TGD bias depended on the receiver type. Jiao Guoqiang held the point of view that the clock offset and TGD bias should be regarded as the ISB (intersystem bias) of BDS-2/3 while solving the combined BDS-2/3 position solution. Some investigators studied the joint BDS-2/3 precise point positioning (PPP), whose precision would be approved by estimating the intersystem bias of BDS-2/3 [42–45]. And the joint BDS-2/3 precise time transfer also benefits from the intersystem bias [46].

In the background above, it is meaningful and necessary to investigate the effect of the signal difference of BDS-2/3 on LEO POD. This paper first analyses the reason for receiver ISB between BDS-2 and BDS-3 and gives the measurement equation. Then, this paper collects the Tianping-1B onboard GPS and BDS-2/3 measurements for the experiments of BDS-3-based POD and the multi-GNSS-based POD.

2. GNSS ISB

The Tianping-1B receiver could receive the dual-frequency code and phase measurements. This paper adopts the ionosphere-free (IF) measurements combined from dual-frequency measurements to correct the ionosphere delay. Therefore, for the onboard GPS and BDS IF measurements, the observation equation is as follows:

$$\begin{cases} P_{IF}^G = \rho^G + c \cdot \left(dt_R^G - dt_s^G\right) + \varepsilon_{P_{IF}}^G \\ P_{IF}^C = \rho^C + c \cdot \left(dt_R^C - dt_s^C\right) + \varepsilon_{P_{IF}}^C \\ L_{IF}^G = \rho^G + c \cdot \left(dt_R^G - dt_s^G\right) + \lambda_{IF}^G N_{IF}^G + \varepsilon_{L_{IF}}^G \\ L_{IF}^C = \rho^C + c \cdot \left(dt_R^C - dt_s^C\right) + \lambda_{IF}^C N_{IF}^C + \varepsilon_{L_{IF}}^C \end{cases} \tag{1}$$

The superscripts G and C denote, respectively, the GPS and BDS; P_{IF} and L_{IF} denote, respectively, the IF code measurements and phase measurements; ρ is the true distance from the receiver to the navigation satellite; dt_R denotes the receiver clock errors of GPS and BDS; dt_S denotes the navigation satellite clock error; λ_{IF} and N_{IF} denotes the dual-frequency IF wavelength and ambiguity value and $\varepsilon_{P_{IF}}$ and $\varepsilon_{L_{IF}}$ denote the IF code and phase noise.

The different GNSS signals have different hardware delay biases of signal transmission in the receiver. In the data processing, the hardware delay bias would be assimilated by the receiver clock error. The receiver clock error estimated is expressed as follows:

$$dt_R = dt_r + dt_{hd} \tag{2}$$

The superscript dt_r denotes the clock's true error, and dt_{hd} denotes the hardware delay bias of the signal transmission in the receiver. The superscript dt_R denotes the GNSS receiver clock error, which contains the clock's true error dt_r and hardware delay bias dt_{hd}. Therefore, the GPS and BDS receiver clock errors are expressed as follows:

$$dt_R^G = dt_r + dt_{hd}^G, \ dt_R^C = dt_r + dt_{hd}^C \tag{3}$$

The difference of dt_{hd}^G and dt_{hd}^C could be considered as the receiver ISB (intersystem bias) of GPS and BDS, recorded as B^{CG}:

$$B^{CG} = dt_{hd}^C - dt_{hd}^G = dt_R^C - dt_R^G \tag{4}$$

Meanwhile, it is worth noting that the magnitude of the hardware delay of the carrier phase measurements is very small, and the carrier phase hardware delay could be absorbed into the ambiguity value [37]. Therefore, the ISBs of multi-GNSS receiver code measurements only are studied in this paper.

In a general way, the method to deal with ISB is to set the hardware delay bias to zero. Equation (4) could be expressed as follows:

$$B^{CG} = dt_R^C - dt_R^G \tag{5}$$

The receiver clock error of BDS could be expressed as follows:

$$dt_R^C = dt_R^G + B^{CG} \tag{6}$$

Therefore, Equation (1) could be expressed as follow:

$$\begin{cases} P_{IF}^G = \rho^G + c \cdot \left(dt_R^G - dt_s^G\right) + \varepsilon_{P_{IF}}^G \\ P_{IF}^C = \rho^C + c \cdot \left(dt_R^C - dt_s^C + B^{CG}\right) + \varepsilon_{P_{IF}}^C \\ L_{IF}^G = \rho^G + c \cdot \left(dt_R^G - dt_s^G\right) + \lambda_{IF}^G N_{IF}^G + \varepsilon_{L_{IF}}^G \\ L_{IF}^C = \rho^C + c \cdot \left(dt_R^C - dt_s^C + B^{CG}\right) + \lambda_{IF}^C N_{IF}^C + \varepsilon_{L_{IF}}^C \end{cases} \tag{7}$$

We find, both Equations (1) and (7) are equivalent to deal with ISB. Therefore, the method to estimate the GPS receiver clock error as one parameter, and to estimate the BDS receiver clock error is another parameter together, equivalent to the way to estimate the GPS receiver clock error as one parameter and estimate the BDS-related ISB as one parameter together. This paper would adopt the first method to deal with the ISBs.

With respect to the receiver, different components and modulations of the BDS-2/3 signal in the B1 band [27] would lead to different B1I hardware delay biases of BDS-2/3. Likewise, the B1I/B3I IF combination signals of BDS-2 and BDS-3 would have different hardware delay biases, and the receiver clock errors estimated of BDS-2 and BDS-3 are different, such as the next equation:

$$dt^{C2}_{HD} \neq dt^{C3}_{HD} \, , \, dt^{C2}_R = dt_r + dt^{C2}_{HD}, \, dt^{C3}_R = dt_r + dt^{C3}_{HD} => \, dt^{C2}_R \neq dt^{C3}_R \tag{8}$$

For the onboard BDS ionosphere-free (IF) measurements from the LEO satellite, the observation equation would be expressed as follows:

$$\begin{cases} P^{C2}_{IF} = \rho^{C2} + c \cdot \left(dt^{C2}_R - dt^{C2}_s\right) + \varepsilon^{C2}_{P_{IF}} \\ P^{C3}_{IF} = \rho^{C3} + c \cdot \left(dt^{C3}_R - dt^{C3}_s\right) + \varepsilon^{C3}_{P_{IF}} \\ L^{C2}_{IF} = \rho^{C2} + c \cdot \left(dt^{C2}_R - dt^{C2}_s\right) + \lambda^{C2}_{IF} N^{C2}_{IF} + \varepsilon^{C2}_{L_{IF}} \\ L^{C3}_{IF} = \rho^{C3} + c \cdot \left(dt^{C3}_R - dt^{C3}_s\right) + \lambda^{C3}_{IF} N^{C3}_{IF} + \varepsilon^{C3}_{L_{IF}} \end{cases} \tag{9}$$

3. POD Strategy

In this paper, a Tianping-1B satellite orbit solution is solved to use the GFZ rapid products of GPS/BDS satellite orbit and clock corrections. The precision of GFZ rapid products is very important due to the LEO orbit precision. We analyzed the BDS and GPS satellite orbit precision provided by GFZ. Then, the dynamic model used and our POD experiment designment were introduced.

3.1. BDS/GPS Onboard Data and Products

The multi-GNSS receiver employed by Tianping-1B continually collects L1/L2 of GPS and B1I/B3I of the BDS-2/3 dual-frequency code and phase measurements [26]. The GPS L1 and L2 signals are from G01~G32, and the BDS-2/3 B1I and B3I signals are from C01~C32. Among BDS-2/3, C01~C05 are Geostationary Equatorial Orbit (GEO) satellites of BDS-2; C06~C10, C13 and C16 are Inclined GeoSynchronous Orbit (IGSO) satellites of BDS-2 and C11, C12 and C14 are Medium Earth Orbit (MEO) satellites of BDS-2. The rest of the BDS satellites are MEO satellites of BDS-3.

Generally, the LEO satellite precise orbit is solved by fixing the GPS/BDS precise products, both orbit and clock correction [7,14]. GPS/BDS precise products are obtained by a precise post-process based on the raw dual-frequency measurements from abundant global monitoring stations, such as L1/L2 for GPS [47]. The multi-GNSS rapid products of GFZ (Geo Forschungs Zentrum) adopt the B1I/B3I dual-frequency code and phase measurements from both BDS-3 and BDS-2, and the GFZ multi-GNSS rapid products treat the BDS-2 B1I/B3I signal the same as that of BDS-3 (ftp.gfz-potsdam.de/GNSS/products/mgex/READ-ME.txt, accessed on 18 February 2022). The orbit accuracy of GFZ rapid products is expressed in the SP3 file header (ftp://igs.ign.fr/pub/igs/igscb/data/format/SP3d.pdf, accessed on 18 February 2022). During the span of POD data processing, we calculated the orbit accuracy average values per BDS and GPS satellite used in the period of DOY 340-363 in 2019.

Figure 1 shows the BDS and each GPS satellite orbit accuracy means during DOY 340–363 in 2019. The orbit accuracy is from 35 cm to 100 cm for the BDS-2 satellites and less than 50 cm for all the BDS-3 satellites. The main GPS satellite orbit accuracy is less than 15 cm, except G06, G17, G21, G25 and G32.

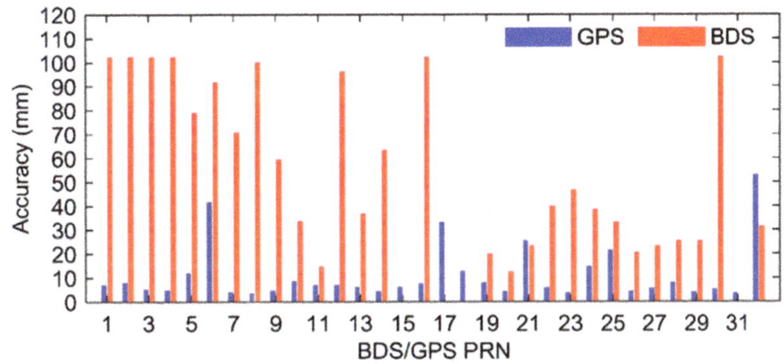

Figure 1. BDS and GPS satellite orbit accuracy.

3.2. POD Model

Table 1 expresses the measurement models, dynamic models and estimated parameters used in this study.

Table 1. Measurement model, dynamic model and estimated parameters.

Measurement Model	Models/Parameters
Observations	IF code measurements (1 m priori sigma), IF phase measurements (1 cm priori sigma)
GNSS precise products	GFZ multi-GNSS rapid products
Ionospheric delay	first-order delay eliminated, high orders neglected
GNSS phase center offset (PCO)	igs14_2086.atx
GNSS phase center variation (PCV)	Neglected
Tianping-1B PCO	Nominal value [26]
Tianping-1B PCV	Neglected
LEO attitude	Nominal model [26]
Smallest elevation angle	8°
Time span/interval	30 h, 10 s
Dynamics	**Models/Parameters**
Earth gravity model	EIGEN_GL04C, 120 × 120 [48]
N-body disturbance	DE421 [49]
Relativistic effect	IERS 2010 [50]
Earth orientation parameter	IERS 2010 [50]
Ocean tide	FES 2004, 10 × 10 [51]
Solid earth and pole tide	IERS 2010 [50]
Solar radiation pressure	Cannonball model [52]
Atmospheric drag	JB2008 [53]
Empirical force	Piecewise periodic estimation of sin and cos coefficients in along and cross directions and not in radial direction
Estimated parameters	**Parameters**
Tianping-1B initial state	Position and velocity at initial state
Receiver clock error	One value estimated per GNSS and per epoch
Ambiguities	One value per GNSS satellite per arc
Solar radiation pressure coefficients	One value per POD arc
Atmospheric drag coefficients	1 per 1.5 h
Empirical force coefficients	1 per 3.0 h

In this paper, the IF code and phase measurements are processed to solve the Tianping-1B orbit solutions with GFZ multi-GNSS rapid products. The a priori noise errors are 1 m for the IF code measurements and 1 cm for the IF phase measurements. The IF measurements eliminate the first-order ionospheric delay and neglect the higher-order delay. The GNSS antenna PCO (phase center offset) is from the igs14_2086.atx, and the Tianping-1B receiver antenna PCO is set to the nominal value $(-0.087, 0.018, -0.23211)$ m [26]. The antenna PCVs of GNSS and receiver are neglected. The nominal attitude of Tianping-1B is a stable zero-yaw attitude mode (refer to [26]). The smallest elevation angle is set to $8°$. The time span of POD arc is set to 30 h, and the interval of measurements is 10 s.

The dynamic models are adopted in the POD experiments, such as earth gravity of the 120×120 EIGEN_GL04C, N-body disturbance of DE421, relativistic effect, ocean tide of 10×10 FES 2004, solid earth and pole tide, solar radiation pressure of the Cannonball model, atmospheric drag of JB2008 and empirical forces.

The least squares algorithm is adopted to solve the unknown variable, such as the satellite initial state, the clock errors, ambiguity parameters and dynamic parameters. The clock errors are estimated per GNSS per epoch. The ambiguity parameter is set to one estimated value per arc per GNSS satellite without cycle slip. The coefficient of the Cannonball model is estimated by one parameter per POD arc. The coefficients of the JB2008 atmospheric drag model are estimated as one parameter per 1.5, and the sin and cos coefficients of the empirical force are estimated as one parameter per 3h in the along and cross directions.

In particular, the receiver ISB is the difference of the receiver clocks of the GPS and the other GNSS. In this paper, the receiver clock error is estimated as one value for GPS per epoch and another value for BDS-2, BDS-3 or BDS-2/3 per epoch, which is the same to estimate the receiver ISB [54].

3.3. Experiment Design

To study the POD potential of the onboard BDS measurements for the LEO satellite, we first solved the LEO satellite high-precision orbit based on the onboard GPS dual-frequency code and phase measurements. Taking the GPS-based POD results as the reference orbit, we evaluated the BDS-based POD results. The BDS-based POD experiments were divided into four parts. The first part was BDS-2-based POD, the second part was BDS-3 POD and the third part was denoted as BDS-2/3 POD, which treated BDS-2 and BDS-3 as one navigation system to solve the LEO satellite orbit. The final part denoted as BDS-2/3-B treated the BDS-2 and BDS-3 signals with different hardware delay biases in the onboard receivers, and BDS-2 and BDS-3 were treated as two navigation systems.

In addition, in order to analyze the performance of multi-GNSS-based POD, this paper preliminarily tried to obtain the orbit solutions of the different combined strategies from GPS, BDS-2 and BDS-3. GPS, BDS-2 and BDS-3 could be viewed as three separate systems for the onboard GNSS receiver. There are three combined POD solutions, the GPS+BDS-2-based POD solution, the GPS+BDS-3-based POD solution and the GPS+BDS-based POD solution. GPS and BDS-2 are treated as two system to solve the receiver ISB in the GPS+BDS-2-based POD solution, and the same thing was used with the GPS+BDS-3-based POD solution. In the GPS+BDS-based POD solution, GPS, BDS-2 and BDS-3 are treated as three systems to solve two ISBs.

4. Results and Analysis

For GPS-based POD orbit, the orbit overlapping consistency is adopted to evaluate the orbit accuracy. The BDS-based POD orbits accuracy is evaluated by comparison with GPS-based POD orbit, and the GPS/BDS-2/3 combined orbit solution is analyzed by the overlapping consistency and comparison with the GPS-based POD orbit.

4.1. GPS-Based Orbit

Figure 2 shows the GPS-based POD residuals of code pseudo-orange and the carrier phase measurements. The residual RMS value of the code pseudo-orange measurements is about 0.76 m, and the residual RMS value of the carrier phase measurement is about 0.96 cm.

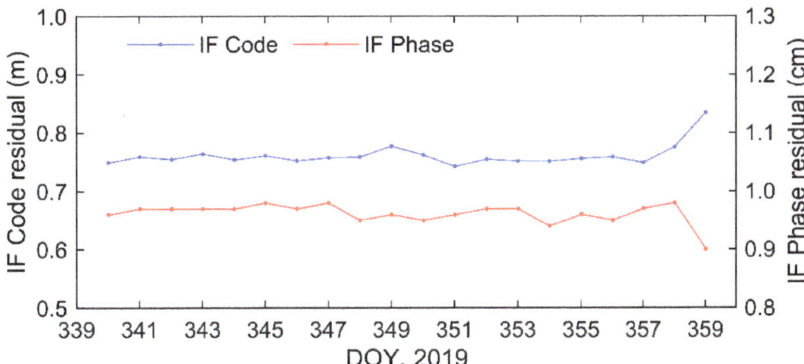

Figure 2. The GPS-based POD residuals.

Figure 3 shows the RMS differences in overlapping consistency of the GPS-based orbits, and Table 2 shows the relevant statistical results. The overlap consistency 3D-RMS average is 1.28 cm, and the standard deviation is less than 0.3 cm. The 3D-RMS maximum value of the overlap consistency is 1.9 cm, and the minimum value of that is 0.8 cm. The radial overlap consistency average is 0.56 cm, and the standard deviation is less than 0.2 cm. The maximum and minimum values of that are, respectively, 0.9 cm and 0.3 cm. The average values of the along-track RMS and cross-track RMS are, respectively, 0.86 cm and 0.71 cm.

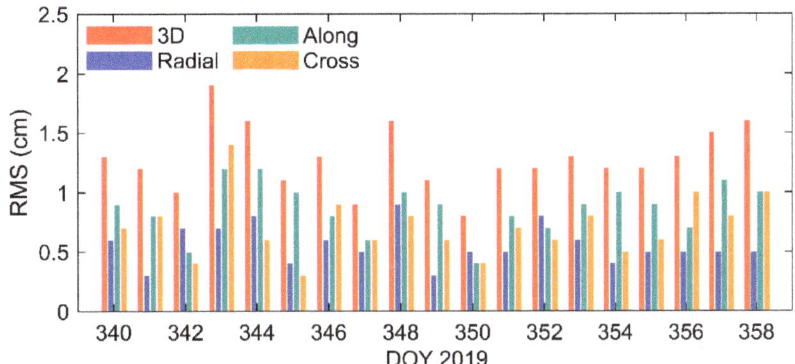

Figure 3. The RMS differences in the overlapping consistency of the GPS-based orbits.

Table 2. The statistics of overlapping consistency of the GPS-based orbits (unit: cm).

Statistics	3D	Radial	Along	Cross
Max	1.9	0.9	1.2	1.4
Min	0.8	0.3	0.4	0.3
Mean	1.28	0.56	0.86	0.71
STD	0.27	0.16	0.22	0.26

It can be seen that the 3D orbital overlap consistency of GPS-based POD is less than 2 cm, and the radial overlap consistency is less than 1 cm. Therefore, the GPS-based orbit solutions could be as a precision product to evaluate BDS-based orbit solutions.

4.2. BDS-3-Based POD

Using the spaceborne BDS-2/3 measurements and GFZ multi-GNSS rapid products, the Tianping1-B orbit is solved with different strategies, as described in Section 3.2.

Figure 4 shows the BDS-based POD residuals with different strategies. For BDS-2-based POD, the IF code residual of BDS-2 is about 1.06 m, and the IF phase residual is about 0.88 cm. For BDS-3-based POD, the IF code and phase residuals are 0.64 m and 0.82 cm, respectively. For BDS-2/3-based POD, the IF code and phase residuals of BDS-2 are 2.16 m and 1.1 cm and that of BDS-3 are 1.85 m and 1.0 cm, respectively. Compared with BDS-2-based and BDS-2/3-based POD, the IF code and phase residuals of both BDS-2 and BDS-3 are larger. For BDS-2/3-B-based POD, the IF code and phase residuals of BDS-2 are 1.07 m and 0.99 cm and that of BDS-3 are 0.64 m and 0.89 cm, respectively. The IF code residual is equivalent to that of BDS-2-based and BDS-3-based POD compared with BDS-2/3-based POD.

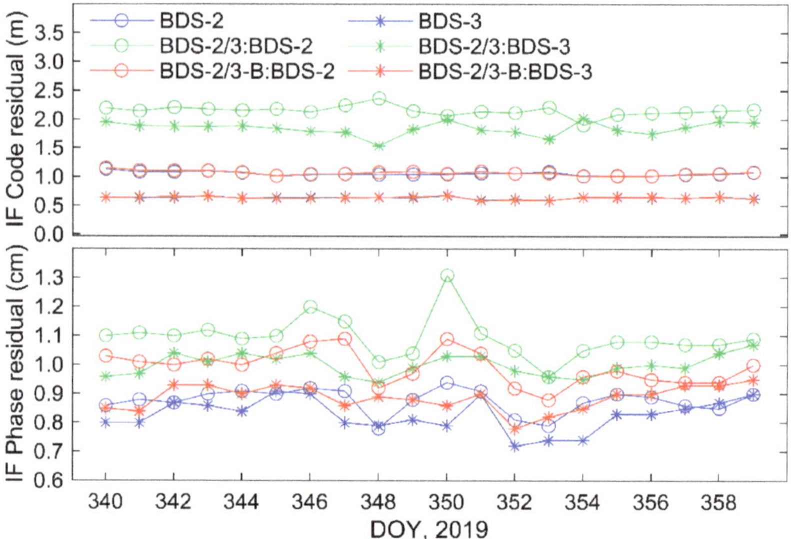

Figure 4. The BDS-based POD residuals with different strategies.

Figure 5 shows the BDS-based orbit RMS difference results compared with the GPS-based orbits, and the RMS difference statistics results are listed in Table 3. As illustrated in Table 3 and Figure 5, the BDS-3-based POD orbit 3D RMS difference is 6.57 cm and is reduced by 56% compared to BDS-2-based POD. This result indicates that the orbit accuracy of LEO POD based on BDS-3 has been greatly improved, compared with BDS-2-based POD.

For BDS-2 and BDS-3 combined POD, the BDS-2/3-B-based orbit 3D RMS difference of 5.37 cm is the lowest. It reduces 64% to be compared to the BDS-2-based orbit and 18% to be compared to the BDS-3-based orbit, which are significant compared to BDS-2/3-based POD. The orbit RMS difference between BDS-2/3-B-based POD and GPS-based POD is 1.96 cm in the radial direction, less than that of BDS-3-based POD and BDS-2/3-based POD. These results show that the receiver ISB estimated in the BDS-2 and BDS-3 combined POD could improve accuracy for the LEO satellite.

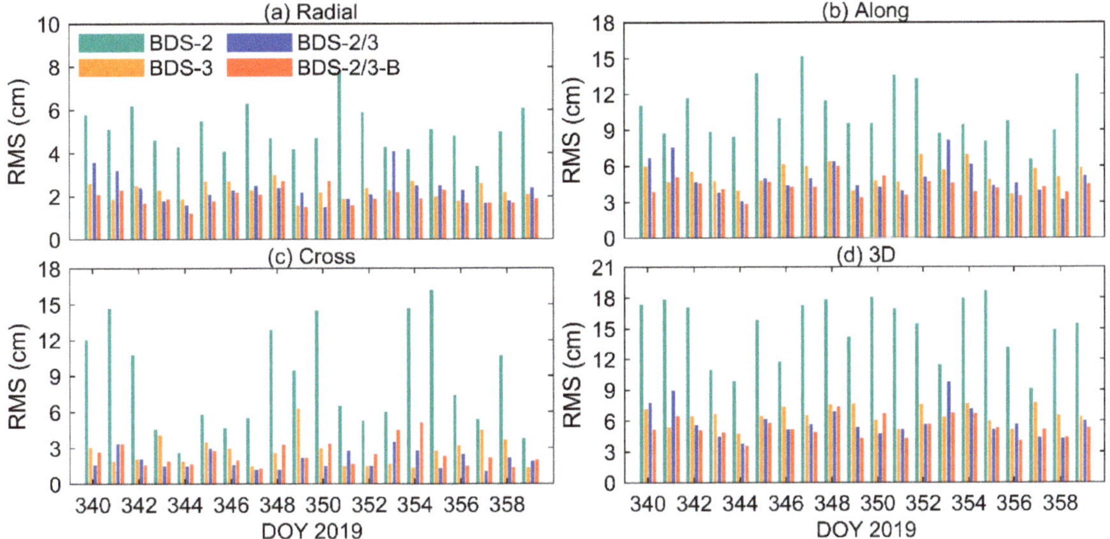

Figure 5. The BDS-based orbit RMS errors results compared with the GPS-based orbits.

Table 3. BDS-based orbit RMS difference statistics results compared with GPS-based orbits (unit: cm).

Test	Reduced Ratio		3D	Radial	Along	Cross
	BDS-2	**BDS-3**				
BDS-2-based POD	/	/	15.10	5.11	10.56	8.69
BDS-3-based POD	56%	/	6.57	2.29	5.36	2.74
BDS-2/3-based POD	61%	10%	5.92	2.35	5.02	2.02
BDS-2/3-B-based POD	64%	18%	5.37	1.96	4.28	2.48

4.3. GPS+BDS-2/3 Combined Orbit Solution

Table 4 presents the IF code and phase residual statistics results of the GPS and BDS joint different strategies, as described in Section 3.2. The BDS-2 IF code and phase residuals are, respectively, about 115 cm and 1 cm for GPS+BDS-2-based POD solution and GPS+BDS-based POD solution. The BDS-3 IF code and phase residuals are, respectively, about 69 cm and 0.92 cm for GPS+BDS-3-based POD solution and GPS+BDS-based POD solution. The GPS IF code and phase residuals are, respectively, about 80 cm and 0.97 cm for each POD solution. These residuals results show that each system IF code residual from the three combined strategies is closed to the IF code residuals of the GPS-based, BDS-2-based and BDS-3-based POD.

Figure 6 shows the results of both the orbit overlapping consistency and the combined orbit RMS errors compared with GPS-based POD, and Table 5 presents, respectively, the orbit RMS difference statistics. The 3D RMS differences for the orbit overlapping consistency are, respectively, 3.15 cm, 2.91 cm and 3.03 cm from the GPS+BDS-based POD solution, GPS+BDS-2-based POD solution and GPS+BDS-3-based POD solution, and the differences between them are less than 3 mm. The radial RMS differences for the orbit overlapping consistency of the three different strategies are, respectively, 0.91 cm, 0.88 cm and 0.91 cm, and the differences between them are less than 1 mm. The along and cross RMS differences for the orbit overlapping consistency of the three different strategies are, respectively, about 2.6 cm and 1.07 cm. It is important to note that the STD values of the GPS+BDS-3-based orbit overlapping consistency from all the POD arcs are smallest in the 3D, radial, along and cross directions.

Table 4. The IF code and phase residual statistics results of the GPS and BDS joint different strategies (unit: cm).

TEST		GPS+BDS-Based POD	GPS+BDS-3-Based POD	GPS+BDS-2-Based POD
BDS-2	**IF code**	114.96	\	115.02
	IF phase	1.00		1.00
BDS-3	**IF code**	68.14	69.03	\
	IF phase	0.91	0.93	
GPS	**IF code**	79.68	80.42	79.30
	IF phase	0.97	0.98	0.96
Entirety	**IF code**	85.58	77.59	89.35
	IF phase	0.96	0.97	0.97

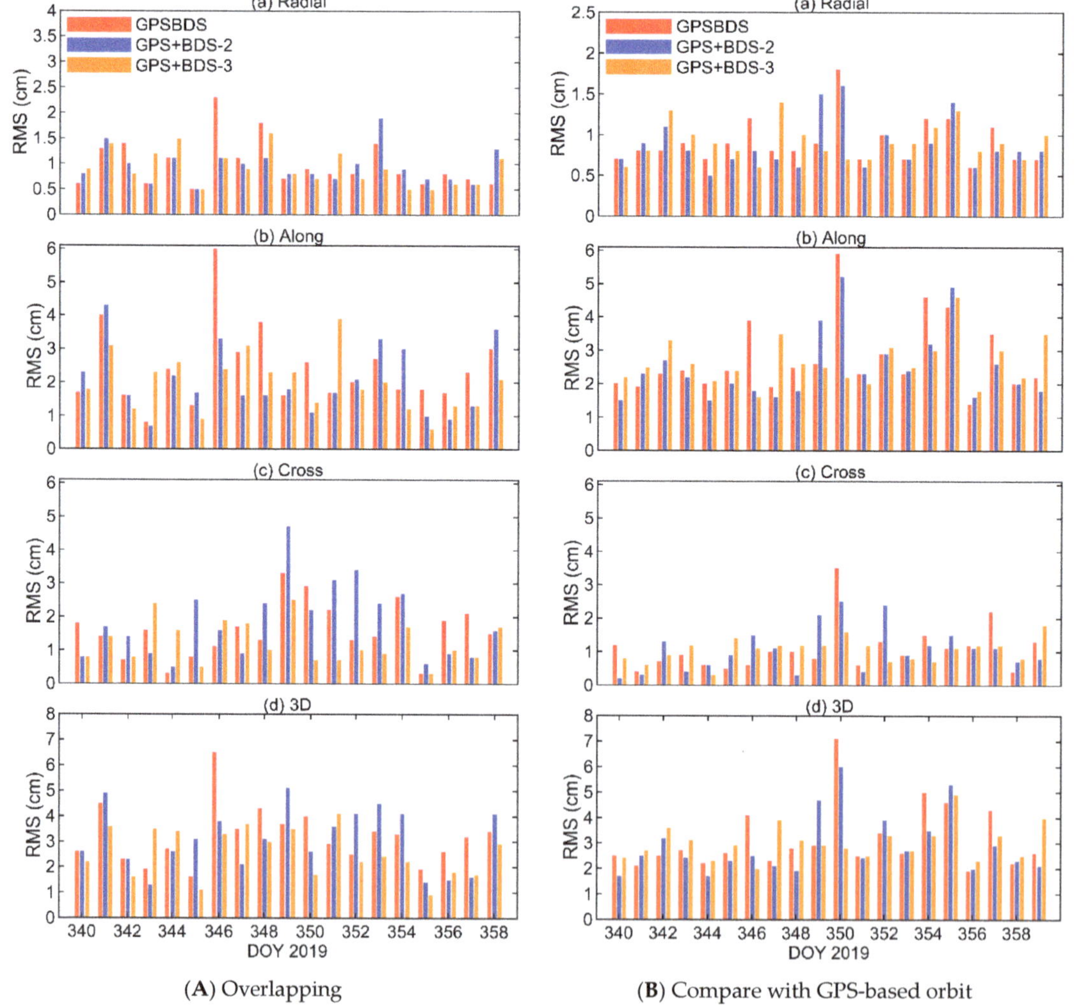

Figure 6. The orbit RMS of the GPS and BDS combined different strategies. (**A**) The orbit RMS in overlapping consistency from GPS+BDS-based POD, GPS+BDS-2-based POD and GPS+BDS-3-based POD. (**B**) The orbit RMS between GPS-based POD and the three GPS and BDS combined POD strategies.

Table 5. The orbit RMS difference statistics of the GPS and BDS combined different strategies (unit: cm).

Test	Statistic	Overlapping				Compared with GPS-Based Orbit			
		3D	Radial	Along	Cross	3D	Radial	Along	Cross
GPS+BDS-based POD	MEAN	3.15	0.91	2.77	1.09	3.20	0.99	2.41	1.59
	STD	1.25	0.27	1.09	0.70	1.13	0.47	1.19	0.81
GPS+BDS-2-based POD	MEAN	2.91	0.88	2.51	1.07	3.07	0.95	2.06	1.85
	STD	1.17	0.30	1.04	0.66	1.21	0.34	1.01	1.12
GPS+BDS-3-based POD	MEAN	3.03	0.91	2.66	1.05	2.57	0.92	1.98	1.24
	STD	0.68	0.22	0.69	0.34	0.95	0.34	0.84	0.63

The orbit RMS differences between the GPS+BDS-based POD solutions and GPS-based POD solutions are, respectively, 3.20 cm, 0.99 cm, 2.41 cm and 1.59 cm in the 3D, radial, along and cross directions. The GPS+BDS-2-based POD orbit RMS differences compared with the GPS-based orbit are, respectively, 3.07 cm, 0.95 cm, 2.06 cm and 1.85 cm in the 3D, radial, along and cross directions. The orbit RMS differences of the GPS and BDS-3 combined POD solutions and GPS-based POD are smaller than that of GPS+BDS-based and GPS+BDS-2-based POD and are, respectively, 2.57 cm, 0.92 cm, 1.98 cm and 1.24 cm in the 3D, radial, along and cross directions. It is also important to note that the STD values of all the POD orbit RMS differences between the GPS+BDS-3-based orbits and GPS-based orbits are the smallest in the 3D, radial, along and cross directions.

5. Conclusions

BDS-3 provides a new option for LEO POD and promotes the application of the multi-GNSS. It is unavoidable to estimate the GNSS receiver ISB in multi-GNSS applications. This paper preliminarily researched the impacts of GNSS receiver ISB for LEO POD in detail through the POD test with Tianping-1B onboard GPS and BDS data.

This paper discussed the reason for the ISB between BDS-2 and BDS-3 and gave the measurement equations. Using GFZ multi-GNSS rapid products, the ISBs were estimated in the Tianping-1B POD based on the BDS-2, BDS-3 and GPS measurements. Using only onboard BDS-3 measurements, the orbit accuracy was 6.57 cm in the 3D direction, improved by 56% from the BDS-2-based POD. When the ISBs of BDS-2 and BDS-3 were estimated in the combined POD, the orbit accuracy of 5.37 cm was 18% better than the BDS-3-based POD, which was nearly two times larger than the improvement ratio of the BDS-2/3-based POD with no ISB.

For the GPS and BDS combined POD, the orbit accuracy of GPS+BDS-3-based POD was higher. The 3D orbit RMS difference between the GPS+BDS-3-based POD solution and GPS-based POD solution was less than 3 cm, and its STD value was the smallest. These results illustrated that BDS-3 is more appropriate to join the GPS to determinate the LEO satellite orbit.

In conclusion, the ISB estimated between BDS-2 and BDS-3 could improve the Tianping-1B orbit accuracy using the onboard BDS-2/3 measurements. GPS and BDS-3 combined POD could obtain a more consistent orbit with a GPS-based orbit. It is meaningful to estimate the ISB to improve the Tianping-1B satellite orbit accuracy when using BDS-2 and BDS-3 measurements together. At the same time, the POD experiments with GPS and BDS provide an important reference for LEO POD with onboard multi-GNSS measurements.

Author Contributions: Conceptualization, X.Z. and X.H.; Formal analysis, X.Z.; Funding acquisition, S.Z.; Methodology, S.Z. and X.Z.; Resources, X.H.; Software, J.C. and Z.W.; Validation, S.Z.; Writing—original draft, X.Z. and S.Z. and Writing—review and editing, X.Z. and J.Y. All authors have read and agreed to the published version of the manuscript.

Funding: This research was funded by the National Natural Science Foundation of China (No. 12173072) and Qian Xuesen Youth Innovation Fund of China Aerospace Science and Technology Corporation In 2021.

Data Availability Statement: The Tianping-1B onboard GPS and BDS-3 data are supported by the Xi'an Satellite Control Center.

Acknowledgments: From the Xi'an Satellite Control Center, Guo Danni and Guo Kai are gratefully acknowledged for providing the onboard GPS and BDS-2/3 data from Tianping-1B. The authors thank GFZ for providing the multi-GNSS rapid products, and the authors are grateful to IGS for providing other necessary products. The authors thank the reviewers and editors for their constructive comments and words that have improved the quality of this manuscript.

Conflicts of Interest: The authors declare no conflict of interest.

References

1. Montenbruck, O.; Hauschild, A.; Langley, R.B.; Siemes, C. CASSIOPE orbit and attitude determination using commercial off-the-shelf GPS receivers. *GPS Solut.* **2019**, *23*, 114. [CrossRef]
2. Hauschild, A.; Montenbruck, O.; Langley, R.B. Flight results of GPS-based attitude determination for the Canadian CASSIOPE satellite. *Navigation* **2020**, *67*, 83–93. [CrossRef]
3. Kahr, E.; Roth, N.; Montenbruck, O.; Risi, B.; Zee, R.E. GPS Relative Navigation for the CanX-4 and CanX-5 Formation-Flying Nanosatellites. *J. Spacecr. Rocket.* **2018**, *55*, 1–14. [CrossRef]
4. Bock, H.; Jäggi, A.; Meyer, U.; Visser, P.; Ijssel, J.V.D.; van Helleputte, T.; Heinze, M.; Hugentobler, U. GPS-derived orbits for the GOCE satellite. *J. Geod.* **2011**, *85*, 807–818. [CrossRef]
5. Hackel, S.; Gisinger, C.; Balss, U.; Wermuth, M.; Montenbruck, O. Long-Term Validation of TerraSAR-X and TanDEM-X Orbit Solutions with Laser and Radar Measurements. *Remote Sens.* **2018**, *10*, 762. [CrossRef]
6. Montenbruck, O.; Hackel, S.; Ijssel, J.V.D.; Arnold, D. Reduced dynamic and kinematic precise orbit determination for the Swarm mission from 4 years of GPS tracking. *GPS Solut.* **2018**, *22*, 79. [CrossRef]
7. Montenbruck, O.; Hackel, S.; Jaggi, A. Precise orbit determination of the Sentinel-3A altimetry satellite using ambiguity-fixed GPS carrier phase observations. *J. Geod.* **2018**, *92*, 711–726. [CrossRef]
8. Guo, J.; Zhao, Q.; Guo, X.; Liu, X.; Liu, J.; Zhou, Q. Quality assessment of onboard GPS receiver and its combination with DORIS and SLR for Haiyang 2A precise orbit determination. *Sci. China Earth Sci.* **2015**, *58*, 138–150. [CrossRef]
9. Xiong, C.; Lu, C.; Zhu, J.; Ding, H. Orbit determination using real tracking data from FY3C-GNOS. *Adv. Space Res.* **2017**, *60*, 543–556. [CrossRef]
10. Cai, Y.; Bai, W.; Wang, X.; Sun, Y.; Du, Q.; Zhao, D.; Meng, X.; Liu, C.; Xia, J.; Wang, D.; et al. In-orbit performance of GNOS on-board FY3-C and the enhancements for FY3-D satellite. *Adv. Space Res.* **2017**, *60*, 2812–2821. [CrossRef]
11. Zhao, Q.; Wang, C.; Guo, J.; Yang, G.; Liao, M.; Ma, H.; Liu, J. Enhanced orbit determination for BeiDou satellites with FengYun-3C onboard GNSS data. *GPS Solut.* **2017**, *21*, 1179–1190. [CrossRef]
12. Li, X.; Zhang, K.; Zhang, Q.; Zhang, W.; Yuan, Y.; Li, X. Integrated Orbit Determination of FengYun-3C, BDS, and GPS Satellites. *J. Geophys. Res. Solid Earth* **2018**, *123*, 8143–8160. [CrossRef]
13. Li, X.; Zhang, K.; Meng, X.; Zhang, Q.; Zhang, W.; Li, X.; Yuan, Y. LEO–BDS–GPS integrated precise orbit modeling using FengYun-3D, FengYun-3C onboard and ground observations. *GPS Solut.* **2020**, *24*, 48. [CrossRef]
14. Li, M.; Li, W.; Shi, C.; Jiang, K.; Guo, X.; Dai, X.; Meng, X.; Yang, Z.; Yang, G.; Liao, M. Precise orbit determination of the Fengyun-3C satellite using onboard GPS and BDS observations. *J. Geod.* **2017**, *91*, 1313–1327. [CrossRef]
15. Chen, X.; Zhao, S.; Wang, M.; Lu, M. Space-borne BDS receiver for LING QIAO satellite: Design, implementation and preliminary in-orbit experiment results. *GPS Solut.* **2016**, *20*, 837–847. [CrossRef]
16. Wang, L.; Xu, B.; Fu, W.; Chen, R.; Li, T.; Han, Y.; Zhou, H. Centimeter-Level Precise Orbit Determination for the Luojia-1A Satellite Using BeiDou Observations. *Remote Sens.* **2020**, *12*, 2063. [CrossRef]
17. Jiang, K.; Li, M.; Zhao, Q.; Li, W.; Guo, X. BeiDou Geostationary Satellite Code Bias Modeling Using Fengyun-3C Onboard Measurements. *Sensors* **2017**, *17*, 2460. [CrossRef]
18. Li, W.; Li, M.; Shi, C.; Fang, R.; Zhao, Q.; Meng, X.; Yang, G.; Bai, W. GPS and BeiDou Differential Code Bias Estimation Using Fengyun-3C Satellite Onboard GNSS Observations. *Remote Sens.* **2017**, *9*, 1239. [CrossRef]
19. Li, X.; Ma, T.; Xie, W.; Zhang, K.; Huang, J.; Ren, X. FY-3D and FY-3C onboard observations for differential code biases estimation. *GPS Solut.* **2019**, *23*, 57. [CrossRef]
20. Lu, C.; Zhang, Q.; Zhang, K.; Zhu, Y.; Zhang, W. Improving LEO precise orbit determination with BDS PCV calibration. *GPS Solut.* **2019**, *23*, 109. [CrossRef]
21. Yang, Y.; Gao, W.; Guo, S.; Mao, Y.; Yang, Y. Introduction to BeiDou-3 navigation satellite system. *Navigation* **2019**, *66*, 7–18. [CrossRef]
22. Liu, L.; Zhang, T.; Zhou, S.; Hu, X.; Liu, X. Improved design of control segment in BDS-3. *Navigation* **2019**, *66*, 37–47. [CrossRef]
23. Chen, J.; Hu, X.; Tang, C.; Zhou, S.; Yang, Y.; Pan, J.; Ren, H.; Ma, Y.; Tian, Q.; Wu, B.; et al. SIS accuracy and service performance of the BDS-3 basic system. *Sci. China-Phys. Mech. Astron.* **2020**, *63*, 269511. [CrossRef]
24. Lv, Y.; Geng, T.; Zhao, Q.; Xie, X.; Zhou, R. Initial assessment of BDS-3 preliminary system signal-in-space range error. *GPS Solut.* **2019**, *24*, 16. [CrossRef]

25. Yang, Y.; Mao, Y.; Sun, B. Basic performance and future developments of BeiDou global navigation satellite system. *Satell. Navig.* **2020**, *1*, 1. [CrossRef]
26. Zhao, X.; Zhou, S.; Ci, Y.; Hu, X.; Cao, J.; Chang, Z.; Tang, C.; Guo, D.; Guo, K.; Liao, M. High-precision orbit determination for a LEO nanosatellite using BDS-3. *GPS Solut.* **2020**, *24*, 102. [CrossRef]
27. Lu, M.; Li, W.; Yao, Z.; Cui, X. Overview of BDS III new signals. *Navig. J. Inst. Navig.* **2019**, *66*, 19–35. [CrossRef]
28. Wang, C.; Cui, X.; Ma, T.; Zhao, S.; Lu, M. Asymmetric Dual-Band Tracking Technique for Optimal Joint Processing of BDS B1I and B1C Signals. *Sensors* **2017**, *17*, 2360. [CrossRef]
29. Gao, Y.; Yao, Z.; Lu, M. A Coherent Processing Technique with High Precision for BDS B1I and B1C Signals. In *China Satellite Navigation Conference*; Springer: Singapore, 2020.
30. Odijk, D.; Nadarajah, N.; Zaminpardaz, S.; Teunissen, P.J.G. GPS, Galileo, QZSS and IRNSS differential ISBs: Estimation and application. *GPS Solut.* **2017**, *21*, 439–450. [CrossRef]
31. Teunissen, P. GNSS Precise Point Positioning. In *Position, Navigation, and Timing Technologies in the 21st Century*; Wiley: Hoboken, NJ, USA, 2021; Chapter 20; pp. 503–528.
32. Tian, Y.; Yuan, L.; Tan, L.; Yan, H.; Xu, S. Regularization and particle filtering estimation of phase inter-system biases (ISB) and the lookup table for Galileo E1-GPS L1 phase ISB calibration. *GPS Solut.* **2019**, *23*, 115. [CrossRef]
33. Mi, X.; Zhang, B.; Odolinski, R.; Yuan, Y. On the temperature sensitivity of multi-GNSS intra- and inter-system biases and the impact on RTK positioning. *GPS Solut.* **2020**, *24*, 112. [CrossRef]
34. Pan, L.; Zhang, Z.; Yu, W.; Dai, W. Intersystem Bias in GPS, GLONASS, Galileo, BDS-3, and BDS-2 Integrated SPP: Characteristics and Performance Enhancement as a Priori Constraints. *Remote Sens.* **2021**, *13*, 4650. [CrossRef]
35. Paziewski, J.; Wielgosz, P. Accounting for Galileo–GPS inter-system biases in precise satellite positioning. *J. Geod.* **2015**, *89*, 81–93. [CrossRef]
36. Odijk, D.; Teunissen, P.J.G. Characterization of between-receiver GPS-Galileo inter-system biases and their effect on mixed ambiguity resolution. *GPS Solut.* **2013**, *17*, 521–533. [CrossRef]
37. Li, X.; Zhang, X.; Ren, X.; Fritsche, M.; Wickert, J.; Schuh, H. Precise positioning with current multi-constellation Global Navigation Satellite Systems: GPS, GLONASS, Galileo and BeiDou. *Sci. Rep.* **2015**, *5*, 8328. [CrossRef] [PubMed]
38. Wang, N.; Li, Z.; Montenbruck, O.; Tang, C. Quality assessment of GPS, Galileo and BeiDou-2/3 satellite broadcast group delays. *Adv. Space Res.* **2019**, *64*, 1764–1779. [CrossRef]
39. Zhang, Y.; Kubo, N.; Chen, J.; Chu, F.-Y.; Wang, A.; Wang, J. Apparent clock and TGD biases between BDS-2 and BDS-3. *GPS Solut.* **2020**, *24*, 27. [CrossRef]
40. Zhang, Y.; Wang, H.; Chen, J.; Wang, A.; Meng, L.; Wang, E. Calibration and Impact of BeiDou Satellite-Dependent Timing Group Delay Bias. *Remote Sens.* **2020**, *12*, 192. [CrossRef]
41. Jiao, G.; Song, S.; Liu, Y.; Su, K.; Cheng, N.; Wang, S. Analysis and Assessment of BDS-2 and BDS-3 Broadcast Ephemeris: Accuracy, the Datum of Broadcast Clocks and Its Impact on Single Point Positioning. *Remote Sens.* **2020**, *12*, 2081. [CrossRef]
42. Jiao, G.; Song, S.; Jiao, W. Improving BDS-2 and BDS-3 joint precise point positioning with time delay bias estimation. *Meas. Sci. Technol.* **2020**, *31*, 025001. [CrossRef]
43. Zhao, W.; Chen, H.; Gao, Y.; Jiang, W.; Liu, X. Evaluation of Inter-System Bias between BDS-2 and BDS-3 Satellites and Its Impact on Precise Point Positioning. *Remote Sens.* **2020**, *12*, 2185. [CrossRef]
44. Cao, X.; Shen, F.; Zhang, S.; Li, J. Time delay bias between the second and third generation of BeiDou Navigation Satellite System and its effect on precise point positioning. *Measurement* **2020**, *168*, 108346. [CrossRef]
45. Song, Z.; Chen, J.; Wang, B.; Yu, C. (Eds.) *Analysis and Modeling of the Inter-System Bias between BDS-2 and BDS-3*; Springer: Singapore, 2020; p. 651.
46. Qin, W.; Ge, Y.; Zhang, Z.; Su, H.; Wei, P.; Yang, X. Accounting BDS3–BDS2 inter-system biases for precise time transfer. *Measurement* **2020**, *156*, 107566. [CrossRef]
47. Villiger, A.; Dach, R. (Eds.) *International GNSS Service Technical Report 2019*; IGS Annual Report; IGS Central Bureau and University of Bern; Bern Open Publishing: Bern, Switzerland, 2020.
48. Flechtner, F.; Dahle, C.; Neumayer, K.H.; König, R.; Förste, C. The release 04 CHAMP and GRACE EIGEN gravity field models. In *System Earth via Geodetic-Geophysical Space Techniques*; Advanced Technologies in Earth Sciences; Springer: Berlin, Germany, 2010; pp. 1643–2190.
49. Folkner, W.; Williams, J.; Boggs, D. *The Planetary and Lunar Ephemeris DE 421*; Interplanetary Network Progress Report; 42–178. Memorandum IOM 343R-08-003; Institute of Technology, Jet Propulsion Laboratory: Pasadena, CA, USA, 2009.
50. Petit, G.; Luzum, B. *IERS Conventions (2010)*; Technical Report; Verlag des Bundesamts fur Kartographie und Geodasie: Avignon, France; Frankfurt am Main, Germany, 2010.
51. Lyard, F.; Lefevre, F.; Letellier, T.; Francis, O. Modelling the global ocean tides: Modern insights from FES2004. *Ocean Dyn.* **2006**, *56*, 394–415. [CrossRef]
52. Rosengren, A.; Scheeres, D. Long-term dynamics of high area-tomass ratio objects in high-Earth orbit. *Adv. Space Res.* **2013**, *52*, 1545–1560. [CrossRef]

53. Bowman, B.; Tobiska, W.K.; Marcos, F.; Huang, C.; Lin, C.; Burke, W. A new empirical thermospheric density model JB2008 using new solar and geomagnetic indices. In Proceedings of the AIAA/AAS Astrodynamics Specialist Conference and Exhibit, Honolulu, HI, USA, 18–21 August 2008.
54. Choy, S.; Zhang, S.; Lahaye, F.; Héroux, P. A Comparison between GPS-only and Combined GPS+GLONASS Precise Point Positioning. *J. Spat. Sci.* **2013**, *58*, 169–190. [CrossRef]

remote sensing

MDPI

Article

The Refined Gravity Field Models for Height System Unification in China

Panpan Zhang [1,2], Zhicai Li [3,4,*], Lifeng Bao [1,2], Peng Zhang [4], Yongshang Wang [4], Lin Wu [1,2] and Yong Wang [1,2]

1   State Key Laboratory of Geodesy and Earth's Dynamics, Innovation Academy for Precision Measurement Science and Technology, Chinese Academy of Sciences, Wuhan 430077, China; zhangpanpan@asch.whigg.ac.cn (P.Z.); baolifeng@asch.whigg.ac.cn (L.B.); linwu@apm.ac.cn (L.W.); ywang@whigg.ac.cn (Y.W.)
2   University of Chinese Academy of Sciences, Beijing 100049, China
3   College of Geoscience and Surveying Engineering, China University of Mining and Technology-Beijing, Beijing 100083, China
4   National Geomatics Center of China, Beijing 100830, China; zhangpeng@ngcc.cn (P.Z.); ys@ngcc.cn (Y.W.)
*   Correspondence: zcli@cumtb.edu.cn; Tel.: +86-13718354006

Abstract: A unified height datum is essential for global geographic information resource construction, ecological environment protection, and scientific research. The goal of this paper is to derive the geopotential value for the Chinese height datum (CNHD) in order to realize the height datum unification in China. The estimation of height datum geopotential value usually depends on high-precision global gravity field models (GFMs). The satellite gravity missions of the Gravity Recovery and Climate Experiment (GRACE) and Gravity field and steady-state Ocean Circulation Exploration (GOCE) provide high-accuracy, medium–long-wavelength gravity field spectra, but satellite-only GFMs are limited to medium–long wavelengths, which will involve omission errors. To compensate for the omission errors in satellite-only GFMs, a spectral expansion approach is used to obtain the refined gravity field models using the EGM2008 (Earth Gravitational Model 2008) and residual terrain model (RTM) technique. The refined GFMs are evaluated by using high-quality GNSS/leveling data, the results show that the quasi-geoid accuracy of the refined DIR_R6_EGM2008_RTM model in China has optimal accuracy and, compared with the EGM2008 model and the DIR_R6 model, this refined model in China is improved by 9.6 cm and 21.8 cm, and the improvement ranges are 35.7% and 55.8%, respectively. Finally, the geopotential value of the Chinese height datum is estimated to be equal to 62,636,853.29 $\mathrm{m^2 s^{-2}}$ with respect to the global reference level defined by $W_0 = 62,636,853.4 \mathrm{\ m^2 s^{-2}}$ by utilizing the refined DIR_R6_EGM2008_RTM model and 1908 high-quality GNSS/leveling datapoints.

Keywords: Chinese height datum; GRACE/GOCE; residual terrain model; spectral expansion approach; height datum geopotential

Citation: Zhang, P.; Li, Z.; Bao, L.; Zhang, P.; Wang, Y.; Wu, L.; Wang, Y. The Refined Gravity Field Models for Height System Unification in China. *Remote Sens.* **2022**, *14*, 1437. https://doi.org/10.3390/rs14061437

Academic Editor: Xiaogong Hu

Received: 26 January 2022
Accepted: 12 March 2022
Published: 16 March 2022

1. Introduction

With the emergence of Global Navigation Satellite System (GNSS), users can obtain consistent ellipsoidal height at global scale. The ellipsoid height relative to a given geocentric ellipsoid can be obtained quickly and accurately by using GNSS. However, the ellipsoidal height is not related to the Earth's gravity field. The height related to the Earth's gravity field usually refers to the orthometric or normal height, which is strictly based on the geopotential number C_P ($C_P = W_0^{LVD} - W_P$, where W_0^{LVD} is the geopotential value for the local vertical datum, and W_P indicates the gravity potential for point P). The local vertical datum refers to the geoid (or quasi-geoid), which is assumed to be coincident with the local mean sea level (MSL). Importantly, even though all local height datums are related to the MSL, the vertical offsets between them may be up to 2 m at global scale [1]. This is due to the fact that the MSL presents geographical and time-dependent variations, but they are a consequence of the natural sea dynamics. Therefore, the vertical datums vary among countries or regions, affecting and restricting the sharing and exchange of global geospatial information, and the unification of the height datums has become a key point in the field of geodesy.

According to the Global Geodetic Observing System (GGOS) of the International Association of Geodesy (IAG), a global vertical datum related to Earth's gravity field should be established. One of the main works of GGOS is to support global geometric and physical heights with centimeter accuracy within a global framework, and to unify all existing physical height systems [2,3]. A global reference system defines constants, conventions, models, and parameters as the necessary basis for the mathematical representation of geometric and physical quantities [4]. According to the IAG resolution No 1, 2015, the gravity potential value for the International Height Reference System (IHRS) is released equal to 62,636,853.4 $m^2 s^{-2}$ [5–7]. According to this resolution, the existing local vertical datum systems can be integrated into the IHRS, which will ensure the consistency of the global height datum systems. The fundamental approach for height datum unification is the geodetic boundary value problem (GBVP) method [8–13]. This method has been widely applied in areas with good coverage of surface gravity data. However, in areas where surface gravity data are poor or restricted, a feasible option for height datum unification is to use high-accuracy GFMs. The GFMs provide expected accuracy at scales from centimeter to decimeter [11]. With the development of high-precision and high-resolution global gravity flied models, the GFMs have become feasible for the unification of vertical datums [14–22].

The GFMs have many advantages for determining the geopotential values of vertical datums, but the results depend largely on the accuracy of the utilized GFMs. The satellite gravity missions such as GRACE [23] and GOCE [24] provide unprecedented information for the medium-wavelength gravity field [25,26], which can improve the accuracy of medium- and long-wavelength geoid. The primary goal of the GOCE satellite is to derive about 1–2 cm geoid accuracy to a target resolution of about 100 km [27]. The GRACE/GOCE-based GFMs have the medium–long wavelength information with high precision; however, these models have certain omission errors. The omission error for geoid (quasi-geoid) height reaches about 32 cm for a GFM up to degree 200, according to Kaula's rule and the variance model of Tscherning and Rapp [28–31]. This omission error cannot be ignored for realizing the height datum unification. In addition to the omission errors of the GFMs, the spectral accuracy of the selected GFM, the uncertainties of GNSS-derived ellipsoidal heights, and the accumulation leveling errors must be considered to determine the vertical datum geopotential with a high accuracy.

The Chinese vertical datum is determined by the MSL observed by the Qingdao tide gauge station during the 27 years from 1952 to 1979. The latest first-order leveling network in China, which is based on the 1985 vertical datum and was completed and put into use in 2017, is the most accurate and practical national modern vertical control network to date [32]. It is used for transmitting normal height at national scale. The aim of the paper is to determine the vertical datum geopotential of China based on the GFM by utilizing the latest normal height results. Thus, the 1985 vertical datum in China can be connected to the IHRS, which can provide a unified height datum for the construction of global geographic information resources, ecological environmental protection, and scientific research. In order to derive more accurate height datum geopotential of China based on GFM, we utilize the spectral expansion method by augmenting the GOCE/GRACE-based GFMs with the EGM2008 model and residual terrain model (RTM) technique [33–37], which can reduce the omission error of the satellite-only GFMs. In addition, considering the large east–west span in China, the systematic accumulated error may occur in the long-distance leveling, and GFM have certain systematic errors between the east and west of China, which has influence on determining the geopotential value of the Chinese height datum. This paper will further analyze the systematic errors influence. Finally, we will combine the refined GFMs and GNSS/leveling to preliminarily derive the geopotential value of the Chinese height datum (CNHD) towards height system unification in China.

The structure of the manuscript is as follows. The materials and method for estimating the Chinese height datum geopotential value are introduced in Section 2. Section 3 provides the results of the Chinese height datum geopotential value and specifically focuses on (a) spectral accuracy of GFMs, (b) the omission errors of the satellite-only GFMs in China, and (c) the determination of the refined GFMs. Finally, discussion and conclusions are provided in Sections 4 and 5, respectively.

2. Materials and Methods

2.1. Materials

This section briefly introduces the materials used in this study. The main data required are (1) GNSS/leveling data; (2) global gravity field models; and (3) topographic data.

2.1.1. GNSS/LEVELING Data

The leveling networks in China contain the first-, second-, third-, and fourth-order leveling networks. The latest first-order leveling network observations, which were completed and placed into use in 2017, are used. This first-order leveling network consists of 148 loops with a length of over 125,746.5 km. Considering the large east–west span in China, the systematic accumulated error may occur in the long-distance leveling; therefore, the leveling data is handled by the least square adjustment in which the length correction of leveling rod, the non-parallel correction of level surface, and the gravity reduction are considered [32]. The maximum error in leveling is only about 3.57 cm (about 6185 km from Qingdao leveling origin) after adjustment. Finally, 1908 high-accuracy and evenly distributed GNSS/leveling datapoints are made available by the National Geomatics Center of China to determine the vertical datum geopotential of China. The GNSS ellipsoidal coordinates are based on ITRF2014 [38], and the GNSS coordinate accuracy reaches the millimeter level; in particular, the accuracy of GNSS ellipsoidal heights is about 5 mm. The distribution for the GNSS/leveling datapoints in China is shown in Figure 1.

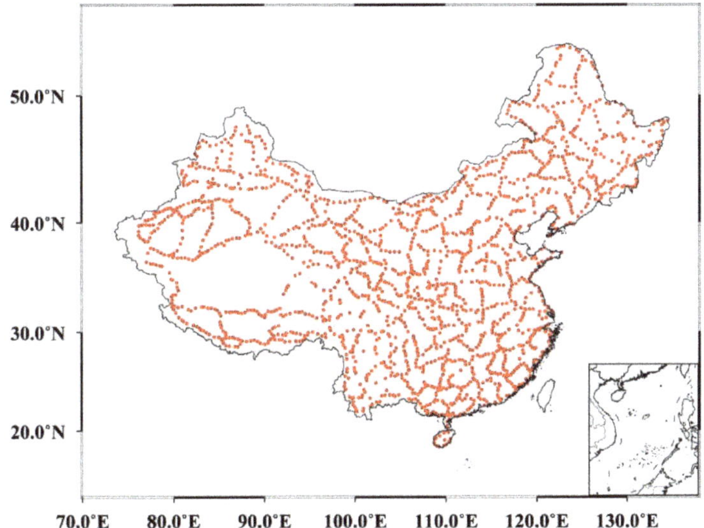

Figure 1. Distribution of GNSS/leveling benchmarks.

Because the GNSS ellipsoidal height is based on a tide-free system, to ensure that the GNSS/leveling-based quasi-geoid height represents a tide-free system, the normal heights are transformed to the tide-free system by the following equation [39]:

$$H^{TF} = H^{MF} + 0.68\left(0.099 - 0.296 \sin^2 \varphi\right) \tag{1}$$

where H^{TF} represents the tide-free normal height, H^{MF} is the mean tide normal height, and φ is the latitude of the GNSS/leveling points.

2.1.2. Global Gravity Field Models (GFMs)

Table 1 shows the GFMs used in this paper. The direct approach is employed to derive the DIR_R5 and DIR_R6 models using GOCE satellite gravity observations, GRACE satellite gravity observations, and satellite laser ranging (SLR). The DIR_R5 model utilizes GOCE data from November 2009–October 2013 and GRACE data from the ten-year period (2003–2012). The DIR_R6 model utilizes GOCE data from October 2009–October 2013 and GRACE data from January 2007–November 2014. The TIM_R5 and TIM_R6 models are derived by the time-wise approach using GOCE observations from November 2009–October 2013 and September 2009–October 2013, respectively. The EGM2008 model combines multi-source gravity observation data to derive a maximum degree of 2190; however, this model is complete to d/o 2159. The gravity data utilized in EGM2008 model mainly refers to global gravity database of $5' \times 5'$. The surface gravity data covering China is composed of two-divisions: one is the gravity data of $5'$ for construction of the EGM2008, without any restrictions, the main coverage areas include eastern, southern, and Central China; another gravity data resource is permitted to use at a resolution of 15 arc-minute, these data cover other regions except for eastern, southern, and Central China.

For the consistency of tide system, the $\overline{C}_{20}$ coefficient of GFM is transformed to tide-free system using the following formula [40]:

$$\overline{C}_{20}^{TF} = \overline{C}_{20}^{ZT} - k_{20} \cdot \langle \Delta \overline{C}_{20} \rangle \tag{2}$$

where $\overline{C}_{20}^{TF}$ and $\overline{C}_{20}^{ZT}$ are the spherical harmonic coefficient under the tide-free system and the zero-tide system, respectively, $k_{20} = 0.30190$ is loading Love number, and $\langle \Delta \overline{C}_{20} \rangle = -1.391412 \cdot 10^{-8}$ represents the value of tidal correction.

Table 1. The details for global gravity field models. The letter "S" in the third column represents satellite data, "G" represents ground observations, "A" represents altimetry observations, and "d/o" represents degree/order.

Models	d/o	Data	Tide System	Reference
EGM2008	2160	S(Grace), G, A	Tide-free	[41]
GO_CONS_GCF_2_TIM_R5	280	S(Goce)	Tide-free	[42]
GO_CONS_GCF_2_TIM_R6	300	S(Goce)	Zero-Tide	[43]
GO_CONS_GCF_2_DIR_R5	300	S(Goce, Grace, Lageos)	Tide-free	[44]
GO_CONS_GCF_2_DIR_R6	300	S(Goce, Grace, Lageos)	Tide-free	[45]

2.1.3. Topographic Data

Topographic data are used to calculate the RTM effect and recover the high-frequency gravity signals missing in the GFMs. The RTM represents the difference between the topographic surface and the long-wavelength reference terrain. The Shuttle Radar Topographic Mission (SRTM) V4.1 data [46] with a spatial resolution of $7.5'' \times 7.5''$ are used to represent the land over China. For the sea area, the SRTM15_PLUS V2 topographic data [47] are used, with a spatial resolution of $15'' \times 15''$. The rock-equivalent topography model RET2012 [48] works as the reference topography, and the reference terrain elevations can be computed by Equation (2) in [49]. In the coastal zone, the mass density of seawater is different from that of the standard topographic mass density. To avoid the necessity of distinguishing density changes in the computational process, the rock-equivalent topography (RET) method can be used to compress the water depth to the equivalent rock height [50].

$$H^* = H(1 - \rho_w/\rho) \tag{3}$$

where H^* and H are the compressed water depth and the original water depth, respectively, ρ_w denotes the density of seawater, and ρ represents the standard topographic mass density.

The SRTM and SRTM15_PLUS data can be merged to obtain the unified topographic data on land and sea. The merging process is mainly divided into two steps: (1) the bicubic interpolation method is employed to interpolate the sea topographic data into a spatial resolution of $7.5'' \times 7.5''$, and Equation (3) is employed to process the sea water depths; (2) the land topographic data are combined with the water depth data obtained by step (1). Figure 2a shows the merged topography. The residual topographic masses (RTM elevations) are the difference between the SRTM merged data (Figure 2a) and the reference terrain. Figure 2b shows the residual terrain model elevations.

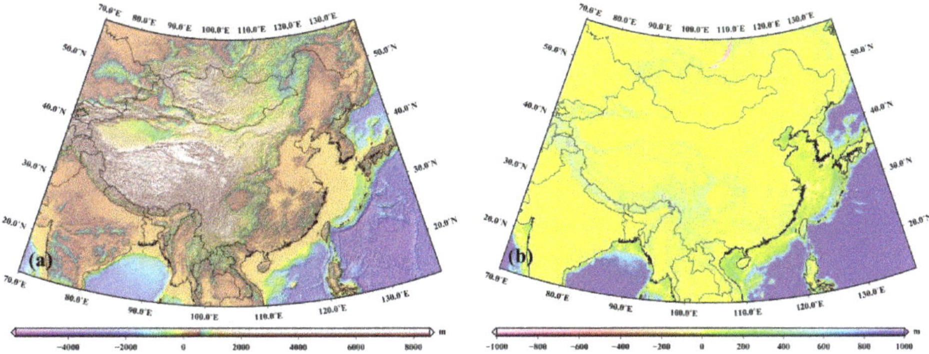

Figure 2. (**a**) Topography based on merged SRTM and (**b**) residual terrain model (RTM) elevations with RET2012 terrain as the reference surface.

2.2. Methods for Determining the Height Datum Geopotential Value

The vertical offset δH between the local height datum and the global geoid can be expressed as

$$\delta H = h - H - \zeta \tag{4}$$

where H represents the normal heights, h is the ellipsoidal heights, and ζ is the height anomalies from GFM (see Figure 3).

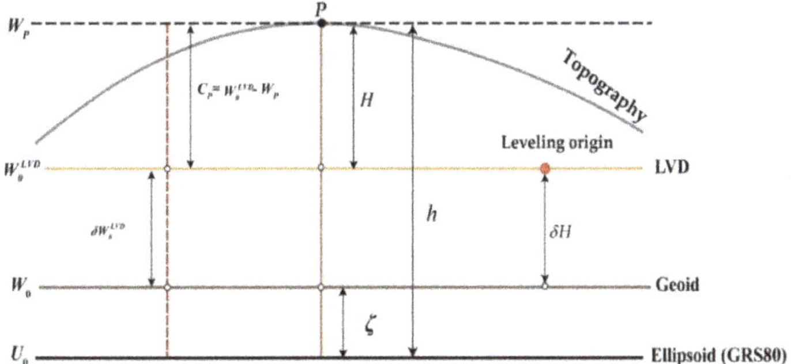

Figure 3. The relations among different reference datum surfaces.

The height anomaly ζ can be calculated by using the satellite-only GFMs, but the resulting omission errors cannot be ignored. Therefore, the EGM2008 model can be used to extend the satellite-only GFMs to obtain the refined GFMs via the spectral expansion approach. The height anomaly ζ_{GGM} determined by the refined GFMs can be expressed as

$$\zeta_{GGM} = \left(\frac{GM - GM_0}{r \cdot \overline{\gamma}} - \frac{W_0 - U_0}{\overline{\gamma}} \right) + \zeta_{GOCE/GRACE}\big|_2^l + \zeta_{EGM2008}\big|_{l+1}^{2160} \tag{5}$$

where $\zeta_{GOCE/GRACE}\big|_2^l$ is the height anomaly determined by the satellite-only GFMs truncated to the degree l, and $\zeta_{EGM2008}\big|_{l+1}^{2160}$ is the height anomaly represented from degrees 201 to 2160 of EGM2008. $GM_0 = 3.986005000 \times 10^{14}$ m^{3}s^{-2} and $U_0 = 62{,}636{,}860.8500$ m^{2}s^{-2} are constants of the gravitational constant and the normal potential value of the GRS80 ellipsoid [51], respectively; r is the geocentric radial for computation point; GM represents geocentric gravitational constant used in the GFM; $\overline{\gamma}$ is the mean normal gravity; and $W_0 = 62{,}636{,}853.4$ m^{2}s^{-2} is the geopotential value of the global geoid [52].

The RTM technology is used to recover the short-scale signal beyond degree 2160 in Equation (5). The RTM represents the difference (residual terrains) between the topographic surface and the reference surface. The gravitational potentials of residual terrains can be expressed as follows [53]:

$$T = G\rho \int_\Omega \frac{1}{r} d\Omega \tag{6}$$

where T is the gravitational effect for the residual terrains, G denotes the gravitational constant, r is the distance between the attraction mass and the computation point, Ω is the volume for the residual terrains, and ρ is the standard topographic mass density.

The residual terrains are decomposed into a set of rectangular-prism mass. Figure 4 shows the coordinate system definition of a rectangular-prism mass body. The computation point P is the origin of this coordinate system; this means that the coordinates defining have to be transformed by a shift (see Equation (1) in [54]). To obtain the gravitational effects of a residual mass distribution, the result of Equation (6) for a single rectangular-prism can be calculated using the following equation:

$$T(P) = G\rho \int_{x_1}^{x_2} \int_{y_1}^{y_2} \int_{z_1}^{z_2} \frac{1}{r} dx\,dy\,dz = G\rho \big|\big|\big| xy \ln(z+r) + yz \ln(x+r) + zx \ln(y+r) -$$
$$\frac{x^2}{2} \tan^{-1}\left(\frac{yz}{xr}\right) - \frac{y^2}{2} \tan^{-1}\left(\frac{xz}{yr}\right) - \frac{z^2}{2} \tan^{-1}\left(\frac{xz}{zr}\right) \big|_{x_1}^{x_2} \big|_{y_1}^{y_2} \big|_{z_1}^{z_2} \tag{7}$$

where $x_1, x_2, y_1, y_2, z_1,$ and z_2 describe the corner coordinates of the prism faces in Figure 4.

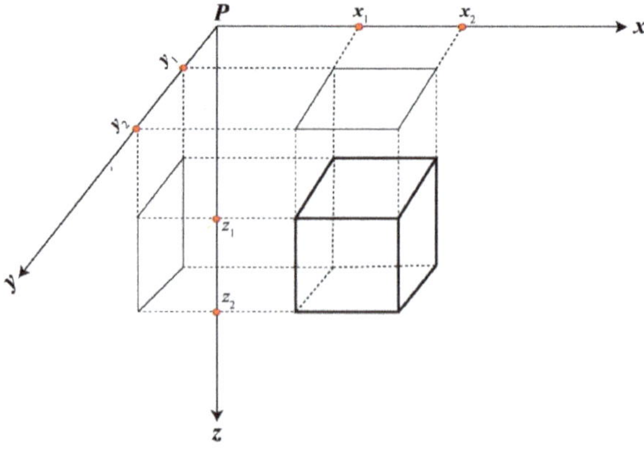

Figure 4. Definition of a prism mass body.

The total gravitational effects of the residual terrains for computation point are derived by a summation of gravitational effects in Equation (7):

$$T(P) = \sum_{i=1}^{M} T(P)_i \tag{8}$$

where M represents the number for prism mass elements.

The height anomaly ζ_{RTM} caused by the residual mass distribution can be expressed by

$$\zeta_{RTM} = \frac{T(P)}{\gamma_P} \tag{9}$$

where γ_P is normal gravity for computation point P.

ζ_{RTM} represents the quasi-geoid height signals of the GFM beyond degree 2160. Then, the height anomaly ζ determined by the refined GFMs can be further expressed as follows:

$$\zeta = \zeta_{GGM} + \zeta_{RTM} = \left(\frac{GM - GM_0}{r\overline{\gamma}} - \frac{W_0 - U_0}{\overline{\gamma}}\right) + \zeta_{GOCE/GRACE}\Big|_2^l + \zeta_{EGM2008}\Big|_{l+1}^{2160} + \zeta_{RTM} \tag{10}$$

The global or local vertical datum is a gravitational equipotential surface. Therefore, the vertical datum offset determined by Equation (4) should theoretically be a fixed constant. However, vertical offsets contain certain discrepancies due to the influences of systematic errors and random errors. Systematic errors may contain possible systematic errors in GFM, accumulated leveling errors and the GNSS height ellipsoidal errors [55–57]. To reduce these random and systematic errors, the vertical datum offset is estimated by applying a planar correction parametric model. For each GNSS/leveling point P, the observation equation can be formulated as follows:

$$l_P = h_P - H_P - \zeta_P = \delta\overline{H} + a_1(B_P - B_0) + a_2(L_P - L_0)\cos B_P \tag{11}$$

where $\delta\overline{H}$ is the unknown height datum offset, (B_P, L_P) are the geodetic coordinates for the computation point, (B_0, L_0) are the geodetic coordinates for leveling origin in the local vertical datum zone, and a_1 and a_2 are the north–south tilt and east–west tilt, respectively. If there are n GNSS/leveling benchmarks in the local vertical datum zone, according to Equation (11), the function model can be expressed as follows:

$$\underbrace{\begin{bmatrix} h_1 - H_1 - \zeta_1 \\ h_2 - H_2 - \zeta_2 \\ \vdots \\ h_n - H_n - \zeta_n \end{bmatrix}}_{l} = \underbrace{\begin{bmatrix} 1 & (B_1 - B_0) & (L_1 - L_0)\cos B_1 \\ 1 & (B_2 - B_0) & (L_2 - L_0)\cos B_2 \\ & \vdots & \\ 1 & (B_n - B_0) & (L_n - L_0)\cos B_n \end{bmatrix}}_{A} \cdot \underbrace{\begin{bmatrix} \delta\overline{H} \\ a_1 \\ a_2 \end{bmatrix}}_{x} \tag{12}$$

The unknown parameter x in Equation (12) can be further determined according to the least square adjustment of the system, denoted by

$$x = \left(A^T P A\right)^{-1} A^T P l \tag{13}$$

where $P = D_{ll}^{-1}$ is the weight matrix, D_{ll} represents the error variance–covariance matrix for the observed values. Assuming the involved terms in Equation (11) are uncorrelated to each other, the D_{ll} can be specified by

$$D_{ll} = D_{hh} + D_{HH} + D_{\zeta\zeta} \tag{14}$$

where D_{hh} and D_{HH} are the error variance–covariance matrix of the ellipsoidal and normal heights, respectively, and $D_{\zeta\zeta}$ represents the error variance–covariance matrix for quasi-geoid heights.

$D_{\zeta\zeta}$ might be determined from the errors of the GFMs and RTM quasi-geoid heights. Voigt and Denker [58] and Grombein et al. [17] showed that the uncertainty of the topography-implied gravity signals is at the sub-mm level; therefore, the errors of RTM quasi-geoid heights can be considered negligible. However, the error variance–covariance matrix of the GFM might generally not be available [17]. In addition, the uncertainties for ellipsoidal heights and normal heights are usually not available. Therefore, it is not possible to obtain D_{ll} in practical cases. Based on the above reasons, we assume $P = I$ in this paper, where I is the identity weight matrix.

After removing the errors effects, we will obtain vertical offsets with considering corrections by

$$\delta H_P = (h_P - H_P - \zeta_P) - a_1(B_P - B_0) - a_2(L_P - L_0)\cos B_P \tag{15}$$

The geopotential value $W_{P_0}^{LVD}$ for point P can be expressed as follows:

$$W_{P_0}^{LVD} = W_0 - \delta H_P \cdot \overline{\gamma}_P \tag{16}$$

where $\overline{\gamma}_P$ denotes the mean normal gravity, which is computed by Equations (4)–(60) in [59].

Finally, we can derive the geopotential value W_0^{LVD} of the local height datum by

$$W_0^{LVD} = W_0 - \frac{\sum\limits_{P=1}^{m} \delta H_P \cdot \overline{\gamma}_P}{m} = W_0 - \delta W_0^{LVD} \tag{17}$$

where m is the number of GNSS/leveling benchmarks.

3. Results

3.1. Spectral Accuracy Evaluation for GFMs

According to the spherical harmonic coefficients and their formal errors, the spectral accuracy of a GFM is evaluated by using the commission errors (the degree error, cumulative degree error, difference degree error, and cumulative difference degree error) and signal-to-noise ratio (SNR). The geoid degree error can be expressed as the square root of the error degree variances, and the cumulative geoid degree error is represented as the square root of the cumulative geoid error degree variances. A relative comparison of satellite-based GFM and EGM2008 can be estimated by the difference degree error and the cumulative difference degree error. These specific calculation formulas can be found in [60].

Figure 5a shows the geoid degree errors for GFMs. We can see from the figure that the geoid degree errors of the satellite-only GFMs before about degree 200 are lower than the degree error of EGM2008. Figure 5b shows the geoid cumulative degree errors for the GFMs. The geoid cumulative degree errors for DIR_R5 and DIR_R6 models are lower than that of the EGM2008 model in the whole spectral domain. The geoid cumulative degree errors for TIM_R5 and TIM_R6 models are lower than that of the EGM2008 model before approximately degree 260. The above analysis shows that the satellite-only GFMs have higher medium–long wavelength accuracy than the EGM2008 model, and the satellite-based GFMs are characterized by noise as increases in degree.

Figure 6a indicates the SNRs for the GFMs. The SNRs of satellite-only GFMs are better than that of the EGM2008 model before about degree 200. The results show that the satellite-only GFMs have a strong geoid signal in the medium–long wavelength band, and the EGM2008 model has a strong geoid signal in the short wavelength band. Figure 6b shows the difference degree error and cumulative difference degree error of the satellite-only GFMs, taking the EGM2008 model as a reference. We can see from the figure that the geoid difference degree errors for DIR_R5 and DIR_R6 are lower than geoid difference degree errors of the TIM_R5 and TIM_R6 before about degree 100, and cumulative geoid difference degree errors of the DIR_R5 and DIR_R6 are lower than cumulative difference degree errors of TIM_R5 and TIM_R6 before about degree 150. This is because the DIR_R5 and DIR_R6 GFM take advantage of GRACE gravity data and LAGEOS satellite laser ranging (SLR) data.

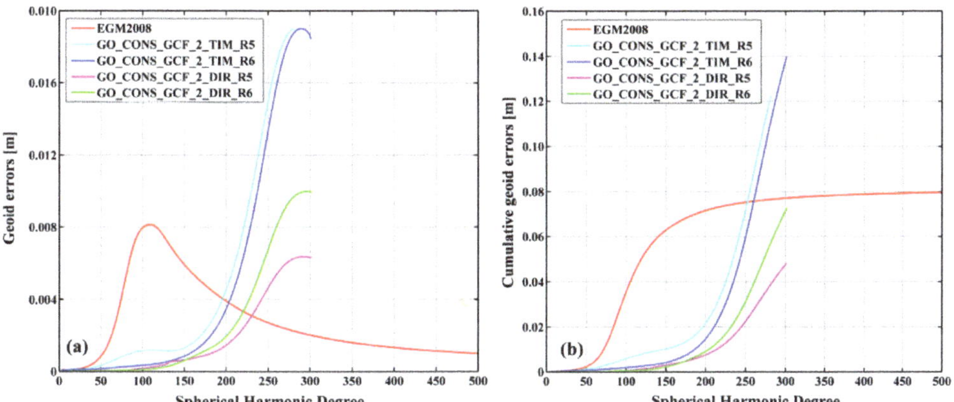

Figure 5. (**a**) Geoid degree error and (**b**) cumulative geoid degree errors for GFMs.

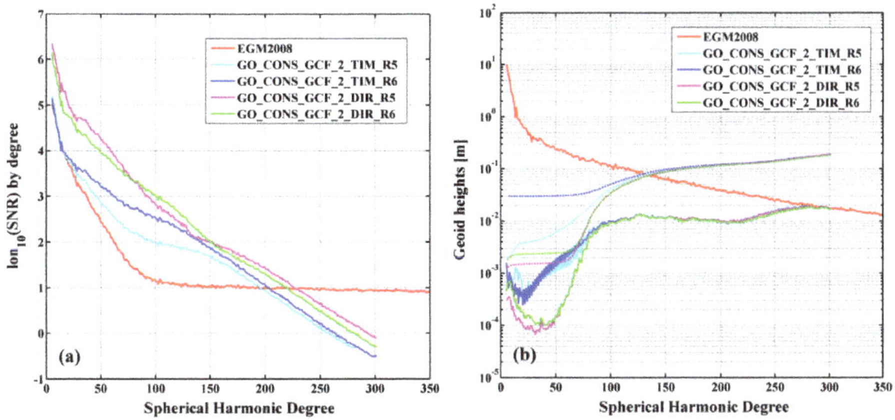

Figure 6. (**a**) Signal to noise ratios (SNRs) of GFMs and (**b**) difference degree errors (solid line) and difference cumulative degree errors (dashed line) between satellite-only GFMs and the EGM2008.

3.2. The Omission Errors for Satellite-Only GFMs

To quantify the magnitudes of these omission errors of satellite-only GFMs, we assume that 200 is the general expansion degree of the satellite-only GFM. The magnitudes of the height anomaly $\zeta_{201\text{-}2160}^{EGM2008}$ using the degrees 201 to 2160 of EGM2008 and the height anomaly ζ_{RTM} inferred from the RTM can indicate the omission errors of the satellite-only GFM. The specific calculation process for ζ_{RTM} can be found in Section 3.3.

Figure 7a indicates the omission error of satellite-only GFM represented by $\zeta_{201\text{-}2160}^{EGM2008}$, Figure 7b shows the omission errors of satellite-only GFM represented by the ζ_{RTM}, and Figure 7c shows the total omission errors obtained from the sum of $\zeta_{201\text{-}2160}^{EGM2008}$ and ζ_{RTM}. Table 2 show statistics for the omission errors of satellite-only GFM in China. We can see from the figure that the omission errors in the rugged regions are larger (mid-west of mainland China) than in other regions. From Table 2, we can see that the omission errors of the satellite-only GFM in mainland China reach the decimeter level, and the largest amplitude is about 350 cm. The omission error signals represented by the RTM are centimeter level in mainland China. Therefore, the omission errors for the satellite-only GFMs must be considered for unification of China vertical datum.

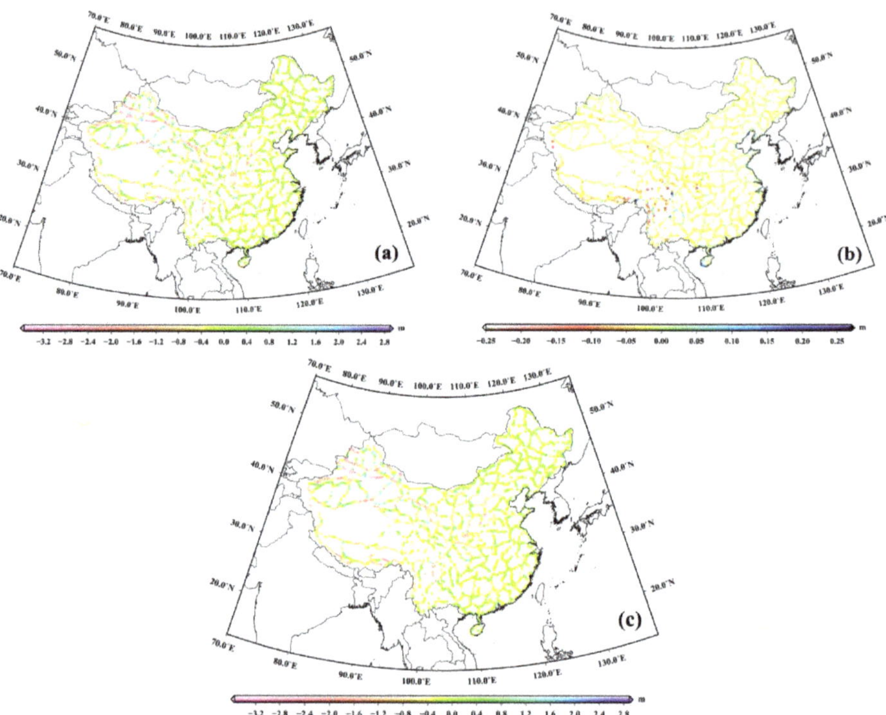

Figure 7. (**a**) Omission errors of satellite-only GFMs represented by $\zeta^{EGM2008}_{201\text{-}2160}$, (**b**) omission errors for satellite-only GFMs represented by ζ_{RTM}, (**c**) total omission errors of satellite-only GFMs.

Table 2. Statistics of the omission errors for satellite-only GFMs in China. Unit: m.

Omission Errors (m)	Max	Min	Mean	Std
$\zeta^{EGM2008}_{201\text{-}2160}$	2.885	−3.529	−0.215	0.601
ζ_{RTM}	0.264	−0.248	−0.006	0.027
$\zeta^{EGM2008}_{201\text{-}2160} + \zeta_{RTM}$	2.892	−3.548	−0.221	0.602

3.3. The Refined GFMs Obtained by the Spectral Expansion Approach

A spectral expansion approach is used to obtain the refined gravity field models using the EGM2008 (Earth Gravitational Model 2008) and residual terrain model (RTM) technique. Because the degree errors of the satellite-only GFMs increase with the increase in degree and order, the noise starts to dominate the signals at high degree and order. Therefore, it is not possible to derive an optimal GFM by combining the maximum degree of satellite-based GFMs and EGM2008. It is not a reasonable strategy to obtain a refined GFM by combining two models directly, and the choice of the optimal degree for the combination is very important. The specific steps used to determine the optimal combination degrees are as follows: (a) the satellite-only GFMs are truncated to degree l (l = 10, 20, 30 . . . . . . N, where N is the maximum degree for satellite-only GFMs) as the 0~l degree of the refined GFMs, and the l~2160 degree of the refined GMFs is supplemented by the corresponding degree of the EGM2008 model; (b) the GNSS/leveling-based height anomaly is used to check the height anomaly determined by the refined GFMs; (c) the combination degree presenting the best accuracy is chosen as the optimal combination degree of the refined GFMs. Figure 8 indicates the standard deviations of the height anomalies differences between GNSS/leveling and the refined GFMs with varying combinations of degree. As seen in Figure 8, the accuracy of the combined GFMs is basically consistent with that of the EGM2008 model before about degree 90. From degree 90 onwards, there are obvious differences in accuracy. We select the optimal combination degree when the standard deviations minimized. Figure 8 shows that 230, 240, 220, and 240 are the optimal combination degrees for obtaining the refined GFMs by combining the TIM_R5, TIM_R6, DIR_R5, and DIR_R6 models, respectively, with the EGM2008 model.

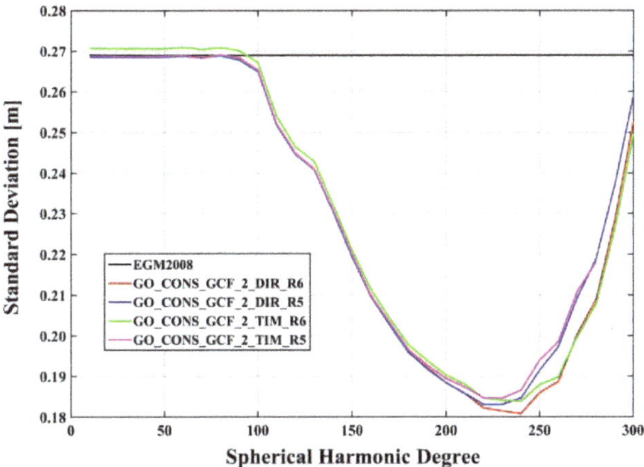

Figure 8. Standard deviations of the height anomalies differences between GNSS/leveling and the refined GFMs with varying expansion d/o.

The RTM technology can further be utilized to obtain higher frequency gravity field signals of the refined GFMs. The RTM gravitational potential is not harmonic when the computation point is below the reference elevation surface. In order to solve the problem of non-harmonic, a harmonic correction can usually be performed by downward continuation through a Bouguer plate [33,61]. Thus, the harmonic correction for quasi-geoid (or geoid) can be considered as zero; however, harmonic correction for gravity can be considered as $4\pi G\rho\Delta H$ (ΔH represents the height difference between the topography and reference terrain). The choice of the RTM integration radius is crucial. For RTM quasi-geoid height, an integration radius of ~200 km is suitable [49]. Therefore, an integration radius of ~200 km is used herein to determine the RTM quasi-geoid heights. According to unified topographic data (in Figure 2), we can obtain RTM quasi-geoid height by using Equations (7)–(9). Figure 9 shows the quasi-geoid contribution of the RTM with a resolution of $7.5'' \times 7.5''$, with maximum, minimum, mean, and standard deviation values of 0.301 m, -0.280 m, 0.001 m, and 0.020 m, respectively.

Table 3 shows the quasi-geoid accuracy statistics of different GFMs in China. From Table 3, without considering the influence of the RTM on the quasi-geoid height, the quasi-geoid accuracies of the refined GFMs in mainland China are better than 18.5 cm. Among them, the DIR_R6_EGM2008 model has the best accuracy, at 18.1 cm. Compared with the EGM2008 model, the quasi-geoid accuracy of the DIR_R6_EGM2008 model is improved by 8.8 cm. On the other hand, these results also indicate that the quasi-geoid medium-long wavelength accuracy of EGM2008 model in China is poor. Considering the influence of the RTM quasi-geoid, the mainland China quasi-geoid accuracies determined by the refined GFMs are better than 17.8 cm, and the DIR_R6_EGM2008_RTM model has the optimal quasi-geoid accuracy with a standard deviation of 17.3 cm. Compared with the EGM2008 model, the quasi-geoid accuracy of the DIR_R6_EGM2008_RTM model is improved by 9.6 cm. Compared with the TIM_R5, TIM_R6, DIR_R5, and DIR_R6 models, the quasi-geoid accuracy of the DIR_R6_EGM2008_RTM model is improved by 0.400 m, 0.391 m, 0.396 m, and 0.391 m to 0.178 m, 0.177 m, 0.176 m, and 0.173 m, respectively.

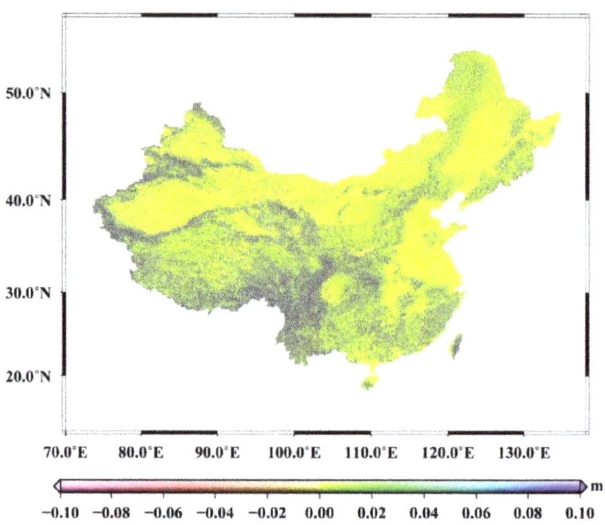

Figure 9. The RTM contribution to quasi-geoid height in China.

Table 3. Statistics of the height anomaly differences between GNSS/leveling and the GFMs. Unit: m.

Models	Max	Min	Mean	STD
EGM2008	1.162	−2.079	0.061	0.269
DIR_R6	2.624	−2.475	−0.114	0.391
DIR_R5	2.536	−2.507	−0.112	0.396
TIM_R6	2.578	−2.457	−0.107	0.391
TIM_R5	2.549	−2.432	−0.114	0.400
TIM_R5_EGM2008	1.044	−1.804	0.045	0.185
TIM_R6_EGM2008	1.099	−1.806	0.050	0.184
DIR_R5_EGM2008	1.007	−1.689	0.048	0.183
DIR_R6_EGM2008	1.072	−1.787	0.044	0.181
EGM2008_RTM	1.256	−1.981	0.068	0.261
TIM_R5_EGM2008_RTM	1.042	−1.803	0.053	0.178
TIM_R6_EGM2008_RTM	1.097	−1.806	0.058	0.177
DIR_R5_EGM2008_RTM	1.005	−1.688	0.056	0.176
DIR_R6_EGM2008_RTM	1.069	−1.787	0.052	0.173

To validate our results of the refined GFMs, the EIGEN-6C4 [62], GECO [63], SGG-UGM-1 [64], SGG-UGM-2 [65], XGM2016 [66], and XGM2019 [67] models are used. These six higher-degree GFMs further improve the medium–long wavelength by adding GOCE data.

Table 4 summarizes the statistics for height anomaly differences between GNSS/leveling and six higher-degree GFMs. These models in Table 4 perform better than the EGM2008 model in China, which can assume the largest impact from the contribution of GOCE solution. The EIGEN-6C4 outperforms GECO, SGG-UGM-1, and SGG-UGM-2 models in China, which is mainly due to the differences in use of GOCE data. Comparing Tables 3 and 4, we can find that the refined GFMs outperform EIGEN-6C4, GECO, SGG-UGM-1, XGM2019, XGM2016, and SGG-UGM-2 in mainland China; the major improvement of these models can be attributed to the GOCE data and topography signals.

Table 4. Statistics of the height anomaly differences between GNSS/leveling and six higher-degree GFMs. Unit: m.

Models	Max	Min	Mean	STD
EIGEN-6C4	1.007	−1.696	0.048	0.187
GECO	1.579	−1.703	0.041	0.223
SGG-UGM−1	1.003	−1.671	0.052	0.194
SGG-UGM−2	1.003	−1.704	0.051	0.191
XGM2019	1.705	−1.737	0.081	0.213
XGM2016	1.016	−1.757	−0.020	0.214

3.4. Determination for the Geopotential Value of Chinese Height Datum

The geopotential $W_0 = 62{,}636{,}853.4 \ \mathrm{m^2s^{-2}}$ adopted as reference level for the IHRS is used herein. Based on the GFM, the vertical datum geopotential value for China is determined. Thus, the vertical datum in China is connected into the IHRS.

In the analysis presented in Section 3.3, we conclude that the DIR_R6_EGM2008_RTM model has optimal quasi-geoid accuracy in China. Therefore, we choose this model to derive the Chinese height datum geopotential value. According to Equation (4), we can obtain vertical offset values of each GNSS/leveling point. Figure 10a represents the height anomaly differences between the GNSS/leveling and the DIR_R6_EGM2008_RTM model in China, thus providing the spatial distribution for the vertical datum offsets. However, it can be found from Figure 10a that the discrepancies exist in mainland China. The vertical offsets have certain discrepancies due to the influence of various factors. Among them, systematic error is a major contributor. Therefore, the planar corrections surface is used further through Equation (11). Then, we can derive vertical offsets after applying the correction surfaces by Equation (15). Figure 10b shows correction surface at GNSS-leveling points. Figure 10c shows vertical offsets after applying the correction surfaces. It can be seen from Figure 10b that there are certain east–west systematic errors in China, and the maximum error is about 12.2 cm. Systematic errors mainly contain errors in the GFM and leveling, The maximum first-order leveling error in China is only about 3.57 cm after height networks adjustment [32]. Therefore, we can conclude that the major systematic effect contributor to the estimation of the height datum geopotential values of China mainly comes from GFM error, and the systematic errors from GFM are more obvious in western China.

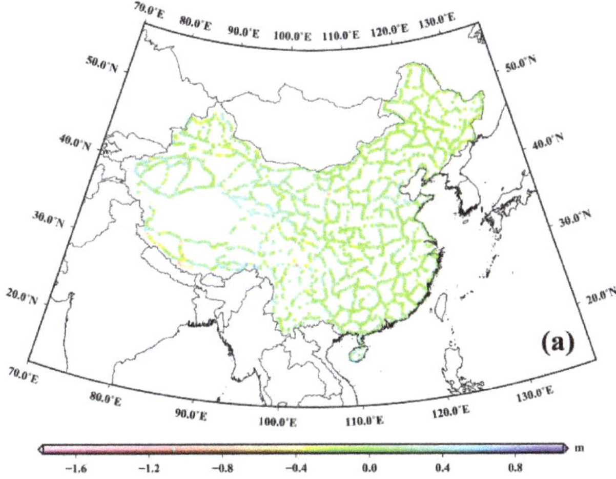

Figure 10. *Cont.*

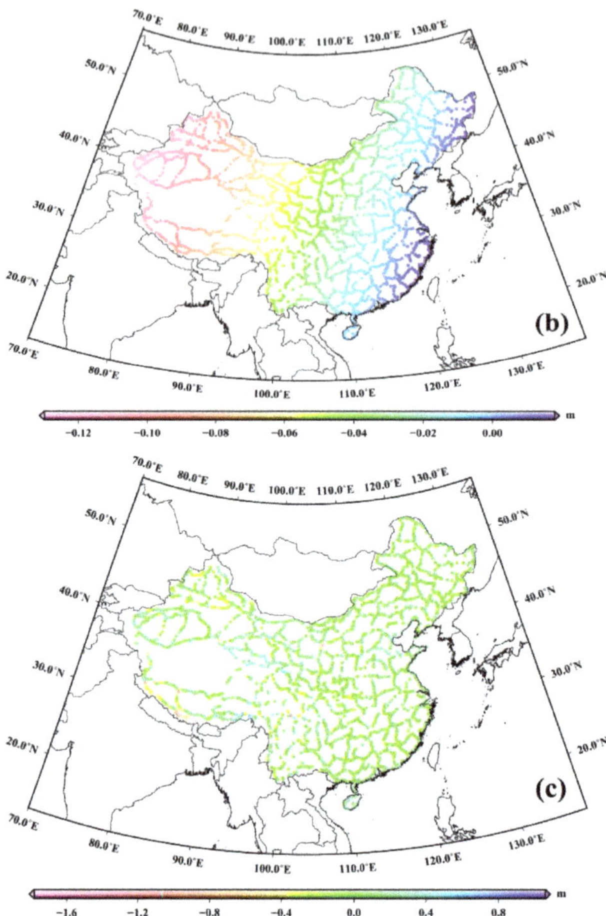

Figure 10. (**a**) Vertical offsets distribution without applying planar correction surface, (**b**) correction surface at GNSS-leveling stations, and (**c**) vertical offsets distribution after applying the planar correction.

The systematic errors influence on the determination of the Chinese height datum geopotential values is evaluated. The results with and without consideration of systematic error effect are compared, as shown in Table 5. Table 5 presents the numerical results of both scenarios based on the DIR_R6_EGM2008_RTM model. As shown in Table 5, we can see that there is a minor improvement of 0.06 m^{2}s^{-2} for DIR_R6_EGM2008_RTM model when considering the planar corrections. Finally, the geopotential value for the Chinese height datum is derived to be equal to 62,636,853.29 m^{2}s^{-2} based on the DIR_R6_EGM2008_RTM model, considering planar corrections.

Table 5. Statistics of the without-planar corrections and with-planar corrections scenarios for the height datum geopotential value in China based on DIR_R6_EGM2008_RTM model. Unit: m^{2}s^{-2}.

Model	Scenarios	Max	Min	Mean
DIR_R6_EGM2008_RTM	Without-planar corrections	62,636,870.86	62,636,842.93	62,636,852.89 ± 1.75
	With-planar corrections	62,636,870.88	62,636,843.31	62,636,853.29 ± 1.69

4. Discussion

The satellite-only GFMs have high spectral accuracy and strong geoid signals. However, the maximum expansion degree of these satellite-based models is limited and there are certain omission errors as a result of the gravity field attenuation at the height of satellite orbit. From Figure 7 and Table 2, we can see that the omission errors of the satellite-only GFM in mainland China reach the decimeter level, and the largest amplitude is about 350 cm. The omission error signals represented by the RTM are centimeter level in mainland China. Therefore, the omission errors for the satellite-only GFMs must be considered for unification of China vertical datum. The satellite-only GFMs have higher spectral accuracy and stronger geoid signals at medium–long wavelength, and EGM2008 model has stronger geoid signals at short wavelength. Therefore, combining the GRACE/GOCE-based GFMs and EGM2008 model to obtain refined GFMs is a feasible strategy.

We combine the GRACE/GOCE-based GFMs and EGM2008 model to derived refined GFMs. The accuracy trend of the refined GFMs in Figure 8 mainly depends on the EGM008 model, the satellite-only GFMs, and GNSS/leveling resources. The GNSS/leveling have a good accuracy and quality; especially, the systematic errors in leveling have been greatly weakened by height network adjustment. Therefore, the results presented in Figure 8 mainly reflect the accuracy of the combined GFMs in China. The accuracy of the combined GFMs relies heavily on the improvement of the satellite-only models. Gruber and Willberg [56] demonstrated that 80% accuracy improvement for high-resolution GFMs compared with EGM2008 reveal the contribution of GOCE solution to medium wavelengths. Therefore, high-quality GNSS/leveling resources and the satellite-only GFMs increase the quality and reliability of the combined GFMs in this study. In the combination process, the satellite-only GFM using GOCE, GRACE, and LAGEOS laser ranging data should be used, which can better meet the quasi-geoid medium–long wavelength signal.

The RTM technology can further be utilized to obtain higher-frequency gravity field signals of the refined GFMs. In the spectral expansion process, the spatial resolution of the refined GFMs obtained by combining the satellite-only GFMs and EGM2008 model is 5′. Because the RET2012 reference topographic model has the same resolution as the refined GFMs, the reference model serves as a high-pass filter that can filter out low-frequency features from the SRTM data. As a result, the residual terrain height can imply gravity field signals at shorter scales than the spatial resolution of the refined GFMs and further compensate for the omission error of the refined GFMs. It can be seen from Figure 9 that larger quasi-geoid signals are strongly correlated with topography. However, it should be noted that the RTM quasi-geoid signals may contain some possible implications due to density anomaly and the uncertainties in the harmonic correction [48].

Finally, we use the refined GFMs to preliminarily derive the geopotential value of the Chinese height datum by utilizing the latest normal height results. The refined GGMs provide obvious accuracy improvement advantage and provide guidance for developing a high-accuracy quasi-geoid in China. Prior to this work, He et al. [20] determined the geopotential value of China's height datum as 62,636,853.47 m^2s^{-2}, which fits well with our value of 62,636,853.29 m^2s^{-2}, considering the standard deviation. The geopotential value of Chinese height datum is determined by utilizing the latest national first-order leveling network and GNSS points in this paper, which are representative to some extent.

We can expect that the refined GFMs can provide a guidance for determining the quasi-geoid or the geopotential value of the Chinese height datum. However, the combination of the satellite-only GFM with the EGM2008 in this study is by a pure complementation of the spherical harmonic coefficients at a specific degree; however, such a procedure might cause a spectral gap between both models, which is different from rigorous combination that is carried out on the basis of the normal equations and covariance by a least-squares. In the next step, the rigorous combination will be considered to derive the refined GFMs. In addition, the well-distributed surface gravity data for determining the geopotential value of the Chinese height datum also play a crucial role and need to be considered further.

5. Conclusions

In this paper, we used a spectral expansion to derive the refined GFMs by combining the EGM2008 and the satellite-only GFMs. The results show that 230, 240, 220, and 240 are the optimal combination degrees for determining the refined GFMs by combining the TIM_R5, TIM _R6, DIR_R5, and DIR_R6 models with the EGM2008 model, respectively. To consider the influence of higher-frequency gravity field signals caused by topography, the RTM is utilized to further compensate for the omission errors in the refined GFMs. The mainland China quasi-geoid accuracies determined by the refined GFMs are better than 17.8 cm, and the DIR_R6_EGM2008_RTM model has an optimal quasi-geoid accuracy with of 17.3 cm. To validate our results of the refined GFMs, the EIGEN-6C4, GECO, SGG-UGM-1, and SGG-UGM-2, XGM2019, and XGM2016 models were used. The results show that the refined GFMs outperform EIGEN-6C4, GECO, SGG-UGM-1, XGM2019, XGM2016, and SGG-UGM-2 in mainland China; the major improvement of these models can be attributed to the GOCE data and topography signals.

The systematic error effects for determining the geopotential value of the Chinese height datum are considered. The results show that the refined GFMs show minor improvements when considering the planar corrections. The major systematic effect contributor for determining the Chinese height datum geopotential values mainly comes from GFM error. Finally, the Chinese height datum geopotential value is derived to be equal to $62,636,853.29 \ \mathrm{m^2 s^{-2}}$ based on the refined DIR_R6_EGM2008_RTM model when considering planar corrections.

Author Contributions: Conceptualization, P.Z. (Panpan Zhang) and L.B.; methodology, P.Z. (Panpan Zhang) and L.B.; software, P.Z. (Panpan Zhang) and Z.L.; validation, P.Z. (Panpan Zhang); formal analysis, P.Z. (Panpan Zhang) and L.B.; investigation, all; resources, Z.L.; writing—original draft preparation, P.Z. (Panpan Zhang); writing—review and editing, L.B., Z.L. and L.W.; visualization, P.Z. (Panpan Zhang); supervision, L.W., L.B. and Z.L.; funding acquisition, L.B. and Z.L. All authors have read and agreed to the published version of the manuscript.

Funding: The research was funded by the National Natural Science Foundation of China (Grant Nos. 42174102, 42192535, and 41931076), the Project supported by the Open Fund of Hubei Luojia Laboratory, the Basic Frontier Science Research Program of Chinese Academy of Sciences (Grant No. ZDBS-LY-DQC028), and the National Key Research and Development Program of China (Grant No. 2016YFB0501405).

Institutional Review Board Statement: Not applicable.

Informed Consent Statement: Not applicable.

Data Availability Statement: The global Earth Models can be downloaded from ICGEMs (http://icgem.gfz-potsdam.de/tom_longtime/, accessed on 30 April 2021) and SRTM data can be derived from NASA (https://srtm.csi.cgiar.org/, accessed on 20 April 2021). In addition, the GNSS/leveling data can be obtained request up on the author.

Acknowledgments: We would thank the International Centre for Global Earth Models and NASA Shuttle Radar Topographic Mission for providing us with the GFMs and the digital terrain model data, respectively.

Conflicts of Interest: The authors declare no conflict of interest.

References

1. Barzaghi, R.; De Gaetani, C.I.; Betti, B. The worldwide physical height datum project. *Rend. Fis. Acc. Lincei.* **2020**, *31*, 27–34. [CrossRef]
2. Sánchez, L.; Ågren, J.; Huang, J.; Wang, Y.M.; Mäkinen, J.; Denker, H.; Ihde, J.; Abd-Elmotaal, H.; Ahlgren, K.; Amos, M.; et al. Advances in the realisation of the International Height Reference System. In Proceedings of the IUGG General Assembly, Rio de Janeiro, Brazil, 12–14 November 2019.
3. Sánchez, L.; Barzaghi, R. Activities and plans of the GGOS Focus Area Unified Height System. In Proceedings of the IUGG XXVII General Assembly, Montreal, QC, Canada, 14 July 2019.

4. Drewes, H. Reference Systems, Reference Frames, and the Geodetic Datum-Basic Considerations. In *Observing Our Changing Earth*; Sideris, M.G., Ed.; Springer: Perugia, Italy, 2009; Volume 133, pp. 3–9.

5. Drewes, H.; Kuglitsch, F.l.; Adám, J.; Rózsa, S. The Geodesist's Handbook 2016. *J. Geod.* **2016**, *90*, 907–1205. [CrossRef]

6. Ihde, J.; Sánchez, L.; Barzaghi, R.; Drewes, H.; Foerste, C.; Gruber, T.; Liebsch, G.; Marti, U.; Pail, R.; Sideris, M. Definition and proposed realization of the international height reference system (IHRS). *Surv. Geophys.* **2017**, *38*, 549–570. [CrossRef]

7. Sánchez, L.; Ågren, J.; Huang, J.; Wang, Y.M.; Mäkinen, J.; Pail, R.; Barzaghi, R.; Vergos, G.S.; Ahlgren, K.; Liu, Q. Strategy for the realisation of the International Height Reference System (IHRS). *J. Geod.* **2021**, *95*, 33. [CrossRef]

8. Gerlach, C.; Rummel, R. Global height system unification with GOCE: A simulation study on the indirect bias term in the GBVP approach. *J. Geod.* **2013**, *87*, 57–67. [CrossRef]

9. Amjadiparvar, B.; Rangelova, E.; Sideris, M.G. The GBVP approach for vertical datum unification: Recent results in North America. *J. Geod.* **2016**, *90*, 45–63. [CrossRef]

10. Ophaug, V.; Gerlach, C. On the equivalence of spherical splines with least-squares collocation and stokes's formula for regional geoid computation. *J. Geod.* **2017**, *91*, 1367–1382. [CrossRef]

11. Sánchez, L.; Sideris, M.G. Vertical datum unification for the International Height Reference System (IHRS). *Geophys. J. Int.* **2017**, *209*, 570–586. [CrossRef]

12. Ebadi, A.; Ardalan, A.; Karimi, R. The Iranian height datum offset from the GBVP solution and spirit-leveling/gravimetry data. *J. Geod.* **2019**, *93*, 1207–1225. [CrossRef]

13. Zhang, P.; Bao, L.; Guo, D.; Wu, L.; Li, Q.; Liu, H.; Xue, Z.; Li, Z. Estimation of Vertical Datum Parameters Using the GBVP Approach Based on the Combined Global Geopotential Models. *Remote Sens.* **2020**, *12*, 4137. [CrossRef]

14. Hayden, T.; Amjadiparvar, B.; Rangelova, E.; Sideris, M.G. Estimating Canadian vertical datum offsets using GNSS/levelling benchmark information and GOCE global geopotential models. *J. Geod. Sci.* **2012**, *2*, 257–269. [CrossRef]

15. Gruber, T.; Gerlach, C.; Haagmans, R. Intercontinental height datum connection with GOCE and GPS-levelling data. *J. Geod. Sci.* **2012**, *2*, 270–280. [CrossRef]

16. Gomez, M.E.; Pereira, R.A.D.; Ferreira, V.G.; Del Cogliano, D.; Luz, R.T.; de Freitas, S.R.C.; Farias, C.; Perdomo, R.; Tocho, C.; Lauria, E.; et al. Analysis of the Discrepancies between the Vertical Reference Frames of Argentina and Brazil. In *IAG 150 Years*; Rizos, C., Willis, P., Eds.; Springer International Publishing: Cham, Switzerland, 2015; Volume 143, pp. 289–295.

17. Grombein, T.; Seitz, K.; Heck, B. On High-Frequency Topography-Implied Gravity Signals for a Height System Unification Using GOCE-based Global Geopotential Models. *Surv. Geophys.* **2016**, *38*, 443–477. [CrossRef]

18. Li, J.; Chu, Y.; Xu, X. Determination of Vertical Datum Offset between the Regional and the global Height Datum. *Acta Geod. Cartogr. Sin.* **2017**, *46*, 1262–1273.

19. Vergos, G.S.; Erol, B.; Natsiopoulos, D.A.; Grigoriadis, V.N.; Tziavos, I.N. Preliminary results of GOCE-based height system unification between Greece and turkey over marine and land areas. *Acta. Geod. Geophys.* **2018**, *53*, 61–79. [CrossRef]

20. He, L.; Chu, Y.; Xu, X.; Zhang, T. Evaluation of the GRACE/GOCE Global Geopotential Model on estimation of the geopotential value for the China vertical datum of 1985. *Chin. J. Geophys.* **2019**, *62*, 2016–2026.

21. Kelly, C.I.; Andam-Akorful, S.A.; Hancock, C.; Laari, P.; Ayer, J. Global gravity models and the Ghanaian Vertical Datum: Challenges of a proper definition. *Surv. Rev.* **2019**, *53*, 44–54. [CrossRef]

22. Zhang, P.; Bao, L.; Guo, D.; Li, Q. Estimation of the height datum geopotential value of Hong Kong using the combined global geopotential models and GNSS/levelling data. *Surv. Rev.* **2021**, 1–11. [CrossRef]

23. Tapley, B.D.; Bettadpur, S.; Watkins, M.; Reigber, C. The gravity recovery and climate experiment: Mission overview and early results. *Geophy. Res. Lett.* **2004**, *31*, L09607. [CrossRef]

24. Drinkwater, M.R.; Floberghagen, R.; Haagmans, R.; Muzi, D.; Popescu, A. GOCE: ESA's first earth explorer core mission. In *Earth Gravity Field from Space—From Sensors to Earth Science*; Beutler, G., Ed.; Kluwer Academic Publishers: Bern, Switzerland, 2003; Volume 108, pp. 419–432.

25. Hirt, C.; Gruber, T.; Featherstone, W.E. Evaluation of the first GOCE static gravity field models using terrestrial gravity, vertical deflections and EGM2008 quasigeoid heights. *J. Geod.* **2011**, *85*, 723–740. [CrossRef]

26. Tziavos, I.N.; Vergos, G.S.; Grigoriadis, V.N.; Tzanou, E.A.; Natsiopoulos, D.A. Validation of GOCE/GRACE Satellite Only and Combined Global Geopotential Models over Greece in the Frame of the GOCESeaComb Project. In *IAG 150 Years*; Rizos, C., Willis, P., Eds.; Springer International Publishing: Cham, Switzerland, 2015; Volume 143, pp. 297–304.

27. Gruber, T.; Visser, P.N.A.M.; Ackermann, C.; Hosse, M. Validation of GOCE gravity field models by means of orbit residuals and geoid comparisons. *J. Geod.* **2011**, *85*, 845–860. [CrossRef]

28. Kaula, W.M. *Theory of Satellite Geodesy*; Blaisdell: Toronto, OH, USA, 1966.

29. Forsberg, R. Modelling of the fine-structure of the geoid: Methods, data requirements and some results. *Surv. Geophys.* **1993**, *14*, 403–418. [CrossRef]

30. Denker, H. Regional Gravity Field Modeling: Theory and Practical Results. In *Sciences of Geodesy—II Innovations and Future Developments*; Xu, G., Ed.; Springer: Heidelberg, Germany, 2013; pp. 185–291.

31. Tscherning, C.C.; Rapp, R.H. *Closed Covariance Expressions for Gravity Anomalies, Geoid Undulations, and Deflections of the Vertical Implied by Anomaly Degree Variance Models*; Reports of the Department of Geodetic Science; The Ohio State University: Columbus, OH, USA, 1974.

32. Wang, W.; Guo, C.; Li, D.; Zhao, H. Elevation change analysis of the national first order leveling points in recent 20 years. *Acta Geod. Cartogr. Sin.* **2019**, *48*, 1–8.

33. Forsberg, R. *A Study of Terrain Reductions, Density Anomalies and Geophysical Inversion Methods in Gravity Field Modelling*; Scientific Report No.5; The Ohio State University: Columbus, OH, USA, 1984.

34. Hirt, C.; Featherstone, W.E.; Marti, U. Combining EGM2008 and SRTM/DTM2006.0 residual terrain model data to improve quasigeoid computations in mountainous areas devoid of Gravity Data. *J. Geod.* **2010**, *84*, 557–567. [CrossRef]

35. Yang, M.; Hirt, C.; Wu, B.; Deng, X.; Tsoulis, D.; Feng, W.; Wang, C.; Zhong, M. Residual Terrain Modelling: The Harmonic Correction for Geoid Heights. *Surv. Geophys.* **2022**. [CrossRef]

36. Hirt, C.; Bucha, B.; Yang, M.; Kuhn, M. A numerical study of residual terrain modelling (RTM) techniques and the harmonic correction using ultra-high degree spectral gravity modelling. *J. Geod.* **2019**, *93*, 1469–1486. [CrossRef]

37. Yang, M.; Hirt, C.; Rexer, M.; Pail, P.; Yamazaki, D. The tree canopy effect in gravity forward modelling. *Geophys. J. Int.* **2019**, *219*, 271–289. [CrossRef]

38. Altamimi, Z.; Rebischung, P.; Métivier, L.; Collilieux, X. ITRF2014: A new release of the International Terrestrial Reference Frame modeling nonlinear station motions. *J. Geophys. Res. Solid Earth* **2016**, *121*, 6109–6131. [CrossRef]

39. Ekman, M. Impacts of geodynamic phenomena on systems for height and gravity. *Bull. Géod.* **1989**, *63*, 281–296. [CrossRef]

40. Gruber, T.; Abrikosov, O.; Hugentobler, U. GOCE Standards. Prepared by the European GOCE Gravity Consortium EGG-C. 2014. Available online: https://earth.esa.int/documents/10174/1650485/GOCE_Standards (accessed on 1 June 2021).

41. Pavlis, N.K.; Holmes, S.A.; Kenyon, S.C.; Factor, J.K. The development and evaluation of the earth gravitational model 2008(EGM2008). *J. Geophys. Res.* **2012**, *117*, B04406. [CrossRef]

42. Brockmann, J.M.; Zehentner, N.; Höck, E.; Pail, R.; Loth, I.; Mayer-Gürr, T.; Schuh, W.-D. EGM_TIM_RL05: An independent geoid with centimeter accuracy purely based on the GOCE mission. *Geophys. Res. Lett.* **2014**, *41*, 8089–8099. [CrossRef]

43. Brockmann, J.M.; Schubert, T.; Schuh, W.D. An Improved Model of the Earth's Static Gravity Field Solely Derived from Reprocessed GOCE Data. *Surv. Geophys.* **2021**, *42*, 277–316. [CrossRef]

44. Bruinsma, S.L.; Förste, C.; Abrikosov, O.; Lemoine, J.-M.; Marty, J.-C.; Mulet, S.; Rio, M.-H.; Bonvalot, S. ESA's satellite-only gravity field model via the direct approach based on all GOCE data. *Geophys. Res. Lett.* **2014**, *41*, 7508–7514. [CrossRef]

45. Förste, C.; Abrykosov, O.; Bruinsma, S.; Dahle, C.; König, R.; Lemoine, J.-M. ESA's Release 6 GOCE Gravity Field Model by Means of the Direct Approach Based on Improved Filtering of the Reprocessed Gradients of the Entire Mission (GO_CONS_GCF_2_DIR_R6). Available online: https://doi.org/10.5880/ICGEM.2019.004 (accessed on 14 May 2021).

46. Jarvis, A.; Reuter, H.I.; Nelson, A.; Guevara, E. Hole-Filled SRTM for the Globe Version 4. CGIAR-SXI SRTM 90 m Database. 2008. Available online: http://srtm.csi.cgiar.org (accessed on 14 April 2021).

47. Tozer, B.; Sandwell, D.T.; Smith, W. Global Bathymetry and Topography at 15 Arc Sec: SRTM15+. *Earth Space Sci.* **2019**, *6*, 1847–1864. [CrossRef]

48. Hirt, C.; Claessens, S.; Fecher, T.; Kuhn, M.; Pail, R.; Rexer, M. New ultrahigh-resolution picture of earth's gravity field. *Geophy. Res. Lett.* **2013**, *40*, 4279–4283. [CrossRef]

49. Hirt, C. RTM gravity forward-modeling using topography/bathymetry data to improve high-degree global geopotential models in the coastal zone. *Mar. Geod.* **2013**, *36*, 183–202. [CrossRef]

50. Hirt, C.; Kuhn, M.; Featherstone, W.E.; Göttl, F. Topographic/isostatic evaluation of new-generation GOCE gravity field models. *J. Geophys. Res.* **2012**, *117*, B05407. [CrossRef]

51. Moritz, H. Geodetic reference system 1980. *J. Geod.* **2000**, *74*, 128–133. [CrossRef]

52. Sánchez, L.; Čunderlík, R.; Dayoub, N.; Mikula, K.; Minarechová, Z.; Šíma, Z.; Vatrt, V.; Vojtíšková, M. A conventional value for the geoid reference potential W0. *J. Geod.* **2016**, *90*, 815–835. [CrossRef]

53. Yang, M.; Hirt, C.; Tenzer, R.; Pail, R. Experiences with the use of mass-density maps in residual gravity forward modelling. *Stud. Geophys. Geod.* **2018**, *62*, 596–623. [CrossRef]

54. Nagy, D.; Papp, G.; Benedek, J. The gravitational potential and its derivatives for the prism. *J. Geod.* **2000**, *74*, 552–560. [CrossRef]

55. Kotsakis, C.; Katsambalos, K.; Ampatzidis, D. Estimation of the zero-height geopotential level W0 in a local vertical datum from inversion of co-located GPS, leveling and geoid heights: A case study in the Hellenic islands. *J. Geod.* **2012**, *86*, 423–439. [CrossRef]

56. Gruber, T.; Willberg, M. Signal and error assessment of GOCE based high resolution gravity field models. *J. Geod. Sci.* **2019**, *9*, 71–86. [CrossRef]

57. Vu, D.T.; Bruinsma, S.; Bonvalot, S.; Remy, D.; Vergos, G.S. A Quasigeoid-Derived Transformation Model Accounting for Land Subsidence in the Mekong Delta towards Height System Unification in Vietnam. *Remote Sens.* **2020**, *12*, 817. [CrossRef]

58. Voigt, C.; Denker, H. Validation of GOCE gravity field models in Germany. In *Assessment of GOCE Geopotential Models*; Huang, J., Reguzzoni, M., Gruber, T., Eds.; Newton's Bulletin: Heidelberg, Germany, 2015; pp. 37–48.

59. Hofmann-Wellenhof, B.; Moritz, H. *Physical Geodesy*; Springer: Wien, WI, USA, 2006.

60. Ustun, A.; Abbak, R. On global and regional spectral evaluation of global geopotential models. *J. Geophys. Eng.* **2010**, *7*, 369. [CrossRef]

61. Forsberg, R.; Tscherning, C.C. Topographic effects in gravity field modelling for BVP. In *Geodetic Boundary Value Problems in View of the One Centimeter Geoid*; Sansó, F., Rummel, R., Eds.; Springer: Heidelberg, Germany, 1997; pp. 241–272.

62. Förste, C.; Bruinsma, S.L.; Abrykosov, O.; Lemoine, J.-M.; Marty, J.C.; Flechtner, F.; Balmino, G.; Barthelmes, F.; Biancale, R. *EIGEN-6C4 The Latest Combined Global Gravity Field Model Including GOCE Data up to Degree and Order 2190 of GFZ Potsdam and GRGS Toulouse*; GFZ Data Services: Potsdam, Germany, 2014. [CrossRef]
63. Gilardoni, M.; Reguzzoni, M.; Sampietro, D. GECO: A global gravity model by locally combining GOCE data and EGM2008. *Stud. Geophys. Geod.* **2016**, *60*, 228–247. [CrossRef]
64. Liang, W.; Xu, X.; Li, J.; Zhu, G. The determination of an ultrahigh gravity field model SGG-UGM-1 by combining EGM2008 gravity anomaly and GOCE observation data. *Acta Geod. Cartogr. Sin.* **2018**, *47*, 425–434.
65. Liang, W.; Li, J.; Xu, X.; Zhang, S.; Zhao, Y. A High-Resolution Earth's Gravity Field Model SGG-UGM-2 from GOCE, GRACE, Satellite Altimetry, and EGM2008. *Engineering* **2020**, *6*, 860–878. [CrossRef]
66. Pail, R.; Fecher, T.; Barnes, D.; Factor, J.F.; Holmes, S.A.; Gruber, T.; Zingerle, P. Short note: The experimental geopotential model xgm2016. *J. Geod.* **2018**, *92*, 443–451. [CrossRef]
67. Zingerle, P.; Pail, R.; Gruber, T.; Oikonomidou, X. The combined global gravity field model XGM2019e. *J. Geod.* **2020**, *94*, 66. [CrossRef]

Article

Analysis of Precise Orbit Determination for the HY2D Satellite Using Onboard GPS/BDS Observations

Hailong Peng [1,2], Chongchong Zhou [3,4,*], Shiming Zhong [3], Bibo Peng [3,4], Xuhua Zhou [5], Haoming Yan [3], Jie Zhang [3], Jinyang Han [3,6], Fengcheng Guo [7] and Runjing Chen [8]

1   National Satellite Ocean Application Service, No. 8 Dahuisi Road, Haidian District, Beijing 100081, China; phl@mail.nsoas.org.cn
2   Key Laboratory of Space Ocean Remote Sensing and Application, MNR, No. 8 Dahuisi Road, Haidian District, Beijing 100081, China
3   State Key Laboratory of Geodesy and Earth's Dynamics, Innovation Academy for Precision Measurement Science and Technology, Chinese Academy of Sciences, No. 340 Xudong Road, Wuhan 430077, China; smzhong@apm.ac.cn (S.Z.); pengbibo@apm.ac.cn (B.P.); yhm@apm.ac.cn (H.Y.); zhangjie@apm.ac.cn (J.Z.); goldensun_han@apm.ac.cn (J.H.)
4   National Geodetic Observatory, Wuhan, Innovation Academy for Precision Measurement Science and Technology, Chinese Academy of Sciences, No. 340 Xudong Road, Wuhan 430077, China
5   Shanghai Astronomical Observatory, Chinese Academy of Sciences, No. 80 Nandan Road, Shanghai 200030, China; xhzhou@shao.ac.cn
6   College of Earth and Planetary Sciences, University of Chinese Academy of Sciences, No. 19A Yuquan Road, Shijingshan District, Beijing 100049, China
7   School of Geography, Geomatic and Planning, Jiangsu Normal University, No. 101 Shanghai Road, Tongshan District, Xuzhou 221116, China; fchguo@jsnu.edu.cn
8   School of Computer and Information Engineering, Xiamen Institute of Technology, No. 600 Ligong Road, Jimei District, Xiamen 361024, China; chenrj@xmut.edu.cn
*   Correspondence: zcc@apm.ac.cn; Tel.: +86-133-9723-4970

Citation: Peng, H.; Zhou, C.; Zhong, S.; Peng, B.; Zhou, X.; Yan, H.; Zhang, J.; Han, J.; Guo, F.; Chen, R. Analysis of Precise Orbit Determination for the HY2D Satellite Using Onboard GPS/BDS Observations. *Remote Sens.* **2022**, *14*, 1390. https://doi.org/10.3390/rs14061390

Academic Editor: Xiaogong Hu

Received: 15 February 2022
Accepted: 10 March 2022
Published: 13 March 2022

Publisher's Note: MDPI stays neutral with regard to jurisdictional claims in published maps and institutional affiliations.

Abstract: High-precision orbits of Low Earth Orbit (LEO) satellites are essential for many scientific applications, such as assessing the change in current global mean sea level, estimating the coefficients of gravity field, and so on. How to determinate the high-precision orbits for LEO satellites has gradually become an important research focus. HY2D is a new altimetry satellite of China, which is equipped with a Global Positioning System (GPS) and the third generations of the BeiDou Global Navigation Satellite System (BDS-3) in order to guarantee the reliability of orbital precision in radar altimetry mission. Therefore, this study adopts one month of spaceborne data to conduct the research of precise orbit determination (POD) for the HY2D satellite. Our analysis results are: (1) The standard deviation of residuals for the HY2D satellite based on spaceborne BDS and GPS data are 9.12 mm and 8.53 mm, respectively, and there are no significant systematic errors in these residuals. (2) The comparison results with Doppler Orbitography and Radio-positioning Integrated by Satellite (DORIS)-derived orbits indicate that the HY2D satellite, using spaceborne BDS and GPS data, can achieve the radial accuracy of 1.4~1.5 cm, and the mean three-dimensional (3D) accuracy are 5.3 cm and 4.3 cm, respectively, which can satisfy high-precision altimetry applications. (3) By means of satellite laser ranging (SLR), the accuracy of Global Navigation Satellite System (GNSS)-derived orbits of HY2D is approximately 3.3 cm, which reflects that the model strategies are reliable.

Keywords: high-precision orbits; HY2D; spaceborne BDS and GPS data; radial accuracy; model strategies

1. Introduction

With the rapid development of space technologies, more and more Low Earth Orbit Satellites (LEOs) have been successfully used in scientific missions [1–7]. Especially, in order to obtain more detailed information about the marine environment and mapping and carry out research about the change in current global mean sea level, tens of ocean altimetry satellites have been launched. For example, there have been successful launches of Seasat, GeoSat, Topex/Poseidon, HY2A/B/C, Jason1, Jason2 and Jason3 altimetry satellites [1,8], which provide a large amount of effective and high-precision data for assessing the change in current global mean sea level. As shown by the successful use of Global Positioning System (GPS) in the precise orbit determination (POD) for Topex/Poseidon satellite [8], the spaceborne GPS technique makes it possible to obtain centimeter-level orbit products and, thus, has been widely used for LEOs [9–15]. Gao et al. [10] adopted DORIS (Doppler Orbitography and Radiopositioning Integrated by Satellite) data of HY2A to analyze POD results, and showed that the radial orbit difference with the CNES (Centre National d'Etudes Spatiales) orbits is about 1.1 cm. Guo et al. [11] conducted POD of HY2A based on GPS and DORIS data, and achieved radial accuracy better than 1.0 cm. Wang et al. [16] adopted three months of GPS observations of HY2C, and achieved the radial accuracy of about 1.2 cm.

China has developed and operated the BeiDou Navigation Satellite System (BDS) independently [17]. Nowadays, there are a total of 34 BDS satellites in orbit, including 15 BDS-2 satellites (six geostationary Earth orbit (GEO) satellites, six inclined geosynchronous orbit (IGSO) satellites, and 3 medium-Earth orbit (MEO) satellites) and 19 BDS-3 satellites (two IGSO satellites and 17 MEO satellites). There are more and more LEO satellites carrying BDS receivers, and numerous studies have been carried out on the POD based on spaceborne BDS data in recent years. In 2013, the FengYun-3C (FY-3C) satellite was launched successfully, which was equipped with BDS-2 and GPS receivers simultaneously. The spaceborne BDS and GPS data of FY-3C satellite provided a great opportunity to analyze the POD performance of LEOs with BDS. Li et al. [18] conducted POD of FY-3C based on BDS-only and BDS/GPS combined; the analysis showed that the BDS-only orbits can reach a three-dimensional (3D) root mean square (RMS) of 8 cm based on the orbit overlap comparison, while the 3D RMS value of combined POD is 3.9 cm. Xiong et al. [19] achieved real-time POD with a precision of 1.24 m for the FY-3C satellite using BDS and GPS pseudo-range observations. Zhao et al. [20] used the derived POD orbits of FY-3C and regional station observations to enhance the BDS orbits and improved the accuracy from 354.3 to 63.1 cm for GEO, 22.70 to 20.0 cm for IGSO, and 20.9 to 16.7 cm for MEO. Based on the FY-3C spaceborne BDS and GPS data from 2013 to 2017, Li et al. [21] found that the combined POD (without GEO) can achieve slightly better precision than the GPS results, which indicates that when high-quality BDS orbit and clock products are used, the combined solution can improve the accuracy of POD for LEOs in comparison with the GPS-only solution. Based on the spaceborne BDS data of FY-3C, Zhang et al. [22] found that the precision of LEO orbit determination and reliability of the solution are improved through the calibration of daily orbit biases in GEO. The measurements of Tianping-1B, launched in 2018, were also collected by Zhao et al. from GPS/BDS-3. Their results indicated that the orbit consistency of the combined BDS-3/GPS solutions was below 3.5 cm [23].

In 2021, China successfully launched a new altimetry, satellite-HaiYang-2D (HY2D), which is a marine operational satellite that can provide precise ocean dynamic environmental information for the warning and forecasting of marine disaster, continuously measuring the sea surface height and sea surface wind and conducting marine scientific research. In order to guarantee the reliability of satellite orbits for radar altimetry mission, HY2D is equipped with a GPS and BDS receiver and carries a laser retro-reflector array for satellite laser ranging (SLR). Therefore, this paper adopts spaceborne GPS and BDS data to conduct research about the POD of HY2D, mainly including the model strategies used in POD processing, Global Navigation Satellite System (GNSS)-derived orbit analysis, and the validation of SLR residuals for HY2D. The relevant research results can lay an important foundation for the development of subsequent altimetry satellites, and that in turn can lay an important foundation for the development of a spaceborne BDS receiver.

This paper is structured as follows. Section 2 introduces materials and mainly consists of general information about the HY2D satellite and data collection. Section 3 mainly presents the POD method and strategies. All results and discussion obtained based on the above methods and strategies are given in Section 4. Finally, conclusions are given in Section 5.

2. Materials

2.1. HY2D Spacecraft

On 19 May 2021, the HY2D satellite was launched at the Jiuquan satellite launch center successfully. The HY2D satellite adopts a non-sun-synchronous orbit with the inclination of 66°, and the orbit altitude of about 960 km. The primary sensors comprise radar altimeter, microwave scatterometer, calibration radiometer, data collection system and ship automatic identification system [24].

Figure 1 shows the HY2D spacecraft and its payloads. It should be pointed out that the GNSS antennas consists of GPS and BDS antennas but cannot receive GPS/BDS signals simultaneously. Thus, the GPS and BDS receivers would switch to each other as needed. The X, Y and Z are the three axes of the satellite flight reference frame (SFF) in the figure, and the +X and +Z axes point toward the direction of flight and nadir, respectively. The +Y axis completes the right-hand orthogonal reference. When the satellite is at zero attitude, the SFF coincides with the satellite body reference frame (SBF), while the SFF is different from the SBF in a non-zero attitude. Table 1 lists the coordinates of the GPS and BDS antenna phase center, laser retro-array (LRA) spherical center, and center of mass in the SBF. With the help of these coordinates, we can perform phase center correction of receivers for HY2D in POD processing.

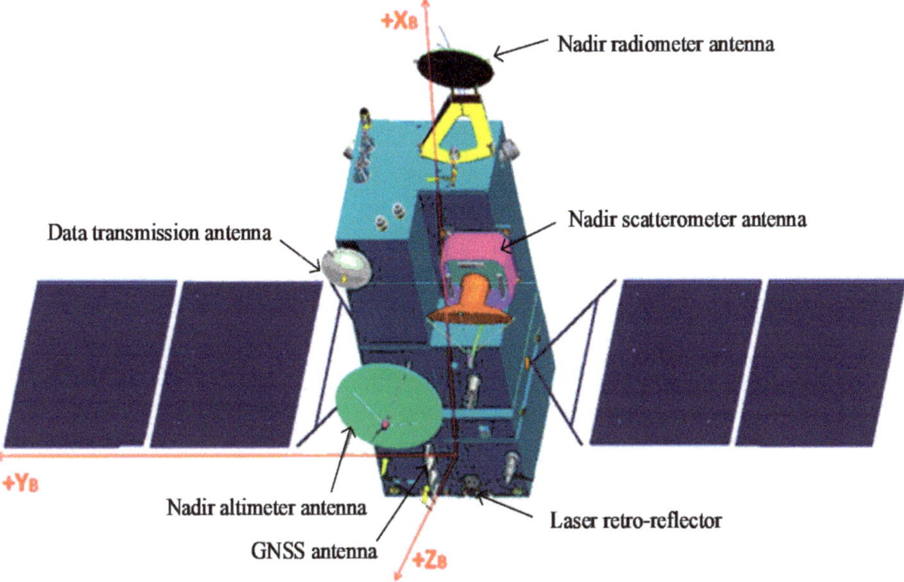

Figure 1. HY2D satellite and its payloads.

Table 1. Coordinates of the center of mass, GPS and BDS antenna phase centers and LRA spherical center in the SBF.

	X (mm)	Y (mm)	Z (mm)
Center of mass	1319.4	−4.7	5.7
GPS phase center (L1)	347.2	−181.9	−1377.5
GPS phase center (L2)	347.4	−181.0	−1396.1
BDS phase center (B1)	427.4	177.7	−1378.5
BDS phase center (B2)	427.7	177.9	−1397.3
LRA spherical center	311.7	−215.5	1060.8

When the satellite is flying, surface forces acting on HY2D are mainly from atmospheric drag, earth radiation pressure and solar radiation pressure [10]. In order to model the forces precisely, having a well knowledge of the characteristics for the HY2D satellite surfaces and radiators is very necessary. Table 2 gives the parameters information of spacecraft surfaces and radiators, including the optical characteristics and projected areas. With help of the information, the solar radiation pressure of HY2D can be described with the Box-Wing model.

Table 2. Projected areas and optical characteristics of HY2D.

			X+	X−	Y+	Y−	Z+	Z−	SA+	SA−
Spacecraft surfaces		Projected area (m^2)	3.62	3.92	5.17	5.46	3.06	6.22	-	-
	Visible	Specular	0.65	0.65	0.65	0.65	0.65	0.65	-	-
		Diffuse	0.00	0.00	0.00	0.00	0.00	0.00		
		Absorbed	0.35	0.35	0.35	0.35	0.35	0.35	-	-
	Infra-red	Specular	0.00	0.00	0.00	0.00	0.00	0.00	-	-
		Diffuse	0.31	0.31	0.31	0.31	0.31	0.31		
		Absorbed	0.69	0.69	0.69	0.69	0.69	0.69	-	-
Radiators and solar arrays		Projected area (m^2)	0.33	0.37	2.61	2.33	4.88	1.72	18.12	18.12
	Visible	Specular	0.87	0.87	0.87	0.87	0.00	0.87	0.10	0.00
		Diffuse	0.00	0.00	0.00	0.00	0.15	0.00	0.00	0.10
		Absorbed	0.13	0.13	0.13	0.13	0.85	0.13	0.90	0.90
	Infra-red	Specular	0.22	0.22	0.22	0.22	0.00	0.22	0.08	0.00
		Diffuse	0.00	0.00	0.00	0.00	0.15	0.00	0.00	0.10
		Absorbed	0.78	0.78	0.78	0.78	0.85	0.78	0.92	0.90

Table 2 shows that the +X and +Z axes point to the directions of flight and nadir, respectively, and -Y axis points to the direction of Sun. As the HY2D satellite belongs to a non-sun-synchronous orbit satellite, the solar arrays are always directed to the sun. Moreover, the HY2D satellite is in a non-zero attitude during the flight, and the SBF has to rotate according to the attitude following the order of yaw-roll-pitch to obtain the SFF. So, it is essential to adopt satellite attitude data to calculate the solar radiation pressure and phase center offset accurately.

Table 3 shows a part of attitude data of HY2D satellite, which is provided by National Satellite Ocean Application Service (NSOAS) at present. As can be shown from Table 3, the attitude data consist of date, time, roll angle, pitch angle, and yaw angle. In addition, the change in yaw angle works well with the increase in time, so, this study needs to use the attitude data of HY2D to correct the phase center offset and solar radiation pressure.

Table 3. The attitude data of HY2D satellite provided by NSOAS.

Date (Year/Month/Day)	Time (Hour/Minute/Second)	Roll (deg.)	Pitch (deg.)	Yaw (deg.)
2021/7/10	19:20:56.322	−0.0770	−0.0385	+88.0330
2021/7/10	19:20:57.346	−0.0770	−0.0385	+88.1155
2021/7/10	19:20:58.370	−0.0770	−0.0385	+88.1980
2021/7/10	19:20:59.394	−0.0770	−0.0385	+88.2805
2021/7/10	19:21:00.418	−0.0770	−0.0385	+88.3630
2021/7/10	19:21:01.442	−0.0770	−0.0385	+88.4455
2021/7/10	19:21:02.466	−0.0825	−0.0385	+88.5225
2021/7/10	19:21:03.490	−0.0825	−0.0385	+88.6050
2021/7/10	19:21:04.514	−0.0825	−0.0385	+88.6875
2021/7/10	19:21:05.538	−0.0825	−0.0385	+88.7700
. . .	. . .	. . .	. . .	. . .
2021/7/10	19:30:47.425	−0.1650	+0.0165	+133.0285
2021/7/10	19:30:48.449	−0.1650	+0.0165	+133.1000
2021/7/10	19:30:49.473	−0.1650	+0.0165	+133.1715
2021/7/10	19:30:50.497	−0.1650	+0.0165	+133.2375
2021/7/10	19:30:51.521	−0.1650	+0.0165	+133.3090
2021/7/10	19:30:52.545	−0.1650	+0.0165	+133.3750
2021/7/10	19:30:53.313	−0.1650	+0.0165	+133.4300
2021/7/10	19:30:54.337	−0.1650	+0.0165	+133.4960

2.2. Data Collection

For evaluating the POD performances of HY2D, we selected spaceborne BDS data from July 6 to July 19, and GPS data from July 20 to August 10 to conduct experiments. Here, Figures 2 and 3 present the number of BDS and GPS satellites observed on a day, respectively.

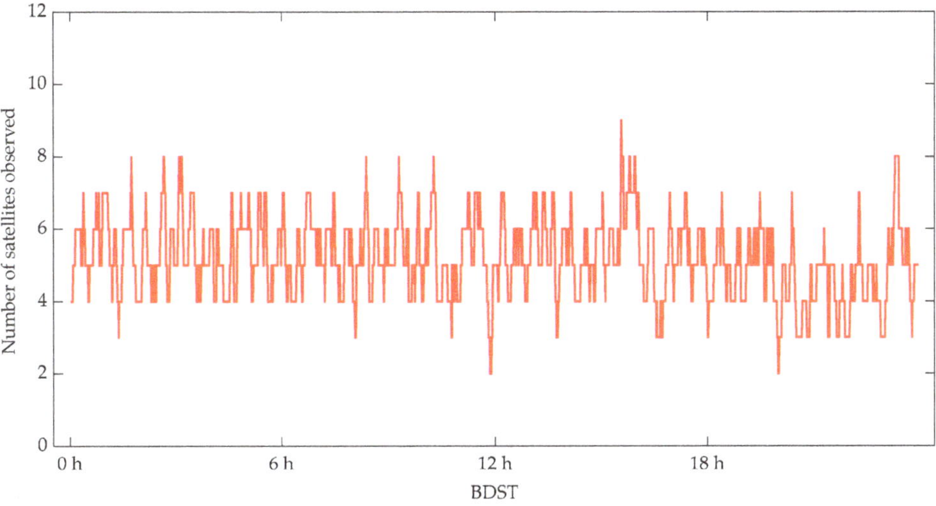

Figure 2. Number of BDS satellites observed on 10 July 2021.

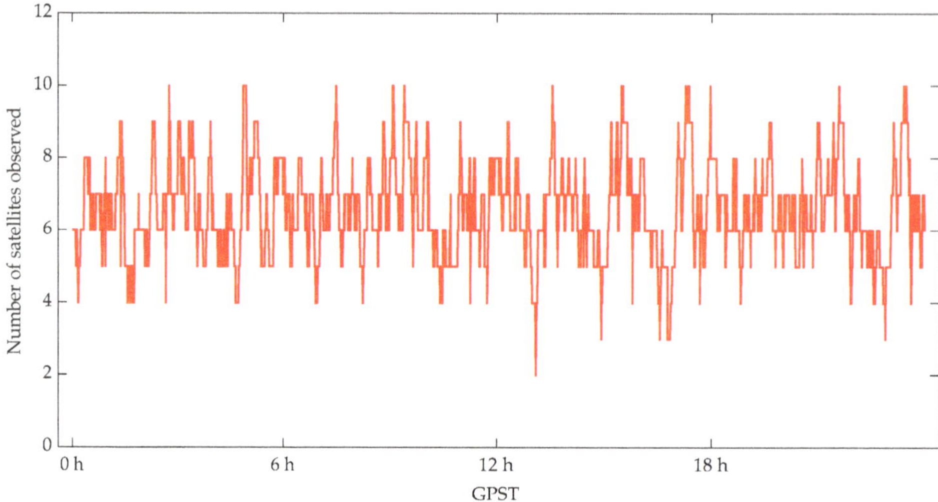

Figure 3. Number of GPS satellites observed on 21 July 2021.

As can be shown from Figure 2, most of the epochs have at least four BDS satellites available, and there are an average of five satellites. However, the smallest is 2, and the biggest is 9. As can be seen from Figure 3, compared to BDS, the usability of satellites for GPS is much better, and most epochs have more than five GPS satellites, and the maximum is up to 10 GPS satellites. This is mainly attributed to the GPS constellation consisting of 32 MEO satellites, more than the current BDS constellation, and the fact that most BDS, GEO and IGSO satellites are located within the Asia-pacific region.

3. POD Method and Strategies

This Section mainly describes the observation and dynamic model used in the POD processing of HY2D. Meanwhile, we also present the detailed dynamic and observation model in the POD for HY2D.

3.1. Observation Model

The GNSS antenna of HY2D satellite only tracks the GPS or BDS signal separately at present, so we selected ionosphere-free combined observation from GPS or BDS to calculate precise orbits. Equation (1) gives ionosphere-free combined observation of GPS and BDS.

$$\begin{cases} P_B = \rho_B - c \cdot \delta t_B^s + c \cdot \delta t_B^t + \varepsilon_{BP} \\ L_B = \rho_B - c \cdot \delta t_B^s + c \cdot \delta t_B^t + N_B \cdot \lambda_B + \varepsilon_{BL} \\ P_G = \rho_G - c \cdot \delta t_G^s + c \cdot \delta t_G^t + \varepsilon_{GP} \\ L_G = \rho_G - c \cdot \delta t_G^s + c \cdot \delta t_G^t + N_G \cdot \lambda_G + \varepsilon_{GL} \end{cases} \tag{1}$$

where P_B and P_G are the code observations of BDS and GPS, respectively, while L_B and L_G are the carrier phase observations of BDS and GPS, respectively; ρ_B and ρ_G are geometrical distance from BDS and GPS signal to receiver, respectively, δt_B^s and δt_G^s stand for satellite clock errors for BDS and GPS, δt_B^t and δt_G^t are clock errors for BDS and GPS receivers, N_B and N_G refer to combined ambiguity parameters of BDS and GPS observations separately, λ_B and λ_G refer to combined wave length of BDS and GPS observations separately, ε_{BP}, ε_{BL}, ε_{GP} and ε_{GL} stand for different type observation noise.

3.2. Dynamic Model

The equation of motion of a single LEO satellite in the inertial frame can be expressed as follows [25]:

$$\ddot{\vec{r}} = -\frac{GM}{\left|\vec{r}\right|^3}\vec{r} + a_r + a_{rtn} \tag{2}$$

where, GM stands for geocentric gravitational constant, $\vec{r}$ and $\ddot{\vec{r}}$ are the position and acceleration of the satellite in the inertial coordinate system, respectively; a_r refers to main perturbation acceleration excluded Earth center gravity, that includes the Sun and Moon perturbation, atmosphere drag, solar radiation pressure, Earth radiation pressure, solid Earth tide, ocean tide and so on. a_{rtn} refers to the periodic radial, tangential, and normal (RTN) perturbation, which can make up for these unmodeled perturbation acceleration errors [26–28]. In this study, the tangential and normal perturbation parameters are estimated for the POD processing.

In order not to affect the calculation efficiency and accuracy of orbit determination, the process of choosing the dynamic and observation model is important. For this purpose, the used dynamic model and measurement model are designed in the POD processing for the HY2D satellite refers to the above descriptions. Furthermore, this study also shows the used model information of the SLR validation for the calculated HY2D orbit. The details are given in Table 4. We adopted the 24 h arc solution based on the dynamical method orbit determination, the atmospheric drag coefficient, the solar radiation pressure coefficient, and RTN perturbation parameters are estimated per 6 h, 24 h, and 24 h, respectively. Moreover, the receiver clock parameter for HY2D is estimated with the gaussian white noise model, and the float solutions of ambiguities are also estimated.

Table 4. Dynamical and observation model employed in the POD for HY2D.

Project	Selection/Description
Dynamic model	
Gravity model	EIGEN-GRGS.RL04. MEAN-FIELD 120 × 120 [29]
Atmosphere drag	MSIS00 density model [30]
Solar radiation pressure	Box-Wing model [31]
Sun and moon ephemeris	JPL DE405 [32]
Earth radiation pressure	Knocke-Ries-Tapley model [33]
Empirical force model	RTN perturbation
Ocean tide [34]	FES2004 [35]
Solid Earth tide [34]	TIDE2000 [34]
Earth orientation parameter	IERS EOP 14 C04 [36], IAU2000A model
Observation model	
Data type	Code and phase observation of ionosphere-free combination
Data interval	30 s
Elevation cutoff	7°
HY2D satellite attitude	Quaternion data
GPS and BDS phase model	IGS14.atx
Orbit determination arc length and integration step	24 h arc dynamic solution, 30 s integral step
GPS and BDS satellite ephemeris and clock	CODE precise products
Atmospheric drag coefficient estimation	Cd/6 h
Solar radiation pressure coefficient estimation	Cr/24 h
RTN perturbation estimation	Tangential and normal/24 h
Receiver clock error estimation	Gaussian white noise
Ambiguities	Float solutions of ionosphere-free combination

Table 4. *Cont.*

Project	Selection/Description
SLR validation for the calculated HY2D orbit	
Cut-off angle	20°
SLR station coordinate	SLRF2014
Troposphere delay correction	Mendes-Pavlis delay model [37]
Relativity correction (propagation path)	IERS Conventions 2010 [34]

4. Results and Discussion

For the sake of analyzing the orbit determination accuracy of the HY2D satellite, we selected spaceborne BDS data from 6 July to 19 July, and GPS data from 20 July to 10 August for determination orbit. Then, we analyzed the residual variation of the POD processing for the HY2D satellite, presented the overlap orbit precision and compared the calculated orbits with precise orbit products provided by CNES, and showed the SLR validation results.

4.1. POD Residuals Analysis

Because the change in daily residuals of POD is basically similar for the HY2D satellite, we present the residual variation of observations for two days of data calculated by the above method and model strategies. Additionally, the selected days are, respectively, July 10 and 21, 2021 for BDS and GPS. Figure 4 shows the residual variation in spaceborne BDS phase observations from HY2D satellite on 10 July 2021. As can be seen from the figure, about 96.41 percent of residual values are located within ±15.00 mm. The average value and standard deviation of residuals are 0.036 mm and 9.12 mm, respectively, which show that these residuals have no significant deviations.

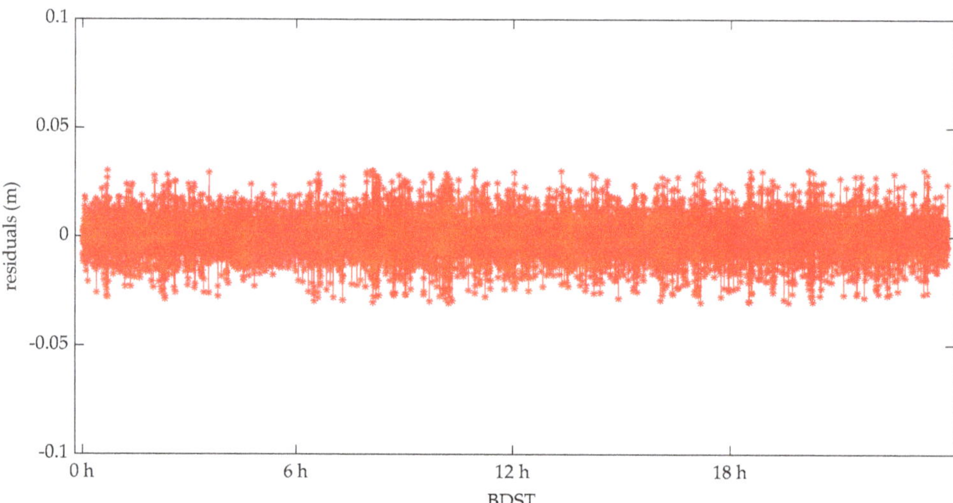

Figure 4. Variation of residuals in the POD for HY2D using spaceborne BDS data.

Moreover, Figure 5 shows the residuals variation in spaceborne GPS phase observations from the HY2D satellite on 21 July 2021. As can be seen from the figure, about 97.21 percent of residuals values are located within ±15.00 mm, similar to BDS. After the statistical calculation of these residuals, their average value and standard deviation are 0.026 mm and 8.53 mm, respectively, and there are also no obvious systematic errors in the residuals of GPS.

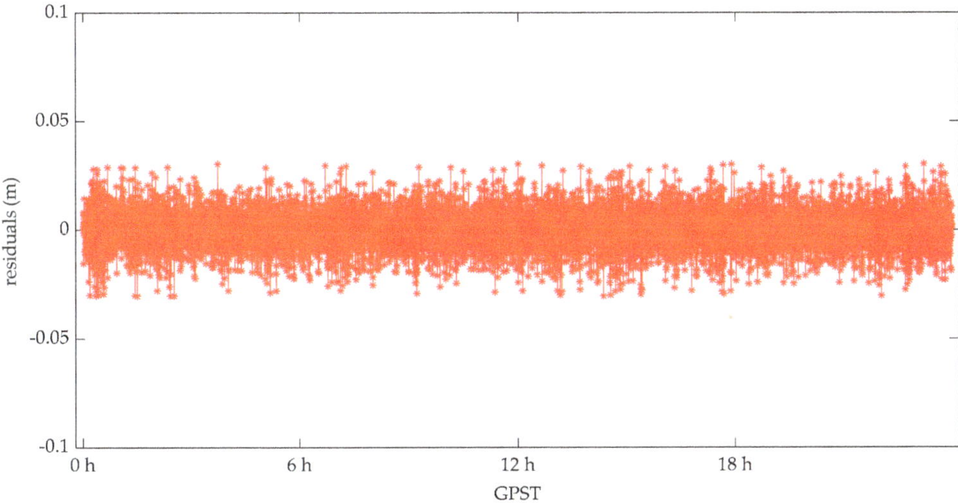

Figure 5. Variation of residuals in the POD for HY2D using spaceborne GPS data.

4.2. POD Precision Analysis for HY2D

In this study, we also compare the calculated orbit using spaceborne BDS and GPS data with the precise Doppler Orbitography and Radio-positioning Integrated by Satellite (DORIS)-derived orbits provided by CNES. Figures 6 and 7, respectively, show the position differences in Radial (R), Tangential (T), and Normal (N) directions between the calculated orbit and the precise orbit.

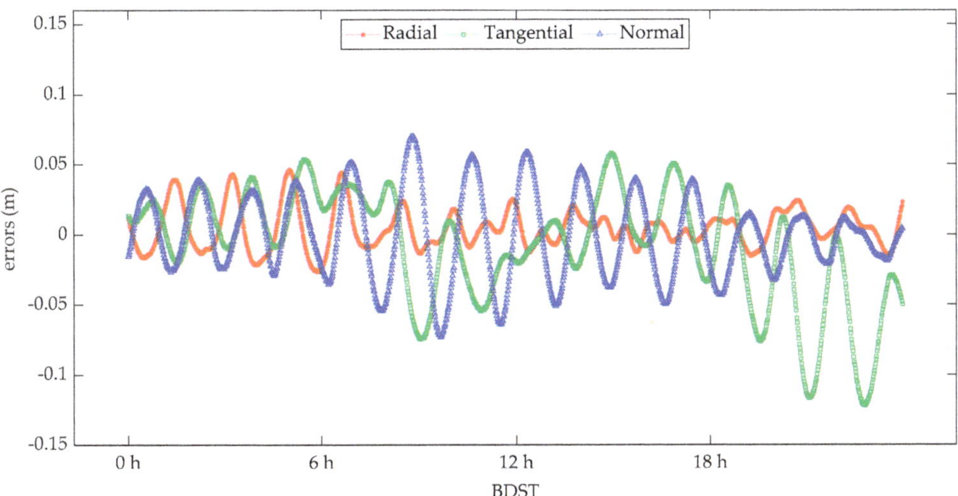

Figure 6. The variation in errors for the HY2D satellite using spaceborne BDS data compared with CNES orbit.

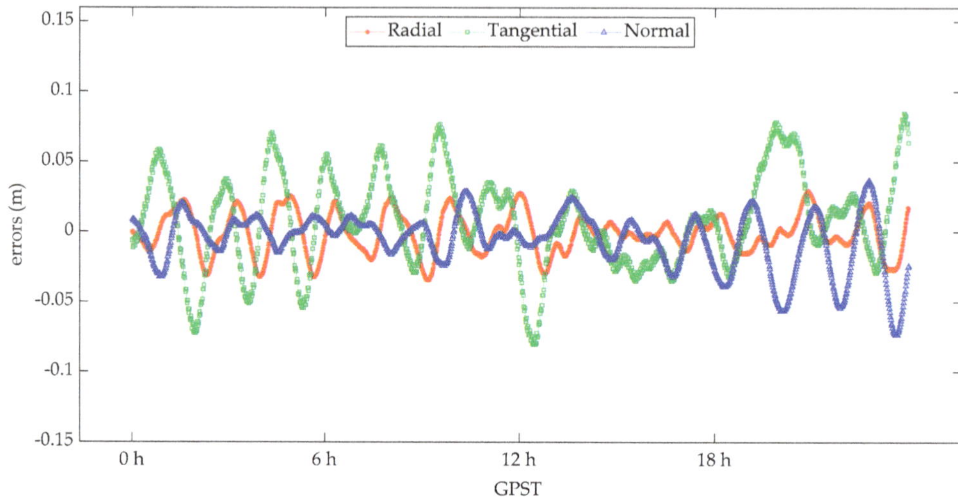

Figure 7. The variation in errors for the HY2D satellite using spaceborne GPS data compared with CNES orbit.

As can be seen from Figures 6 and 7, the variations in the radial direction have smaller fluctuation, which are basically located within ±0.03 m. However, the fluctuation in the tangential direction is relatively bigger, and this may be related to the used models of atmospheric drag and solar pressure, which are hard to estimate exactly. It should be noted that the normal fluctuations of GPS are relatively smaller than BDS; this is related to the number of GPS satellites, which are more than the BDS satellites and the orbit precision of GEO and IGSO satellites is lower. To analyze the precision visually, Table 5 gives the accuracy statistics of the HY2D satellite using spaceborne BDS and GPS data. We can find that the mean values are close to zero, which illustrate that these error values are basically unbiased and have no obvious systematic errors. For spaceborne BDS data, the radial, tangential, normal and 3D accuracy can achieve 1.5 cm, 4.1 cm, 3.0 cm, and 5.3 cm, respectively, and the radial, tangential, normal and 3D accuracy are, respectively 1.5 cm, 3.5 cm, 2.0 cm, and 4.3 cm for spaceborne GPS data. However, because of the fewer satellites available, the consistency the BDS-derived orbits are slightly worse than the GPS-derived orbits.

Table 5. Accuracy statistics of POD for HY2D using spaceborne BDS and GPS data (unit: cm).

		Mean		RMS
	Radial	0.4		1.5
BDS	Tangential	−0.8		4.1
	Normal	0.1		3.0
	3DRMS		5.3	
	Radial	−0.1		1.5
GPS	Tangential	0.8		3.5
	Normal	−0.5		2.0
	3DRMS		4.3	

For validating the reliability of the used method and model strategies, we also utilize more spaceborne BDS and GPS data to conduct POD solutions for HY2D satellite, and the results of comparison with precise orbit provided by CNES are shown in Figures 8 and 9 in radial, tangential, normal and 3D directions.

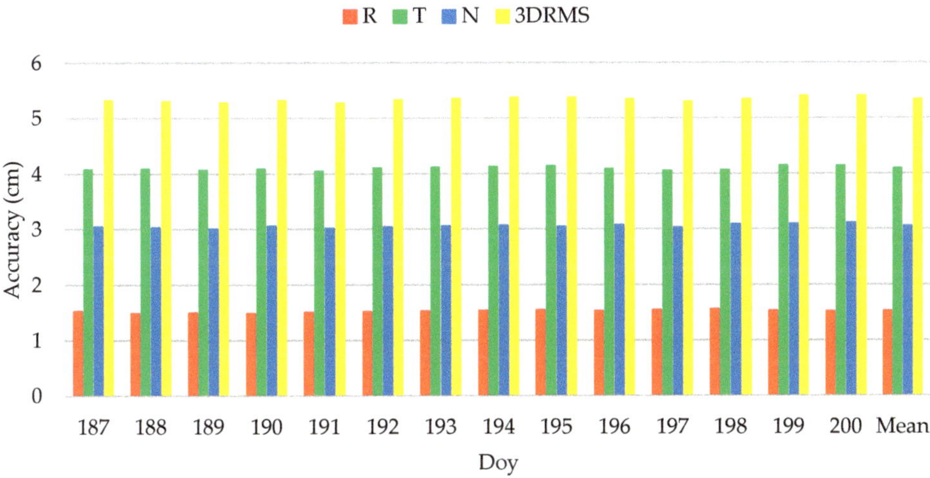

Figure 8. The RMS values for HY2D satellite using spaceborne BDS data.

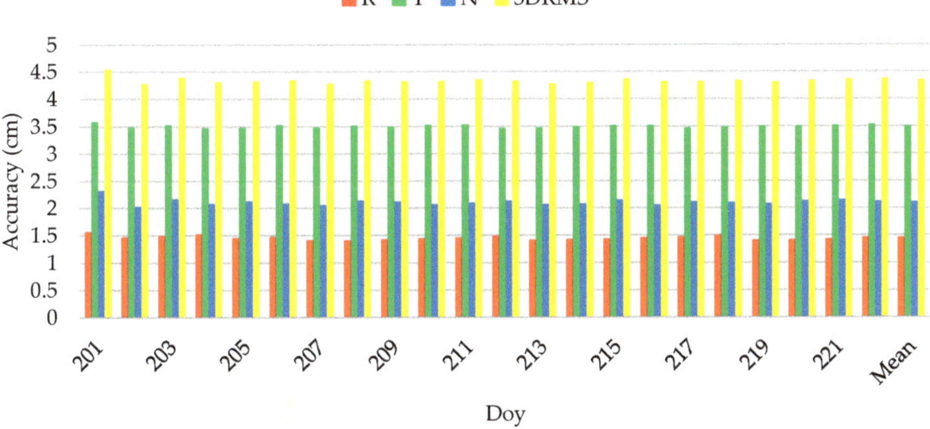

Figure 9. The RMS values for HY2D satellite using spaceborne GPS data.

As can be shown from Figures 8 and 9, the radial RMS values are basically 1.4~1.5 cm for BDS and GPS data, and the tangential and normal accuracies are slightly poorer compared to the radial direction. Overall, the precision based on spaceborne GPS data are slightly better than that of spaceborne BDS data, and the results are in agreement with the previous analysis, which shows that the used method and model strategies have a certain reliability. Moreover, Tables 6 and 7 show the accuracy statistics of HY2D. As can be shown from Tables 6 and 7, the average radial accuracy based on spaceborne BDS and GPS data are, respectively, 1.5 cm and 1.4 cm, the average three-dimensional accuracy are 5.3 cm and 4.3 cm, respectively. These results illustrate that the HY2D satellite, using spaceborne BDS and GPS data, can achieve the radial precision of 1.4~1.5 cm, the 3D position precision better than 5.5 cm, and the radial precision can satisfy high-precision altimetry applications.

Table 6. The precision statistics of HY2D satellite using spaceborne BDS data (unit: cm).

Doy	R	T	N	3DRMS
187	1.5	4.1	3.1	5.3
188	1.5	4.1	3.0	5.3
189	1.5	4.1	3.0	5.3
190	1.5	4.1	3.1	5.3
191	1.5	4.1	3.0	5.3
192	1.5	4.1	3.0	5.3
193	1.5	4.1	3.1	5.4
194	1.5	4.1	3.1	5.4
195	1.6	4.1	3.1	5.4
196	1.5	4.1	3.1	5.3
197	1.5	4.1	3.1	5.3
198	1.6	4.1	3.1	5.3
199	1.5	4.2	3.1	5.4
200	1.5	4.1	3.1	5.4
Mean	1.5	4.1	3.1	5.4

Table 7. The statistics of accuracy for HY2D satellite using spaceborne GPS data (unit: cm).

Doy	R	T	N	3DRMS
201	1.6	3.6	2.3	4.5
202	1.5	3.5	2.0	4.3
203	1.5	3.5	2.2	4.4
204	1.5	3.5	2.1	4.3
205	1.5	3.5	2.1	4.3
206	1.5	3.5	2.1	4.3
207	1.4	3.5	2.1	4.3
208	1.4	3.5	2.1	4.3
209	1.4	3.5	2.1	4.3
210	1.4	3.5	2.1	4.3
211	1.5	3.5	2.1	4.4
212	1.5	3.5	2.1	4.3
213	1.4	3.5	2.1	4.3
214	1.4	3.5	2.1	4.3
215	1.4	3.5	2.2	4.4
216	1.5	3.5	2.1	4.3
217	1.5	3.5	2.1	4.3
218	1.5	3.5	2.1	4.3
219	1.4	3.5	2.1	4.3
220	1.4	3.5	2.1	4.3
221	1.4	3.5	2.1	4.4
222	1.5	3.5	2.1	4.4
Mean	1.4	3.5	2.1	4.3

4.3. SLR Validation for the POD of HY2D

SLR is an independent measurement technique compared to GNSS, and thus SLR validation can validate objectively the accuracy and reliability of GNSS-derived orbits [11,38]. Here, we select the SLR normal point data from the corresponding periods of the HY2D satellite (download: ftp://edc.dgfi.tum.de/pub/slr/data/npt_crd/, accessed on 10 February 2022). The SLR residual results cannot verify the accuracies of the components in each direction or orbital position directly, but they denote that the distance precision between the given SLR station and the validated satellite. When the higher cut-off angle is set, the SLR residuals can better reflect the radial accuracy. Because HY2D is an altimetry satellite, its radial precision is our primary concern. In this study, we set the cut-off angles as 20, and give the time series of SLR residuals in Figure 10.

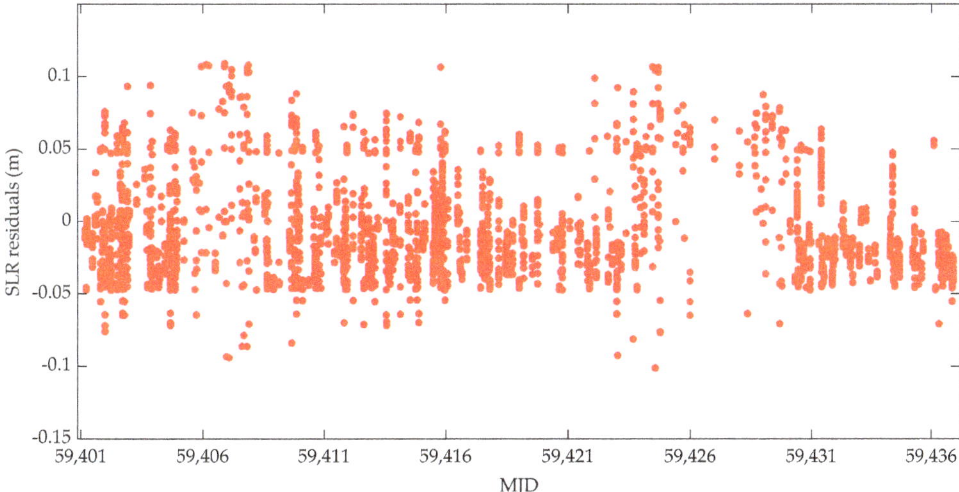

Figure 10. The variation of SLR residuals for HY2D satellite.

As can be seen from Figure 10, most of the residual values are located within ±8.0 cm, and have no obvious abnormal value. The orbits of the first 14 days are calculated using spaceborne BDS data, and the next 22 days orbits are calculated using spaceborne GPS data. The range of SLR residuals for BDS-derived orbits is slightly bigger than that of GPS-derived orbits, which is related to that the precision based on spaceborne GPS data are slightly better than that of spaceborne BDS data. After the statistical calculation of these residuals, their average value and standard deviation are, respectively, −0.7 cm and 3.3 cm, which illustrate that these residuals have no significant deviations. The results also indicate that the orbital accuracy of HY2D satellite can satisfy the current high-precision altimetry requirements.

5. Conclusions

Based on the spaceborne BDS and GPS data of the newly launched HY2D satellite, this paper conducts research into POD using one month of data. The model strategies are developed to solve the precise orbits of the HY2D satellite, and this study selects a month of real onboard GPS and BDS data to validate the developed model strategies. The main conclusions of the study are given in the following:

(1)   The standard deviation of the POD residuals based on spaceborne BDS and GPS data are, respectively, 9.12 mm and 8.53 mm, and these residual variations show no significant deviations or systematic errors.

(2)   The comparison results with DORIS-derived orbits show that the average radial RMS value of spaceborne BDS and GPS are 1.5 cm and 1.4 cm, respectively, and the corresponding 3D RMS accuracy are 5.3 cm and 4.3 cm, respectively. Overall, these results indicate that the POD processing of the HY2D satellite using spaceborne BDS and GPS data can achieve 1 cm radial precision and satisfy the current high-precision altimetry applications.

(3)   According to the SLR validation results, it is shown that the standard deviation of residuals is 3.3 cm, which indicates that the orbital accuracy of the HY2D satellite is approximately 3.3 cm, and the used model strategy is also reliable.

As an example of the HY2D satellite, this study adopts spaceborne BDS and GPS data to conduct POD research. Eventually, the model strategies are developed to solve the precise orbit of the HY2D satellite. Meanwhile, we use the external DORIS-derived precise orbit of the HY2D satellite and SLR normal point data to validate the calculated orbit, and the results show that the HY2D satellite can achieve 1 cm radial precision.

Author Contributions: C.Z. provided the initial idea for this research; H.P., C.Z., S.Z. and X.Z. collected the experimental data and conducted the experiment; C.Z., H.P., S.Z., B.P., H.Y. analyzed the results of the experiment; C.Z., J.Z., J.H., F.G. and R.C. wrote the paper. All authors have read and agreed to the published version of the manuscript.

Funding: This work was supported by the National Key Research Program of China "Collaborative Precision Positioning Project" (No. 2016YFB0501900), the National Natural Science Foundation of China (Grant No. 42174222, 41904165, 62101219, 41804019), the State Key Laboratory of Geodesy and Earth's Dynamics self-deployment project (No. S21L6101, S21L8101), the Natural Science Foundation of Hubei Province (No. 2017CFB372), the Natural Science Foundation of Jiangsu Province (No. BK20210921), the Natural Science Foundation of Fujian Province (No. 2018J01480).

Institutional Review Board Statement: Not applicable.

Informed Consent Statement: Not applicable.

Data Availability Statement: Not applicable.

Acknowledgments: The authors acknowledge the satellite information of HY2D provide by NSOAS. Our sincere thanks to the NSOAS for providing space-borne GPS and BDS data for HY2D; CODE for providing GPS satellite orbits, clocks and Earth rotation parameters; the CNES for providing precise orbits for HY2D; the ILRS for providing SLR data of the HY2D satellite. Meanwhile, we would like to thank the anonymous reviewers for their valuable comments.

Conflicts of Interest: The authors declare that they have no known competing financial interests or personal relationships that could have appeared to influence the work reported in this paper.

References

1. Cerri, L.; Berthias, J.P.; Bertiger, W.I.; Haines, B.J.; Lemoine, F.G.; Mercier, F.; Ries, J.C.; Willis, P.; Zelensky, N.P.; Ziebart, M. Precision Orbit Determination Standards for the Jason Series of Altimeter Missions. *Mar. Geodesy* **2010**, *33*, 379–418. [CrossRef]
2. D'Amico, S.; Ardaens, J.-S.; Larsson, R. Spaceborne Autonomous Formation-Flying Experiment on the PRISMA Mission. *J. Guid. Control. Dyn.* **2012**, *35*, 834–850. [CrossRef]
3. Zhang, Y.; Vincent, T. HY2-ICD-0-00009-CNES. Technical Document for HY2A Ground Segment. 2011.
4. Kang, Z.; Tapley, B.; Bettadpur, S.; Ries, J.; Nagel, P.; Pastor, R. Precise orbit determination for the GRACE mission using only GPS data. *J. Geodesy* **2006**, *80*, 322–331. [CrossRef]
5. Bock, H.; Jäggi, A.; Meyer, U.; Visser, P.; Ijssel, J.V.D.; van Helleputte, T.; Heinze, M.; Hugentobler, U. GPS-derived orbits for the GOCE satellite. *J. Geodesy* **2011**, *85*, 807–818. [CrossRef]
6. Bock, H.; Jäggi, A.; Švehla, D.; Beutler, G.; Hugentobler, U.; Visser, P. Precise orbit determination for the GOCE satellite using GPS. *Adv. Space Res.* **2007**, *39*, 1638–1647. [CrossRef]
7. Ijssel, J.V.D.; da Encarnacao, J.T.; Doornbos, E.; Visser, P. Precise science orbits for the Swarm satellite constellation. *Adv. Space Res.* **2015**, *56*, 1042–1055. [CrossRef]
8. Tapley, B.; Ries, J.C.; Davis, G.W.; Eanes, R.J.; Schutz, B.E.; Shum, C.K.; Watkins, M.M.; Marshall, J.A.; Nerem, R.S.; Putney, B.H.; et al. Precision orbit determination for TOPEX/POSEIDON. *J. Geophys. Res. Earth Surf.* **1994**, *99*, 24383–24404. [CrossRef]
9. Zhou, X.; Wang, X.; Zhao, G.; Peng, H.; Wu, B. The Precise Orbit Determination for HY2A Satellite Using GPS, DORIS and SLR Data. *Geomat. Inf. Sci. Wuhan Univ.* **2015**, *40*, 1000–1005.
10. Gao, F.; Peng, B.; Zhang, Y.; Evariste, N.H.; Liu, J.; Wang, X.; Zhong, M.; Lin, M.; Wang, N.; Chen, R.; et al. Analysis of HY2A precise orbit determination using DORIS. *Adv. Space Res.* **2015**, *55*, 1394–1404. [CrossRef]
11. Guo, J.; Zhao, Q.; Guo, X.; Liu, X.; Liu, J.; Zhou, Q. Quality assessment of onboard GPS receiver and its combination with DORIS and SLR for Haiyang 2A precise orbit determination. *Sci. China Earth Sci.* **2014**, *58*, 138–150. [CrossRef]
12. Haines, B.; Bar-Sever, Y.; Bertiger, W.; Desai, S.; Willis, P. One-Centimeter Orbit Determination for Jason-1: New GPS-Based Strategies. *Mar. Geodesy* **2004**, *27*, 299–318. [CrossRef]
13. Lemoine, F.G.; Zelensky, N.; Chinn, D.; Pavlis, D.; Rowlands, D.; Beckley, B.; Luthcke, S.; Willis, P.; Ziebart, M.; Sibthorpe, A.; et al. Towards development of a consistent orbit series for TOPEX, Jason-1, and Jason-2. *Adv. Space Res.* **2010**, *46*, 1513–1540. [CrossRef]
14. Liu, M.; Yuan, Y.; Ou, J.; Chai, Y. Research on Attitude Models and Antenna Phase Center Correction for Jason-3 Satellite Orbit Determination. *Sensors* **2019**, *19*, 2408. [CrossRef] [PubMed]
15. Wu, S.C.; Yunck, T.P.; Thornton, C.L. Reduced-dynamic technique for precise orbit determination of low earth satellites. *J. Guid. Control. Dyn.* **1991**, *14*, 24–30. [CrossRef]

16. Wang, Y.; Li, M.; Jiang, K.; Li, W.; Zhao, Q.; Peng, H.; Lin, M. Precise orbit determination of the Haiyang 2C altimetry satellite using attitude modeling. *GPS Solutions* **2022**, *26*, 1–14. [CrossRef]

17. Min, L.; Qu, L.; Zhao, Q.; Jing, G.; Xing, S.; Li, X. Precise Point Positioning with the BeiDou Navigation Satellite System. *Sensors* **2014**, *14*, 927–943.

18. Li, M.; Li, W.; Shi, C.; Jiang, K.; Guo, X.; Dai, X.; Meng, X.; Yang, Z.; Yang, G.; Liao, M. Precise orbit determination of the Fengyun-3C satellite using onboard GPS and BDS observations. *J. Geodesy* **2017**, *91*, 1313–1327. [CrossRef]

19. Xiong, C.; Lu, C.; Zhu, J.; Ding, H. Orbit determination using real tracking data from FY3C-GNOS. *Adv. Space Res.* **2017**, *60*, 543–556. [CrossRef]

20. Zhao, Q.; Wang, C.; Guo, J.; Yang, G.; Liao, M.; Ma, H.; Liu, J. Enhanced orbit determination for BeiDou satellites with FengYun-3C onboard GNSS data. *GPS Solutions* **2017**, *21*, 1179–1190. [CrossRef]

21. Li, X.; Zhang, K.; Meng, X.; Zhang, W.; Zhang, Q.; Zhang, X.; Li, X. Precise Orbit Determination for the FY-3C Satellite Using Onboard BDS and GPS Observations from 2013, 2015, and 2017. *Engineering* **2019**, *6*, 904–912. [CrossRef]

22. Zhang, Q.; Guo, X.; Qu, L.; Zhao, Q. Precise Orbit Determination of FY-3C with Calibration of Orbit Biases in BeiDou GEO Satellites. *Remote Sens.* **2018**, *10*, 382. [CrossRef]

23. Zhao, X.; Zhou, S.; Ci, Y.; Hu, X.; Cao, J.; Chang, Z.; Tang, C.; Guo, D.; Guo, K.; Liao, M. High-precision orbit determination for a LEO nanosatellite using BDS-3. *GPS Solutions* **2020**, *24*, 1–14. [CrossRef]

24. The General Information of HY-2 Satellites. Available online: https://ilrs.gsfc.nasa.gov/missions/satellite_missions/current_missions/hy2d_general.html (accessed on 10 February 2022).

25. Montenbruck, O.; Gill, E.; Lutze, F. Satellite Orbits: Models, Methods, and Applications. *Appl. Mech. Rev.* **2002**, *55*, B27–B28. [CrossRef]

26. Rim, H.; Kang, Z.; Nagel, P.; Yoon, S.; Bettadpur, S.; Schutz, B.; Tapley, B. CHAMP precision orbit determination. Advances in the Astronautical Sciences. 2002.

27. Rim, H.J.; Yoon, S.P.; Schutz, R.E. Effect of GPS orbit accuracy on CHAMP precision orbit determination. In Proceedings of the AAS/AIAA Space Flight Mechanics Meeting, San Antonio, TX, USA, 27–30 January 2002; 2002; 2, pp. 1411–1418.

28. Zhao, Q. *Research on Precision Orbit Determination Theory and Software of both GPS Navigation Constellation and LEO Satellites*; Wuhan University: Wuhan, China, 2004.

29. Lemoine, J.M.; Biancale, R.; Reinquin, F.; Bourgogne, S.; Gégout, P. CNES/GRGS RL04 Earth gravity field models, from GRACE and SLR data. *GFZ Data Serv.* **2019**. [CrossRef]

30. Picone, J.M.; Hedin, A.E.; Drob, D.P.; Aikin, A.C. NRLMSISE-00 empirical model of the atmosphere: Statistical comparisions and scientific issues. *J. Geophys. Res.* **2002**, *107*, SIA15-1–SIA15-16. [CrossRef]

31. Rodriguez-Solano, C.J.; Hugentobler, U.; Steigenberger, P. Adjustable box-wing model for solar radiation pressure impacting GPS satellites. *Adv. Space Res.* **2012**, *49*, 1113–1128. [CrossRef]

32. Standish, E.M. JPL Planetary and Lunar Ephemerides, DE405/LE405, Jet Propulsion Laboratory Interoffice Memoran-Dum. 312.F-98-048. 1998.

33. Knocke, P.; Ries, J.; Tapley, B. Earth radiation pressure effects on satellites. In Proceedings of the Astrodynamics Conference, Stowe, Vermont, 15–17 August 1988. [CrossRef]

34. Petit, G.; Luzum, B. IERS conventions. *IERS Technical Note No.36*; IERS Convention Centre, 2010. Available online: https://www.iers.org/IERS/EN/Publications/TechnicalNotes/tn36.html (accessed on 10 February 2022).

35. Lyard, F.; Lefevre, F.; Letellier, T.; Francis, O. Modelling the global ocean tides: Modern insights from FES2004. *Ocean Dyn.* **2006**, *56*, 394–415. [CrossRef]

36. Bizouard, C.; Lambert, S.; Gattano, C.; Becker, O.; Richard, J.-Y. The IERS EOP 14C04 solution for Earth orientation parameters consistent with ITRF 2014. *J. Geodesy* **2018**, *93*, 621–633. [CrossRef]

37. Mendes, V.B.; Pavlis, E.C. High-accuracy zenith delay prediction at optical wavelengths. *Geophys. Res. Lett.* **2004**, *31*, 189–207. [CrossRef]

38. Pearlman, M.; Degnan, J.; Bosworth, J. The International Laser Ranging Service. *Adv. Space Res.* **2002**, *30*, 135–143. [CrossRef]

 *remote sensing*

Article

High-Rate One-Hourly Updated Ultra-Rapid Multi-GNSS Satellite Clock Offsets Estimation and Its Application in Real-Time Precise Point Positioning

Guoqiang Jiao [1,2] and Shuli Song [1,*]

[1] Shanghai Astronomical Observatory, Chinese Academy of Sciences, Shanghai 200030, China; jiaoguoqiang@shao.ac.cn
[2] University of Chinese Academy of Sciences, Beijing 100049, China
* Correspondence: slsong@shao.ac.cn

Abstract: The requirement of timeliness is increasing while obtaining precise tempo-spatial information with the development of global navigation satellite systems (GNSSs). Due to the poor network environment and communication conditions in some regions or application scenarios, it is difficult for users to receive real-time (RT) precise products. The hourly updated ultra-rapid products with low latency and high accuracy are of great interest in GNSS real-time and near-real-time fields. However, it is difficult to achieve the high-rate one-hourly updated precise clock estimation (PCE); since many ambiguity parameters need to be estimated, the computation is time-consuming. At present, the highest time resolution of ultra-rapid clock offsets is 15 min. The low samplings affect the prediction accuracy of clock offsets and the precise point positioning (PPP) performances. To meet these requirements, we proposed an efficient method and design a new framework for high-rate one-hourly updated ultra-rapid PCE. We modified the epoch-difference (ED) PCE model in the parameter estimation. According to the characteristics of the modified ED PCE model, the Open Multi-Processing (OpenMP) and Intel Math Kernel Library (MKL) technologies are used to construct a parallel system to realize the parallelism among satellites, epochs, and stations. The comprehensive assessment in the precision of clock offsets and PPP performances is conducted. The result demonstrates that the one-hourly updated multi-GNSS clock offsets with 30 s sampling can be obtained within 20 min. The estimated clock offsets accuracy increases with the improvement of the time resolution. The STD and RMS are improved by (0.97 to 9.09% and 0.12 to 5.56%) in the observation session, (2.82 to 23.08% and 0.95 to 9.09%) in the first hour of the prediction session, and (0.11 to 3.85% and 0.12 to 4.19%) in the second hour of the prediction session compared with low-rate products, respectively. The high-rate one-hourly updated ultra-rapid clock offsets significantly improves the RT-PPP performances. The positioning accuracy can be improved by 1.52~25.74%, and the convergence time can be improved by 21.96~65.75%. The RT-PPP performances are basically the same as GeoForschungsZentrum Potsdam (GFZ) rapid products and slightly better than the Center National d'Etudes Spatiales (CNES) RT products (CLK93). The one-hourly updated ultra-rapid products with low latency, high accuracy, and not limited by network conditions can be well applied to real-time or near real-time applications and research.

Keywords: high-rate; one-hourly updated ultra-rapid clock offsets; precise clock estimation (PCE); epoch-difference (ED); parallel processing; precise point positioning (PPP)

Citation: Jiao, G.; Song, S. High-Rate One-Hourly Updated Ultra-Rapid Multi-GNSS Satellite Clock Offsets Estimation and Its Application in Real-Time Precise Point Positioning. *Remote Sens.* **2022**, *14*, 1257. https://doi.org/10.3390/rs14051257

Academic Editor: Yunbin Yuan

Received: 30 December 2021
Accepted: 2 March 2022
Published: 4 March 2022

Publisher's Note: MDPI stays neutral with regard to jurisdictional claims in published maps and institutional affiliations.

1. Introduction

Precise satellite orbits and clock offsets are essential for obtaining precise tempo-spatial information in precise point positioning (PPP), which plays an important role in precise orbit determination (POD) for low Earth orbit (LEO) satellites, atmospheric retrieval, precise positioning, timing, and so on. The International GNSS Service (IGS) has officially provided final, rapid, and ultra-rapid products since 5 November 2000 [1]. As summarized in IGS product introductions (https://igs.org/products/#about, accessed on 1 March 2022), the IGS final products with highest precision are delayed for 12~18 days; the rapid products are delayed for 17~41 h [2]. With the development of global navigation satellite systems (GNSSs) and the increasing requirements for positioning accuracy and low latency, the final and rapid products can hardly meet the real-time and near-real-time user needs. The ultra-rapid products with low latency and high accuracy are of great interest in GNSS real-time and near-real-time fields. IGS integrated ultra-rapid GPS-only products from eight analysis centers (AC), Center for Orbit Determination in Europe (CODE), Natural Resources Canada (NRCan), and so on, which are updated every six hours [3,4]. At present, the latency and time resolution of the ultra-rapid products released by most ACs are 6 h and 15 min, respectively. The ultra-rapid orbits and clock offsets consist of observation and prediction sessions. The high latency means that we need to predict the satellite orbit and clock offsets for a longer time. With the increase in prediction time, the accuracy of satellite orbits and clock offsets will be reduced correspondingly. Especially for satellite clock offsets prediction, the accuracy of satellite clock offsets will be reduced with the longer prediction time. Therefore, with the six-hourly updated ultra-rapid GPS-only products, it is difficult to meet the high-precision real-time applications due to the low sampling and accuracy loss of clock offsets prediction.

To support better real-time services, IGS established the Real Time Working Group (RTWG) and provided the real-time service (RTS) to the GNSS community [5]. The Center National d'Etudes Spatiales (CNES) adopted the Kalman Filter (KF) estimator for satellite precise clock estimation (PCE) based on the raw observations [6]. The EPOS-RT and PANDA software used the Square Root Information Filter (SRIF) to obtain the real-time clock offsets based on the mixed-differenced model [7–9]. Whether in the raw observation receiving or real-time product broadcasting, a good network is essential for the real-time PCE. However, the available observations for real-time product solution will be affected by the network environment, and its corresponding product accuracy will be decreased [5]. In addition, it is difficult for users with poor network environment and communication ability to receive real-time precise products. On the contrary, the ultra-rapid products estimation is not expected to require high network conditions. It is not necessary for server and client to receive data continuously. All the same, the high latency and low samplings for ultra-rapid products are the problems that cannot be ignored. Therefore, it is important to provide the ultra-rapid orbit and clock products with low latency, high precision, and high samplings. To reduce the latency, Li et al. [10] used multi-thread technology and decreased the sampling rate (5 to 10 min) to estimate one-hourly updated multi-GNSS products based on undifferenced (UD) ionosphere-free (IF) models. Similarly, Chen et al. [11] updated data processing strategy and optimized software to obtain one-hourly updated multi-GNSS products with 5 min sampling. Wuhan University (WHU) GNSS AC released one-hourly updated multi-GNSS products with 5 min sampling [12,13]. Although, the latency and the samplings of ultra-rapid clock offsets has been improved from 6 to 1 h and 15 to 5~10 min, respectively. However, the samplings are not adequate for some applications. The time resolution of satellite clock offsets will affect the PPP in terms of convergence time and positioning accuracy, especially for the kinematic PPP applications [7,14,15]. To meet real-time needs and improve the positioning performance, it is meaningful to develop the high-rate one-hourly ultra-rapid PCE.

Furthermore, the single GPS is gradually developing to multi-constellation GNSS [16]. Multi-GNSS observations can improve the positioning accuracy, convergence time, and the stability and reliability of GNSS services [17]. However, the development of multi-GNSS has brought severe challenges to GNSS data processing. Figure 1 depicts the number of estimated parameters for the different PCE models with the increase in the number of the tracking stations. Due to the large number of estimation parameters (including ambiguities) in multi-GNSS, the computation is time-consuming. Bock et al. [18] proposed the epoch-difference (ED) PCE model to achieve high-rate GPS-only PCE for final products.

In addition, most GNSS ACs used this model to obtain high-rate rapid and final products. Although the ED model greatly reduces the time-consuming, the current ED model and data processing strategy are difficult to realize the high-rate one-hourly updated ultra-rapid PCE. The research on high-rate one-hourly updated ultra-rapid multi-GNSS PCE is slightly insufficient. To meet the real-time and near-real-time requirements for multi-GNSS PPP and ensure the stability and reliability of GNSS precise service, it is necessary to investigate the multi-GNSS high-rate one-hourly updated ultra-rapid multi-GNSS PCE.

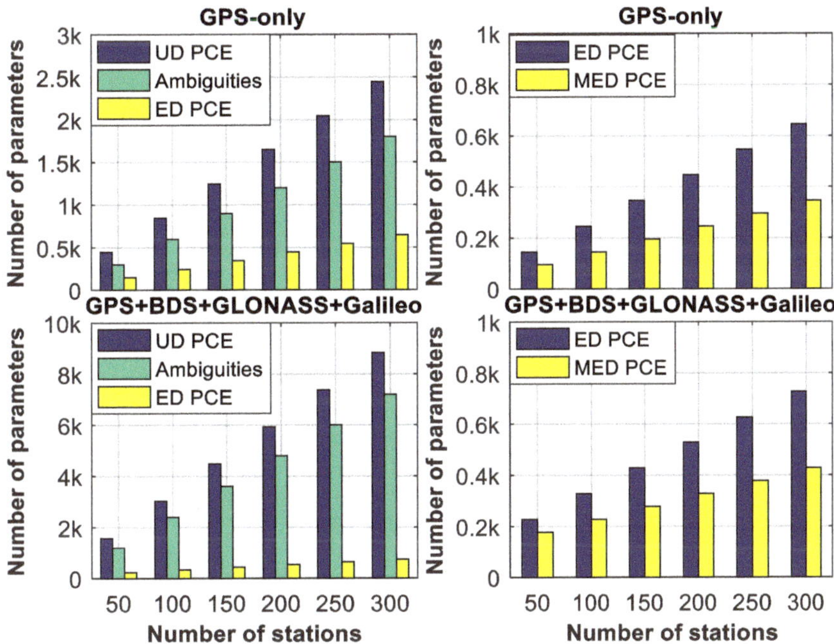

Figure 1. The number of estimated parameters for the different PCE models with the increase in the number of the tracking stations, in which UD PCE, ED PCE, and MED PCE denote undifferenced ionosphere-free PCE model, epoch-difference PCE model, and modified epoch-difference PCE, respectively. Suppose a station can track six satellites for each constellation in one epoch.

With this background, we introduce an efficient model and design a new framework for high-rate one-hourly updated ultra-rapid PCE. We investigate the influence of tropospheric delay on ED clock offsets based on the traditional ED model and further propose the modified ED PCE model to reduce the latency and improve calculation efficiency. Furthermore, Open Multi-Processing (OpenMP) and Intel Math Kernel Library (Intel MKL) technologies are used to improve the PCE processing efficiency and improve the time resolution. First, we present the mathematical models of multi-GNSS modified ED PCE and the methods of restoring ED clock offsets to absolute clock offsets in Section 2. Then, the data processing strategies of multi-GNSS modified ED PCE and new data processing framework are depicted in Section 3. In the following section, we verify the accuracy of the estimated clock offsets. Furthermore, the real-time PPP (RT-PPP) performance using real-time products and high-rate one-hourly updated products are also compared. Finally, we summarize the contributions and give some recommendations.

2. Methods

This section begins with the single-frequency observation models. Then, the modified ED models using GPS, BDS-2, BDS-3, GLONASS, and Galileo observations are developed in detail. Additionally, the methods of restoring ED clock offsets to absolute clock offsets are introduced.

The linearized equations of single-frequency observation equation can be expressed as [19]:

$$
\begin{cases}
P_{r,j}^s = R_r^s + c \cdot \left(\delta t_r + b_j^r + \widetilde{b}_j^r \right) - c \cdot \left(\delta t^s + b_j^s + \widetilde{b}_j^s \right) + g_w \cdot \tau_w + I_{r,j}^s + \varepsilon_{P_j} \\
\Phi_{r,j}^s = R_r^s + c \cdot \left(\delta t_r + B_j^r + \widetilde{B}_j^r \right) - c \cdot \left(\delta t^s + B_j^s + \widetilde{B}_j^s \right) + g_w \cdot \tau_w - I_{r,j}^s + \lambda_j \cdot N_{r,j}^s + \varepsilon_{\Phi_j}
\end{cases}
\tag{1}
$$

where r, s, and j depict the receiver, satellite, and the frequency band, respectively; P and Φ are the range and phase observations; R_r^s is the geometrical range between satellite to receiver; c is the lightspeed; δt_r and δt^s denote the receiver and satellite clock offsets; b_j^r and b_j^s represent the receiver and satellite constant time-invariant uncalibrated code delays (UCDs); B_j^r and B_j^s depict the corresponding constant time-invariant uncalibrated phase delays (UPDs); $\widetilde{b}_j^r$, $\widetilde{b}_j^s$, $\widetilde{B}_j^r$, and $\widetilde{B}_j^s$ represent time-varying part; τ_w and g_w depict the zenith wet delay (ZWD) and the corresponding wet mapping function; $I_{r,j}^s$ is the slant ionospheric delay; $N_{r,j}^s$ denote the integer ambiguity with its wavelength λ_j; ε_{P_j} and ε_{Φ_j} are the range and phase observation noises containing multipath and unmodeled error.

Most IGS and International GNSS Monitoring and Assessment System (iGMAS) ACs adopted undifferenced IF model to obtain the satellite clock offsets [1], which can be expressed as:

$$
\begin{cases}
P_{r,IF_{i,j}}^s = R_r^s + c \cdot \delta t_{IF_{i,j}}^r - c \cdot \delta t_{IF_{i,j}}^s + g_w \cdot \tau_w + \varepsilon_{P_{IF_{i,j}}} \\
\Phi_{r,IF_{i,j}}^s = R_r^s + c \cdot \delta t_{IF_{i,j}}^r - c \cdot \delta t_{IF_{i,j}}^s + g_w \cdot \tau_w + \lambda_{IF_{i,j}} \cdot N_{r,IF_{i,j}}^s + \varepsilon_{\Phi_{IF_{i,j}}}
\end{cases}
\tag{2}
$$

The IF combination for observations and hardware delays can be described as [19]:

$$
\begin{cases}
\alpha_{i,j} = \frac{f_i^2}{f_i^2 - f_j^2} ; \beta_{i,j} = \frac{-f_j^2}{f_i^2 - f_j^2} \\
(\cdot)_{IF_{i,j}} = \alpha_{i,j} \cdot (\cdot)_i + \beta_{i,j} \cdot (\cdot)_j \\
(\cdot) = P_r^s, \Phi_r^s, b^r, b^s, B^r, B^s, \widetilde{b}^r, \widetilde{b}^s, \widetilde{B}^r, \widetilde{B}^s \\
\delta t_{IF_{i,j}}^r = \delta t_r + b_{IF_{i,j}}^r + \widetilde{B}_{IF_{i,j}}^r \\
\delta t_{IF_{i,j}}^s = \delta t^s + b_{IF_{i,j}}^s + \widetilde{B}_{IF_{i,j}}^s \\
\lambda_{IF_{i,j}} \cdot N_{r,IF_{i,j}}^s = \left(B_{IF_{i,j}}^r - b_{IF_{i,j}}^r \right) - \left(B_{IF_{i,j}}^s - b_{IF_{i,j}}^s \right) + \alpha_{i,j} \cdot \lambda_j \cdot N_{r,j}^s + \beta_{i,j} \cdot \lambda_i \cdot N_{r,i}^s
\end{cases}
\tag{3}
$$

where $\alpha_{i,j}$ and $\beta_{i,j}$ are frequency factors; f_i and f_j depict the ith and jth frequency; $P_{r,IF_{i,j}}^s$, $\Phi_{r,IF_{i,j}}^s$, $b_{IF_{i,j}}^r$, $b_{IF_{i,j}}^s$, $B_{IF_{i,j}}^r$, $B_{IF_{i,j}}^s$, $\widetilde{b}_{IF_{i,j}}^r$, $\widetilde{b}_{IF_{i,j}}^s$, $\widetilde{B}_{IF_{i,j}}^r$, and $\widetilde{B}_{IF_{i,j}}^s$ are IF combinations for the corresponding observations and hardware delays, respectively; N_{IF} denote the float ambiguity; $\varepsilon_{P_{IF_{i,j}}}$ and $\varepsilon_{\Phi_{IF_{i,j}}}$ are IF observations noises for pseudorange and carrier phase, and $\varepsilon_{P_{IF_{i,j}}}$ will absorb $\widetilde{b}_{IF_{i,j}}^r - \widetilde{B}_{IF_{i,j}}^r + \widetilde{B}_{IF_{i,j}}^s - \widetilde{b}_{IF_{i,j}}^s$.

As mentioned above, it is difficult to achieve the high-rate one-hourly updated PCE because of many estimated parameters for the UD model. To ensure the accuracy, stability, and computational efficiency of one-hourly ultra-rapid PCE, the ED model is applied to perform PCE, which can be expressed as:

$$
\begin{cases}
\Delta P_{r,IF_{i,j}}^s(t_{k+1}, t_k) = c \cdot \Delta \delta t_{IF_{i,j}}^r(t_{k+1}, t_k) - c \cdot \Delta \delta t_{IF_{i,j}}^s(t_{k+1}, t_k) + \Delta g_w(t_{k+1}, t_k) \cdot \tau_w(t_{k+1}) + \varepsilon_{\Delta P_{IF_{i,j}}} \\
\Delta \Phi_{r,IF_{i,j}}^s(t_{k+1}, t_k) = c \cdot \Delta \delta t_{IF_{i,j}}^r(t_{k+1}, t_k) - c \cdot \Delta \delta t_{IF_{i,j}}^s(t_{k+1}, t_k) + \Delta g_w(t_{k+1}, t_k) \cdot \tau_w(t_{k+1}) + \varepsilon_{\Delta \Phi_{IF_{i,j}}}
\end{cases}
\tag{4}
$$

where Δ is the ED operator; $s = GPS, BDS, GLONASS, Galileo$; for example, the ED clock offsets are $\Delta \delta t_{IF_{i,j}}^r(t_{k+1}, t_k) = \delta t_{IF_{i,j}}^r(t_{k+1}) - \delta t_{IF_{i,j}}^r(t_k)$. The ambiguity can be removed by ED operation when there are no cycle slips in carrier phase observations. The inter-system bias is quite stable over a day and can be eliminated by ED [20]. Therefore, only ZWDs and the ED clock offsets of receiver and satellite remain. One receiver equipped high-precision atomic clock is selected as reference clock to avoid the singularity between receiver and satellite clock offsets.

Although the traditional ED model and the corresponding data processing strategy are difficult to achieve the high-rate one-hourly updated ultra-rapid PCE, the ED model greatly reduces the number of estimated parameters and reduces the consuming time. Based on the characteristic of traditional ED model and the corresponding data processing strategy, we investigated the influence of tropospheric delays on ED PCE. Generally, the tropospheric delays cannot be ignored in the ED PCE. The systematic bias caused by mapping function and systematic variation of tropospheric delays will be absorbed by absolute clock offsets [7,21]. The ZWD are generally solved once an hour. If ZWD can be eliminated by ED, as shown in Figure 1, the estimated parameters will be further reduced. In addition, there is no correlation between epochs on the premise of estimating ED clock offsets as white noise. The parallel processing between epochs can be realized to further improve the solution efficiency. The modified ED model without ZWD estimations can be expressed as:

$$
\begin{bmatrix} \Delta P^s_{r,IF_{i,j}} \\ \Delta \Phi^s_{r,IF_{i,j}} \\ clk_{ref} \end{bmatrix} = \begin{bmatrix} e_n^T \otimes e_2 \otimes e_m & e_k^T \otimes e_1 \otimes e_m \\ e_1^T \otimes e_1 \otimes e_l \end{bmatrix} \begin{bmatrix} \Delta \delta t^r_{IF_{i,j}} \\ \Delta \delta t^s_{IF_{i,j}} \end{bmatrix} + \begin{bmatrix} \varepsilon_{\Delta P_{IF_{i,j}}} \\ \varepsilon_{\Delta \Phi_{IF_{i,j}}} \end{bmatrix} \tag{5}
$$

where n is the number of stations, m represents the number of observations, k depicts the number of the satellites, l is the number of observations at reference station. e_n, e_m, e_k, and e_l are n-row, m-row, k-row, and l-row vector in which all values are 1; clk_{ref} is the reference clock. For simplicity, SHA1 and SHA2 denote the modified ED PCE model (ignoring ZWD) and the traditional ED PCE model (estimating ZWD), respectively.

The estimated ED clock offsets must be restored to the absolute clock offsets before it can be used normally. The low-rate clock offsets are used as control points to obtain absolute clock offsets, which can be written as [18]:

$$
\begin{bmatrix} \widetilde{\delta t}_{fix}(t_1) \\ \Delta \delta t(t_2,t_1) \\ \Delta \widetilde{\delta t}(t_3,t_2) \\ \cdots \\ \Delta \delta t(t_i,t_{i-1}) \\ \widetilde{\delta t}_{fix}(t_i) \\ \Delta \delta t(t_{i+1},t_i) \\ \cdots \\ \Delta \delta t(t_n,t_{n-1}) \\ \widetilde{\delta t}_{fix}(t_n) \end{bmatrix} = \begin{bmatrix} \delta t(t_1) \\ -\delta t(t_1) & \delta t(t_2) \\ & -\delta t(t_2) & \delta t(t_3) \\ & & \cdots & \vdots \\ & & & -\delta t(t_{i-1}) & \delta t(t_i) \\ & & & & \delta t(t_i) \\ & & & & -\delta t(t_i) & \delta t(t_{i+1}) \\ & & & & & \cdots & \vdots \\ & & & & & & -\delta t(t_{n-1}) & -\delta t(t_n) \\ & & & & & & & -\delta t(t_n) \end{bmatrix} \tag{6}
$$

where $\widetilde{\delta t}_{fix}(t_1)$, $\widetilde{\delta t}_{fix}(t_i)$, and $\widetilde{\delta t}_{fix}(t_n)$ are the low-rate clock offsets; $\Delta \delta t(t_{i+1},t_i)$ is the estimated ED clock offsets; n denotes the number of epochs of the low-rate clock offsets.

3. Datasets and Processing Strategies

3.1. Datasets

Approximately 120 stations are used to estimate the high-rate one-hourly ultra-rapid clock offsets. The hourly updated observations from IGS MGEX and iGMAS are used to perform the PCE. The selected GNSS stations are shown in Figure 2. To verify the performance of high-rate hourly updated ultra-rapid clock offsets, the GNSS stations without PCE are selected for RT-PPP. The stations marked red are used for high-rate PCE, and the other GNSS stations marked blue are used for RT-PPP.

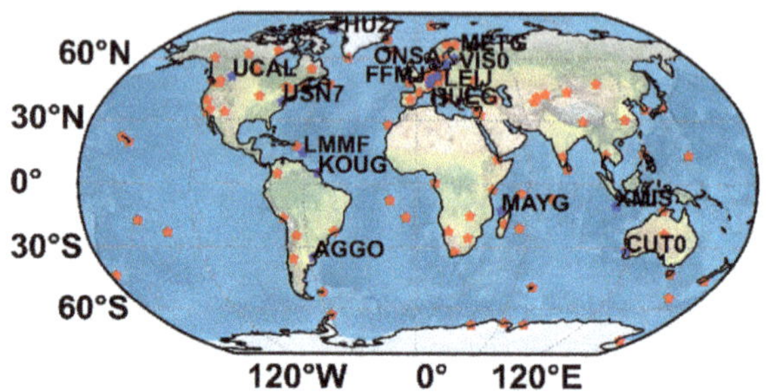

Figure 2. Distribution of the selected GNSS tracking stations for PCE and PPP, in which the stations marked red are used for PCE and the stations marked blue are used for RT-PPP.

3.2. Processing Strategies

The one-hourly updated ultra-rapid multi-GNSS PCE processing strategies are illustrated in Table 1. The dual-frequency observations of L1 (1575.42 MHz) and L2 (1227.60 MHz) for GPS, G1 (1602+k*9/16 MHz) and G2 (1246+k*7/16 MHz) for GLONASS, E1 (1575.42 MHz) and E5a (1176.45 MHz) for Galileo, B1I (1561.098 MHz) and B3I (1268.52 MHz) for BDS-2 and BDS-3 are used to generate high-rate hourly updated ultra-rapid multi-GNSS clock offsets. The satellite orbit and station coordinates are fixed to the SHAO one-hourly updated ultra-rapid solutions [11]. The reference frame and time system of SHAO (Shanghai Astronomical Observatory) precise products are ITRF 2014 and GPS time (GPST), and the satellite orbit and clock offsets is 5 min sampling [11]. As for the noteworthy clock, one receiver equipped high-accuracy atomic clock is selected as the reference clock to avoid the singularity between receiver and satellite ED clock offsets. Because the low-rate clock offsets are used as control points to obtain absolute clock offsets, the reference clock is consistent with SHAO to avoid additional bias. The dry tropospheric delay is corrected using the modified Hopfield model and Global Pressure and Temperature 3 (GPT3), and the Vienna mapping functions 3 (VMF3) is used to obtain the mapping functions of both dry and wet parts [22]. Regarding the wet part, the two schemes mentioned above are proposed to investigate the effect of troposphere estimation on PCE. As for phase center offset (PCO) and phase center variations (PCV), GPS, GLONASS, Galileo satellite antennas are corrected using the antenna file data provided by IGS MGEX [23,24]. BDS-2 and BDS-3 are corrected using the BeiDou official data (http://www.beidou.gov.cn/, accessed on 1 March 2022).

Table 1. The data processing strategy for precise clock estimation (PCE).

Items	Strategies				
System	GPS	GLONASS	Galileo	BDS-2	BDS-3
Frequency	L1/L2	G1/G2	E1/E5a	B1I/B3I	B1I/B3I
Observations	Pseudorange and carrier phase observations				
Elevation cutoff	7 degrees				
Observation weighting	Elevation weight [sin(elevation)]				
Satellite orbit	Fixed by SHAO one-hourly updated ultra-rapid precise orbit products				
Satellite ED clock offsets	Estimated				

Table 1. *Cont.*

Items	Strategies	
Satellite absolute clock offsets	Estimated by using satellite ED clock offsets and low-rate absolute clock offsets	
Receiver coordinate	Fixed by SHAO station coordinates	
Reference clock	Receiver clock	
Receiver ED clock	Ordinary station: EstimatedReference station: Fixed	
Tropospheric delay	SHA1: Modified Hopfield for dry and wet part SHA2: Modified Hopfield for dry part and estimated for wet part (10^{-9} m^2/s)	
Ionospheric delay	Eliminated first order by IF observations	
Satellite antenna	IGS MGEX values	BDS official
Receiver antenna	IGS MGEX values	
Phase windup effect	Corrected [25]	
Relativistic effect	Corrected [26]	
Earth rotation	Corrected [26]	
Tide effect	Solid Earth, Pole, and Ocean tide	
Ambiguity	Eliminated by ED model	

Figure 3 illustrates the slide window for one-hourly updated ultra-rapid orbit and clock offsets products generation. To achieve one-hourly updated ultra-rapid orbit and clock offsets, the computation time of POD and high-rate PCE is within one hour. SHAO can solve low-rate ultra-rapid multi-GNSS orbit and clock offsets in 35 to 40 min based on the Lenovo SR650 server (4*Inter (R) Xeon (R) Gold 6244 CPU @3.60 GHz), which means that the high-rate PCE needs to be solved in about 20 min. Moreover, the latency of one-hourly ultra-rapid multi-GNSS satellite products is one hour, we need to predict it to meet the real-time applications. The orbit prediction method is integrating the prediction orbit by the fitted orbit, from 24 h, and adjacent afore 24 h orbits (observation session) [11]. The clock offsets prediction model is modified Auto Regressive Integrated Moving Average (ARIMA) [27].

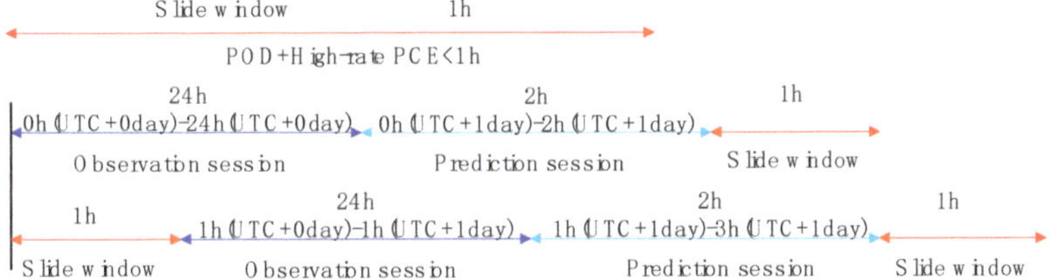

Figure 3. The slide window for one-hourly updated ultra-rapid orbit and clock offsets products generation.

According to the characteristics of the modified ED PCE model, we propose a corresponding new data processing flow, as shown in Figure 4, to achieve high-rate one-hourly updated ultra-rapid PCE. First, the IGS real-time stream and one-hourly updated observations obtained from IGS MGEX and iGMAS are merged and preprocessed. After that, the POD is performed to obtain one-hourly updated ultra-rapid precise products including orbit, low-rate clock offsets, Earth rotation parameters (ERP), and station coordinates. Then, the one-hourly updated ultra-rapid precise products and observations are imputed into high-rate PCE system marked red in Figure 4. In the processing of calculating correction information, there is no correlation between stations, the OpenMP can be used for parallel processing of multiple stations. In the processing of calculating ED clock offsets, the Intel MKL is used to accelerate the matrix solution. Because the SHA1 model does not need to estimate ZWD, the OpenMP can be used for parallel processing of multiple epochs. However, the OpenMP cannot be used for SHA2 parallel processing of multiple epochs. In the process of obtaining absolute clock offsets, the OpenMP can be used for parallel processing of multiple satellites and Intel MKL can be used to accelerate the matrix solution. This high-rate PCE system has strong flexibility, and it can be connected to the ultra-rapid, rapid, final, and real-time POD system to obtain high-rate clock offsets.

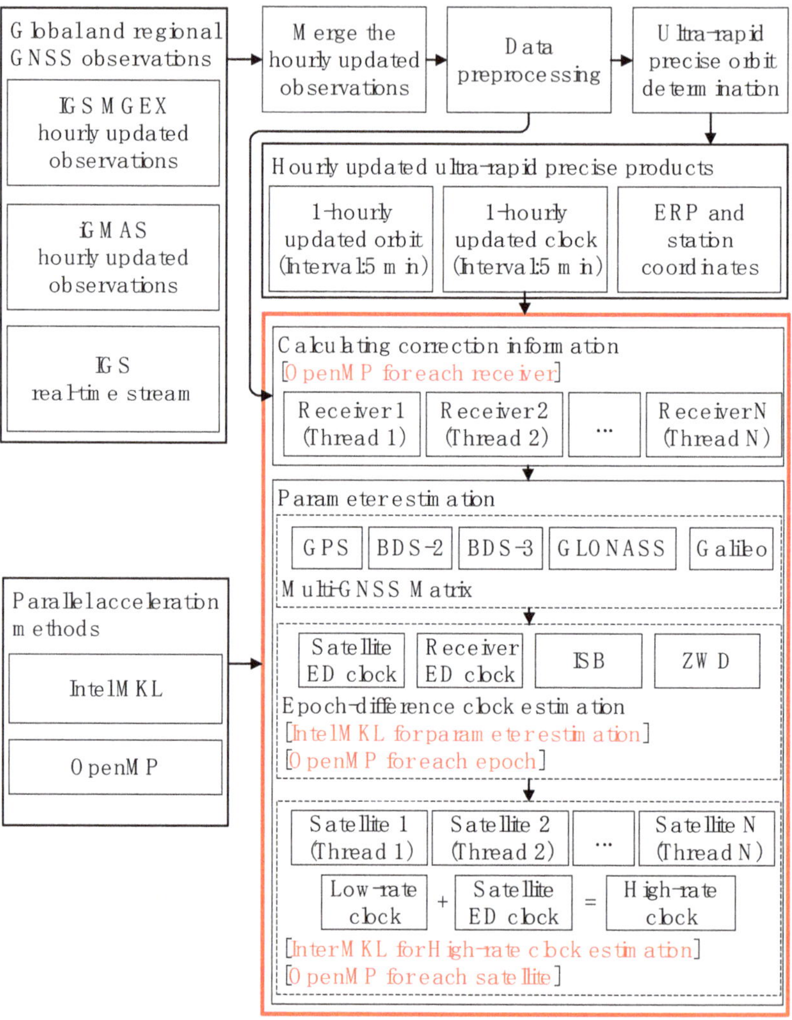

Figure 4. The data processing flow for one-hourly updated ultra-rapid orbit and clock offsets products generation.

4. Validation and Results

This part begins with the validation of the estimated clock offsets. Because the high-rate one-hourly updated ultra-rapid products are mainly oriented to real-time and near-real-time users, we compared the estimated high-rate clock offsets and real-time clock offsets provided by CNES (CLK93) in terms of the clock offsets accuracy and PPP performances. Finally, the performance of the estimated clock offsets is discussed.

4.1. The ED Clock Offsets

The time series of estimated ED clock offsets and its comparison with GFZ rapid (GBM) and CODE final (COD) multi-GNSS products are depicted in Figure 5. To show ED clock offsets more clearly, the ED clock offsets are shifted by the same amount (1 ns) to avoid overlapping. Figure 6 shows their differences. Because BDS constellations are composed of Geostationary Earth Orbit (GEO), Inclined Geosynchronous Satellite Orbit (IGSO), and Medium Earth Orbit (MEO) satellites, different orbit types of BDS satellites have different accuracy in POD [28,29]. Therefore, the corresponding clock offsets have different accuracy for different orbit types. In order to analyze BDS-2 objectively, we discuss it separately. From Figures 5 and 6, we can find that the difference between GBM and the estimated ED clock offsets for SHA1 and SHA2 is very small, and they are basically distributed on the diagonal with slope of 1. From Figures 5 and 7, we can draw a similar conclusion that the estimated ED clock offsets are basically consistent with COD. This means that the performance of the estimated ED clock is basically consistent with GBM and COD. Certainly, we can also find that the performance of BDS-2 is slightly worse than other systems, especially for GEO satellites. There are three reasons for the difference. First, BDS-2 is a regional navigation system, and the stations tracked BDS-2 signals are basically located in Asia and Europe. There are less observations for BDS-2 POD and PCE because of the unevenly distributed stations. Secondly, BDS-2 contains five GEO and seven IGSO satellites. The orbit accuracy of BDS-2 GEO and IGSO satellites is poor due to the force models and the observation geometry. Therefore, the estimated ED clock offsets calculated by fixed GEO orbit is slightly poor. Last, but most important, there are some differences between GFZ and SHAO in BDS data processing strategies in terms of the precise attitude model, the solar radiation pressure model, and DCB corrections [11,30], which leads to the poor consistency of the results. In order to more accurately and objectively explain the accuracy of the estimated ED clock offsets compared with GBM and COD, the mean, STD, and RMS of the estimated ED clock offsets between GBM, COD and SHA1, SHA2 for multi-GNSS satellites are listed in Table 2. Our calculation is basically consistent with the results of GBM and COD. The difference in ED clock offsets between GBM, COD and SHA1, SHA2 for GPS and Galileo satellites range from 6 to 7 ps, and the difference for BDS and Galileo range from 10 to 16 ps. The results in Table 2 confirm the findings of Figures 5–7.

Table 2. The statistics of means, STD, and RMS for GPS, BDS-2 GEO, BDS-2 IGSO, BDS-2 MEO, BDS-3, GLONASS, and Galileo satellites ED clock (unit: ps).

Type	System	GPS	BDS-2 (GEO)	BDS-2 (IGSO)	BDS-2 (MEO)	BDS-3	GLONASS	Galileo
SHA1 (GBM)	Mean	−0.26	0.96	−0.47	−0.30	−0.37	−0.20	−0.25
	STD	6.45	21.31	15.19	14.36	11.04	12.64	7.17
	RMS	6.46	21.33	15.20	14.36	11.05	12.65	7.17
SHA2 (GBM)	Mean	−0.13	0.94	−0.43	−0.25	−0.24	−0.13	−0.18
	STD	6.29	21.25	15.54	14.56	11.03	12.71	7.16
	RMS	6.32	21.27	15.55	14.57	11.03	12.71	7.16
SHA1 (COD)	Mean	−0.21	Non	−0.27	−0.22	−0.25	−0.11	−0.22
	STD	6.73	Non	14.93	14.50	11.22	12.61	6.87
	RMS	6.82	Non	15.01	14.52	11.24	12.71	6.90
SHA2 (COD)	Mean	−0.09	Non	−0.15	−0.17	−0.13	−0.03	−0.12
	STD	6.54	Non	15.46	13.90	11.46	13.26	7.32
	RMS	6.57	Non	15.50	14.00	11.48	13.30	7.34

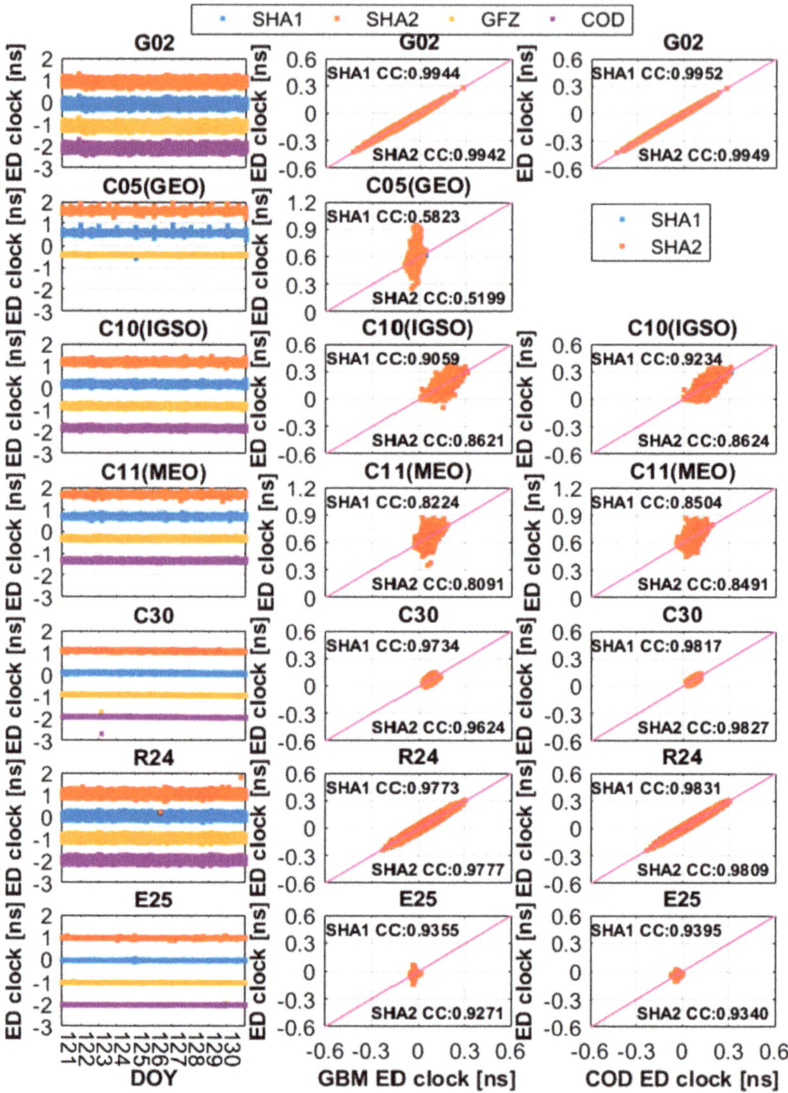

Figure 5. The time series of estimated ED clock offsets and its comparison with GFZ rapid and COD final clock products, in which CC denotes the correlation coefficient.

Now, turn to the influence of tropospheric delays on ED clock offsets estimation. Figures 5–7 not only show the difference between GBM, COD, and the estimated ED clock offsets but also depict the difference between SHA1 and SHA2. As shown in Figure 5, SHA1 and SHA2 are highly overlapped, and it is difficult to distinguish the difference. From Figures 6 and 7 and Table 2, we can see that the difference between SHA1 and SHA2 is very small, ranging from 0 to 0.35 ps in terms of the statistics of mean, STD, and RMS for GPS, BDS-2 GEO, BDS-2 IGSO, BDS-2 MEO, BDS-3, GLONASS, and Galileo satellites, which can be ignored. To prove this point more favorably and reflect the situation of each satellite clearly, the relationship between SHA1 and SHA2 in terms of the mean, STD, and RMS of each satellite is depicted in Figure 8. The mean of each satellite for SHA2 is closer to 0 than SHA1, but the difference is very tiny. In addition, the STD and RMS of each satellite are basically distributed on the diagonal with slope of 1. In general, the ED clock of SHA1 and SHA2 are basically consistent both in mean, STD, and RMS. However, from a calculation time-consuming point of view, adding the ZWD parameter will increase the computation complexity. The corresponding time-consuming will increase. As shown in Figure 9, the mean time-consuming of SHA1 is about 16.5 min, and that of SHA2 is about 23.5 min, based on Lenovo SR650 server (4*Inter (R) Xeon (R) Gold 6244 CPU @3.60GHz) using OpenMP and Intel MKL. In case of poor data quality, the time-consuming of SHA1 is about 20 to 23 min. In normal conditions, SHA1 takes less than 20 min.

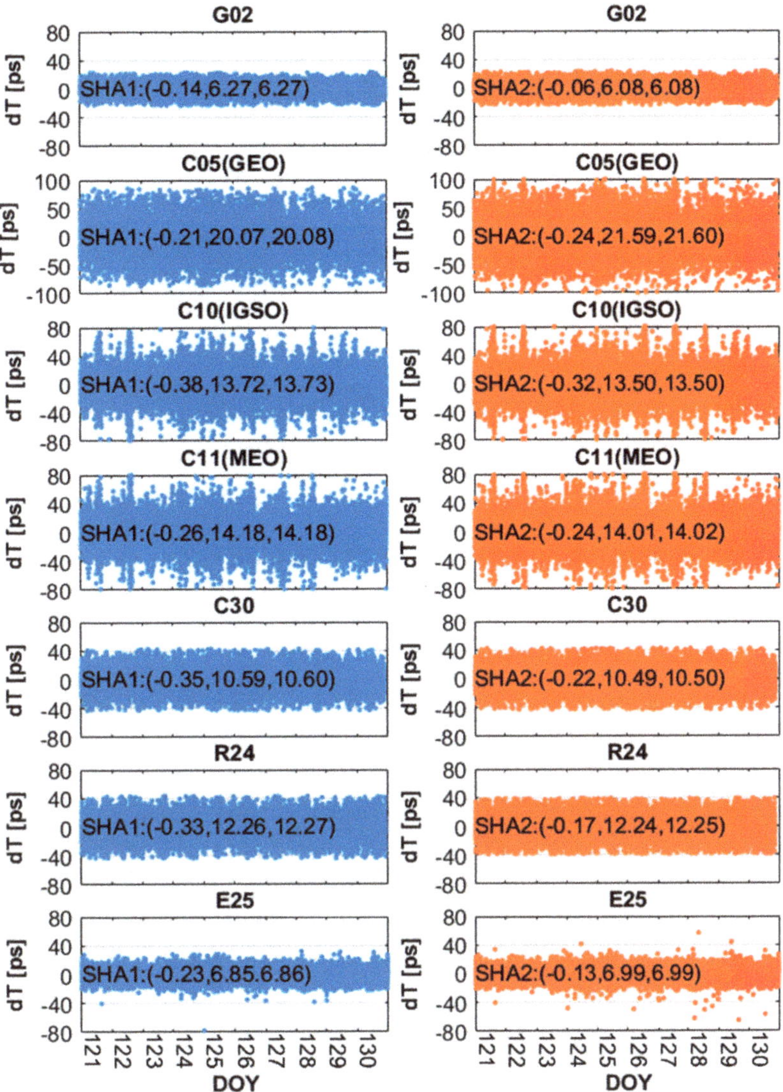

Figure 6. The difference between GFZ rapid clock offsets and estimated ED clock offsets, in which (a, b, c) denotes (mean, STD, and RMS).

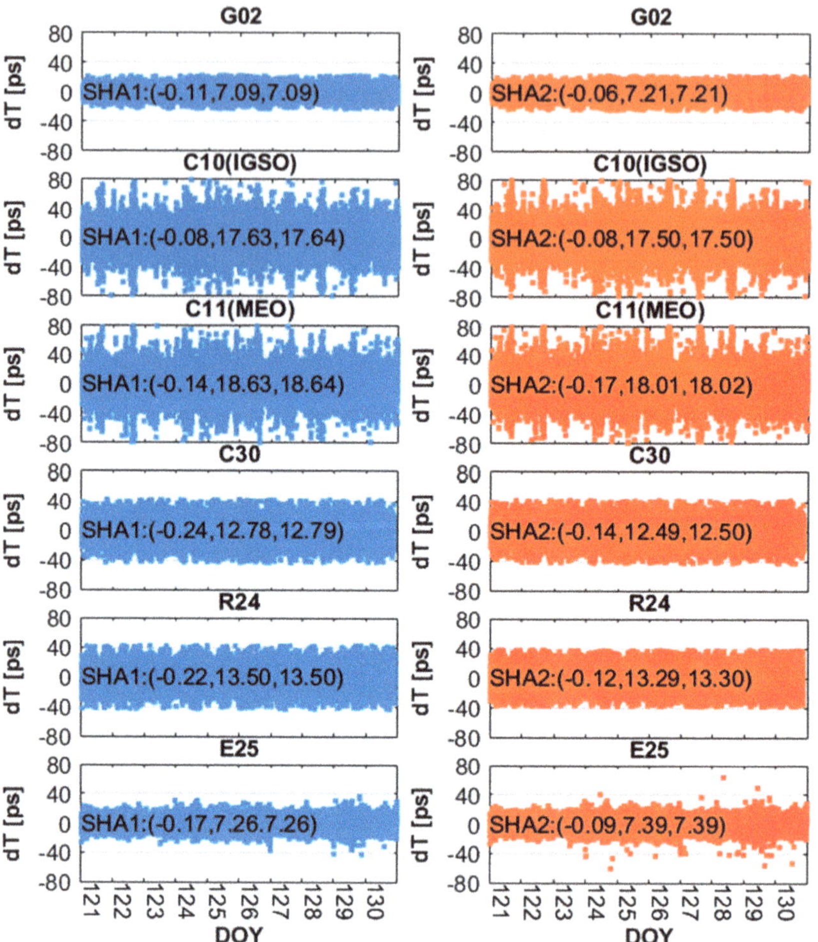

Figure 7. The difference between CODE final clock offsets and estimated ED clock offsets, in which (a, b, c) denotes (mean, STD, and RMS).

As mentioned above, the time-consuming of POD and high-rate PCE must be less than 1 h to achieve one-hourly updated orbits and clocks. Because the difference between SHA1 and SHA2 is very tiny, and its difference is less than 0.35 ps in terms of mean, STD, and RMS. As stated earlier, SHA1 can use the OpenMP for parallel processing of multiple stations, multiple epochs, and multiple satellites, while SHA2 can only use the OpenMP for parallel processing of multiple stations and multiple satellites. The computational efficiency of SHA1 is significantly higher than that of SHA2. Considering the calculation time and accuracy, SHA1 as the optimal strategy is applied to obtain one-hourly updated ultra-rapid clock.

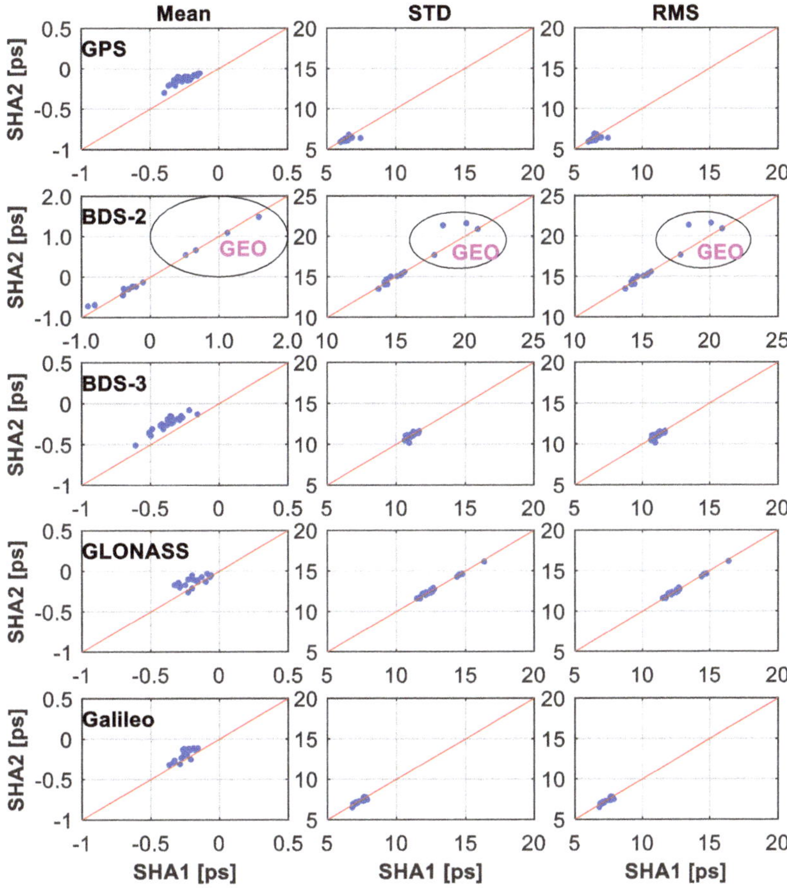

Figure 8. The mean, STD, and RMS of each satellite for GPS, BDS-2, BDS-3, GLONASS, and Galileo using two tropospheric processing strategies.

4.2. The Absolute Clock Offsets

The ED clock offsets calculated by the SHA1 strategy are restored to the absolute clock offsets by using Equation (6). The double-difference (DD) method of selecting one satellite to eliminate the clock datum [31] is used to assess the absolute clock. GPS PRN01 (G01), BDS-2 PRN07 (C07), BDS-3 PRN19 (C19), GLONASS PRN24 (R24), and Galileo PRN01 (E01) are selected as the reference satellite to assess the clock of other GNSS satellites. Figures 10 and 11 show the STD and RMS of high-rate and low-rate one-hourly updated GNSS ultra-rapid clock for observation. In addition, the ultra-rapid products are more targeted to real-time and near real-time users; we added the real-time stream products CLK93 released by CNES to discuss the precision of estimated clock and CLK93. When users use the low-rate clock, they need to use mathematical interpolation algorithm to interpolate the low-rate clock, which causes the loss of clock accuracy [14]. In order to reflect the actual applications of low-rate clock, the clock interpolated by the ninth order Lagrange interpolation algorithm was also evaluated.

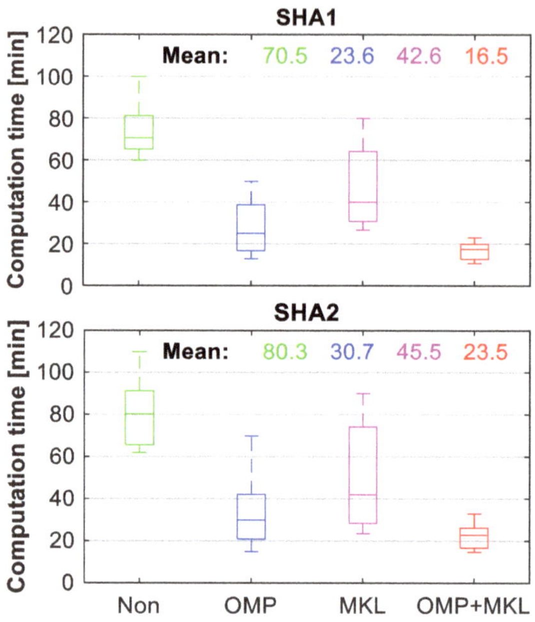

Figure 9. The computation time for 1-day high-rate multi-GNSS PCE using 120 GNSS tracking stations, in which Non, OMP, and MKL denote no acceleration method, OpenMP, and Intel MKL, respectively.

It can be seen from Figure 10 that the STD of the estimated high-rate clock (SHA30s (PCE)), low-rate clock (SHA300s), and clock offsets interpolated by mathematical interpolation algorithm (SHA30s (Math)) for GPS, BDS-2 GEO, BDS-2 non-GEO, BDS-3, GLONASS, and Galileo satellites are basically better than 0.2, 1.2, 0.3, 0.3, 0.4, and 0.2 ns, respectively. From SHA30s (PCE), SHA300s, and SHA30s (Math), we can find that SHA30s (PCE) has a certain improvement compared with SHA30s (Math), but the improvement is not obvious compared with SHA300s. There are three reasons for the little increase in SHA30s (PCE) compared with SHA300s. Firstly, the weight of the control points is higher than that of the ED clock offsets when the ED clock offsets are restored to the absolute clock offsets by using low-rate clock offsets as control points. Secondly, the accuracy of SHA300s is obtained by evaluating the clock offsets on time node (300s intervals), and there is no precision loss caused by the interpolation algorithm. Finally, the accuracy of SHA300s provide by SHAO is already good.

It is worth mentioning that the STD of the estimated clock offsets is slightly better than CLK93. There are two possible explanations: the clock offsets precision is inevitably disturbed by the stability of real-time observation streams, more commonly, the available observation data for real-time product solution are less than ultra-rapid solutions. In other words, one-hourly updated ultra-rapid orbit and clock offsets products can ensure better accuracy in the case of poor network environment. Whether it is used as backup data or practical data, one-hourly updated ultra-rapid orbit and clock offsets products are very reliable.

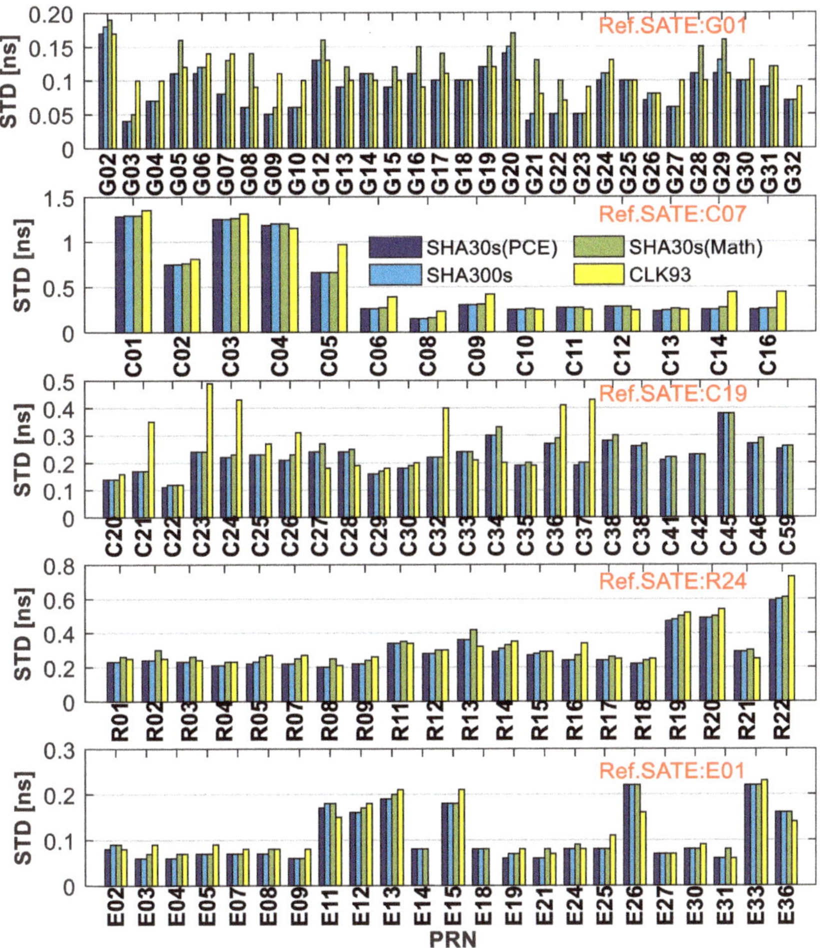

Figure 10. The STD of multi-GNSS ultra-rapid estimated clock for observation session, in which SHA30s (PCE), SHA300s, SHA30s (Math), and CLK93 depict the estimated high-rate clock, low-rate clock, clock interpolated by mathematical interpolation algorithm, and CNES real-time clock, respectively. The Ref.SATE means the reference satellite.

Note that RMS, the RMS of SHA30s (PCE), SHA300s, and SHA30s (Math) for GPS, BDS-2, BDS-3, GLONASS, and Galileo satellites are basically better than 0.5, 3.0, 0.9, 1.2, and 0.4 ns. The RMS of satellites such as G04, G14, G18, G23, C11, C12, and C14 are especially large compared with other satellites. Because the different ACs adopt different processing strategies such as the precise attitude model, the solar radiation pressure model, and DCB corrections [11,30,32], these inconsistencies lead to the differences between SHAO and GBM on some satellites. Moreover, the RMS of GLONASS satellites is worse than other GNSSs, which is due to the possible bias caused by the float ambiguity solutions in POD and frequency division multiple access (FDMA) signal transmit mechanism. From the comparison between SHA30s (PCE) and CLK93, we can find that the RMS of SHA30s (PCE) is better than that of CLK93. As commented earlier, the main reasons are as follows: on the one hand, different ACs have differences in data processing strategies, on the other hand, the available observation data for real-time product solution is not as stable as that of ultra-rapid solutions.

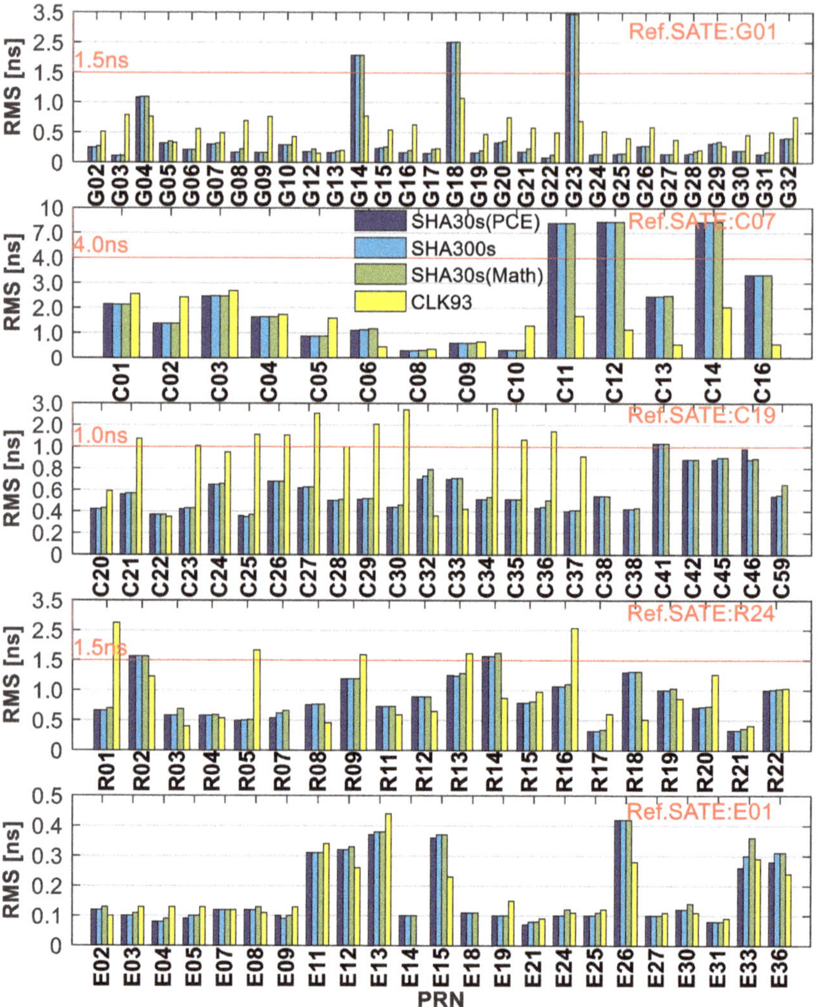

Figure 11. The RMS of high-rate and low-rate one-hourly updated GNSS ultra-rapid clock for observation session, in which SHA30s (PCE), SHA300s, SHA30s (Math), and CLK93 depict the estimated high-rate clock, low-rate clock, clock interpolated by mathematical interpolation algorithm, and CNES real-time clock, respectively.

Now, pay attention to the mean STD and RMS for the observation session and prediction session. Combined with Figures 10–12, we can obtain four key findings. In terms of the precision of the estimated high-rate clock offsets and SHAO low-rate clock offsets, GPS and Galileo are the best, BDS-3 is the second, and GLONASS and BDS-2 are relatively worse. Secondly, SHA30s (PCE) is optimal in both the observation and prediction session. The absolute clock of the prediction and observation sessions has been modified by the improvement of the time resolution. The STD and RMS are improved by (0.97 to 9.09% and 0.12 to 5.56%) in the observation session, (2.82 to 23.08% and 0.95 to 9.09%) in the first hour of the prediction session, and (0 to 3.85% and 0.12 to 4.19%) in the second hour of the prediction session, respectively. Thirdly, the precision of SHA30 (PCE) is slightly better than CLK93 in the observation section and basically the same as CLK93 in the prediction section. Finally, there are some differences between SHAO, GFZ, and CNES in the data processing strategy, and the clock offsets of different ACs show differences in RMS.

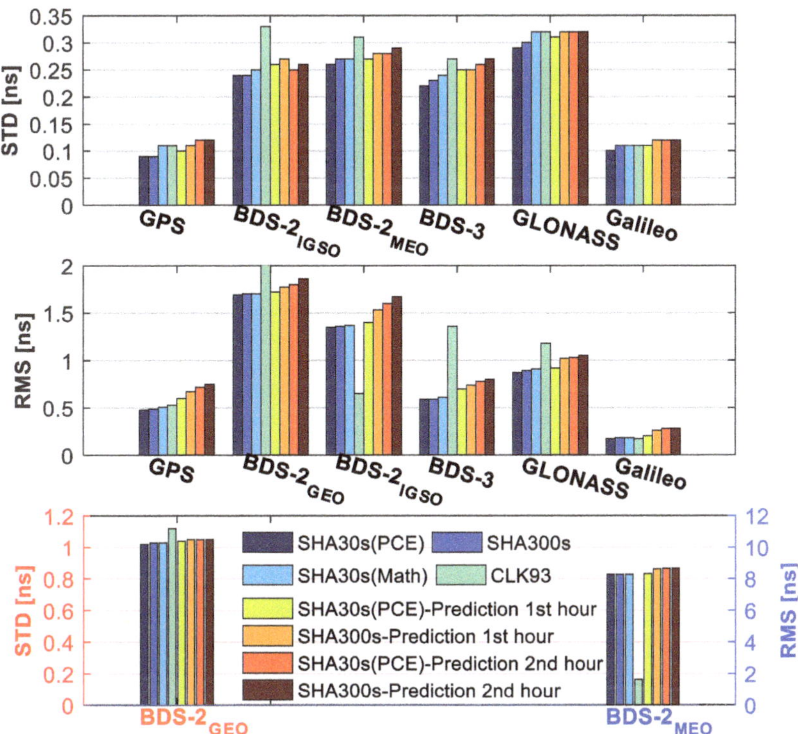

Figure 12. Average STD and RMS of high-rate and low-rate one-hourly updated GNSS ultra-rapid clock for observation session and prediction session.

4.3. Real-Time PPP Validation

The RT-PPP experiments will be carried out to verify the practicability and reliability of the estimated clocks. The PPP solutions using GBM products (30 s interval) is taken as the reference. Figure 13 depicts the PPP convergence at HUEG for GPS-only, BDS-only, GLONASS-only, Galileo-only, and multi-GNSS (GPS + BDS + GLONASS + Galileo). To show the PPP convergence clearly, only the first 2 h are shown in Figure 13. From Figure 13, we can see that the convergence time for SHA30s (PCE) is obviously improved, and its performance is basically same as that of GBM products. Figure 14 shows the average convergence time. During the observation session, SHA30s (PCE) can improve the convergence time by (48.16, 31.55, 37.06%) for GPS-only PPP, (32.88, 21.96, 22.60%) for BDS-only PPP, (40.78, 31.15, 35.05%) for GLONASS-only PPP, (40.49, 23.49, 25.57%) for Galileo-only PPP, and (64.75, 60.28, 55.07%) for multi-GNSS PPP, compared with SHA300s in north, east, and up directions, respectively. In the first hour of the prediction session, SHA30s (PCE) can improve the convergence time by (49.84, 32.47, 37.63%) for GPS-only PPP, (33.62, 24.53, 28.12%) for BDS-only PPP, (44.16, 27.69, 28.69%) for GLONASS-only PPP, (44.82, 25.03, 23.95%) for Galileo-only PPP, and (65.75, 60.98, 56.21%) for multi-GNSS PPP, respectively. In the second hour of the prediction session, SHA30s (PCE) can improve the convergence time by (57.27, 34.85, 36.03%) for GPS-only PPP, (39.76, 27.80, 31.31%) for BDS-only PPP, (43.99, 28.29, 26.30%) for GLONASS-only PPP, (45.74, 26.30, 28.29%) for Galileo-only PPP, and (62.12, 55.51, 57.14%) for multi-GNSS PPP, respectively. The estimated high-rate one-hourly ultra-rapid clock greatly shortens the PPP convergence time, which is basically the same as GBM and slightly better than CLK93 in the observation session. In the prediction session, SHA30s (PCE) performs good, and the convergence time of SHA30s (PCE) is at the same level as CLK93.

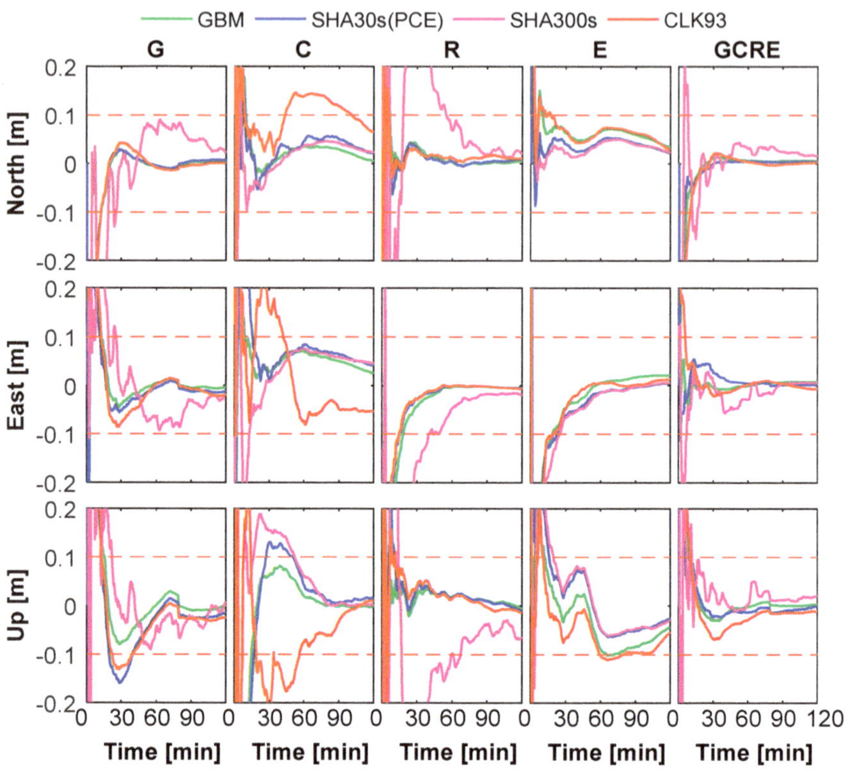

Figure 13. PPP convergence at HUEG for GPS-only, BDS-only, GLONASS-only, Galileo-only, and multi-GNSS (GPS + BDS + GLONASS + Galileo) from 0:00 to 2:00 on DOY 121, 2021.

As for positioning accuracy, the final convergence accuracy of SHA30s (PCE) and SHA300s is basically the same in the observation section. However, the positioning accuracy of SHA300s decreases slightly in the prediction session, while the performance of SHA30s (PCE) is still good. From the Figures 15 and 16, it is obvious that the positioning accuracy will decrease with the increase in prediction time. However, using high-rate clock offsets can appropriately avoid the decrease in positioning accuracy. The positioning accuracy of SHA30s (PCE) for GPS-only, BDS-only, GLONASS-only, Galileo-only, and multi-GNSS PPP is improved by 14.90, 21.68, 13.38, 9.06, and 25.74% in the observation session, 12.34, 1.52, 13.37, 8.66, and 12.99% in the first hour of the prediction session, and 7.97, 9.86, 7.76, 22.45, and 11.65% in the second hour of the prediction session, respectively. Similarly, SHA30s (PCE) improves the PPP positioning accuracy, which is slightly better than CLK93 in the observation session. In the prediction session, the positioning accuracy of SHA30s (PCE) is at the same level as CLK93. Even in the case of poor network environment, the estimated high-rate one-hourly ultra-rapid clock can ensure the better PPP performance.

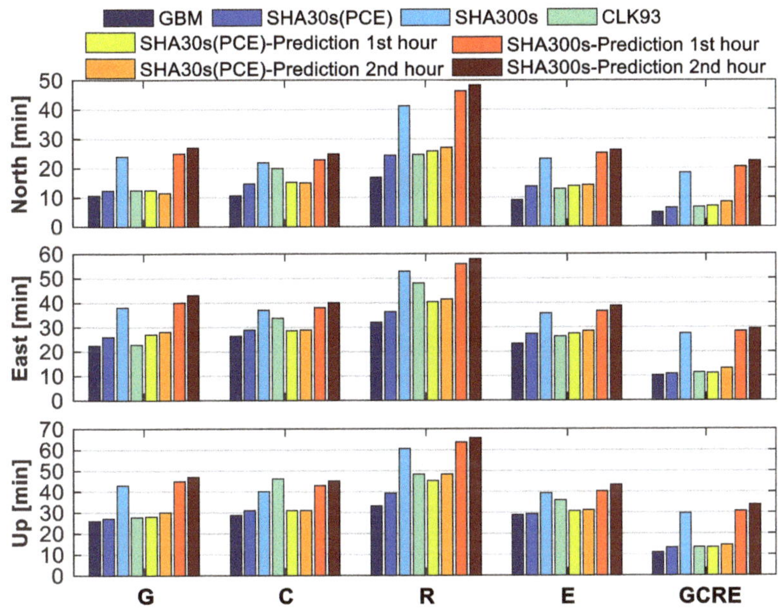

Figure 14. The average convergence time for GPS-only, BDS-only, GLONASS-only, Galileo-only, and multi-GNSS PPP from DOY 121 to 130, 2021.

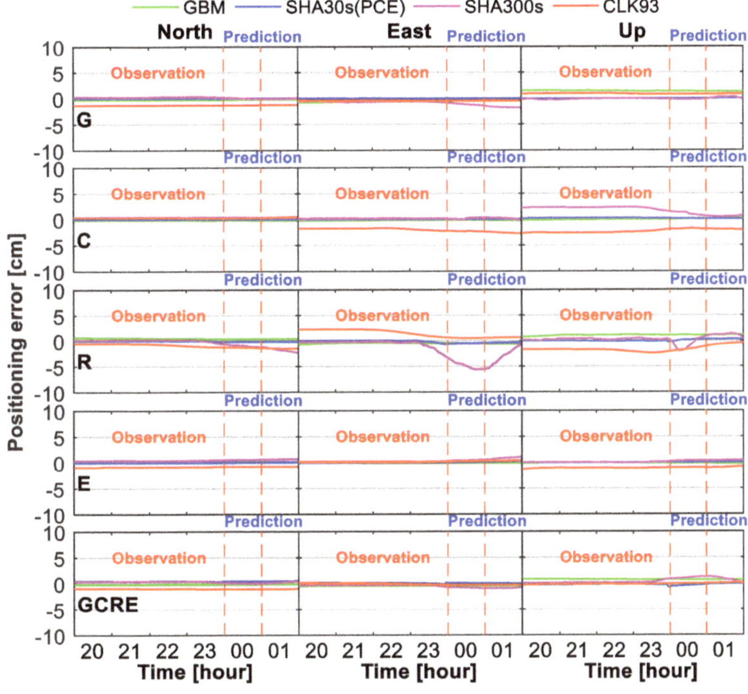

Figure 15. PPP positioning error at HUEG for GPS-only, BDS-only, GLONASS-only, Galileo-only, and multi-GNSS from 20:00 on DOY 121 to 2:00 on DOY 122, 2021.

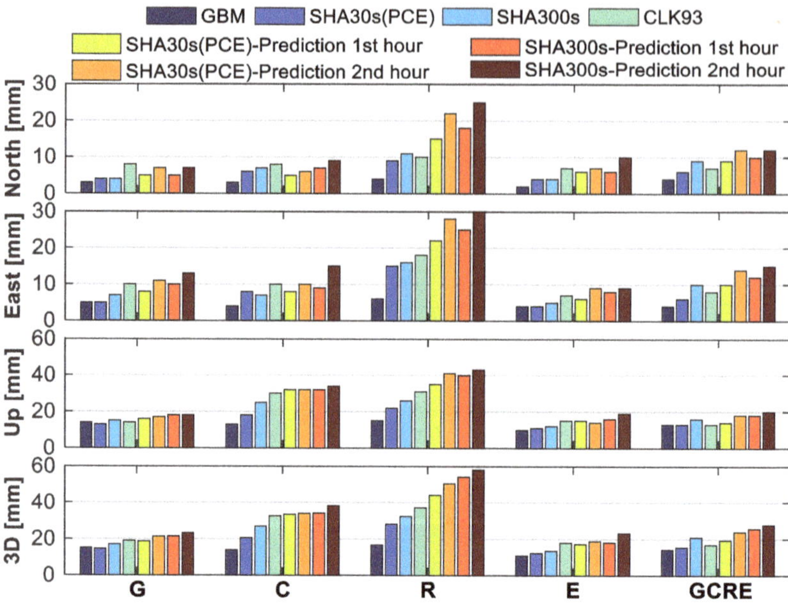

Figure 16. The average RMS for GPS-only, BDS-only, GLONASS-only, Galileo-only, and multi-GNSS PPP from DOY 121 to 130, 2021.

In summary, the estimated high-rate one-hourly ultra-rapid clock can improve the PPP performance in terms of positioning accuracy and convergence time and can be well applied to real-time PPP applications.

5. Conclusions

This contribution focused on the high-rate one-hourly updated ultra-rapid PCE and its application in RT-PPP. We modified the ED PCE model by investigating the influence of tropospheric delays on ED PCE and proposed a new framework to obtain high-rate one-hourly updated ultra-rapid clock. The OpenMP and Intel MKL are used to construct a parallel processing system to improve computation efficiency. Through the analysis and discussion, we can draw the following conclusions.

As for the influence of tropospheric delays on ED PCE, the difference between SHA1 and SHA2 is very tiny, ranging from 0 to 0.35 ps in terms of the statistics of mean, STD, and RMS for multi-GNSS satellites, which can be ignored. From a calculation time-consuming point of view, adding the ZWD parameter will increase the computation complexity. The corresponding time-consuming will increase. The time-consuming of SHA1 is about 16.5 min, and that of SHA2 is about 23.5 min based on the Lenovo SR650 server (4*Inter (R) Xeon (R) Gold 6244 CPU @3.60GHz). Therefore, SHA1 as the optimal strategy with low time-consuming and high accuracy is recommended to obtain one-hourly updated ultra-rapid clock offsets. From the precision of the clock offsets point of view, improving the time resolution can improve the precision of the absolute clock offsets in both prediction and observation sessions. The STD and RMS are improved by (0.97 to 9.09% and 0.12 to 5.56%) in the observation session, (2.82 to 23.08% and 0.95 to 9.09%) in the first hour of the prediction session, and (0.11 to 3.85% and 0.12 to 4.19%) in the second hour of the prediction session, respectively.

Similarly, the estimated clock offsets perform well in RT-PPP. The estimated clock offsets improve the PPP positioning accuracy and greatly shorten the convergence time. The positioning accuracy can be improved by 9.06~25.74% in the observation session, 1.52~13.37% in the first hour of the prediction session, and 7.76~22.45% in the second hour of the prediction session, respectively. The convergence time can be significantly improved by 21.96~64.75, 23.95~65.75, and 26.30~62.12% for the observation session, the first hour of the prediction session, and the second hour of the prediction session compared with low-rate products, respectively. The RT-PPP performance of SHA30s (PCE) is basically the same as GFZ rapid products and slightly better than CLK93. Whether in the prediction session or in the observation session, the estimated clock shows good performance. The estimated high-rate one-hourly ultra-rapid clock can ensure the better PPP performance in the case of poor network environment.

Synthesizing the analysis and discussion above, the estimated high-rate one-hourly updated ultra-rapid precise clock offsets (URL: ftp://igsdepot.ign.fr/pub/igs/products/mgex/, accessed on 1 March 2022) have excellent performance, which can be well applied to real-time or near real-time applications and research.

Author Contributions: Conceptualization, G.J. and S.S.; methodology, G.J.; software, G.J.; validation, G.J. and S.S.; formal analysis, G.J.; investigation, G.J. and S.S.; resources, S.S.; data curation, G.J. and S.S.; writing—original draft preparation, G.J.; writing—review and editing, G.J. and S.S.; visualization, G.J.; supervision, S.S.; project administration, S.S.; funding acquisition, S.S. All authors have read and agreed to the published version of the manuscript.

Funding: This research was funded by the National Natural Science Foundation of China grant number No. 41730109 and No. 12073063.

Institutional Review Board Statement: Not applicable.

Informed Consent Statement: Not applicable.

Data Availability Statement: The datasets analyzed during this study are available in the IGS MGEX repository (ftp://cddis.gsfc.nasa.gov/pub, accessed on 1 March 2022). The multi-GNSS precise orbit and clock products are available in the GFZ (ftp://ftp.gfz-potsdam.de/GNSS/products/mgex/, accessed on 1 March 2022), CODE (ftp://igs.ign.fr/pub/igs/, accessed on 1 March 2022), and CNES (http://www.ppp-wizard.net/products/REAL_TIME/, accessed on 1 March 2022) ACs.

Acknowledgments: The authors would like to thank Ke Su for his valuable suggestions.

Conflicts of Interest: The authors declare no conflict of interest.

References

1. Kouba, J. *A Guide to Using International GNSS Service (IGS) Products*; Jet Propulsion Laboratory: Pasadena, CA, USA, 2009; p. 34. Available online: http://igscb.jpl.nasa.gov/igscb/resource/pubs/GuidetoUsingIGSProducts.pdf (accessed on 1 March 2022).
2. Dow, J.M.; Neilan, R.E.; Rizos, C. The International GNSS Service in a changing landscape of Global Navigation Satellite Systems. *J. Geod.* **2009**, *83*, 191–198. [CrossRef]
3. Choi, K.K.; Ray, J.; Griffiths, J.; Bae, T.-S. Evaluation of GPS orbit prediction strategies for the IGS Ultra-rapid products. *GPS Solut.* **2012**, *17*, 403–412. [CrossRef]
4. Lutz, S.; Beutler, G.; Schaer, S.; Dach, R.; Jäggi, A. CODE's new ultra-rapid orbit and ERP products for the IGS. *GPS Solut.* **2014**, *20*, 239–250. [CrossRef]
5. Elsobeiey, M.; Al-Harbi, S. Performance of real-time Precise Point Positioning using IGS real-time service. *GPS Solut.* **2015**, *20*, 565–571. [CrossRef]
6. Laurichesse, D.; Cerri, L.; Berthias, J.P.; Mercier, F. Real Time Precise GPS Constellation and Clocks Estimation by Means of a Kalman Filter. In Proceedings of the 26th International Technical Meeting of the Satellite Division of The Institute of Navigation (ION GNSS + 2013), Nashville, TN, USA, 16–20 September 2013; pp. 1155–1163.
7. Ge, M.; Chen, J.; Douša, J.; Gendt, G.; Wickert, J. A computationally efficient approach for estimating high-rate satellite clock corrections in realtime. *GPS Solut.* **2011**, *16*, 9–17. [CrossRef]
8. Ge, M.; Gendt, G.; Dick, G.; Zhang, F.P.; Rothacher, M. A New Data Processing Strategy for Huge GNSS Global Networks. *J. Geod.* **2006**, *80*, 199–203. [CrossRef]
9. Zuo, X.; Jiang, X.; Li, P.; Wang, J.; Ge, M.; Schuh, H. A square root information filter for multi-GNSS real-time precise clock estimation. *Satell. Navig.* **2021**, *2*, 28. [CrossRef]
10. Li, X.; Chen, X.; Ge, M.; Schuh, H. Improving multi-GNSS ultra-rapid orbit determination for real-time precise point positioning. *J. Geod.* **2018**, *93*, 45–64. [CrossRef]
11. Chen, Q.; Song, S.; Zhou, W. Accuracy Analysis of GNSS Hourly Ultra-Rapid Orbit and Clock Products from SHAO AC of iGMAS. *Remote Sens.* **2021**, *13*, 1022. [CrossRef]

12. Guo, J.; Xu, X.; Zhao, Q.; Liu, J. Precise orbit determination for quad-constellation satellites at Wuhan University: Strategy, result validation, and comparison. *J. Geod.* **2015**, *90*, 143–159. [CrossRef]
13. Ma, H.; Zhao, Q.; Xu, X. A New Method and Strategy for Precise Ultra-Rapid Orbit Determination. In Proceedings of the China Satellite Navigation Conference (CSNC), Shanghai, China, 23–25 May 2017; Lecture Notes in Electrical Engineering. Springer: Singapore, 2017; pp. 191–205.
14. Guo, F.; Zhang, X.; Li, X.; Cai, S. Impact of sampling rate of IGS satellite clock on precise point positioning. *Geo-Spat. Inf. Sci.* **2010**, *13*, 150–156. [CrossRef]
15. Zhang, X.; Li, X.; Guo, F. Satellite clock estimation at 1 Hz for realtime kinematic PPP applications. *GPS Solut.* **2010**, *15*, 315–324. [CrossRef]
16. Hein, G.W. Status, perspectives and trends of satellite navigation. *Satell. Navig.* **2020**, *1*, 22. [CrossRef] [PubMed]
17. Bahadur, B.; Nohutcu, M. Comparative analysis of MGEX products for post-processing multi-GNSS PPP. *Measurement* **2019**, *145*, 361–369. [CrossRef]
18. Bock, H.; Dach, R.; Jäggi, A.; Beutler, G. High-rate GPS clock corrections from CODE: Support of 1 Hz applications. *J. Geod.* **2009**, *83*, 1083–1094. [CrossRef]
19. Ye, S.; Zhao, L.; Song, J.; Chen, D.; Jiang, W. Analysis of estimated satellite clock biases and their effects on precise point positioning. *GPS Solut.* **2017**, *22*, 16. [CrossRef]
20. Zhang, W.; Lou, Y.; Gu, S.; Shi, C.; Haase, J.S.; Liu, J. Joint estimation of GPS/BDS real-time clocks and initial results. *GPS Solut.* **2015**, *20*, 665–676. [CrossRef]
21. Zhang, Q.; Moore, P.; Hanley, J.; Martin, S. Auto-BAHN: Software for near real-time GPS orbit and clock computations. *Adv. Space Res.* **2007**, *39*, 1531–1538. [CrossRef]
22. Landskron, D.; Bohm, J. VMF3/GPT3: Refined discrete and empirical troposphere mapping functions. *J. Geod.* **2018**, *92*, 349–360. [CrossRef]
23. Montenbruck, O.; Steigenberger, P.; Prange, L.; Deng, Z.; Zhao, Q.; Perosanz, F.; Romero, I.; Noll, C.; Stürze, A.; Weber, G.; et al. The Multi-GNSS Experiment (MGEX) of the International GNSS Service (IGS)–Achievements, prospects and challenges. *Adv. Space Res.* **2017**, *59*, 1671–1697. [CrossRef]
24. Rebischung, P.; Schmid, R. IGS14/igs14.atx: A new framework for the IGS products. In Proceedings of the AGU Fall Meeting 2016, San Francisco, CA, USA, 12–16 December 2016; Volume 121, pp. 6109–6131.
25. Wu, J.-T.; Wu, S.C.; Hajj, G.A.; Bertiger, W.I.; Lichten, S.M. Effects of antenna orientation on GPS carrier phase. In Proceedings of the Astrodynamics 1991, San Diego, CA, USA, 19–22 August 1991; pp. 1647–1660.
26. Petit, G.; Luzum, B. *IERS Conventions(2010)*; IERS Technical Note No. 36; Verlag des Bundesamts für Kartographie und Geodäsie: Frankfurt am Main, Germany, 2010.
27. Zhou, W.; Huang, C.; Song, S.; Chen, Q.; Liu, Z. Characteristic Analysis and Short-Term Prediction of GPS/BDS Satellite Clock Correction. In Proceedings of the China Satellite Navigation Conference (CSNC) 2016 Proceedings: Volume III, Changsha, China, 27 April 2016; pp. 187–200. [CrossRef]
28. Yang, Y.; Gao, W.; Guo, S.; Mao, Y.; Yang, Y. Introduction to BeiDou-3 navigation satellite system. *Navigation* **2019**, *66*, 7–18. [CrossRef]
29. Yang, Y.; Li, J.; Wang, A.; Xu, J.; He, H.; Guo, H.; Shen, J.; Dai, X. Preliminary assessment of the navigation and positioning performance of BeiDou regional navigation satellite system. *Sci. China Earth Sci.* **2013**, *57*, 144–152. [CrossRef]
30. Deng, Z.; Ge, M.; Uhlemann, M.; Zhao, Q. Precise orbit determination of BeiDou satellites at GFZ. In Proceedings of the IGS Workshop, Vienna, Austria, 2 May 2014; pp. 23–27.
31. Yao, Y.; He, Y.; Yi, W.; Song, W.; Cao, C.; Chen, M. Method for evaluating real-time GNSS satellite clock offset products. *GPS Solut.* **2017**, *21*, 1417–1425. [CrossRef]
32. Uhlemann, M.; Gendt, G.; Ramatschi, M.; Deng, Z. GFZ global multi-GNSS network and data processing results. In *IAG 150 Years*; Springer: Berlin/Heidelberg, Germany, 2015; pp. 673–679.

Article

Performance of BDS B1 Frequency Standard Point Positioning during the Main Phase of Different Classified Geomagnetic Storms in China and the Surrounding Area

Junchen Xue [1,*], **Sreeja Vadakke Veettil** [2], **Marcio Aquino** [2], **Xiaogong Hu** [1], **Lin Quan** [3], **Dun Liu** [4], **Peng Guo** [1] **and Mengjie Wu** [1]

[1] Shanghai Astronomical Observatory, Chinese Academy of Sciences, Shanghai 200030, China; hxg@shao.ac.cn (X.H.); gp@shao.ac.cn (P.G.); mjwu@shao.ac.cn (M.W.)
[2] Nottingham Geospatial Institute, University of Nottingham, Nottingham NG7 2TU, UK; v.sreeja@gmail.com (S.V.V.); Marcio.Aquino@nottingham.ac.uk (M.A.)
[3] Beijing Institute of Tracking and Telecommunication Technology, Beijing 100094, China; quanlin@BITTT.cn
[4] China Research Institute of Radiowave Propagation, Qingdao 266107, China; liud@crirp.ac.cn
[*] Correspondence: jcxue@shao.ac.cn

Abstract: Geomagnetic storms are one of the space weather events. The radio signals transmitted by modern navigation systems suffer from the effects of magnetic storms, which can degrade the performance of the whole system. In this study, the performance of the BeiDou Navigation Satellite System (BDS) B1 frequency standard point positioning (SPP) in China and the surrounding area during different classes of storm was investigated for the first time. The statistical analysis of the results revealed that the accuracy of the BDS-2 B1 frequency SPP deteriorated during the storms. The probability of the extrema of the positioning error statistics was largest during strong storms, followed by moderate and weak storms. The positioning accuracy for storms of a similar class was found not to be at the same level. The root mean square error in positioning for the different classes of storm could be at least tens of centimeters in the east, north and up directions. The findings in this study could contribute toward the error constraint of BDS positioning accuracy during different classes of geomagnetic storm and be beneficial to other systems, such as BDS-3, as well.

Keywords: BDS; SPP; main phase; geomagnetic storms

Citation: Xue, J.; Veettil, S.V.; Aquino, M.; Hu, X.; Quan, L.; Liu, D.; Guo, P.; Wu, M. Performance of BDS B1 Frequency Standard Point Positioning during the Main Phase of Different Classified Geomagnetic Storms in China and the Surrounding Area. *Remote Sens.* **2022**, *14*, 1240. https://doi.org/10.3390/rs14051240

Academic Editor: Yunbin Yuan

Received: 25 January 2022
Accepted: 1 March 2022
Published: 3 March 2022

Publisher's Note: MDPI stays neutral with regard to jurisdictional claims in published maps and institutional affiliations.

1. Introduction

A geomagnetic storm is defined as a period when the ring current becomes intense enough to exceed the key threshold of the disturbance storm time (Dst) index. Geomagnetic storms are induced by the intense and continuing interplanetary convection electric field and energy injection into the magnetosphere–ionosphere system [1]. The enhanced interplanetary convection electric field is motivated by a constant southward interplanetary magnetic field (IMF) [2]. Solar wind carries the coronal magnetic field out into the entire heliosphere, thus forming the IMF [3]. Based on the signatures in the magnetic field, a geomagnetic storm can be divided into three phases: initial, main and recovery. The main phase is the principal characteristic of a geomagnetic storm [1,4].

The largest global atmospheric effects can be activated by geomagnetic storms [5]. Storms can generate disturbances in the ionosphere, which varies with the location of the region under consideration, the local time (LT) of the geomagnetic storm onset and other parameters [6]. The Equatorial Ionization Anomaly also responds to the effects of storms [7]. The disturbed condition of the ionosphere during geomagnetic storms is known as an ionospheric perturbation, which can have great effects on radio propagation-dependent applications, especially for Global Navigation Satellite System (GNSS) single frequency users. These effects can usually be corrected by ionospheric models. In the case of BeiDou

Navigation Satellite System (BDS-2) single frequency users, a Klobuchar-style ionospheric navigation model is applied for the corrections [8]. The mean correction precision of the model is better than 65% and it performs better for middle latitudes than low latitudes [9]. For BDS-3 single frequency users, the BeiDou global broadcast ionospheric delay correction model (BDGIM) is proposed for the ionospheric delay correction. Ionospheric errors can be mitigated by 80.9% in the China region and 77.6% at the global scale [10].

There are case studies demonstrating the adverse effects of geomagnetic storms on GNSS positioning. During storms, carrier phase cycle slips in Global Positioning System (GPS) signals may occur, which result from loss of lock (LOL) events [11]. Astafyeva et al. [12] demonstrated that the density of GPS LOL events can increase to 0.25% on the L1 band and 3% on the L2 band during super storms (Dst ≤ -250 nT) and 0.15% on L1 and 1% on L2 during intense storms (-250 nT $<$ Dst ≤ -100 nT). Specifically, the tracking performance of GPS receivers in the high latitudes was investigated for the 2015 St. Patrick's Day strong storm. The significant scintillation caused by the storm contributed to a LOL on the GPS L2 band but had little influence on the tracking of the GPS L1 signal [13]. Kinematic GPS positioning could also be degraded during ionospheric disturbances induced by geomagnetic storms. The repeatability of kinematic positioning, which was estimated using a two-step approach based on double difference L3 phase measurements, reached 12.8, 8.1 and 26.1 cm in the north (N), east (E) and up (U) directions, respectively, during the 2003 Halloween storm [14]. The accuracy of real-time kinematic (RTK) positioning could also be deteriorated during a strong geomagnetic storm [15] and even during a weak storm at high latitudes [16]. Furthermore, positioning errors in network RTK and precise point positioning (PPP) techniques increased rapidly during the 2015 St. Patrick's Day strong storm [17]. An investigation was also performed into the effects of moderate and weak geomagnetic storms on the performance of GNSS–SBAS in low-latitude African regions using an SBAS emulator to simulate specific EGNOS-like messages. The SBAS performance in equatorial African regions showed a non-linear relationship with the geomagnetic storm indices [18]. Additionally, GPS instrumental biases, including receiver and satellite biases, are routinely estimated using a dual frequency geometry-free combination. These computations can also be affected during geomagnetic storms [19].

Even though previous studies have revealed the effects of individual, or several, geomagnetic storms on the Earth's upper atmosphere and GNSS applications, few papers have studied the performance of BDS-based applications during geomagnetic storms, especially during the different classes of storm. GNSS single frequency users are supposed to be more obviously affected compared to other positioning modes, such as PPP, during those periods. In this study, the effects of different types of storm on the performance of BDS-2 single frequency standard point positioning (SPP) are investigated comprehensively, especially for the most commonly used B1 frequency. In addition, the differences between the effects of separate storms are studied.

2. Methodology

Dst index can be used as a criterion to classify geomagnetic storms [4]. In this study, Dst indices were extracted from combined OMNI files that were obtained from the NASA database (https://omniweb.gsfc.nasa.gov, accessed on 20 January 2022). All storms in solar cycle 24 were analyzed and divided into three classes: strong, moderate and weak. Table 1 states the threshold conditions that were applied in the classification of storms (see [1,20]).

Table 1. The thresholds implemented in the classification of geomagnetic storms. Type refers to the storm classes, Dst is in nT and ΔT is in hours.

Type	Dst (nT)	ΔT (h)
Strong	-100	3
Moderate	-50	2
Weak (typical substorm)	-30	1

The basic strategy for selecting storms was that the Dst should be as minimal as possible and the duration of each storm should be more than 12 h. To ensure that each storm was independent and not influenced by another storm, a condition was applied that the Dst index for the ten days before and after the main phase day must be greater than the minimum value for each individual class of storm. Finally, five cases with noticeable main phases were chosen for each class of storm from 2015 to 2018 [20]. The principal characteristic of a geomagnetic storm is the main phase [4]. The details of the main phases of the storms, including the related Dst values, the start and end epoch and the duration, are presented in Table 2 (see [20]); Dates with a suffix of "0" refer to the start epoch while a suffix of "1" represents the end epoch. The same meanings are applied to the suffix used for the Dst values.

The earliest period for collecting BDS-2 observations from the Multi-GNSS Experiment (MGEX) network [21] in this study was 2015. BDS-2 observations were obtained for the chosen stations on the dates listed in Table 2. Those stations from the MGEX network, namely DAEJ, GMSD, JFNG, LHAZ, HKWS and HKSL, are distributed throughout China and the surrounding area (see Figure 1). The sampling interval was 30s. The related information, such as geodetic coordinates, receiver and antenna version, is shown in Table 3. The last two columns show the dates that the hardware, such as the receivers or antennae, was changed or updated.

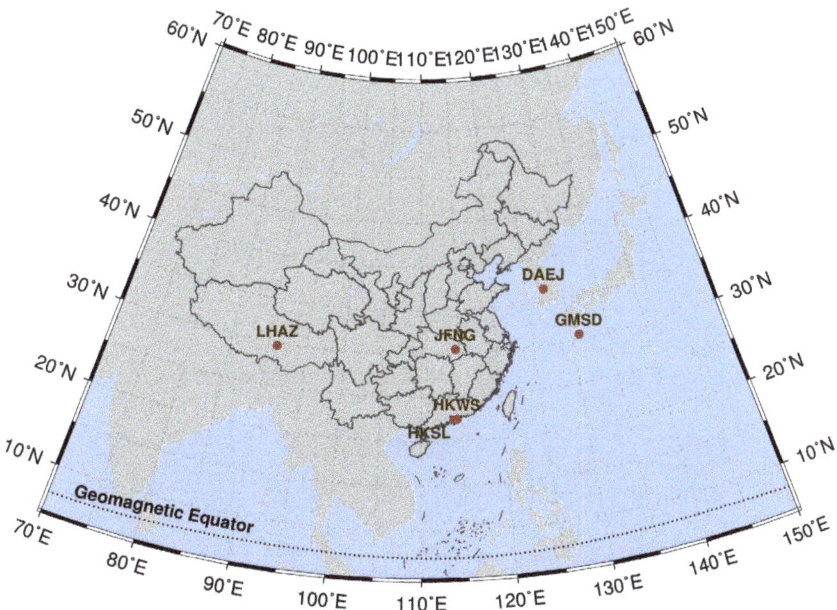

Figure 1. The distribution of the selected MGEX stations throughout China and the surrounding area: a red dot indicates the location a station; the black dotted line indicates the geomagnetic equator. The latitude and longitude are in degrees).

Table 2. The main phases of the different classes of geomagnetic storm from 2015 to 2018: Type, storm classes; MJD, modified Julian date; MON, month; DOY, day of the year; Duration, the period of the main phase; STR, strong storm; MED, moderate storm; MNM, weak storm. Dst is in nT and duration is in hours.

TYPE	MJD0	YEAR0	MON0	DAY0	DOY0	HOUR0	Dst0 (nT)	MJD1	YEAR1	MON1	DAY1	DOY1	HOUR1	Dst1 (nT)	Duration (h)
STR	57,098	2015	3	17	76	5	56	57,098	2015	3	17	76	22	−223	17
	57,195	2015	6	22	173	6	13	57,196	2015	6	23	174	4	−204	22
	57,302	2015	10	7	280	2	−9	57,302	2015	10	7	280	22	−124	20
	57,375	2015	12	19	353	22	43	57,376	2015	12	20	354	22	−155	24
	58,355	2018	8	25	237	8	19	58,356	2018	8	26	238	6	−174	22
MED	57,180	2015	6	7	158	19	24	57,181	2015	6	8	159	8	−73	13
	57,273	2015	9	8	251	20	−2	57,274	2015	9	9	252	12	−98	16
	57,406	2016	1	19	19	19	15	57,407	2016	1	20	20	16	−93	21
	57,838	2017	3	26	85	22	15	57,839	2017	3	27	86	14	−74	16
	58,064	2017	11	7	311	4	25	58,065	2017	11	8	312	1	−74	21
MNM	57,544	2016	6	5	157	8	32	57,545	2016	6	6	158	6	−44	22
	57,716	2016	11	24	329	5	−12	57,717	2016	11	25	330	5	−46	24
	57,784	2017	1	31	31	11	−5	57,785	2017	2	1	32	9	−45	22
	57,920	2017	6	16	167	7	30	57,920	2017	6	16	167	23	−31	16
	58,269	2018	5	31	151	21	5	58,270	2018	6	1	152	19	−39	22

Table 3. Information regarding the longitude, latitude and receiver and antenna versions of the stations. Site refers to the station name and the latitude and longitude are in degrees.

SITE	LATITUDE	LONGITUDE	RECEIVER	ANTENNA	YEAR	DOY
DAEJ	36.40	127.37	TRIMBLE NETR9	TRM59800.00	2017	087
				TRM59800.00	2015	075
GMSD	30.56	131.02	TRIMBLE NETR9	TRM41249.00	2017	311
				TRM59800.00	2018	151
JFNG	30.52	114.49	TRIMBLE NETR9	TRM59800.00	2015	075
LHAZ	29.66	91.10	LEICA GR25	LEIAR25.R4	2016	157
HKWS	22.43	114.34	LEICA GR25	LEIAR25.R4	2015	353
			LEICA GR50		2017	031
HKSL	22.37	113.93	LEICA GR25	LEIAR25.R4	2015	353
			LEICA GR50		2016	329

The data were processed in the kinematic mode of SPP using BDS-2 single frequency pseudorange observations. Considering the dispersive nature of the ionosphere, only the B1 pseudorange was used here. The pseudorange observation equation is illustrated as follows:

$$B_1 = \rho + dt_r - dt^s + T + I_1 + db_{r1} - db^{s1} + \varepsilon \tag{1}$$

where B_1 is the BDS-2 B1 pseudorange observation, ρ is the geometric range, dt_r is the receiver clock error, dt^s is the satellite clock error, T is the tropospheric delay, I_1 is the ionospheric delay, db_{r1} is the receiver differential code bias (DCB), db^{s1} is the satellite DCB and ε is the noise error.

A conventional option was set for the SPP program. The satellite orbit and clock were computed from IGS navigation data. The tropospheric delays were derived using the Saastamoinen model. The ionospheric delays were calculated using the broadcasted BDS-2 navigation ionospheric model. Time group delays in the broadcast ephemeris were extracted, converted and utilized to compute the satellite DCB. For each epoch, the station coordinates and receiver clock error were estimated using the Gauss–Newton least square method. The weight was set with the satellite elevation angle. The elevation mask angle was set to $10°$.

As a result, the station coordinates in the Cartesian coordinate system were compared to the precise solutions from SINEX files obtained from MGEX products. Positioning errors were obtained in the east (E), north (N) and up (U) directions of the local station coordinate framework. The related statistics were performed for the main phase periods using indices such as minimum (MIN), maximum (MAX), bias and root mean square error (RMSE). The MIN and MAX represent the minimum and maximum of the positioning errors for the three directions. The bias and RMSE were computed from the positioning errors for each component as well. The formulas are demonstrated as follows:

$$\begin{aligned}
MIN &= minimum\{\Delta POS_i\} \\
MAX &= maximum\{\Delta POS_i\} \\
BIAS &= <\Delta POS_i> \\
RMSE &= \sqrt{<\Delta POS_i^2>} \\
\Delta POS_i &= POS_{ref,i} - POS_{est,i}, i = 1, n
\end{aligned} \tag{2}$$

where $<>$ is the average of the variable, $POS_{ref,i}$ is the precise solution from the SINEX files, $POS_{est,i}$ is the solution obtained from this processing and n is the total number of samples.

3. Results and Discussion

The accuracy of BDS-2 B1 frequency SPP positioning during the main phase of different classes of storm is analysed in this section. First, the positioning errors in the three directions for all stations during a period of 3 days' representative of each class of storm are presented in Figures 2–4. The periods shown in each of the figures cover the three days before and after the main phases of the individual storms. In each of the figures, the Dst time series is applied to indicate the activities of the storms. The slant total electron content (TEC) of each station was computed and converted into slant time delays (DION) in the B1 frequency. The positioning errors of the selected GNSS stations in the E, N and U directions are shown below. The main phases of the geomagnetic storms are indicated with the red dash–dot lines. The recovery phase is the period after the epoch indicated by the right-hand dash–dot line. From the figures, we can see that the positioning errors during the main phases were clearly different from those in other periods. This implies that the positioning errors were influenced during the main phases of the storms. In addition, the degradation of the positioning errors could occur at random epochs of the main phase with the decrease in Dst values. This could last until the recovery phase, as shown in Figure 2. Furthermore, the variations of positioning errors in the U direction were more significant than in the other directions.

For the strong storm shown in Figure 2, the positioning errors for the selected GNSS stations in the E, N and U directions represent the fluctuations during the main phase of the storm, especially for the U direction in which errors could reach approximately 10 m. The fluctuations of the positioning errors for the different stations varied. In general, the positioning errors of the stations could reach about 2 m in the E direction, about 5 m in the N direction and about 11 m in the U direction. The characteristics of the ionospheric delays were changed during the main phase. The maximum delays and the related epochs were different from other periods. Moreover, in this event, a larger degradation of the positioning errors occurred at the beginning of the recovery phase. The reason for this can be attributed to the fact that the Dst values were lower, the geomagnetic disturbance remained intense throughout and the ionospheric error correction was insufficient.

It is worth noting that there were some sharp increases in the positioning errors of the LHAZ station in the E, N and U directions during the main phase of the storm. Nonetheless, there were no similar increases observed for the HKWS and HKSL stations, which are also located at lower latitudes. The DIONs shown in Figure 2 stayed at a low level, thus suggesting that the correction of ionospheric error could be sufficient. The increases were observed at nearly 1 LT and lasted for about 30 min, thus indicating that this occurred locally and temporarily. The time series of F10.7 cm radio flux was also checked for the same period to discover whether there could be sudden ionospheric disturbances. However, no sudden variations were observed and the values were under 75 sfu (1 sfu = 10^{-22} w/m^2/hz), thus implying that there were no abrupt changes in solar activity. Solar radio burst (SRB) events can also affect GNSS positioning [22] and, therefore, the data from the NOAA websites (see ftp://ftp.swpc.noaa.gov/pub/indices/events, accessed on 20 January 2022) were checked to detect any possible occurrences of SRB events. It was noticed that no SRB events occurred during that epoch.

Further, BKG Ntrip Client (BNC) software [23] was used for the data quality check at the LHAZ station. The number of satellites and the position dilution of precision (PDOP) were computed and are shown in Figure 5. It was found that there were no variations in the number of satellites during the periods that sharp increases in the PDOP values were observed. This indicates that the degradation was directly caused by the poor geometrical structure of the constellations. The orbits of most of the observed satellites were geostationary earth orbits (GEOs) and inclined geosynchronous orbits (IGSOs). There was only one medium earth orbit (MEO) satellite, named C14, which was at a high elevation during this time. These factors might have contributed to the sharp jumps in the PDOP. However, frequent jumps in the number of satellites were observed near the end of the whole jump period. On checking the PDOP values at other stations, no such jumps were observed in the same period. The reason for this could be attributed to receiver signal tracking issues during the storm. In addition, it can be seen that the PDOP values varied and most values were greater than 5 during the whole storm period, thus suggesting that the influence of this storm on PDOP was strong.

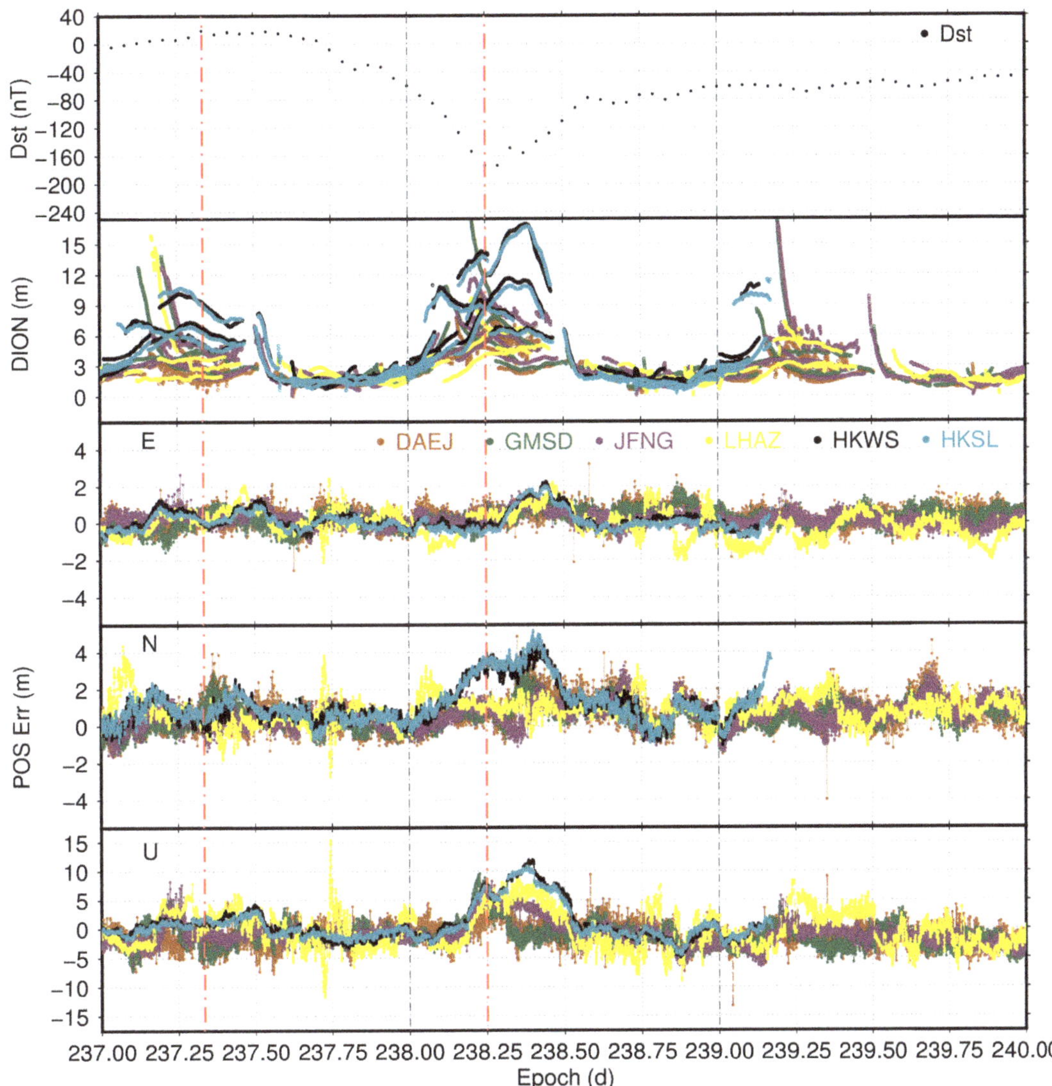

Figure 2. The time series of the positioning errors for BDS-2 B1 frequency during a strong storm around DOY 238, 2018. The times series from top to bottom are for Dst, DION and the positioning errors (POS Err) in E, N and U directions, respectively. The brown dotted line is for DAEJ, the green dotted line is for GMSD, the purple dotted line is for JFNG, the yellow dotted line is for LHAZ, the black dotted line is for HKWS and the cyan dotted line is for HKSL. The red dash–dot lines indicate the borders of the main phase, the left-hand line is the start epoch and the right-hand line is the end epoch. The gray dash–dot lines indicate the borders of the days. The *X*-axis is epoch (DOY) in GPST and the *Y*-axis is Dst in nT. DION and POS Err are in meters.

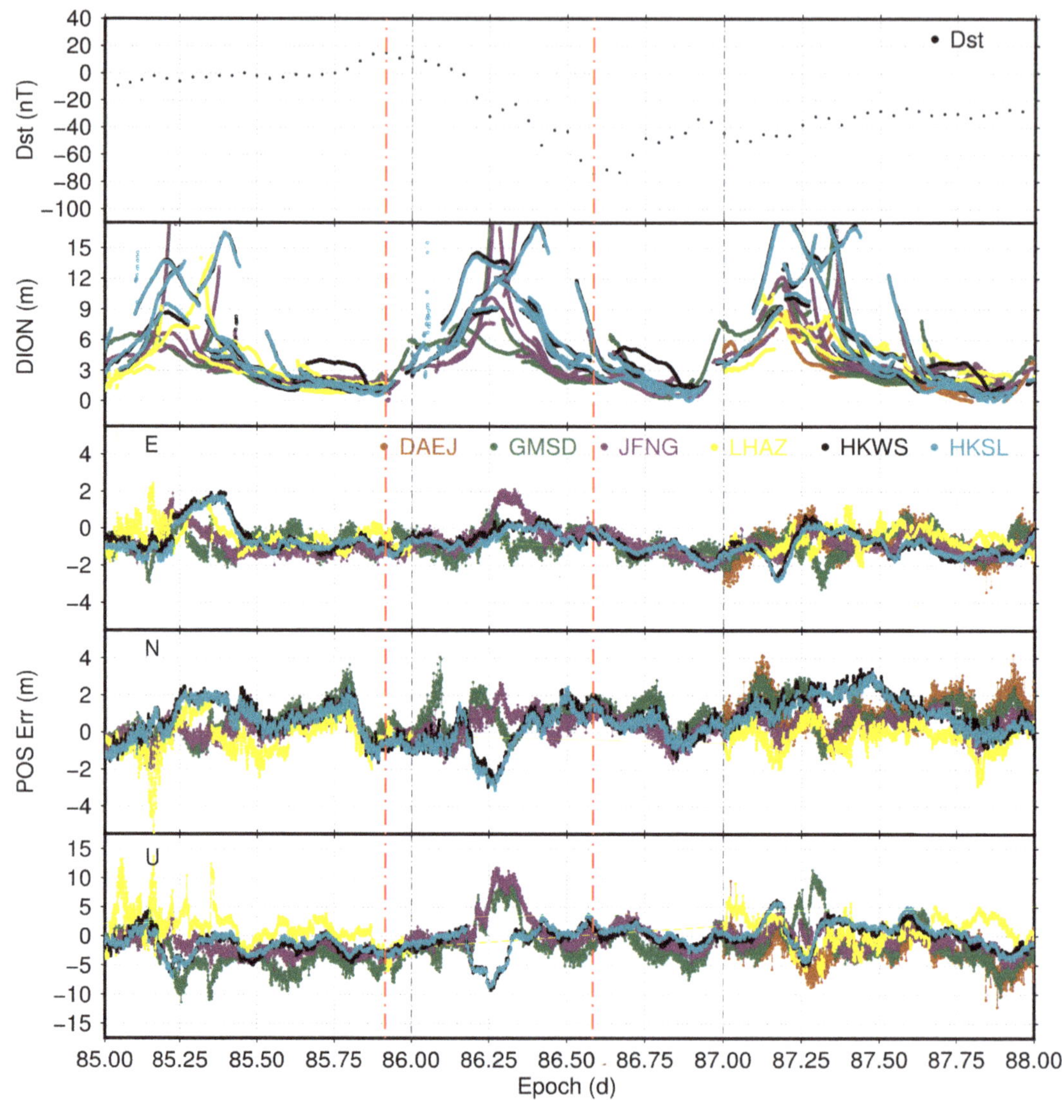

Figure 3. The time series of the positioning errors for BDS-2 B1 frequency during a moderate storm around DOY 086, 2017. The times series from top to bottom are for Dst, DION and the positioning errors (POS Err) in E, N and U directions, respectively. The brown dotted line is for DAEJ, the green dotted line is for GMSD, the purple dotted line is for JFNG, the yellow dotted line is for LHAZ, the black dotted line is for HKWS and the cyan dotted line is for HKSL. The red dash–dot lines indicate the borders of the main phase, the left-hand line is the start epoch and the right-hand line is the end epoch. The gray dash–dot lines indicate the borders of the days. The *X*-axis is epoch (DOY) in GPST and the *Y*-axis is Dst in nT. DION and POS Err are in meters.

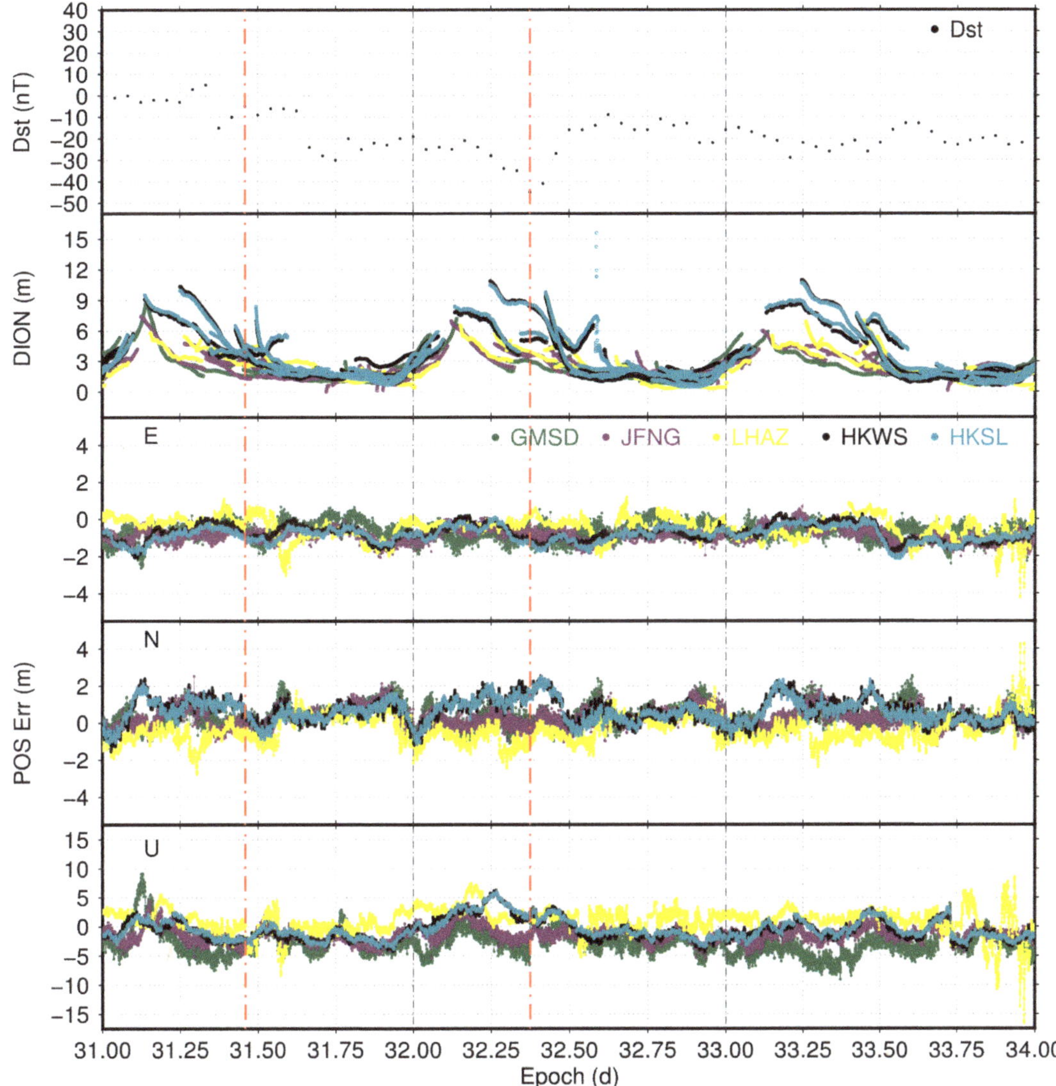

Figure 4. The time series of the positioning errors for BDS-2 B1 frequency during a weak storm around DOY 032, 2017. The times series from top to bottom are for Dst, DION and the positioning errors (POS Err) in E, N and U directions, respectively. The brown dotted line is for DAEJ, the green dotted line is for GMSD, the purple dotted line is for JFNG, the yellow dotted line is for LHAZ, the black dotted line is for HKWS and the cyan dotted line is for HKSL. The red dash–dot lines indicate the borders of main phase, the left-hand line is the start epoch and the right-hand line is the end epoch. The gray dash–dot lines indicate the borders of the days. The *X*-axis is epoch (DOY) in GPST and the *Y*-axis is Dst in nT. DION and POS Err are in meters).

The moderate storm event indicated by the Dst time series is shown in Figure 3. The positioning errors of most of the stations varied in all directions during the main phase in comparison to other periods. The positioning errors were degraded to some extent during the main phase. The degradation in the U direction was more obvious than in the other two directions. The maxima of the positioning errors were about 2 m in the E direction,

4 m in the N direction and 10 m in the U direction. When comparing the maxima of the positioning errors from the selected moderate storm and strong storm, the former were lower than the latter. It was found that the characteristics of the positioning errors of HKWS and HKSL were different from those of the other stations. This could be attributed to the different version of receiver hardware used at those two stations (LEICA version, see Table 3). The ionospheric delays were slightly enhanced during the main phase compared to other periods. In addition, there were also changes in the positioning errors along the E, N and U directions during the recovery phase. This suggests the there were influences of the geomagnetic storm on the positioning errors when the Dst index was still at the lower level.

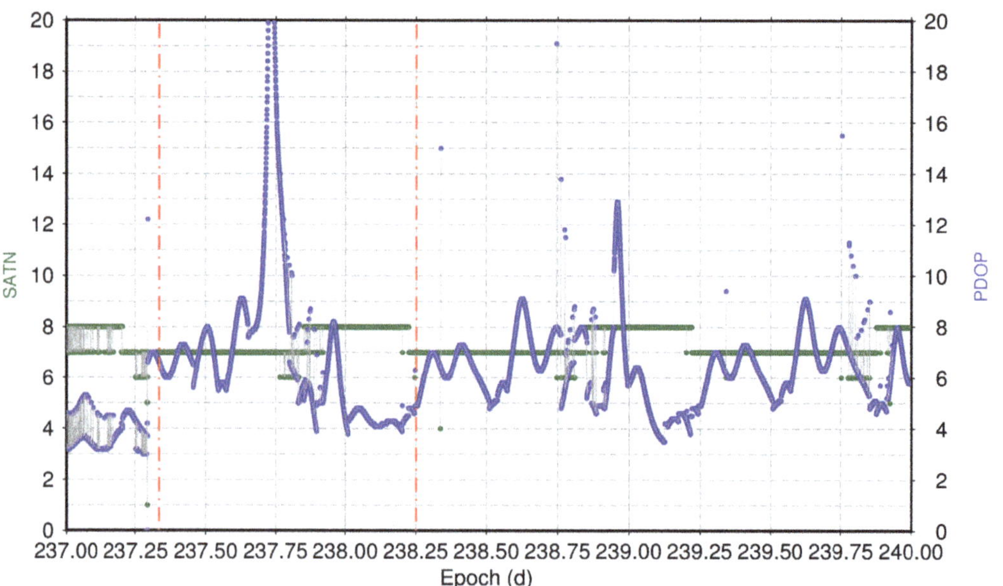

Figure 5. The number of satellites and the PDOP for the LHAZ station during a strong storm around DOY 238, 2018. The green dots indicate the number of satellites and the blue dots indicate the PDOP. The red dash–dot lines indicate the borders of the main phase, the left-hand line is the start epoch and the right-hand line is the end epoch. The gray dash–dot lines indicate the borders of the days. The X-axis is epoch (DOY) in GPST, the left Y-axis is the number of satellites and the right Y-axis is PDOP.

The time series of the positioning errors during the selected weak storm are shown in Figure 4. From the figure, we can see that the activity of this storm was at a low level in terms of the Dst time series. The ionospheric delays were at a reduced status as well. There were slight variations in the positioning errors in the three directions during the main phase compared to the other periods. The positioning errors could reach about 2 m in the E direction, 2 m in the N direction and 8 m in the U direction. The statistical results seemed lower than those from the strong and moderate storms. This suggests that the influence of the weak storm on the positioning errors was less than those of the strong and moderate storms. It can also be seen that the variation in the U direction was greater than in the other directions. The variations in the U direction were stronger near the end of the main phase when the Dst was at its minimum. Additionally, there were also fluctuations in the positioning errors along the U direction during the recovery phase.

Tables 4–6 show the statistics for the BDS-2 B1 frequency positioning errors during the main phases of the independent storms. From the statistical results, the probability of the extrema in the four statistical indices was largest during strong storms, overall followed by moderate and weak storms. The difference in positioning errors between different classes of storm was greatest in the U direction.

Table 4. The statistical indices for the BDS-2 B1 frequency positioning errors during the main phases of strong storms: MJD, modified Julian date; SITE, station name; MIN, MAX, BIAS and RMSE, statistical indices. The three columns for each index are for results in the E, N and U directions, respectively. The last two rows are the mean and median of the statistics in each column. All indices are in meters).

MJD	SITE	MIN			MAX			BIAS			RMSE		
57,098	GMSD	−4.95	−2.62	−3.75	0.66	3.01	16.84	−1.08	−0.21	1.68	1.45	0.88	3.37
	JFNG	−3.69	−3.53	−4.04	1.39	4.45	10.21	−1.20	−0.19	1.75	1.41	1.33	2.97
57,196	GMSD	−2.91	−2.34	−9.22	0.31	1.81	7.59	−1.08	0.07	−0.29	1.23	0.81	3.59
	JFNG	−2.26	−3.51	−5.23	1.10	1.90	8.03	−0.52	−0.19	1.87	0.91	0.97	3.26
57,302	GMSD	−4.71	−5.17	−12.09	0.69	7.01	17.00	−1.32	−0.12	−1.08	1.52	2.26	5.29
	JFNG	−2.52	−3.95	−6.42	1.28	4.29	10.31	−0.88	−0.16	0.01	1.14	1.60	2.85
57,376	GMSD	−4.18	−2.90	−7.59	1.20	31.76	33.54	−0.93	1.11	2.89	1.34	3.25	6.45
	JFNG	−2.50	−2.54	−4.61	3.66	4.49	22.24	−0.20	0.93	4.69	1.46	1.79	7.92
58,356	DAEJ	−2.54	−1.31	−7.65	2.15	3.94	4.87	0.25	0.43	−0.98	0.54	0.80	1.98
	GMSD	−1.37	−0.97	−5.89	1.17	2.83	9.57	0.09	0.39	−0.66	0.40	0.71	2.34
	JFNG	−1.07	−1.33	−4.17	1.43	2.06	7.54	0.17	0.38	−0.98	0.41	0.61	1.83
	LHAZ	−2.12	−2.74	−11.67	2.44	3.83	15.37	0.10	0.63	−0.40	0.71	1.02	2.95
	HKWS	−0.77	−0.51	−3.30	1.33	3.60	7.93	0.09	0.95	0.23	0.41	1.22	2.20
	HKSL	−0.79	−0.71	−2.73	1.03	3.77	7.47	−0.01	0.93	0.10	0.37	1.25	1.93
MEAN		−2.60	−2.44	−6.31	1.42	5.63	12.75	−0.47	0.35	0.63	0.95	1.32	3.50
MEDIAN		−2.51	−2.58	−5.56	1.24	3.80	9.89	−0.36	0.39	0.06	1.03	1.12	2.96

From Table 4, we can see that the maximum positioning error during the strong storms was up to 33 m, the minimum was nearly −12 m, the bias approached 5 m and the RMSE for the E, N and U directions reached 1.52, 3.25 and 7.92 m, respectively. The mean of all RMSEs was 0.95, 1.32 and 3.50 m in the E, N and U directions, respectively, whilst the median was 1.03, 1.12 and 2.96 m for the E, N and U directions, respectively. Furthermore, the positioning accuracy was not comparable between the different strong storms; the accuracy during some storms was better than during others. This indicates that not all strong storms had a similar influence on the positioning accuracy. The same feature was observed during the moderate and weak storms as well.

From Table 5, we can see that the maximum positioning error during moderate storms reached 12 m, the minimum was close to −11 m, the bias approximated 4 m and the RMSE could be up to 1.83, 1.87 and 5.40 m for the E, N and U directions, respectively. The mean and median for the RMSEs of the E, N and U directions were 0.94, 1.09, 3.21 m and 0.90, 1.07, 3.06 m, respectively.

Table 5. The statistical indices for the BDS-2 B1 frequency positioning errors during the main phases of moderate storms: MJD, modified Julian date; SITE, station name; MIN, MAX, BIAS and RMSE, statistical indices. The three columns for each index are for results in the E, N, U directions, respectively. The last two rows are the mean and median of the statistics in each column. All indices are in meters.

MJD	SITE	MIN			MAX			BIAS			RMSE		
57,181	GMSD	−1.66	−2.05	−10.40	0.42	1.40	2.41	−0.46	0.15	−3.84	0.63	0.50	5.01
	JFNG	−2.26	−2.90	−5.98	2.23	2.29	9.07	−0.67	−0.15	1.09	1.09	1.15	3.82
57,274	GMSD	−4.13	−2.02	−14.74	1.47	5.17	6.92	−1.42	0.83	−3.69	1.76	1.71	5.40
	JFNG	−2.64	−2.25	−10.34	0.60	2.98	7.24	−1.02	0.15	−1.61	1.32	1.07	3.44
57,407	GMSD	−3.26	−4.09	−8.82	0.35	3.86	4.35	−1.70	0.04	−1.54	1.83	1.27	2.76
	JFNG	−2.57	−2.99	−5.18	0.31	2.41	5.55	−1.30	−0.06	0.05	1.42	0.96	2.03
	HKWS	−2.11	−2.66	−5.93	1.59	3.89	9.51	−0.69	1.13	2.28	1.16	1.86	3.86
	HKSL	−2.22	−2.97	−4.11	1.55	3.99	11.98	−0.70	1.16	2.97	1.24	1.87	4.26
57,839	GMSD	−2.29	−1.30	−7.93	1.58	4.06	9.35	−0.72	0.73	−0.39	0.94	1.07	3.67
	JFNG	−1.87	−1.65	−4.70	2.13	2.69	11.78	−0.19	0.40	1.41	0.90	0.85	4.01
	LHAZ	−1.47	−0.89	−5.93	−0.02	0.60	−0.18	−0.92	−0.24	−2.47	0.98	0.37	2.78
	HKWS	−1.59	−2.99	−9.33	0.63	2.32	3.23	−0.43	−0.09	−1.39	0.64	1.17	3.06
	HKSL	−1.61	−3.16	−8.81	0.33	2.47	3.93	−0.56	−0.25	−1.22	0.73	1.22	3.05
58,065	DAEJ	−1.30	−2.07	−10.97	1.84	4.20	4.34	0.33	0.83	−2.57	0.66	1.27	3.55
	GMSD	−1.48	−2.00	−6.87	2.01	3.13	4.18	0.27	0.34	−1.28	0.61	0.84	2.33
	JFNG	−1.55	−1.56	−5.70	1.76	2.53	2.80	0.05	0.50	−1.98	0.48	0.90	2.48
	LHAZ	−1.68	−1.26	−4.74	1.71	2.16	7.08	−0.18	0.17	−0.78	0.70	0.65	2.20
	HKWS	−1.86	−1.72	−3.89	1.28	3.72	2.53	0.06	0.37	−0.78	0.44	1.00	1.62
	HKSL	−1.86	−1.76	−3.87	1.12	3.84	2.54	−0.07	0.35	−0.76	0.42	1.02	1.70
MEAN		−2.07	−2.23	−7.28	1.20	3.04	5.72	−0.54	0.33	−0.87	0.94	1.09	3.21
MEDIAN		−1.86	−2.05	−5.98	1.47	2.98	4.35	−0.56	0.34	−1.22	0.90	1.07	3.06

From Table 6, we can see that the positioning errors during weak storms were generally lower than those observed during moderate and strong storms. However, it can be noticed from Table 6 that the LHAZ, HKWS and HKSL stations presented an irregular behavior on MJD 57717 (DOY 329, 2016). Among the three stations, the maximum positioning error reached 50.08 m in the N direction of HKSL, while the minimum reached −93.21 m in the N direction of HKWS. The corresponding bias and RMSE were also large. There were no such big changes for the GMSD and JFNG stations, although they lie in higher latitudes. The reasons for this irregular behavior in the positioning errors of the three stations are analyzed in details later. Except for the anomalous results, the maximum was up to 10 m, the minimum was −20 m, bias was approximately 3 m and the RMSE could reach 2.15, 1.95 and 4.87 m for the E, N and U directions, respectively.

The corresponding time series of the irregular positioning errors during the selected weak storm are showed in Figure 6. The ionospheric delays were slightly depressed during the main phase. The positioning errors of all stations showed large fluctuations at the beginning of the storm, which were more noticeable in the N and U directions. There were also minor fluctuations in the E direction. The LT corresponding to the start epoch of the storm was around 14 h, which is when the ionospheric activity is the most intense during a day. There were large jumps in the positioning errors, especially in the N and U directions. Furthermore, there were no positioning estimations in any direction for many epochs. The strongest jumps occurred at 20 LT and lasted until 8 LT the next morning. Moreover, there were other jumps at around 20 LT in the recovery phase. These jumps were clearly visible in all three directions.

Table 6. The statistical indices for the BDS-2 B1 frequency positioning errors during the main phases of weak storms: MJD, modified Julian date; SITE, station name; MIN, MAX, BIAS and RMSE, statistical indices. The three columns for each index are for results in the E, N and U directions, respectively. The last two rows are the mean and median of the statistics in each column. All indices are in meters.

MJD	SITE	MIN			MAX			BIAS			RMSE		
	GMSD	−2.28	−3.36	−9.62	1.98	3.89	4.97	−0.42	0.32	−2.04	0.95	1.50	3.17
	JFNG	−2.32	−3.30	−6.13	0.62	4.24	5.24	−0.67	0.29	−0.58	0.82	1.35	2.18
57,545	LHAZ	−4.40	−3.98	−19.90	2.46	7.16	9.04	0.42	0.11	1.40	0.98	1.27	3.60
	HKWS	−2.63	−4.36	−7.30	0.65	4.83	9.12	−0.82	0.23	1.41	0.99	1.95	3.07
	HKSL	−2.89	−4.64	−6.25	0.78	4.87	9.27	−0.88	0.23	1.56	1.10	1.95	3.00
	GMSD	−3.89	−2.87	−13.36	−0.09	1.96	6.45	−2.00	−0.23	−2.67	2.15	0.86	4.87
	JFNG	−3.07	−2.89	−6.26	−0.58	2.03	4.75	−1.92	−0.56	−0.38	1.98	0.94	2.09
57,717	LHAZ	−3.01	71.06	−39.24	3.07	29.86	16.22	−0.99	−3.19	0.11	1.43	10.71	5.59
	HKWS	−3.06	93.21	−39.35	1.54	14.19	10.45	−1.26	−12.34	−3.34	1.76	23.58	9.12
	HKSL	−2.97	84.15	−28.16	1.49	50.08	23.53	−1.20	−13.47	−3.46	1.69	30.80	11.09
	GMSD	−1.97	−1.36	−6.58	0.67	2.25	2.89	−0.63	0.44	−2.49	0.80	0.77	3.03
	JFNG	−2.06	−1.55	−5.41	0.08	2.13	4.08	−0.81	0.33	−1.32	0.89	0.66	2.03
57,785	LHAZ	−3.03	−2.44	−8.28	0.73	1.77	7.53	−0.57	−0.32	1.10	0.83	0.84	2.61
	HKWS	−1.67	−1.20	−4.29	0.20	2.28	6.27	−0.68	0.67	−0.19	0.79	0.92	2.45
	HKSL	−1.57	−1.14	−3.91	0.14	2.15	6.10	−0.77	0.67	−0.02	0.85	0.90	2.44
	DAEJ	−2.80	−2.02	−8.70	1.06	3.21	3.15	−0.91	0.55	−2.09	1.05	1.05	3.12
	GMSD	−2.77	−1.91	−10.43	0.85	2.12	2.43	−0.98	0.27	−2.46	1.14	0.80	3.29
57,920	JFNG	−2.10	−1.88	−3.52	0.51	2.04	3.48	−0.60	0.17	0.19	0.76	0.74	1.40
	LHAZ	−1.71	−4.44	−0.80	0.61	7.71	8.47	−0.63	0.23	3.72	0.80	1.47	4.11
	HKWS	−2.00	−3.11	−1.73	0.28	2.93	4.95	−0.58	−0.50	0.98	0.76	1.11	1.76
	HKSL	−1.96	−3.21	−1.81	0.53	1.73	4.89	−0.44	−0.54	1.00	0.73	1.07	1.74
	DAEJ	−2.43	−2.66	−9.93	1.42	2.95	9.50	−0.18	0.30	−1.83	0.61	0.77	2.58
	GMSD	−1.72	−1.99	−7.34	1.28	2.06	1.10	−0.17	0.11	−2.75	0.52	0.60	3.15
58,270	JFNG	−2.00	−2.05	−5.99	1.13	2.25	2.46	−0.15	0.38	−2.10	0.61	0.83	2.52
	LHAZ	−1.96	−1.51	−5.99	1.56	2.03	7.14	−0.46	0.70	−1.44	0.82	0.97	2.71
	HKWS	−0.83	−0.89	−4.83	1.33	2.42	1.86	0.23	0.88	−1.44	0.51	1.08	1.95
	HKSL	−0.93	−0.65	−4.18	1.32	2.07	1.65	0.17	0.81	−1.32	0.53	0.97	1.81
MEAN		−2.37	11.40	−9.97	0.95	6.19	6.56	−0.66	−0.87	−0.76	0.99	3.35	3.35
MEDIAN		−2.28	−2.66	−6.26	0.78	2.28	5.24	−0.63	0.23	−1.32	0.83	0.97	2.71

It was initially supposed that the irregular behavior could be related to the fact that ionospheric activity at low latitudes is more intense than at high latitudes [24]. However, the influence of ionospheric activity could not reach such a level (tens of meters) after the correction of ionospheric model (see Figure 6 and reference [25]). This can also be proved by the positioning errors at other epochs of the main phase. Further, the number of satellites and the PDOP for the three stations are shown in Figure 7. It can be seen that there were large jumps in the PDOP. The maximum could be at the level of several hundred counts. Combined with the number of satellites, these jumps were directly caused by a loss of satellite tracking. A similar feature was observed on the next day (DOY 330, 2016) as well. There are many reasons for the failure of tracking satellites, such as issues with receiver hardware or software, signal strength, space weather, etc. In this study, the main reason for the jumps in the positioning errors for the three stations (LHAZ, HKWS and HKSL) could be attributed to a comprehensive effect. The issues with that specific receiver (LEICA version, see Table 3) might be the most possible reason, which caused similar jumps for these three stations. In addition, the time series of the related space weather indices were checked. The southward IMF Bz had high-frequency variations during the storm period. The solar wind speed increased before the end of the main phase and subsequently decreased. There were continual and consistent variations in Kp and Dst during that period. The AE had large jumps near the epoch of the minimum Dst. There were no sharp changes in F10.7 or the solar wind pressure time series, nor any occurrence of SRBs events. The variations of the space weather indices imply that the storm activity became complex during this period.

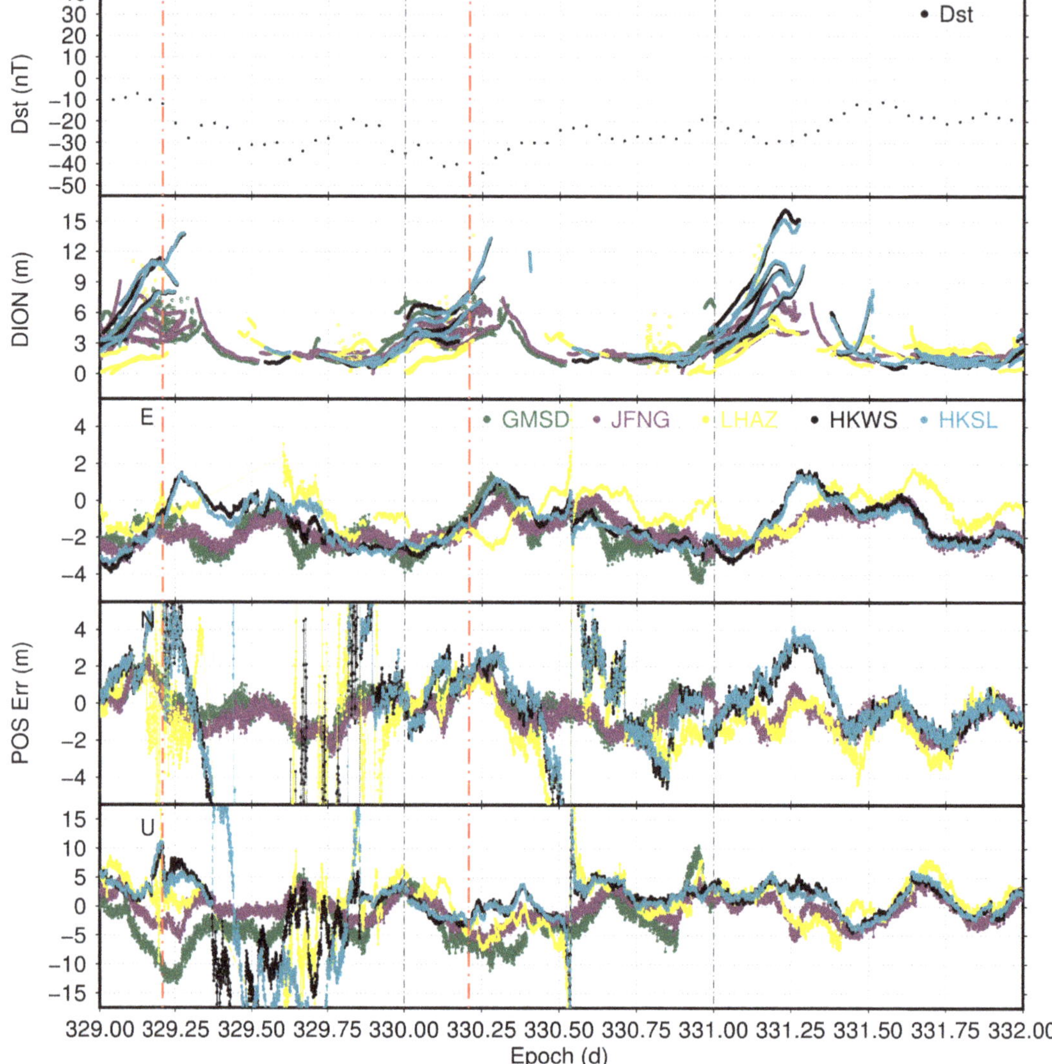

Figure 6. The time series of the positioning errors for BDS-2 B1 frequency during the main phase around DOY 330, 2016. The times series from top to bottom are for Dst, DION and the positioning errors (POS Err) in E, N and U directions, respectively. The green dotted line is for GMSD, the purple dotted line is for JFNG, the yellow dotted line is for LHAZ, the black dotted line is for HKWS and the cyan dotted line is for HKSL. The red dash–dot lines indicate the borders of the main phase, the left-hand line is the start epoch and the right-hand line is the end epoch. The gray dash–dot lines indicate the borders of the days. The *X*-axis is epoch (DOY) in GPST and the *Y*-axis is Dst in nT. DION and POS Err are in meters.

After removing these three stations, the final statistics are illustrated in Table 7. As shown in the table, the mean of the RMSEs for the B1 positioning errors in the three directions was 0.92, 1.06 and 2.70 m and the median was 0.82, 0.96 and 2.60 m. The statistics were lower than those for strong and moderate storms.

Futhermore, the 3D RMSEs of each station were computed around the selected events. By comparing the 3D RMSEs of the pre-storm period and main phase, the decrease in positioning performance was derived. The mean relative percentages of the 3D RMSEs were 5.38%, 1.53% and 0.59% for the strong, moderate and weak storms in this study, respectively.

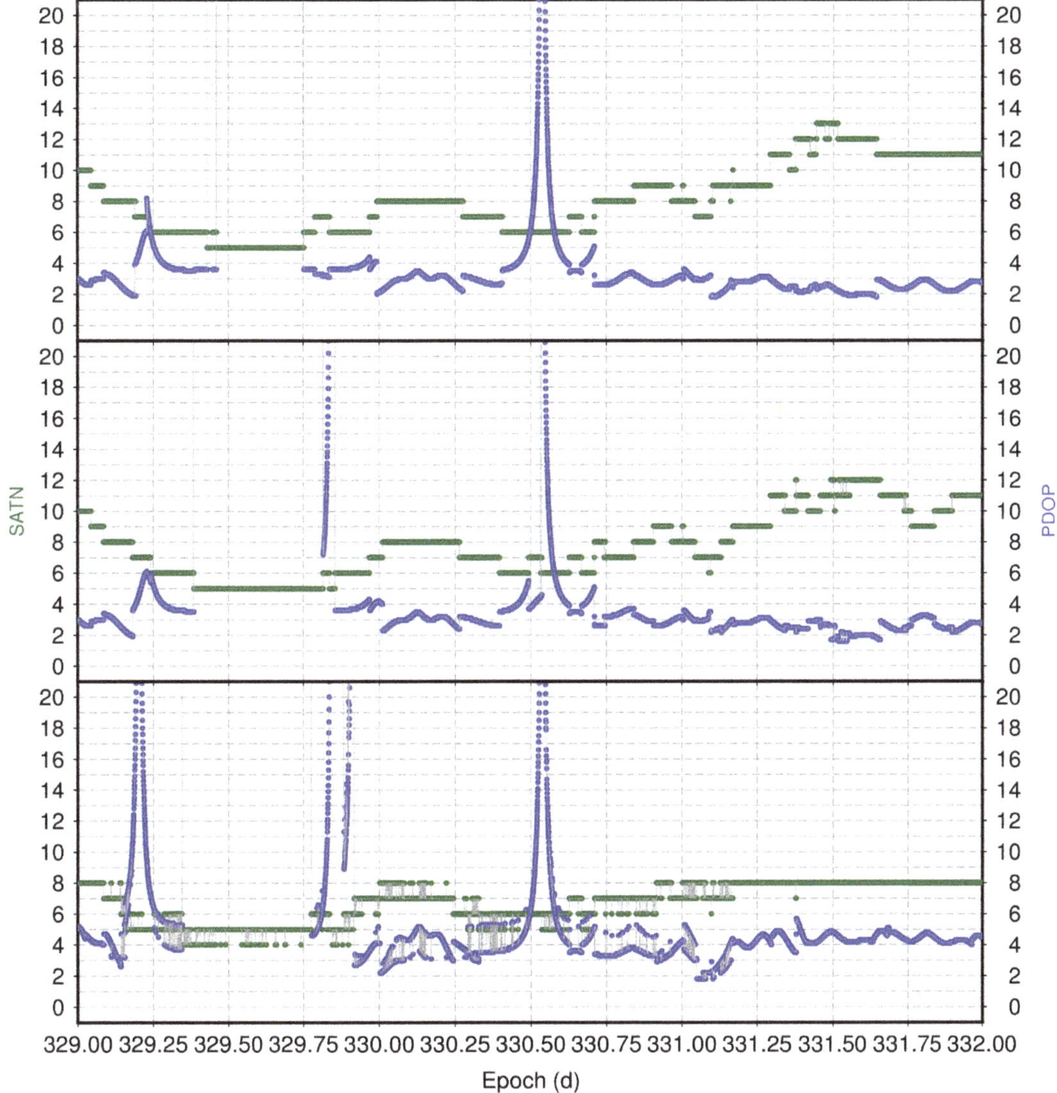

Figure 7. The number of satellites and the PDOP for the HKSL, HKWS and LHAZ stations (top to bottom) around DOY 330, 2016. The green dots indicate the number of satellites and the blue dots indicate the PDOP. The *X*-axis is GPST, the left *Y*-axis is the number of satellites and the right *Y*-axis is PDOP.

Table 7. The statistical indices for the BDS-2 B1 positioning errors during the main phases of weak storms without the anomalous stations: (MJD, modified Julian date; MIN, MAX, BIAS and RMSE, statistical indices. The three columns for each index are for results in the E, N and U directions, respectively. The last two rows are the mean and median of the statistics. All indices are in meters).

MJD	MIN			MAX			BIAS			RMSE		
MEAN	−2.29	−2.48	−6.77	0.81	3.05	5.28	−0.60	0.23	−0.57	0.92	1.06	2.70
MEDIAN	−2.08	−2.25	−6.19	0.70	2.25	4.96	−0.62	0.28	−0.95	0.82	0.96	2.60

4. Conclusions

In this study, a statistical analysis of BDS-2 B1 frequency SPP positioning errors during the main phases of different classes of storm in China and the surrounding area was conducted. From the results, it was observed that the positioning accuracy was affected to different degrees during the storms. Some relevant conclusions can be drawn from the analyses. Firstly, the probability of the extrema of the positioning error statistics was greatest during strong storms, followed by moderate and weak storms. Secondly, during the same class of storm, the positioning accuracy could vary. Thirdly, the positioning accuracy could be influenced even during the recovery phase of a storm.

The findings of this study will hopefully contribute toward the error constraint of BDS positioning accuracy during different strengths of geomagnetic storms. Additionally, the influence of storms could be comparable to other GNSS systems; thus, the findings could also be beneficial to those systems. However, it should be noted that the analysis in this study could have been limited by the uncertainties of the error models of SPP, although the further study of those uncertainties would be beneficial to the improvement of SPP during these storms. In addition, since the study period was in the descending phase of solar cycle 24, the effects on the positioning accuracy might not be entirely apparent. Thus, the study needs to be extended to the beginning of solar cycle 25 and with the addition of more storm events. Investigations into BDS-3 applications, such as the BDGIM model [10], could be performed with more BDS observations as well. Further work could also be focused on case studies of the effects and analyses should be performed with combinations of geophysical indices, drivers and sources, etc.

Author Contributions: Conceptualization, J.X.; data curation, J.X.; formal analysis, J.X., S.V.V., M.A., X.H., L.Q. and D.L.; funding acquisition, J.X.; investigation, J.X., S.V.V. and M.A.; methodology, J.X.; project administration, J.X.; resources, J.X.; software, J.X.; supervision, S.V.V. and M.A.; validation, J.X., S.V.V. and M.A.; visualization, J.X.; writing—original draft, J.X.; writing—review & editing, J.X., S.V.V., M.A., X.H., L.Q., D.L., P.G. and M.W. All authors have read and agreed to the published version of the manuscript.

Funding: This research was funded by the National Natural Science Foundation of China, grant number 11703066, and the China Scholarship Council, grant number 201704910002.

Data Availability Statement: The datasets analysed during this study are available in the IGS MGEX repository (ftp://cddis.gsfc.nasa.gov/pub, accessed on 20 January 2022), NASA OMNI (https://omniweb.gsfc.nasa.gov, accessed on 20 January 2022) and the NOAA database (ftp://ftp. swpc.noaa.gov/pub/indices/events, accessed on 20 January 2022).

Acknowledgments: The authors would like to thank the anonymous referees for their valuable suggestions.

Abbreviations

The following abbreviations are used in this manuscript:

BDS	BeiDou Navigation Satellite System
SPP	Standard Point Positioning
RMSE	Root Mean Square Error
Dst	Disturbance Storm Time
IMF	Interplanetary Magnetic Field
GNSS	Global Navigation Satellite System
LT	Local Time
GPS	Global Positioning System
LOL	Loss of Lock
RTK	Real-Time Kinematic
PPP	Precise Point Positioning
DCB	Differential Code Bias
SRB	Solar Radio Burst
BNC	BKG Ntrip Client
PDOP	Position Dilution of Precision
GEO	Geostationary Earth Orbit
IGSO	Inclined Geosynchronous Orbit
MEO	Medium Earth Orbit
MGEX	Multi-GNSS Experiment
TEC	Total Electron Content

References

1. Gonzalez, W.D.; Joselyn, J.A.; Kamide, Y.; Kroehl, H.W.; Rostoker, G.; Tsurutani, B.T.; Vasyliunas, V.M. What is a geomagnetic storm? *J. Geophys. Res. Space Phys.* **1994**, *99*, 5771–5792. [CrossRef]
2. Hori, T.; Lui, A.T.Y.; Ohtani, S.; C:son Brandt, P.; Mauk, B.H.; McEntire, R.W.; Maezawa, K.; Mukai, T.; Kasaba, Y.; Hayakawa, H. Storm-time convection electric field in the near-Earth plasma sheet. *J. Geophys. Res. Space Phys.* **2005**, *110*. [CrossRef]
3. Owens, M.J.; Forsyth, R.J. The Heliospheric Magnetic Field. *Living Rev. Sol. Phys.* **2013**, *10*, 5. [CrossRef]
4. Loewe, C.A.; Prölss, G.W. Classification and mean behavior of magnetic storms. *J. Geophys. Res. Space Phys.* **1997**, *102*, 14209–14213. [CrossRef]
5. Lastovicka, J. Effects of geomagnetic storms in the lower ionosphere, middle atmosphere and troposphere. *J. Atmos. Terr. Phys.* **1996**, *58*, 831–843. [CrossRef]
6. Danilov, A.; Lastovicka, J. Effects of geomagnetic storms on the ionosphere and atmosphere. *Int. J. Geomagn. Aeron.* **2001**, *2*, 209–224.
7. Sreeja, V.; Devasia, C.V.; Ravindran, S.; Pant, T.K.; Sridharan, R. Response of the equatorial and low-latitude ionosphere in the Indian sector to the geomagnetic storms of January 2005. *J. Geophys. Res. Space Phys.* **2009**, *114*. [CrossRef]
8. BeiDou ICD. Beidou Navigation Satellite System Signal In Space Interface Control Document Open Service Signal (Version 2.0). 2013. Available online: http://www.beidou.gov.cn (accessed on 20 January 2022).
9. Wu, X.; Hu, X.; Wang, G.; Zhong, H.; Tang, C. Evaluation of COMPASS ionospheric model in GNSS positioning. *Adv. Space Res.* **2013**, *51*, 959–968. [CrossRef]
10. Yuan, Y.; Wang, N.; Li, Z.; Huo, X. The BeiDou global broadcast ionospheric delay correction model (BDGIM) and its preliminary performance evaluation results. *Navigation* **2019**, *66*, 55–69. [CrossRef]
11. Rama Rao, P.V.S.; Gopi Krishna, S.; Vara Prasad, J.; Prasad, S.N.V.S.; Prasad, D.S.V.V.D.; Niranjan, K. Geomagnetic storm effects on GPS based navigation. *Ann. Geophys.* **2009**, *27*, 2101–2110. [CrossRef]
12. Astafyeva, E.; Yasyukevich, Y.; Maksikov, A.; Zhivetiev, I. Geomagnetic storms, super-storms, and their impacts on GPS-based navigation systems. *Space Weather* **2014**, *12*, 508–525. [CrossRef]
13. Jin, Y.; Oksavik, K. GPS Scintillations and Losses of Signal Lock at High Latitudes During the 2015 St. Patrick's Day Storm. *J. Geophys. Res. Space Phys.* **2018**, *123*, 7943–7957. [CrossRef]
14. Bergeot, N.; Bruyninx, C.; Defraigne, P.; Pireaux, S.; Legrand, J.; Pottiaux, E.; Baire, Q. Impact of the Halloween 2003 ionospheric storm on kinematic GPS positioning in Europe. *GPS Solut.* **2011**, *15*, 171–180. [CrossRef]
15. Jacobsen, K.S.; Schäfer, S. Observed effects of a geomagnetic storm on an RTK positioning network at high latitudes. *J. Space Weather Space Clim.* **2012**, *2*, A13. [CrossRef]
16. Andalsvik, Y.L.; Jacobsen, K.S. Observed high-latitude GNSS disturbances during a less-than-minor geomagnetic storm. *Radio Sci.* **2014**, *49*, 1277–1288. [CrossRef]
17. Jacobsen, K.S.; Andalsvik, Y.L. Overview of the 2015 St. Patrick's day storm and its consequences for RTK and PPP positioning in Norway. *J. Space Weather. Space Clim.* **2016**, *6*, A9. [CrossRef]

18. Abe, O.E.; Paparini, C.; Ngaya, R.H.; Otero Villamide, X.; Radicella, S.M.; Nava, B. The storm-time assessment of GNSS-SBAS performance within low latitude African region using a testbed-like platform. *Astrophys. Space Sci.* **2017**, *362*, 1–19. [CrossRef]
19. Zhang, W.; Zhang, D.H.; Xiao, Z. The influence of geomagnetic storms on the estimation of GPS instrumental biases. *Ann. Geophys.* **2009**, *27*, 1613–1623. [CrossRef]
20. Xue, J.; Aquino, M.; Veettil, S.V.; Hu, X.; Quan, L. Performance of BDS Navigation Ionospheric Model During the Main Phase of Different Classified Geomagnetic Storms in China Region. *Radio Sci.* **2020**, *55*, e2019RS007033. [CrossRef]
21. Montenbruck, O.; Steigenberger, P.; Prange, L.; Deng, Z.; Zhao, Q.; Perosanz, F.; Romero, I.; Noll, C.; Stürze, A.; Weber, G.; et al. The Multi-GNSS Experiment (MGEX) of the International GNSS Service (IGS)—Achievements, prospects and challenges. *Adv. Space Res.* **2017**, *59*, 1671–1697. [CrossRef]
22. Sreeja, V.; Aquino, M.; de Jong, K. Impact of the 24 September 2011 solar radio burst on the performance of GNSS receivers. *Space Weather* **2013**, *11*, 306–312. [CrossRef]
23. Weber, G.; Mervart, L.; Stürze, A. *BKG Ntrip Client (BNC): Version 2.12*; Verlag des Bundesamtes für Kartographie und Geodäsie: 2016. Available online: https://software.rtcm-ntrip.org/export/7214/ntrip/trunk/BNC/src/bnchelp.html (accessed on 1 January 2022).
24. Biqiang, Z.; Weixing, W.; Libo, L.; Tian, M. Morphology in the total electron content under geomagnetic disturbed conditions: results from global ionosphere maps. *Ann. Geophys.* **2007**, *25*, 1555–1568. [CrossRef]
25. Quan, L.; Xue, J.; Hu, X.; Li, L.; Liu, D.; Wang, D. Performance of GPS single frequency standard point positioning in China during the main phase of different classified geomagnetic storms. *Chin. J. Geophys.* **2021**, *64*, 3030–3047.

Article

Toward an Optimal Selection of Constraints for Terrestrial Reference Frame (TRF)

Shize Song [1,2]**, Zhongkai Zhang** [3,4,5,]*** and Guangli Wang** [1,2]

[1] Shanghai Astronomical Observatory, Chinese Academy of Sciences, No. 80 Nandan Road, Shanghai 200030, China; ssz@shao.ac.cn (S.S.); wgl@shao.ac.cn (G.W.)
[2] University of Chinese Academy of Sciences, 19A Yuquanlu, Beijing 100049, China
[3] College of Geospatial Information, Information Engineering University, Zhengzhou 450000, China
[4] Henan Industrial Technology Academy of Spatio-Temporal Big Data, Zhengzhou 450046, China
[5] The College of Geography and Environmental Science, Henan University, Kaifeng 475004, China
* Correspondence: zhang.astro@foxmail.com

Abstract: Given that the observations from current space geodetic techniques do not carry all the necessary datum information to realize a Terrestrial Reference System (TRS), and each of the four space geodetic techniques has limits, for instance: Very Long Baseline Interferometry (VLBI) ignores the center of mass and satellite techniques lack the TRS orientation, additional constraints have to be added to the observations. This paper reviews several commonly used constraints, including inner constraints, internal constraints, kinematic constraints, and minimum constraints. Moreover, according to their observation equations and normal equations, the similarities and differences between them are summarized. Finally, we discuss in detail the influence of internal constraints on the scale of VLBI long-term solutions. The results show that there is a strong correlation between the scale parameter and the translation parameter introduced by the combination model at the Institut National de l'Information Géographique et Forestière (IGN), and internal constraints force these two groups of parameters to meet certain conditions, which will lead to the coupling of scale and translation parameters and disturbing the scale information in VLBI observations. The minimum or kinematic constraints are therefore the optimum choices for TRF.

Keywords: TRF; constraints; correlation; optimal selection; combination model

Citation: Song, S.; Zhang, Z.; Wang, G. Toward an Optimal Selection of Constraints for Terrestrial Reference Frame (TRF). *Remote Sens.* **2022**, *14*, 1173. https://doi.org/10.3390/rs14051173

Academic Editor: José Fernández

Received: 29 January 2022
Accepted: 25 February 2022
Published: 27 February 2022

Publisher's Note: MDPI stays neutral with regard to jurisdictional claims in published maps and institutional affiliations.

1. Introduction

The International Terrestrial Reference System (ITRS) definition fulfills the following conditions [1]:

1. It is geocentric, and its origin is the center of mass for the whole Earth, including oceans and atmosphere;

2. The unit of length is the meter (Le Système International d'Unités (SI)). The scale is consistent with the geocentric coordinate time (TCG) time coordinate for a geocentric local frame, in agreement with the International Astronomical Union (IAU) and the International Union of Geodesy and Geophysics (IUGG) (1991) resolutions. (The mean rate of the coordinate time TCG coincides with the mean rate of the proper time of an observer situated at the geocenter (with the Earth removed), whereas the mean rate of the terrestrial time (TT) coincides with the mean rate of the proper time of an observer situated on the geoid. $TCG - TT \approx 0.7$ ppb [2].) This is obtained by appropriate relativistic modeling;

3. Its orientation was initially given by the Bureau International de l'Heure (BIH) orientation at 1984.0;

4. The time evolution of the orientation is ensured by using a no-net-rotation (NNR) condition with regards to horizontal tectonic motions over the whole Earth.

By defining the above datum definition, ITRS is implemented. The International Terrestrial Reference Frame (ITRF) is a long-term, linear reference frame, as defined by the International Earth Rotation and Reference Systems Service (IERS) Conventions (2010). Thus, the ITRF datum information includes the origin, scale, orientation, and corresponding rates. As the ITRS realization, ITRF is the combination of the four space geodetic techniques (Global Navigation Satellite Systems (GNSS), Satellite Laser Ranging (SLR), VLBI, and Doppler Orbitography and Radio-positioning Integrated by Satellite (DORIS). As an unbiased-ranging technology, SLR uniquely determines the ITRF origin, which is related to the focal point of the satellite orbits [3,4]. Due to VLBI measuring extremely precise baseline lengths involving the speed of light [5] and the unambiguous nature of SLR measurements [6], the scale of the ITRF is provided by both VLBI and SLR. In order to ensure continuity, the orientation of ITRF is conventionally aligned with the BIH earth orientation parameter (EOP) series at 1984.0 [1,7].

However, observations from any space geodesy do not contain all the necessary datum information to completely define a TRS. For example, VLBI ignores the Earth center of mass, and satellite techniques including GNSS, SLR, and DORIS lack the orientation datum information. Therefore, for the realization of the ITRS, constraints have to be added to the observations to make the normal Equation (NEQ) invertible. The constraints are required to complete the rank deficiency of NEQ without causing the station network distortion and affecting the datum information of the observations, such as the scale of VLBI [8]. Several constraints commonly used in ITRF calculations include: inner constraints [9], minimum constraints [8], internal constraints [10], and kinematic constraints [9,11]. These constraints are reviewed in this work. Furthermore, according to their observation equations and normal equations, the similarities and differences between them are summarized.

In this paper, we only discuss constraints on the VLBI scale information in the VLBI intra-technique combination, i.e., stacking the time series. According to different intra-technique combination models, the corresponding datum constraint is selectable. IERS contains three ITRS combination centers (CCs): IGN, Deutsches Geodätisches Forschungsinstitut (DGFI-TUM) and Jet Propulsion Laboratory (JPL). The ITRFs provided by IGN and DGFI-TUM are secular reference frames, and the one from JPL is based on a Kalman filter approach producing time series of weekly solutions [12]. In this work, we assume that the ITRF model is long-term and linear within the scope of the IERS Conventions (2010) [1].

The computation strategy of DGFI-TUM is based on the combination at the NEQ level, while the ITRS CC at IGN is at the solution level. Before the combination, time series analysis of the input data is needed to improve the accuracy of the linear frame, which includes outlier detection, discontinuities, velocity changes, and estimation of seasonal signals. Outliers are detected and eliminated in the process of stacking time series and producing long-term solutions [13,14]. If the normalized residual exceeded a threshold of 3, the observations would be eliminated as outliers. After several iterations, all outliers are removed. The discontinuities and velocity variations in the time series of station positions are estimated by considering equipment changes from station log files [15] and earthquake information [16–18]. In this paper, we directly use the discontinuities and velocity change information provided by IGN (available at https://itrf.ign.fr/ITRF_solutions/2014/computation_strategy.php?page=2 (accessed on 22 January 2022)).For ITRF2014, seasonal signals of the stations were estimated during the stacking [13]. For DGFI-TUM's ITRS realization 2014 (DTRF2014), time series of atmospheric and hydrological non-tidal loading corrections provided by Tonie van Dam was used to reduce the influence of seasonal signals before the stacking (available at https://doi.pangaea.de/10.1594/PANGAEA.864046?format=html#download (accessed on 22 January 2022)). In this paper, seasonal signals are not estimated or corrected because it is not expected that the seasonal signals affect the ITRF datum information and the velocities of stations with more than 2.5 years of observations [13,19]. Before the stacking, postseismic deformation (PSD) for sites mainly caused by major earthquakes was modeled (corresponding data and subroutines available at https://itrf.ign.fr/ITRF_solutions/2014/ITRF2014_files.php (accessed on 22 January 2022)).

For the IGN combination, in order to provide the origin and orientation datum, minimum constraints are applied to the VLBI 24h sessions provided under the form of normal equations before stacking the time series. The addition of minimum constraints on the datum of orientation and origin to the VLBI free normal equations not only preserves its physical scale parameter but also allows its inversion [10]. The IGN intra-technique combination model is based on the seven-parameter similarity transformation [9,20]. In addition to calculating station coordinates and EOPs, the time series of the seven transformation parameters between each daily minimum constrained solution and the corresponding long-term stacked solution was also estimated at IGN. Since the transformation parameters were introduced to the accumulated NEQ by the combination model at IGN, two constraints could be selected in the intra-technique combination (the time series stacking), where the minimum constraints [8] were imposed over the station coordinates, and internal constraints [9,10] were applied over the time series of each of the seven transformation parameters. The DGFI model is based on the combination of the NEQs free from additional constraints. The VLBI Solution Independent Exchange (SINEX) files contain normal equation systems free from datum constraints and resulting from a combination of the Analysis Centers' (ACs) contributions at the NEQ level [21]. Therefore, DGFI-TUM directly stacks VLBI time series of NEQs to generate a multi-year normal equation system [14,22]. As the stacked normal equation does not contain the time series of the transformation parameters, DGFI-TUM can only impose conditions over station positions and velocities.

Since the constrained parameters are different (the internal constraints act on the time series of each of the seven transformation parameters, while the minimum constraints act on the station coordinates), it is necessary to compare the two constraints by means of specific calculations. Altamimi et al. compared the calculation results realized by internal and minimum constraints, respectively, [10]: (1) The post-fit residuals derived from the two methods are still the same; (2) the Helmert parameters between these two corresponding accumulated solutions (intra-technique combination) are different, which is estimated by the 14-parameter similarity transformation. By analyzing the observation equations and implicit conditions of these two constraints, the rationality of similar results (case 1) can be verified. However, the difference of the obtained transformation parameters (case 2) indicates that the datum of the long-term solutions corresponding to the two constraints is inconsistent. By calculating the scales at epoch 2010.0 and scale rates of DTRF2014 DORIS, GNSS, SLR, and VLBI with respect to ITRF2014 [23], there is a non-negligible linear trend between the long-term solutions of the same technique. Although the ITRF2014 scale is defined by the average of the VLBI and SLR scale [13], the scale of the multi-technique combination solutions is determined by the long-term solution (intra-technique combination). Therefore, the scale between the same technologies of DTRF2014 and ITRF2014 should be a constant offset rather than a linear trend. By calculating scale offsets and their rates calculated from the 14-parameter Helmert transformation of the VLBI and SLR single-technique solutions of (intra-technique combination) DGFI-TUM and IGN (transformation epochs are at 2000.0 and 2010.0) [24], the results show that there is a linear trend between the scales of single-technique solutions (the intra-technique combination) of IGN and DGFI-TUM. Considering the nearly 40 years worth of observation history for VLBI and SLR, this linear trend is not negligible. According to the preliminary calculations of ITRF2020, the scale difference between SLR and VLBI is about 3 mm [25], versus 8.7 mm in ITRF2014 [13]. However, IGN has not provided the modified combination method. In DTRF2014, which used minimum constraints instead of internal constraints, the scales of SLR and VLBI are considered to be statistically equal ($\pm$3.3 mm) [22]. In algebraic terms, minimum constraints can complete the NEQ rank deficiency of long-term solutions and not more [8]. Moreover, when we check the VLBI minimum constraint solution (input solutions to ITRF2014 [21] and the long-term solutions produced in this work), the origin and orientation are indeed expressed in an a priori reference frame. Therefore, the minimum constraints do not affect the scale information of VLBI observations.Based on the above analysis, we believe that internal constraints may affect the datum of long-term solutions obtained from the intra-technical combination. This paper will describe how internal constraints affect the scale datum of VLBI technology, considering the intra-technique combination model of IGN and DGFI-TUM, respectively.

The main objective of this article is to review and compare several commonly used constraints and select the optimal ones for TRF.

Section 2 introduces the combination model of IGN and DGFI-TUM, reviews kinematic constraints, minimum constraints, internal and inner constraints, and summarizes the relationship between these constraints. Section 3 gives the results of four VLBI long-term solutions with different constraints and intra-technique combination models. Section 4 discusses how internal constraints affect the VLBI long-term solution. Section 5 concludes.

2. Materials and Methods

The scale difference between the VLBI long-term solutions of IGN and DGFI-TUM [24] may occur for two reasons: the combination model and constraints. This section briefly introduces the combination models of IGN and DGFI-TUM. The ITRF combination procedure can be divided into two main steps: the intra-technique and inter-technique combination [13,14]. We review several commonly used constraints, including inner constraints, internal constraints, kinematic constraints, and minimum constraints. Moreover, according to their observation equations and normal equations, the similarities and differences between them are summarized. The research on constraints is quite well-established. Based on existing results (see Section 1), this section summarizes the relationship between these constraints. For this purpose, we review these constraints and study the relationship between them by comparing their preconditions, observation equations, and normal equations. In order to avoid the interference of other factors on the VLBI long-term solutions, the same time series analysis method was used.

2.1. Combination Model at IGN

Since the computation strategy of IGN is based on the combination of solutions, all input data are converted to the minimum constraint solutions before the combination. The combination model of IGN can be summarized as [20]:

$$
\begin{cases}
X_s^i = X_c^i + (t_s^i - t_0)\dot{X}_c^i + T_k + D_k X_c^i + R_k X_c^i \\
\quad + (t_s^i - t_k)[\dot{T}_k + \dot{D}_k X_c^i + \dot{R}_k X_c^i] \\
\dot{X}_s^i = \dot{X}_c^i + \dot{T}_k + \dot{D}_k X_c^i + \dot{R}_k X_c^i
\end{cases}
\tag{1}
$$

$$
\begin{cases}
x_s^p & = x_c^p + R_{yk} \\
y_s^p & = y_c^p + R_{xk} \\
UT_s & = UT_c - \dfrac{1}{f} R_{zk} \\
\dot{x}_s^p & = \dot{x}_c^p \\
\dot{y}_s^p & = \dot{y}_c^p \\
LOD_s & = LOD_c
\end{cases}
\tag{2}
$$

where positions (at epoch t_s^i) and velocities of each station i of a single-technique solution (s = VLBI, SLR, GNSS or DORIS) are represented by X_s^i and $\dot{X}_s^i$, respectively, and those of the combined solution c by X_c^i at reference epoch t_0 and $\dot{X}_c^i$. For each solution s expressed in the frame k at epoch t_k, T_k is the translation vector including three origin components (T_x, T_y, T_z), R_k the rotation matrix composed of 3 rotation parameters (R_x, R_y, and R_z), and D_k the scale factor. The dotted parameters are their derivatives with respect to time. The EOPs contain pole coordinates (x_s^p, y_s^p) and universal time (UT_s), and $\dot{x}_s^p$, $\dot{y}_s^p$, and LOD_s are their rates. The conversion factor f from universal time into sidereal time is equal to 1.002737909350795.

Since the daily or weekly solution X_s^i does not involve station velocity $\dot{X}_s^i$ and transformation parameter rates, all of Equation (2) and only the first line of Equation (1) are used in the intra-technique combination. In the second step of the inter-technique combination, the two equations above the four long-term solutions are combined.

2.2. Combination Model at DGFI-TUM

The DGFI-TUM model is based on the combination of the NEQs free from additional constraints. Its combination model can be described by Equations (3)–(7) [14].

$$
\begin{cases}
\hat{x} = N^{-1}y \\
\hat{\sigma}^2 = \dfrac{l^T Pl - y^T \hat{x}}{n - u}
\end{cases}
\tag{3}
$$

where n and u represent the number of observations and estimates, respectively, N the matrix of NEQ, $\hat{x}$ the estimates, $\hat{\sigma}^2$ the posteriori variance factor, $y = A^T Pl$ the product, $l^T Pl$ the square sum of the observed vector minus the computed vector (O-C). A, l, and P are the coefficient matrix, the observation vector, and the weight matrix of the observations, respectively. The covariance matrix $C_{\hat{x}\hat{x}}$ of the estimates is described by Equation (4).

$$
C_{\hat{x}\hat{x}} = \hat{\sigma}^2 N^{-1}
\tag{4}
$$

The final NEQ (Equation (3)) is obtained in two steps. In the intra-technique combination, time series of NEQs from the Technique Centres (TCs) are stacked to one accumulated NEQ $\hat{x}_i = N_i^{-1}y_i$ (i is one of the four space geodetic technologies). In the second step, the accumulated NEQs (obtained from step 1) of the different techniques are combined:

$$
N = \sum_i \lambda_i N_i
\tag{5}
$$

$$
y = \sum_i \lambda_i y_i
\tag{6}
$$

$$
l^T Pl = \sum_i \lambda_i (l^T Pl)_i
\tag{7}
$$

where λ_i represents the weight factors estimated for the techniques.

2.3. Kinematic Constraints

Kinematic constraints have physical meanings [9]: (a) with respect to the origin by imposing constant reference coordinates for the barycenter of the station network, (b) with respect to orientation by imposing zero relative angular momentum for the network stations with the assumption that mass points (stations) have equal masses, and (c) with respect to the scale by imposing a constant mean quadratic size (related to the distances from stations to their barycenter). The observation equations of NNR and no-net-translation (NNT) conditions are given below. For kinematic constraints related to scale, please refer to [9,11].

(1) NNT

$$
\sum_{i=1}^{N} \delta x_i = 0
\tag{8}
$$

$$
\sum_{i=1}^{N} \delta v_i = 0
\tag{9}
$$

(2) NNR

$$
\sum_{i=1}^{N} [x_i^{ap} \times] \delta x_i = 0
\tag{10}
$$

$$\sum_{i=1}^{N} [x_i^{ap} \times] \delta v_i = 0 \tag{11}$$

$$[x_i^{ap} \times] = \begin{bmatrix} 0 & z_i^0 & -y_i^0 \\ -z_i^0 & 0 & x_i^0 \\ y_i^0 & -x_i^0 & 0 \end{bmatrix} \tag{12}$$

where δx_i is the correction of the station position x_i, and its prior value is x_i^{ap}. δv_i is the correction of the station velocity v_i, and its prior value is v_i^{ap}.

For simplicity, we only give NEQs of NNT and NNR with respect to the station position (corresponding to Equations (8) and (10)).

The matrix form of NNT is:

$$A_{NNT} \cdot \delta x = \underbrace{\begin{bmatrix} 1 & 0 & 0 & \cdots & 1 & 0 & 0 & \cdots & 1 & 0 & 0 \\ 0 & 1 & 0 & \cdots & 0 & 1 & 0 & \cdots & 0 & 1 & 0 \\ 0 & 0 & 1 & \cdots & 0 & 0 & 1 & \cdots & 0 & 0 & 1 \end{bmatrix}}_{3 \times 3N} \underbrace{\begin{bmatrix} \delta x_1 \\ \delta y_1 \\ \delta z_1 \\ \vdots \\ \delta x_i \\ \delta y_i \\ \delta z_i \\ \vdots \\ \delta x_N \\ \delta y_N \\ \delta z_N \end{bmatrix}}_{3N \times 1} = \underset{3 \times 1}{\mathbf{0}} \tag{13}$$

where N is the number of stations.

The NEQ of NNT is:

$$\underset{3N \times 3N}{\mathbf{N}_{NNT} \delta x} = (A_{NNT}^T \mathbf{P}_x A_{NNT}) \cdot \delta x = 0 \tag{14}$$

where the diagonal matrix $\mathbf{P}_x$ is the weight matrix. $\mathbf{P}_x$ is equivalent to a coefficient p_x in Equation (14). The coefficient matrix $\mathbf{N}_{NNT}$ of NEQ is:

$$\mathbf{N}_{NNT} = p_x \begin{bmatrix} 1 & 0 & 0 & \cdots & 1 & 0 & 0 & \cdots & 1 & 0 & 0 \\ 0 & 1 & 0 & \cdots & 0 & 1 & 0 & \cdots & 0 & 1 & 0 \\ 0 & 0 & 1 & \cdots & 0 & 0 & 1 & \cdots & 0 & 0 & 1 \\ \vdots & \vdots & \vdots & \ddots & \vdots & \vdots & \vdots & \ddots & \vdots & \vdots & \vdots \\ 1 & 0 & 0 & \cdots & 1 & 0 & 0 & \cdots & 1 & 0 & 0 \\ 0 & 1 & 0 & \cdots & 0 & 1 & 0 & \cdots & 0 & 1 & 0 \\ 0 & 0 & 1 & \cdots & 0 & 0 & 1 & \cdots & 0 & 0 & 1 \end{bmatrix}$$

The matrix form of NNR is:

$$A_{NNR} \cdot \delta x = \underset{3 \times 3N}{\left[\ [x_1^{ap} \times]\ \cdots\ [x_i^{ap} \times]\ \cdots\ [x_N^{ap} \times]\ \right]} \underset{3N \times 1}{\begin{bmatrix} \delta x_1 \\ \delta y_1 \\ \delta z_1 \\ \vdots \\ \delta x_i \\ \delta y_i \\ \delta z_i \\ \vdots \\ \delta x_N \\ \delta y_N \\ \delta z_N \end{bmatrix}} = \underset{3 \times 1}{\mathbf{0}} \tag{15}$$

The NEQ of NNR is:

$$\underset{3N \times 3N}{N_{NNR} \delta x} = (A_{NNR}^T P_x A_{NNR}) \cdot \delta x = 0 \tag{16}$$

The coefficient matrix N_{NNR} of NEQ is:

$$N_{NNR} = P_x \left[\ [x_1^{ap} \times]^T [x_1^{ap} \times]\ \cdots\ [x_i^{ap} \times]^T [x_i^{ap} \times]\ \cdots\ [x_N^{ap} \times]^T [x_N^{ap} \times]\ \right]$$

2.4. Minimum Constrains

We review the minimum constraints [8] and give the corresponding normal equations. Before giving the minimum constraint, the similarity transformation of 14 Helmert parameters is introduced. Assuming that any solution X_s only includes the station position and velocity, then X_s can be transformed into the ITRF solution X by the 14 Helmert parameters, and the linearized matrix form is directly given, as follows:

$$X = X_s + A\theta \tag{17}$$

where $\theta = (T1, T2, T3, D, R1, R2, R3, \dot{T}1, \dot{T}2, \dot{T}3, \dot{D}, \dot{R}1, \dot{R}2, \dot{R}3)$ are 14 Helmert parameters, and $T1, T2$, and $T3$; D; $R1, R2$, and $R3$ represent three translation parameters; one scale parameter; and three rotation parameters, respectively, and the partial derivative symbol represents their corresponding rates. A is the design matrix derived from 14 to parameter similarity transformation, where N is the number of stations in the solution X_s, and $(\ldots, x_i^0, \ldots)$ represnts approximate positions of the station i:

$$A = \begin{bmatrix} \cdot & \cdot & \cdot & \cdot & \cdot & \cdot & \cdot & \cdot & \cdot & \cdot & \cdot & \cdot \\ 1 & 0 & 0 & x_i^0 & 0 & z_i^0 & -y_i^0 & & & & & \\ 0 & 1 & 0 & y_i^0 & -z_i^0 & 0 & x_i^0 & & \approx 0 & & & \\ 0 & 0 & 1 & z_i^0 & y_i^0 & -x_i^0 & 0 & & & & & \\ & & & & & & & 1 & 0 & 0 & x_i^0 & 0 & z_i^0 & -y_i^0 \\ & & \approx 0 & & & & & 0 & 1 & 0 & y_i^0 & -z_i^0 & 0 & x_i^0 \\ & & & & & & & 0 & 0 & 1 & z_i^0 & y_i^0 & -x_i^0 & 0 \\ \cdot & \cdot & \cdot & \cdot & \cdot & \cdot & \cdot & \cdot & \cdot & \cdot & \cdot & \cdot \end{bmatrix} \tag{18}$$

Using an unweighted least-squares adjustment, θ can be solved:

$$\theta = (A^T A)^{-1} A^T (X - X_S) \tag{19}$$

The θ in Equation (19) is regarded as the virtual observation value, and the error equation is given:

$$v_\theta = B(X - X_s) \tag{20}$$

where $B = (A^T A)^{-1} A^T$, and v_θ is the residual of θ. Because of the regularity of the matrix $A^T A$, it is even possible to use $B = A^T$ [26]. In geodetic analysis, the TRF difference between X and X_s is very small, so a prior value of the virtual observation value θ can be set to zero.

The NEQ of the minimum constraints derived from Equation (20) is:

$$(A\Sigma_\theta^{-1}A^T)(X - X_s) = 0 \tag{21}$$

where a diagonal matrix composed of small empirical variances corresponding to the 14 Helmert parameters is represented by Σ_θ.

For comparison with kinematic constraints, Equation (21) is simplified to:

$$p_x AA^T \delta x = 0 \tag{22}$$

where the diagonal matrix Σ_θ^{-1} can be simplified as a coefficient p_x.

2.5. Equivalence between Minimum Constraints and Kinematic Constraints

When only considering the origin datum (i.e., columns 1, 2, and 3 of the matrix A in Equation (18)), its normal equation of minimum constraints (see Equation (23)) is equivalent to that of kinematic constraints (see Equation (14)):

$$p_x AA^T \delta x = p_x \begin{bmatrix} 1 & 0 & 0 & \cdots & 1 & 0 & 0 & \cdots & 1 & 0 & 0 \\ 0 & 1 & 0 & \cdots & 0 & 1 & 0 & \cdots & 0 & 1 & 0 \\ 0 & 0 & 1 & \cdots & 0 & 0 & 1 & \cdots & 0 & 0 & 1 \\ \vdots & \vdots & \vdots & \ddots & \vdots & \vdots & \vdots & \ddots & \vdots & \vdots & \vdots \\ 1 & 0 & 0 & \cdots & 1 & 0 & 0 & \cdots & 1 & 0 & 0 \\ 0 & 1 & 0 & \cdots & 0 & 1 & 0 & \cdots & 0 & 1 & 0 \\ 0 & 0 & 1 & \cdots & 0 & 0 & 1 & \cdots & 0 & 0 & 1 \end{bmatrix} = 0 \tag{23}$$

Similarly, the orientation constraints implemented by these two constraints are also equivalent, only considering columns 5, 6, and 7 of the matrix A. We checked the VLBI minimum constraint solution (input solutions to ITRF2014 [21] and the long-term solutions produced in this work), with the results showing that the origin and orientation datum defined by the minimum constraint conform to the kinematic constraints.

The NEQs of the datum rates constraints between the minimum constraints and the kinematic constraints are also equivalent.

2.6. Internal Constraints

In the intra-technique combination at IGN, internal constraints are used to define the datum of ITRF (scale and origin) [10,13,20]. The intra-technique combination model (see Equation (1)) produces time series of 7 Helmert transformation parameters between the input and a single-technique combination solution. With the assumption that the linear time evolution for the station positions and transformation parameters, we can write for each of the 7 transformation parameters P_k (at epoch t_k) [10]:

$$P_k = P_k(t_0) + (t_k - t_0)\dot{P}_k \tag{24}$$

where t_0 represents the selected reference epoch of the combination solution.

Then, the least-squares adjustment can yield the following normal equation system of Equation (24):

$$\begin{pmatrix} K & \sum_{k \in K}(t_k - t_0) \\ \sum_{k \in K}(t_k - t_0) & \sum_{k \in K}(t_k - t_0)^2 \end{pmatrix} \begin{pmatrix} P_k(t_0) \\ \dot{P}_k \end{pmatrix} = \begin{pmatrix} \sum_{k \in K} P_k \\ \sum_{k \in K}(t_k - t_0)P_k \end{pmatrix} \tag{25}$$

The minimal/intrinsic constraints are used to impose some conditions on these transformation parameters:

$$\begin{cases} P_k(t_0) & = 0 \\ \dot{P}_k & = 0 \end{cases} \tag{26}$$

Imposing the two conditions implied in Equation (26) to NEQ (25), the right term of NEQ must be zero to ensure a unique zero solution. Internal constraints are derived:

$$\begin{cases} \sum_{k \in K} P_k & = 0 \\ \sum_{k \in K} \dfrac{P_k}{(t_k - t_0)^{-1}} & = 0 \end{cases} \tag{27}$$

2.7. Inner Constraints

Altamimi and Dermanis discuss and derive inner constraints and kinematic constraints [9] and conclude that the partial inner constraints for station parameters may coincide with the kinematical constraints, while partial inner constraints related to transformation parameters are equivalent to internal constraints. Inner constraints are given directly (see literature for detailed derivation):

$$\sum_{i=1}^{N} \delta x_{0i} - \sum_{k=1}^{M} d_k = 0 \tag{28}$$

$$\sum_{i=1}^{N} \delta v_i - \sum_{k=1}^{M} (t_k - t_0) d_k = 0 \tag{29}$$

$$\sum_{i=1}^{N} [x_{0i}^{ap} \times] \delta x_{0i} + \sum_{k=1}^{M} \theta_k = 0 \tag{30}$$

$$\sum_{i=1}^{N} [x_{0i}^{ap} \times] \delta v_i + \sum_{k=1}^{M} (t_k - t_0) \theta_k = 0 \tag{31}$$

$$\sum_{i=1}^{N} (x_{0i}^{ap})^T \delta x_{0i} - \sum_{k=1}^{M} s_k = 0 \tag{32}$$

$$\sum_{i=1}^{N} (x_{0i}^{ap})^T \delta v_i - \sum_{k=1}^{M} (t_k - t_0) s_k = 0 \tag{33}$$

where d_k, θ_k, and s_k represent translation, rotation, and scale transformation parameters, respectively.

The partial inner constraints for station parameters (the left parts of Equations (28)–(33)) are the exact kinematical constraints (see Section 2.3), while the partial inner constraints for transformation parameters are equivalent to internal constraints (see Section 2.6).

2.8. Relationship between Constraints

According to Sections 2.3–2.5, we can conclude that the minimum constraints and kinematic constraints are equivalent. Inner constraints unify internal constraints and kinematic constraints (see Section 2.3, 2.6 and 2.7).

Compared with the minimum constraints imposed over station parameters, the internal constraints act on the time series of Helmert parameters between each daily or weekly frame k and the intra-technique combination solution (see the first line of Equation (1), i.e., intra-technique combination). The internal constraint contains a condition that the conversion parameter is 0 (see Equation (26)), while the a priori value of θ, the virtual observation value of the minimum constraints, is also 0 (see Section 2.3). Variances of the transformation parameters θ are small in minimum constraints, while the transformation parameters in internal constraints have no statistical significance (or can be seen as infinitesimal). Therefore, the minimum constraints and internal constraints have both similarities and differences. The long-term solutions realized by these two constraints are compared at IGN, with the results showing that the post-fit residuals of the two methods are still the same and the transformation parameters between the two corresponding long-term accumulated solutions (intra-technique combination) are different, which is estimated by the 14-parameter similarity transformation.

However, only comparing the forms and results between the minimum constraints and internal constraints cannot explain the reasons for these differences. A detailed analysis of the process of intra-technique combination helps us understand how the internal constraints affect the datum of long-term solutions.

3. Results

Firstly, according to the VLBI intra-technique combination, we compare the scale of the long-term solutions realized by the internal constraints and the minimum constraints. Three years (2004.1.5–2006.12.29) of the time series of 24 h session data are available, which are provided by the International VLBI Service for Geodesy and Astrometry (IVS) for the ITRF2014 in the SINEX format [21]. Except for estimating seasonal signals of station position, the input data were analyzed using the same time series analysis method as ITRF2014 to eliminate the effects of discontinuity, velocity variation, PSD, and outliers [13]. We performed four stacking tests. Test 1, 2, and 3 use the IGN combination model (see Section 2.1), while test 4 uses the DGFI-TUM combination model (see Section 2.2). Before the stacking, minimum constraints are applied to inputs (NEQs) of tests 1, 2, and 3 to determine the origin and orientation datum. We extend the constrained NEQs of tests 1, 2, and 3 by 6 or 7 parameters of a similarity transformation (see Table 1).

Table 1. Extending NEQs by transformation parameters.

Parameters	Test 1	Test 2	Test 3	Test 4
Translation	Yes	Yes	Yes	No
Rotation	Yes	Yes	Yes	No
Scale	Yes	No	Yes	No

Tests 1 and 3 are extended by seven transformation parameters. Test 2 is extended by six parameters, corresponding to translation and rotation.

After extending the obtained NEQs by the station velocity parameter, the time series of NEQs are combined to one normal equation system. Minimum constraints or internal ones are applied to the accumulated normal equations of the four tests to complete the rank deficiency. For test 1, we choose internal constraints for the translation and scale components and the minimum constraint approach to define the orientation datum. For test 2, since scale parameters are not extended, we only choose minimum constraints for the translation and rotation components. For test 3, we choose minimum constraints for the translation and rotation components and the internal constraint approach to define the scale datum. For test 4, we choose minimum constraints for the translation and rotation components.

The scale parameters of four long-term solutions with respect to the ITRF2014 VLBI solution are shown in Table 2.

Table 2. The scale parameters of four long-term solutions with respect to the ITRF2014 VLBI solution.

Epoch	Parameter	Test 1	Test 2	Test 3	Test 4
2004.1.5		0.4268	0.2777	0.3485	0.2757
2005.1.1	Scale (ppb)	0.4203	0.4347	0.4612	0.4357
2006.12.29		0.4072	0.7500	0.6878	0.7570

Since we discuss linear frames in this work, the scale parameters at two epochs can reflect the scale rate.

The scale parameters of tests 2 and 4 are equal (see Table 2), and the difference between their station coordinates is negligible, indicating that the combination model does not affect the scale datum of VLBI. Compared with test 1, there is a linear trend (0.16 ppb/yr) on the scale of test 4. By estimating a 14-parameter similarity transformation, Altamimi et al. compared DTRF2014 solutions to the ITRF2014, with the result showing that there is a linear trend term between the VLBI solutions of DTRF2014 and ITRF2014 (see Figure 1) [23].

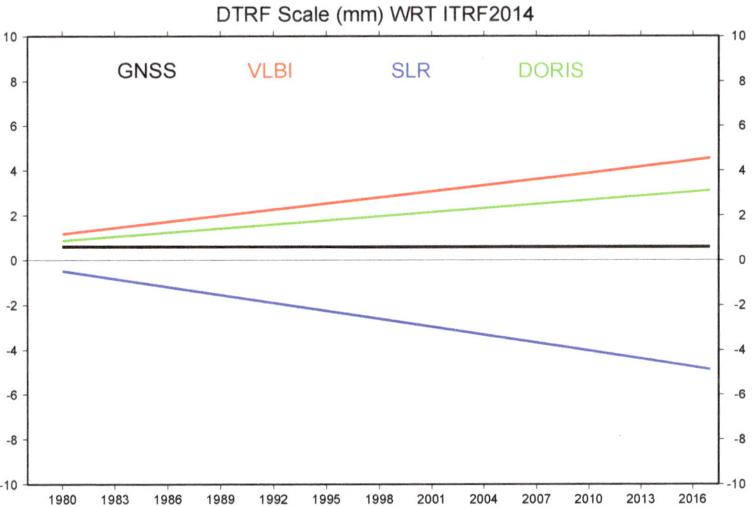

Figure 1. DTRF2014 scale with respect to ITRF2014.

Angermann et al. estimated the difference of VLBI single-technique solutions (intra-technique combination) of DGFI-TUM and IGN by 14 Helmert parameters, and the results showed that there is a linear trend between scales of these two solutions (see Figure 2) [24]. In order to minimize the scale impact for these two techniques, the scale of the ITRF2014 is defined in such a way that there is a zero scale factor (at epoch 2010.0) and a zero scale rate with respect to the average of the explicit scales and scale rates of the VLBI and SLR solutions [13,20]. Therefore, there is a constant offset in Figure 1 with respect to Figure 2.

For tests 1 and 3, the combination model of the two tests is the same, and the scale datum is realized by internal constraints. However, the scales of the two tests are inconsistent. Therefore, we have reason to suspect that there is a correlation between the time series of transformation parameters (especially scale and translation) introduced by the intra-technique combination. Moreover, in order to satisfy the internal constraint conditions, the scale and translation parameters may be mutually absorbed. In other words, the linear trend in VLBI scales is caused by internal constraints.

Although the VLBI inputs submitted to ITRF2014 are unconstrained normal equations, tests 1 and 3 are combined on the solution level, that is, all NEQs are applied with minimum constraints before superposition. Therefore, it is necessary to check the singularity of NEQs after setting up seven transformation parameters and station velocity to avoid the influence of over-constraints on the datum. Without loss of generality, the EOP and Helmert parameters are eliminated from the NEQs, and only the station coordinate parameters (including position and velocity) are retained. $n_1, n_2, \ldots, n_m$ (m is the number of station coordinate parameters) are the row or column vectors of NEQs. From a geometric point of view, the lack of the NEQ datum information leading to the rank deficiency of the NEQ is based on the fact that the columns of the normal matrix are orthogonal to the corresponding columns of the transformation matrix (see Equation (18)) [27]. The column vectors of the

14-Helmert transformation matrix are expressed as $g_1, g_2, \ldots, g_{14}$. The rank deficiency of NEQs can be calculated by Equation (34):

$$
< n, \, g >= \begin{bmatrix} \dfrac{n_1^T g_1}{\|n_1\| \cdot \|g_1\|} & \dfrac{n_1^T g_2}{\|n_1\| \cdot \|g_2\|} & \cdots & \dfrac{n_1^T g_{14}}{\|n_1\| \cdot \|g_{14}\|} \\ \dfrac{n_2^T g_1}{\|n_2\| \cdot \|g_1\|} & \dfrac{n_2^T g_2}{\|n_2\| \cdot \|g_2\|} & \cdots & \dfrac{n_2^T g_{14}}{\|n_2\| \cdot \|g_{14}\|} \\ \vdots & \vdots & \cdots & \vdots \\ \dfrac{n_m^T g_1}{\|n_m\| \cdot \|g_1\|} & \dfrac{n_m^T g_2}{\|n_m\| \cdot \|g_2\|} & \cdots & \dfrac{n_m^T g_{14}}{\|n_m\| \cdot \|g_{14}\|} \end{bmatrix}
\tag{34}
$$

where $< n, \, g >$ is the cosine value of the angle between the two vectors.

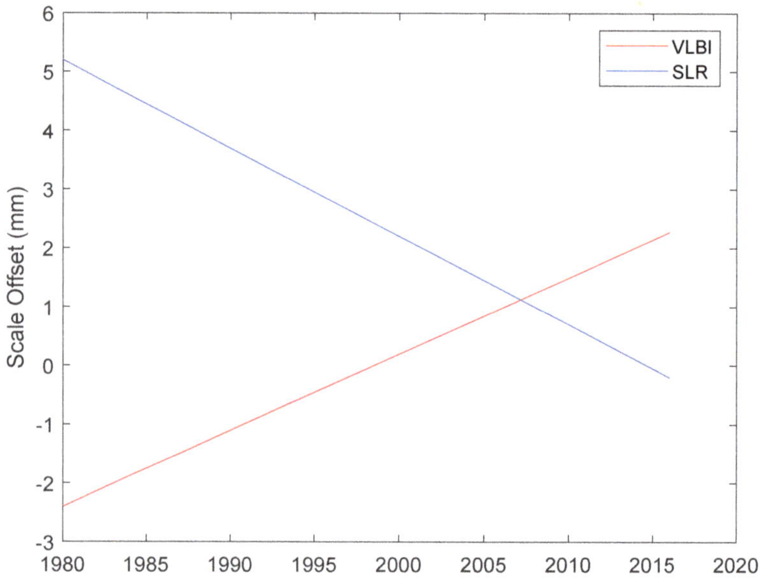

Figure 2. The linear trend between scales of the VLBI (red) and SLR (blue) long-term solutions from DGFI-TUM and IGN.

The singularity of NEQs can also be analyzed by calculating the orthogonality between the eigenvector of the coefficient matrix of NEQs and the similarity transformation matrix [27]. According to the above two singular analyses of NEQs, we can conclude that the long-term solutions obtained in this paper are not over-constrained. In other words, the scale datum of tests 1 and 3 is implemented only by internal constraints.

4. Discussion

In this section, we will check the correlation between the Helmert parameters of the session-wise NEQs. Firstly, the minimum constrained solutions of NEQs are substituted into the seven-parameter similarity transformation model. Then, the Pearson correlation coefficient between the transformation parameters in each normal equation of the similarity transformation is calculated. Figure 3 shows a strong correlation between scale and translation parameters and a weak correlation between scale and rotation parameters.

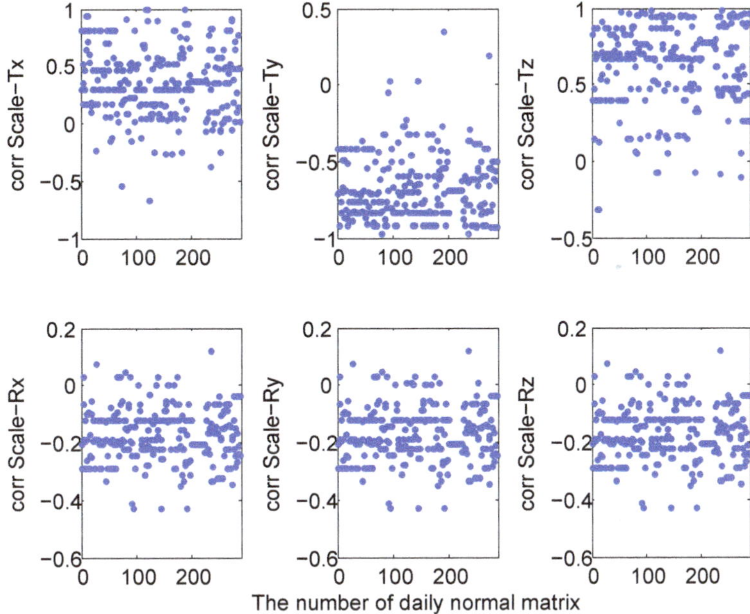

Figure 3. Correlation between scale parameters and other Helmert parameters.

Figure 4 shows the time series of translation parameters with respect to the long-term solutions of tests 1 (blue points on the left) and 2 (red points on the right).

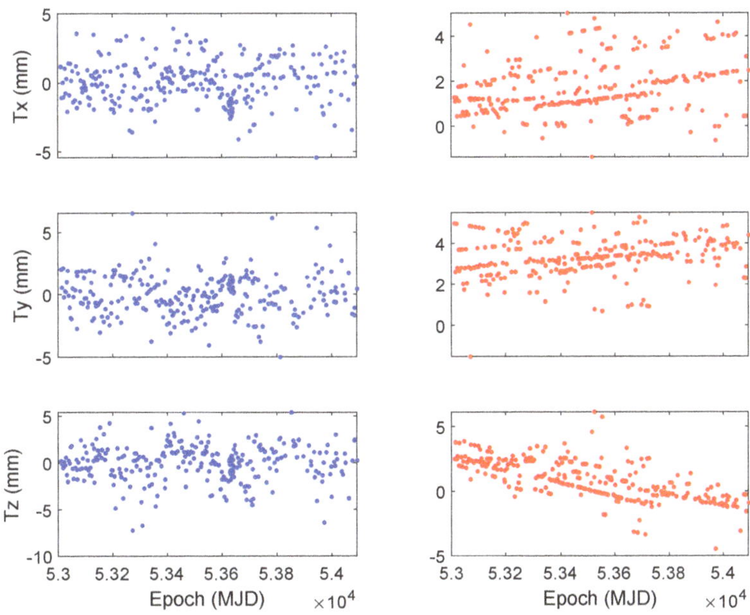

Figure 4. Time series of translation parameters with respect to the long-term solutions of tests 1 (blue) and 2 (red). MJD is Modified Julian Day.

These two results above confirm our previous assumption that there is a strong correlation between scale and translation parameters, and the internal constraint makes scale and translation parameters absorb each other, thus interfering with the scale of VLBI.

Figure 5 shows a weak correlation between translation and rotation parameters. The difference between the rotation parameters of tests 1 and 2 is at the μas level, that is, the orientation of the two frames can be considered the same.

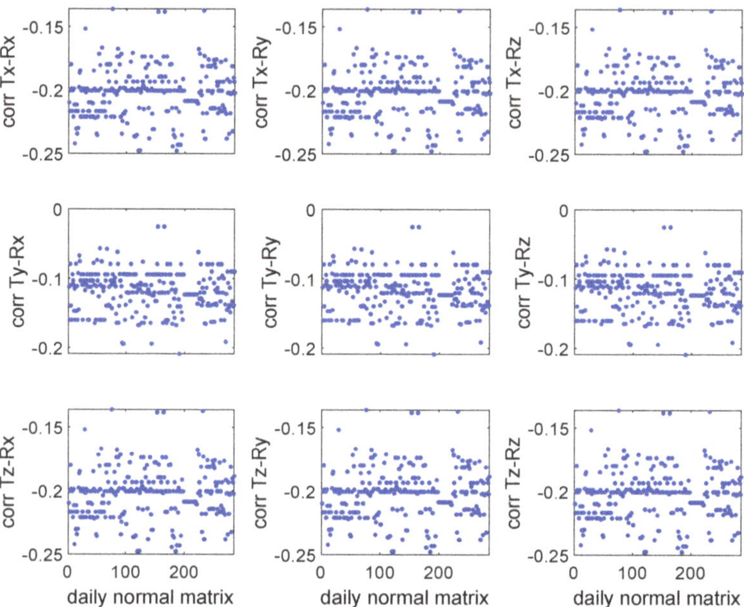

Figure 5. Correlation between translation and rotation parameters.

It should be noted that the internal constraints have strict requirements on stations' coverage of IVS sessions. The poor station coverage of IVS sessions leads to a strong correlation between rotation and other transformation parameters. ITRF2014 has to exclude the IVS sessions without more than four stations and those with poor coverage [13], although the excluded data are not outliers. Figures 6 and 7 show the station distribution of the two excluded sessions, corresponding to a session of no more than four stations and one with regional coverage, respectively.

Since the minimum constraints act on the station coordinates, the effect of these excluded sessions on tests 2 and 4 is not significant. Therefore, DGFI-TUM did not exclude VLBI data [28].

A comparison between solutions of the intra-technique combination [23] (see Figure 1) or inter-technique combination [24] (see Figure 2) shows that the scale of VLBI has a significant linear trend with respect to SLR in ITRF2014. With several tests showing that the scale offset between SLR and VLBI is less than 3.3 mm at 2000.0 [29], the scales of SLR and VLBI in the DTRF2014 was assumed to be statistically equal [22]. According to the preliminary calculations in the ITRF2020, the scale difference between SLR and VLBI is about 3 mm [25], versus 8.7 mm in ITRF2014 [13]. ITRF2020 does not provide a corresponding combination method. However, according to our calculations, internal constraints affect the scale of VLBI long-term solutions. The influence of the intra-technique combination model on the VLBI long-term solution is not significant. The influence of internal constraints on the long-term solution of SLR and that of the inter-technique combination model on the datum needs to be further studied.

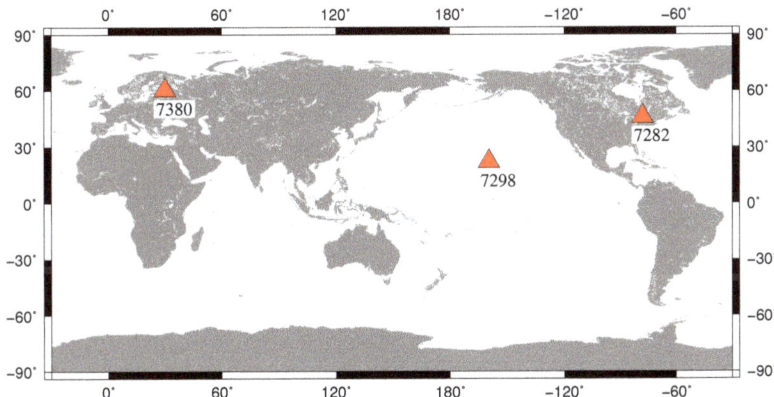

Figure 6. Station distribution of a session of less than five stations. The red triangle represents the VLBI station and its station code in the white box.

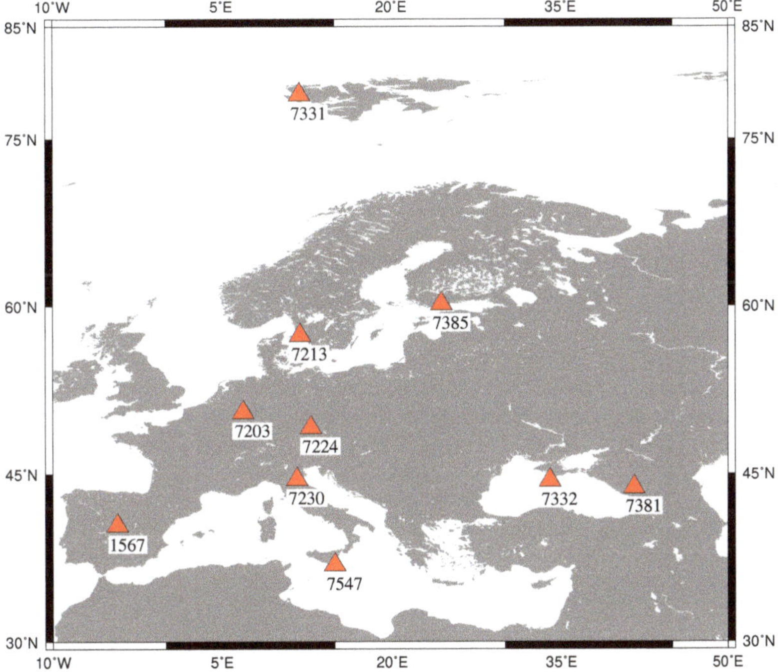

Figure 7. Station distribution of a session with regional coverage. The red triangle represents the VLBI station and its station code in the white box.

5. Conclusions

Different groups of scientists performed extensive serious research on constraints, against which our work is benchmarked. Inner constraints include internal constraints and kinematic constraints, and kinematic constraints are equivalent to minimum constraints. There is a strong correlation between the scale and translation parameters introduced by the VLBI intra-technique combination at IGN. Therefore, in order to satisfy the conditions of internal constraints, the scale and translation parameters absorb each other, thus interfering with the scale of VLBI. However, when the minimum constraints are applied, the VLBI long-term solution derived from the intra-technique combination model at IGN is equivalent to that of DGFI-TUM. That is, the intra-technique combination model does not affect the scale of VLBI.

Different from internal constraints imposed on the transformation parameters, minimum constraints act on the station coordinates, complete the rank deficiency of the singular NEQs, and ensure that the accumulated solution is expressed in the identical TRF with the reference solution. Therefore, minimum constraints do not interfere with the inherent datum of space geodetic technology. Compared with the minimum constraint, there is a linear trend in the scale of VLBI long-term solutions realized by the internal constraint. According to the comparison between ITRF2014 and DTRF2014 (the time span of VLBI input data is between 1980.0 and 2015.0), this linear trend leads to a maximum offset of more than 4 mm. However, compared to DTRF2014, ITRF SLR has a negative scale rate (also leading to a maximum offset of more than 4 mm), which is the opposite of the VLBI rate. Therefore, the maximum difference between the VLBI and SLR scales is intensified. In future work, we will analyze the influence of the internal constraints on the SLR datum information and investigate whether the inter-technique combination model can affect the scale datum of VLBI and SLR.

Author Contributions: Conceptualization, S.S., G.W. and Z.Z.; methodology, S.S., G.W. and Z.Z.; software, S.S.; validation, S.S., G.W. and Z.Z.; formal analysis, S.S., G.W. and Z.Z.; resources, S.S., G.W. and Z.Z.; data curation, S.S., G.W. and Z.Z.; writing—original draft preparation, S.S., G.W. and Z.Z.; writing—review and editing, S.S., G.W. and Z.Z.; visualization, S.S., G.W. and Z.Z.; supervision, S.S., G.W. and Z.Z.; project administration, S.S., G.W. and Z.Z.; funding acquisition, G.W. and Z.Z. All authors have read and agreed to the published version of the manuscript.

Funding: This research was funded by the National Natural Science Foundation of China (CN) grant number 11873077 and the Natural Science Foundation of Henan Province of China.

Institutional Review Board Statement: Not applicable.

Informed Consent Statement: Not applicable.

Data Availability Statement: Publicly available datasets were analyzed in this study. These data can be found here: [https://cddis.nasa.gov/ (accessed on 22 January 2022)].

Acknowledgments: I'd like to thank Tianhe Xu from Shandong University, Weihai, and Naifeng Fu from School of Marine Science and Technology, Tianjin University for their help to this work. The authors would like to thank IVS for providing high-quality VLBI data.

Conflicts of Interest: The authors declare no conflict of interest.

Abbreviations

The following abbreviations are used in this manuscript:

TRF	Terrestrial Reference Frame
TRS	Terrestrial Reference System
VLBI	Very Long Baseline Interferometry
IGN	Institut National de l'Information Géographique et Forestière
ITRS	The International Terrestrial Reference System
SI	Le Système International d'Unités
TCG	Geocentric Coordinate Time
TT	terrestrial time
IAU	The International Astronomical Union
IUGG	The International Union of Geodesy and Geophysics
BIH	Bureau International de l'Heure
NNR	no net rotation
ITRF	The International Terrestrial Reference Frame
IERS	The International Earth Rotation and Reference Systems Service
GNSS	Global Navigation Satellite Systems
SLR	Satellite Laser Ranging
DORIS	Doppler Orbitography and Radio-positioning Integrated by Satellite

EOP	earth orientation parameter
NEQ	the normal equation
CC	combination center
DGFI-TUM	Deutsches Geodätisches Forschungsinstitut
JPL	Jet Propulsion Laboratory
DTRF2014	DGFI-TUM's ITRS realization 2014
PSD	postseismic deformation
SINEX	Solution Independent Exchange
ACs	the Analysis Centres
NNT	no-net-translation
MJD	Modified Julian Day

References

1. Petit, G.; Luzum, B. (Eds.) *IERS Conventions (2010), IERS Technical Note No. 36*; Verlag des Bundesamts für Kartographie und Geodäsie: Frankfurt am Main, Germany, 2010.
2. Altamimi, Z.; Sillard, P.; Boucher, C. ITRF2000: A new release of the International Terrestrial Reference Frame for earth science applications. *J. Geophys. Res. Solid Earth* **2002**, *107*, ETG 2-1–ETG 2-19. Available online: https://agupubs.onlinelibrary.wiley.com/doi/pdf/10.1029/2001JB000561 (accessed on 25 January 2022). [CrossRef]
3. Altamimi, Z. The SLR Contribution to the ITRF. In Proceedings of the Science Session and Full Proceedings CD-ROM, 2003; Noomen, R., Klosko, S., Noll, C., Pearlman, M., Eds.; NASA/CP-2003-212248; Institut Géographique National 6–8 Avenue Blaise Pascal, Cites Descartes: Champs-sur-Marne, France, 2003.
4. Pavlis, E.C.; Kuzmicz-Cieslak, M. Variations in the Realization of the Origin of the ITRF From Satellite Laser Ranging. In Proceedings of the EGU General Assembly Conference Abstracts, Vienna, Austria, 22–27 April 2012; p. 6915.
5. Wahl, D.; Heinkelmann, R.; Schuh, H. Investigation of scale effects in the TRF determined by VLBI. In Proceedings of the EGU General Assembly Conference Abstracts, Vienna, Austria, 22–27 April 2017; p. 9547.
6. Pavlis, E.C. The Global SLR Network and the Origin and Scale of the TRF in the GGOS Era. In Proceedings of the 15th International Workshop on Laser Ranging, Canberra, Australia, 16–20 October 2006.
7. Boucher, C.; Altamimi, Z. (Eds.) *Evolution of the Realizations of the Terrestrial Reference System Done by the BIH and IERS (1984–1988), IERS Technical Note No. 04*; Central Bureau of IERS—Observatoire de Paris: Paris, France, 1990.
8. Altamimi, Z.; Boucher, C.; Sillard, P. New trends for the realization of the international terrestrial reference system. *Adv. Space Res.* **2002**, *30*, 175–184. [CrossRef]
9. Altamimi, Z.; Dermanis, A. The Choice of Reference System in ITRF Formulation. In Proceedings of the International Association of Geodesy Symposia, Buenos Aires, Argentina, 31 August–4 September 2009; Volume 137. [CrossRef]
10. Altamimi, Z.; Collilieux, X.; Legrand, J.; Garayt, B.; Boucher, C. ITRF2005: A new release of the International Terrestrial Reference Frame based on time series of station positions and Earth Orientation Parameters. *J. Geophys. Res. Solid Earth* **2007**, *112*. [CrossRef]
11. Angermann, D.; Drewes, H.; Seitz, M.; Meisel, B.; Gerstl, M.; Kelm, R.; Müller, H.; Seemüller, W.; Tesmer, V. ITRS Combination Center at DGFI: A terrestrial reference frame realization 2003. Verlag der Bayerischen Akademie der Wissenschaften: München, Germany, 2004; ISBN 3-7696-8593-8.
12. Abbondanza, C.; Chin, T.M.; Gross, R.S.; Heflin, M.B.; Parker, J.W.; Soja, B.S.; van Dam, T.; Wu, X. JTRF2014, the JPL Kalman filter and smoother realization of the International Terrestrial Reference System. *J. Geophys. Res. Solid Earth* **2017**, *122*, 8474–8510. [CrossRef]
13. Altamimi, Z.; Rebischung, P.; Métivier, L.; Collilieux, X. ITRF2014: A new release of the International Terrestrial Reference Frame modeling nonlinear station motions. *J. Geophys. Res. Solid Earth* **2016**, *121*, 6109–6131. [CrossRef]
14. Seitz, M.; Angermann, D.; Bloßfeld, M.; Drewes, H.; Gerstl, M. The 2008 DGFI realization of the ITRS: DTRF2008. *J. Geod.* **2012**, *86*, 1097–1123. [CrossRef]
15. Rebischung, P.; Altamimi, Z.; Ray, J.; Garayt, B. The IGS contribution to ITRF2014. *J. Geod.* **2016**, *90*, 611–630. [CrossRef]
16. Métivier, L.; Collilieux, X.; Lercier, D.; Altamimi, Z.; Beauducel, F. Global coseismic deformations, GNSS time series analysis, and earthquake scaling laws. *J. Geophys. Res. (Solid Earth)* **2014**, *119*, 9095–9109. [CrossRef]
17. Dziewonski, A.M.; Chou, T.A.; Woodhouse, J.H. Determination of earthquake source parameters from waveform data for studies of global and regional seismicity. *J. Geophys. Res. Solid Earth* **1981**, *86*, 2825–2852. [CrossRef]
18. Ekström, G.; Nettles, M.; Dziewoński, A. The global CMT project 2004–2010: Centroid-moment tensors for 13,017 earthquakes. *Phys. Earth Planet. Inter.* **2012**, *200–201*, 1–9. [CrossRef]
19. Collilieux, X.; Altamimi, Z.; Coulot, D.; van Dam, T.; Ray, J. Impact of loading effects on determination of the International Terrestrial Reference Frame. *Adv. Space Res.* **2010**, *45*, 144–154. [CrossRef]
20. Altamimi, Z.; Collilieux, X.; Laurent, M. ITRF2008: An improved solution of the international terrestrial reference frame. *J. Geod.* **2011**, *85*, 457–473. [CrossRef]
21. Bachmann, S.; Thaller, D.; Roggenbuck, O.; Lösler, M.; Messerschmitt, L. IVS contribution to ITRF2014. *J. Geod.* **2016**, *90*, 631–654. [CrossRef]

22. Seitz, M.; Bloßfeld, M.; Angermann, D.; Seitz, F. DTRF2014: DGFI-TUM's ITRS realization 2014. *Adv. Space Res.* **2021**, *69*, 2391–2420. [CrossRef]
23. Altamimi, Z.; Dick, W.R. *ITRS Center Evaluation of DTRF2014 and JTRF2014 with Respect to ITRF2014*; Presented at the IERS Technical Note No. 40; Verlag des Bundesamts für Kartographie und Geodäsie, Frankfurt am Main, Germany, 2020.
24. Angermann, D.; Bloßfeld1, M.; Seitz1, M.; Rudenko1, S. Comparison of latest ITRS Realizations: ITRF2014, DTRF2014 and JTRF2014; Presented at the IERS Technical Note No. 40; Verlag des Bundesamts für Kartographie und Geodäsie: Frankfurt am Main, Germany, 2020.
25. Zuheir Altamimi. The International Terrestrial Reference Frame: An Update. 2021. Available online: https://www.unoosa.org/documents/pdf/icg/2021/ICG15/19.pdf (accessed on 25 January 2022).
26. Dach, R.; Lutz, S.; Walser, P.; Fridez, P. Minimum Constraint Conditions. In *Bernese GNSS Software Version 5.2*; Dach, R., Lutz, S., Walser, P., Fridez, P., Eds.; Astronomical Institute, University of Bern: Bern, Switzerland, 2015; pp. 221–222.
27. Kotsakis, C.; Chatzinikos, M. Rank defect analysis and the realization of proper singularity in normal equations of geodetic networks. *J. Geod.* **2017**, *91*, 527–652. [CrossRef]
28. Seitz, M.; Bloßfeld, M.; Angermann, D.; Schmid, R.; Gerstl, M.; Seitz, F. *The New DGFI-TUM Realization of the ITRS: DTRF2014 (Data)*; Deutsches Geodätisches Forschungsinstitut: Munich, Germany, 2016. [CrossRef]
29. Bloßfeld, M.; Angermann, D.; Seitz, M. DGFI-TUM Analysis and Scale Investigations of the Latest Terrestrial Reference Frame Realizations. In *International Symposium on Advancing Geodesy in a Changing World*; Freymueller, J.T., Sánchez, L., Eds.; Springer International Publishing: Cham, Switzerland, 2019; pp. 3–9.

Relative Kinematic Orbit Determination for GRACE-FO Satellite by Jointing GPS and LRI

Zhouming Yang [1], Xin Liu [1,*], Jinyun Guo [1], Hengyang Guo [1], Guowei Li [2], Qiaoli Kong [1] and Xiaotao Chang [3]

[1] College of Geodesy and Geomatics, Shandong University of Science and Technology, Qingdao 266590, China; yangzhouming21@mails.ucas.ac.cn (Z.Y.); guojy@sdust.edu.cn (J.G.); xiaoguo@sdust.edu.cn (H.G.); qiaolikong@sdust.edu.cn (Q.K.)
[2] Shandong Provincial Institute of Land Surveying and Mapping, Jinan 250102, China; ligw@shandong.cn
[3] Land Satellite Remote Sensing Application Center of Ministry of Natural Resource, Beijing 100048, China; cxt@sasmac.cn
* Correspondence: skd994268@sdust.edu.cn

Abstract: As the first in-orbit formation satellites equipped with a Laser Ranging Interferometer (LRI) instrument, Gravity Recovery and Climate Experiment Follow-on (GRACE-FO) satellites are designed to evaluate the effective ability of the new LRI ranging system applied to satellite-to-satellite tracking. To evaluate the application of LRI in GRACE-FO, a relative kinematic orbit determination scheme for formation satellites integrating Kalman filters and GPS/LRI is proposed. The observation equation is constructed by combining LRI and spaceborne GPS data, and the intersatellite baselines of GRACE-FO formation satellites are calculated with Kalman filters. The combination of GPS and LRI techniques can limit the influence of GPS observation errors and improve the stability of orbit determination of the GRACE-FO satellites formation. The linearization of the GPS/LRI observation model and the process of the GPS/LRI relative kinematic orbit determination are provided. Relative kinematic orbit determination is verified by actual GPS/LRI data of GRACE-FO-A and GRACE-FO-B satellites. The quality of relative kinematic orbit determination is evaluated by reference orbit check and K-Band Ranging (KBR) check. The result of the reference orbit check indicates that the accuracy of GRACE-FO relative kinematic orbit determination along X, Y, and Z (components of the baseline vector) directions is better than 2.9 cm. Compared with the relative kinematic orbit determination by GPS only, GPS/LRI improves the accuracy of the relative kinematic orbit determination by approximately 1cm along with X, Y and Z directions, and by about 1.8 cm in 3D directions. The overall accuracy of relative kinematic orbit determination is improved by 25.9%. The result of the KBR check indicates that the accuracy of the intersatellite baseline determination is about +/−10.7 mm.

Keywords: GRACE-FO; formation satellites; spaceborne GPS; Laser Ranging Interferometer; relative kinematic orbit determination

Citation: Yang, Z.; Liu, X.; Guo, J.; Guo, H.; Li, G.; Kong, Q.; Chang, X. Relative Kinematic Orbit Determination for GRACE-FO Satellite by Jointing GPS and LRI. *Remote Sens.* **2022**, *14*, 993. https://doi.org/10.3390/rs14040993

Academic Editor: Xiaogong Hu

Received: 7 December 2021
Accepted: 15 February 2022
Published: 17 February 2022

Publisher's Note: MDPI stays neutral with regard to jurisdictional claims in published maps and institutional affiliations.

1. Introduction

Gravity Recovery and Climate Experiment Follow-on (GRACE-FO) is regarded as a new mission of gravity formation satellites launched jointly by the National Aeronautics and Space Administration (NASA) and Helmholtz-Centre Potsdam, German Research Centre for Geosciences (GFZ), with a design life of 5 years. Having been launched successfully at Vandenberg Air Force Base in California on May 22, 2018, GRACE-FO satellites attempt to replace GRACE satellites that had orbited for 15 years and retired in June, 2017 [1–3]. GRACE-FO satellites carry the same equipment as GRACE, including GPS, a K-Band Ranging (KBR) System, a satellite accelerometer, and star sensor [4,5]. Similarly, GRACE-FO satellites also adopt an orbit design similar to that of GRACE satellites, with an orbit height of about 500 ± 10 km, an orbit eccentricity of less than 0.005, and an orbit inclination of about 89° [5,6]. GRACE-FO satellites, mainly used to accurately measure and invert the time-varying Earth's gravitational field, are regarded as an important mission for collecting

Earth's gravitational field data of high resolution [2]. The high quality satellite orbit and relative position can ensure the high quality processing of gravity data, which contributes to the estimation of gravity field models with a monthly resolution [7,8]. Therefore, the study of the precise orbit determination of GRACE-FO satellites is considered crucial. To ensure a stable relative distance in formation flight, GRACE satellites are equipped with a KBR system. In addition, GRACE-FO satellites have on-board a Laser Ranging Interferometer (LRI) [9–12]. LRI can accurately measure the intersatellite distance of GRACE-FO formation satellites and provide distance data of high quality [13].

At present, several formation satellites have been performing in orbit, including GRACE, GRACE-FO, TanDEM-X, PRISMA, SJ-9 and TechSat21 [14–21]. The existing formation satellites mainly realize relative positioning based on a GPS technique [19]. Generally, the relative kinematic orbit determination of LEO satellites is based on carrier-phase differential observations (CDGPS) [14]. The differential technique promises to eliminate and weaken some common observation errors, such as receiver clock bias, satellite clock bias and ionospheric delay errors [19–30]. Gu et al. [31] combined single and double differential techniques to jointly solve the orbit of formation satellites. A comparison with satellite laser ranging (SLR) indicates that the accuracy of the orbit was improved by 25%. Van Barneveld [32] analyzed the influence of ionospheric delay on the calculation of intersatellite long baselines (>100 km) and compared existing ionospheric delay models.

The new generation GRACE-FO satellites are equipped with an LRI system, which makes it possible to use LRI observations to enhance the quality and stability of formation satellites orbit. Currently, the intersatellite distance are generally used to verify the intersatellite baseline, and to maintain the relative state of formation satellites. GRACE-FO satellites can be equipped with dual ranging system (LRI and KBR). Concerning the LRI ranging system, this manuscript describes the following activities: firstly, the inter-satellite distance observed by LRI is combined with the GPS observation data of the GRACE-FO satellites to form the GPS/LRI observations. Secondly, the estimated orbits are compared to reference orbits and KBR measurements. Thirdly, this study analyzed the relative kinematic orbit determination results of GPS/LRI and GPS only.

2. Mathematical Model of GPS and LRI

LEO satellites' orbit determination generally takes pseudo range and carrier phase of GPS as the main observed values. The pseudo range and carrier phase observation models are as follows:

$$P_{r,j}^s = \rho_r^s + c(dt_r - dt^s) + I_{r,j}^s + \varepsilon_P^s \tag{1}$$

$$L_{r,j}^s = \rho_r^s + c(dt_r - dt^s) - I_{r,j}^s + \lambda_j N_{r,j}^s + \lambda_j \varepsilon_L^s, \tag{2}$$

where $P_{r,j}^s$ is the observed code pseudo range; ρ_r^s is the geometric distance between the GRACE-FO satellite to the GPS satellite; c is the velocity of light; dt_r and dt^s are receivers and satellite clock offsets, respectively; $I_{r,j}^s$ is the ionospheric delay; $L_{r,j}^s$ is the observed value of carrier phase; $\lambda_j N_{r,j}^s$ is the ambiguity; and $\lambda_j \varepsilon_L^s$ and ε_P^s are observation noise. To eliminate ionospheric delay, ionospheric-free combinations are usually adopted:

$$P_{r,IF}^s = \rho_r^s + c(dt_r - dt^s) + \varepsilon_{P,IF}^s \tag{3}$$

$$L_{r,IF}^s = \rho_r^s + c(dt_r - dt^s) + \lambda_{IF} N_{r,IF}^s + \varepsilon_{L,IF}^s, \tag{4}$$

where $P_{r,IF}^s$ is the pseudo range of ionospheric-free combinations; $L_{r,IF}^s$ is the carrier phase of ionospheric-free combinations; λ_{IF} is the wavelength of carrier phase L_{IF}, $N_{r,IF}^s$ is the ambiguity of ionospheric-free combinations, and $\varepsilon_{r,IF}^s$ and $\varepsilon_{P,IF}^s$ are observation noise of pseudo range and carrier phase ionospheric-free combinations. The equations of ionospheric-free combinations are as follows:

$$L_{IF} = \frac{f_1^2}{f_1^2 - f_2^2} L_1 - \frac{f_2^2}{f_1^2 - f_2^2} L_2, \tag{5}$$

where f_1 and f_2 are the frequencies of carrier phases L_1 and L_2. Ionospheric-free combinations can eliminate most of the ionospheric delay, and loses integer characteristic.

To build double-difference observation equations, inter-satellite single-difference equations should first be constructed. The single-difference models are as follows:

$$\begin{cases} \nabla P_{r,IF}^{s_0,s} = \rho_r^s - \rho_r^{s_0} - c(dt^s - dt^{s_0}) \\ \nabla L_{r,IF}^{s_0,s} = \rho_r^s - \rho_r^{s_0} - c(dt^s - dt^{s_0}) + \lambda_{IF}(N_{r,IF}^s - N_{r,IF}^{s_0}) \\ \nabla N_{r,IF}^{s_0,s} = N_{r,IF}^s - N_{r,IF}^{s_0} \end{cases} \tag{6}$$

where $\nabla L_{r,IF}^{s_0,s}$ and $\nabla P_{r,IF}^{s_0,s}$ are single-difference observed values of the carrier phase and pseudo range. The satellite s_0 with the largest altitude angle is selected as the reference satellite. The double-difference models are as follows:

$$\begin{cases} \Delta\nabla P_{r_0,r,IF}^{s_0,s} = \nabla\rho_r^{s_0,s} - \nabla\rho_{r_0}^{s_0,s} \\ \Delta\nabla L_{r_0,r,IF}^{s_0,s} = \nabla\rho_r^{s_0,s} - \nabla\rho_{r_0}^{s_0,s} + \lambda_{IF}(\nabla N_{r,IF}^{s_0,s} - \nabla N_{r_0,IF}^{s_0,s}) \\ \Delta\nabla N_{r_0,r,IF}^{s_0,s} = \nabla N_{r,IF}^{s_0,s} - \nabla N_{r_0,IF}^{s_0,s} \end{cases} \tag{7}$$

where double-difference observed values of $\Delta\nabla P_{r_0,r,IF}^{s_0,s}$, $\Delta\nabla L_{r_0,r,IF}^{s_0,s}$ and $\Delta\nabla N_{r_0,r,IF}^{s_0,s}$ are double-difference pseudo range, carrier phase and ambiguity, respectively.

For the first time, GRACE-FO formation satellites are equipped with a LRI ranging system. The model of LRI is as follows [5]:

$$\Psi = c * (-\psi_M(t) + R_T(t-\tau)) / (2f), \tag{8}$$

where Ψ is the LRI measurement (unit: m), $\psi_M(t)$ is the observed phase value of LRI at t, $R_T(t-\tau)$ is the time delay correction of the LRI acquisition system, τ is the time of signal transmission of LRI, and f is the LRI-frequency of the observed phase. The ground nominal value of f in GRACE-FO-A is 2.81616393e14 Hz, and the ground nominal value of f in GRACE-FO-B is 2.81615684e14 Hz. The original phase observed values (LRI1A) of LRI have glitches [33]. Glitches of LRI mainly occurs during thruster firings. JPL laboratory detects glitches of LRI by three methods. Firstly, differences in phase observed values are performed, and then glitches of LRI are detected by comparing the observed differential values with the tolerance. Secondly, small glitches of LRI are fitted by a phase filter model [33]. Thirdly, glitches are further detected by residuals of Two-Way Range [11]. In order to simplify experiments, this study adopts secondary-treated Level-1B data of JPL laboratory, and all glitches of LRI were detected and repaired.

Based on the principle of LRI and product description published by GFZ, LRI observations deliver a biased distance N_{LRI}. To obtain the actual intersatellite distance, the biased distance must be gained in advance. This study proposes a simple and feasible calculation process for the biased LRI distance.

(1) The intersatellite distance is resolved by using relative kinematic orbit determination from spaceborne GPS data. Assuming that there are n epochs, the intersatellite distance between two LEO satellites $\rho_{r_0,r}^i (i = 0 \cdots n)$ is obtained according to the relative kinematic orbit determination of spaceborne GPS;

(2) LRI range minus the intersatellite distance $\rho_{r_0,r}^i$ at the corresponding epoch, and the initial value $N_{LRI,0}^i$ of LRI biased distance can be introduced;

(3) The mean value $\overline{N}_{LRI}$ of $N_{LRI,0}^i (i = 0 \cdots n)$ should be calculated;

(4) The difference between the initial value ($N_{LRI,0}^i (i = 0 \cdots n)$) and the mean value ($\overline{N}_{LRI}$) is calculated, and epochs with a difference greater than a threshold are marked as outliers. This threshold is defined as follows:

$$\left| N_{LRI,0}^i - \overline{N}_{LRI} \right| < 3\sigma_{rel}, \tag{9}$$

where σ_{rel} is the precision of relative kinematic orbit determination of spaceborne GPS. For the convenience of the experiment, σ_{rel} was set to 0.15 m ($\sigma_{rel} = 0.15$ m);

(5) The marked observed values of LRI should be eliminated, and then steps from 2 to 4 are repeated until all $N^i_{LRI,0}$ pass the inspection formula;

(6) Taking into account the LRI bias, ranging values of each epoch are corrected to obtain the intersatellite distance of LEO satellites based on LRI ranging.

To evaluate the precision of LRI (actually LRI biased), the LRI biased distances for 4 months (days 121 to 242 of 2019) are calculated. The statistical results are shown in Table 1, confirming the good quality of LRI distances. More than 98.5% of LRI observations can be used to solve LRI bias.

Table 1. RMS of LRI bias (unit: m) and percentage of LRI outliers (unit: %).

DAY	RMS of LRI Bias	Percentage of LRI Outliers
121−151	0.0492	1.21
151−182	0.0480	0.73
182−212	0.0535	0.98
212−242	0.0544	1.85
Average	0.0513	1.19

3. GPS/LRI Observation Equation and Linearization

As observed distance values, LRI can establish the observation of the LEO satellite by combining with GPS. The GPS/LRI observation equation in matrix form reads as follows:

$$V = BX - l, \tag{10}$$

where l is obtained by subtracting the calculated values from actual observed values, V is the correction of observed values, and B is the linearized parameter coefficient matrix to be solved. B is expressed as follows:

$$B = \begin{bmatrix} a_x^{s_0,s_1} & a_y^{s_0,s_1} & a_z^{s_0,s_1} & \lambda & 0 & \cdots & 0 \\ a_x^{s_0,s_2} & a_y^{s_0,s_2} & a_z^{s_0,s_2} & 0 & \lambda & 0 & 0 \\ \vdots & \vdots & \vdots & 0 & 0 & \ddots & 0 \\ a_x^{s_0,s_{n-1}} & a_y^{s_0,s_{n-1}} & a_z^{s_0,s_{n-1}} & 0 & 0 & 0 & \lambda \\ a_x^{s_0,s_1} & a_y^{s_0,s_1} & a_z^{s_0,s_1} & 0 & 0 & 0 & 0 \\ \vdots & \vdots & \vdots & \vdots & \vdots & \vdots & \vdots \\ a_x^{s_0,s_{n-1}} & a_y^{s_0,s_{n-1}} & a_z^{s_0,s_{n-1}} & 0 & 0 & 0 & 0 \\ a_x^{LRI} & a_y^{LRI} & a_z^{LRI} & 0 & 0 & 0 & 0 \end{bmatrix}, \tag{11}$$

where $a_x^{s_0,s} = -\frac{x^{s_0}-x_r}{\rho_r^{s_0}} + \frac{x^s-x_r}{\rho_r^s}$; $a_y^{s_0,s} = -\frac{y^{s_0}-y_r}{\rho_r^{s_0}} + \frac{y^s-y_r}{\rho_r^s}$; $a_z^{s_0,s} = -\frac{z^{s_0}-z_r}{\rho_r^{s_0}} + \frac{z^s-z_r}{\rho_r^s}$;
$a_x^{LRI} = \frac{x_r-x_{r_0}}{\rho_{r_0,r}}$; $a_y^{LRI} = \frac{y_r-y_{r_0}}{\rho_{r_0,r}}$; $a_z^{LRI} = \frac{z_r-z_{r_0}}{\rho_{r_0,r}}$; $\rho_{r_0,r} = \sqrt{(x_r-x_{r_0})^2 + (y_r-y_{r_0})^2 + (z_r-z_{r_0})^2}$;
$\rho_r^{s_0}$ and ρ_r^s are the geometric distances from GRACE-FO to different GPS satellites. x_r, y_r, z_r, $x^{s_0}, y^{s_0}, z^{s_0}$ and x^s, y^s, z^s are Cartesian Coordinates. The absolute vector l is given by:

$$l = \begin{bmatrix} l_{r_0,r}^{s_0,s_1} \cdots l_{r_0,r}^{s_0,s_{n-1}}, p_{r_0,r}^{s_0,s_1} \cdots p_{r_0,r}^{s_0,s_{n-1}}, v_{LRI} \end{bmatrix} \tag{12}$$

$$\begin{aligned} l_{r_0,r}^{s_0,s} &= \rho_r^s - \rho_r^{s_0} - \rho_{r_0}^s + \rho_{r_0}^{s_0} - \Delta\nabla L_{r,IF}^{s_0,s} \\ p_{r_0,r}^{s_0,s} &= \rho_r^s - \rho_r^{s_0} - \rho_{r_0}^s + \rho_{r_0}^{s_0} - \Delta\nabla P_{r,IF}^{s_0,s} \\ v_{LRI} &= \rho_{r_0,r} - \Psi \end{aligned} \tag{13}$$

After the observation equation is linearized, a Kalman filter process can be used for iterative processing.

4. Kalman Filter Theory Applied to Relative Kinematic Orbit Determination of LEO Satellites

Relative Kinematic orbit determination of LEO satellites aims to obtain coordinates of satellites. Under the condition of double-difference orbit determination, the parameters to be obtained not only include the coordinates of satellites, but also the ambiguities of double-difference carrier observed values. The parameters of LEO satellites are to be calculated by setting n double-difference carrier observed values:

$$X = [\Delta x, \Delta y, \Delta z, \Delta\nabla N_{r_0,r,IF}^{s_0,s_1} \cdots \Delta\nabla N_{r_0,r,IF}^{s_0,s_{n-1}}],\tag{14}$$

where $\Delta x, \Delta y, \Delta z$ are components of the baseline vector between the GRACE-FO satellites (Earth-fixed reference frame) and $\Delta\nabla N_{r_0,r,IF}^{s_0,s_1} \cdots \Delta\nabla N_{r_0,r,IF}^{s_0,s_{n-1}}$ are ambiguities of the observed carrier values. The Kalman filter equations of LEO satellites are as follows:

$$\begin{cases} J = D_0 B^T \left(B D_0 B^T + R\right)^{-1} \\ X = X_0 + Jl \\ D = \left(D_0^{-1} + B^T R^{-1} B\right)^{-1} \end{cases},\tag{15}$$

where D_0 is the prior variance matrix of parameters; R is the observation variance matrix; J is the gain matrix; X_0 and X are the prior value and calculated value of parameters to be solved, respectively; l is the observed minus computed (O-C); and D is the posterior variance matrix of parameters.

Pseudo-range single point positioning is used to replace the state-transition matrix to generate prior coordinates of LEO satellites (prediction coordinates), with an accuracy of about 10 m, which requires continuous refinement of LEO satellites' coordinates according to observed carrier values and their corresponding variance information. The variance processing of parameters is regarded as an important basis for Kalman filter calculation. Therefore, this manuscript provides the detailed settings for the corresponding parameter variance matrix and observation variance matrix in the Kalman filter. The variance matrix formula of observed values R is as follows:

$$\underset{(2n-2)\times(2n-2)+1}{R} = \begin{bmatrix} R_L^{s_0} + R_L^{s_1} & R_L^{s_0} & R_L^{s_0} & 0 & 0 & 0 \\ \vdots & \ddots & \vdots & \vdots & \vdots & \vdots \\ R_L^{s_0} & R_L^{s_0} & R_L^{s_0} + R_L^{s_{n-1}} & 0 & 0 & 0 \\ 0 & 0 & 0 & R_P^{s_0} + R_L^{s_1} & R_P^{s_0} & R_P^{s_0} \\ \vdots & \vdots & \vdots & \vdots & \ddots & \vdots \\ 0 & 0 & 0 & R_P^{s_0} & R_P^{s_0} & R_P^{s_0} + R_L^{s_{n-1}} \\ 0 & 0 & 0 & 0 & 0 & 0 \end{bmatrix}\tag{16}$$

$$\begin{cases} R_L^s = 2 \times \left(a_L^2 + \dfrac{b_L^2}{\sin^2 el^s}\right) \\ R_P^s = 2 \times \left(a_P^2 + \dfrac{b_P^2}{\sin^2 el^s}\right) \end{cases}.\tag{17}$$

R_L^s and R_P^s are the phase prior variances and pseudo-range prior variances, el^s is the elevation angle of GPS satellites, $a_L = 0.003$ and $b_L = 0.003$ are the error factors of phase observation (unit: m), $a_P = 0.3$ and $b_P = 0.3$ are the error factors of pseudo-range observation (unit: m), and R_{LRI} is the variance of LRI's observed values. In the experiment of this manuscript, $R_{LRI} = 0.01^2$ (unit: m^2).

In this work, Kalman filter orbit determination does not take dynamic information into consideration, and pseudo-range single point positioning is used to generate a priori coordinate X_0. The prior value of the ambiguity parameter adopts the floating-point value resolved in the previous epoch. The initial-state parameter variance matrix D_0 of LEO satellites is assigned as follows:

$$\underset{(n+3)\times(n+3)}{D_0} = \begin{bmatrix} 30^2 & & \\ & \ddots & \\ & & 30^2 \end{bmatrix},$$

(18)

where the variance corresponding to the coordinate parameter is reset as 30^2(unit: m^2) in the next epoch, while the variance corresponding to the ambiguity parameter directly adopts the posterior variance of previous epochs.

5. Relative Kinematic Orbit Determination Process of GPS/LRI

The essence of relative kinematic orbit determination of spaceborne GPS/LRI is to regard LRI as a directly observed value, and build the observation equation with GPS double-difference observations for resolving the intersatellite baseline. The relative kinematic orbit determination process of GPS/LRI is shown in Figure 1.

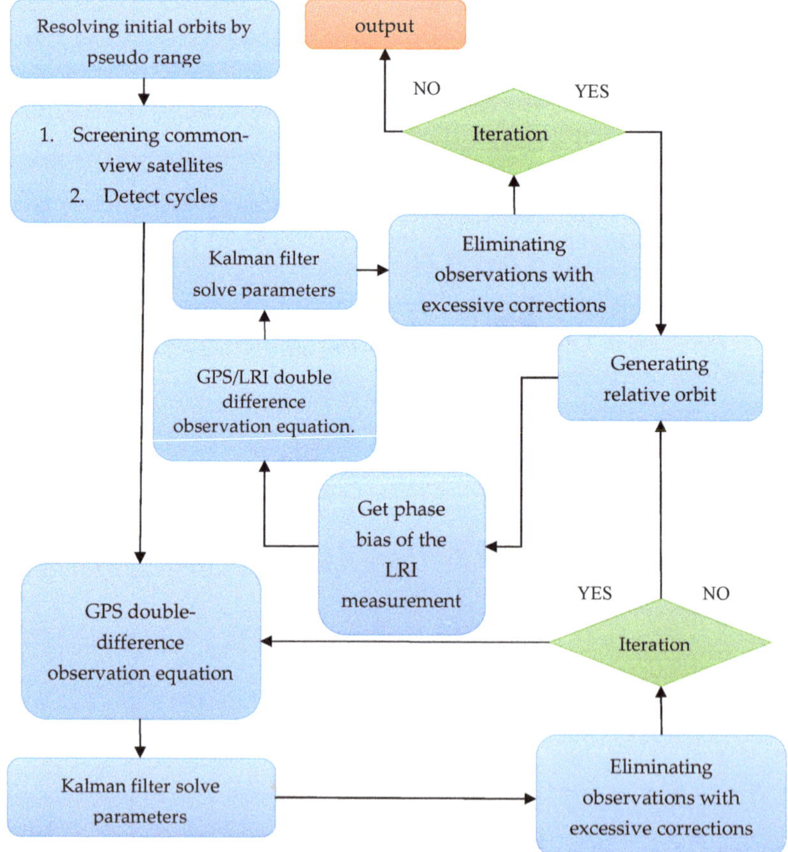

Figure 1. Orbit determination process.

The specific process of kinematic orbit determination is shown below:
(1) Resolving initial coordinates of orbits by the ionospheric-free pseudo range;
(2) Screening common-view satellites of GRACE-FO-A and GRACE-FO-B;

(3) Detecting cycle slips according to Melbourne–Wübbena (MW) and geometry-free (GF) combination [34,35]. The models of MW and GF are provided below:

$$L_{MW} = (f_1 L_{r,1}^s - f_2 L_{r,2}^s)/(f_1 - f_2) - (f_1 P_{r,1}^s + f_2 P_{r,2}^s)/(f_1 + f_2) \tag{19}$$

$$L_{GF} = L_{r,1}^s - L_{r,2}^s. \tag{20}$$

Equations (19) and (20) are usually used to jointly detect cycle slips. In this paper, the threshold of cycle slips detection between epochs (i and $i + 1$) is set as:

$$\begin{cases} N_w^{i+1} - N_w^i < 5 \\ L_{GF}^{i+1} - L_{GF}^i < 0.05 \end{cases} (unit : m). \tag{21}$$

The detection for cycle slips is regarded as data preprocessing;

(4) Constructing double-difference observation equation according to Equations (1) to (7);

(5) Updating the parameters to be resolved by Kalman filter;

(6) Correcting phase center offsets (PCO) and Sensor Offsets of LEO satellite. Taking GRACE-FO-A and GRACE-FO-B as examples, Tables 2 and 3 show the three components of PCO and Sensor Offsets in the body- fixed coordinate of GRACE-FO satellites. As for LEO satellites, their solar panel must always aim at the sun, so the satellite gesture will change according to time. Therefore, several steps should be taken in calculating PCO and Sensor Offsets correction of LEO satellites with spaceborne GPS:

(a) The coordinates of LEO satellites without being corrected by PCO and Sensor Offset are converted from Earth-fixed coordinate system to inertial coordinate system.

(b) According to the attitude of LEO satellites, the three components of PCO and Sensor offset in the body- fixed coordinate system are transferred to the inertial coordinate system.

(c) The corrections of PCO and Sensor Offset in the inertial system are added to the coordinates of LEO satellites.

(d) The coordinates of LEO satellites are transferred from the inertial coordinate system to the Earth-fixed coordinate system;

Table 2. PCO of GRACE-FO satellites (unit: mm).

Scheme	Frequency	PCO		
		North	East	Up
GRACE-FO-A	L1	1.49	0.60	−7.01
	L2	0.96	0.86	22.29
GRACE-FO-B	L1	1.49	0.60	−7.01
	L2	0.96	0.86	22.29

Table 3. Sensor Offset of GRACE-FO satellites (unit: mm).

Satellite	Sensor Offset		
	North	East	Up
GRACE-FO-A	−261.8	−0.8	−531.6
GRACE-FO-B	−260.0	0.5	−530.6

(7) The intersatellite baseline is recalculated according to the correction of mass center, and the relative kinematic orbit determination results are generated.

6. Case Study and Analysis

To verify the influence of LRI on the relative kinematic orbit determination of GRACE-FO formation satellites, a control experiment is designed according to actual observation data of GRACE-FO satellites. The experimental group achieves relative kinematic orbit determination by adopting GPS/LRI combinations, and the control group only uses GPS. The paper adopts reference orbit comparison and KBE check to analyze and compare the experimental results. The observation data of GPS/LRI and KBR in GRACE-FO satellites are provided by GFZ (https://isdc.gfz-potsdam.de/, accessed on 9 June 2020), collected from day 121 to day 242 in 2019, a total of 121 days. Samples of GPS data are collected at the interval of 10 s, and those of LRI and KBR data are collected at the interval of 2 s and 5 s, respectively. Then, synchronous GPS, LRI and KBR data should be used for check. The precise ephemeris, satellite clock offsets and Earth rotation parameter files are provided by the Center for Orbit Determination in Europe (CODE), and broadcast ephemeris are provided by the International GNSS Service (IGS). More details about these data are listed in Table 4.

Table 4. Data description.

Data Type	Source	Detail
GPS measurements	GFZ	Sampling rate 1 s; observed values of P1, P2, L1 and L2
LRI measurements	GFZ	Sampling rate 2 s; including observation time and intersatellite distance
GPS precise ephemeris GPS precise clock	CODE	Final precise ephemeris; Sampling rate 900 s; Final precise clock error products; Sampling rate 30 s
Earth rotation parameters	CODE	121-day values
Post-processed science orbit	GFZ	Reduced dynamic orbit
K-Band Ranging measurements	GFZ	Sampling rate 5 s
Broadcast ephemeris	IGS	BRDC station

6.1. Reference Orbit Check

The orbit published by GFZ is regarded as the reference value. The relative orbit determination results of GPS/LRI and GPS are separately compared with the reference orbit. Figures 2–6 show the comparison results of days 121–242's reference orbits. As is shown, there are no systematic errors in the GRACE-FO satellites' relative kinematic orbit determination. The main reason is that the relative kinematic orbit determination method used in this paper essentially belongs to a geometric orbit determination without being affected by the dynamic satellite model. However, the residual errors of GPS/LRI joint relative kinematic orbit determination result in X, Y, and Z (components of the baseline vector) directions are obviously smaller than those of GPS only. Compared with relative kinematic orbit determination by GPS only, the distribution of orbit residuals calculated by GPS/LRI is closer, and the obvious spikes in the figure are weakened. The main reason is that on-board GPS is quite different from ground stations in that the number of satellites observed by the on-board GPS receiver in some arcs is smaller than five. Meanwhile, the travel speed of LEO satellites can reach 6–7 km per second, so the observed GPS satellites change frequently. Usually, a satellite can be observed for only 10 to 30 min. Therefore, the observation data quality is poor for a small part of the arc, reducing the stability of the geometric method for relative kinematic orbit determination. The Geometric Dilution Precision (GDOP) is related to the orbit determination accuracy. We analyzed the GDOP of GRACE-FO satellites for 121 to 242 days. When the value of GDOP increases, the orbit accuracy will decrease. However, GPS / LRI can limit the impact of GDOP increase on the orbit accuracy. Therefore, combined GPS/LRI observations can better eliminate GPS observations of poor quality and reduce the impact of GPS observation quality on relative kinematic orbit determination. Additionally, the results of 121 days' relative kinematic orbit determination show that introducing GPS/LRI observations is obviously better than using GPS only.

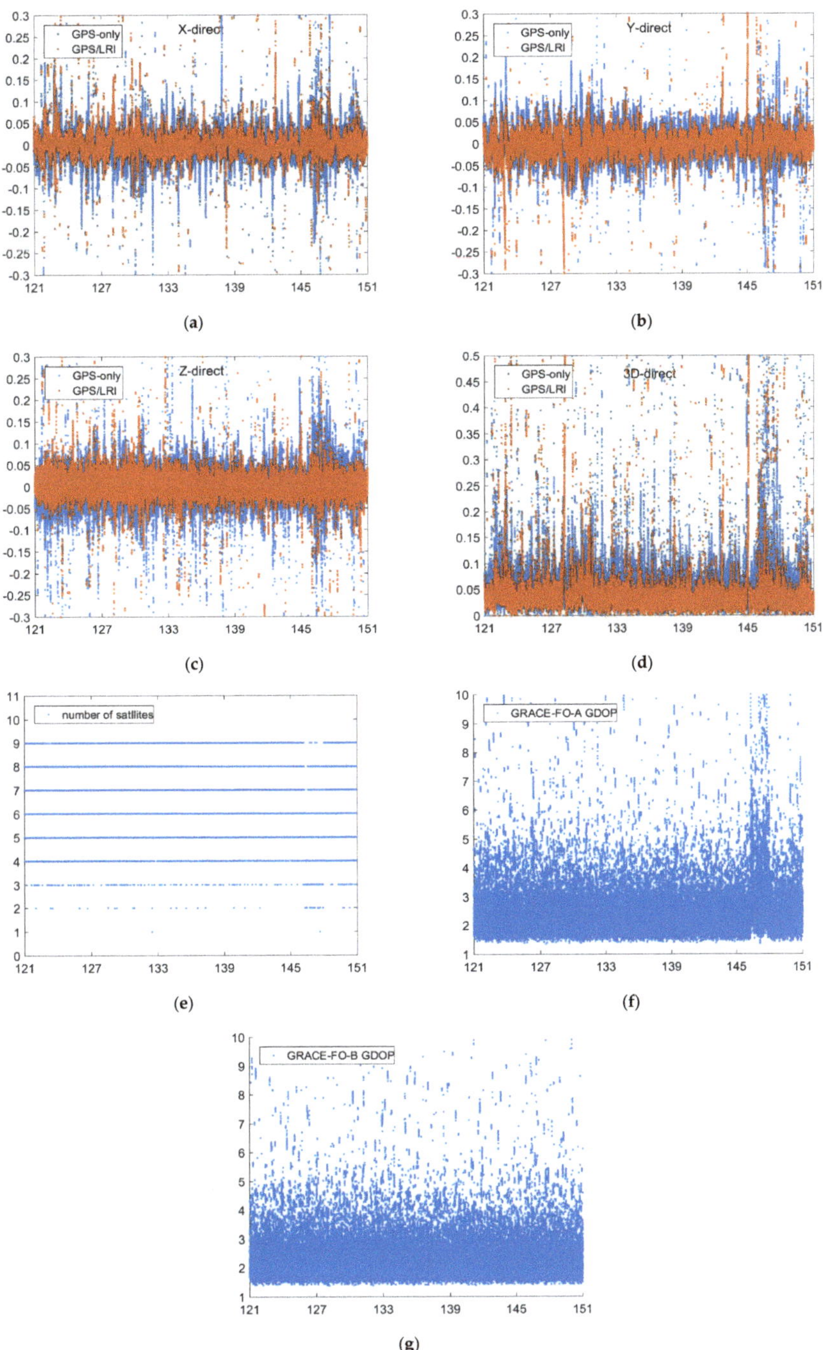

Figure 2. Reference orbit check for GRACE-FO in the X, Y, Z, 3D (X, Y, Z are components of the baseline vector, unit: m, (**a**–**d**)) directions (days 121–151), number of satellites for each epoch (**e**) and GDOP (**f**,**g**).

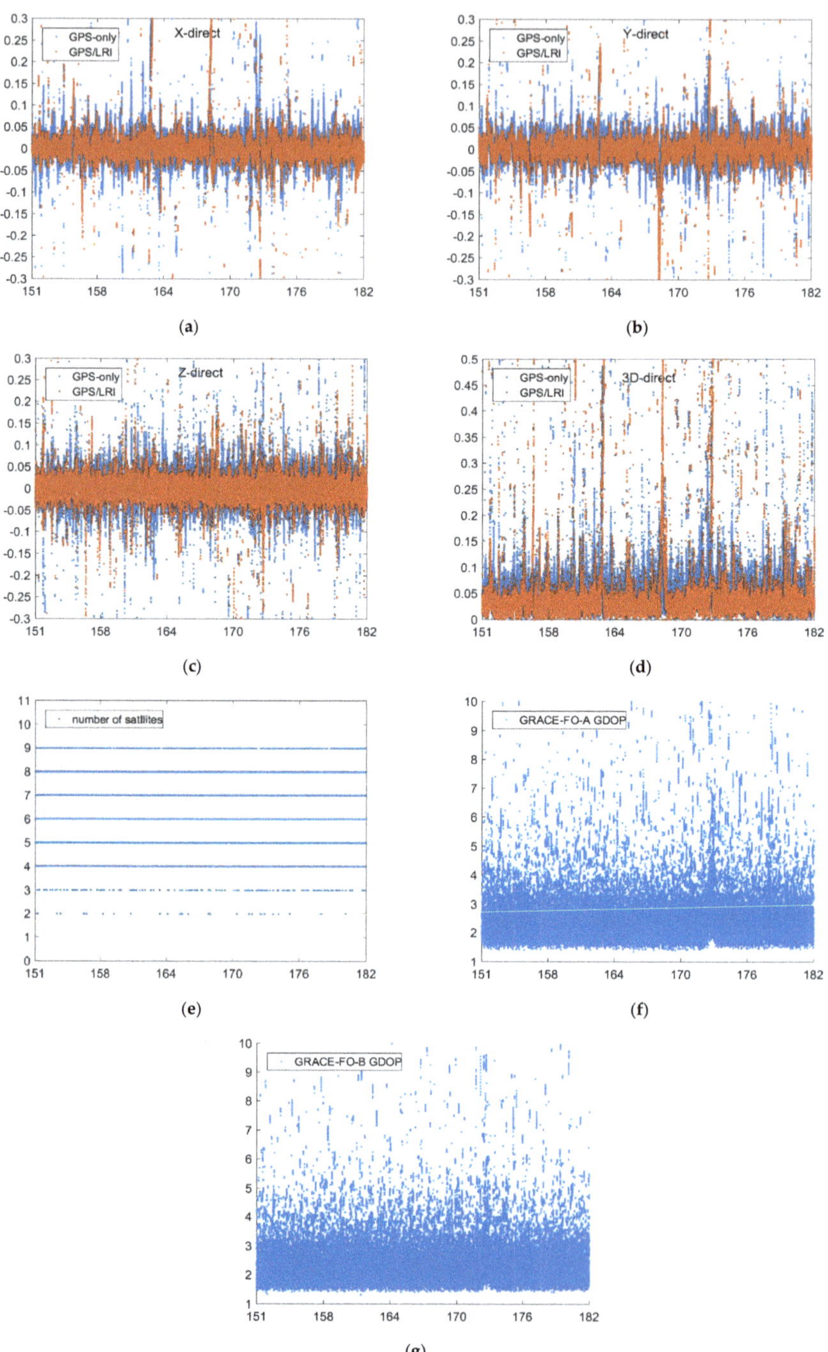

Figure 3. Reference orbit check for GRACE-FO in the X, Y, Z, 3D (X, Y, Z are components of the baseline vector, unit: m, (**a**–**d**)) directions (days 151–182), number of satellites for each epoch (**e**) and GDOP (**f**,**g**).

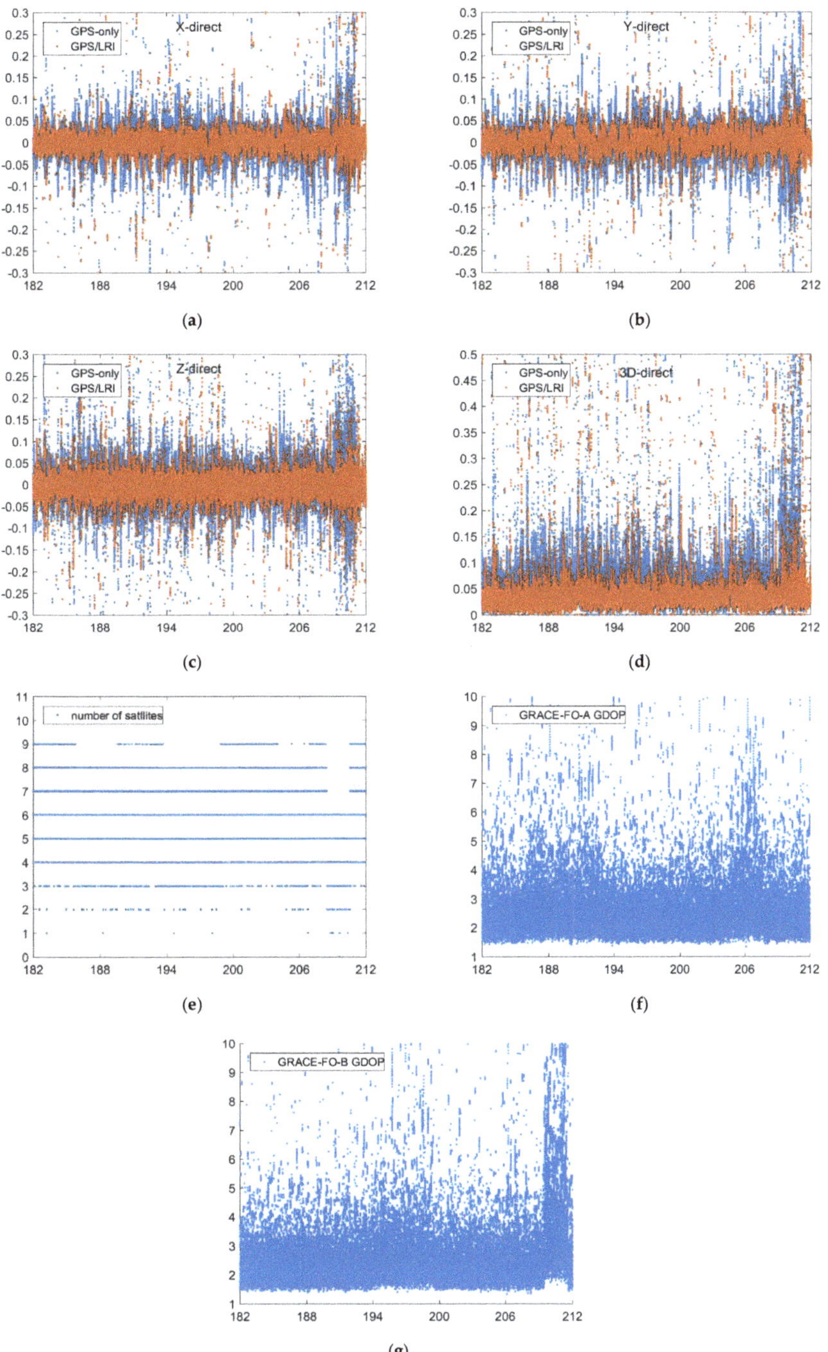

Figure 4. Reference orbit check for GRACE-FO in the X, Y, Z, 3D (X, Y, Z are components of the baseline vector, unit: m, (**a–d**)) directions (days 182–212), number of satellites for each epoch (**e**) and GDOP (**f,g**).

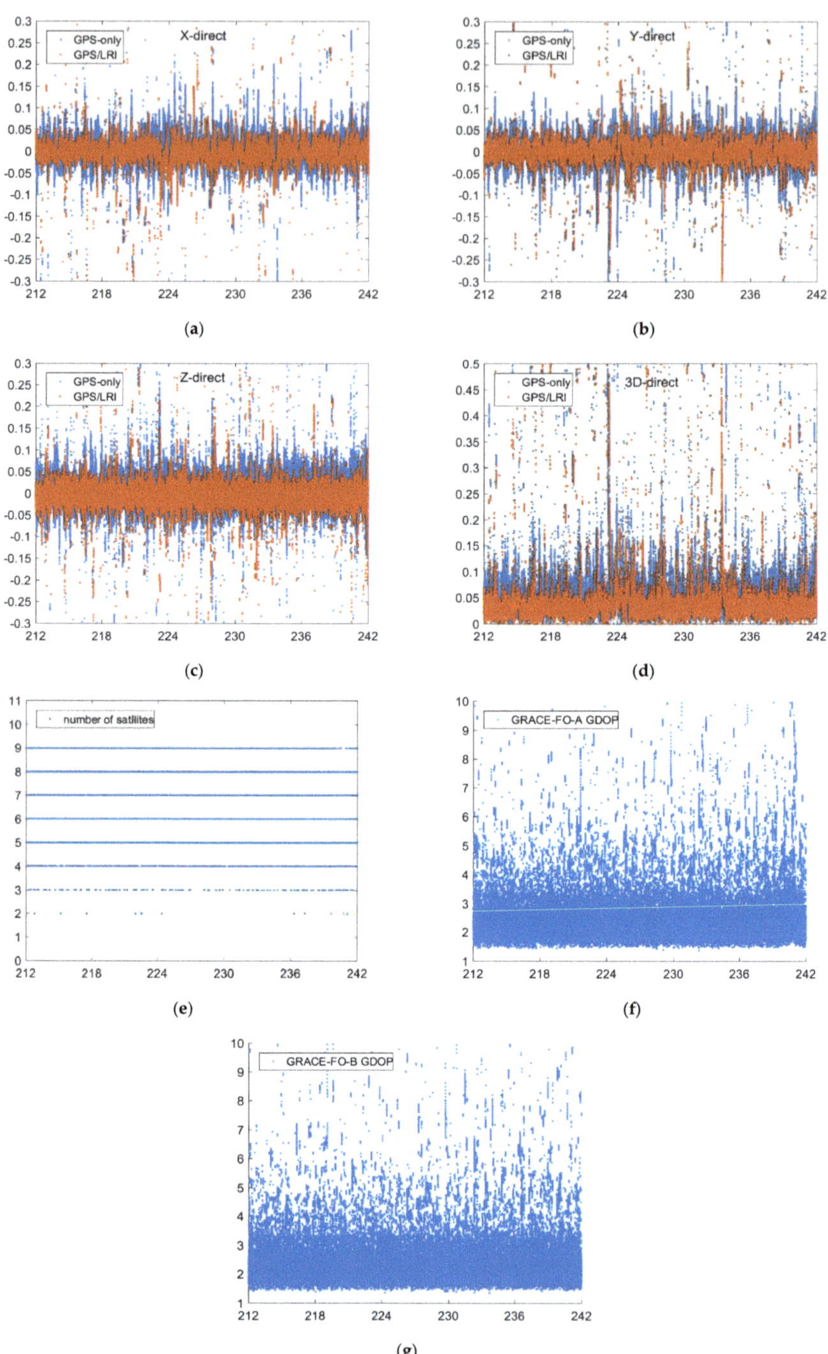

Figure 5. Reference orbit check for GRACE-FO in the X, Y, Z, 3D (X, Y, Z are components of the baseline vector, unit: m, (**a**–**d**)) directions (days 212–242), number of satellites for each epoch (**e**) and GDOP (**f**,**g**).

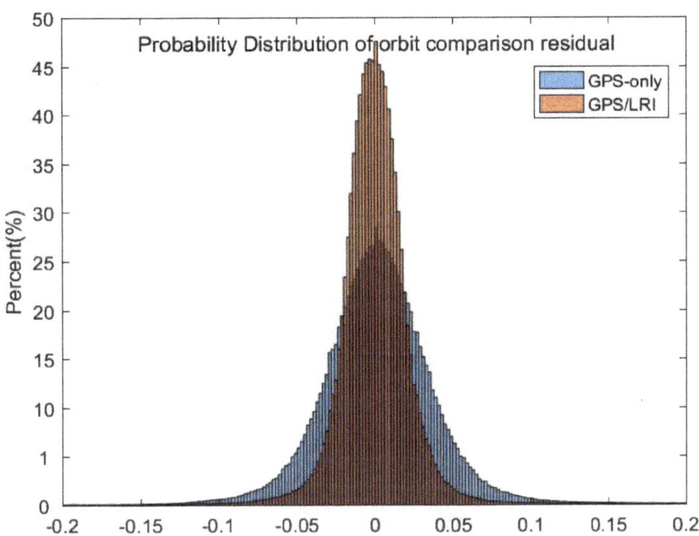

Figure 6. Days 121–242 orbit comparison residuals (in the X, Y and Z directions, unit: m) probability distribution for the GPS/LRI and GPS relative orbit.

To obtain more specific results of relative kinematic orbit determination, statistics over 121 days are carried out in this part. X, Y, and Z refer to the baseline vectors of the two satellites in Earth-fixed coordinates. It is shown in Table 5 that along the X direction, the 121 days accuracy of GPS/LRI relative kinematic orbit determination is improved by 12.4 mm, and the relative kinematic orbit determination accuracy along the X direction reaches 25.4 mm, with an accuracy increase of 32.8%; along the Y direction, the accuracy of GPS/LRI relative kinematic orbit determination is improved by 8.5 mm, and the accuracy along the Y direction reaches 28.6 mm, with an accuracy increase of 22.9%; along the Z direction, the accuracy of GPS/LRI relative kinematic orbit determination is improved by 15.2 mm, and the accuracy along the Z direction reaches 21.9 mm, with an accuracy increase of 41.0%; along the 3D direction, the accuracy of GPS/LRI relative kinematic orbit determination is improved by 17.5 mm, and the accuracy along the 3D direction reaches 50.1 mm, with an accuracy increase of 25.9%. LRI observed values are added to increase the number of redundant observations, enhancing the geometric strength of observed values, so the accuracy of GPS/LRI is greatly improved compared with that of only GPS for formation orbit determination. Statistical results show that the joint GPS/LRI data can effectively improve the overall accuracy of relative kinematic orbit determination of GRACE-FO formation satellites and limit orbit determination errors due to GPS observation quality, improving the stability of relative kinematic orbit determination.

Table 5. Statistics of reference orbit check (days 121–242) in Earth-fixed coordinate (unit: m).

Type	GPS-Only				GPS/LRI			
	X	Y	Z	3D	X	Y	Z	3D
MEAN	0	0.0019	0.0017	—	0	0	0.0009	—
MEDIAN	0	0.0025	0.0013	—	0	0.0008	0.0012	—
RMS	0.0378	0.0371	0.0371	0.0676	0.0254	0.0286	0.0219	0.0501

To analyze the orbital residual, we set GRACE-FO-B as the reference orbit and solve the orbit of GRACE-FO-A according to the relative kinematic orbit determination results. The orbit of the GRACE-FO-A satellite and the orbit released by GFZ are compared by using the ORBCMP module of Bernese 5.2 software. Under the radial, along–track and out–of–plane (RSW) decomposition, we analyze the residual results. Figures 7–10 show the comparison results of day 121–242's reference orbits. Figure 11 shows the probability distribution of satellite orbit residuals. In the R and W directions, the residual distributions of GPS-only and GPS/LRI are similar. However, in the S direction, GPS/LRI can significantly improve the relative kinematic orbit determination accuracy of GRACE-FO-A satellite. For further analysis, we calculated the MEAN, MEDIAN and RMS of orbital residuals in R, S and W directions. It is shown in Table 6 that along the R direction, the 121 days accuracy of GPS/LRI relative kinematic orbit determination is improved by 10.0 mm, and the relative kinematic orbit determination accuracy along the R direction reaches 29.0 mm, with an accuracy increase of 34.48%; along the S direction, the accuracy of GPS/LRI relative kinematic orbit determination is improved by 31.7 mm, and the accuracy along the S direction reaches 8.9 mm, with an accuracy increase of 78.8%; along the W direction, the accuracy of GPS/LRI relative kinematic orbit determination is not significantly improved. The RMS of the 3D vector is decreased by 18 mm which corresponds to an improvement of 26.3%.

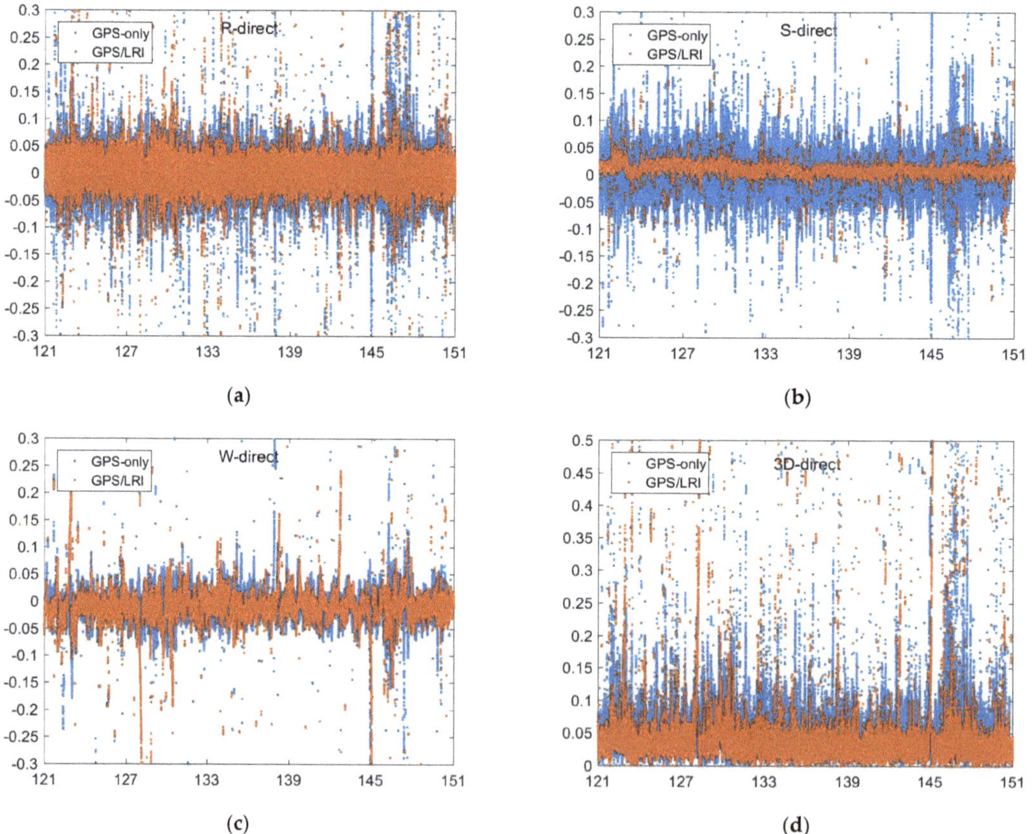

Figure 7. Reference orbit check for GRACE-FO-A in the R, S, W, 3D directions (days 121–151, (**a**–**d**)).

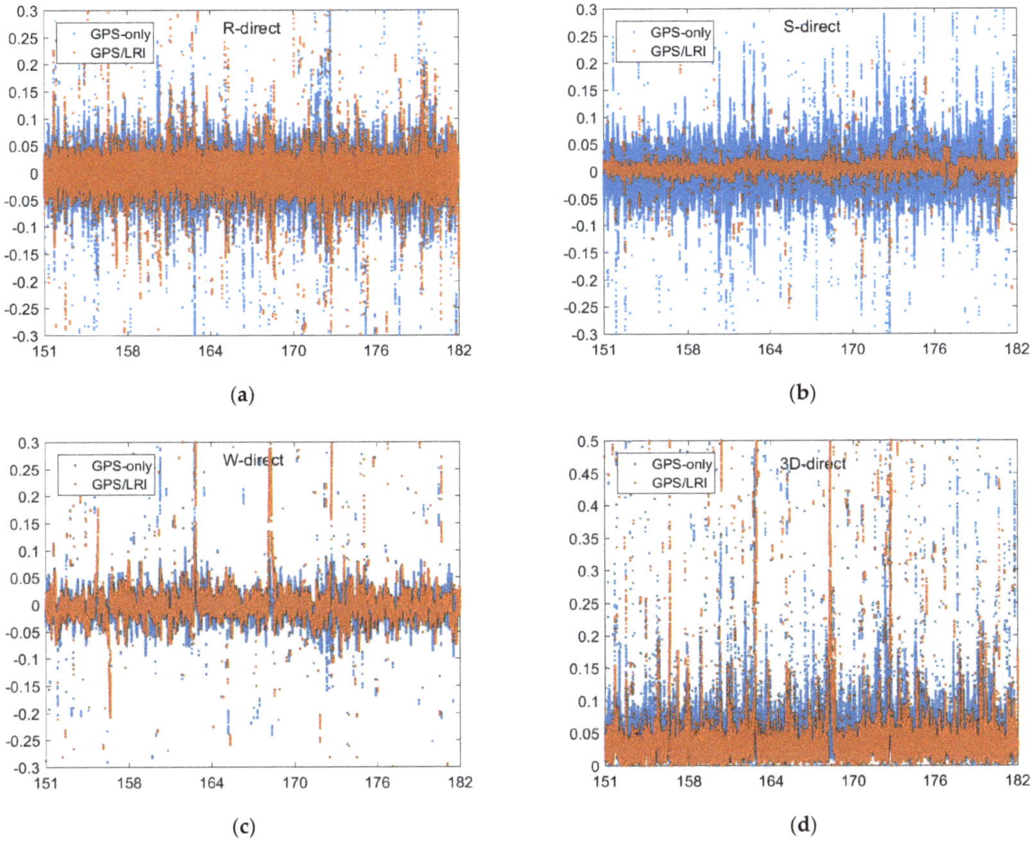

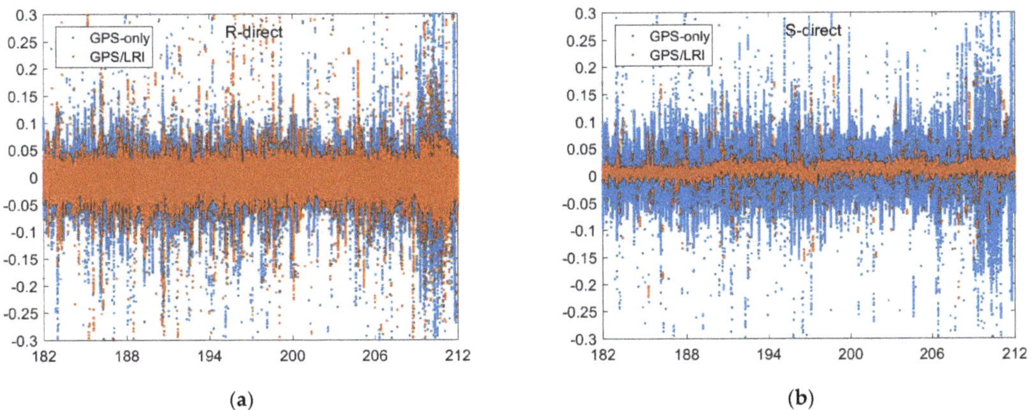

Figure 8. Reference orbit check for GRACE-FO-A in the R, S, W, 3D directions (days 151–182, (**a**–**d**)).

Figure 9. *Cont.*

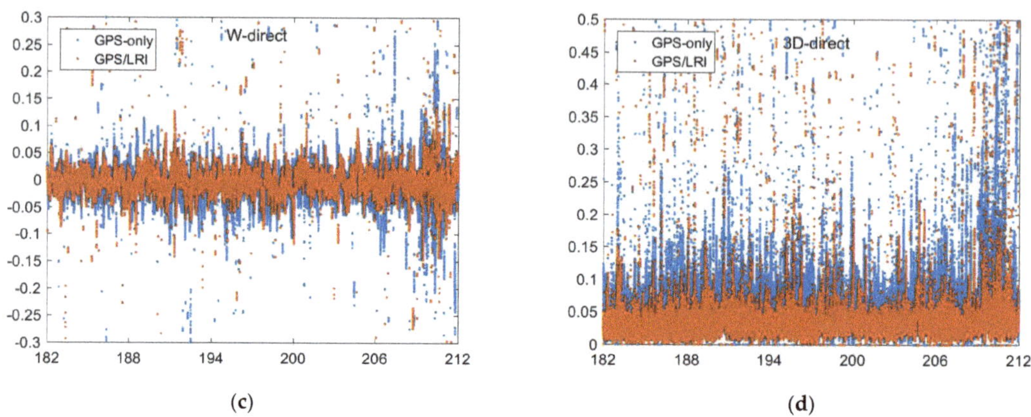

(c)   (d)

Figure 9. Reference orbit check for GRACE-FO-A in the R, S, W, 3D directions (days 182–212, (**a**–**d**)).

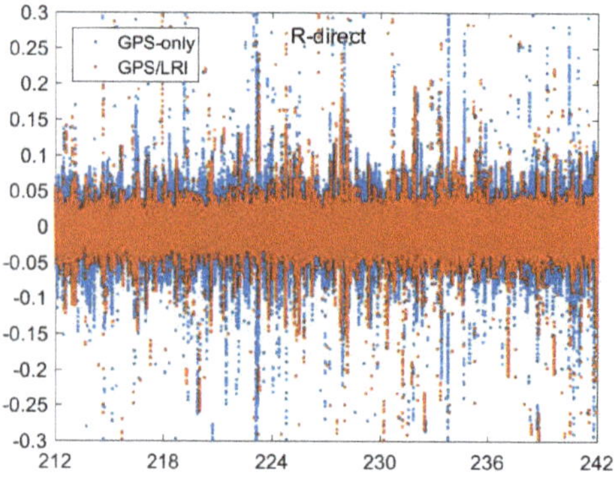

(a) 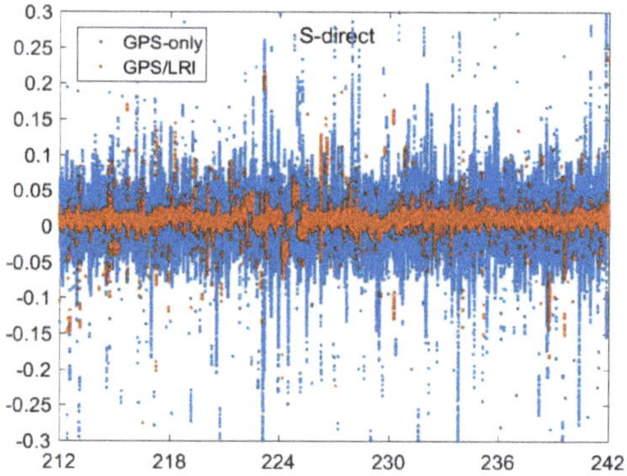   (b)

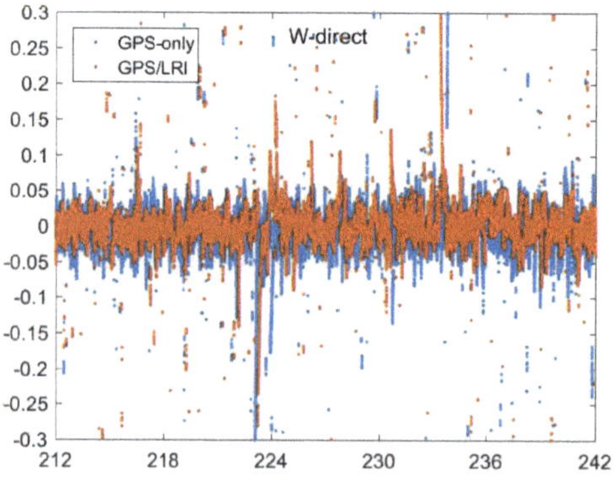

(c) 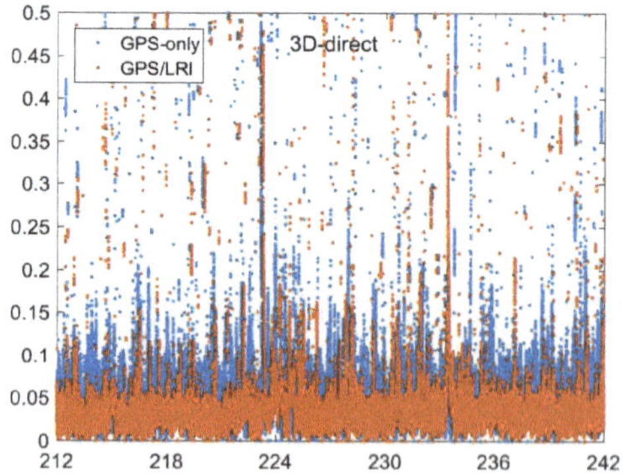   (d)

Figure 10. Reference orbit check for GRACE-FO-A in the R, S, W, 3D directions (days 212–242, (**a**–**d**)).

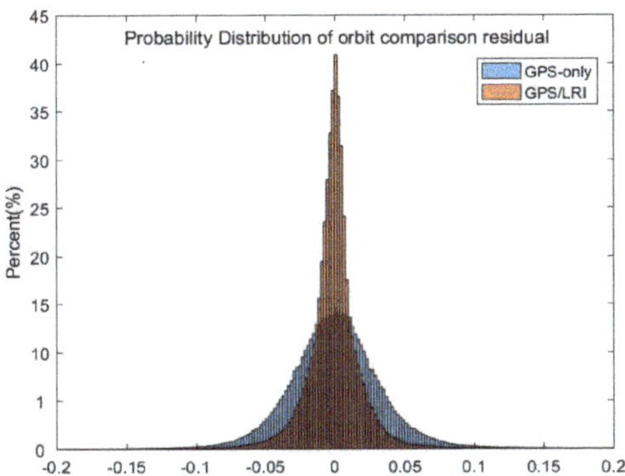

Figure 11. Days 121–242 orbit comparison residuals (in the R, S and W directions, unit: m) probability distribution for the GPS/LRI and GPS relative orbit.

Table 6. Statistics of reference orbit check (days 121–242) in RSW (unit: m).

Type	GPS-Only				GPS/LRI			
	R	S	W	3D	R	S	W	3D
MEAN	−0.0010	0	−0.0005	–	−0.0013	0	0	–
MEDIAN	−0.0009	0	0	–	−0.0008	0	0	–
RMS	0.0390	0.0406	0.0330	0.0678	0.0290	0.0089	0.0305	0.0500

To evaluate the GPS/LRI under sunlight and solar eclipse, we use the ORBGEN module of Bernese 5.2 software to mark the epochs in solar eclipse. GRACE-FO satellites were under solar eclipse for 25% of the total period. Figure 12 show the comparison results of day 121–242's reference orbits in the sunlight and solar eclipse. In the case of sunlight and solar eclipse, there is no significant difference in GPS/LRI orbit residuals. In both cases, the accuracy of GPS/LRI orbit determination is higher than the GPS-only. To further analyze the impact of solar eclipse on GPS/LRI, we calculated the RMS, MEAN and MEDIAN of orbital residuals. It is shown in Tables 7 and 8 that the RMS values differ by 2.2 mm and 2.1 mm in the 3D-directions. Along the X, Y, Z and 3D direction the 121-days accuracy of GPS/LRI relative kinematic orbit determination, under the conditions of sunlight and solar eclipse, is not a significant difference.

Table 7. Statistics of X, Y, Z and 3D residual (unit: m, days 121–242) in Sunlight and Solar eclipse.

Type	GPS/LRI in Sunlight				GPS/LRI in Solar Eclipse			
	X	Y	Z	3D	X	Y	Z	3D
MEAN	0	0.0007	−0.0009	–	0	−0.0007	−0.0009	–
MEDIAN	0	0.0008	−0.0011	–	0	0	−0.0014	–
RMS	0.0255	0.0282	0.0216	0.0437	0.0260	0.0302	0.0228	0.0459

Table 8. Statistics of X, Y, Z and 3D residual (unit: m, days 121–242) in Sunlight and Solar eclipse.

Type	GPS-Only in Sunlight				GPS-Only in Solar Eclipse			
	X	Y	Z	3D	X	Y	Z	3D
MEAN	0.0012	0.0018	0.0014	–	0.0007	0.0021	0.0023	–
MEDIAN	0.0007	0.0025	0.0011	–	0	0.0022	0.0017	–
RMS	0.0382	0.0375	0.0374	0.0652	0.0368	0.0362	0.0363	0.0631

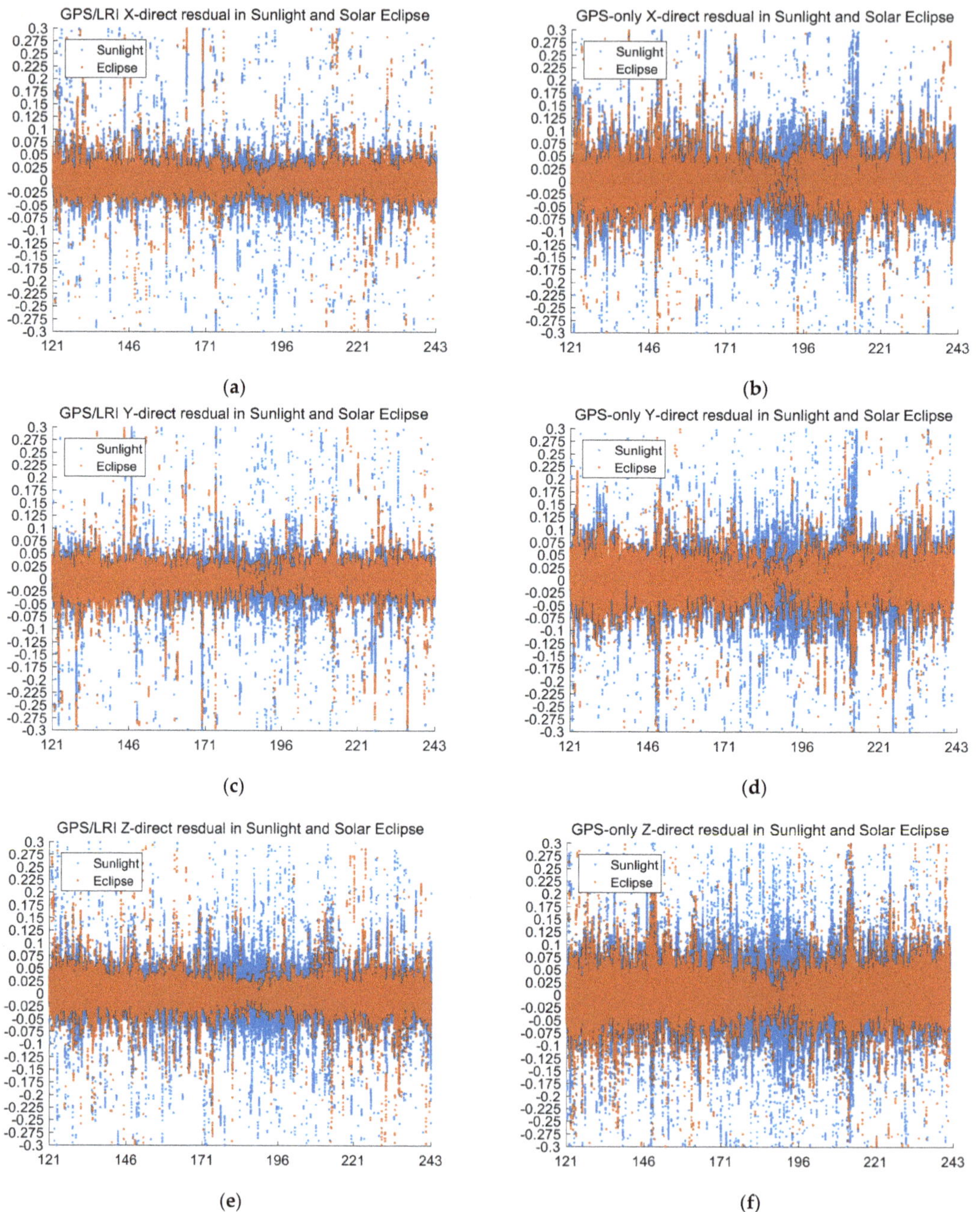

Figure 12. D (X, Y, Z are components of the baseline vector, unit: m, (**a**–**d**)) directions (days 121–242).

To evaluate the GPS/LRI in satellite orbit maneuver, we analyzed the orbit eccentricity of GRACE-FO satellites for 121 days. Landerer points out that regular orbit maintenance maneuvers will be performed throughout the GRACE-FO mission duration [36]. Figure 13 shows the orbital eccentricity of GRACE-FO satellites and their hourly variation.

We marked the time of orbital maneuver in the figure with circles. Table 9 shows the date of the satellite orbit maneuver. Figure 14 shows the comparison results of reference orbits in the orbit maneuver. Table 10 shows the RMS, MEAN and MEDIAN of orbital residuals during orbital maneuver.

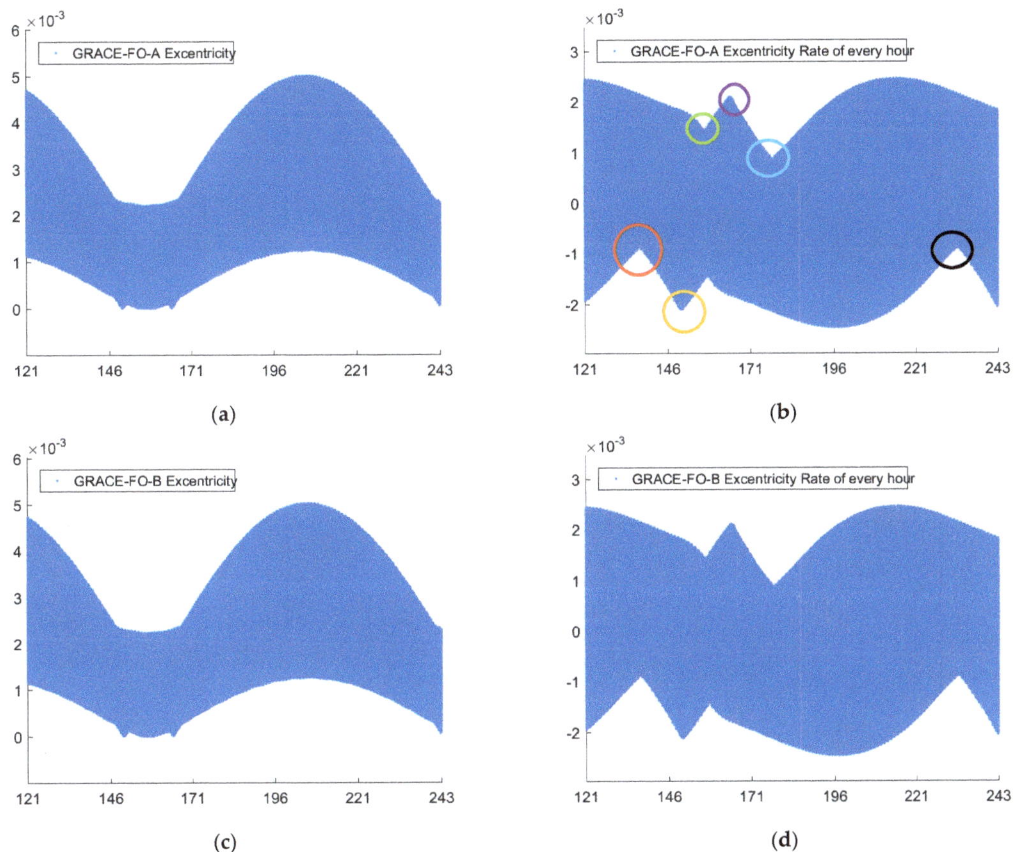

Figure 13. Eccentricity and eccentricity variation of GRACE-FO satellites (hourly variation).

Table 9. Days of Orbital Maneuver.

Circle	Days
red	135−136
yellow	149−150
green	155−156
purple	163−164
blue	175−176
black	230−231

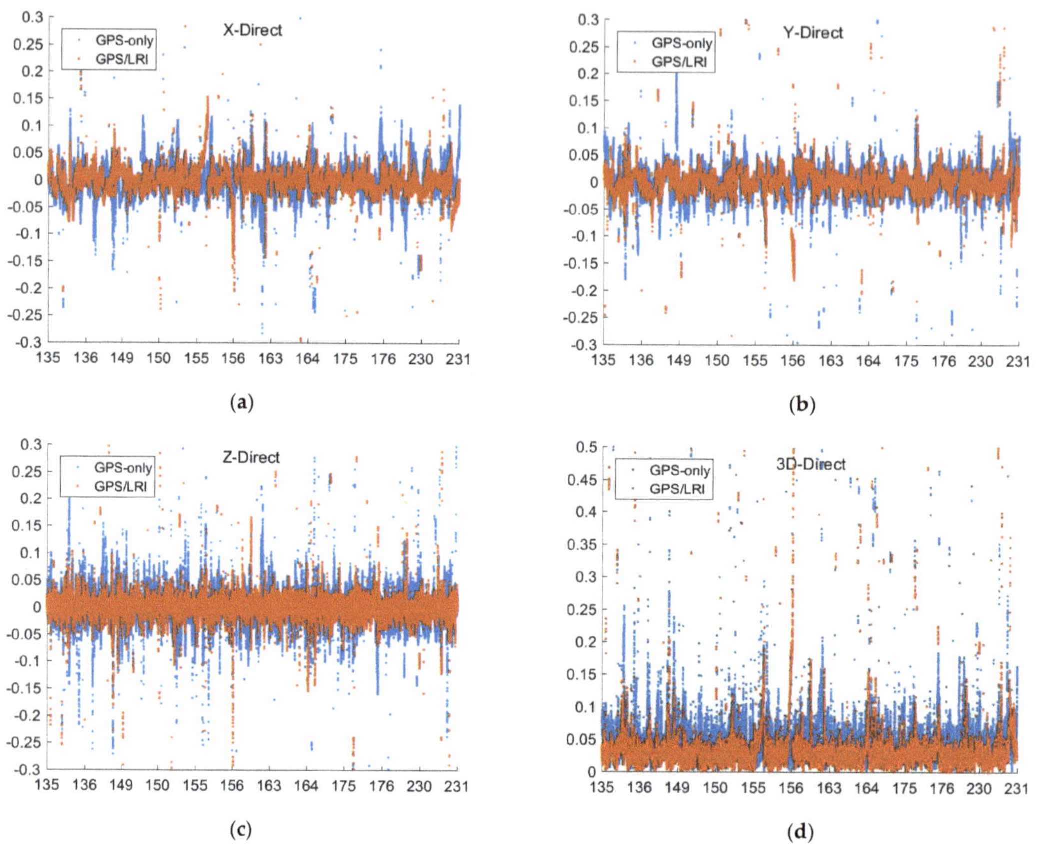

Figure 14. Reference orbit check for Orbital Maneuver in the X, Y, Z, 3D (X, Y, Z are components of the baseline vector, unit: m, (**a–d**)) directions (days 121–242).

Table 10. Statistics of X, Y, Z and 3D residual (unit: m, days 121–242) in Orbital Maneuver.

Type	GPS-Only				GPS/LRI			
	X	**Y**	**Z**	**3D**	**X**	**Y**	**Z**	**3D**
MEAN	−0.0018	0	0	—	−0.0010	0	0	—
MEDIAN	−0.0023	0	0	—	−0.0011	0	0	—
RMS	0.0367	0.0339	0.0381	0.0604	0.0291	0.0258	0.0228	0.0466

6.2. KBR Validation

The KBR system is one of the key scientific instruments carried by GRACE-FO satellites with a sampling interval of 5 s, ranging accuracy of 10 um, and range-changing rate accuracy of 1 um/s [37]. Actually, KBR and LRI are distinctly separate intersatellite ranging systems. Therefore, KBR can be used as an important external condition to verify GPS/LRI relative kinematic orbit determination results. The basic principle of KBR check is to obtain the difference between KBR-observed values and intersatellite baseline, and then evaluate the quality of the orbit difference between LEO satellites. Figures 15–19 show the residual distribution of KBR from day 121 to 242 in 2019. It shows that the relative kinematic orbit determination residuals by GPS only are mainly distributed within ±0.05 m, and the fluctuation of orbit determination residuals presents random distribution with some residual spikes. The residual values of a few epochs exceed ± 0.075 m. However, the relative kinematic orbit determination results of joint GPS/LRI are quite smooth with the residual fluctuation range within ± 0.01 m, and only very few epochs have KBR residual values greater than ± 0.025 m. Compared with relative kinematic orbit determination by GPS, relative kinematic orbit determination by joint GPS/LRI reduces the orbital spikes of relative kinematic orbit determination caused by GPS observation quality, and improves the orbital accuracy of relative kinematic orbit determination.

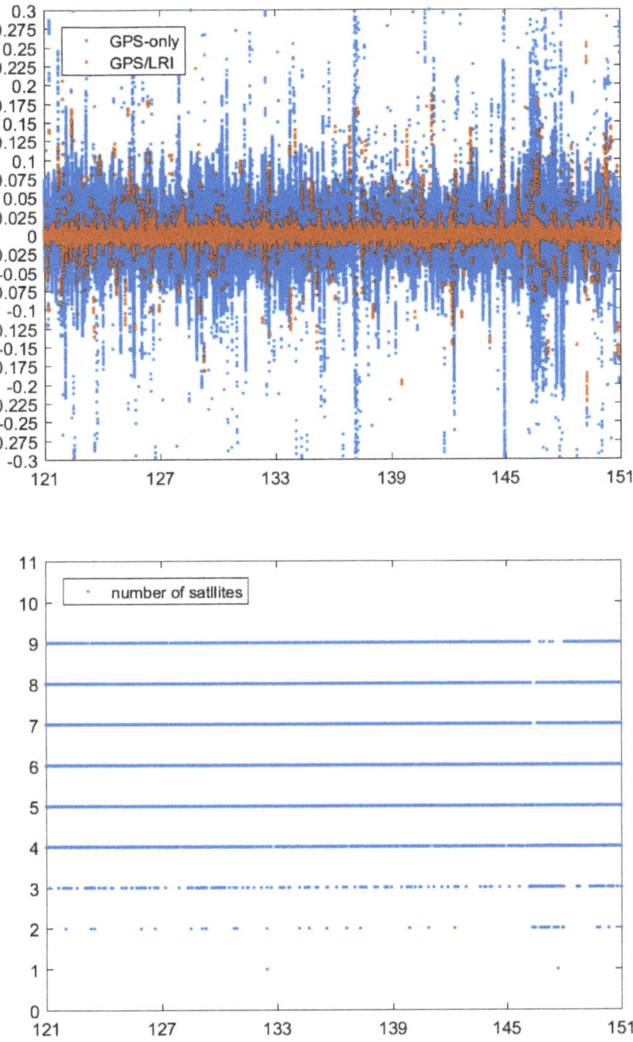

Figure 15. Day 121–151 KBR residuals (unit: m) for the GPS/LRI and GPS-only relative orbit and number of satellites.

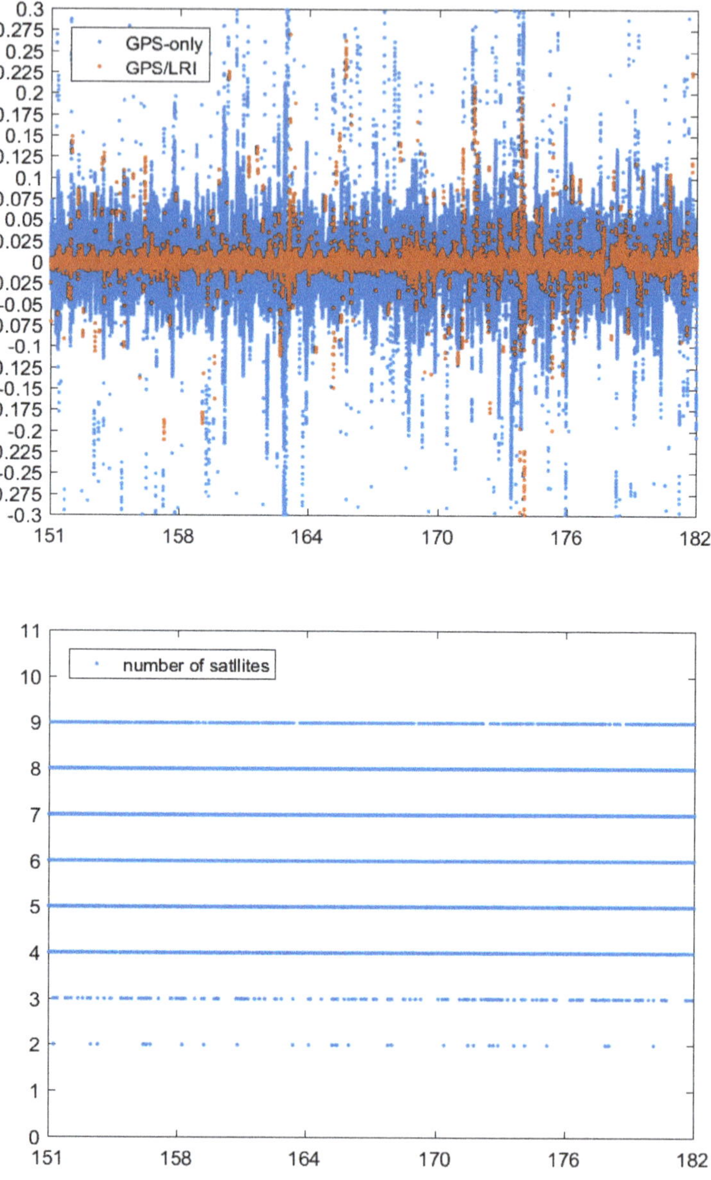

Figure 16. Day 151–182 KBR residuals (unit: m) for the GPS/LRI and GPS-only relative orbit and number of satellites.

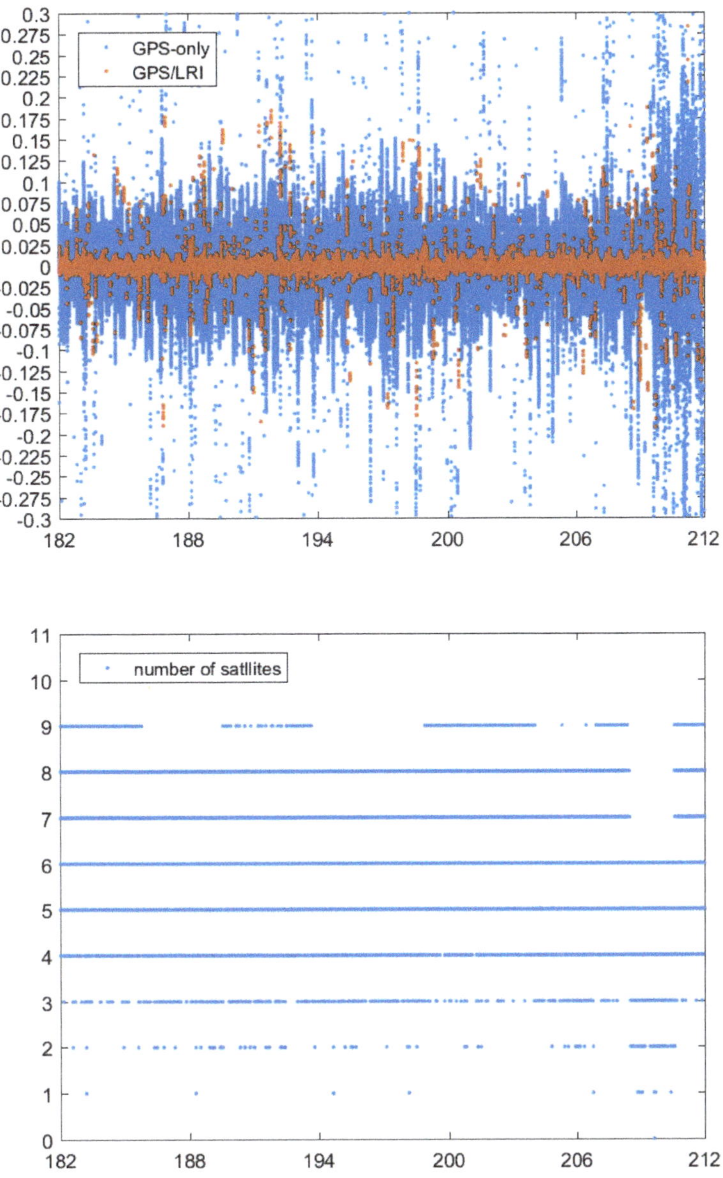

Figure 17. Day 182–212 KBR residuals (unit: m) for the GPS/LRI and GPS-only relative orbit and number of satellites.

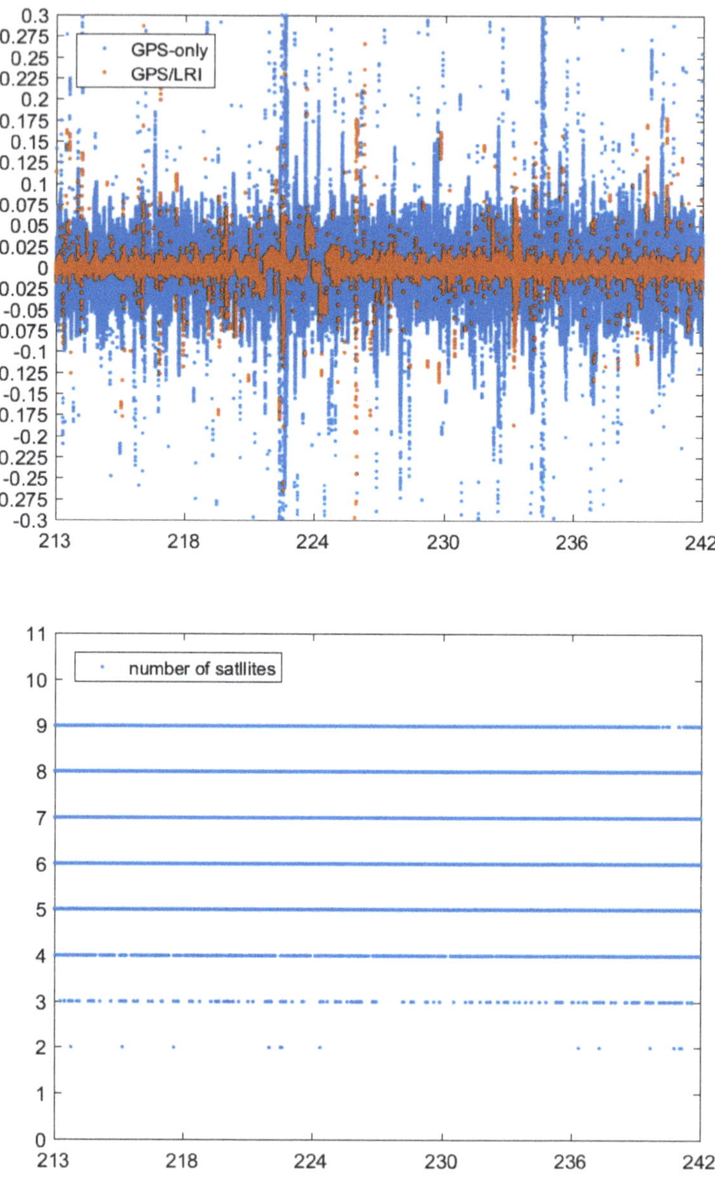

Figure 18. Day 213–242 KBR residuals (unit: m) for the GPS/LRI and GPS-only relative orbit and number of satellites.

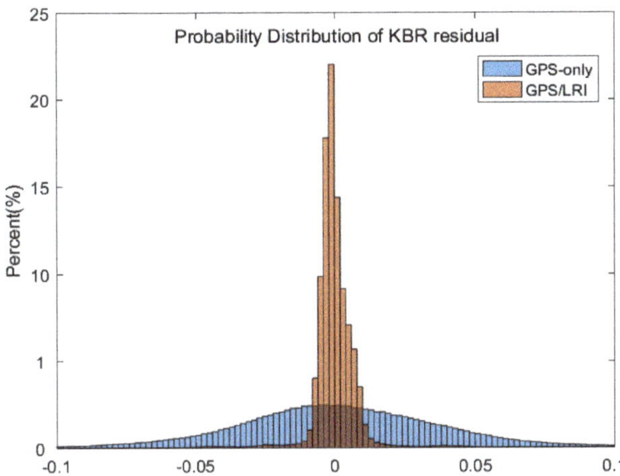

Figure 19. Day 121–242 KBR residuals (unit: m) probability distribution for the GPS/LRI and GPS-only relative orbit.

To obtain more detailed statistical information, we have conducted for all 121 days a comparison between KBR ranges and the relative kinematic orbit determination results. RMS, MEAN and MEDIAN values of 121 days KBR residuals are listed in Table 11. The accuracy of 121 days' relative kinematic orbit determination based on GPS only approximates 42.8 mm while that calculated from joint GPS/LRI data approximates 10.7 mm. It indicates that joint GPS/LRI improves the quality of relative kinematic orbit determination, and confirms the results of reference orbit verification. The mean value of KBR residuals in 121 days is less than 0.5 mm, indicating no obvious systematic errors in joint GPS/LRI and GPS relative kinematic orbit determination, and the results of relative kinematic orbit determination comply with the statistical law.

Table 11. Statistics of KBR check (unit: m).

Day	GPS-Only			GPS\LRI		
	RMS	**MEAN**	**MEDIAN**	**RMS**	**MEAN**	**MEDIAN**
121–151	0.0415	0	0	0.0112	0	−0.0007
151–182	0.0407	0	0	0.0093	0	−0.0006
182–212	0.0458	0	0	0.0086	0	−0.0007
212–242	0.0433	0	0	0.0135	0	−0.0007
Average	0.0428	0	0	0.0107	0	−0.0007

KBR can be used to evaluate the ability of GPS/LRI under sunlight and eclipse conditions. Figure 20 shows the KBR residuals of GPS/LRI and GPS-only relative kinematic orbit determination. There is no significant difference in the KBR residuals of GPS/LRI relative kinematic orbit determination under sunlight and solar eclipse. Table 12 shows the RMS, MEAN and MEDIAN of KBR residuals for GPS/LRI and GPS-only. Under the sunlight and solar eclipse, the KBR residuals RMS corresponding to GPS/LRI are 9.3 mm and 9.7 mm, and their differences are very slight. The RMS values of KBR residuals corresponding to GPS-only are 42.4 mm and 42.2 mm. Their differences are also small.

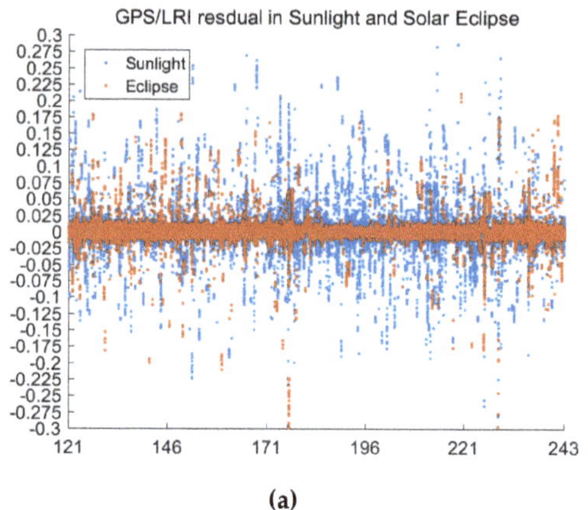

(a)

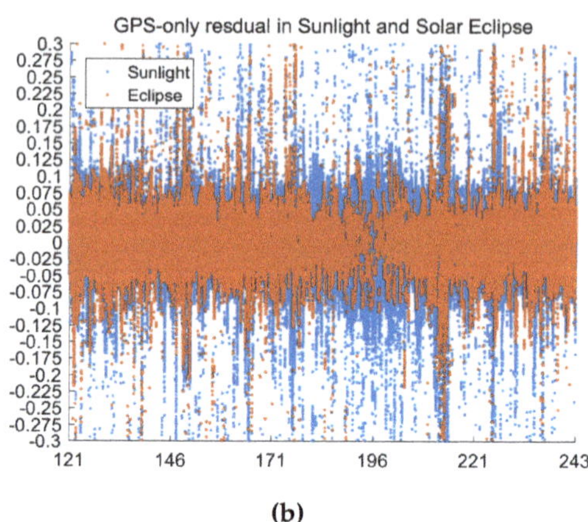

(b)

Figure 20. Days 121–242 KBR residuals (unit: m) for the GPS/LRI (**a**) and GPS-only (**b**) relative orbit in the Sunlight and Solar eclipse).

Table 12. Statistics of KBR residual (unit: m, days 121–242) in Sunlight and Solar eclipse.

Type	GPS/LRI		GPS-Only	
	Sunlight	Solar Eclipse	Sunlight	Solar Eclipse
MEAN	0	0	−0.0007	−0.0006
MEDIAN	0	0.0007	0	0
RMS	0.0093	0.0097	0.0424	0.0422

7. Conclusions

In this manuscript, we studied the LRI ranging system carried by GRACE-FO formation satellites, achieving high-quality relative kinematic orbit determination based on joint GPS/LRI data and the Kalman filter. Additionally, we introduced the LRI ranging systems as well as the LRI observation equation, and described the process of relative kinematic orbit determination of GPS/LRI in detail. Ultimately, relative kinematic orbit determination results of GPS/LRI were verified and analyzed by comparing to a reference orbit and verifying KBR. Accordingly, we draw two conclusions: 1. Compared with relative kinematic orbit determination by GPS only, kinematic orbit determination by GPS/LRI achieves more robust results, which can weaken some orbits' error and significantly improve the accuracy of relative kinematic orbit determination. The accuracy of relative kinematic orbit determination by GPS/LRI is improved by 17.5 mm in 3D directions, and the accuracy of relative kinematic orbit determination is improved by 25.9%, reaching 50.1 mm (compared with the reference orbit released by GFZ). 2. The results of KBR validation are ideal. The residual distribution of KBR is smooth without obvious fluctuation. The consistency between the relative kinematic orbit determination and the KBR measurements is at the +/−10.7 mm level.

Author Contributions: Conceptualization, writing—original draft preparation, writing—review and editing, and methodology, Z.Y., X.L. and J.G.; software, Z.Y.; validation, H.G., G.L. and Q.K.; data curation, X.C.; formal analysis, H.G.; funding acquisition, J.G. All authors have read and agreed to the published version of the manuscript.

Funding: This research was funded by The National Natural Science Foundation of China (42174041, 41774001), the Independent and Controllable Project of National Surveying and Mapping (816-517) and the SDUST Research Fund (2014TDJH101).

Institutional Review Board Statement: Not applicable for studies not involving humans or animals.

Remote Sens. **2022**, 14, 993

Informed Consent Statement: Not applicable for studies not involving humans.

Data Availability Statement: Our sincere thanks go to the German Research Centre Geosciences(GFZ) for providing space-borne GPS data for GRACE-FO (ftp://isdcftp.gfz-potsdam.de/grace-fo, accessed on 9 June 2020); CODE and IGS for providing GPS satellite orbits, clocks and Earth rotation parameters; the GFZ for providing precise orbits, KBR and LRI data for GRACE-FO.

Acknowledgments: This research is supported by National Natural Science Foundation of China (grant numbers 42174041, 41774001) and SDUST Research Fund (grant number2014TDJH101).

Conflicts of Interest: The authors declare no conflict interest.

References

1. Dragon, K. *GRACE Follow-On Mission Plan*; NASA Jet Propulsion Laboratory/California Institute of Technology: Pasadena, CA, USA, 2015.
2. Kornfeld, R.P.; Arnold, B.W.; Gross, M.A.; Dahya, N.T.; Klipstein, W.M.; Gath, P.F.; Bettadpur, S. GRACE-FO: The Gravity Recovery and Climate Experiment Follow-On Mission. *J. Spacecr. Rocket.* **2019**, *56*, 931–951. [CrossRef]
3. Xia, Y.; Liu, X.; Guo, J.; Yang, Z.; Qi, L.; Ji, B.; Chang, X. On GPS data quality of GRACE-FO and GRACE satellites: Effects of phase center variation and satellite attitude on precise orbit determination. *Acta Geod. Geophys.* **2020**, *56*, 93–111. [CrossRef]
4. Case, K.; Kruizinga, G.; Wu, S.-C. *GRACE Level 1B Data Product User Handbook*; JPL Publication D-22027; NASA Jet Propulsion Laboratory/California Institute of Technology: Pasadena, CA, USA, 2002.
5. Wen, H.Y.; Kruizinga, G.; Paik, M.; Landerer, F.; Bertiger, W.; Sakumura, C.; Bandikova, T.; Mccullough, C. *Gravity Recovery and Climate Experiment Follow-On (GRACE-FO) Level-1 Data Product User Handbook*; NASA Jet Propulsion Laboratory/California Institute of Technology: Pasadena, CA, USA, 2019.
6. Bertiger, W.; Desai, S.D.; Haines, B.; Harvey, N.; Moore, A.W.; Owen, S.; Weiss, J.P. Single receiver phase ambiguity resolution with GPS data. *J. Geodesy* **2010**, *84*, 327–337. [CrossRef]
7. Allende-Alba, G.; Montenbruck, O. Robust and precise baseline determination of distributed spacecraft in LEO. *Adv. Space Res.* **2015**, *57*, 46–63. [CrossRef]
8. Bettadpur, S. *Gravity Recovery and Climate Experiment Product Specification Document (Rev 4.5–February 20, 2007)*; The University of Texas at Austin, Center for Space Research: Austin, TX, USA, 2007.
9. Dubovitsky, S. *Definition of Time Specification Relevant to LRI Operations*; NASA Jet Propulsion Laboratory/California Institute of Technology: Pasadena, CA, USA, 2016.
10. Bertiger, W.; Harvey, N.; Spero, R. *An Alternate Formulation of The LRI Measurement*; NASA Jet Propulsion Laboratory/California Institute of Technology: Pasadena, CA, USA, 2019.
11. Abich, K.; Abramovici, A.; Amparan, B.; Baatzsch, A.; Okihiro, B.B.; Barr, D.C.; Bize, M.P.; Bogan, C.; Braxmaier, C.; Burke, M.J.; et al. In-Orbit Performance of the GRACE Follow-on Laser Ranging Interferometer. *Phys. Rev. Lett.* **2019**, *123*, 031101. [CrossRef] [PubMed]
12. Sodnik, Z.; Cugny, B.; Karafolas, N.; Nicklaus, K.; Herding, M.; Baatzsch, A.; Dehne, M.; Diekmann, C.; Voss, K.; Heinzel, G.; et al. Laser ranging interferometer on Grace follow-on. In *International Conference on Space Optics—ICSO 2016*; International Society for Optics and Photonics: Bellingham, DC, USA, 2017.
13. Turyshev, S.; Sazhin, M.V.; Toth, V. General relativistic laser interferometric observables of the GRACE-Follow-On mission. *Phys. Rev. D* **2014**, *89*, 105029. [CrossRef]
14. Kroes, R.; Montenbruck, O.; Bertiger, W.; Visser, P. Precise GRACE baseline determination using GPS. *GPS Solut.* **2005**, *9*, 21–31. [CrossRef]
15. Jäggi, A.; Hugentobler, U.; Bock, H.; Beutler, G. Precise orbit determination for GRACE using undifferenced or doubly differenced GPS data. *Adv. Space Res.* **2007**, *39*, 1612–1619. [CrossRef]
16. Krieger, G.; Moreira, A.; Fiedler, H.; Hajnsek, I.; Werner, M.; Younis, M.; Zink, M. TanDEM-X: A Satellite Formation for High-Resolution SAR Interferometry. *IEEE Trans. Geosci. Remote Sens.* **2007**, *45*, 3317–3341. [CrossRef]
17. Van Barneveld, P. *Orbit Determination of Satellite Formations*; Delft University of Technology: Delft, The Netherlands, 2012. [CrossRef]
18. D'Amico, S.; Ardaens, J.-S.; De Florio, S. Autonomous formation flying based on GPS—PRISMA flight results. *Acta Astronaut.* **2013**, *82*, 69–79. [CrossRef]
19. Chen, P.; Shu, L.; Ding, R.; Han, C. Kinematic single-frequency relative positioning for LEO formation flying mission. *GPS Solut.* **2014**, *19*, 525–535. [CrossRef]
20. Ju, B.; Gu, D.; Herring, T.A.; Allende-Alba, G.; Montenbruck, O.; Wang, Z. Precise orbit and baseline determination for maneuvering low earth orbiters. *GPS Solut.* **2015**, *21*, 53–64. [CrossRef]
21. Guo, H.; Guo, J.; Yang, Z.; Wang, G.; Qi, L.; Lin, M.; Peng, H.; Ji, B. On Satellite-Borne GPS Data Quality and Reduced-Dynamic Precise Orbit Determination of HY-2C: A Case of Orbit Validation with Onboard DORIS Data. *Remote Sens.* **2021**, *13*, 4329. [CrossRef]

22. Blewitt, G. Carrier phase ambiguity resolution for the Global Positioning System applied to geodetic baselines up to 2000 km. *J. Geophys. Res. Earth Surf.* **1989**, *94*, 10187–10203. [CrossRef]
23. Maosheng, Z.; Liu, X.; Guo, J.; Qinghua, Z.; Zhimin, L.; Ji, B.; Yongzhong, O. Epoch-by-epoch phase difference method to evaluate GNSS single-frequency phase data quality. *Rev. Int. Métodos Numér. Cálc. Diseño Ing.* **2020**, *36*, 42. [CrossRef]
24. Qi, L.; Guo, J.; Xia, Y.; Yang, Z. Effect of Higher-Order Ionospheric Delay on Precise Orbit Determination of GRACE-FO Based on Satellite-Borne GPS Technique. *IEEE Access* **2021**, *9*, 29841–29849. [CrossRef]
25. Shi, J.; Huang, Y.; Ouyang, C. A GPS relative positioning quality control algorithm considering both code and phase observation errors. *J. Geodesy* **2019**, *93*, 1419–1433. [CrossRef]
26. Montenbruck, O.; Kroes, R. In-flight performance analysis of the CHAMP BlackJack GPS Receiver. *GPS Solut.* **2003**, *7*, 74–86. [CrossRef]
27. Montenbruck, O.; Gill, E. Ionospheric Correction for GPS Tracking of LEO Satellites. *J. Navig.* **2002**, *55*, 293–304. [CrossRef]
28. Banville, S.; Langley, R.B. Mitigating the impact of ionospheric cycle slips in GNSS observations. *J. Geod.* **2012**, *87*, 179–193. [CrossRef]
29. Hwang, C.; Tseng, T.-P.; Lin, T.-J.; Švehla, D.; Hugentobler, U.; Chao, B.F. Quality assessment of FORMOSAT-3/COSMIC and GRACE GPS observables: Analysis of multipath, ionospheric delay and phase residual in orbit determination. *GPS Solut.* **2009**, *14*, 121–131. [CrossRef]
30. Kim, J.; Lee, Y.J. Using ionospheric corrections from the space-based augmentation systems for low earth orbiting satellites. *GPS Solut.* **2014**, *19*, 423–431. [CrossRef]
31. Gu, D.; Ju, B.; Liu, J.; Tu, J. Enhanced GPS-based GRACE baseline determination by using a new strategy for ambiguity resolution and relative phase center variation corrections. *Acta Astronaut.* **2017**, *138*, 176–184. [CrossRef]
32. Van Barneveld, P.; Montenbruck, O.; Visser, P. Epochwise prediction of GPS single differenced ionospheric delays of formation flying spacecraft. *Adv. Space Res.* **2009**, *44*, 987–1001. [CrossRef]
33. Grace, S.R. *Follow-On Laser Ranging Interferometer Phase Jump Removal*; NASA Jet Propulsion Laboratory/California Institute of Technology: Pasadena, DC, USA, 2018.
34. Blewitt, G. An Automatic Editing Algorithm for GPS data. *Geophys. Res. Lett.* **1990**, *17*, 199–202. [CrossRef]
35. Liu, Z. A new automated cycle slip detection and repair method for a single dual-frequency GPS receiver. *J. Geodesy* **2010**, *85*, 171–183. [CrossRef]
36. Landerer, F.W.; Flechtner, F.M.; Save, H.; Webb, F.H.; Bandikova, T.; Bertiger, W.I.; Bettadpur, S.V.; Byun, S.H.; Dahle, C.; Dobslaw, H.; et al. Extending the Global Mass Change Data Record: GRACE Follow-On Instrument and Science Data Performance. *Geophys. Res. Lett.* **2020**, *47*. [CrossRef]
37. Kang, Z.; Tapley, B.; Bettadpur, S.; Ries, J.; Nagel, P.; Pastor, R. Precise orbit determination for the GRACE mission using only GPS data. *J. Geodesy* **2006**, *80*, 322–331. [CrossRef]

Article

Local Enhancement of Marine Gravity Field over the Spratly Islands by Combining Satellite SAR Altimeter-Derived Gravity Data

Yihao Wu [1], Junjie Wang [1], Adili Abulaitijiang [2], Xiufeng He [1], Zhicai Luo [3], Hongkai Shi [1,*], Haihong Wang [4] and Yuan Ding [1]

[1] School of Earth Sciences and Engineering, Hohai University, Nanjing 211100, China; yihaowu@hhu.edu.cn (Y.W.); junjie2020@hhu.edu.cn (J.W.); xfhe@hhu.edu.cn (X.H.); dingyuanhhu@hhu.edu.cn (Y.D.)

[2] Institute of Geodesy and Geoinformation, University of Bonn, 53012 Bonn, Germany; adili@space.dtu.dk

[3] MOE Key Laboratory of Fundamental Physical Quantities Measurement, School of Physics, Huazhong University of Science and Technology, Wuhan 430074, China; zcluo@hust.edu.cn

[4] School of Geodesy and Geomatics, Wuhan University, Wuhan 430072, China; hhwang@sgg.whu.edu.cn

* Correspondence: shk@hhu.edu.cn

Citation: Wu, Y.; Wang, J.; Abulaitijiang, A.; He, X.; Luo, Z.; Shi, H.; Wang, H.; Ding, Y. Local Enhancement of Marine Gravity Field over the Spratly Islands by Combining Satellite SAR Altimeter-Derived Gravity Data. *Remote Sens.* **2022**, *14*, 474. https://doi.org/10.3390/rs14030474

Academic Editor: Xiaogong Hu

Received: 29 November 2021
Accepted: 14 January 2022
Published: 20 January 2022

Publisher's Note: MDPI stays neutral with regard to jurisdictional claims in published maps and institutional affiliations.

Abstract: The marine gravity field recovery close to land/island is challenging owing to the scarcity of measured gravimetric observations and sorely contaminated satellite radar altimeter-derived data. The satellite missions that carried the synthetic aperture radar (SAR) altimeters supplied data with improved quality compared to that retrieved from the conventional radar altimeters. In this study, we combine the satellite altimeter-derived gravity data for marine gravity field augmentation over island areas; in particular, the feasibility for regional augmentation by incorporating the SAR altimeter-derived gravity data is investigated. The gravity field modeling results over the Spratly Islands demonstrate that the marine gravity field is augmented by the incorporation of newly published satellite altimeter-derived gravity data. By merging the gravity models computed with the Sentinel-3A/B SAR altimetry data, the quasi-geoid and mean dynamic topography are dramatically improved, by a magnitude larger than 4 cm around areas close to islands, in comparison with the results directly derived from a combined global geopotential model alone. Further comparison of regional solutions computed from heterogeneous gravity models shows that the ones modeled from the SAR-based gravity models have better performances, the errors of which are reduced by a magnitude of 2~4 cm over the regions close to islands, in comparison with the solutions modeled with the gravity models developed without SAR altimetry data. These results highlight the superiority of using the SAR-based gravity data in marine gravity field recovery, especially over the regions close to land/island.

Keywords: marine gravity field refinement; satellite altimetry; synthetic aperture radar altimeter; Sentinel-3A/B; quasi-geoid; mean dynamic topography

1. Introduction

High-resolution gravity field determination at seas is a basic task in geodesy. Thanks to the Gravity Recovery and Climate Experiment (GRACE) [1,2] and Gravity Field and Steady-State Ocean Circulation Explorer (GOCE) missions [3,4], the global gravity field has been prominently strengthened at long wavelength up to hundreds of kilometers [5–8]. On the other hand, by incorporating ground-based data at short-wavelength, the derived Global Geopotential Models (GGMs) (known as high-degree or combined GGMs) can map the gravity signals at a mean spatial resolution of 5 arc-minutes (~10 km) in global scale [9–11].

Despite the tremendous progresses made in global geopotential model computation over decades, the lack of globally distributed gravity data is still a major barrier to improve the combined GGMs. For regions inland, such as most areas in Asia and Africa, the measured land/airborne gravimetric observations are limited; accessible and fill-in measurements were involved in the computation of combined GGMs [9,12,13], whereas satellite altimetric gravity data was usually used over oceans as shipborne/airborne gravimetric measurements were usually inaccessible or confidentially kept due to political reasons [14]. However, the satellite altimeter-derived data is notoriously known to be of low quality close to land/island, owing to the contamination of radar altimeter waveforms and degraded quality of geophysical models used for data corrections [15–18]. Inevitably, the errors in satellite altimeter-derived gravimetric data were spread to the combined GGMs. As a result, the errors in the combined GGMs reached 6~10 centimeters over most oceans [9], and even a magnitude of decimeter level or larger over polar areas and coastal regions [19,20]. The errors in a GGM may bring about dramatic disturbances in the investigation of the mean ocean state in detail [21,22].

The improved satellite altimetry techniques result in the augmentation of global marine gravity field, by a factor of 2~4, in comparison with the gravity field models developed with old altimeter-derived records [23–25]. In particular, the CryoSat-2 and Sentinel-3A/B missions that carried synthetic aperture radar (SAR) altimeters derived precise records of sea surface height up to several kilometers from the coast [17,26]. The CryoSat-2 applies the low-resolution mode (LRM), SAR mode, and SAR interferometry (SARIn) mode over different areas [27–29]. Sentinel-3A/B inherited the SAR altimeter from CryoSat-2, and it retrieved observations with denser spatial coverage in comparison with the conventional altimetry (or operated in the LRM mode). Moreover, the data derived from the SAR waveforms have higher signal-to-noise ratio and lower speckle noise [30–32], and the accuracy of SAR altimetry data reaches several centimeters to decimeter level close to coast and lakes [33–35]. By incorporating recent altimetry observations, the derived gravimetric products have improved precision versus the ones developed without these observations [36].

The launch of the satellite altimetry missions carrying the SAR altimeter provides solid basis for local gravity field recovery close to land/island; however, little attention has been paid to gravity field enhancement over coastal or island areas by combing the SAR altimetry data, especially for the use of recently released data from Sentinel-3A/B. To our best knowledge, no existing literature has investigated and quantified the additional signals introduced from the SAR-based data retrieved from Sentinel-3A/B on regional gravity field modeling. This study aims at strengthening the marine gravity field over regions close to land/island on a regional scale based on gravity data derived from satellite altimeters. In particular, we study the feasibility for local augmentation based on SAR-based gravity data. Moreover, we compare the performances of different altimetric gravity models in marine gravity field recovery. In the following, the study area and data sources are included in Section 2. Section 3 displays the gravity field modeling results, where the added signals retrieved from the newly published satellite altimeter-derived gravity data are quantified and validated; in particular, the possibility for local enhancement based on SAR-based gravity data is investigated. Section 4 contains a summary of the study and the main conclusions.

2. Study Area and Data

We chose the study area as the Spratly Islands located in the southern part of the South China Sea (SCS) (see the area inside the red rectangle in Figure 1a), and the bathymetry is retrieved from the General Bathymetric Chart of the Oceans (GEBCO) [37]. The Spratly Islands are the largest archipelago in the SCS and are also the southernmost archipelago, where small islands, sandbanks, shoals, and coral reefs are the predominant structures [38] (see Figure 1b). The Spratly Islands archipelago is rich in natural resources, and is also a disputed archipelago with complicate governances. As a result, it is hard to implement shipborne/airborne gravimetric surveys over this area. The lack of measured gravimetric observations brings about the difficulty for marine gravity field recovery over the Spratly Islands, and the current combined GGMs were computed by using satellite altimetric gravity data. Moreover, small islands, atolls, sandbanks, shoals, and reefs are in abundance in the Spratly Islands, and the return waveforms from radar altimeters have been severely contaminated; as such, there exist multiple challenges in local marine gravity refinement. However, this offers a chance to research the feasibility of using SAR-based gravimetric data in regional augmentation.

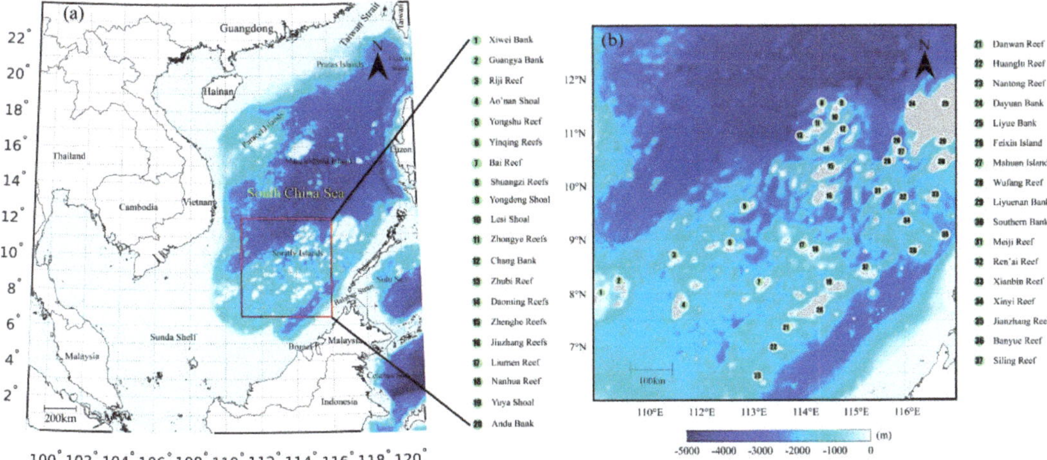

Figure 1. (**a**) Study area and (**b**) the primary islands of the Spratly Islands. The geographical coordinates are expressed in the Plate Carrée projection.

As there were no measured gravimetric observations available over the Spratly Islands, we use the satellite altimeter-derived data for marine gravity field modeling. For this purpose, several recently released altimetric gravity models are used. First, the widely used models developed at the Technical University of Denmark (DTU) space and Scripps Institution of Oceanography (SIO) are introduced. For the series of models developed at DTU space, the latest model, namely DTU21GRA, and several predecessors, i.e., DTU13GRA, DTU15GRA, and DTU17GRA [36,39], are used. Typically, for the models that were developed in the DTU space, the updated gravity models had improved accuracy versus the previous versions, which were mainly due to the use of newly released altimeter data and the updated data preprocessing methods. For instance, the comparison with an airborne survey over North Greenland showed that the standard deviation (SD) of the misfits between DTU17GRA and the airborne data was 3.78 mGal. This value increased to 8.81, 5.91, and 5.45 mGal when DTU10GRA, DTU13GRA, and DTU15GRA were evaluated, respectively. The major improvement of DTU17GRA over DTU15GRA is that the computation of DTU17GRA contained more CryoSat-2 data and SARAL/AltiKa data from 2016 to 2017 in the geodetic phase, while the improvement of DTU21GRA over DTU17GRA is that the former was computed by combining 5 years of Sentinel-3A and 3 years of Sentinel-3B and reprocessed Cryosat-2 data (processed with the SAMOSA+ physical retracker).

For the models developed in SIO, the latest model, called SIO V31.1, and its previous version, namely, SIO V30.1 (hereinafter referred to as SIO31 and SIO30, respectively) [23,24,40], are introduced. In addition, two recently published regional models over the SCS, i.e., SCSGA V1.0 [41] and HY-2A V1.0 [42], are used. In short, we call these two regional models SCSGA and HY2A, respectively. All these satellite altimetric gravity models have the spatial resolutions of one arc-minute, and the datasets used in these models' development and other associated information are shown in Table 1.

Table 1. Description of the selected satellite altimeter-derived marine gravity field models.

Model	Year	Reference Gravity Field and Reference Mean Dynamic Topography	Data Used in Model Development	Is This Model Used for Regional Gravity Field Enhancement?
DTU13GRA	2013	EGM2008+DOT2008	The modeling of DTU13GRA additionally involved data from CryoSat-2 and Jason-1, compared to its predecessor, i.e., DTU10GRA, which was derived by combining data from Topex/Poseidon (T/P), Geosat, ICESat, Jason-1, GFO, Envisat, and retracked data from ERS-1.	Yes
DTU15GRA	2015	EGM2008+DOT2008	Added more data from Geosat, ERS-1, Cryosat-2, and Jason-1.	Yes
DTU17GRA	2017	EGM2008+DOT2008	Added more CryoSat-2 data and SARAL/AltiKa data from 2016 to 2017 in the geodetic phase.	Yes
DTU21GRA	2021	EGM2008+DOT2008	The major improvement of DTU21GRA over DTU17GRA is that 5 years of Sentinel-3A and 3 years of Sentinel-3B and reprocessed Cryosat-2 data (processed with the SAMOSA+ physical retracker) were added.	Yes
SIO V23.1	2013	EGM2008+DOT2008	This model was derived by incorporating data from T/P, Geosat, Envisat, Jason-1, ERS-1/2, and CryoSat-2.	No
SIO V28.1	2019	EGM2008+DOT2008	Involved more data from Jason-2 and Cryosat-2 and added data from SARAL/Altika.	No
SIO V29.1	2019	EGM2008+DOT2008	Included 2 years data from Sentinel-3A/B.	No
SIO V30.1	2020	EGM2008+DOT2008	Involved data from SARAL/Altika as well as more data from Cryosat-2 and Sentinel-3A/B.	Yes
SIO V31.1	2021	EGM2008+DOT2008	Included more data from Cryosat-2, SARAL/Altika, and Sentinel-3A/B.	Yes
SCSGA V1.0	2020	EGM2008+CNES-CLS13MDT	Included data from T/P, GFO, ERS-1/2, Envisat, Jason-1/2, HY-2A, CryoSat-2, and SARAL/AltiKa.	Yes
HY-2A V1.0	2020	EGM2008+DOT2008	This model was developed by incorporating data from HY-2A and data from T/P, Geosat, Jason-1, Envisat, ERS-1/2, CryoSat-2, and SARAL/AltiKa.	Yes

For the models developed at the DTU space, it is noticeable that although DTU13GRA/DTU15GRA/DTU17GRA was computed with CryoSat-2 data, CryoSat-2 operated in the LRM over the SCS, and no SAR altimetry data was used in the computation of these models. For the same reason, SCSGA and HY2A are the models that were computed without combining SAR altimetry data, whereas the development of SIO30, SIO31, and DTU21GRA incorporated Sentinel-3A/B SAR altimeter measurements over the SCS, and the altimeter in the Sentinel-3A/B mission operated in the SAR mode all around the world.

The comparisons with an airborne survey over North Greenland and the marine gravity data from the NGA showed that DTU17GRA had improved accuracy, compared to the previous versions, e.g., DTU10GRA, DTU13GRA, and DTU15GRA [36]. Moreover, the validation with independent shipborne gravity data over the SCS displayed that the SCSGA model had comparable precision to SIO V27.1 but had better performance than DTU13GRA/DTU17GRA [41], whereas HY2A had degraded precision compared with SIO V27.1 and DTU17GRA [42].

3. Results and Discussions

3.1. Local Refinement with SAR-Based Altimetric Gravity Data

The remove–compute–restore approach is applied for modeling [43,44], and we use the residual terrain model (RTM) to reduce the high-frequency topographical signals [45]. XGM2019e_2159, with a full degree and order (d/o) of 2190/2159, is chosen as the reference model [46]. A satellite-only GGM, namely, GOCO06S, was used to represent the long-wavelength component of XGM2019e_2159, and the gravimetric data sources from National Geospatial-Intelligence Agency (NGA) and DTU13GRA were used to compute the short-wavelength component of this model [46]. The validation against local airborne gravimetric observations demonstrated that XGM2019e_2159 had improved precision, by a magnitude of ~1 mGal, compared to the GGMs that have similar expansion degrees [47].

We model the residual gravity field by using Poisson wavelets, and the long- and short-wavelength components of gravity field are recovered from the reference model and RTM, respectively [48,49]. To compute the residual gravity data, the gravity anomalies synthesized from XGM2019e_2159 up to degree 2190 and the associated RTM corrections are subtracted from the satellite altimeter-derived gravity data. Then, we parameterized the residual gravity field based on Poisson wavelets, and the function that links the gravity anomaly to the disturbing potential is seen in Klee et al. (2008) [50]. The area extends 7° N to 12° N in the latitudinal direction, and 111.0° E to 116° E in the longitudinal direction is chosen as the computational region. We locate the Poisson wavelets 5 km beneath the terrain, and we set the mean resolution of Poisson wavelets as 10 km. Moreover, we compute the quasi-geoid model with a spatial resolution of ~2 km, in correspondence with the mean resolution of the satellite altimeter-derived gravity models.

To highlight the use of SAR altimetry data in local augmentation, we investigate the performances of various solutions modeled with different altimetric gravity models that are computed with and without SAR altimetry data. To do this, four different versions of altimetric gravity models that were developed at DTU space, i.e., DTU13GRA, DTU15GRA, DTU17GRA, and DTU21GRA, are used. It is of note that only the DTU21GRA model was developed with the SAR altimetry data retrieved from Sentinel-3A/B over the Spratly Islands, while the other three models were developed without SAR altimetry data. The residual quasi-geoids computed with various satellite altimeter-derived gravity models are seen in Figure 2 and reach a magnitude of several centimeters (see the statistics in Table 2). The magnitude of the contents retrieved from DTU13GRA is slightly larger than 3 cm, and the standard deviation value is ~0.6 cm. Although the reference model in local gravity field modeling, i.e., XGM2019e_2159, was computed based on the DTU13GRA data over oceans, there may still be the additional contents in the DTU13GRA data that were unresolved in XGM2019e_2159, whereas the minimum/maximum value changes to −3.0/4.0 cm (−3.4/4.8 cm) when DTU15GRA (DTU17GRA) is applied for modeling. The additional signals derived from these satellite altimeter-derived gravity models concentrate over the areas close to islands, e.g., close to the Zhongye Reefs (11.1° N, 114.3° E), Daoming Reefs (10.8° N, 114.5° E), Zhenghe Reefs (10.4° N, 114.6° E), and Jiuzhang Reefs (9.8° N, 114.5° E). The added signals retrieved from DTU17GRA have stronger structures than those derived from the DTU13GRA data, which may be due to that the computation of DTU17GRA involved more high-quality data, including more data from Jason-1, CryoSat-2, and SARAL/AltiKa, as well as the updated data preprocessing strategies. As a result, DTU17GRA had improved accuracy compared to DTU13GRA [36].

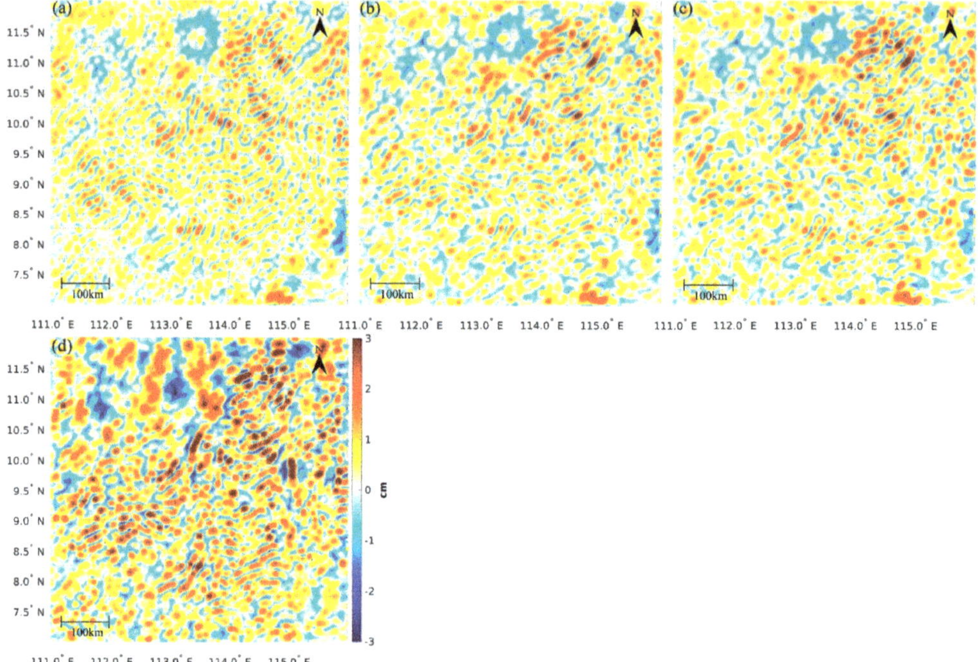

Figure 2. The added signals in terms of quasi-geoid heights derived from (**a**) DTU13GRA, (**b**) DTU15GRA, (**c**) DTU17GRA, and (**d**) DTU21GRA. The geographical coordinates are expressed in the Plate Carrée projection. Note: only DTU21GRA was computed with the SAR altimetry data derived from Sentinel-3A/B over the Spratly Islands.

Table 2. Statistics of the residual quasi-geoid heights derived from various satellite altimeter-derived gravity models developed at DTU space (units: cm).

	Max	**Min**	**Mean**	**SD**
DTU13GRA	3.4	−3.2	0.0	0.6
DTU15GRA	4.0	−3.0	0.0	0.6
DTU17GRA	4.8	−3.4	0.0	0.7
DTU21GRA	5.3	−5.2	−0.0	1.0

In contrast, the added signals range from –5.2 to 5.3 cm (the SD value is ~1.0 cm) when DTU21GRA is used. The contents retrieved from DTU21GRA have stronger structures than those derived from the three models above, particularly over the northern part and the regions close to islands. The DTU21GRA data computed with the SAR altimetry data retrieved from Sentinel-3A/B may further contribute to the marine gravity field augmentation, especially over island areas, compared to the altimetric gravity models computed without SAR altimeter data, as the SAR waveforms have the relatively high signal-to-noise ratio and can alleviate the well-known coast problem. Moreover, the SAR altimeters supply observations with denser spatial coverage than conventional radar altimeters, and the incorporation of DTU21GRA is beneficial to recover the short-wavelength component of local gravity field.

Further, the geodetic mean dynamic topography (MDT) solutions based on the quasi-geoids derived from different altimetric gravity data are computed and compared. The MDT represents the departure of the MSS from the geoid [51–54], and we use the quasi-geoid rather than the geoid, due to fact that the quasi-geoid coincides with the geoid at seas [55]. We choose DTU21MSS as the MSS model. For the derivation of DTU21MSS, more than 5 years of Sentinel-3A data and 2 years of Sentinel-3B data were combined [56]. Moreover, an updated waveform retracker, i.e., the SAMOSA+ physical retracker, was used to preprocess the Cryosat-2 data. This further improves the quality of the MSS model over coastal regions compared to its previous version, e.g., DTU18MSS [57].

The raw MDTs are seen in Figure 3, and we find that the dramatic oscillations exist in the model derived from XGM2019e_2159 over island areas. The errors of the quasi-geoid computed from a combined GGM spread to the MDT, which possibly brings about errors larger than several centimeters at seas [9], and even to decimeter level over coastal regions and polar areas [19,20]. In comparison, smaller disturbances are seen in the solutions computed by using the quasi-geoids that are augmented by incorporating the satellite altimeter-derived gravity data (see Figure 3b–e). For instance, see the patterns around the Shuangzi Reefs (11.4° N, 114.4° E), Zhongye Reefs (11.1° N, 114.3° E), Daoming Reefs (10.8° N, 114.5° E), Zhenghe Reefs (10.4° N, 114.6° E), and Jiuzhang Reefs (9.8° N, 114.5° E). By using the newly published satellite altimeter-derived gravity data, the quasi-geoid and mean dynamic topography are augmented, compared to the GGM-derived solutions. The mutual comparisons demonstrate that the MDT computed with the quasi-geoid enhanced by DTU21GRA shows relatively smooth structures, especially in the regions close to islands. This is probably owing to that the computation of DTU21GRA involved the Sentinel-3A/3B SAR altimetry data, which brings about regional augmentation.

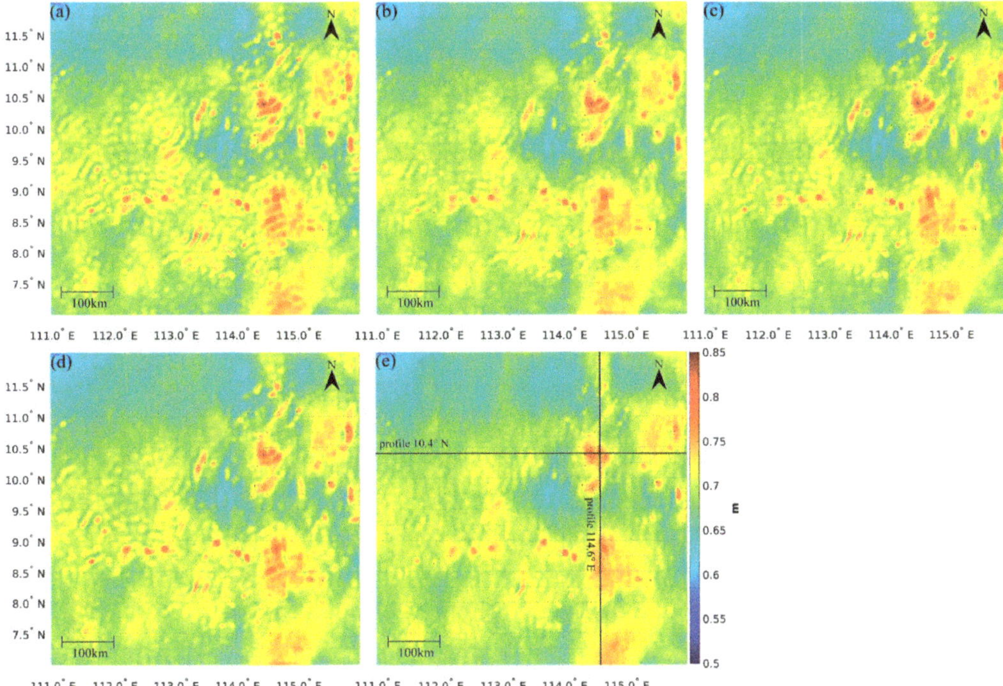

Figure 3. MDT modeled with the quasi-geoid computed from (**a**) XGM2019e_2159, (**b**) DTU13GRA, (**c**) DTU15GRA, (**d**) DTU17GRA, and (**e**) DTU21GRA. The geographical coordinates are expressed in the Plate Carrée projection. Note: only DTU21GRA was computed with the SAR altimetry data derived from Sentinel-3A/B over the study area.

The raw MDTs are further investigated along latitude 10.4° N and longitude 114.6° E, and the profiles are seen in Figure 3e. In Figure 4a, we see that the MDT derived from the quasi-geoid augmented by merging DTU21GRA has smoother properties than the one directly derived from XGM2019e_2159 and the solutions computed from other altimetric gravity models. This is particularly the case when this profile crosses the Zhenghe Reefs (114.6° E) and Wufang Reef (115.7° E), where the spike-like patterns appear in the MDTs. These spikes are usually identified as the errors due to the degraded quality of satellite altimeter-derived data close to land/island [58]. Over these regions, the XGM2019e_2159-derived solution has the most prominent disturbances; and the inconsistencies between this model and the one strengthened by combining DTU21GRA is greater than 3 cm close to the Zhenghe Reefs. The mutual comparison shows that the MDTs computed from DTU13GRA, DTU15GRA, and DTU17GRA almost have consistent structures, although the computation of DTU17GRA incorporated more CryoSat-2 and SARAL/AltiKa data that was not included in the derivation of DTU13GRA/DTU15GRA. In contrast, the application of the DTU21GRA data reduces these spike-like errors, by a magnitude exceeding 2 cm close to the Zhenghe Reefs and Wufang Reef, compared to the associated MDTs modeled with the altimetric gravity data computed without SAR altimetry data. This also corresponds with the results shown in Figure 3, where the MDT modeled with the DTU21GRA data has relatively smooth structures. These results demonstrate the superiority of using the SAR-based gravity data in the determination of marine quasi-geoid and mean dynamic topography over the areas close to land/island.

Figure 4. Profiles of various MDTs computed from different gravity field models along latitude 10.4° N (**a**) and longitude 114.6° E (**b**). Note: only DTU21GRA was computed with the SAR altimeter data over the study area.

The properties of the MDTs along longitude 114.6° N are seen in Figure 4b. We also find that the mean dynamic topography computed with the augmentation of DTU21GRA has relatively small variations, especially when this profile crosses the Zhenghe Reefs (10.4° E) and Lesi Shoal (11.35° E). Moreover, the discrepancy between the XGM2019e_2159-derived MDT and the one strengthened by using the DTU21GRA data exceeds 3 cm (4 cm) over the Zhenghe Reefs (Lesi Shoal). Similarly, compared to the MDT modeled with the altimetric gravity model computed without SAR altimeter data, i.e., DTU13GRA, DTU15GRA, and DTU17GRA, the use of DTU21GRA data can reduce these spike-like errors by a magnitude exceeding 2 cm (3 cm) over the Zhenghe Reefs (Lesi Shoal).

3.2. Performances of Heterogeneous Altimetric Gravity Models

Moreover, we investigate the performances of heterogeneous gravity models released from different institutes in marine gravity field modeling. To achieve this, five representative models, i.e., HY2A, SCSGA, SIO30, SIO31, and DTU21GRA, are selected for the case study. The magnitudes of the added signals in terms of quasi-geoid heights derived from different gravity models are not consistent (see Figure 5). The maximum/minimum of the additional signals retrieved from HY2A is −5.5/7.0 cm (see Table 3). In comparison, the signals derived from SCSGA have weaker patterns, and the maximum/minimum value is −4.2/4.4 cm, whereas the added signals computed from the altimetric gravity data modeled with the Sentinel-3A/B SAR altimeter data, i.e., SIO30/SIO31/DTU21GRA, reach a magnitude greater than 5 cm. The differences among different MDTs result from the different data preprocessing techniques and weighting methods, as well as the different datasets used in model development.

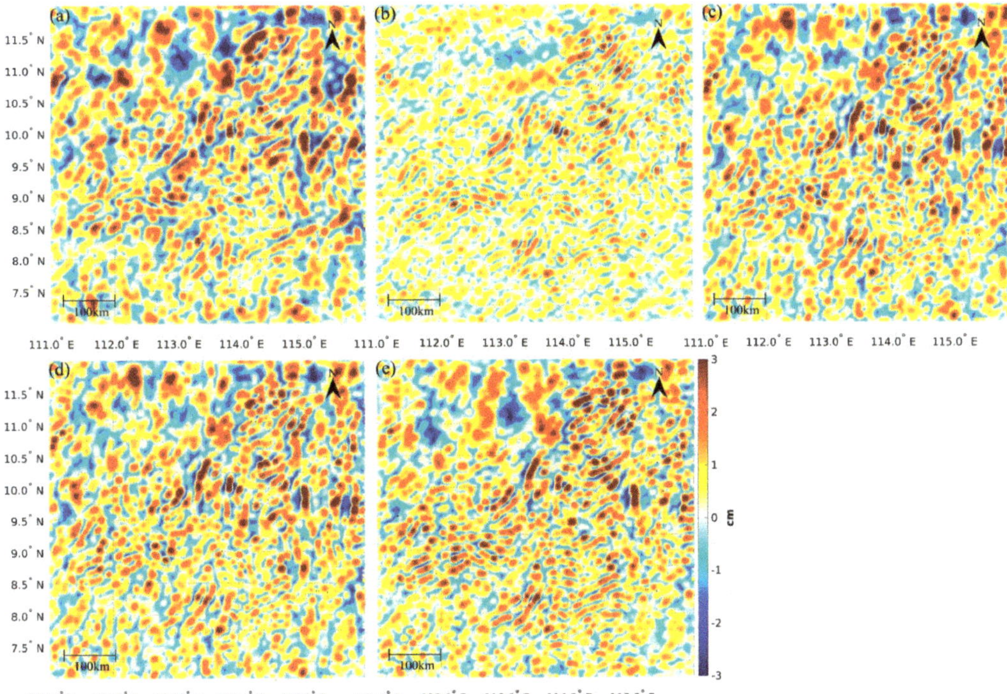

Figure 5. The added signals in terms of quasi-geoid heights modeled with (**a**) HY2A, (**b**) SCSGA, (**c**) SIO30, (**d**) SIO31, and (**e**) DTU21GRA. The geographical coordinates are expressed in the Plate Carrée projection. Note: only SIO30, SIO31, and DTU21GRA were computed with the SAR altimetry data derived from Sentinel-3A/B over the study area.

Table 3. Statistics of the residual quasi-geoid signals modeled with heterogeneous gravity models (units: cm).

	Max	Min	Mean	SD
HY2A	5.5	−7.0	−0.0	1.2
SCSGA	4.4	−4.2	0.0	0.6
SIO30	6.0	−5.4	−0.0	1.0
SIO31	6.2	−5.6	−0.0	1.0
DTU21GRA	5.3	−5.2	−0.0	1.0

Further, the quasi-geoids computed from different altimetric gravity models are assessed in MDT modeling. Figure 6 displays the raw MDTs, and the comparison of the MDTs computed from various gravity models shows that the structures of different MDTs are heterogeneous. For the MDTs augmented by HY2A and SCSGA, the variations are still significant over the island areas. In comparison, the models enhanced by SIO30, SIO31, and DTU21GRA have smaller disturbances, especially over the regions close to the Shuangzi Reefs (11.4° N, 114.4° E), Zhongye Reefs (11.1° N, 114.3° E), Daoming Reefs (10.8° N, 114.5° E), Zhenghe Reefs (10.4° N, 114.6° E), Jiuzhang Reefs (9.8° N, 114.5° E), and Wufang Reef (10.4° N, 115.7° E). The heterogeneous data preprocessing strategies, as well as data combination methods used in model development, partially cause these discrepancies among different MDTs. However, the most important reason may be due to that the computation of SIO30, SIO31, and DTU21GRA involved the Sentinel-3A/B SAR altimeter data over the study area. These results agree with the validation of heterogeneous gravity models against independent gravimetric observations over the SCS, which showed that HY2A had degraded quality versus SIO V27.1/DTU17GRA [42].

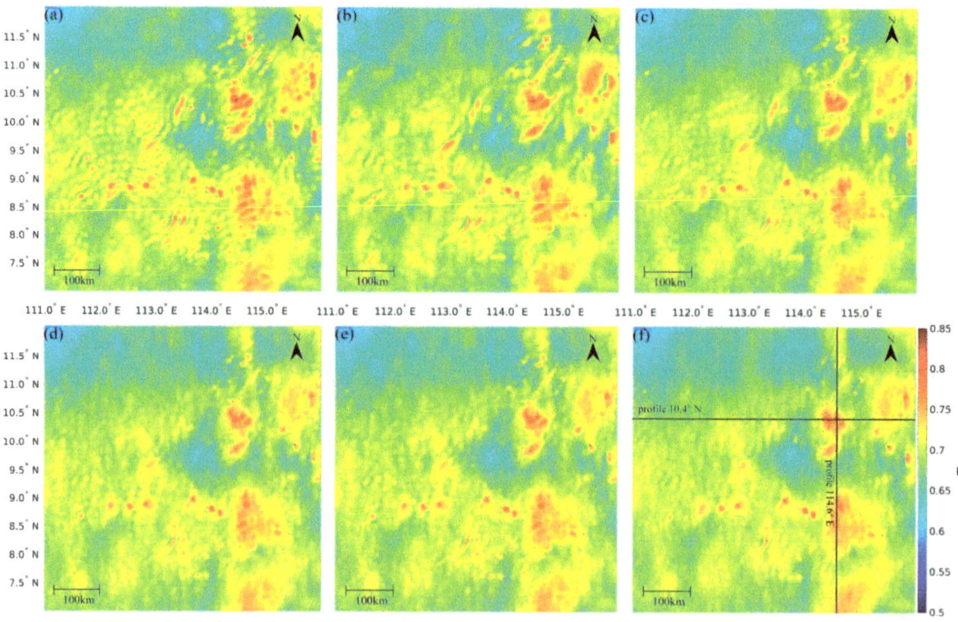

Figure 6. MDTs derived from the quasi-geoid model computed from (**a**) XGM2019e_2159, (**b**) HY2A, (**c**) SCSGA, (**d**) SIO30, (**e**) SIO31, and (**f**) DTU21GRA. The geographical coordinates are expressed in the Plate Carrée projection. Note: only SIO30, SIO31, and DTU21GRA were computed with the SAR altimetry data derived from Sentinel-3A/B over the study area.

Figure 7a shows the different MDTs along latitude 10.4° N. Although the MDT derived from the quasi-geoid enhanced by HY2A/SCSGA has smaller turbulences than that derived from XGM2019e_2159, the spike-like errors are still dramatic close to the Zhenghe Reefs (114.6° E) and Wufang Reef (115.7° E). The spikes are reduced in the solutions strengthened by using SIO31/DTU21GRA, with a reduction greater than 2 cm over the Zhenghe Reefs and 3 cm over the Wufang Reef, compared to the MDT modeled with the HY2A data. It agrees well with the modeling results in Figure 6, where the MDT modeled with the SIO31/DTU21GRA data has relatively smooth structures close to islands. The MDTs computed with the SIO30, SIO31, and DTU21GRA data almost show consistent structures along this profile, and the mutual discrepancies among these three solutions are within 2 cm. The properties of different MDTs along longitude 114.6° N show similar results to the profile along latitude 10.4° E (see Figure 7b). The MDTs from the altimetric gravity data computed with the SAR altimetry data, i.e., SIO30, SIO31, and DTU21GRA, have relatively small variations compared with the one derived from XGM2019e_2159 alone and the solution derived from the HY2A/SCSGA data. The application of the SAR altimetry data reduces the spike-like errors by a magnitude larger than 2 cm (4 cm) over the Zhenghe Reefs (Lesi Shoal), compared with the MDT modeled with the HY2A data. These results show that the performances of heterogeneous gravity models are not consistent, and the SAR-based models may be preferable.

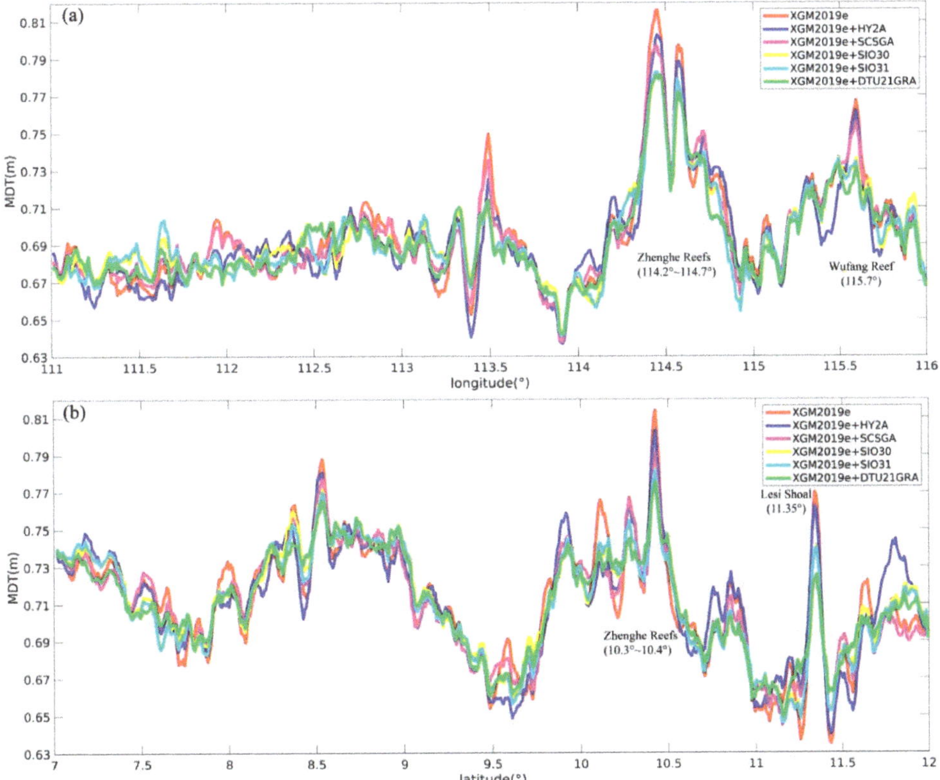

Figure 7. Profiles of various MDTs modeled with different altimetric gravity models along latitude 10.4° N (**a**) and longitude 114.6° E (**b**).

3.3. The MDTs Modeled with Different Reference Models

In addition, we research the performances of various satellite altimeter-derived gravity models in marine gravity field modeling based on heterogeneous reference models, where three combined models that have similar expansion degrees to XGM2019e_2159, i.e., EGM2008, SGG-UGM-1, and GECO, are introduced. EGM2008 (d/o 2190/2159) was computed by merging GRACE observations with ground-based gravity data [9]. Moreover, two models computed by combining GOCE gravity gradients, namely GECO (d/o 2190/2159) [11] and SGG-UGM-1 (d/o 2159/2159) [59], are used. The application of GOCE gradients enhanced the global gravity field at low-frequency bands, approximately from degree 30 to 220 [60].

Three representative altimetric gravity models, i.e., HY2A, SIO31, and DTU21GRA, are used, where HY2A represents the altimetric gravity model computed by only using the conventional radar altimeter data, and SIO31 and DTU21GRA represent the SAR-based altimetric gravity models. Figure 8 shows the MDTs modeled from different reference models; for all the MDTs directly modeled from the GGMs alone, we find dramatic variations over areas close to islands. The associated solution augmented by using HY2A demonstrates smaller turbulences in comparison with the MDTs directly derived from these GGMs. This is mainly due to the incorporation of heterogeneous satellite altimeter-derived gravity data in model development, where EGM2008/SGG-UGM-1/GECO was computed by involving DNSC07GRA data at short-wavelength over the SCS. However, limited high-quality altimeter data was used in developing DNSC07GRA, where no SARAL/AltiKa or CryoSat-2 data was included. In comparison, five years of CryoSat-2 data, two years of SARAL/AltiKa, and data from the HY-2A geodetic mission were included in the computation of the HY2A model, and, consequently, the HY2A model had improved accuracy versus DTU10GRA/DNSC07GRA [42]. The MDTs computed from the SIO31/DTU21GRA data further reduce these disturbances over island areas, by a magnitude of several centimeters, compared with the solutions modeled from the HY2A data. Moreover, the inconsistencies among the MDTs modeled from different reference models are prominent over the northern and eastern study area, owing to the heterogeneous data sources and modeling approaches adopted in those reference models' computation.

Figures 9 and 10 show the MDTs computed from different reference models along latitude 10.4° N and longitude 114.6° E, respectively. For all the MDTs, we see that the GGM-derived MDT has the largest variations, whereas the MDTs modeled additionally with the HY2A data reduces these disturbances: see the Zhenghe Reefs and Wufang Reef along latitude 10.4° N and the Zhenghe Reefs and Lesi Shoal along longitude 114.6° E. Moreover, the MDTs modeled from the SIO31/DTU21GRA data computed with the SAR altimetry data have less oscillations than those computed from the HY2A data. This is especially a case close to the Lesi Shoal along longitude 114.6° E, where the use of DTU21GRA data reduces the spikes significantly, by a magnitude larger than 10 cm and 4 cm, compared to the results computed from these GGMs alone and ones modeled by additionally combining the HY2A data, respectively. This also demonstrates that the use of altimetric gravity data computed with the SAR altimetry data can significantly enhance the marine gravity field, no matter which global geopotential model is applied for modeling.

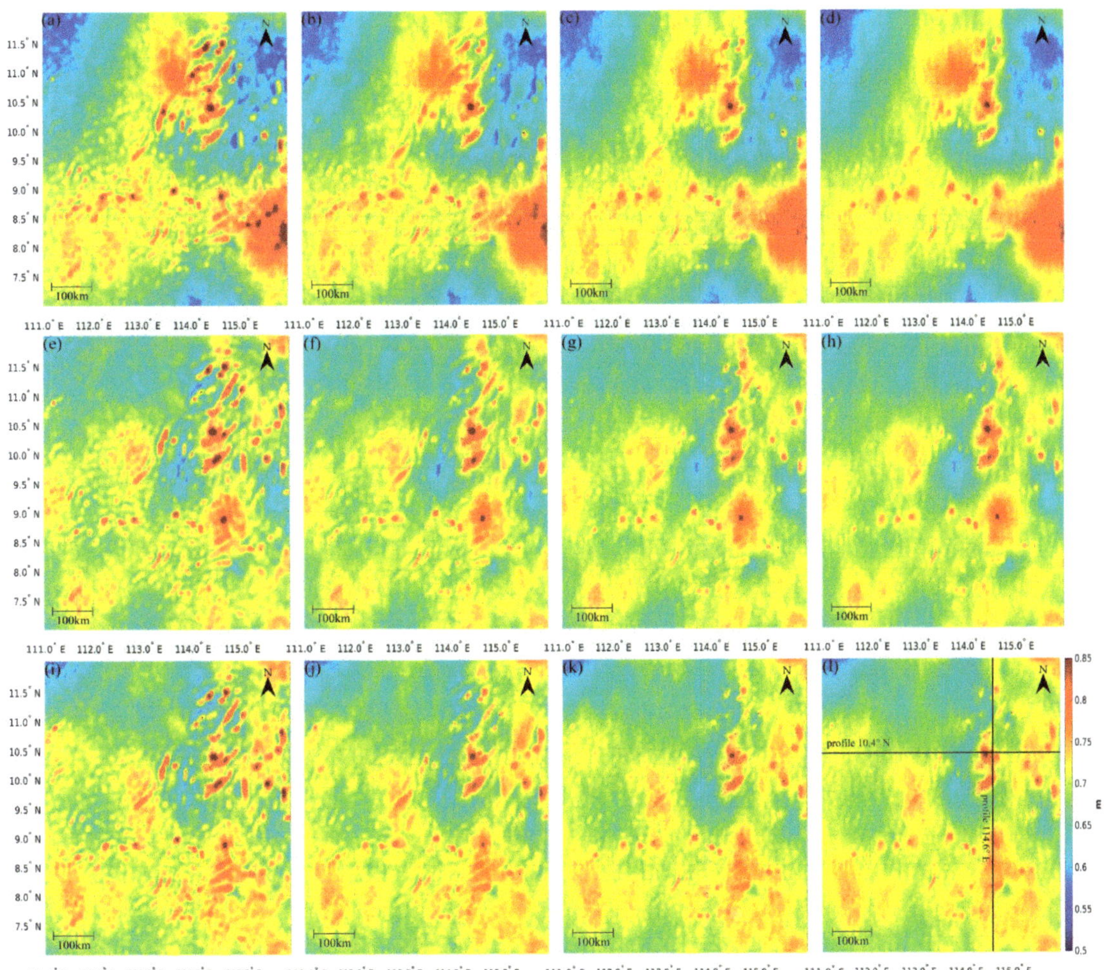

Figure 8. MDTs derived from various satellite altimeter-derived gravity models when different global geopotential models are used as the reference models. The first, second, and third rows are the solutions modeled when EGM2008, GECO, and SGG-UGM-1 are used as reference models, respectively. The first column (**a,e,i**) demonstrates the solutions modeled with different global geopotential models (up to the maximal d/o), the second column (**b,f,j**) manifests the solutions enhanced by HY2A, the third column (**c,g,k**) displays the solutions strengthened by SIO31, and the fourth column (**d,h,l**) shows the solutions augmented by DTU21GRA. The geographical coordinates are expressed in the Plate Carrée projection.

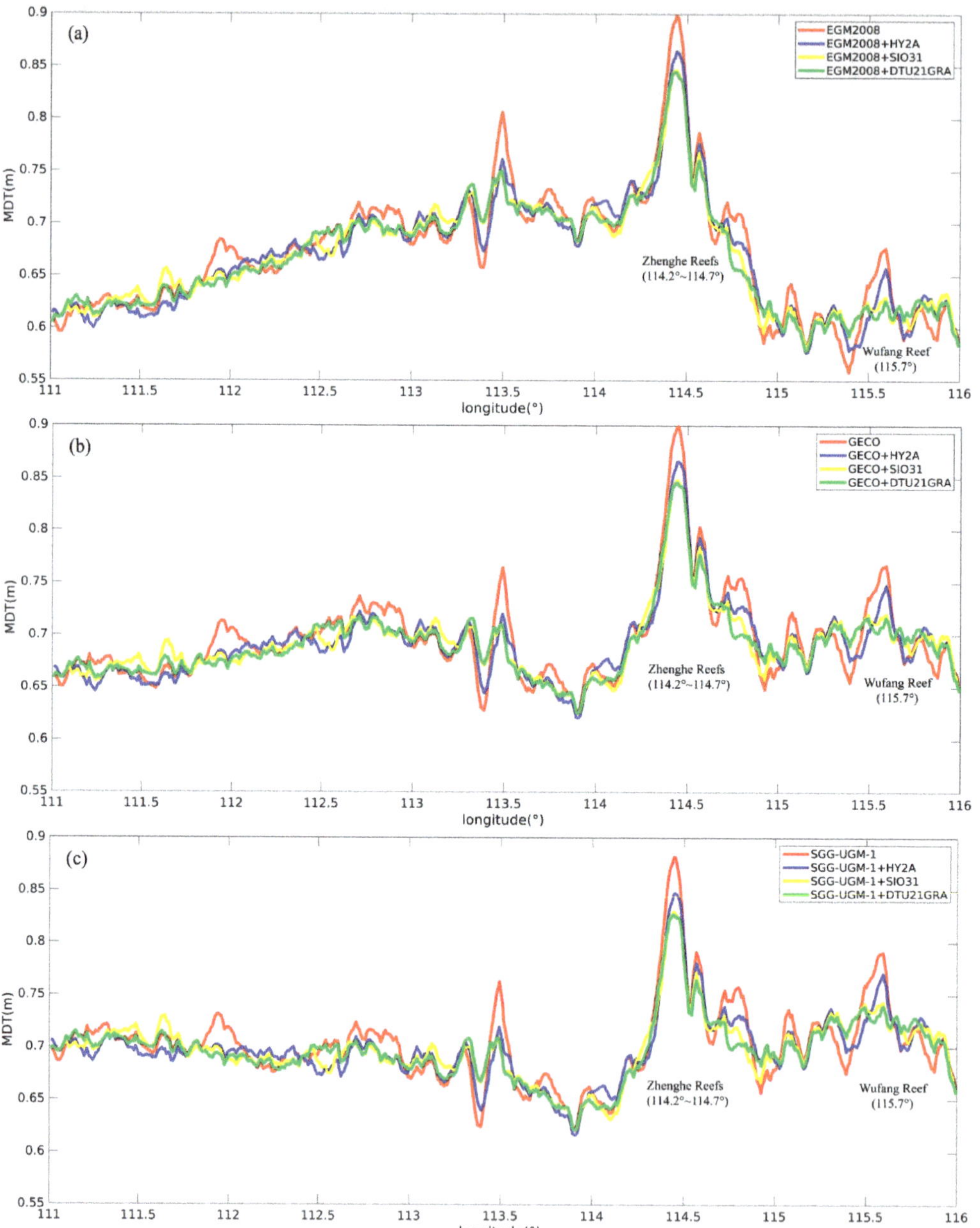

Figure 9. Profiles of the MDTs modeled with (**a**) EGM2008, (**b**) GECO, and (**c**) SGG-UGM-1 based on different altimetric gravity models along latitude 10.4° N.

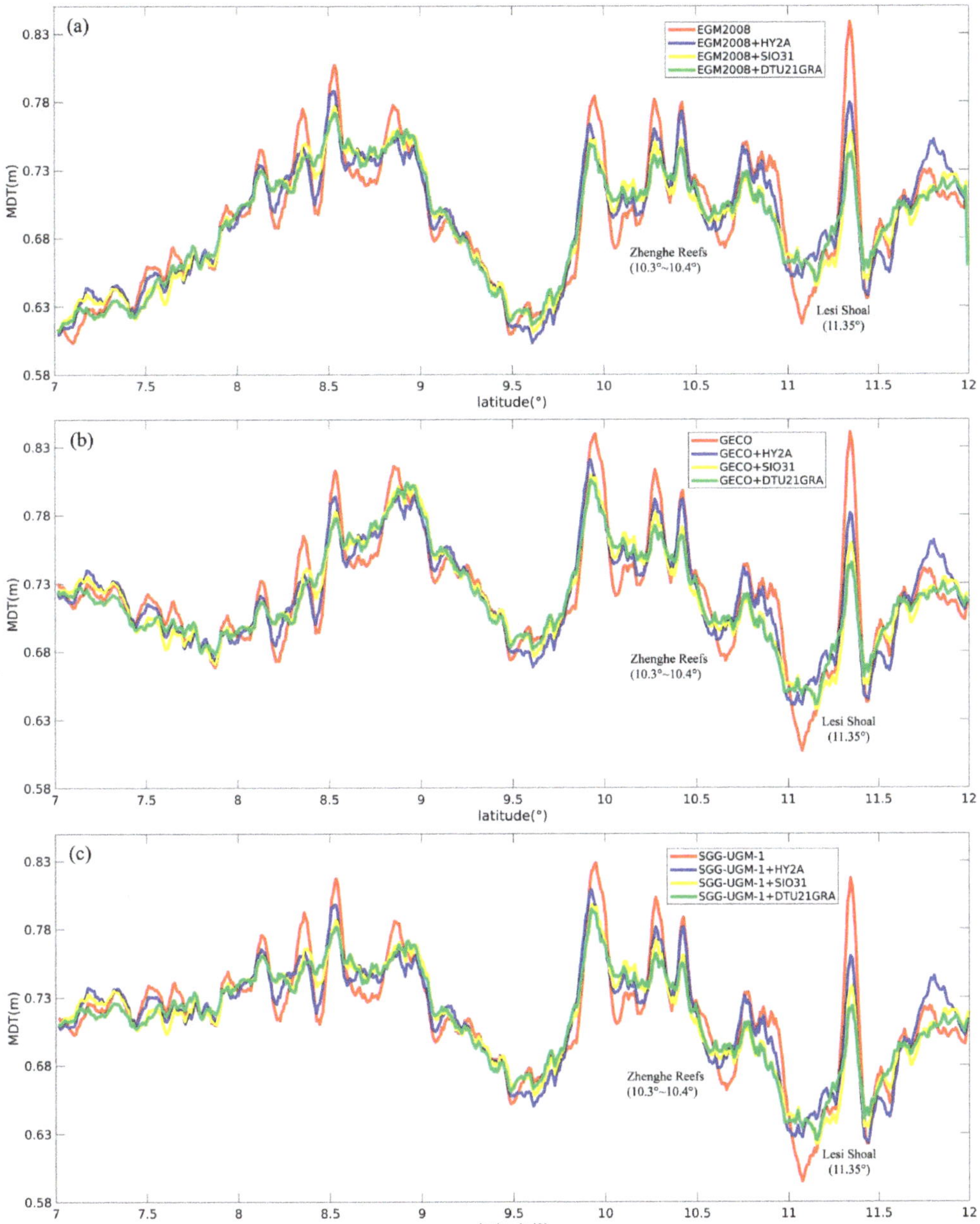

Figure 10. Profiles of the MDTs modeled with (**a**) EGM2008, (**b**) GECO, and (**c**) SGG-UGM-1 based on different altimetric gravity models along longitude 114.6° N.

4. Conclusions

In this study, the satellite altimeter-derived gravity models are used in marine gravity field augmentation over island areas; in particular, we study the feasibility of regional enhancement by using SAR-based gravity models. The gravity field modeling results in the Spratly Islands demonstrate that the incorporation of the newly published satellite altimeter-derived gravity observations strengthens the marine gravity field. By involving the gravity data computed with the Sentinel-3A/B SAR altimetry data, the accuracies of regional solutions (including quasi-geoid and MDT models) are dramatically improved, by a magnitude larger than 4 cm around areas close to islands, in comparison with the results directly derived from a newly released high-degree GGM, namely, XGM2019e_2159. This suggests that the newly published altimetric gravity models contain additional signals that were unresolved in the currently available GGMs.

Moreover, we model local gravity field based on heterogeneous combined GGMs. For all the solutions derived from the GGMs alone, significant variations are observed, which may bring about errors that are greater than 10 cm. These results show that the application of a GGM alone is inadequate in MDT modeling with centimeter-level accuracy over the areas close to land/island.

Further comparison of the solutions modeled from heterogeneous gravity models show that the performances of different models are heterogeneous, where the quasi-geoid/MDT solutions augmented with the SAR-based gravity models have better results. The use of the SAR-based gravity models in local augmentation improves the quasi-geoid/MDT solutions, by a magnitude of 2~4 cm, versus the solutions computed from the gravity models developed without SAR altimetry data. These results highlight the superiority of using the SAR-based altimetric gravity data in marine gravity field recovery, especially over the regions close to land/island. Moreover, the use of SAR altimetry data can alleviate the well-known coast problem, which also allays the issue of data scarcity in narrow/shallow waters. It is noticeable that the use of SAR-based altimetry data cannot fully overcome the coast problem; moreover, it also cannot replace the role of ground-based gravity data in gravity field modeling. By combing the ground-based gravity data, the local gravity field can be further improved.

Future work involves the further improvement of local gravity field model and the application of the model computed in this study in research fields such as geodesy and oceanography. By merging SAR altimeter records from the missions such as Sentinel-6 and using the improved waveform retracking methods, the local gravity field may be improved. Moreover, the quasi-geoid and mean dynamic topography models computed with the SAR altimetric gravity data are beneficial in understanding sea level change, ocean currents, and water exchanges with the surrounding regions close to the Spratly Islands. More importantly, these models are useful for establishing a unified vertical height datum over the Spratly Islands, where extensive coastlines and many islands exist, which make the traditionally used methods, such as the hydrostatic leveling, oceanic dynamic leveling, and trigonometric leveling methods, difficult to use for height datum unification.

Author Contributions: Conceptualization, Y.W. and A.A.; methodology, Y.W. and J.W.; software, Y.W.; validation, Y.W. and A.A.; formal analysis, Y.W.; writing—original draft preparation, Y.W.; writing—review and editing, Y.W. and H.S.; visualization, H.W. and H.S.; supervision, X.H. and Z.L.; project administration, Y.W. and Y.D.; funding acquisition, Y.W. All authors have read and agreed to the published version of the manuscript.

Funding: This research was funded by the National Natural Science Foundation of China, grant numbers 42004008, 41830110, 41931074, and 41974016; the Natural Science Foundation of Jiangsu Province, China, grant numbers BK20190498 and BK20190495; the Fundamental Research Funds for the Central Universities, grant number B210201013; the State Scholarship Fund from Chinese Scholarship Council, grant number 201306270014; and the Guangxi Key Laboratory of Spatial Information and Geomatics, grant number 19-185-10-06.

Institutional Review Board Statement: Not applicable.

Informed Consent Statement: Not applicable.

Data Availability Statement: The satellite altimetric gravity models and mean sea surface models that developed at DTU space are available on https://ftp.space.dtu.dk/pub, accessed on 28 November 2021, and the altimetric gravity models computed by the Scripps Institution of Oceanography are freely accessible at https://topex.ucsd.edu/pub/global_grav_1min, accessed on 28 November 2021.

All the global geopotential models can be publicly accessed from http://icgem.gfz-potsdam.de/home, accessed on 28 November 2021.

Acknowledgments: The authors would like to give our sincerest thanks to the three anonymous reviewers for their constructive suggestions and comments, which are of great value for improving the manuscript. The authors also thank the Editor for the kind assistances and beneficial comments. The authors are grateful for the kind support from the editorial office. We gratefully acknowledge the funders of this study.

Conflicts of Interest: The authors declare no conflict of interest.

References

1. Tapley, B.D.; Chambers, D.P.; Bettadpur, S.; Ries, J.C. Large scale ocean circulation from the GRACE GGM01 geoid. *Geophys. Res. Lett.* **2003**, *30*, 2163. [CrossRef]
2. Tapley, B.D.; Bettadpur, S.; Watkins, M.; Reigber, C. The gravity recovery and climate experiment: Mission overview and early results. *Geophys. Res. Lett.* **2004**, *31*, L09607. [CrossRef]
3. Pail, R.; Bruinsma, S.; Migliaccio, F.; Förste, C.; Goiginger, H.; Schuh, W.D.; Höck, E.; Reguzzoni, M.; Brockmann, J.M.; Abrikosov, O.; et al. First GOCE gravity field models derived by three different approaches. *J. Geod.* **2011**, *85*, 819–843. [CrossRef]
4. Pail, R.; Goiginger, H.; Schuh, W.-D.; Höck, E.; Brockmann, J.M.; Fecher, T.; Gruber, T.; Mayer-Gürr, T.; Kusche, J.; Jäggi, A.; et al. Combined satellite gravity field model GOCO01S derived from GOCE and GRACE. *Geophys. Res. Lett.* **2010**, *37*, L20314. [CrossRef]
5. Bruinsma, S.L.; Förste, C.; Abrikosov, O.; Marty, J.C.; Rio, M.H.; Mulet, S.; Bonvalot, S. The new ESA satellite-only gravity field model via the direct approach. *Geophys. Res. Lett.* **2013**, *40*, 3607–3612. [CrossRef]
6. Brockmann, J.M.; Schubert, T.; Schuh, W.D. An improved model of the Earth's static gravity field solely derived from reprocessed GOCE Data. *Surv. Geophys.* **2021**, *42*, 277–316. [CrossRef]
7. Brockmann, J.M.; Zehentner, N.; Höck, E.; Pail, R.; Loth, I.; Mayer-Gürr, T.; Schuh, W.-D. EGM_TIM_RL05: An independent geoid with centimeter accuracy purely based on the GOCE mission. *Geophys. Res. Lett.* **2014**, *41*, 8089–8099. [CrossRef]
8. Kvas, A.; Brockmann, J.M.; Krauss, S.; Schubert, T.; Gruber, T.; Meyer, U.; Mayer-Gürr, T.; Schuh, W.D.; Jäggi, A.; Pail, R. GOCO06s—A satellite-only global gravity field model. *Earth Syst. Sci. Data* **2021**, *13*, 99–118. [CrossRef]
9. Pavlis, N.K.; Holmes, S.A.; Kenyon, S.C.; Factor, J.K. The development and evaluation of Earth Gravitational Model (EGM2008). *J. Geophys. Res. Solid Earth* **2012**, *117*, B04406, Erratum in *J. Geophys. Res. Solid Earth.* **2013**, *118*, 2633. [CrossRef]
10. Förste, C.; Bruinsma, S.L.; Abrikosov, O.; Lemoine, J.M.; Schaller, T.; Götze, H.J.; Ebbing, J.; Marty, J.C.; Flechtner, F.; Balmino, G.; et al. EIGEN-6C4 The latest combined global gravity field model including GOCE data up to degree and order 2190 of GFZ Potsdam and GRGS Toulouse. In Proceedings of the 5th GOCE User Workshop, Paris, France, 25–28 November 2014.
11. Gilardoni, M.; Reguzzoni, M.; Sampietro, D. GECO: A global gravity model by locally combining GOCE data and EGM2008. *Stud. Geophys. Geod.* **2015**, *60*, 228–247. [CrossRef]
12. Fecher, T.; Pail, R.; Gruber, T. GOCO05c: A new combined gravity field model based on full normal equations and regionally varying weighting. *Surv. Geophys.* **2017**, *38*, 571–590. [CrossRef]
13. Zingerle, P.; Pail, R.; Gruber, T.; Oikonomidou, X. The combined global gravity field model XGM2019e. *J. Geod.* **2020**, *94*, 66. [CrossRef]
14. Schwabe, J.; Scheinert, M. Regional geoid of the Weddell Sea, Antarctica, from heterogeneous ground-based gravity data. *J. Geod.* **2014**, *88*, 821–838. [CrossRef]
15. Deng, X.; Featherstone, W.E. A coastal retracking system for satellite radar altimeter waveforms: Application to ERS2 around Australia. *J. Geophys. Res. Oceans* **2006**, *111*, C06012. [CrossRef]
16. Andersen, O.B.; Scharroo, R. Range and geophysical corrections in coastal regions: And implications for mean sea surface determination. In *Coastal Altimetry*; Vignudelli, S., Ed.; Springer: Berlin/Heidelberg, Germany, 2011.
17. Abulaitijiang, A.; Andersen, O.B.; Stenseng, L. Coastal sea level from inland CryoSat-2 interferometric SAR altimetry. *Geophys. Res. Lett.* **2015**, *42*, 1841–1847. [CrossRef]
18. Idžanović, M.; Ophaug, V.; Andersen, O.B. The coastal mean dynamic topography in Norway observed by CryoSat-2 and GOCE. *Geophys. Res. Lett.* **2017**, *44*, 5609–5617. [CrossRef]
19. McAdoo, D.C.; Farrell, S.L.; Laxon, S.; Ridout, A.; Zwally, H.J.; Yi, D. Gravity of the Arctic Ocean from satellite data with validations using airborne gravimetry: Oceanographic implications. *J. Geophys. Res. Oceans* **2013**, *118*, 917–930. [CrossRef]
20. Wu, Y.; Abulaitijiang, A.; Featherstone, W.E.; McCubbine, J.C.; Andersen, O.B. Coastal gravity field refinement by combining airborne and ground-based data. *J. Geod.* **2019**, *93*, 2569–2584. [CrossRef]
21. Farrell, S.L.; McAdoo, D.C.; Laxon, S.W.; Zwally, H.J.; Yi, D.; Ridout, A.; Giles, K. Mean dynamic topography of the Arctic Ocean. *Geophys. Res. Lett.* **2012**, *39*, L01601. [CrossRef]
22. Skourup, H.; Farrell, S.L.; Hendricks, S.; Ricker, R.; Armitage, T.W.K.; Ridout, A.; Anderson, O.B.; Haas, C.; Baker, S. An assessment of state-of-the-art mean sea surface and geoid models of the Arctic Ocean: Implications for sea ice freeboard retrieval. *J. Geophys. Res. Oceans* **2017**, *122*, 8593–8613. [CrossRef]

23. Sandwell, D.T.; Garcia, E.; Soofi, K.; Wessel, P.; Smith, W.H.F. Towards 1mGal global marine gravity from CryoSat-2, Envisat, and Jason-1. *Lead. Edge* **2013**, *32*, 892–899. [CrossRef]
24. Sandwell, D.T.; Müller, R.D.; Smith, W.H.F.; Garcia, E.; Francis, R. New global marine gravity model from CryoSat-2 and Jason-1 reveals buried tectonic structure. *Science* **2014**, *346*, 65–67. [CrossRef] [PubMed]
25. Garcia, E.S.; Sandwell, D.T.; Smith, W.H.F. Retracking CryoSat-2, Envisat and Jason-1 radar altimetry waveforms for improved gravity field recovery. *Geophys. J. Int.* **2014**, *196*, 1402–1422. [CrossRef]
26. Aldarias, A.; Gomez-Enri, J.; Laiz, I.; Tejedor, B.; Cipollini, P. Validation of Sentinel-3A SRAL Coastal Sea Level Data at High Posting Rate: 80 Hz. *IEEE Trans. Geosci. Remote Sens.* **2020**, *58*, 3809–3821. [CrossRef]
27. Wingham, D.J.; Francis, C.R.; Baker, S.; Bouzinac, C.; Cullen, R.; de Chateau-Thierry, P.; Laxon, S.W.; Mallow, U.; Mavrocordatos, C.; Phalippou, L.; et al. CryoSat: A mission to determine the fluctuations in Earth's land and marine ice fields. *Adv. Space Res.* **2006**, *37*, 841–871. [CrossRef]
28. Calafat, F.M.; Cipollini, P.; Bouffard, J.; Snaith, H.; Féménias, P. Evaluation of new cryosat-2 products over the ocean. *Remote Sens. Environ.* **2017**, *191*, 131–144. [CrossRef]
29. García, P.; Martin-Puig, C.; Roca, M. SARin mode, and a window delay approach, for coastal altimetry. *Adv. Space Res.* **2018**, *62*, 1358–1370. [CrossRef]
30. Boy, F.; Desjonquères, J.D.; Picot, N.; Moreau, T.; Raynal, M. CryoSat-2 SAR-mode over oceans: Processing methods, global assessment, and benefits. *IEEE Trans. Geosci. Remote Sens.* **2017**, *55*, 148–158. [CrossRef]
31. Dinardo, S.; Fenoglio-Marc, L.; Buchhaupt, C.; Becker, M.; Scharroo, R.; Fernandes, M.J.; Benveniste, J. Coastal SAR and PLRM altimetry in German Bight and west Baltic Sea. *Adv. Space Res.* **2018**, *62*, 1371–1404. [CrossRef]
32. Peng, F.; Deng, X. Validation of Sentinel-3A SAR mode sea level anomalies around the Australian coastal region. *Remote Sens. Environ.* **2020**, *237*, 111548. [CrossRef]
33. Nielsen, K.; Stenseng, L.; Andersen, O.B.; Villadsen, H.; Knudsen, P. Validation of cryosat-2 sar mode based lake levels. *Remote Sens. Environ.* **2015**, *171*, 162–170. [CrossRef]
34. Cipollini, P.; Calafat, F.M.; Jevrejeva, S.; Melet, A.; Prandi, P. Monitoring sea level in the coastal zone with satellite altimetry and tide gauges. *Surv. Geophys.* **2017**, *38*, 33–57. [CrossRef]
35. Kleinherenbrink, M.; Naeije, M.; Slobbe, C.; Egido, A.; Smith, W. The performance of CryoSat-2 fully-focussed SAR for inland water-level estimation. *Remote Sens. Environ.* **2020**, *237*, 111589. [CrossRef]
36. Andersen, O.B.; Knudsen, P. The DTU17 Global Marine Gravity Field: First Validation Results. In *International Association of Geodesy Symposia*; Springer: Berlin/Heidelberg, Germany, 2019.
37. Weatherall, P.; Marks, K.M.; Jakobsson, M.; Schmitt, T.; Tani, S.; Arndt, J.E.; Rovere, M.; Chayes, D.; Ferrini, V.; Wigley, R. A new digital bathymetric model of the world's oceans. *Earth Space Sci.* **2015**, *2*, 331–345. [CrossRef]
38. Dong, Y.; Liu, Y.; Hu, C.; Xu, B. Coral reef geomorphology of the spratly islands: A simple method based on time-series of landsat-8 multi-band inundation maps. *ISPRS J. Photogramm. Remote Sens.* **2019**, *157*, 137–154. [CrossRef]
39. Andersen, O.B.; Knudsen, P.; Kenyon, S.; Factor, J.K.; Holmes, S. The DTU13 Global marine gravity field—First evaluation. In Proceedings of the OSTST Meeting, Boulder, CO, USA, 8–11 October 2013.
40. Sandwell, D.T.; Harper, H.; Tozer, B.; Smith, W.H.F. Gravity field recovery from geodetic altimeter missions. *Adv. Space Res.* **2021**, *68*, 1059–1072. [CrossRef]
41. Zhu, C.; Guo, J.; Gao, J.; Liu, X.; Hwang, C.; Yu, S.; Yuan, J.; Ji, B.; Guan, B. Marine gravity determined from multi-satellite GM/ERM altimeter data over the South China Sea: SCSGA V1.0. *J. Geod.* **2020**, *94*, 50. [CrossRef]
42. Zhang, S.; Andersen, O.B.; Kong, X.; Li, H. Inversion and validation of improved marine gravity field recovery in south china sea by incorporating HY-2A altimeter waveform data. *Remote Sens.* **2020**, *12*, 802. [CrossRef]
43. Omang, O.C.D.; Forsberg, R. How to handle topography in practical geoid determination: Three examples. *J. Geod.* **2000**, *74*, 458–466. [CrossRef]
44. Featherstone, W.E.; McCubbine, J.C.; Brown, N.J.; Claessens, S.J.; Filmer, M.S.; Kirby, J.F. The first Australian gravimetric quasigeoid model with location-specific uncertainty estimates. *J. Geod.* **2018**, *92*, 149–168. [CrossRef]
45. Forsberg, R. *A Study of Terrain Reductions, Density Anomalies and Geophysical Inversion Methods in Gravity Field Modelling*; Report No. 355; Department of Geodetic Science and Surveying, The Ohio State University: Colombus, OH, USA, 1984.
46. Zingerle, P.; Pail, R.; Gruber, T.; Oikonomidou, X. *The Experimental Gravity Field Model XGM2019e*; GFZ Data Services: Potsdam, Germany, 2019. [CrossRef]
47. Wu, Y.; He, X.; Luo, Z.; Shi, H. An Assessment of Recently Released High-Degree Global Geopotential Models Based on Heterogeneous Geodetic and Ocean Data. *Front. Earth Sci.* **2021**, *9*, 749611. [CrossRef]
48. Wu, Y.; Zhou, H.; Zhong, B.; Luo, Z. Regional gravity field recovery using the GOCE gravity gradient tensor and heterogeneous gravimetry and altimetry data. *J. Geophys. Res. Solid Earth* **2017**, *122*, 6928–6952. [CrossRef]
49. Wu, Y.; Luo, Z.; Chen, W.; Chen, Y. High-resolution regional gravity field recovery from Poisson wavelets using heterogeneous observational techniques. *Earth Planets Space* **2017**, *69*, 1–15. [CrossRef]
50. Klees, R.; Tenzer, R.; Prutkin, I.; Wittwer, T. A data-driven approach to local gravity field modelling using spherical radial basis functions. *J. Geod.* **2008**, *82*, 457–471. [CrossRef]
51. Becker, S.; Brockmann, J.M.; Schuh, W.D. Mean dynamic topography estimates purely based on GOCE gravity field models and altimetry. *Geophys. Res. Lett.* **2014**, *41*, 2063–2069. [CrossRef]

52. Bingham, R.J.; Knudsen, P.; Andersen, O.B.; Pail, R. An initial estimate of the North Atlantic steady-state geostrophic circulation from GOCE. *Geophys. Res. Lett.* **2011**, *38*, L01606. [CrossRef]
53. Rio, M.H.; Guinehut, S.; Larnicol, G. New CNES-CLS09 global mean dynamic topography computed from the combination of GRACE data, altimetry, and in situ measurements. *J. Geophys. Res. Oceans* **2011**, *116*, C07018. [CrossRef]
54. Rio, M.H.; Mulet, S.; Picot, N. Beyond GOCE for the ocean circulation estimate: Synergetic use of altimetry, gravimetry, and in situ data provides new insight into geostrophic and Ekman currents. *Geophys. Res. Lett.* **2014**, *41*, 8918–8925. [CrossRef]
55. Tenzer, R.; Foroughi, I. Effect of the Mean Dynamic Topography on the Geoid-to-Quasigeoid Separation Offshore. *Mar. Geod.* **2018**, *41*, 368–381. [CrossRef]
56. Andersen, O.B.; Abulaitijiang, A.; Zhang, S.; Rose, S.K. A new high resolution Mean Sea Surface (DTU21MSS) for improved sea level monitoring. In Proceedings of the EGU General Assembly (EGU21-16084), Vienna, Austria, 19–30 April 2021. [CrossRef]
57. Andersen, O.B.; Knudsen, P.; Stenseng, L. A New DTU18 MSS Mean Sea Surface–Improvement from SAR Altimetry. In Proceedings of the 25 Years of Progress in Radar Altimetry Symposium, Ponta Delgada, Portugal, 24–29 September 2018.
58. Wu, Y.; Abulaitijiang, A.; Andersen, O.B.; He, X.; Luo, Z.; Wang, H. Refinement of mean dynamic topography over island areas using airborne gravimetry and satellite altimetry data over the northern of South China Sea. *J. Geophys. Res. Solid Earth* **2021**, *126*, e2021JB021805. [CrossRef]
59. Liang, W.; Xu, X.; Li, J.; Zhu, G. The determination of an ultra-high gravity field model SGG-UGM-1 by combining EGM2008 gravity anomaly and GOCE observation data. *Acta Geod. Cartogr. Sin.* **2018**, *47*, 425–434. [CrossRef]
60. Gruber, T.; Rummel, R.; Abrikosov, O.; Hees, V.R. GOCE Level 2 Product Data Handbook, GO-MA-HPF-GS-0110. 2014. Issue 4.2. Available online: https://earth.esa.int/documents/10174/1650485/GOCE_Product_Data_Handbook_Level-2 (accessed on 28 November 2021).

remote sensing

MDPI

Article

Measuring Height Difference Using Two-Way Satellite Time and Frequency Transfer

Peng Cheng [1], Wenbin Shen [1,2,*], Xiao Sun [1], Chenghui Cai [1], Kuangchao Wu [1] and Ziyu Shen [3]

[1]   Time and Frequency Geodesy Center, School of Geodesy and Geomatics, Wuhan University, Wuhan 430079, China; pengcheng97@whu.edu.cn (P.C.); xsun@whu.edu.cn (X.S.); chcai@whu.edu.cn (C.C.); Kcwu@whu.edu.cn (K.W.)

[2]   State Key Laboratory of Information Engineering in Surveying, Mapping and Remote Sensing, Wuhan University, Wuhan 430079, China

[3]   School of Resource, Environmental Science and Engineering, Hubei University of Science and Technology, Xianning 437100, China; zyshen@hbust.edu.cn

*   Correspondence: wbshen@sgg.whu.edu.cn

Abstract: According to general relativity theory (GRT), the clock at a position with lower geopotential ticks slower than an identical one at a position with higher geopotential. Here, we provide a geopotential comparison using a non-transportable hydrogen clock and a transportable hydrogen clock for altitude transmission based on the two-way satellite time and frequency transfer (TWSTFT) technique. First, we set one hydrogen clock on the fifth floor and another hydrogen clock on the ground floor, with their height difference of 22.8 m measured by tape, and compared the time difference between these two clocks by TWSTFT for 13 days. Then, we set both clocks on the ground floor and compared the time difference between the two clocks for seven days for zero-baseline calibration (synchronization). Based on the measured time difference between the two clocks at different floors, we obtained the height difference 28.0 ± 5.4 m, which coincides well with the tape-measured result. This experiment provides a method of height propagation using precise clocks based on the TWSTFT technique.

Keywords: TWSTFT; satellite; geopotential; altitude transmission

Citation: Cheng, P.; Shen, W.; Sun, X.; Cai, C.; Wu, K.; Shen, Z. Measuring Height Difference Using Two-Way Satellite Time and Frequency Transfer. *Remote Sens.* **2022**, *14*, 451. https://doi.org/10.3390/rs14030451

Academic Editor: Xiaogong Hu

Received: 15 December 2021
Accepted: 16 January 2022
Published: 18 January 2022

Publisher's Note: MDPI stays neutral with regard to jurisdictional claims in published maps and institutional affiliations.

1. Introduction

The gravity potential (geopotential) plays a significant role in geodesy. It is essential for defining the geoid and measuring orthometric height. The conventional method of determining the geopotential is combining leveling and gravimetry, but there are shortcomings: with the increase in measurement length, the error accumulates and becomes larger and larger, and it is impossible or difficult to transfer the orthometric height between two points separated by oceans.

To overcome the shortcomings existing in the conventional method, time–frequency comparison methods based on general relativity theory (GRT) [1] were proposed in recent decades [2–7]. The basic idea is that by comparing the time elapsed or frequency shift between two remote clocks, the geopotential difference between the two sites where the clocks are located could be determined.

In the clock-transportation method, the most critical conditions are the clock's precision and time transfer. Precise clocks generally include microwave-atomic clocks (MACs) and optical-atomic clocks (OACs). MACs play an important role in time service research. However, with the development of a high-precision clock, its precision could not match that of the OAC. The concept of OAC was first proposed by Nobel laureate Dehmelt (1973) in the late 1970s. He used the energy level transition of a single ion to realize an ultra-high-precision optical clock. In recent years, the precision of OAC has reached a total systematic uncertainty of 9.4×10^{-19} and frequency stability of $1.2 \times 10^{-15} / \sqrt{\tau}$ [8]. However, OACs are often bulky and can only work in laboratory environments, which significantly limits their application scope and makes it difficult to conduct a clock-transportation experiment. It has always been the wish of scientists to realize a transportable, reliable, and quasi-continuous high-precision OAC, but it is also challenging. To broaden the application scope of OAC, many research groups in the worldwide have been devoted to the development of transportable optical-atomic clocks (TOCs). In 2014, a group reported a TOC based on laser-cooled strontium atoms trapped in an optical lattice, and this TOC fits within a volume of <2 m^3, and its relative uncertainty is 7×10^{-15} [9]. Three years later, a group from Physikalisch-Technische Bundesanstalt (PTB) reported a TOC with ^{87}Sr, with its characterization against an OAC resulting in a systematic uncertainty of 7.4×10^{-17} [10]. With the development of TOC, many scientists started to conduct clock-transportation experiments. Grotti et al. (2018) reported the first field measurement campaign with a ^{87}Sr TOC with an uncertainty of 1.8×10^{-16} and a ^{171}Yb OAC with an uncertainty of 1.6×10^{-16}. They used these clocks and fiber link to determine the geopotential difference between the middle of a mountain and a location 90 km away with a height difference of 1000 m. Their experimental result of potential difference of 10,034(174) m^{2}s^{-2} agrees well with value of 10,032.1(16) m^{2}s^{-2} determined independently by the conventional geodetic approach [3].

The clock-transportation experiments mentioned above used optical fiber to transfer frequency signals. Although optical fiber has very high accuracy, distance still limits its application. By comparison, though GNSS common-view technique and TWSTFT have lower accuracy than optical fiber frequency signal transfer [11–15], they can realize long-distance time–frequency signal transmission. In TWSTFT, one uses a geostationary satellite as 'bridge' to transmit time–frequency signals from one station to another one, and the time elapse recorded at one station is compared with that at another one. Because using TWSTFT only requires the stations in a place where the geostationary satellite can receive and transmit signals, there is hardly any limit on the positions of stations. Most errors in the signal-transmission process are offset because of the approximate symmetry of the signal transmission path of TWSTFT technology. Therefore, this symmetry causes the high precision of this technology [16]. In August 1962, the USNO (U.S. Naval Observatory) launched the first communication satellite, Telstar I, to transmit telephone and high-speed data signals. Then, the USNO and the NPL (National Physical Laboratory) collaborated in an experiment using this satellite to relate the precise clocks at the USNO and the RGO (Royal Greenwich Observatory). This is considered to be the first two-way satellite time transfer experiment, and the accuracy of the experiment was 20 μs [17]. In 1992, some satellite systems and modems adapted for TWSTFT were commercialized. About ten coordinated universal time (UTC) laboratories have equipped with the modems and other equipment for the clock comparisons with TWSTFT [18]. Later, many TWSTFT experiments were conducted, and they almost exchanged timing information via the communication satellite; paired ground stations transmit and receive pseudo-random noise (PN) coded signals in TWSTFT links. The TWSTFT technique became promising using geostationary satellites for high-accuracy time and frequency transfer [16,19–21]. The accuracy of TWSTFT has been further improved to around 0.2 ns, and its improvement has been seriously limited by the chip rate of the coded signal [22]. Therefore, further improvements should come from the use of carrier phase information, because the resolution of the carrier phase is 100 to 1000 times more accurate than that of the code [23]. In 2016, an experiment using carried-phase TWSTFT was performed between the two stations of NICT (National Institute of Information and Communications Technology) and KRISS (Korea Research Institute of Standards and Science) with Sr and Yb OAC, and the instability for a frequency transfer at the 10^{-16} level after 12 h was achieved [24]. Riedel et al. (2020) conducted a 26-day comparison of five simultaneously operated OACs and six MACs located at SYRTE, NPL, INRIM, LNE, and PTB by using TWSTFT and GPSPPP. Considering the correlations and gaps of measurement data, they improved the statistical analysis procedure; combined overall uncertainties in the range of 1.8×10^{-16} to 3.5×10^{-16} for the OAC comparisons were found [25]. To investigate the feasibility of transportable atomic clock comparison using TWSTFT, we conducted a MAC comparison experiment at the Beijing Institute of Radio Metrology and Measurement (BIRMM), Beijing [26,27].

In Section 2, we introduce the principle of measuring the height difference by the TW-STFT, and in Section 3, we discuss the error corrections of TWSTFT. Section 4 demonstrates our experiment and data processing. In Section 5, we provide the results. Conclusions and discussions are placed in Section 6.

2. Methods

2.1. Height Measurement Based on Time Difference

According to general relativity theory (GRT), a clock at a position with lower height (stronger geopotential) ticks slower than an identical at a position with higher height (weaker geopotential) [5,28]. Inversely, we can determine the geopotential difference between two points A and B by measuring the elapsed time difference between two clocks located at A and B, expressed as (accurate to $1/c^2$) [7,29–31]

$$\frac{\Delta t_{AB}}{T} = \frac{t_B - t_A}{T} = -\frac{W_B - W_A}{c^2} = -\frac{\Delta W_{AB}}{c^2} \tag{1}$$

where t_A and t_B denote the times at sites A and B, respectively, after a standard time period of T; W_A and W_B are the geopotential at sites A and B, respectively (note that we apply the definition of geopotential given by geodetic community); and c is the speed of light in the vacuum. From Equation (1), we can determine the geopotential difference $\Delta W_{AB} = W_B - W_A$ based on $\Delta t_{AB}/T$.

As shown in Figure 1, given the height of point A and the geopotential difference ΔW_{AB} between A and B, one can measure the orthometric height of point B, expressed as [32,33]

$$H_B = H_A - \frac{\Delta W_{AB}}{\overline{g}} = H_A + \frac{\Delta t_{AB}}{T} \cdot \frac{c^2}{\overline{g}} \tag{2}$$

where H_A and H_B are the orthometric heights of point A and B, respectively, and $\overline{g}$ is the 'mean value' between g_B at point B and g_{O_B} at point O_B

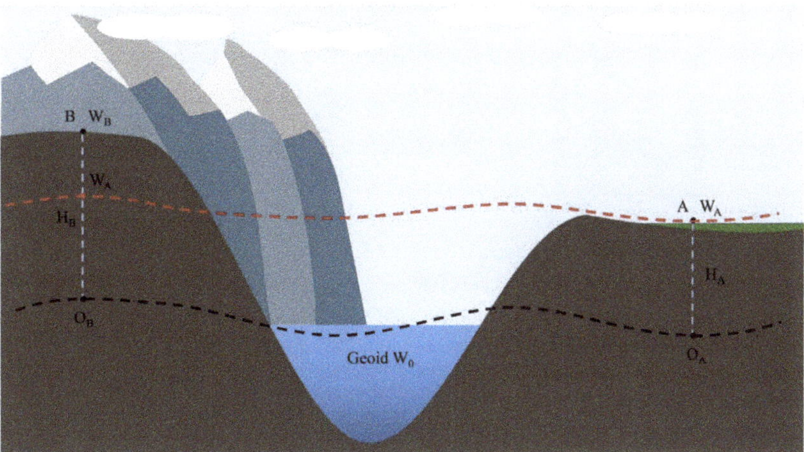

Figure 1. Principle of determining the orthometric heights. W_A and W_B are the geopotentials at point A and B, O_A and O_B are the projection points on the geoid (bold dashed curve) corresponding to point A and B along the plumb lines (light-blue dashed curves), red dashed curve denotes equipotential surface passing through point A, W_0 is the geopotential on the geoid.

2.2. One Pulse Per Second (1 PPS) Signal

The 1 PPS signal provides precise clock synchronization [34]. The original signal is the frequency signal (with typical frequency 10 MHz) output by the atomic clock. As an analog signal, in distant transmission, the information carried by the original frequency signal is easily distorted by various interferences [35]. Therefore, it needs to be converted into a digital signal, 1 PPS, to complete the time-information transmission. The process of converting frequency signal (say 10 MHz) into 1 PPS signal is shown in Figure 2. Suppose the frequency of the reference signal is 10 MHz, 10^7 cycles of the reference signal are one second. The generated 1 PPS signal rises from a low electrical level at the beginning of one reference signal cycle, and keeps the high electrical level for a short time (generally the pulse width is 20 µs), then declines to a low electrical level and keeps the site until the end of the 10^7 cycles (calculated from the first rise). This is a complete cycle (1 s) of 1 PPS signal. When the 1 PPS signal is used for time signal comparison, the rising edge of the signal will be used as the point to trigger the timer. Usually, the electrical level of the trigger is a predetermined value between the lowest electrical level and the highest electrical level. Ideally, when the electrical level rises to this predetermined value, the switch is triggered immediately, and the timing starts. However, there is a trigger delay during the process of the trigger switch. In fact, the switch can only be triggered when the actual electrical level is slightly higher than the predetermined value, so the trigger time will be within the time corresponding to the red oblique line [36]. Hence, the rising time duration δt is very important for precise time synchronization. Usually, δt should be smaller than 10 ns, since if δt is too large, the rising edge's slope (the blue slope line at the bottom of Figure 2) will become too small, which will increase the uncontrollable duration (red line) and reduce the time synchronization precision.

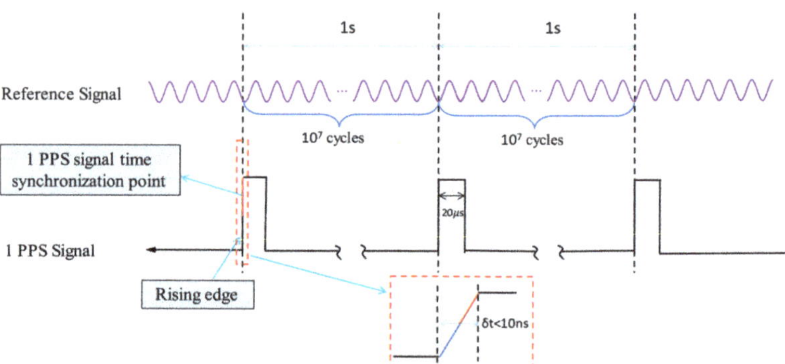

Figure 2. Generation of the one pulse per second (1 PPS) signal.

2.3. Transmission of 1 PPS Signal

Figure 3 shows the transmission process of 1 PPS signal in the TWSTFT. When the 1 PPS signal is generated by a clock, a part of the 1 PPS signal enters time interval counter (TIC) as trigger gate open pulse. Another part of the 1 PPS signal is transmitted to a satellite, through the modulation and emitter. The satellite uses a transparent transponder to transmit it to another ground station. After the signal from the satellite is received, the receiver and demodulation recover the 1 PPS signal from the received signal. The recovered 1 PPS signal enters the TIC as a pulse to trigger the gate close. The above processes are conducted between each of the two stations.

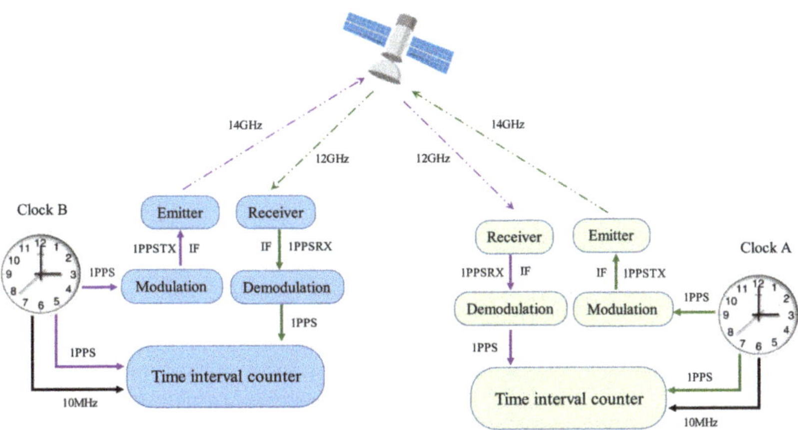

Figure 3. Principle of the two-way satellite time and frequency transfer (TWSTFT). There are two same TWSTFT observation systems located at two sites A and B. Every system include clock, emitter, receiver, etc. (see text) (modified after ITU-R 2015).

If the 1 PPS signal is directly transmitted to another station, it is difficult to compare the 1 PPS signals generated by two clocks at two sites synchronously. Hence, using the PN code to accompany the 1 PPS signal for time marking is necessary. The maximal length linear feedback shift register sequences (m-sequences) are the PN code usually used in TWSTFT. The m-sequence is the largest code that can be generated by a given shift register or a delay element of a given length [37]. The code has good autocorrelation characteristics and can be easily generated (Enge et al., 1987). At the same time, to reduce the impact of the bit error on the final time comparison, the frame synchronization bits (FSB) will be inserted at the rising edge of the 1 PPS signal. The bit error means that in signal transmission, decay causes the signal to be damaged, so the originally transmitted signal is '1', but the received signal is '0' (it should also be '1'), and the FSB is a specific set of bits at the beginning of each frame of signal [38]. The spreading spectrum consists of 1 PPS and PN, which is completed when the 1 PPS signal combines with the synchronized PN code by way of modulo-2 sum in the code generator (CG). This spread spectrum signal (SSS) is the baseband signal with wide bandwidth and decays rapidly during transmission. If it is directly used for long-distance baseband transmission, due to quick attenuation, the signal at the receiver will have a too-low signal-to-noise ratio (SNR) to be identified [39]. Therefore, the SSS needs to be modulated on a proper carrier. Generally, the carrier is a cosine wave, with its amplitude, frequency, and phase being known, and is used to transmit the information (signal) by changing the carrier's amplitude, or frequency, or phase, or the combinations of these entities.

The SSS is modulated on the carrier in the modulator with the binary phase-shift keying (BPSK) [40]. The BPSK changes the initial phase of the carrier to transmit the binary digital information of the SSS, while the carrier's amplitude and frequency remain unchanged (in the modulation process). In this modulation technology, the initial phase of the carrier has only two values of '0' and 'π', which correspond to the digits '1' and '0' of the 1 PPS signal, respectively. When transmitting signal '1', the modem generates a carrier with initial phase 0, and when transmitting signal '0', the modem generates a carrier with the initial phase π. The modulated signal is an intermediate frequency (IF) signal, which is named 1 PPSTX.

Referring to Figure 3, the frequency of the 1 PPSTX signal generated at station A is about 70 MHz, which is not suitable for long-distance signal transmission. Therefore, the frequency of the 1 PPSTX signal is amplified to 14 GHz through the multiplier in the up-converter (UC) and transmitted to the geostationary satellite through the transmitter.

After it receives the up-link signal, the satellite transparent transponder (STT) converts the signal's frequency to 12 GHz and transmits the signal (as a down-link signal) to station B. When the down-link signal arrives at station B, the signal's frequency is reduced to about 70 MHz through the multiplier in the down-converter (DC) (the function of which is similar to UC, but in the opposite way). At the same time, the received signal will also be digitized to facilitate subsequent processing. Usually, when this signal is converted to the IF range, a frequency shift will impact the signal acquisition. The reason is that the carrier used to demodulate the IF signal to the baseband signal needs to have the same frequency as the IF signal. When capturing the received signal, the synchronization information of the signal is searched, and the known PN code is used to synchronize the received signal and PN code. Then, the local carrier frequency and code phase are almost identical to the received signal. After the capture is implemented, the signal will be tracked in the delay-locked loop (DDL), which can fully synchronize the local carrier and the PN code with the received signal, thus removing the carrier and PN code from the received signal and recovering the 1 PPS signal. Thus far, we have completed the signal demodulating and despreading and obtained the transmitting 1 PPS signal generated by station A [41]. However, due to the existence of bit error, it is necessary to detect the synchronization code in the 1 PPS signal to ensure that the correct synchronization point of the 1 PPS signal is generated [38]. Since it has been synchronized by PN code, the recovered 1 PPS signal needs to be input to TIC to compare with the local 1 PPS signal. Each rising edge of the local 1 PPS signal will be used as the opening point of the timing gate, and each corresponding rising edge of the recovered 1 PPS signal will be used as the closing point of the timing gate. In the TIC, the 10 MHz signal output by the local clock is used as the time base reference, measuring the time length between the opening and the closing each time, and then the TIC outputs the time difference as the observed value. Therefore, at both clock sites, the time signals are transmitted nominally at the same instant. Each clock site receives the signal from the other clock site, and its arrival time is measured. After exchanging the measured data, the time–frequency difference between the two clocks is calculated.

2.4. Time Difference Calculation in TWSTFT

The TWSTFT is based on the exchange of timing signals through geostationary telecommunication satellites. Any one of the clocks at site A and B generates a pulse signal every second (1 PPS signal). If the two clocks run at the same rate, the time difference between the two pulses holds the same at every second. Therefore, measuring the time difference between the two pulses, the running rate difference between the two clocks can be calculated. The task of TWSTFT technology is how to accurately compare the two time pulses from two stations.

It is assumed that the clocks of the two stations A and B have been synchronized in advance. At the appointed time, the clock C_A and C_B at A and B simultaneously generate the pulse signals P_A and P_B. A part of P_A enters TIC(A) as trigger gate open pulse. Another part of P_A is transmitted to station B. After a short time, the pulse P_B is coming and entering the TIC(A) as a pulse to trigger the gate close. The TIC(A) will calculate and record the time from gate opening to gate closing. The above process is repeated every second.

The TIC reading at station A is expressed as:

$$\tau_I^A = \tau^A - \tau^B + \tau_t^B + \tau_{pu}^B + \tau_{su}^B + \tau_s^B + \tau_{pd}^A + \tau_{sd}^A + \tau_r^A \tag{3}$$

and that at station B is expressed as:

$$\tau_I^B = \tau^B - \tau^A + \tau_t^A + \tau_{pu}^A + \tau_{su}^A + \tau_s^A + \tau_{pd}^B + \tau_{sd}^B + \tau_r^B \tag{4}$$

Combining Expressions (3) and (4), the time scale difference could be expressed as:

$$\tau^A - \tau^B = \tfrac{1}{2}[(\tau_I^A - \tau_I^B) + (\tau_s^A - \tau_s^B) - (\tau_{sd}^A - \tau_{su}^A) + (\tau_{sd}^B - \tau_{su}^B) + \left(\tau_{pu}^A - \tau_{pd}^A\right)$$
$$- \left(\tau_{pu}^B - \tau_{pd}^B\right) + (\tau_t^A - \tau_r^A) - (\tau_t^B - \tau_r^B)] \tag{5}$$

where τ^k denotes the local time-scale, and k means station k (k = A, B); τ_I^k is the TIC reading; τ_{su}^k and τ_{sd}^k are the Sagnac effect delay in the up-link and down-link, respectively; τ_{pu}^k and τ_{pd}^k the signal path up-link and down-link delays, respectively; τ_s^k is the satellite path delay through the transponder, τ_t^k the transmitter delay, and τ_r^k the receiver delay [42].

3. Error Analysis and Corrections

The error sources in TWSTFT observations mainly come from three aspects: (1) equipment delay errors; (2) signal-propagation-path-delay errors; and (3) Sagnac effect errors.

3.1. Equipment Delay Error

Equipment delay error mainly includes TIC measurement error, modem errors, satellite transparent transponder delay error, and the emitting and receiving delay error of Earth's station equipment. The TIC measurement error and modems error, which are caused by their own measurement accuracy errors, are about a dozens of picoseconds and 100 ps [42], respectively. The satellite transparent transponder delay error, mainly due to the signal from A to B and the signal from B to A through different forwarding channels of the transparent transponder, is hard to control, and is generally in the range of 100 ps [43]. The emitting and receiving equipment delay error of the ground station is around 0.2~0.5 ns, including cable delay, the transmission and receiving system error, and temperature variation.

It may not be accurate enough to measure the delay error of the signal after passing through each piece of equipment alone. Therefore, the zero-baseline measurement could be a better way to measure all equipment-delay errors. Zero-baseline measurement is when two atomic clocks are placed at the same place simultaneously, and then a TWSTFT experiment is conducted.

3.2. Propagation Path Delay Error

The signal's propagation path delay error is caused by the satellite's signal path delay errors of up-link and down-link, mainly including the delay propagation path geometry error and tropospheric delay error, ionospheric delay error, and delay errors of the different distances between two stations and the satellite.

The propagation path geometry delay is related to the coordinates of satellites and ground stations. For the signal's arriving and receiving time, if there is an error for ground stations or satellite position, the error will directly affect the time delay between the satellite and ground station. However, the clock difference calculation formula includes the difference between the up-link τ_{pu}^k and down-link path τ_{pd}^k; therefore, through the difference between two paths, one can eliminate the error with only a few picoseconds left. That means, at the same time, coordinate errors of satellite and ground stations do not affect the calculation of the clock difference.

Tropospheric delay is also called the tropospheric refraction error. The troposphere, through changing the propagation path of the signal, causes time delay. Many factors can affect tropospheric delay, such as ground climate, atmospheric pressure, temperature, and humidity TEC. Tropospheric delay can use the tropospheric delay model to correct; common models are the Hopfield model, Saastamoinen model, EGNOS model, etc. [44–46]. The differences between each model are mainly in the low-altitude angle; at the zenith direction, the difference is very small. Moreover, the up-link and down-link paths are symmetrical, hence, the tropospheric delay can be mostly cancelled. The remaining time delay is less than 10ps, and it can be neglected in the present study [42].

The ionospheric delay error is mainly caused by charged particles in the ionosphere. The charged particles can change the speed and path of the signal. The ionospheric delay depends on the total electron content (TEC) and the signal's frequency. Because the up-link and down-link signals at each station differ in carrier frequency, the following formula is used to correct the ionospheric delay error [47]:

$$\tau_{pu}^{k}\Big|_{\text{ion}} - \tau_{pd}^{k}\Big|_{\text{ion}} = \frac{40.28\,e_{ks}}{c}\left(\frac{1}{f_U^2} - \frac{1}{f_D^2}\right) \tag{6}$$

where f_u and f_D are the up-link and down-link frequencies of the signals, respectively; e_{ks} is the TEC along the signal-propagation path between ground station k and satellite S, and c is the speed of light in vacuum. Thus, we have the following equation to correct the ionospheric delay:

$$\frac{1}{2}\left[\tau_{pu}^{A}\Big|_{\text{ion}} - \tau_{pd}^{A}\Big|_{\text{ion}}\right] - \frac{1}{2}\left[\tau_{pu}^{B}\Big|_{\text{ion}} - \tau_{pd}^{B}\Big|_{\text{ion}}\right] = \frac{20.14(e_{As} - e_{Bs})}{c}\left(\frac{1}{f_U^2} - \frac{1}{f_D^2}\right) \tag{7}$$

Obviously, after the two-way signal transfer, the total effect of ionospheric delay could be further reduced. However, to better eliminate the error, we need to know the $\Delta e = e_{As} - e_{Bs}$. Usually, the global ionospheric TEC map provided by the IGS (International GNSS Service) every two hours can be used to eliminate or reduce the delay error. Taking the typical value of TEC, it can be estimated that the ionospheric delay is about 0.1 ns.

There is an error caused by the distances between the two stations and the satellite. The distance between the two stations and satellite is different, which means the signal arrival time from A to S and B to S is different. There is the problem that, when the signal of one station arrives at the satellite, another signal is on the way. Because the geostationary satellite is not strictly relative static with the Earth, the satellite's location is different between the first-arriving and the late-arriving. It will cause the signal path from A to B and B to A to be asymmetrical, seriously influencing the advantages of symmetrical paths of TWSTFT in eliminating errors. Approximately, when the distance converted from the longitude difference between the two ground stations and the satellite longitude is 300 km, the maximum error is about 30 ps. To reduce signal arrival time difference, we can slightly adjust the transmission delay to make them arrive at the satellite simultaneously. If the time difference between the two signals arriving at the satellite is less than 5 ms, the effect in TWSTFT will be less than 1 ps [43].

3.3. Sagnac Effect Error

When we transmitted the signal from the station to the satellite, the signal's route is not time-varying [45]. Since the satellite and ground stations are moving, this motion causes the signal-propagation change; the corresponding influence is referred to as the Sagnac effect [48]. After the two-way link difference, the error caused by the Sagnac effect is about 100-200 ns and needs further correction. The Sagnac effect correction for a one-way route from the ground station to satellite is given in a model that provides sufficient accuracy by the following expression [42,48,49]:

$$\tau_{sd}^{k} = \frac{\Omega}{c^2}R\left\{a\cos\left[\tan^{-1}\left(\tan\varphi^k - f\tan\varphi^k\right)\right] + H^k\cos\varphi^k\right\}\sin\left(\lambda^k - \lambda^s\right) \tag{8}$$

where Ω is the Earth's rotation rate, R is the distance from the satellite to the geocenter, a is Earth's equatorial average radius, f is the flattening of the Earth ellipsoid, H^k (k = A or B) is the height of the station above the ellipsoid, λ^s is the longitude of the satellite, and φ^k and λ^k are the latitude and longitude of the station, respectively.

After the model correction, only about a 10–100 picosecond error will be left. The reason for this residual is that, for the ground observer, the position of a geostationary satellite is not completely fixed. There will be a slight periodic movement with a daily period around a central point. As shown in Figure 4, in terrestrial reference frame, the position of satellite relative to station changes over time. That means the λ^s and R in Equation (8) will change over time. This leads to Sagnac effect error, which is no longer a constant but varies with the orbital period of the satellite in a Earth-fixed system, and its maximum peak to peak amplitude is hundreds of ps. In our study, the influence magnitude is 10–100 ps. At the current accuracy level, it has almost no effect on our experimental results, but it may be considered if higher accuracy is needed. Various error sources and their magnitudes of influence are listed in Table 1.

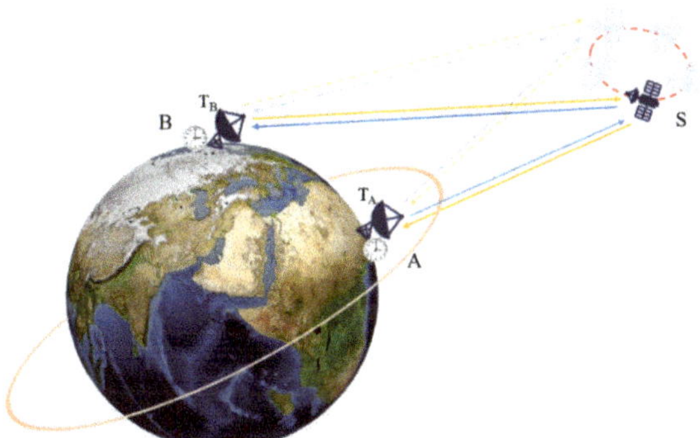

Figure 4. The influence of small periodic movement of satellite on TWSTFT. Two antennas T_A and T_B located, respectively, at station A and B, are connected with clocks. S is the geostationary satellite.

Table 1. Influences of various errors on TWSTFT.

Error Sources	Error Magnitude/ps (Two-Way)	Correction Model	Residual Error/ps
Time interval counter delay	10~100	Zero-baseline calibration	5
Modems delay	100	Zero-baseline calibration	10
Satellite transparent transponder delay	80	Zero-baseline calibration	10
Transmission and receiving system delay	200~500	Zero-baseline calibration	30
Propagation path geometry delay	<10	Neglected	<10
Tropospheric delay	10	Neglected	10
Ionospheric delay	100	Model correction	<10
Asymmetry of station and satellite position delay	30	Delay transmission compensation	<1
Sagnac effect delay	1~2×10^5	Model correction	10~100

4. Experiments and Data Processing

4.1. Experiments

To verify the feasibility of this method, we need clocks with high stability to maintain a stable frequency standard and a reliable time-transfer technique to measure the time difference between two positions. Therefore, we used two hydrogen atomic clocks in the experiment; a non-transportable clock C_A (H-MASER VCH-1003A) and a transportable clock C_B (H-MASER BM2101-01); both of their relative nominal frequency stabilities are 5×10^{-15} in one day. The TWSTFT was used as the time-transfer technique, one of the most accurate time-transfer methods than other GNSS-related techniques [50]. Time transfer by satellite does not have higher stability than fiber link, but it has better performance for long-distance time transfer.

We conducted the experiment in a building at the BIRMM, Beijing. The experiment we conducted was a geopotential difference measurement from 15 December 2016 to 27 December 2016 for a total of 13 days. During this process, C_B was moved to the fifth floor, while C_A was placed on the ground floor (see Figure 5a). The height difference between the fifth floor and ground floor is 22.8 m. After the experiment was conducted, we carried out a zero-baseline measurement from 27 December 2016 to 3 January 2017 for a total of seven days to calculate the clock drift and equipment error for calibration. As Figure 5b shows, clocks C_A and C_B were both placed on the ground floor. The equipment was temperature-stabilized and controlled during the experiment. At every site (the fifth floor or ground floor), through connection cables, one hydrogen clock, integrated control cabinet (ICC), and a satellite antenna were connected (see Figure 5).

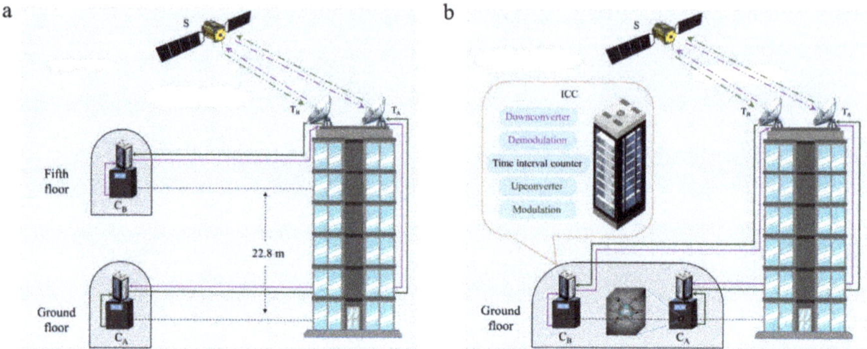

Figure 5. Time difference measurement. (**a**) Geopotential difference measurement. (**b**) Zero-baseline measurement. T_A and T_B are the antennas connected to clock C_A and clock C_B, respectively. ICC is integrated control cabinet, which includes modulation, upconvertor, downconvertor, demodulation, time interval counter, etc. S is the geostationary satellite used in the experiments.

4.2. Data Processing

In the experiments, the data we collected were the time difference between clock C_A and C_B corresponding to time. The sampling rate of data is 1 Hz. According to Equation (1), we just want to calculate $\frac{\Delta t_{AB}}{t}$, which is the slope of data, so it is insignificant whether the two clocks are synchronized in the beginning. What we are concerned with is the running rates of the two clocks.

There are some defects because of an equipment sampling error in the row data (Figure 6a,b). There are accidental data jumps, outliers, and some missing data. To improve the quality of the data, the following procedures are adopted:

(1) Through the data analysis with a large magnitude of change, we found that the data had some jumps, so we used fitting to restore them to the correct positions to ensure the continuity of all data.

(2) To avoid the influence of outliers on the calculation results, we adopted 3σ criterion (PauTa criterion) to identify and eliminate outliers. The occasional outliers might be due to the fact that during its propagation, the radio frequency signal will suffer from various influences, which causes its distortion. Therefore, in the demodulation process, the sampling decision device cannot accurately reproduce the original 1 PPS signal, leading to outliers.

(3) During continuous observation, some missing data may be caused by accidental failure of touch elements in TIC. We used linear interpolation to supplement the missing data.

(4) We used singular spectrum analysis (SSA) [51] to remove periodic terms. This allows better extraction of trend items.

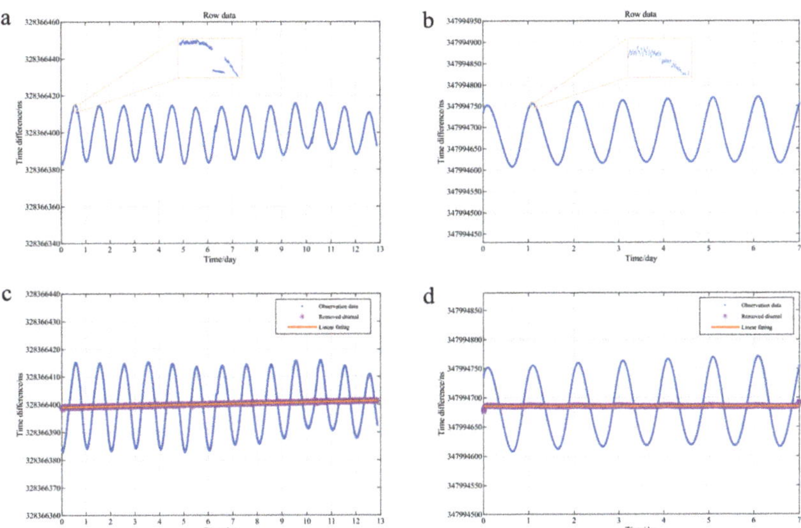

Figure 6. Comparison between two clocks. (**a**) Row data of two clocks on different floors. (**b**) Row data of two clocks on the same ground floor. (**c**) Residual data of two clocks on different floors after processing. (**d**) Residual data of two clocks on the same ground floor after processing.

After these procedures, we obtained the 'valid' data. Then, we corrected some errors in time transfer, as discussed in Section 3. In the experiment, we just want to calculate the time interval difference ($\frac{\Delta t_{AB}}{f}$) caused by geopotential difference. In other words, we do not need to consider the synchronization accuracy of time, and we only need to consider the change rate of the time difference between the two sites. Therefore, the errors we need to consider are the changes over time rather than the fixed values over time in equipment delay, ionospheric delay, and the Sagnac effect. Among equipment delays, the time-varying part is the frequency drift of the clock. The output frequency of the atomic clock has a linear drift with time, which can be found in the zero-baseline experiment (Figure 6c), and could be corrected by the Zero-baseline experiment. In the ionospheric delay correction model, the TEC value changes in real-time. However, because the TEC values on the signal path are all the same (two antennas in one building), the influences on the result can be ignored. The time delay caused by the Sagnac effect is a fixed value because both the variation of the Earth's rotation speed and satellite orbits is minimal, which can be considered as constants under the current experimental accuracy [48]. Hence, the time delay caused by the Sagnac effect will not influence the experimental result at the present accuracy requirement.

Then, we obtained the data with an apparent linear trend and used a linear function to fit the data. As shown in Figure 6c and d, the blue curves denote original observations, the purple curves denote the data after removing the diurnal terms from the original data noted as residual data, and orange lines are linear fittings of the purple curves. In Figure 6c, the data show relatively intense jumps in the geopotential measurement. The reason for the jumps is that the temperature-control equipment was not particularly stable, which influenced the accuracy of the observations. Nevertheless, on the whole, this is in good agreement with the linear fitting line. This trend mainly includes the frequency drift of the clock, equipment error, and the time interval difference caused by geopotential differences. Therefore, we only need to eliminate the frequency drift of the clock and equipment error by zero-baseline measurement to obtain the results. In Figure 6d, in the whole zero-baseline comparison observation period, the data are relatively smooth and have a strong trend, which is conducive to the calculation of results.

5. Results

After calculation, the slopes of the geopotential comparison experiment and zero-baseline experiment are $k_{geo} = 2.11639 \times 10^{-15}$ and $k_{zero} = -0.93617 \times 10^{-15}$, respectively. The slope of the zero-baseline measurement is the constant system shift; therefore, subtracting it from that of the geopotential comparison experiment could determine the difference of the clock running rates between C_A and C_B. To calculate the difference of the clock running rates ($\frac{\Delta t_{AB}}{T}$), we differenced the two slopes:

$$\frac{\Delta t_{AB}}{T} = k_{geo} - k_{zero} = 3.0526 \times 10^{-15} \tag{9}$$

where k_{remote} and k_{zero} denote the slopes of the geopotential comparison experiment and zero-baseline experiment, respectively.

Based on Formulas (1) and (2), the measured height difference is $\Delta H = \frac{\Delta t_{AB}}{T} \frac{c^2}{g} = 28.00$ m. The residual standard deviation of the regression equation is expressed as [52]:

$$S = \sqrt{\frac{\sum_{i=1}^{n}(y_i - \hat{y}_i)^2}{n - 2}} \tag{10}$$

where n is the total number of the sampling interval of the whole time series, y_i denotes the ith observation, and $\hat{y}_i$ denotes the ith fitting value. The uncertainty of the slope is given by [49]

$$u = \frac{S}{\sqrt{\sum_{i=1}^{n}(x_i - \bar{x})^2}} \tag{11}$$

where S is the residual standard deviation of the regression equation, x_i is the time of the ith observation, and $\bar{x}$ is the mean value of x.

Based on Formulas (10) and (11), we obtained the uncertainties of the slopes of the geopotential comparison experiment and zero-baseline experiment, written as $u_{geo} = 0.26 \times 10^{-15}$ and $u_{zero} = 0.52 \times 10^{-15}$, respectively. On the basis of the propagation law of errors, the uncertainty of the difference of the clock running rates ($\frac{\Delta t_{AB}}{T}$) is

$$u_D = \sqrt{u_{remote}^2 + u_{zero}^2} = 0.58 \times 10^{-15} \tag{12}$$

Finally, substituting the uncertainty values into Equations (1) and (2), we obtained the accuracy of the measurements, 5.4 m.

The relevant results are shown in Table 2.

Table 2. Results of geopotential comparison and zero-baseline comparison.

	Geopotential Comparison	Zero-Baseline
Slope	2.11639×10^{-15}	-0.93617×10^{-15}
The uncertainty of the slope	0.26×10^{-15}	0.52×10^{-15}
Measured height difference between A and B (m)	28.0 ± 5.4	
True value (m)	22.8	
Deviation (m)	5.2 ± 5.4	

6. Conclusions

The experimental results in period 1 (geopotential comparison measurement) provide an uncertainty of the slope 0.26×10^{-15}, which has better accuracy than period 2 (zero-baseline comparison measurement) 0.52×10^{-15}. The reason for this result may be that the observation time of period 1 is longer than period 2; a longer observation time is helpful to weaken the influence of observation noise on experimental results.

Based on TWSTFT observations, we determined the height difference between A and B as 28.0 ± 5.4 m. The bias between the measured value and the corresponding true value of 22.8 m is 5.2 m. Our current experimental accuracy is 5.4 m, which is not very high. The main reason is due to the accuracy limit of the hydrogen atomic clocks used, the stability of which is 5×10^{-15} and matches the final measurement accuracy. The current result is only exploratory research on this method, which may not be directly applied, but this result proves the feasibility of the method. Although the results reached expectations, there are still some problems worth further exploration. We cannot explain why there is strong periodicity in the observed data, but it is conjectured that these are connected with the atomic clock's performances and the relative motion of the satellite. The temperature could significantly influence the running rate of the atomic clock [53,54]. Due to the limitation of the experiment conditions, the ambient temperature of the atomic clock is not completely constant, and there is a certain fluctuation, which may be the reason for the jump in the observed data. In addition, the change of satellite orbit may also have an impact on the results. As described in Section 3.3, the satellite's orbit with a daily period, which is not fixed and changes every day.

Here, we employed two hydrogen MACs and TWSTFT technology to measure the height difference. The results of the experiment indicate that the accuracy of TWSTFT used in this experiment is sufficient for future applications. Our experimental results are preliminary. Further studies and experiments are needed to advance this research.

Author Contributions: Initiation and experiment design, W.S.; Conducting experiments, X.S., C.C., K.W., Z.S., and W.S.; Methodology, P.C. and W.S.; Software, P.C.; Data processing, P.C.; Validation, P.C., X.S., and C.C.; Formal analysis, P.C., W.S., and K.W.; Investigation, P.C. and W.S.; Original draft preparation, P.C.; Writing—review and editing, P.C., W.S., X.S., C.C., K.W. and Z.S.; Visualization, P.C.; Supervision, W.S.; Project administration, W.S. All authors have read and agreed to the published version of the manuscript.

Funding: This research was funded by National Natural Science Foundation of China (Grant nos. 41721003, 42030105, 41631072, 41804012, 41874023, 41974034), Space Station Project (2020)228 and Natural Science Foundation of Hubei Province (grant No. 2019CFB611).

Institutional Review Board Statement: Not applicable.

Informed Consent Statement: Not applicable.

Data Availability Statement: The data obtained in the study are available from the corresponding author upon reasonable request.

Acknowledgments: Thanks are given to the Beijing Institute of Radio Metrology and Measurement (BIRMM) for providing the experiment platform, and especially to the Frequency Measurement Laboratory of BIRMM for providing satellite two-way time transfer system. We would like to express our sincere thanks to three anonymous reviewers for their valuable comments and suggestions, which greatly improved the manuscript.

Conflicts of Interest: The authors declare no conflict of interest. The funders had no role in the design of the study; in the collection, analyses, or interpretation of data; in the writing of the manuscript, or in the decision to publish the results.

References

1. Einstein, A. *Die Feldgleichungen der Gravitation*; Sitzungsberichte der Preussischen Akademie der Wissenschaften: Berlin, Germany, 1915; pp. 844–847.
2. Takamoto, M.; Ushijima, I.; Ohmae, N.; Yahagi, T.; Kokado, K.; Shinkai, H.; Katori, H. Test of general relativity by a pair of transportable optical lattice clocks. *Nat. Photonics* **2020**, *14*, 411–415. [CrossRef]
3. Grotti, J.; Koller, S.; Vogt, S.; Häfner, S.; Sterr, U.; Lisdat, C.; Denker, H.; Voigt, C.; Timmen, L.; Rolland, A.; et al. Geodesy and metrology with a transportable optical clock. *Nat. Phys.* **2018**, *14*, 437–441. [CrossRef]
4. Bondarescu, R.; Bondarescu, M.; Hetényi, G.; Boschi, L.; Jetzer, P.; Balakrishna, J. Geophysical applicability of atomic clocks: Direct continental geoid mapping. *Geophys. J. Int.* **2012**, *191*, 78–82. [CrossRef]
5. Shen, W.B.; Ning, J.; Chao, D.; Liu, J. A proposal on the test of general relativity by clock transportation experiments. *Adv. Space Res.* **2009**, *43*, 164–166. [CrossRef]

6. Shen, W.; Chao, D.; Jin, B. On relativistic geoid. *Boll Geod. Sci. Affini.* **1993**, *52*, 207–216.
7. Bjerhammar, A. On a relativistic geodesy. *Bull. Géodésique* **1985**, *59*, 207–220. [CrossRef]
8. Chen, J.S.; Hankin, A.M.; Clements, E.R.; Chou, C.W.; Wineland, D.J.; Hume, D.B.; Leibrandt, D.R.; Brewer, S.M. An ^{27}Al$^+$ Quantum-Logic Clock with a Systematic Uncertainty below 10^{-18}. *Phys. Rev. Lett.* **2019**, *123*, 33201.
9. Poli, N.; Schioppo, M.; Vogt, S.; Falke, S.; Sterr, U.; Lisdat, C.; Tino, G.M. A transportable strontium optical lattice clock. *Appl. Phys. B* **2014**, *117*, 1107–1116. [CrossRef]
10. Koller, S.B.; Grotti, J.; Vogt, S.; Al-Masoudi, A.; Dörscher, S.; Häfner, S.; Sterr, U.; Lisdat, C. Transportable Optical Lattice Clock with 7×10^{-17} Uncertainty. *Phys. Rev. Lett.* **2017**, *118*, 73601. [CrossRef] [PubMed]
11. Rovera, G.D.; Abgrall, M.; Courde, C.; Exertier, P.; Fridelance, P.; Guillemot, P.; Laas-Bourez, M.; Martin, N.; Samain, E.; Sherwood, R.; et al. A direct comparison between two independently calibrated time transfer techniques: T2L2 and GPS Common-Views. *J. Physics. Conf. Ser.* **2016**, *723*, 12037. [CrossRef]
12. Siu, S.; Wang, J.-L.; Tseng, W.-H.; Liao, C.-S.; Hu, H.-F. Primary reference time clocks performance monitoring using GNSS common-view technique in telecommunication networks. In Proceedings of the 2016 18th Asia-Pacific Network Operations and Management Symposium (APNOMS), Kanazawa, Japan, 5–7 October 2016; pp. 1–4.
13. Hang, Y.; Hongbo, W.; Shengkang, Z.; Haifeng, W.; Fan, S.; Xueyun, W. Remote time and frequency transfer experiment based on BeiDou Common View. In Proceedings of the 2016 European Frequency and Time Forum (EFTF), York, UK, 4–7 April 2016; pp. 1–4.
14. Petit, G.; Defraigne, P. The performance of GPS time and frequency transfer: Comment on 'A detailed comparison of two continuous GPS carrier-phase time transfer techniques'. *Metrologia* **2016**, *53*, 1003–1008. [CrossRef]
15. Lee, S.W.; Schutz, B.E.; Lee, C.; Yang, S.H. A study on the Common-View and All-in-View GPS time transfer using carrier-phase measurements. *Metrologia* **2008**, *45*, 156–167. [CrossRef]
16. Kirchner, D. Two-way time transfer via communication satellites. *Proc. IEEE* **1991**, *79*, 983–990. [CrossRef]
17. Steele, J.M.; Markowitz, W.; Lidback, C.A. Telstar Time Synchronization. *IEEE Trans. Instrum. Meas.* **1964**, *IM-13*, 164–170. [CrossRef]
18. Jiang, Z.; Konaté, H.; Lewandowski, W. Review and preview of two-way time transfer for UTC generation—From TWSTFT to TWOTFT. In Proceedings of the 2013 Joint European Frequency and Time Forum & International Frequency Control Symposium (EFTF/IFC), Prague, Czech Republic, 21–25 July 2013; pp. 501–504.
19. Bauch, A. Time and frequency comparisons using radiofrequency signals from satellites. *Comptes Rendus Phys.* **2015**, *16*, 471–479. [CrossRef]
20. Hanson, D.W. Fundamentals of two-way time transfers by satellite. In Proceedings of the 43rd Annual Symposium on Frequency Control, Denver, CO, USA, 31 May–2 June 1989; pp. 174–178.
21. Imae, M.; Hosokawa, M.; Imamura, K.; Yukawa, H.; Shibuya, Y.; Kurihara, N.; Fisk, P.; Lawn, M.A.; Li, Z.; Li, H. Two-way satellite time and frequency transfer networks in Pacific Rim region. *IEEE Trans. Instrum. Meas.* **2002**, *50*, 559–562. [CrossRef]
22. Fujieda, M.; Gotoh, T.; Nakagawa, F.; Tabuchi, R.; Aida, M.; Amagai, J. Carrier-phase-based two-way satellite time and frequency transfer. *IEEE Trans. Ultrason. Ferroelectr. Freq. Control* **2012**, *59*, 2625–2630. [CrossRef] [PubMed]
23. Jing, W.; Wang, J.; Zhao, D.; Lu, X.; Wu, J. A measurement method of the GEO satellite local oscillator error. In Proceedings of the 2013 Joint European Frequency and Time Forum & International Frequency Control Symposium (EFTF/IFC), Prague, Czech Republic, 21–25 July 2013; pp. 339–342.
24. Fujieda, M.; Yang, S.; Gotoh, T.; Hwang, S.; Hachisu, H.; Kim, H.; Lee, Y.K.; Tabuchi, R.; Ido, T.; Lee, W.; et al. Advanced Satellite-Based Frequency Transfer at the 10^{-16} Level. *IEEE Trans. Ultrason. Ferroelectr. Freq. Control* **2018**, *65*, 973–978. [CrossRef]
25. Riedel, F.; Al-Masoudi, A.; Benkler, E.; Dörscher, S.; Gerginov, V.; Grebing, C.; Häfner, S.; Huntemann, N.; Lipphardt, B.; Lisdat, C.; et al. Direct comparisons of European primary and secondary frequency standards via satellite techniques. *Metrologia* **2020**, *57*, 45005. [CrossRef]
26. Wu, K.; Shen, Z.; Shen, W.B.; Sun, X.; Wu, Y. A preliminary experiment of determining the geopotential difference using two hydrogen atomic clocks and TWSTFT technique. *Geod. Geodyn.* **2020**, *11*, 229–241. [CrossRef]
27. Shen, W.; Sun, X.; Cai, C.; Wu, K.; Shen, Z. Geopotential determination based on a direct clock comparison using two-way satellite time and frequency transfer. *Terr. Atmos. Ocean. Sci.* **2019**, *30*, 21–31. [CrossRef]
28. Weinberg, S. Gravitation and cosmology: Principles and Applications of the General Theory of Relativity. *Am. J. Phys.* **1972**, *41*, 598–599. [CrossRef]
29. Pavlis, N.; Weiss, M. The relativistic redshift with 3×10^{-17} uncertainty at NIST, Boulder, Colorado, USA. *Metrologia* **2003**, *40*, 66. [CrossRef]
30. Dittus, H.; Lämmerzahl, C.; Peters, A.; Schiller, S. OPTIS—A Satellite test of Special and General Relativity. *Adv. Space Res.* **2007**, *39*, 230–235. [CrossRef]
31. Vessot, R.; Levine, M.; Mattison, E.; Blomberg, E.; Hoffman, T.; Nystrom, G.; Farrel, B.; Decher, R.; Eby, P.; Baugher, C. Test of Relativistic Gravitation with a Space-Borne Hydrogen Maser. *Phys. Rev. Lett.* **1981**, *45*, 2081–2084. [CrossRef]
32. Shen, Z.; Shen, W.; Peng, Z.; Liu, T.; Zhang, S.; Chao, D. Formulation of Determining the Gravity Potential Difference Using Ultra-High Precise Clocks via Optical Fiber Frequency Transfer Technique. *J. Earth Sci. China* **2018**, *30*, 422–428. [CrossRef]
33. Heiskanen, W.A.; Moritz, H. *Physical Geodesy*; Freeman and Company: San Francisco, CA, USA, 1967.

34. Saburi, Y.; Yamamoto, M.; Harada, K. High-precision time comparison via satellite and observed discrepancy of synchronization. *IEEE Trans. Instrum. Meas.* **1976**, *IM-25*, 473–477. [CrossRef]
35. Asami, K. An Algorithm to Improve the Performance of M-Channel Time-Interleaved A-D Converters. *IEICE Trans. Fundam. Electron. Commun. Comput. Sci.* **2007**, *E90-A*, 2846–2852. [CrossRef]
36. Ridders, C. Accurate determination of threshold voltage levels of Schmitt trigger. *IEEE Trans. Circuits Syst.* **1985**, *32*, 969–970. [CrossRef]
37. Dinan, E.H.; Jabbari, B. Spreading codes for direct sequence CDMA and wideband CDMA cellular networks. *IEEE Commun. Mag.* **1998**, *36*, 48–54. [CrossRef]
38. Celano, T.P.; Francis, S.P.; Gifford, G.A. Continuous satellite two-way time transfer using commercial modems. In Proceedings of the 2000 IEEE/EIA International Frequency Control Symposium and Exhibition (Cat. No.00CH37052), Kansas City, MO, USA, 9 June 2000; pp. 607–611.
39. Barrow, B.B. Error Probabilities for Telegraph Signals Transmitted on a Fading FM Carrier. *Proc. IRE* **1960**, *48*, 1613–1629. [CrossRef]
40. Piester, D.; Bauch, A.; Becker, J.; Staliuniene, E.; Schlunegger, C. On measurement noise in the European TWSTFT network. *IEEE Trans. Ultrason. Ferroelectr. Freq. Control* **2008**, *55*, 1906–1912. [CrossRef] [PubMed]
41. Goldsmith, A. *Wireless Communications*; Cambridge University Press: Cambridge, UK, 2005.
42. ITU-R. The Operational Use of Two-Way Satellite Time and Frequency Transfer Employing Pseudorandom Noise Codes. TF Series: Time Signals and Frequency Standards Emissions, Recommendation ITU-R TF.1153- 3 (08/2015), International Telcomunication Union. 2015. Available online: https://www.itu.int/rec/R-REC-TF.1153-4-201508-I/en (accessed on 5 September 2021).
43. Lin, H.T.; Huang, Y.J.; Tseng, W.H.; Liao, C.S.; Chu, F.D. The TWSTFT links circling the world. In Proceedings of the 2014 IEEE International Frequency Control Symposium (FCS), Taipei, Taiwan, 19–22 May 2014; pp. 1–4.
44. Hopfield, H.S. Tropospheric effect on electromagnetically measured range: Prediction from surface weather data. *Radio Sci.* **1971**, *6*, 357–367. [CrossRef]
45. Sastamoinen, J. Atmospheric correction for troposphere and stratosphere in radio ranging of satellites, in the Use of Artifical Satellites for Geodesy. *Geophys. Monogr. Ser.* **1972**, *15*, 247–251.
46. Penna, N.; Dodson, A.; Chen, W. Assessment of EGNOS Tropospheric Correction Model. *J. Navigation.* **2001**, *54*, 37–55. [CrossRef]
47. Rose, J.A.R.; Watson, R.J.; Allain, D.J.; Mitchel, N.C. Ionospheric corrections for GPS time transfer. *Radio Sci.* **2014**, *49*, 196–206. [CrossRef]
48. Tseng, W.; Feng, K.; Lin, S.; Lin, H.; Huang, Y.; Liao, C. Sagnac Effect and Diurnal Correction on Two-Way Satellite Time Transfer. *IEEE Trans. Instrum. Meas.* **2011**, *60*, 2298–2303. [CrossRef]
49. Bidikar, B.; Rao, G.; Laveti, G. Sagnac Effect and SET Error Based Pseudorange Modeling for GPS Applications. *Procedia Comput. Sci.* **2016**, *87*, 172–177. [CrossRef]
50. Hackman, C.; Levine, J. A Long-Term Comparison of GPS Carrierphase Frequency Transfer and Two-Way Satellite Time/Frequency Transfer. In Proceedings of the 38th Annual Precise Time and Time Interval (PTTI) Meeting, Reston, VA, USA, 7–9 December 2006; pp. 1–15.
51. Vautard, R.; Yiou, P.; Ghil, M. Singular-spectrum analysis: A toolkit for short, noisy chaotic signals. *Phys. D Nonlinear Phenom.* **1992**, *58*, 95–126. [CrossRef]
52. Hocking, R. Methods and Applications of Linear Models. In *Regression and the Analysis of Variance*; John Wiley & Sons, Inc.: Hoboken, NJ, USA, 2005.
53. Ascarrunz, F.G.; Jefferts, S.R.; Parker, T.E. Environmental effects on errors in two-way time transfer. In Proceedings of the 20th Biennial Conference on Precision Electromagnetic Measurements, Braunschweig, Germany, 17–21 June 1996; pp. 518–519.
54. Ascarrunz, F.G.; Jefferts, S.R.; Parker, T.E. Earth station errors in two-way time and frequency transfer. *IEEE Trans. Instrum. Meas.* **1997**, *46*, 205–208. [CrossRef]

 *remote sensing*

Communication

Preliminary Analysis and Evaluation of BDS-3 RDSS Timing Performance

Rui Guo [1,*], Dongxia Wang [1], Nan Xing [2], Zhijun Liu [1], Tianqiao Zhang [1], Hui Ren [1] and Shuai Liu [1]

[1] Beijing Satellite Navigation Center, Beijing 100094, China; wdx2008abc@163.com (D.W.); liuzhijun1111@163.com (Z.L.); zhangtqee@aliyun.com (T.Z.); m18601961081@163.com (H.R.); liushuai0810@sina.com (S.L.)

[2] Department of Astronomy, Beijing Normal University, Beijing 100875, China; xingnan@bnu.edu.cn

* Correspondence: shimbarsalon@163.com

Abstract: Radio determination satellite service (RDSS) is one of the characteristic services of Beidou navigation satellite system (BDS), and also distinguishes with other GNSS systems. BDS-3 RDSS adopts new signals, which is compatible with BDS-2 RDSS signals in order to guarantee the services of old users. Moreover, the new signals also separate civil signals and military signals which are modulated on different carriers to improve their isolation and RDSS service performance. Timing is an important part of RDSS service, which has been widely used in the field of the power, transportation, marine and others. Therefore, the timing accuracy, availability and continuity is an important guarantee for RDSS service. This paper summarizes the principle of one-way and two-way timing, and provides the evaluation method of RDSS timing accuracy, availability and continuity. Based on BDS-3 RDSS signal measurements of system, the performance of one-way timing and two-way timing is analyzed and evaluated for the first time. The results show that: (1) the accuracy of one-way timing and two-way timing is better than 30 ns and 8 ns respectively, which are better than the official claimed accuracy; (2) the RMS of one-way timing accuracy is 5.45 ns, which is 20% smaller than BDS-2, and the availability and continuity are 100%; (3) the RMS of two-way timing accuracy is 3.59 ns, which is 34% smaller than one-way timing, and both of the availability and continuity are 100%; (4) the orbit maneuver of GEO satellite make the one-way timing has 7.68 h recovery, but has no affection on the two-way timing.

Keywords: BDS-3; RDSS; one-way timing; two-way timing; accuracy; availability; continuity

Citation: Guo, R.; Wang, D.; Xing, N.; Liu, Z.; Zhang, T.; Ren, H.; Liu, S. Preliminary Analysis and Evaluation of BDS-3 RDSS Timing Performance. *Remote Sens.* **2022**, *14*, 352. https://doi.org/10.3390/rs14020352

Academic Editor: Jose Moreno

Received: 20 November 2021
Accepted: 8 January 2022
Published: 13 January 2022

Publisher's Note: MDPI stays neutral with regard to jurisdictional claims in published maps and institutional affiliations.

1. Introduction

According to the three-step development strategy [1,2], the Beidou navigation satellite system (BDS) has experienced BDS-1 (1990–2003), BDS-2 (2003–2012) and BDS-3 (2016–2020). BDS-3 has started to run formally for the whole world on 31 July 2020, providing the Position, Navigation and Timing service (PNT), Radio Determination Satellite Service (RDSS), Global Short Message Communication (GSMC), Satellite-Based Augmentation Service (SBAS), Precise Point Positioning (PPP), Search and Rescue (SAR), and so on. Among them, RDSS is an important component and a distinctive feature of BDS that has advantages and a competitiveness which are different from Global Positioning Systems (GPS), Global Navigation Satellite Systems (GLONASS) and the Galileo system [3–5].

BDS-2 is composed of fourteen satellites and has provided services since 2012. Among the satellites are five geosynchronous earth orbit (GEO) satellites, five inclined geosynchronous orbit (IGSO) satellites and four medium earth orbit (MEO) satellites. Five GEO satellites can broadcast RDSS signals, and each satellite has two wide beams. BDS-2 RDSS can provide a timing service. The official claimed accuracy of BDS-2 RDSS is 50 ns for the one-way timing service, and 20 ns for the two-way timing service [6,7].

BDS-3 is composed of thirty satellites and has provided services since 2020. Among the satellites are three GEO satellites, three IGSO satellites and twenty-four MEO satellites.

Three GEO satellites can broadcast RDSS signals. In order to improve the capacity of the system, BDS-3 RDSS adopts narrow beams, and each satellite has six narrow beams to provide inbound and outbound links. At the same time, BDS-3 RDSS separates civil signals and military signals, which are modulated on different carriers to improve their isolation, and to realize the security and compatibility. The official claimed accuracy of BDS-3 RDSS is 50 ns for the one-way timing service, and 10 ns for the two-way timing service [6,7].

As a spatial information infrastructure, the availability and continuity of satellite navigation system services are important. Up to now, there are rare existing studies on BDS-3 RDSS timing performances. Based on BDS-3 RDSS signal measurements, this paper analyzed and evaluated the performance of the one-way timing service and two-way timing service for the first time, which provides technical support for RDSS timing application.

2. Materials and Methods

2.1. One-Way Timing Principle

This section describes the basic principle of the RDSS one-way timing service. In the one-way timing process, the user starts a timer at the beginning of a second (the local 1-pps). At the same time, it receives the n-th frame of the outbound signal (which carries a time stamp) transmitted by the central control system (CCS) through the satellite, and uses the time stamp of the n-th frame to stop the timer. Then, the timer measures the time interval Δ_1 [8,9]. The schematic diagram of the one-way timing principle is illustrated in Figure 1.

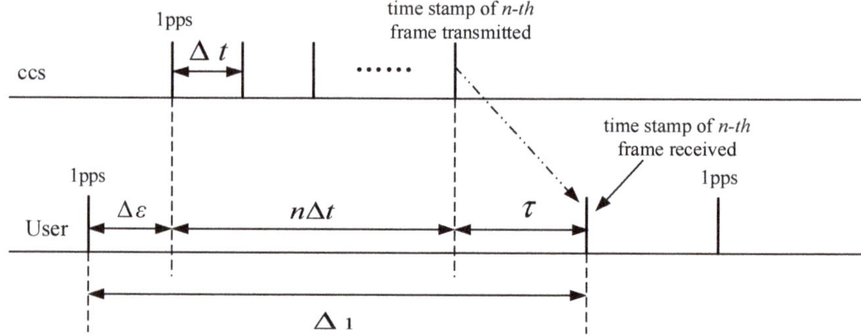

Figure 1. The one-way timing principle.

The clock difference between the user and CCS, denoted by $\Delta\varepsilon$, can be obtained as:

$$\Delta\varepsilon = \Delta_1 - \tau - n\Delta t \tag{1}$$

where Δt = 125 ms is the frame period according to the interface control document (ICD) of the BDS-3 RDSS outbound signal, and τ is the forward propagation delay of the signal, which can be calculated as:

$$\tau = t_{equ} + t_{R01} + t_{R\tau1} \tag{2}$$

where t_{equ} is the one-way time delay of the equipment, which is stored in the user. $t_{R01} = R_{01}/c + t_{R01-pro}$ is the propagation delay from CCS to the satellite; $R01$ is the geometrical distance between CCS and the satellite; c is the light speed; $t_{R01} = R_{01}/c + t_{R01-pro}$ is the ionosphere and troposphere delay, which can be accurately calculated using the parameters broadcast by CCS. $t_{R\tau1} = R_{\tau1}/c + t_{R\tau1-pro}$ is the propagation delay from the satellite to the user; $R_{\tau1}$ is the geometrical distance between the satellite and the user; $t_{R\tau1} = R_{\tau1}/c + t_{R\tau1-pro}$ is the ionosphere and troposphere delay, which is calculated by the user according to the satellite position, user position and broadcast correction parameters [8,9].

From the above Equations (1) and (2), it can be seen that the main factors affecting the accuracy of RDSS one-way timing include GEO satellites' ephemeris errors, ionosphere and troposphere delay correction errors, various equipment delay errors, user position errors, and so on.

2.2. Two-Way Timing Principle

This section describes the basic principle of the RDSS two-way timing service. In a two-way timing process, the user also starts and stops a timer at the start of the local second and at the time stamp of the n-th frame received. What differs is that the forward propagation delay is calculated by CCS and is transmitted to the user. As shown in Figure 2, when the user receives the n-th frame, it sends an inbound signal to CCS immediately. CCS receives the inbound signal, measures the interval Δ_2, and calculates the forward propagation delay τ [8–10], based on Equation (3).

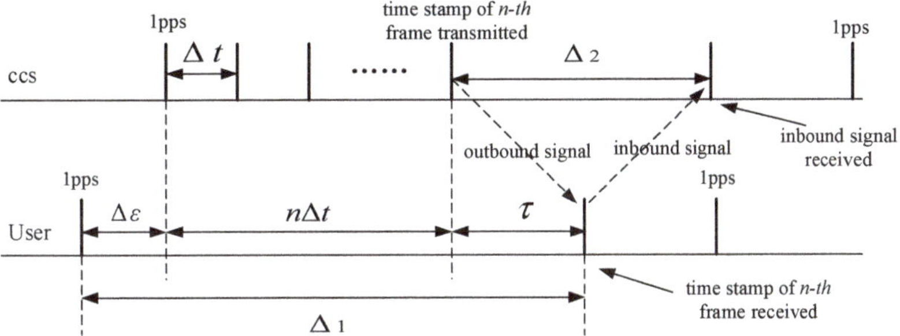

Figure 2. The two-way timing principle.

Assuming that the signal drift of satellite in the process is ignored, the forward propagation delay can be obtained as:

$$\tau = t_{equ\_one-way} + \left(\Delta_2 - t_{equ\_two-way} - t_+ - t_- \right)/2 \tag{3}$$

where $t_{equ\_one-way}$ and $t_{equ\_two-way}$ refer to the equipment time delay of the one-way and two-way (including the time delay of CCS, satellite transmitters and users), Δ_2 refers to the signal round-trip time measured by CCS, t_+ refers to the ionosphere and troposphere delay from CCS to the user through the i-th satellite and t_- refers to the ionosphere and troposphere delay from the user to CCS [8–10].

From the above Equation (3), it can be seen that the main factors affecting the accuracy of RDSS two-way timing include ionosphere and troposphere delay correction errors, various equipment delay errors, and so on. Therefore, the two-way timing accuracy is higher than the one-way timing accuracy. It is worth noting that the signal drift of the satellite in the process is ignored when the forward propagation delay is calculated in Equation (3). However, the drift is related to the speed of motion, which means that periodic fluctuations can also appear in the two-way timing.

3. Results

3.1. Analysis and Evaluation Method

The performance index of BDS-3 RDSS timing service includes timing accuracy, availability and continuity. The analysis methods of each index are described briefly below.

3.1.1. Timing Accuracy

An RDSS receiver, of which its position is known precisely, is used to receive the outbound RDSS signal and then calculate the one-way timing accuracy. Meanwhile, the same receiver is used to send inbound two-way timing signal. CCS will calculate the timing accuracy and compare the results with the known reference 1 pps, which provide a time-frequency reference for the receivers, and measure the timing error of each epoch. The mean value, standard deviation and root mean square (RMS) can be obtained as [8]:

$$\bar{\tau} = \frac{1}{n} \sum_{i=1}^{n} \tau_i \tag{4}$$

$$\text{STD} = \sqrt{\frac{1}{n-1} \sum_{i=1}^{n} |\tau_i - \bar{\tau}|^2} \tag{5}$$

$$\tau_{RMS} = \sqrt{\frac{1}{n} \sum_{i=1}^{n} |\tau_i|^2} \tag{6}$$

where τ_i is the discrete point of timing accuracy, n is the number of sample points, $\bar{\tau}$ is the mean value which reflects the constant deviation of timing results, STD is the standard deviation, which reflects the stability and variation range of timing results, τ_{RMS} is RMS, which is the comprehensive result of mean value and standard deviation, and reflects comprehensively the deviation and stochastic error of timing results.

3.1.2. Availability

The availability of the timing service is the time percentage of when the timing accuracy meets the requirements, during the specified period, conditions and service area [6]. For BDS-3 RDSS one-way timing and two-way timing, the accuracy requirements are 50 ns and 10 ns respectively.

Taking the timing error sequence as the evaluation object, the available time percentage of timing service is calculated as:

$$Ava_l = \frac{\sum\limits_{t=t_{start},inc=T}^{t_{end}} bool\{EPEk \leq f_{Acc}\}}{1 + \frac{t_{end} - t_{start}}{T}} \tag{7}$$

where t_{start} and t_{end} is the start time and end time respectively, T is the sampling interval, $EPEk$ is the timing error at a certain time, f_{Acc} is the timing accuracy threshold and $bool\{*\}$ is the Boolean function.

3.1.3. Continuity

RDSS messages are broadcast very second, and the continuity is evaluated without considering the message type. If the receiver receives a message at the current time but it is lost at the next time, it would be recorded as an interrupt, and the interruption duration is the time that receivers cannot receive the message [6].

In order to ensure that the statistical results reflected the messages continuity strictly, three different receivers were used in the experiment. This indicates that an interruption happened only when all three receivers lost the message at the time.

It should be known that the continuity shoud exclude the planned interrupted exception time, which will be published with the warning announcement on the BDS official website, subsequently.

3.2. One-Way Timing Performance Analysis

In order to evaluate the performance of BDS-3 RDSS one-way timing, the following analyses were conducted by the real measured data from the Beijing RDSS receiver. The assessment scenarios included two different states: normal state and satellite orbit maneuver.

3.2.1. Normal State

From 14 February 2021 to 16 February 2021, the receiver collected data from beam two of the BDS-3 C59 satellite. During this period, the satellite was in the normal state, without satellite orbit maneuver or other abnormal status. In this evaluation, the receiver collected timing results for 3 days successively with a 1 s sampling interval, and the one-way timing accuracy curve can be seen in Figure 3.

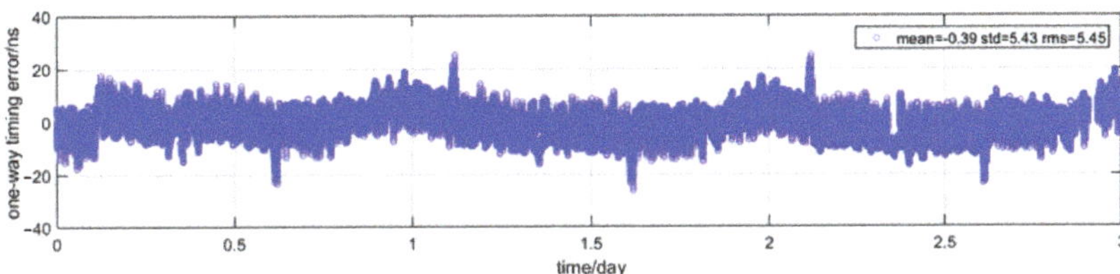

Figure 3. RDSS one-way time service error (normal state).

It can be seen from Figure 3 that: (1) the one-way timing error of all epoch points is less than 30 ns, and the RMS is 5.45 ns, which is 20% smaller than that of the BDS-2 one-way timing error of 6.81 ns [11]; (2) The mean value of one-way timing error is 0.39 ns, which means that there is no significant deviation between the reference 1 pps and the timing results, the equipment delay was calibrated accurately and both ionosphere and troposphere delays were corrected properly; (3) The standard deviation of one-way timing error is 5.43 ns, which actually reflects the white noise in the signal. Moreover, the orbital errors, ionosphere and troposphere delay residual errors always appear as bias.

Furthermore, Figure 3 shows that one-way timing results had a period of 1 day, of which the characteristic was related to the GEO satellite orbit characteristic. In order to further analyze the characteristics of one-way timing error, the spectrum of the one-way timing is shown in Figure 4 and the error removing orbit period is shown in Figure 5.

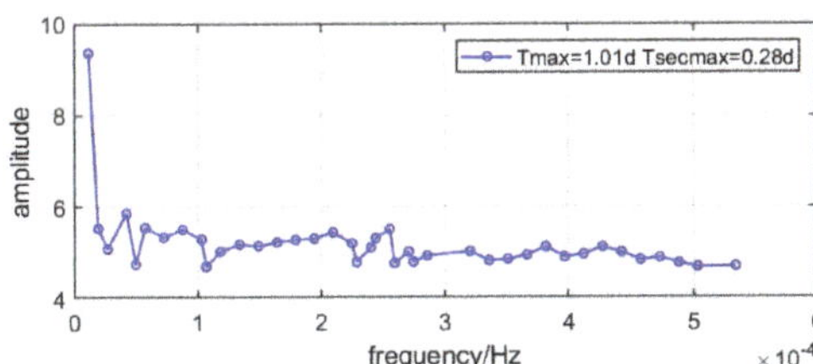

Figure 4. Spectrum analysis results of RDSS time service error.

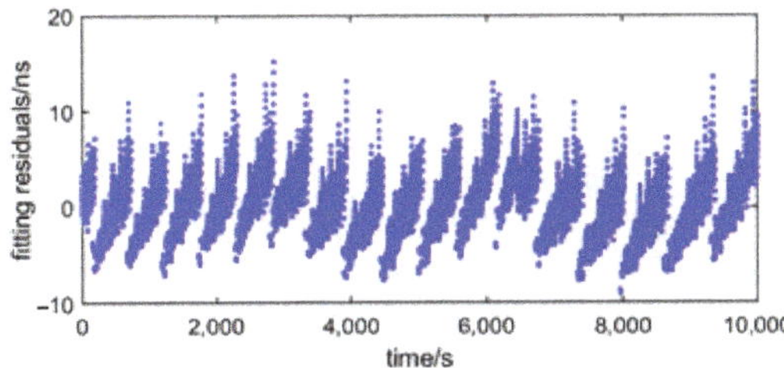

Figure 5. RDSS one-way timing error (remove orbit period).

It can be seen from Figure 4 that the maximum spectrum is 1.01 days (about 24 h 14 m 24 s), which is basically consistent with the orbital period of the GEO Satellite, of 23 h 56 m, and the reason for the inconsistency may be the observation data size and the measurement error. When compared with Figure 3, It can be seen from Figure 5 that the characteristics of the GEO satellite orbital period have been eliminated. However, there is still a periodic jump of the timing error, of which its magnitude is about 10 ns. This phenomenon is related to the RDSS timing principle and the receiver's algorithm. According to the Interface Control Document (ICD), GEO satellite updates the ephemeris every 6 s. Receivers need to interpolate the ephemeris to obtain the satellite position at the timing epoch, and then calculate the timing results based on the fixed point. The interpolation algorithm is the main factor causing the timing error jump.

In order to further evaluate the performance under normal state, the availability and continuity results are given in Table 1. The availability and continuity of the RDSS one-way timing service are 100%.

Table 1. The availability and continuity of one-way timing service (normal state).

Time	14 February 2021	15 February 2021	16 February 2021
Availability	100%	100%	100%
Continuity	100%	100%	100%

3.2.2. Orbit Maneuver State

The ephemeris accuracy of GEO satellites is lower than that of MEO and IGSO satellites due to the orbital characteristics. The orbit of the GEO satellite needs to be adjusted and maintained regularly [12,13]. One-way timing accuracy is mainly limited by orbit determination accuracy during GEO satellite orbit maneuver, because it is related to satellite orbital error. Therefore, the data of one-way timing services were used for analysis and evaluated the orbit maneuver state and recovery of the C59 satellite from 8 February 2021 to 10 February 2021, where the maneuver period was from 11:30 to 13:30 on 8 February 2021.

It can be seen from Figure 6 that: (1) during the orbit maneuver state, the accuracy recovery time of one-way timing is 7.68 h. Specifically, from the start of orbit maneuver, the timing error is within 100 ns about 2.18 h, then the timing error increases to 1000 ns at about 5.08 h and then the timing error gradually decreases and finally returns to the level of 50 ns about 0.52 h. (2) Affected by orbit maneuver, the RMS of the RDSS one-way timing error is 15.33 ns, and its standard deviation is 15.31 ns. This is because in the process of RDSS one-way timing, receivers extrapolate the orbit data according to the ephemeris broadcast by satellites, which is greatly affected by the ephemeris errors of GEO satellites.

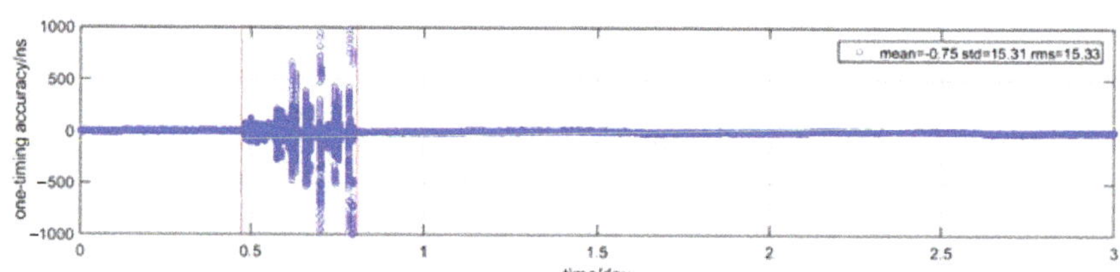

Figure 6. BDS-3 RDSS one-way timing accuracy (orbit maneuver state).

In order to further evaluate the orbit maneuver, the timing accuracy results are given in Table 2. The RMS of normal 8 February, and the recovery on 9 and 10 February are 5.52 ns, 5.49 ns and 5.46 ns, respectively.

Table 2. The accuracy of the RDSS one-way timing service (orbit maneuver state, RMS, unit: ns).

Time	14 February 2021		9 February 2021	10 February 2021
	Normal	Maneuver		
Timing accuracy	5.52	23.21	5.49	5.46

3.3. Two-Way Timing Performance Analysis

Similar to the one-way timing, the performance analysis of two-way timing also used the signal measurements from the Beijing RDSS receiver [13]. The assessment scenarios included two different states: normal state and satellite orbit maneuver.

3.3.1. Normal State

From February 14 to 16, 2021, the receiver collected data from beam two of the BDS-3 C59 satellite. During this period, the satellite was in the normal state, without satellite orbit maneuver or other abnormal status. In this evaluation, the receiver adopted timing results for 3 days successively, with z 1 min sampling interval, and the one-way timing accuracy curve can be seen in Figure 7.

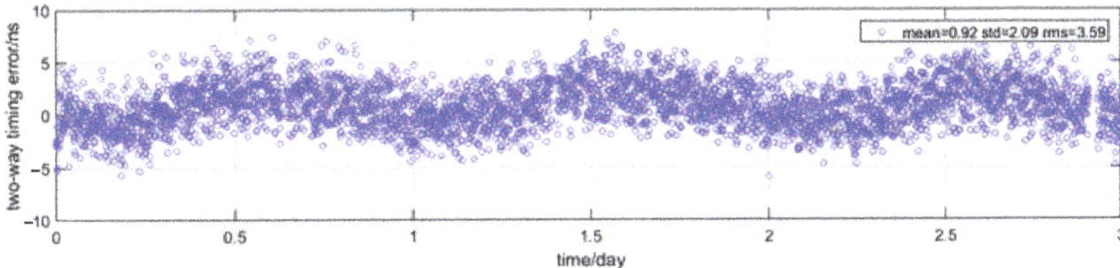

Figure 7. RDSS two-way timing error (normal state).

It can be seen from Figure 7 that: (1) the two-way timing error of all epochs is less than 8 ns, and the RMS is 3.59 ns, which is 34% smaller than that of the one-way timing error of 5.45 ns; (2) the mean value of the two-way timing error is 0.92 ns, which is not much larger than that of the one-way timing of 0.39 ns. This indicates that the equipment delay was calibrated accurately, and both the ionosphere and troposphere delays were corrected properly; (3) The standard deviation is 2.09 ns, which is reduced by 62% when compared with the one-way timing, indicating that GEO satellites orbital errors, ionosphere and troposphere delay errors are greatly reduced in the two-way timing process.

Furthermore, Figure 7 shows that two-way timing results also have a period which is connected to ionosphere and troposphere parameters, which are related to the sun zenith angle, since the GEO satellite is located in geostationary orbit. The spectrum of the two-way timing is shown in Figure 8 and the error removing the orbit period is shown in Figure 9.

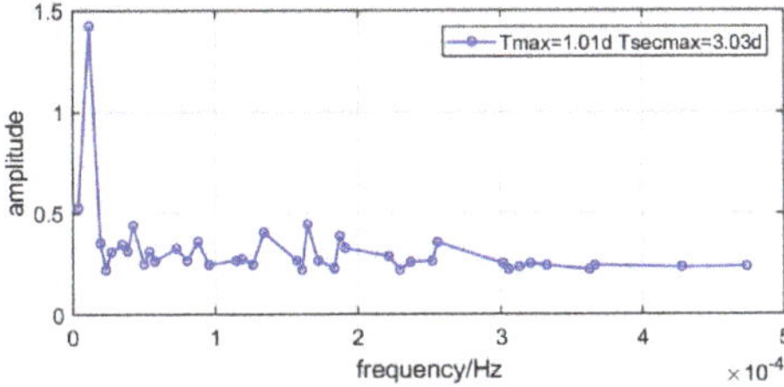

Figure 8. Spectrum analysis results of RDSS timing error.

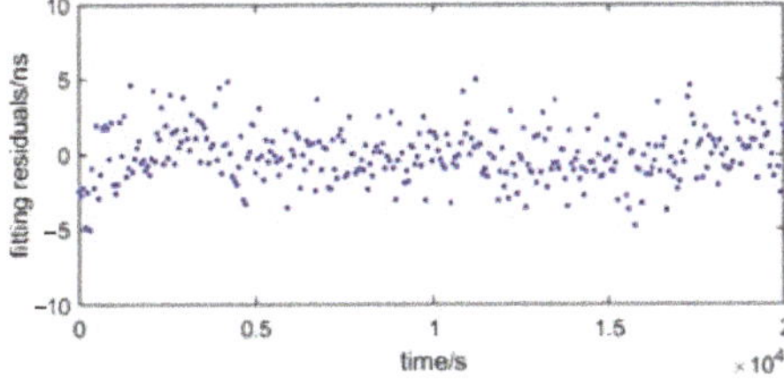

Figure 9. RDSS two-way timing error (remove orbit period).

It can be seen from Figure 8 the maximum spectrum is 1.01 days (about 24 h 14 m 24 s), which is consistent with the analysis results of the one-way timing error series, indicating that the two-way timing is still affected by the GEO satellite orbital error. It is worth noting that the drift is related to the speed of motion, which means that periodic fluctuations can also appear in the two-way timing. When compared with Figure 7, it can be seen from Figure 9 that the characteristics of the GEO satellite orbit period have been eliminated, and there is a 5 ns timing error, which is a comprehensive reflection of the ionosphere and troposphere correction residual errors.

In order to further evaluate the performance under normal state, the availability and continuity results are given in Table 3. The availability and continuity of the RDSS two-way timing service are 100%.

Table 3. The availability and continuity of two-way timing service (normal state).

Time	14 February 2021	15 February 2021	16 February 2021
Availability	100%	100%	100%
Continuity	100%	100%	100%

3.3.2. Orbit Maneuver State

In order to comprehensively analyze the performance of BDS-3 two-way timing, we also analyzed and evaluated the two-way timing results under the orbit maneuvering state. The data was also from 8 February 2021 to 10 February 2021, where the maneuver period was from 11:30 to 13:30 on 8 February 2021.

It can be seen from Figure 10 that the two-way timing under orbit maneuver is consistent with that of the normal state, which means the GEO satellite ephemeris errors were eliminated by the two-way subtraction. Therefore, RDSS two-way timing accuracy has a certain impact on orbit residuals, but it is not obvious. Specifically, the two-way timing error of all epochs during orbit maneuver is smaller than 8 ns, and the RMS, mean value and standard deviation are 3.60 ns, 0.95 ns and 2.07 ns, respectively, which are very close with the results of the normal state. It is worth noting that whether it is maneuvering or not, the GEO satellite moves in the geostationary orbit, and the two-way timing still appears in the periodic fluctuation.

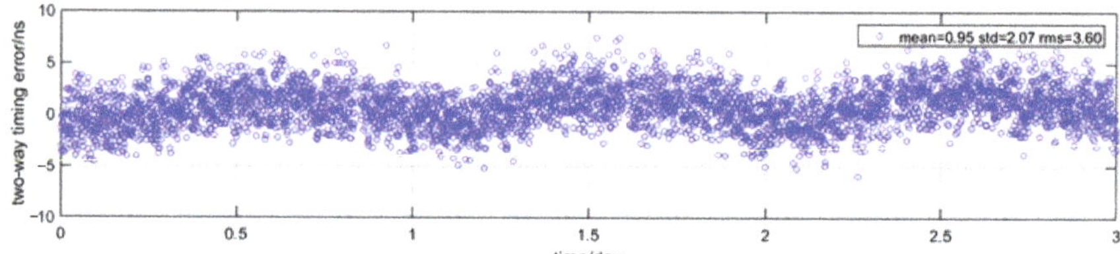

Figure 10. RDSS two-way timing accuracy (orbit maneuver state).

In order to further evaluate the performance of the orbit maneuver state, the availability and continuity results are given in Table 4. The availability and continui ty of theRDSS two-way timing service are 100%.

Table 4. The availability and continuity of two-way timing service (orbit maneuver state).

Time	8 February 2021	9 February 2021	10 February 2021
Availability	100%	100%	100%
Continuity	100%	100%	100%

4. Preliminary Conclusions

This paper discusses the principles of one-way timing and two-way timing, and provides analysis and evaluation methods of timing accuracy, availability and continuity indexes. Using RDSS signal measurements, the performance of BDS-3 RDSS one-way timing and two-way timing were evaluated for the first time under both the normal state and orbit maneuver state. The preliminary conclusions are obtained as follows:

(1) Both one-way timing and two-way timing results are periodical, and the spectrum analysis shows that the period is 1.01 days.

(2) The accuracy of one-way timing is better than 30 ns, and for two-way timing it is better than 8ns.

(3) One-way timing is affected by satellite orbit, while two-way timing is not affected.

(4) The availability and continuity of one-way timing in the normal state are 100%, which cannot be guaranteed in the orbit maneuver state.

(5) The availability and continuity of two-way timing both in normal state and orbit maneuver state are 100%.

In conclusion, the preliminary evaluation results show that the timing performance of BDS-3 is better than the officially claimed one, which can provide technical reference for BDS-3 RDSS users in the normal state and orbit maneuver state.

Author Contributions: Conceptualization, R.G. and D.W.; methodology, Z.L.; software, N.X.; validation, T.Z. and H.R.; formal analysis, R.G.; resources, S.L.; data curation, Z.L.; writing—original draft preparation, R.G. and D.W.; writing—review and editing, D.W.; funding acquisition, R.G. All authors have read and agreed to the published version of the manuscript.

Funding: This research was funded by the National Natural Science Foundation of China, grant number 41874043, 61603397, 41704037.

Data Availability Statement: The data sources are supported by Beijing Satellite Navigation Center.

Acknowledgments: Thanks for the data sources support of Beijing Satellite Navigation Center, and thanks for the fund support of the National Natural Science Foundation of China (Nos. 41874043, 61603397, 41704037).

Conflicts of Interest: The authors declare no conflict of interest.

References

1. Yang, Y. Progress, Contribution and Challenges of Compass/Beidou Satellite Navigation System. *Acta Geod. Cartogr. Sin.* **2010**, *39*, 1–6.

2. Yang, Y.; Li, J.; Wang, A.B.; Xu, J.; He, H.; Guo, H.R.; Shen, J.F.; Dai, X. Preliminary assessment of the navigation and positioning performance of BeiDou regional navigation satellite system. *Sci. China Earth Sci.* **2014**, *57*, 144–152. [CrossRef]

3. Tan, S. Innovative development and forecast of BeiDou system. *Acta Geod. Cartogr. Sin.* **2017**, *46*, 1284–1289.

4. Tan, S. Theory and application of comprehensive RDSS position and report. *Acta Geod. Cartogr. Sin.* **2009**, *38*, 1–5.

5. Zhao, J.; Ou, J.; Yuan, H. A new ambiguity resolution method using combined RNSS-RDSS of BeiDou. *Acta Geod. Cartogr. Sin.* **2016**, *45*, 404.

6. China Satellite Navigation Office. *The Application Service Architecture of BeiDou Navigation Satellite System*; China Satellite Navigation Office: Beijing, China, 2019.

7. China Satellite Navigation Office. *BeiDou Navigation Satellite System Open Service Performance Standard*, version 3.0; China Satellite Navigation Office: Beijing, China, 2021.

8. Tan, S. *The Engineering of Satellite Navigation and Positioning*, 2nd ed.; National Defend Industry Press: Beijing, China, 2010.

9. Li, B.D.; Liu, L.; Ju, X.M. *Geostationary Satellite Positioning*; The People's Liberation Army Press: Beijing, China, 1992.

10. Li, B.D.; Liu, L.; Ju, X.M.; Shi, X.; Zhu, L.F. Accuracy analysis of satellite bidirectional timing. *J. Time Freq.* **2010**, *33*, 129–133. [CrossRef]

11. Wang, D.; Guo, R.; Zhang, T.; Hu, X. Timing performance evaluation of Radio Determination Satellite Service (RDSS) for Beidou system. *Acta Astronaut.* **2019**, *156*, 125–133. [CrossRef]

12. Guo, R.; Zhou, J.H.; Hu, X.G.; Liu, L.; Bo, T.; Li, X.J.; Shan, W. Precise orbit determination and rapid orbit recovery supported by time synchronization. *Adv. Sp. Res.* **2015**, *55*, 2889–2898. [CrossRef]

13. Tang, C.; Hu, X.; Zhou, S.; Guo, R.; He, F.; Liu, L.; Zhu, L.; Li, X.; Wu, S.; Zhao, G. Improvement of orbit determination accuracy for Beidou Navigation Satellite System with Two-way Satellite Time Frequency Transfer. *Adv. Space Res.* **2016**, *58*, 1390–1400. [CrossRef]

 *remote sensing* 

Article

A New Mapping Function for Spaceborne TEC Conversion Based on the Plasmaspheric Scale Height

Mengjie Wu [1,2], Peng Guo [3,*], Wei Zhou [4], Junchen Xue [1], Xingyuan Han [5], Yansong Meng [5] and Xiaogong Hu [1,2]

[1] Shanghai Astronomical Observatory, Chinese Academy of Sciences, Shanghai 200030, China; mjwu@shao.ac.cn (M.W.); jcxue@shao.ac.cn (J.X.); hxg@shao.ac.cn (X.H.)
[2] Shanghai Key Laboratory for Space Positioning and Navigation, Shanghai 200030, China
[3] Advance Research Institute, Taizhou University, Taizhou 318000, China
[4] Beijing Institute of Tracking and Telecommunications Technology (BITTT), Beijing 100094, China; zhouwei_0611@163.com
[5] China Academy of Space Technology (Xi'an), Xi'an 710100, China; hanxy@cast504.com (X.H.); mengys@cast504.com (Y.M.)
*   Correspondence: gp@shao.ac.cn

Abstract: The mapping function is crucial for the conversion of slant total electron content (TEC) to vertical TEC for low Earth orbit (LEO) satellite-based observations. Instead of collapsing the ionosphere into one single shell in commonly used mapping models, we defined a new mapping function assuming the vertical ionospheric distribution as an exponential profiler with one simple parameter: the plasmaspheric scale height in the zenith direction of LEO satellites. The scale height obtained by an empirical model introduces spatial and temporal variances into the mapping function. The performance of the new method is compared with the mapping function F&K by simulating experiments based on the global core plasma model (GCPM), and it is discussed along with the latitude, seasons, local time, as well as solar activity conditions and varying LEO orbit altitudes. The assessment indicates that the new mapping function has a comparable or better performance than the F&K mapping model, especially on the TEC conversion of low elevation angles.

Keywords: radio occultation; TEC; mapping function; plasmaspheric scale height; GCPM

Citation: Wu, M.; Guo, P.; Zhou, W.; Xue, J.; Han, X.; Meng, Y.; Hu, X. A New Mapping Function for Spaceborne TEC Conversion Based on the Plasmaspheric Scale Height. *Remote Sens.* **2021**, *13*, 4758. https://doi.org/10.3390/rs13234758

Academic Editor: Chung-yen Kuo

Received: 27 October 2021
Accepted: 19 November 2021
Published: 24 November 2021

Publisher's Note: MDPI stays neutral with regard to jurisdictional claims in published maps and institutional affiliations.

1. Introduction

Benefitting from the accomplishment of the global navigation satellite system (GNSS) and many low Earth orbit (LEO) satellite missions, ionospheric exploration is greatly facilitated by various spaceborne measurements. Several successful LEO missions, such as Constellation Observing System for Meteorology, Ionosphere, and Climate (COSMIC), contribute greatly to the ionospheric modeling and data assimilation system [1–5]. One significant product provided by these satellites is the sounding measurements that contain the total electron density content (TEC) along the signal paths. To obtain the absolute TEC from the raw GNSS/LEO observations, data analysis centers and scientific researchers have devoted great efforts into the main procedures, including the cycle slip detection and correction, carrier-phase to pseudorange leveling, multipath effect correction, and the differential code bias (DCB) estimation [6–10]. The conversion between the slant TEC (STEC) along the ray path and the vertical TEC (VTEC) in the zenith is an essential procedure during the data processing. An obliquity factor called mapping function (MF) is commonly used to do the conversion. MF is a crucial parameter in the estimation of receiver DCB by assuming the simultaneous VTECs transformed from two GNSS slant observations of one LEO antenna are equal [8]. Furthermore, the mapping function plays a significant role in ionospheric and plasmaspheric TEC modeling due to the fact that the accuracy of the MF is highly correlated with the estimated VTEC and DCB [11].

Several mapping functions are widely used in the TEC conversion, such as the thin layer model (TLM) [12], the simple "geometric" model proposed by Foelsche and Kichengast (called F&K hereafter) [13], and some extended investigations [14–18]. The vertical structure of the ionosphere is usually assumed concentrating on one single thin layer. The height of the layer is called the ionospheric effective height (IEH) or shell height, which is calculated by the centroid method or integral median method or simplified as a constant value. The performance of mapping functions differs when receivers are located at different orbit heights or applying different IEH selections. How to choose the optimized IEH for the LEO-based TEC conversion needs systematical investigation. According to Zhong et al. [19] and Huang & Yuan [20], the IEH selection becomes more significant for the mapping function with an increasing zenith angle. Xiang & Gao [16] reviewed several mapping functions and proposed a mapping function that utilizes the ionospheric varying height assisted by the International Reference Ionosphere (IRI) model [21]. The study by Gulyaeva et al. [22] explored the center-of-mass of the ionosphere as the effective varying shell height produced by the ionospheric equivalent slab thickness. Zhong et al. [19] examined the applicability of three mapping functions for LEO-based GNSS observations: the thin layer model, the F&K model, and the Lear model, and found that the F&K geometric mapping function together with the IEH from the centroid method is more suitable to convert the LEO-based TEC measurements. The optimized IEH for the F&K can be approximately expressed as $(2.5h_{LEO} + 110)$ km under medium solar activity (MSA) conditions. As a consequence, the COSMIC-based TEC conversion (with satellites running at ~800 km) will get minimum mapping errors when the IEH is set to be around 2,100 km during MSA years. The current data processing in the COSMIC data analysis and archive center (CDAAC) applied the F&K mapping model, and the IEH is fixed at several hundreds or thousands of kilometers to simplify the uses [3,23]. However, one common drawback of these thin-layer mapping functions is that they ignore the vertical structure of the ionosphere and plasmasphere, and the IEH without spatial and temporal variations will affect the accuracy of the TEC conversion. One option to improve the TEC mapping is to estimate the 3D ionospheric electron density distribution by tomography or data assimilation based on the dense measuring network of sufficient spatial and temporal resolution and deduce realistic and persistent estimates of the ionospheric effective height [24]. Another approach is to assume multiple layers in the ionosphere implemented with specific mapping functions [17,20]. To further consider the vertical distribution of the ionosphere, Hoque & Jakowski [15] proposed a multi-layer mapping function according to the typical structure described by the Chapman layer. Nevertheless, the above methods are dependent on either extensive data coverage or accurate ionospheric models and parameters that are difficult to obtain in the global scope. Therefore, more advanced ionospheric mapping functions with simple free parameters deserve continuous attention and potential development.

The scale height in the plasmasphere (H_P) is an important factor to present the dynamic nature and variations of the plasmasphere [25]. In Wu et al. [26], we have developed a scale height model depending on parameters, including the month of the year, local time, geographic latitude, and the solar radio flux index F10.7. This study is aimed to propose a new mapping function taking advantage of the former scale height model (named as the scale-height-based mapping function) to give an accurate obliquity factor to convert LEO-based slant TEC to vertical TEC and vice versa. The ionosphere is no longer collapsed into any shell but assumes an exponential vertical structure. The scale height introduces realistic spatial and temporal variations in the ionospheric and plasmaspheric electron density into the TEC conversion. With the proposed method, we expect to solve the problem of optimized IEH specification and neglect of ionospheric physical variations in the current single layer mapping models and, finally, facilitate the TEC and DCB estimations of LEO satellites. Section 2 introduces the mathematical solutions of the new mapping function, and Section 3 gives the distribution of plasmaspheric scale height and the associated mapping function values; in Section 4, global assessments of the scale-height-based mapping function and the F&K model are performed according to different seasons, locations, solar activity levels, and orbit altitudes, and the conclusion is drawn in Section 5.

2. Materials and Methods

The scale height is one of the most important characteristics in the ionosphere–plasmasphere system due to the role in shaping the electron density profile [25,27–29]. There are various definitions of scale heights in published literature, including the theoretical plasma scale height, the vertical scale height, and the effective scale height. Alessio Pignalberi et al. [25] discussed the possible relationships among the different definitions of scale height. In this study, the plasmaspheric scale height adopts the definition of effective scale height, in which the analytical formulation of fitting the electron profile is chosen as exponential function. The new scale-height-based mapping function is not associated with specific IEH or thin-layer assumption but relates to the plasmaspheric scale height deduced from the exponential layer ionospheric model proposed by Stankov et al. [30]; i.e., the vertical electron density profile can be represented as:

$$N_e(h) = N_e(h_0) \cdot exp\left(-\frac{h - h_0}{H_P}\right)$$

(1)

where $N_e(h_0)$ is the base electron density of plasmasphere (here set to be the electron density at the receiver height h_0), and H_P is the plasmaspheric scale height. Thus, the mapping function can be written as the ratio of slant and vertical integral of electron density along the slant ray and the vertical path:

$$M(z) = \frac{STEC}{VTEC} = \frac{\int_{\vec{r_0}}^{\vec{r_c}} N_e\left(\vec{r}\right) d\vec{r}}{\int_{h_0}^{h_c} N_e(h) dh}$$

(2)

where $\vec{r_0}$ and $\vec{r_c}$ represent the locations of LEO and GNSS satellites from the Earth center, respectively; h_0 and h_c are the corresponding satellite heights above Earth's surface; z is the zenith angle of the ray at the LEO receiver. Figure 1 shows the geometric schematic of the mapping function. If the ionospheric spherical symmetry assumption is adopted, together with Equation (1), the numerical expression of the scale-height-based mapping function is

$$M(z) = \frac{\int_{r_0}^{r_c} exp\left(-\frac{r - r_0}{H_P}\right) \cdot \frac{r}{\sqrt{r^2 - r_v^2}} dr}{\int_{h_0}^{h_c} exp\left(-\frac{h - h_0}{H_P}\right) dh}$$

(3)

where $r_v = r_0 \cdot sin(z)$; r_0 represents the radius of LEO satellite orbit from the Earth center; z is the zenith angle of the ray at the LEO receiver. $M(z)$ can be obtained by numerical integral if H_P is known. Since the plasmaspheric scale height varies with the solar activity, season, local time, and location [26], this mapping function is embedded with temporal and spatial variations. It should be mentioned that the scale-height-based mapping function implicitly assumes the exponential distribution of the ionosphere and plasmasphere. Considering the general orbit range of spaceborne satellites, the assumption is reasonable when dealing with LEO-based TEC conversion.

Besides the numerical expression, an analytical solution for the mapping function can be written as:

$$M(z) = \frac{\sqrt{2r_0/H_P}}{sin(z)} exp(I^2) \cdot \frac{\sqrt{\pi}}{2} erfc(I)$$
$$erfc(I) = \frac{2}{\sqrt{\pi}} \int_I^\infty exp(-y^2) dy$$

(4)

where $I = \sqrt{\frac{r_0}{2H_P} cot(z)}$ and $erfc(\ )$ is the complementary error function. The process to obtain Equation (4) is shown in Appendix A. There are differences between the numerical and analytical expression of $M(z)$ because the analytical solution adopts some approximations. We noticed that Hoque & Jakowski [15,31] presented a multi-layer mapping function approach for potential space-based augmentation system (SBAS) service under the assumption of Chapman layer distribution of the vertical electron density depending on the peak ionization height and atmospheric scale height. Here, we adopt the hypothesis of an exponential scale height to deal with the LEO-based slant to vertical TEC conversion.

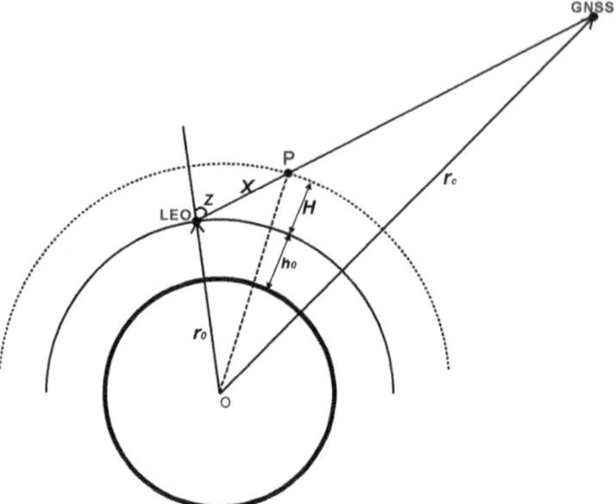

Figure 1. Geometric schematic of LEO-based GNSS TEC mapping model; LEO—Low Earth orbit.

The scale-height-based mapping function is compared with the F&K method in this study. The F&K mapping function was originally proposed to describe the dependence of the hydrostatic path delay with only one free parameter: "effective height". When applied to the LEO-based TEC conversion, the IEH is assumed at the height of a thick spherical layer above the receiver, hypothetically concentrating the whole impact of the ionosphere. The F&K mapping function is written as

$$M(z) = \frac{1 + R_{shell}/R_{orbit}}{\sqrt{(R_{shell}/R_{orbit})^2 - sin^2(z)} + cos(z)}$$

(5)

where R_{shell} and R_{orbit} are the radii of the assumed layer and LEO satellite, respectively; z is the zenith angle. The mapping factor may differ a lot with diverse IEH specifications.

3. Results

We calculate a mapping factor grid of zenith angle z varying between 5° and 80° with 5° resolution, and H_P varying in (100, 6000) km with a 50 km step according to Equations (3) and (4). The obliquity factor of arbitrary TEC conversion can be obtained by interpolation with specified z and H_P. The grid map along with zenith angle and H_P retrieved from the numerical and analytical solutions are demonstrated in Figure 2 (panel (a) and (c)). As an example, the spaceborne receiver is assumed at 800 km and the transmitter is assumed at 20,000 km. Moreover, the F&K model with the shell height varying from 100 km to 6000 km is displayed in panel (b). The blank in panel (b) is caused by the unreasonable result of the F&K model when the IEH is lower than the assumed LEO satellite height at high zenith angles. Similarly, the analytical resolution is not available when the ray path is approaching the zenith direction. We see the mapping factor increases when the zenith angle gets greater, and the difference brought by varying the H_P value or shell height selection becomes significant when the zenith angle is larger than 40°. The mapping factor has similar variations along with the scale height or IEH. The numerical and analytical results differ more with a large zenith angle and higher H_P values (shown in panel (d)).

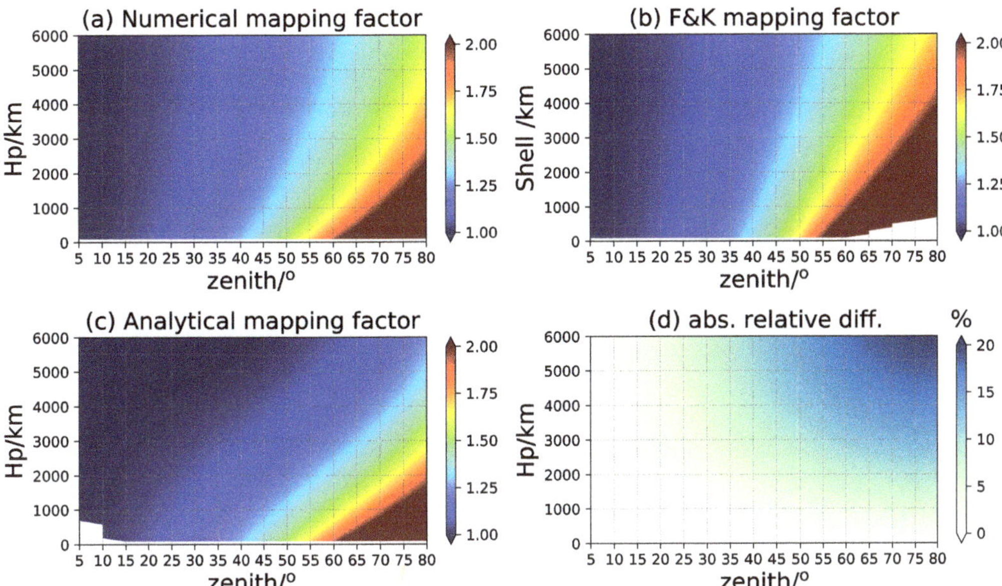

Figure 2. The mapping factor grid of zenith angle and the plasmaspheric scale height H_P; panel (**a,c**) are the numerical and analytical results of scale-height-based mapping function, respectively; (**b**) is those for F&K model; (**d**) is the absolute relative difference between (**a,c**).

The only free parameter in the scale-height-based mapping function is the plasmaspheric scale height. In Wu et al. [26], we have calculated the plasmaspheric scale height from the COSMIC measurements from 2007 to 2014 and proposed an empirical monthly climate model that has reasonable accuracy validated by the scale height retrieved from the International Satellites for Ionospheric Studies (ISIS) observations during its mission period. This H_P model is suitable to provide the essential parameter in the scale-height-based mapping function. Besides the COSMIC-derived H_P, the scale height obtained by the global core plasma model (GCPM) provides an alternative option [32]. The GCPM provides empirically derived core plasma density and ion composition (H^+, H_e^+, and O^+) as a function of the geomagnetic and solar conditions throughout the inner magnetosphere. The model merges with the IRI model at low altitudes and is composed of separate models for the plasmasphere, plasmapause, trough, and polar cap. The H_P is calculated from the overhead vertical TEC simulated with the GCPM electron density field and the base electron density at the assumed receiver altitude since the integral in Equation (1) could be approximately written as $TEC \approx N_e(h_0) \cdot H_P$. The H_P retrieved from the observations and empirical model (represented by H_{P_COSMIC} and H_{P_GCPM}, respectively) are shown according to four seasons, local time, and dipole geomagnetic latitude in the low solar activity (LSA) year 2008 (Figure 3) and high solar activity (HSA) year 2013 (Figure 4), respectively. The four seasons are represented by March equinox (March, April, May), June solstice (June, July, August), September equinox (September, October, November), and December solstice (January, February, December) in this work. The H_{P_COSMIC} is generally greater than the GCPM simulations, but they have generally similar variations along with the latitude, local time, season, and solar activity. The nighttime scale height is larger than that in the daytime and usually achieves the maximum before the sunrise, which is consistent with the conclusion drawn in Wu et al. [26]. The seasonal variations are more predominant for COSMIC scale heights, with higher values in the winter hemisphere and lower at the summer half. Under high solar activity conditions, the scale height decreases for both the observational and empirical models. The obliquity factor will differ a bit when applying H_{P_COSMIC} and H_{P_GCPM} in the mapping function. At the high latitudes ($\pm 60° \sim \pm 90°$), the electron density is quite small and rarely changed, so the scale height value can be regarded as a constant. The H_P derived from satellite observations is assumed more reliable when dealing with the realistic mapping issues, but, in the following study, the H_{P_GCPM} is discussed because we are aiming at the comprehensive validation of the scale-height-based mapping function in a simulation way.

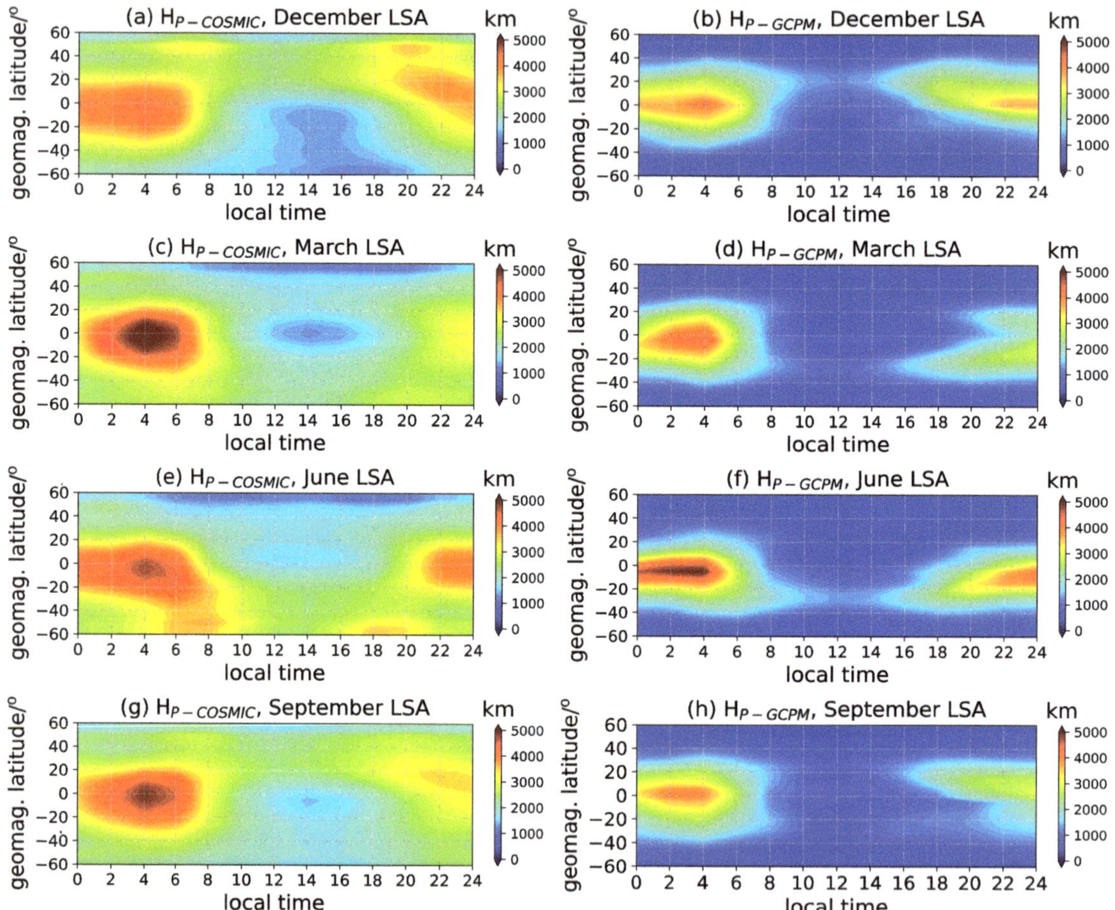

Figure 3. The distribution of the COSMIC and GCPM retrieved H_P (represented by 'H_{P_COSMIC}' and 'H_{P_GCPM}') along with the local time, geomagnetic latitude, and season in the LSA year; panels (**a,c,e,g**) are for H_{P_COSMIC}, panels (**b,d,f,h**) are for H_{P_GCPM}.

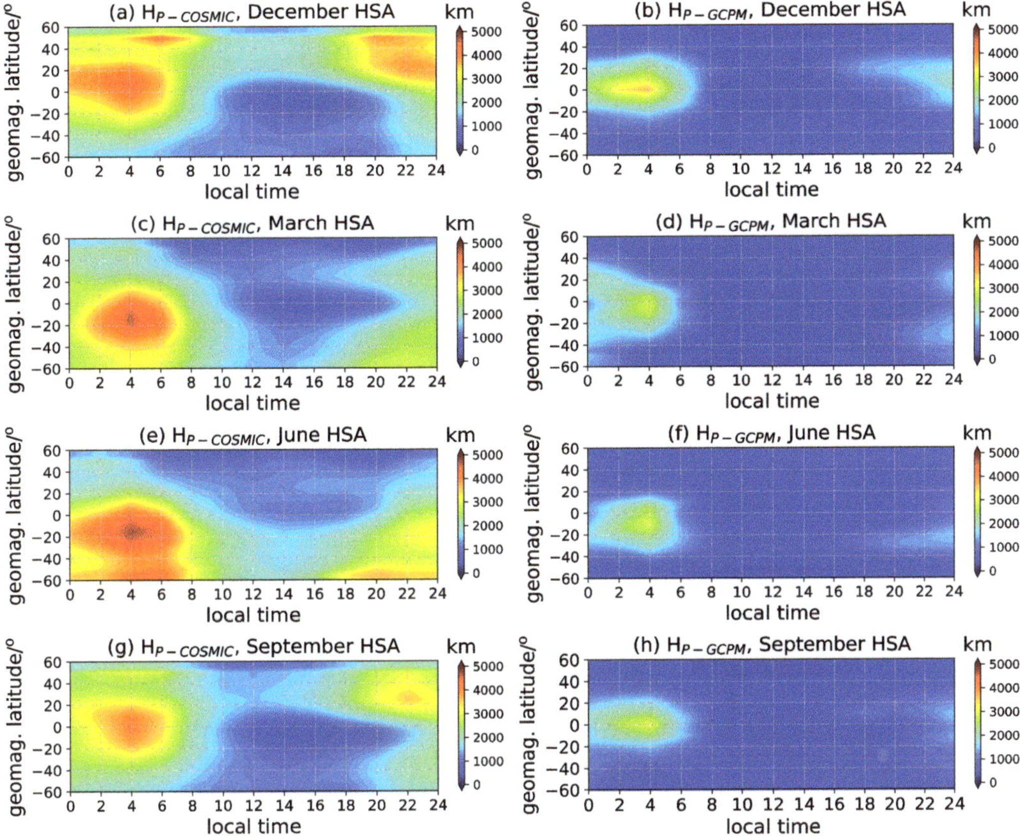

Figure 4. Same as Figure 3 but for the HSA year.

4. Discussion

4.1. Assessment by Global Statistics

To assess the performance of the new mapping function, the GCPM is used to generate simulating LEO-based TEC observations and examine the retrieved errors of both mapping models. The LEO satellite orbit is chosen at 800 km, which is the typical orbit altitude of the COSMIC-1 constellation that made a great contribution to the ionosphere and plasmasphere exploration. The signal transmitter is at 20,000 km, referring to the GNSS satellites. It should be noted that both the scale-height-based and the F&K mapping function assume the spherical symmetric distribution of the ionosphere, which means the electron density on the same sphere is identical everywhere. Thus, the simulated STEC of a different zenith angle is actually equal to the integral of vertical electron density along the slant ray path with a 50 km step. The VTEC retrieved from the slant TEC and mapping function is compared to the VTEC 'truth' integrated directly from the GCPM background (regarded as $VTEC_{mod}$). The relative root mean square (RMS) of each zenith angle z is calculated as the assessment criteria:

$$RMS(z) = \sqrt{\frac{\sum_{i=1}^{N}\left[\left(\frac{STEC_i(z)}{M_i(z)} - VTEC_{mod}\right)/VTEC_{mod}\right]^2}{N}} \times 100\% \tag{6}$$

where N is the TEC observation number of global grids at the zenith angle z. Given the difference between the numerical and analytical representation of mapping function, we checked the performance of the scale-height-based mapping function of both solutions, denoted as 'H_{P_N}' and 'H_{P_A}', respectively, hereafter. To introduce independent reference into the validation, the F&K mapping function is also considered with adjustable IEH represented as a function of the LEO orbit height and solar flux proxy F10.7 (F_{107}) according to Zhong et al. [19]:

$$IEH = (0.0027F_{107} + 1.79)h_{LEO} - 5.52F_{107} + 1350 \tag{7}$$

Thus, the IEH specifications take different satellite altitudes and solar activity levels into consideration, and the F&K mapping model is optimized by choosing a constant IEH. The zenith angle of the simulated slant ray path varies from $0°$ to $80°$ in steps of $5°$, and the geophysical location of the receiver changes with horizontal resolution of latitudinal $5°$ from $-90°$ to $90°$ and longitudinal $10°$ from $-180°$ to $180°$. The simulating experiments are executed on one day in four seasons with a temporal resolution of 2 h under both the low and high solar activity conditions. The mean retrieved RMS errors for each season are calculated according to the zenith angles and shown in Figure 5. The F&K mapping function depends on the fixed IEH globally; thus, the retrieved errors are closely associated with the ionospheric variations of the GCPM background. According to Figure 5, the behavior of the numerical and analytical mapping functions differs from each other while indicating better retrieved results than the F&K model, especially with increasing zenith angles greater than $40°$. In the LSA years, the retrieved errors of the numerical mapping function are slightly less than those of the analytical solution at lower zeniths. When the solar activity is more active, the mapping errors of all the methods decrease to some extent, and the analytical mapping function achieves the best performance throughout the year. The difference between the two scale-height-based models is associated with the variation in the H_P value under different solar activity conditions and the zenith angles. From the statistics, we conclude that, with the LEO satellite located at the 800 km, the H_P-based mapping function is a competitive approach in comparison with the F&K model, especially at low elevation angles; i.e., at high zenith angles.

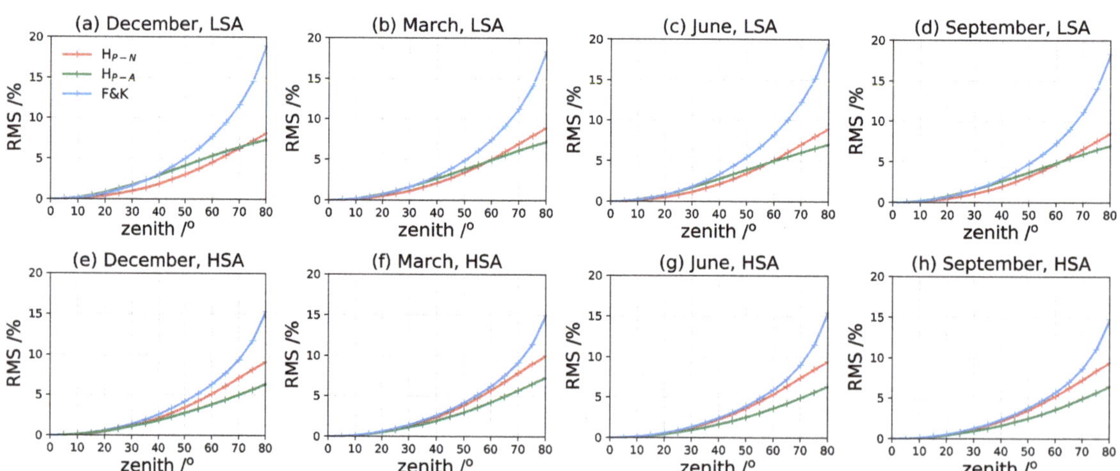

Figure 5. The relative RMS statistics of the scale-height-based mapping function with numerical and analytical solutions (denoted as 'H_{P_N}' and 'H_{P_A}', respectively), and the F&K model ('F&K') in four seasons under low and high solar activity conditions (represented by 'LSA' and 'HSA'); panels (**a–d**) are for the LSA year, and (**e–h**) are for HSA year.

4.2. Global Variations of Mapping Errors

To further analyze the time and spatial dependence of the mapping functions, the global maps of mean retrieved error with the zenith angle fixed at 40° are displayed in Figures 6 and 7 of the LSA and HSA years, respectively. The scale-height-based mapping function of the numerical or analytical solutions and the F&K model exhibit quite different characteristics with the local time, geographical latitude, and seasons. Corresponding to the predominant seasonal asymmetric features of H_P, the VTEC retrieved by the new method is overestimated in the nighttime winter hemisphere, especially when using the analytical mapping factor. Both the scale-height-based models underestimate the VTEC during the daytime at low and equatorial latitudes. Consistent with the statistics in Figure 5, the general performance of the numerical mapping function is fairly good, with the mapping error varying between −5% and 5%. The latitudinal and temporal variations of the mapping errors reflect the reliability of the exponential assumption in the GCPM model. It proves that the variations in the ionosphere and plasmasphere are more complicated and nonuniform at low latitudes. For the F&K model, the relative error maps indicate the inhomogeneous variations of the ionospheric effective height of the background. The VTEC is significantly underestimated during the whole day at low latitudinal areas, while it is overestimated at higher latitudes. The performance of all the methods is remarkably improved under higher solar activity, as shown in Figure 7. Generally, the scale-height-based mapping function achieves better conversion results than the F&K method, especially in the equatorial anomaly regions between −30° and 30°.

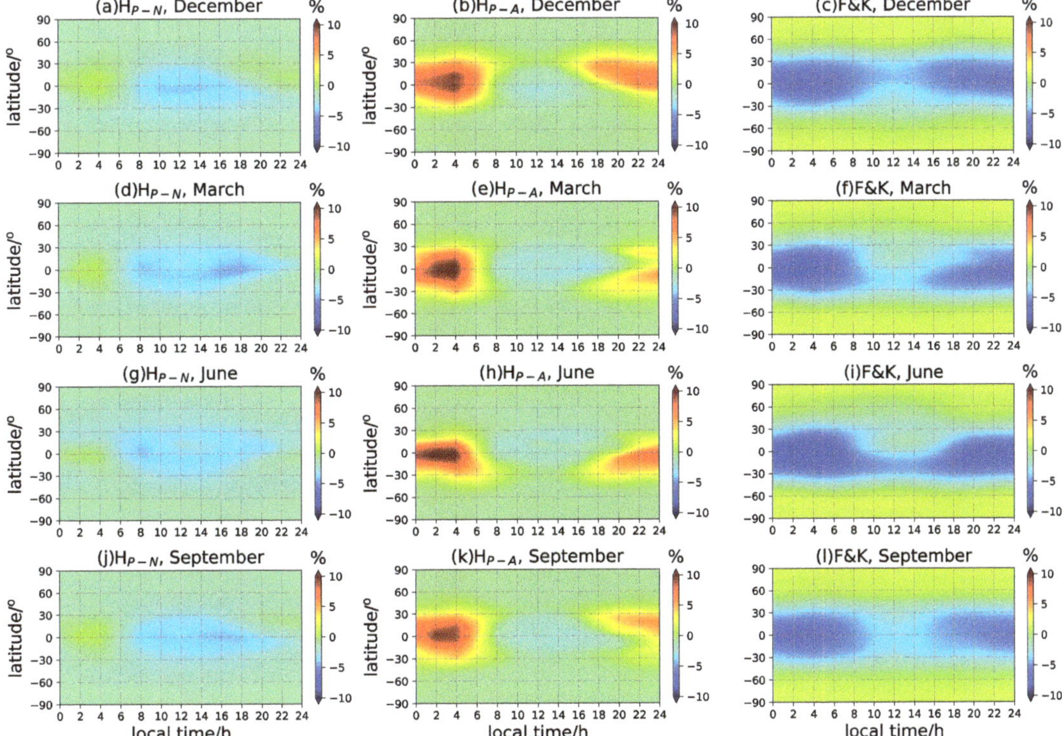

Figure 6. The local time and latitude-dependent variations of the retrieved VTEC mean deviation for the mapping experiments in different seasons in LSA years. The zenith angle is fixed at 40°. The 'H_{P_N}', 'H_{P_A}', and 'F&K' represent the numerical and analytical mapping functions and the F&K model, respectively; panels (**a,d,g,j**) are H_{P_N}-based mapping errors; (**b,e,h,k**) are H_{P_A}-based mapping errors; (**c,f,i,l**) are F&K mapping errors.

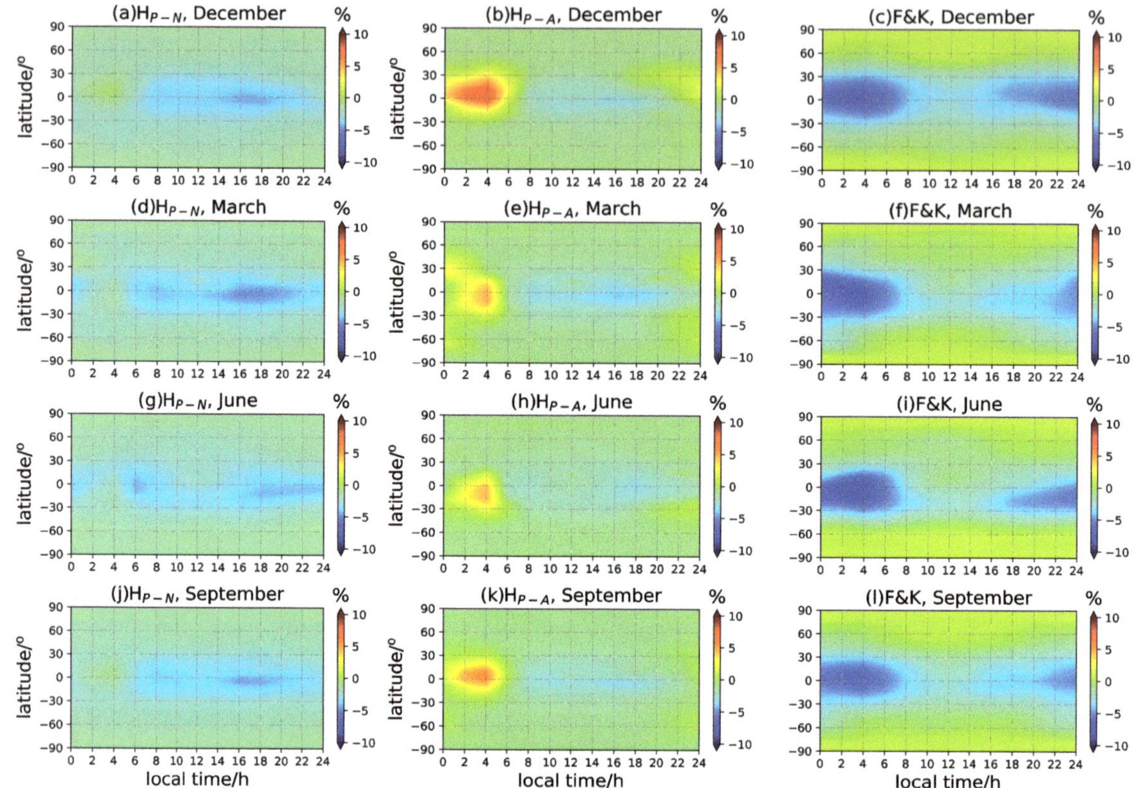

Figure 7. Same as Figure 6 but for HSA year.

Figures 8 and 9 demonstrate the mean deviation of the VTEC along with the latitude and longitude of three chosen zenith angles, 20°, 40°, and 60°, in the December solstice. The VTEC at low latitudinal areas beside the geomagnetic equator rather underestimated when applying the numerical mapping function and the F&K model. At higher elevation angles, the TEC conversion errors brought by various mapping functions are negligible with minor differences. However, the residuals increase pronouncedly when the elevation angle is lower, especially for the F&K model. The performance of the analytical solution is generally the best at mid- and high latitudes, especially in a high solar activity year. The winter hemispheric overestimation along the geomagnetic equator is associated with the seasonal variation in scale height. As mentioned earlier, the geographical error distribution of the F&K model actually reflects the inhomogeneity of the IEH and the ionospheric background model. The results depend greatly on the IEH selection and the ionospheric model used to specify the IEH. The scale-height-based mapping model considers the vertical structure and variations in the ionosphere by involving the plasmaspheric scale height, and it is more convenient and reliable to provide realistic TEC conversion results. Moreover, the degradation in the performance is minor for scale-height-based mapping along with an increasing zenith, so it is especially suitable to be used on observations of low elevation angles.

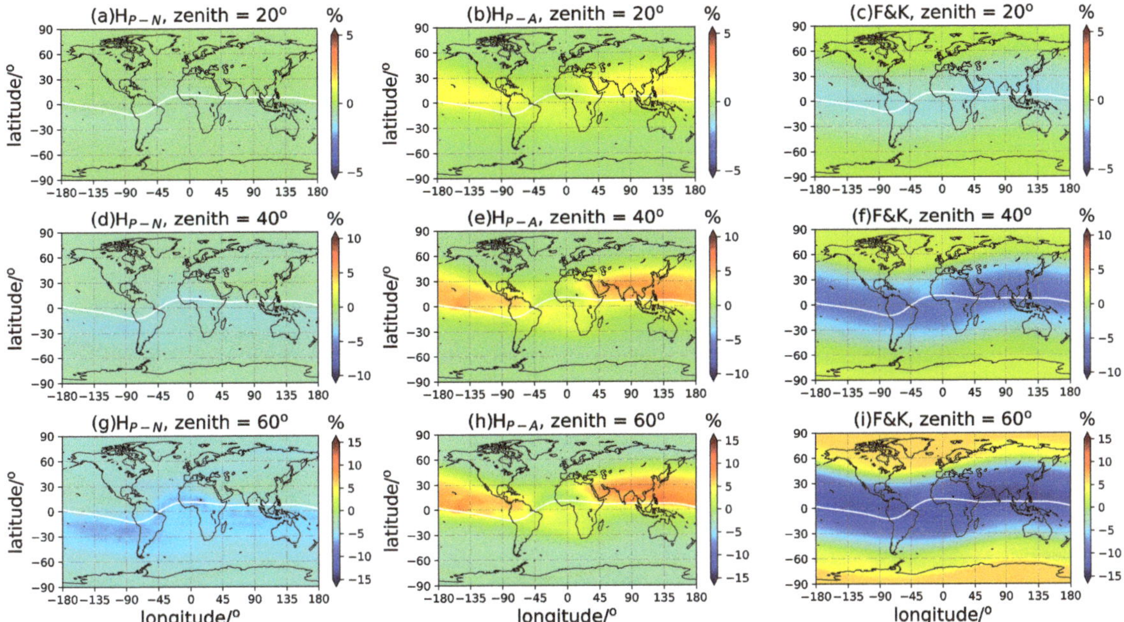

Figure 8. The longitude- and latitude-dependent variations of the retrieved VTEC mean deviation in the December solstice in LSA years. The zenith angle is chosen at 20° (panels (**a**–**c**)), 40° (panels (**d**–**f**)), and 60° (panels (**g**–**i**)), respectively. The 'H_{P_N}', 'H_{P_A}', and 'F&K' represent the numerical and analytical mapping functions and the F&K model, respectively.

4.3. Influence of LEO Orbit Altitudes

Given the diversity of the orbit altitudes of different LEO satellites, we set up another two scenarios for more detailed discussion: the receiver being located at 500 km and 1400 km, respectively. The performances of the scale-height-based mapping function and the F&K model with the IEH specified according to Equation (7) are investigated in the same way as in Section 4.1. In Figure 10, when the LEO orbit altitude is 500 km, the mapping errors increase. The reason is that, at 500 km of altitude, the reliability of considering an exponential vertical profile for the electron density (with the associated scale height) is quite low [33–35]. This hypothesis adopted in the scale-height-based mapping will produce higher errors than at 1400 km of altitude. In an LSA year, both the numerical and analytical mapping function achieve fewer RMS errors than the F&K model, and the analytical solution is slightly better than the numerical one. Under an HSA condition, the numerical mapping factor results in similar or even a bit worse results than the F&K model below a zenith angle of 60°, while the analytical mapping function is promising to obtain the least mapping errors. The situation is quite different with the LEO satellite higher at 1400 km. According to Figure 11, the numerical mapping function remarkably improves the mapping errors in both LSA and HSA years. The analytical solution has a relatively poor peformance at high elevation compared to the F&K method but still functions well with greater zenith angles.

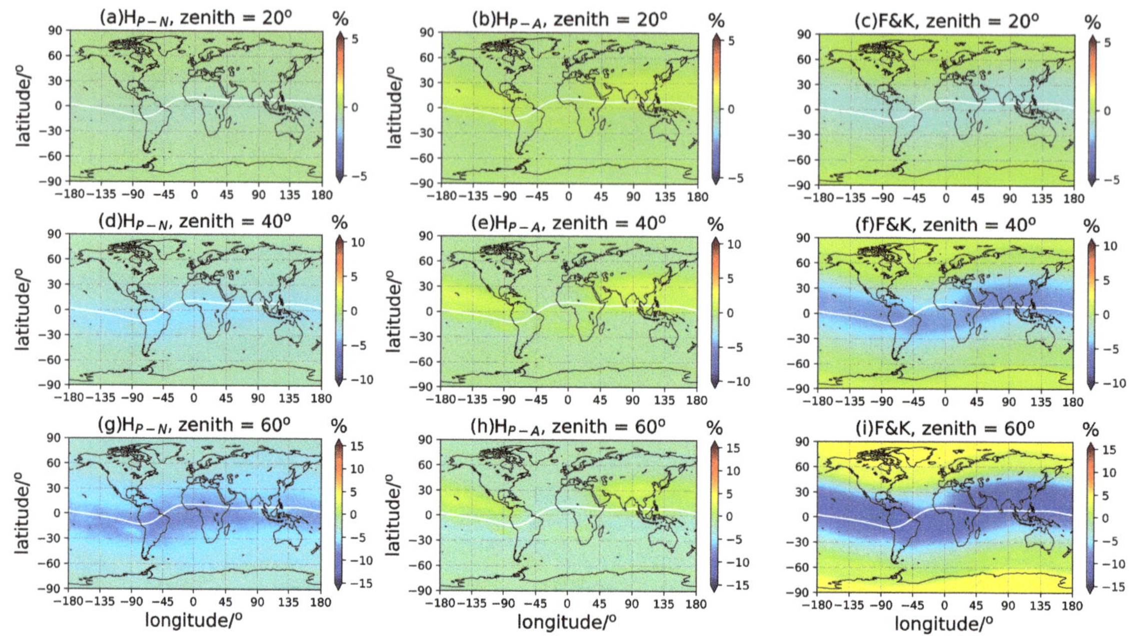

Figure 9. Same as Figure 8 but for HSA year.

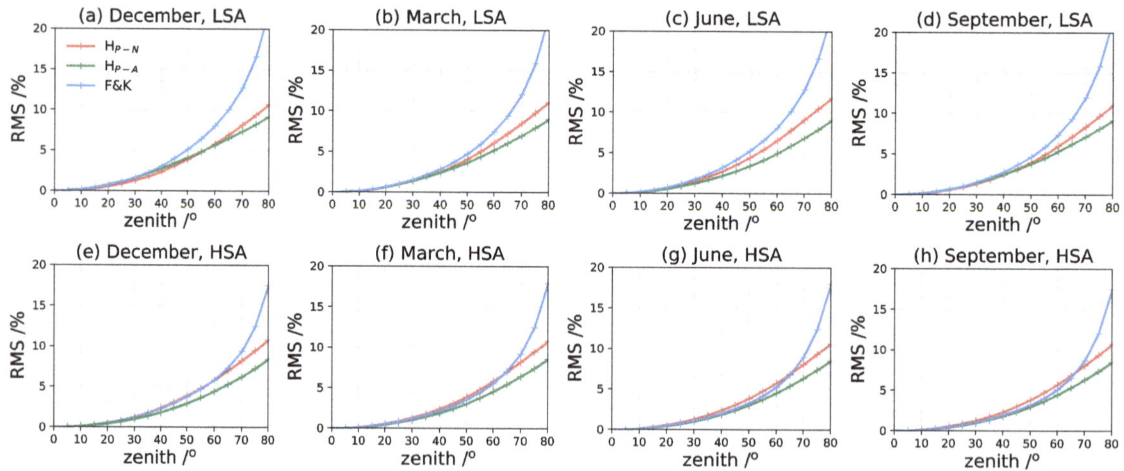

Figure 10. Same as Figure 5 but for LEO satellite at 500 km.

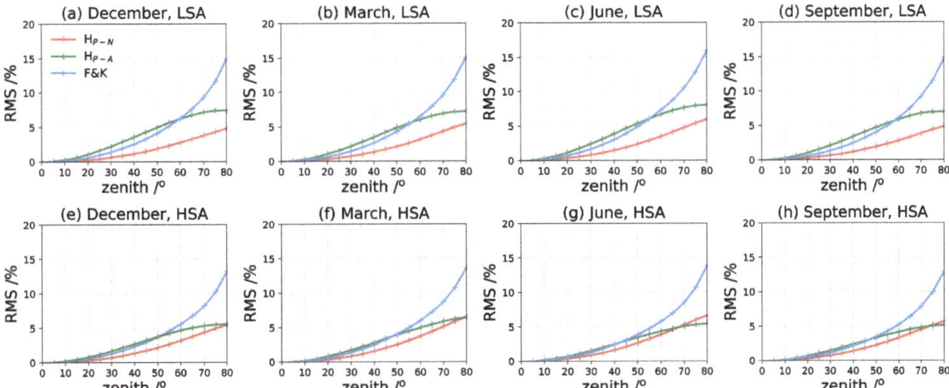

Figure 11. Same as Figure 5 but for LEO satellite at 1400 km.

To understand the distinctive performance of the scale-height-based mapping function refering to different LEO orbit height, it should be noted that the H_P obtained in this work is basically the averaged scale height under the assumption that the ionosphere and plasmasphere are varying roughly in an exponential trend above the satellite orbit. In fact, the actual scale height is varying with the altitude gradually below the O^+-H^+ transition height, which peaks at 700 km and achieves an average value of 862 km [36], and approaching constant when the dominant ion becomes H^+ or H_e^+ in the plasmasphere [26]. Therefore, the averaged H_P has a larger deviation with the varying scale height, especially when the satellite is locating below the transition height, such as in the 500-km-scenario. The retrieved VTEC is generally less than the truth since the mapping factor is overestimated because of the averaged H_P. It agrees with the underestimation of the VTEC of the numerical solution in Figures 6–9. Along with the increase in the orbit height, the averaged H_P is closer to the realistic scale height and leads to fewer conversion errors. That accounts for the improvement in the performance of the numerical solution with the orbit varying from 500 km to 1400 km. Due to the assumption and approximation adopted in the analytical mapping function, a higher H_P value leads to a smaller mapping factor (Figure 2). Therefore, the analytical solution basically compensates the overestimation of $M(z)$, sometimes presenting a better performance than the numerical one in an HSA year. Given the overall performances of the three methods, the scale-height-based mapping function with the numerical solution is the most stable approach under varying solar activity conditions and in changing LEO orbits, and it improves the F&K model significantly, especially at lower elevation angles.

5. Conclusions

This paper proposed a new model of an LEO-based TEC mapping function based on a prior model of the plasmaspheric scale height. The numerical and analytical forms of the mapping function are illustrated, and the mapping factor grid is obtained as a function of the scale height and zenith angle. The scale height and mapping factor are accessible to the public at the database: https://www.researchgate.net/publication/353703384_mapping_factor_grid (accessed on 1 August 2021) and https://www.researchgate.net/publication/353704297_plasmapsheric_scale_height (accessed on 1 August 2021).

The new mapping model is driven by the only free parameter, H_P, obtained either from realistic observations or the empirical model. The performance of this mapping function is assessed by the simulated TEC conversion experiments based on the GCPM electron density field. This method is promising to be applied to LEO-based TEC conversion as a comparable or better alternative to the F&K model according to the assessments under various spatial and temporal specifications. Instead of collapsing the ionosphere into a thin shell, the new formulation considers the exponential layer assumption of the electron density distribution and applies the plasmaspheric scale height to introduce seasonal, local time, and geolocation-dependent variations into the mapping model. The performance of the F&K model is highly correlated with the IEH specification at different latitudes during varying solar activity periods, as well as the orbit altitudes of LEO satellites. The scale-height-based mapping model has advantages in dealing with the spaceborne TEC conversion, especially with low-elevation-angle observations of higher orbit height.

We recognize that the major errors remaining in the TEC conversion are associated with the horizontal inhomogeneous distribution of the ionosphere. Therefore, a comprehensive mapping model involving the azimuth angle variation of the signal ray paths is now under development based on this investigation.

Author Contributions: Conceptualization, M.W. and J.X.; methodology, W.Z.; software, M.W. and J.X.; validation, X.H.(Xingyuan Han), Y.M.; formal analysis, P.G. and M.W. ; investigation, M.W.; resources, X.H.(Xiaogong Hu); data curation, X.H.(Xiaogong Hu); writing—original draft preparation, M.W.; writing—review and editing, P.G., W.Z. and X.H.(Xingyuan Han); supervision, P.G.; project administration, X.H.(Xiaogong Hu); funding acquisition, M.W. and Y.M. All authors have read and agreed to the published version of the manuscript.

Funding: This research was funded by the National Key R&D Program of China, grant number 2020YFA0713501; the National Natural Science Foundation of China, grant number 11903064, U1831116; the stability support fund project of national key laboratory, 2020SSFNKLSMT-06; the stability support fund project of Science and Industry Bureau, grant number HTKJ2020KL504006; the opening project of Shanghai Key Laboratory of Space Navigation and Positioning Techniques, NO. KFKT201906.

Data Availability Statement: The data presented in this study are openly available in ResearchGate at https://www.researchgate.net/publication/353703384_mapping_factor_grid (accessed on 1 August 2021) and https://www.researchgate.net/publication/353704297_plasmapsheric_scale_height (accessed on 1 August 2021).

Acknowledgments: The authors are very thankful for the open access provided by CDAAC for all the COSMIC data involved and the national aeronautics and space administration for GCPM sources codes. The processed data and figure datasets used for this paper are available at the site: https://doi.org/10.5281/zenodo.5205255 (accessed on 1 August 2021).

Conflicts of Interest: The authors declare no conflict of interest.

Abbreviation List

TEC	total electron content
STEC	slant TEC
VTEC	vertical TEC
LEO	low Earth orbit
GCPM	global core plasma model
GNSS	global navigation satellite system
COSMIC	Constellation Observing System for Meteorology, Ionosphere, and Climate
DCB	differential code bias
MF	mapping function
TLM	thin layer model
IEH	ionospheric effective height
IRI	International Reference Ionosphere
MSA	medium solar activity
LSA	low solar activity
HAS	high solar activity
CDAAC	COSMIC data analysis and archive center
SBAS	space-based augmentation system
ISIS	International Satellites for Ionospheric Studies
RMS	root mean square

Appendix A

Assuming an arbitrary point P on the ray path in Figure 1, the relationship between the slant ray X and the vertical projection H to the receiver is approximately written as

$$H \approx Xcos(z) + \frac{X^2 sin^2(z)}{2r_0} \tag{A1}$$

where r_0 represents the radius of LEO satellite orbit from the Earth center; z is the zenith angle of the ray path. Under the assumption of ionospheric spherical symmetry, the slant TEC is obtained by integrating the electron density along the ray path,

$$STEC = \int_0^\infty N_0 exp\left(-\frac{H}{H_P}\right)dX = N_0 \cdot \int_0^\infty exp\left[-\left(\frac{Xcos(z)}{H_P} + \frac{X^2 sin^2(z)}{2r_0 H_P}\right)\right]dX \tag{A2}$$

If $a = \frac{sin^2(z)}{2r_0 H_P}$ and $b = \frac{cos(z)}{2H_P}$, then

$$\begin{aligned} STEC &= N_0 \cdot \int_0^\infty exp\left[-\left(2bX + aX^2\right)\right]dX \\ &= N_0 exp\left(\frac{b^2}{a}\right) \cdot \int_0^\infty exp\left[-a\left(X + \frac{b}{a}\right)^2\right]dX \\ &= \frac{N_0}{\sqrt{a}} exp\left(\frac{b^2}{a}\right) \cdot \int_{\frac{b}{\sqrt{a}}}^\infty exp(-y^2)dy \end{aligned} \tag{A3}$$

N_0 is the electron density at the receiver. The vertical TEC is easy to obtain with integration as $VTEC \approx N_0 \cdot H_P$; therefore, the analytic solution of $M(z)$ relates to the scale height H_P, zenith angle z, and the complementary error function ($erfc(I)$):

$$\begin{aligned} M(z) &= \frac{\sqrt{2r_0/H_P}}{sin(z)} exp\left(I^2\right) \cdot \frac{\sqrt{\pi}}{2} erfc(I) \\ erfc(I) &= \frac{2}{\sqrt{\pi}} \int_I^\infty exp(-y^2)dy \end{aligned} \tag{A4}$$

where $I = \sqrt{\frac{r_0}{2H_P}} cot(z)$.

References

1. Schreiner, W.; Rocken, C.; Sokolovskiy, S.; Syndergaard, S.; Hunt, D. Estimates of the precision of GPS radio occultations from the COSMIC/FORMOSAT-3 mission. *Geophys. Res. Lett.* **2007**, *34*, 1–5. [CrossRef]
2. Yue, X.; Schreiner, W.S.; Kuo, Y.H.; Hunt, D.C.; Wang, W.; Solomon, S.C.; Burns, A.G.; Bilitza, D.; Liu, J.Y.; Wan, W.; et al. Global 3-D ionospheric electron density reanalysis based on multisource data assimilation. *J. Geophys. Res. Space Phys.* **2012**, *117*, 1–17. [CrossRef]
3. Lin, J.; Yue, X.; Zhao, S. Estimation and analysis of GPS satellite DCB based on LEO observations. *GPS Solut.* **2016**, *20*, 251–258. [CrossRef]
4. Wu, M.J.; Guo, P.; Fu, N.F.; Hu, X.G.; Hong, Z.J. Improvement of the IRI Model Using F2 Layer Parameters Derived From GPS/COSMIC Radio Occultation Observations. *J. Geophys. Res. Space Phys.* **2018**, *123*, 9815–9835. [CrossRef]
5. Fu, N.; Guo, P.; Wu, M.; Huang, Y.; Hu, X.; Hong, Z. The two-parts step-by-step ionospheric assimilation based on ground-based/spaceborne observations and its verification. *Remote Sens.* **2019**, *11*, 1172. [CrossRef]
6. Mannucci, A.J.; Wilson, B.D.; Yuan, D.N.; Ho, C.H.; Lindqwister, U.J.; Runge, T.F. A global mapping technique for GPS-derived ionospheric total electron content measurements. *Radio Sci.* **1998**, *33*, 565–582. [CrossRef]
7. Syndergaard, S. A new algorithm for retrieving GPS radio occultation total electron content. *Geophys. Res. Lett.* **2002**, *29*, 55-1–55-4. [CrossRef]
8. Yue, X.; Schreiner, W.S.; Hunt, D.C.; Rocken, C.; Kuo, Y.H. Quantitative evaluation of the low Earth orbit satellite based slant total electron content determination. *Space Weather* **2011**, *9*. [CrossRef]
9. Yuan, L.; Jin, S.; Hoque, M. Estimation of LEO-GPS receiver differential code bias based on inequality constrained least square and multi-layer mapping function. *GPS Solut.* **2020**, *24*, 57. [CrossRef]
10. Noja, M.; Stolle, C.; Park, J.; Lühr, H. Long-term analysis of ionospheric polar patches based on CHAMP TEC data. *Radio Sci.* **2013**, *48*, 289–301. [CrossRef]
11. Yuan, L.; Hoque, M.; Jin, S. A new method to estimate GPS satellite and receiver differential code biases using a network of LEO satellites. *GPS Solut.* **2021**, *25*, 71. [CrossRef]

12. Klobuchar, J.A. Ionospheric Time-Delay Algorithm for Single-Frequency GPS Users. *IEEE Trans. Aerosp. Electron. Syst.* **1987**, *AES-23*, 325–331. [CrossRef]
13. Foelsche, U.; Kirchengast, G. A simple "geometric" mapping function for the hydrostatic delay at radio frequencies and assessment of its performance. *Geophys. Res. Lett.* **2002**, *29*, 1473. [CrossRef]
14. Schaer, S. Mapping and predicting the Earth's ionosphere using the Global Positioning System. Ph.D. TThesis, University of Bern, Bern, Switzerland, 1999.
15. Hoque, M.M.; Jakowski, N. Mitigation of ionospheric mapping function error. In Proceedings of the 26th International Technical Meeting of the Satellite Division of the Institute of Navigation, ION GNSS+ 2013, Nashville, Tennessee, 16–20 September 2013; Volume 3, pp. 1848–1855.
16. Xiang, Y.; Gao, Y. An enhanced mapping function with ionospheric varying height. *Remote Sens.* **2019**, *11*, 1497. [CrossRef]
17. Lyu, H.; Hernández-Pajares, M.; Nohutcu, M.; García-Rigo, A.; Zhang, H.; Liu, J. The Barcelona ionospheric mapping function (BIMF) and its application to northern mid-latitudes. *GPS Solut.* **2018**, *22*, 1–13. [CrossRef]
18. Su, K.; Jin, S.; Jiang, J.; Hoque, M.; Yuan, L. Ionospheric VTEC and satellite DCB estimated from single-frequency BDS observations with multi-layer mapping function. *GPS Solut.* **2021**, *25*, 67. [CrossRef]
19. Zhong, J.; Lei, J.; Dou, X.; Yue, X. Assessment of vertical TEC mapping functions for space-based GNSS observations. *GPS Solut.* **2016**, *20*, 353–362. [CrossRef]
20. Huang, Z.; Yuan, H. Analysis and improvement of ionospheric thin shell model used in SBAS for China region. *Adv. Space Res.* **2013**, *51*, 2035–2042. [CrossRef]
21. Bilitza, D.; Altadill, D.; Truhlik, V.; Shubin, V.; Galkin, I.; Reinisch, B.; Huang, X. International Reference Ionosphere 2016: From ionospheric climate to real-time weather predictions. *Space Weather* **2017**, *15*, 418–429. [CrossRef]
22. Gulyaeva, T.L.; Nava, B.; Stanislawska, I. Modeling Center-of-Mass of the Ionosphere From the Slab-Thickness. *Radio Sci.* **2021**, *56*, 1–17. [CrossRef]
23. Pedatella, N.M.; Zakharenkova, I.; Braun, J.J.; Cherniak, I.; Hunt, D.; Schreiner, W.S.; Straus, P.R.; Valant-Weiss, B.L.; Vanhove, T.; Weiss, J.; et al. Processing and Validation of FORMOSAT-7/COSMIC-2 GPS Total Electron Content Observations. *Radio Sci.* **2021**, *56*, 1–15. [CrossRef]
24. Hernàndez-Pajares, M.; Juan, J.M.; Sanz, J.; Garcia-Fernàndez, M. Towards a more realistic ionospheric mapping function. In Proceedings of the XXVIII URSI General Assembly, Delhi, India, 23–29 October 2005.
25. Pignalberi, A.; Pezzopane, M.; Nava, B.; Coïsson, P. On the link between the topside ionospheric effective scale height and the plasma ambipolar diffusion, theory and preliminary results. *Sci. Rep.* **2020**, *10*, 17541. [CrossRef] [PubMed]
26. Wu, M.; Xu, X.; Li, F.; Guo, P.; Fu, N. Plasmaspheric scale height modeling based on COSMIC radio occultation data. *J. Atmos. Solar-Terr. Phys.* **2021**, *217*, 105555. [CrossRef]
27. Liu, L.; Le, H.; Wan, W.; Sulzer, M.P.; Lei, J.; Zhang, M.L. An analysis of the scale heights in the lower topside ionosphere based on the Arecibo incoherent scatter radar measurements. *J. Geophys. Res. Space Phys.* **2007**, *112*. [CrossRef]
28. Liu, L.; Wan, W.; Zhang, M.L.; Ning, B.; Zhang, S.R.; Holt, J.M. Variations of topside ionospheric scale heights over Millstone Hill during the 30-day incoherent scatter radar experiment. *Ann. Geophys.* **2007**, *25*, 2019–2027. [CrossRef]
29. Luan, X.; Liu, L.; Wan, W.; Lei, J.; Zhang, S.R.; Holt, J.M.; Sulzer, M.P. A study of the shape of topside electron density profile derived from incoherent scatter radar measurements over Arecibo and Millstone Hill. *Radio Sci.* **2006**, *41*, 1–11. [CrossRef]
30. Stankov, S.M.; Jakowski, N.; Heise, S.; Muhtarov, P.; Kutiev, I.; Warnant, R. A new method for reconstruction of the vertical electron density distribution in the upper ionosphere and plasmasphere. *J. Geophys. Res. Space Phys.* **2003**, *108*. [CrossRef]
31. Hoque, M.M.; Jakowski, N.; Berdermann, J. A new approach for mitigating ionospheric mapping function errors. In Proceedings of the 27th International Technical Meeting of the Satellite Division of the Institute of Navigation, Tampa, FL, USA, 8–12 September 2014.
32. Gallagher, D.L.; Craven, P.D.; Comfort, R.H. Global core plasma model. *J. Geophys. Res. Space Phys.* **2000**, *105*, 18819–18833. [CrossRef]
33. Fonda, C.; Coïsson, P.; Nava, B.; Radicella, S.M. Comparison of analytical functions used to describe topside electron density profiles with satellite data. *Ann. Geophys.* **2009**, *48*. [CrossRef]
34. Pignalberi, A.; Pezzopane, M.; Rizzi, R. Modeling the Lower Part of the Topside Ionospheric Vertical Electron Density Profile Over the European Region by Means of Swarm Satellites Data and IRI UP Method. *Space Weather* **2018**, *16*, 304–320. [CrossRef]
35. Stankov, T.; Verhulst, S.M. Evaluation of ionospheric profilers using topside sounding data. *Radio Sci.* **2014**, *49*, 181–195. [CrossRef]
36. Kutiev, I.; Marinov, P. Topside sounder model of scale height and transition height characteristics of the ionosphere. *Adv. Space Res.* **2007**, *39*, 759–766. [CrossRef]

Article

On Satellite-Borne GPS Data Quality and Reduced-Dynamic Precise Orbit Determination of HY-2C: A Case of Orbit Validation with Onboard DORIS Data

Hengyang Guo [1], Jinyun Guo [1,*], Zhouming Yang [1], Guangzhe Wang [1], Linhu Qi [1], Mingsen Lin [2,3], Hailong Peng [2,3] and Bing Ji [4]

1   College of Geodesy and Geomatics, Shandong University of Science and Technology, Qingdao 266590, China; xiaoguo@sdust.edu.cn (H.G.); yangzhouming21@mails.ucas.ac.cn (Z.Y.); wangguangzhe@sdust.edu.cn (G.W.); QiLinhu@sdust.edu.cn (L.Q.)
2   National Satellite Ocean Application Service, Beijing 100081, China; mslin@mail.nsoas.org.cn (M.L.); phl@mail.nsoas.org.cn (H.P.)
3   Key Laboratory of Space Ocean Remote Sensing and Application, MNR, Beijing 100081, China
4   Department of Navigation Engineering, Naval University of Engineering, Wuhan 430033, China; jbing1978@126.com
*   Correspondence: guojy@sdust.edu.cn

Citation: Guo, H.; Guo, J.; Yang, Z.; Wang, G.; Qi, L.; Lin, M.; Peng, H.; Ji, B. On Satellite-Borne GPS Data Quality and Reduced-Dynamic Precise Orbit Determination of HY-2C: A Case of Orbit Validation with Onboard DORIS Data. *Remote Sens.* **2021**, *13*, 4329. https://doi.org/10.3390/rs13214329

Academic Editor: Xiaogong Hu

Received: 16 September 2021
Accepted: 25 October 2021
Published: 28 October 2021

Publisher's Note: MDPI stays neutral with regard to jurisdictional claims in published maps and institutional affiliations.

Abstract: Haiyang-2C (HY-2C) is a dynamic, marine-monitoring satellite that was launched by China and is equipped with an onboard dual-frequency GPS receiver named HY2_Receiver, which was independently developed in China. HY-2C was successfully launched on 21 September 2020. Its precise orbit is an important factor for scientific research applications, especially for marine altimetry missions. The performance of the HY2_Receiver is assessed based on indicators such as the multipath effect, ionospheric delay, cycle slip and data utilization, and assessments have suggested that the receiver can be used in precise orbit determination (POD) missions involving low-Earth-orbit (LEO) satellites. In this study, satellite-borne GPS data are used for POD with a reduced-dynamic (RD) method. Phase centre offset (PCO) and phase centre variation (PCV) models of the GPS antenna are established during POD, and their influence on the accuracy of orbit determination is analysed. After using the PCO and PCV models in POD, the root mean square (RMS) of the carrier-phase residuals is around 0.008 m and the orbit overlap validation accuracy in each direction reaches approximately 0.01 m. Compared with the precise science orbit (PSO) provided by the Centre National d'Etudes Spatiales (CNES), the RD orbit accuracy of HY-2C in the radial (R) direction reaches 0.01 m. The accuracy of satellite laser ranging (SLR) range validation is better than 0.03 m. Additionally, a new method is proposed to verify the accuracy of the RD orbit of HY-2C by using space-borne Doppler orbitography and radiopositioning integrated by satellite (DORIS) data directly. DORIS data are directly compared to the result calculated using the accurate coordinates of beacons and the RD orbit, and the results indicate that the external validation of HY-2C RD orbit has a range rate accuracy of within 0.0063 m/s.

Keywords: HY-2C; Satellite-borne GPS; precise orbit determination; reduced-dynamic orbit; SLR; DORIS

1. Introduction

The Haiyang-2C (HY-2C) satellite was successfully launched on 21 September 2020; it was the third satellite launched by China to monitor the dynamic marine environment and the second satellite in the marine dynamic satellite series established as part of China's national civil space infrastructure [1]. The main goal is to obtain high-precision and high-resolution real-time observations of the ocean, such as the sea surface height and significant wave height [2]. The mass of HY-2C is 1677 kg, the Semi major axis is 7328.583 km, the Eccentricity is 0.000012, the average height is 957 km at an inclination of 66°, and the expected lifetime is 3 years. The satellite is connected to HY-2B and HY-2D to form an all-weather, high-frequency, global, marine dynamic environment-monitoring system that covers large and medium scales.

Precise orbit determination (POD) is an important prerequisite for HY-2C to perform altimetry missions, so this satellite requires a high orbit accuracy. To ensure a high-precision orbit, the HY-2C satellite is equipped with three independent payloads, namely, China's self-developed, satellite-borne, dual-frequency global positioning system (GPS) receiver, named HY2_Receiver [1], a Doppler orbitography and radiopositioning integrated by satellite (DORIS) receiver named DGXX and a laser reflector array (LRA) for precise orbit determination by Satellite Laser Ranging (SLR) [3].

The POD of low-Earth-orbit (LEO) satellites using SLR is affected by the number of stations and the meteorological conditions, and the amount of range data is small [4]. Therefore, SLR can only be used for orbit determination with dynamic methods [5]. DORIS POD relies solely on the dynamic method. In contrast GPS POD can be based on the dynamic approach as well as reduced-dynamic (RD) and kinematic methods [6,7].

Since Bertiger et al. successfully applied the satellite-borne GPS POD technique to TOPEX/Poseidon for the first time, GPS receivers have been employed in combination with many LEO satellites, and the satellite-borne GPS POD technique has become increasingly mature as one of the main ways to determine the precise orbit of LEO satellites [8]. The dynamic method of POD relies on an accurate dynamic model and must adjust the corresponding model parameters in the orbit determination process; however, this approach is cumbersome and involves complicated dynamics theory [9]. Consequently, this method is rarely used to determine LEO satellite orbits [10]. A kinematic method of orbit determination only requires geometric information constructed based on satellite-borne GPS observations, and this information is used to calculate the orbit [11]; this approach also has high requirements regarding the geometric configuration and data continuity of GPS satellites. The RD method achieves a balance between dynamic modelling and the use of geometric information [12]. The accuracy of orbit with the RD method is also higher than that based on the dynamic method and the kinematic method [13,14].

Satellite-borne GPS POD technology has been successfully applied in conjunction with the CHAMP, GRACE, Jason-1, GOCE, COSMIC and other LEO satellites performing geodetic and oceanographic surveys. These satellites have strict orbit accuracy requirements [15–20]. Jäggi et al. have researched the application of pseudorandom pulse parameters in orbit determination and successfully determined the RD orbit of CHAMP [21]. Guo et al. determined the precise orbit of HY-2A by using satellite-borne GPS observations [22]. Lin et al. calculate the precise orbit of HY-2A based on GPS data, and the accuracy of the orbit in the radial (R) direction was better than 3 cm [1]. Gong et al. propose a progressive method to determine the POD of LEO satellites, which leads to an improved orbit accuracy [23].

The capability of the GPS receiver directly affects the quality of observations and the accuracy of orbit determination. The quality of observations reflects the performance of the onboard GPS receiver. The quality of GPS data can be analysed in terms of the multipath effect, ionospheric delay (IOD), elevation angle and utilization of GPS observations. Montenbruck et al. use the GPS elevation angle and multipath effects to evaluate the performance of the BlackJack receiver carried by the CHAMP satellite [24]. Hwang et al. have found that the proportion of the kinematic orbits of FORMOSAT-3/COSMIC that can be used for gravity research is less than 30%, and the factors that limit orbit use mainly include multipath effects, cycle slips, IOD and an insufficient number of visible GPS satellites [25,26]. Hwang et al. analyse the performance of GPS receivers carried by FORMOSAT/COSMIC and GRACE in orbit determination through multipath, IOD and phase residual methods [27].

To analyse the performance of the HY2_Receiver and the accuracy of the RD orbit of HY-2C, the quality of GPS observations is analysed, and internal and external validations are performed. A new external method is proposed to validate the accuracy of the RD orbit using onboard DORIS data.

DORIS is a highly accurate tracking system developed by France, with approximately 60 beacons globally for one-way accurate measurements. DORIS contributes to the International Terrestrial Reference Frame (ITRF), and ITRF2014 benefits from improved analysis strategies of the seven contributing International DORIS Service (IDS) analysis centres [28]. The CNES/CLS analysis centre contributes to the geodetic and geophysical research activity through DORIS data analysis [29]. In 1990, the SPOT-2 satellite, first equipped with a DORIS receiver, received a centimetre orbit [30]. DORIS instruments are presently flying aboard SPOT-4 and SPOT-5, Jason-1 and Jason-2, and Envisat. DIODE, developed by Centre National d'Etudes Spatiales (CNES) is successfully applied to Jason-1 [31]. Mercier et al. utilize DORIS phase and pseudorange data to precisely determine the orbits of Jason-2 [32]. Zelensky et al. determine the precise orbit of Envisat by using satellite-borne DORIS observations [33].

Mercier et al. have proposed a POD strategy based on RINEX DORIS 3.0 format with an accuracy of 1~3 cm [32]. Due to higher frequency, shorter wavelength, and higher measurement accuracy, the accuracy of high-frequency phase data of DORIS data (at frequency 2036.25 MHz) is at the millimetre level. Converting the phase data into a format of average range rate, the nominal accuracy reaches 0.5 mm/s [34].

Considering that SLR can serve as an external validation method for orbit determination by the GPS POD method, the new method proposed in this paper also innovates the use of onboard DORIS as an external validation method.

The primary aim of this paper is to assess the performance of the HY2_Receiver and the accuracy of HY-2C RD orbits, and to estimate the impact of PCO and PCV models on the orbit of HY-2C simultaneously. The rest of this article is arranged as follows. Section 2 introduces the HY-2C satellite information, data required for orbit determination and orbit quality assessment, introduces the method and strategy of HY-2C orbit determination, proposes a new external method to assess LEO orbits with onboard DORIS data directly and discusses the reliability of the new external method to validate orbits with onboard DORIS data directly. Section 3 analyses the quality of GPS data, estimates PCO and PCV models and studies the influence of these models in POD; based on phase residuals, differences between overlapping orbits are determined, comparisons with precise scientific orbits (PSO) are made, independent SLR validation is performed to assess the accuracy of HY-2C RD orbits and an example of HY-2C RD orbits and PSO validations with onboard DORIS data are given. Finally, the conclusions are presented.

2. Materials and Methods

2.1. Materials

2.1.1. HY-2C Spacecraft

Based on the Ziyuan series satellite platform, the load compartment structure of the Haiyang-2 series of satellites was reconfigured to meet the requirements of additional payloads. HY-2C has the same size and structure as HY-2A and HY-2B, but onboard equipment has been updated [7].

HY-2C is mainly equipped with a radar altimeter, a microwave scatterometer, and a microwave radiometer. The equipment used for POD includes an onboard DORIS receiver DGXX, a space-borne dual-frequency GPS receiver named HY2_Receiver that was developed in China, and a laser reflector array (LRA) [7].

A satellite-navigation body reference coordinate system (SNCS) is used [35], and a view of the satellite in this coordinate system is shown in Figure 1. Notably, the scatterometer antenna, data transmission antenna, altimeter antenna, DORIS antenna, and LRA are parallel to the Z axis.

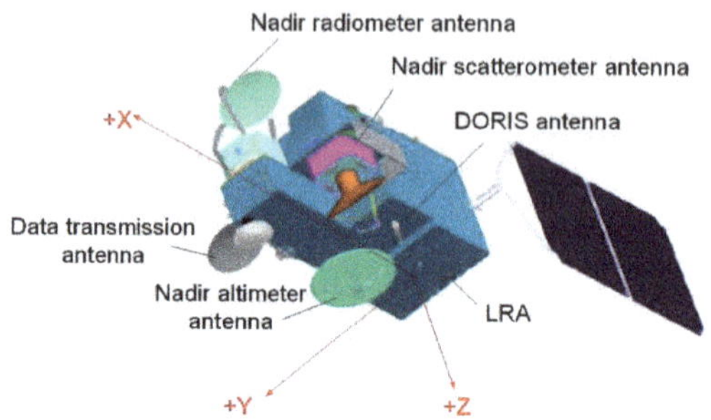

Figure 1. Satellite View in the SNCS [https://osdds.nsoas.org.cn (accessed on 2 June 2021)].

The National Satellite Ocean Application Service (NSOAS) provide some parameters, such as the correction from the antenna phase centre to the satellite centre of mass (Sensor Offset). The Centre National d'Etudes Spatiales (CNES) provides the centre of mass (COM) of the satellite [36], as shown in Table 1. The DORIS antenna and LRA are set in the positive direction of the Z axis, and the GNSS antenna is set in the negative direction of the Z axis [36].

Table 1. Antenna and COM parameters of HY-2C.

Parameters	Satellite Mass (kg)	dX (m)	dY (m)	dZ (m)
Sensor offset		0.3492	−0.1794	−1.3671
COM (Beginning of life)	1677.0	1.3320	0.0086	0.0034
COM (End of life)	1591.0	1.3755	0.0090	0.0036

In order to facilitate analysis of specific perturbation, the force acting on the satellite is specified in a satellite orbit coordinate system (radial, along-track, cross-track; RTN). This RTN system is well suited for easy analyses of the POD quality.

The POD team from the CNES provides the correction parameters for the DORIS antenna phase centre relative to the COM of the satellite. Relevant documentation can be found at: ftp://ftp.ids-doris.org/pub/ids/satellites/DORISSatelliteModels.pdf (accessed on 24 June 2021). The DORIS system operates at two frequencies, 2 GHz and 400 MHz, and the positions of these antennas in the SNCS are listed in Table 2.

Table 2. Positions of the DORIS antennas in the SNCS.

Antenna	X (m)	Y (m)	Z (m)
2 GHz	0.7100	−0.8010	1.3190
400 MHz	0.7100	−0.8010	1.1500

2.1.2. Data for GPS POD and Validation

Satellite-borne GPS observations are released by NSOAS.

Precise GPS satellite ephemeris data are provided by the Center for Orbit Determination in Europe (CODE) [37] with a sampling interval of 15 min. Precise GPS satellite clock offsets are provided by CODE at a sampling interval of 30 s. Earth rotation parameters are also provided by CODE. Since the precise ephemeris data are provided by CODE every 15 min and the sampling interval of observations is 30 s during orbit determination, it is necessary to interpolate the precise ephemeris data [38]. Bernese 5.2 uses the ninth-order Lagrange interpolation method to interpolate the precise GPS ephemeris data [39].

PSO data for orbit accuracy analysis are released by CNES, and these data are obtained with a dynamic method using onboard DORIS observations [40].

SLR tracking data are provided by the International Laser Ranging Service (ILRS) [41].

DORIS observations are provided by IDS in RINEX DORIS 3.0 format. RINEX has been developed by the Astronomical Institute of the University of Berne for the easy exchange of the GPS data to be collected during the large European GPS campaign, EUREF 89. The RINEX format can also be easily adapted to contain DORIS data as it makes little assumption about the actual content of the data file, but only constrains the formatting of the data. Thus, it can be easily adapted to contain data other than GNSS. DORIS data includes pseudorange and phase, which is the same as the format of GNSS data. For a detailed description of the RINEX DORIS 3.0 format, please refer to the documents provided by the IDS: https://ids-doris.org/about-doris-rinex-format.html (accessed on 25 June 2021). When using DORIS data to check an orbit [42], it is necessary to convert the phase in 3.0 format to the range rate, i.e., to convert the 3.0 data format to the 2.2 data format. A summary of the data sources is provided in Table 3.

Table 3. Sources of data.

Data	Organization	Address
GPS observations	NSOAS	https://osdds.nsoas.org.cn (accessed on 2 June 2021)
GPS precise ephemeris	CODE	ftp://ftp.aiub.unibe.ch/CODE (accessed on 2 June 2021)
GPS precise clock offset	CODE	ftp://ftp.aiub.unibe.ch/CODE (accessed on 2 June 2021)
Earth rotation parameter	CODE	ftp://ftp.aiub.unibe.ch/CODE (accessed on 2 June 2021)
PSO of HY-2C	CNES	https://ids-doris.org (accessed on 2 June 2021)
SLR tracking data	ILRS	https://cddis.nasa.gov/archive/slr (accessed on 2 June 2021)
DORIS data	IDS	ftp://doris.ign.fr/pub/doris/data/h2c (accessed on 24 June 2021)

2.2. Method

2.2.1. Reduced-Dynamic Method

When Bernese 5.2 GNSS software [39] is applied, an a priori dynamic orbit is used for GPS clock synchronization and then for the preprocessing of phase data. The cycle slips in the phase observations are marked, and for each cycle slip, a new ambiguity parameter is set in the parameter's estimation process [43]. The "cleaned" phase observations are used in the GPSEST module to determine the dynamic satellite parameters. These parameters include the initial state vector (6 Keplerian elements), 9 solar radiation coefficients [44] and 3 pseudostochastic pulse parameters in the R, T and N directions. Arc lengths of 24 h were selected, and the pulses are estimated every 6 min [45].

2.2.2. Orbit Determination Strategies for HY-2C

The models and parameters used together with satellite-borne GPS observations to achieve RD orbit determination are shown in Table 4.

Table 4. RD orbit determination strategies of HY-2C.

Model/Parameters	Description
Mean earth gravity	EGM2008_SMALL
Ocean tides	FES2004
Solid-earth tides	TIDE2000
N-body	JPL DE405
Relativity	IERS2010XY
GPS phase model	IGS14.atx
Pseudostochastic pulses	Estimate every 6 min
Elevation cutoff	5°
Sampling interval	30 s

2.2.3. Validation of Orbits with DORIS Data

Suppose that the DORIS phase observations corresponding to the two frequencies f_1 and f_2 at time t_i are $\varphi_1(t_i)$ and $\varphi_2(t_i)$, respectively. After ionospheric correction, the corresponding phase value at f_1 is:

$$\varphi_1'(t_i) = \varphi_1(t_i) + \frac{f_2}{\left(f_2^2 - f_1^2\right)} \times \left[f_1 \cdot \varphi_2(t_i) - f_2 \cdot \varphi_1(t_i)\right] \tag{1}$$

If $t_{i+1} = t_i + \Delta t$ (where $9\ s < \Delta t < 11\ s$), the average distance change rate within Δt is:

$$d\Delta\varphi(t_i) = \frac{\left[\varphi_1'(t_{i+1}) - \varphi_1'(t_i)\right]}{\Delta t} \tag{2}$$

The time system used by DORIS is International Atomic Time (TAI), while the RD orbit relates to GPS time (GPST). Therefore, the time offset between both scales has to be considered before validation is performed.

Onboard DORIS observations are used to check the satellite orbit, and they can be obtained based on the average distance change rate:

$$d\Delta\varphi_s\left(\Delta t_{jk}\right) = \frac{\Delta\rho_{jk} + \Delta\rho_{rel} + \Delta\rho_{erp} + \Delta\rho_{tide} + \Delta\rho_{trop} + \Delta\rho_{com} + \varepsilon}{\Delta t_{jk}} \tag{3}$$

where Δt_{jk} represents the time difference between t_k and t_j; $\Delta\rho_{jk}$ represents the distance of satellite movement in the Δt_{jk} time interval; $\Delta\rho_{jk} = \rho_k - \rho_j$ and $\rho_i = \sqrt{(x_i - x_b)^2 + (y_i - y_b)^2 + (z_i - z_b)^2}$. When $i = k$, (x_k, y_k, z_k) represents the position of the satellite at time t_k; when $i = j$, (x_j, y_j, z_j) represents the position of the satellite at time t_j. (x_b, y_b, z_b) represents the precise location of the DORIS beacon stations. The beacon station location information can be obtained from the IDS website: https://ids-doris.org/doris-system/tracking-network/site-logs.html (accessed on 24 June 2021). $\Delta\rho_{rel}$ represents the effect of relativity in the time interval Δt_{jk}; $\Delta\rho_{erp}$ represents the correction of the Earth rotation parameters in the time interval Δt_{jk}; $\Delta\rho_{tide}$ represents the correction of the tides in the time interval Δt_{jk}; $\Delta\rho_{trop}$ represents the correction of the tropospheric delay in the time interval Δt_{jk}; $\Delta\rho_{com}$ represents the correction of the DORIS antenna phase centre; and ε represents random error.

The difference between the average range rate obtained by Equation (2) and the average range rate obtained by Equation (3) is calculated to obtain the DORIS validation residuals:

$$res = d\Delta\varphi_s\left(\Delta t_{jk}\right) - d\Delta\varphi(t_i) \tag{4}$$

When using DORIS data to validate the RD orbit of HY-2C, we need to take into account some errors, such as the DORIS phase-centre correction and the impact of tidal effects. CNES has released the correction data of DORIS phase centre, we only need to correct it to the signal propagation path between HY-2C and the beacon stations. When calculating tide corrections, it is necessary to consider solid-earth tide, ocean tide and polar tide (although there are few DORIS stations at high latitudes) corrections. For tropospheric delay correction, the Saastamoinen model is used to calculate the zenith delay [46], and then the mapping function Niell Mapping Function (NMF) is used to map it to the signal propagation path [47].

According to the law of error propagation, the median error of the range rate is calculated using the satellite positions corresponding to two epochs with an interval of 10 s. The range rate function is:

$$v = \frac{dD}{dt} = \frac{\sqrt{(x_i - x_{i+1})^2 + (y_i - y_{i+1})^2 + (z_i - z_{i+1})^2}}{t} \tag{5}$$

where v represents the range rate in the time interval t; (x_i, y_i, z_i) and $(x_{i+1}, y_{i+1}, z_{i+1})$ represent the satellite positions in two adjacent epochs in the time interval t, respectively; and D represents the distance of satellite movement in the time interval t.

The total differential of the position parameters in Equation (5) is calculated to obtain the medium error relation:

$$m_v = \frac{\sqrt{2a_x{}^2 m_x{}^2 + 2a_y{}^2 m_y{}^2 + 2a_z{}^2 m_z{}^2}}{t} \tag{6}$$

where $a_x = \frac{(x_i - x_{i+1})}{D}$, $a_y = \frac{(y_i - y_{i+1})}{D}$, and $a_z = \frac{(z_i - z_{i+1})}{D}$; m_x, m_y, and m_z respectively represent the errors of the satellite position in three directions in an Earth-Centred and Earth-Fixed coordinate system (ECEF).

If the satellite has a 2 cm error in three directions, according to Equation (6), the median error of the range rate is 0.0052 m/s. Considering other random errors when using DORIS data to check orbits, the error is doubled to v = 0.0104 m/s as the threshold.

The law of error propagation verifies that the method proposed to check satellite orbits using space-borne DORIS data is reasonable and can be used to establish a precise external data validation method

3. Result and Analysis

3.1. Quality Assessment of HY-2C GPS Observations

In this section, the GPS data are preprocessed with G-Nut/Anubis software [48] and the performance of the HY2_Receiver is assessed by using indicators such as multipath effects, the IOD, the cycle slip and the utilization of GPS observations. G-Nut/Anubis is the third GNSS software developed by the Research Institute of Geodesy, Topography and Cartography in the Czech Republic; it can perform quality inspection and analysis operations based on observations and supports the RINEX 3 file format.

3.1.1. Multipath Effect

Theoretically, the influence of multipath effects on the pseudorange can reach up to tens of metres, and the influence on the carrier phase is only one quarter of its wavelength. Therefore, when calculating multipath effects, the influence of the phase multipath error is usually ignored. Anubis uses a linear combination of pseudorange and carrier phase information to calculate the pseudorange multipath $MP1$ and $MP2$ [48]:

$$MP1 = M_1 - \left(1 + \frac{2}{\alpha - 1}\right)\lambda_1 N_1 + \left(\frac{2}{\alpha - 1}\right)\lambda_2 N_2 - \left(1 + \frac{2}{\alpha - 1}\right)m_1 + \left(\frac{2}{\alpha - 1}\right)m_2 \tag{7}$$

$$MP2 = M_2 - \left(\frac{2\alpha}{\alpha - 1}\right)\lambda_1 N_1 + \left(\frac{2\alpha}{\alpha - 1} - 1\right)\lambda_2 N_2 - \left(\frac{2\alpha}{\alpha - 1}\right)m_1 + \left(\frac{2\alpha}{\alpha - 1} - 1\right)m_2 \tag{8}$$

where m_i is the multipath phase noise; M_i is the pseudorange multipath effect on both frequencies; λ_i and N_i represent the wavelength of L_i and the integer ambiguity of L_i, respectively; and $\alpha = \frac{f_1{}^2}{f_2{}^2}$, where f_i represents the frequency of different bands.

To specifically analyse the relationship between multipath effects and the elevation angle, the GPS satellite numbered G09 on DOY 348 in 2020 is taken as an example. The period from 20:25 to 21:05 (UTC) is selected. The pseudorange multipath as well as the elevation angle of G09 as observed from HY-2C are calculated and shown in Figure 2. Figure 2a illustrates a statistical assessment of $MP1$ and the elevation angle, and Figure 2b gives a statistical assessment of $MP2$ and the elevation angle.

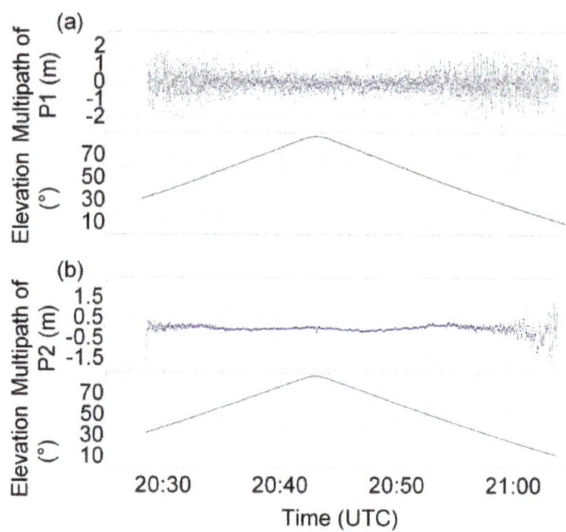

Figure 2. Multipath effects of HY-2C and G09 satellite on DOY 348 in 2020 (**a**) MP1, (**b**) MP2.

Figure 2 shows that *MP1* considerably fluctuates when the elevation angle is less than 30°, and the range of fluctuations is between −2 m and 2 m; when the elevation angle is larger than 30°, the fluctuations are between −1 m and 1 m. The fluctuation of *MP2* is between −1.5 m and 1.5 m, but when the elevation angle is larger than 30°, the fluctuations range between −0.5 m and 0.5 m. By comparing Figure 2a with Figure 2b, it can be found that the *MP1* value is more sensitive to low elevation angles than the *MP2* value.

Table 5 summarizes the report of Anubis regarding the GPS data obtained between DOY 348 and DOY 354 in 2020. Cycle slip represents the ratio of the actual number of epochs observed in a certain period of time to the number of epochs in which a cycle slip occurs. The data utilization represents the ratio of the actual number of epochs to the theoretical number of epochs observed by the GPS receiver. In reality, due to the effects of cycle slips and elevation angle changes, observations will be interrupted, leading to missing data. The cycle slip and data utilization are mainly used to reflect the antijamming capability of satellite-borne GPS receivers. These two indicators can be used to evaluate the performance of the HY2_Receiver.

Table 5. Data quality assessment of HY-2C.

DOY	MP1/cm	MP2/cm	Cycle Slip	Data Utilization/%
348	18.0	12.9	1/33	98.26
349	19.0	12.6	1/40	96.41
350	19.0	12.5	1/43	99.98
351	18.6	12.1	1/39	99.98
352	18.4	14.1	1/34	99.99
353	20.0	17.0	1/48	99.99
354	18.7	13.1	1/51	99.98

In Table 5, the *MP1* values are all smaller than 20.0 cm, and the *MP2* values are all smaller than 17.0 cm. Additionally, the variations in *MP2* are smaller than those in *MP1*, which indicates that the observations of the L_1 band are more susceptible to the influence of multipath effects. As shown in Table 5, the utilization of HY-2C satellite-borne GPS observations is higher than 98% on days except DOY 349 in 2020. The utilization of satellite-borne GPS observations is higher than 99.9% over these 5 days, starting from DOY 350 in 2020. The utilization indicates that the HY2_Receiver operates normally in orbit. When the completeness of observations is normal, the cycle slip is small, which generally indicates that the multipath error and IOD error of GPS observations are large [49].

3.1.2. Ionospheric Delay

A geometry-free combined LEO satellite carrier phase can be expressed as [50]:

$$\lambda_1 L_1(t_j) - \lambda_2 L_2(t_j) = -\Delta I(t_j) + \lambda_1 N_1(t_j) - \lambda_2 N_2(t_j) \tag{9}$$

where $\Delta I(t_j) = I_1(t_j) - I_2(t_j), j = 1, 2, \cdots, n$; I_i is IOD, $i = 1, 2$; L_i represents the observed value of the carrier phase; and λ_i and N_i represent the wavelength of L_i and integer ambiguity of L_i, respectively.

The differences between the combined observations of two consecutive epochs can be obtained by:

$$
\begin{aligned}
D(t_j) &= \lambda_1 L_1(t_j) - \lambda_2 L_2(t_j) - \lambda_1 L_1(t_{j-1}) + \lambda_2 L_2(t_{j-1}) \\
&= -\Delta I(t_j) + \lambda_1 N_1(t_j) - \lambda_2 N_2(t_j) + \Delta I(t_{j-1}) - \lambda_1 N_1(t_{j-1}) + \lambda_2 N_2(t_{j-1})
\end{aligned} \tag{10}
$$

When there is no cycle slip, the difference in ambiguity between two consecutive epochs is 0, and Equation (10) is called the ionospheric residual combination. Under normal circumstances, the IOD residual curve is smooth, but when the ionospheric residual error changes significantly in a short period of time, cycle slip may occur [13]. The corresponding expression of the rate of IOD change is:

$$\dot{D}(t_j) = \frac{D(t_j)}{\Delta t_j} \tag{11}$$

where $\Delta t_j = t_j - t_{j-1}$.

Equation (10) is used to calculate the IOD residuals, and Equation (11) is used to calculate the rate of IOD change. A 24-hour window will make the picture too dense to clearly show the variation, so the time period (20:00–21:20) is chosen on DOY 348 in 2020 to highlight the variations. Figure 3a shows the IOD residuals in the selected time span, and Figure 3b shows the rate of IOD change in the selected time span. Figure 3a shows that most of the IOD residuals remain at low levels. However, from 20:20 to 20:40, the residual values of the G13, G14, G28 and G30 satellites are large, and from 20:30~21:00, the residuals of the G06 and G09 satellites are large. The colour of the rate of IOD change significantly darkens at the end of a period of data collection. Studies have shown that the discontinuity of carrier-phase data and the occurrence of abnormal values will cause this phenomenon [27].

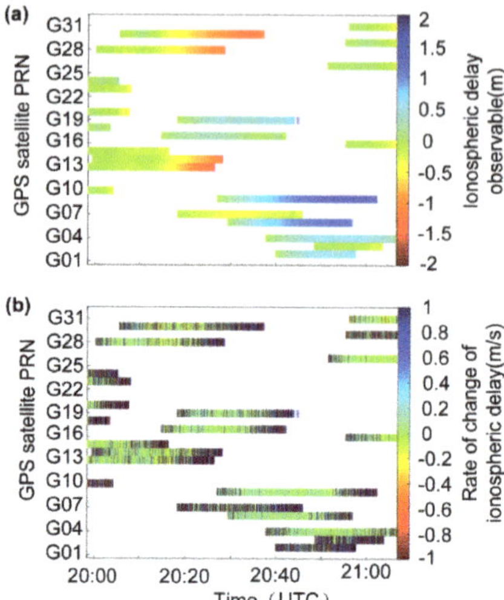

Figure 3. IOD residuals and rate of IOD change on DOY 348 in 2020 (**a**) IOD residuals, (**b**) rate of IOD change.

3.2. Assessment of Orbit Accuracy

Before performing satellite orbit determination, it is necessary to consider issues related to the phase centre of LEO satellite antennas. Onboard GPS observations are based on the distance from the instantaneous phase centre of the GPS satellite antenna to the instantaneous antenna phase centre at the time the signal from the LEO satellite is received; additionally, the reference point of the precise orbit of the LEO satellite is the centre of mass. Generally, the mean antenna phase centre (MAPC) and the antenna reference point (ARP) do not coincide, and the deviation is the PCO. Due to antenna manufacturing deviations and because LEOs are affected by the environment during operation, the GPS signal centre will change. The deviation between the instantaneous phase centre and the average phase centre is PCV, which is a function of the elevation and azimuth angle between the LEO and GPS satellites [51].

Carrier-phase residual analysis, overlapping orbit comparison, PSO comparison, and SLR range validation are used to validate the accuracy of the RD orbit of HY-2C. The first two methods are used to establish internal validation, and the latter are used to establish external validation.

3.2.1. Effect of PCV and PCO on Orbits

In this study, satellite-borne GPS data are selected to estimate PCO and PCV in orbit over 7 days, starting from DOY 348 in 2020 [25]. Due to the high coupling of PCO and PCV, they cannot be estimated at the same time. PCO is substituted into the observation equation as an unknown parameter and the orbital parameters are estimated at the same time [52]. According to the principle of least squares, PCO is calculated for multiple days, and the average is taken as the final corrected value [53]. Similarly, the PCV values are estimated at each grid point. This method is not susceptible to the influence of factors such as receiver clock error and ambiguity among parameters [54,55].

To systematically validate the impacts of PCO and PCV on orbit determination, three schemes are designed: Scheme 1 does not use PCO and PCV information; Scheme 2 only uses PCO information; and Scheme 3 uses PCO and $10° \times 10°$ PCV information. Figure 4 shows the $10° \times 10°$ PCV model estimated in the satellite GPS antenna coordinate system.

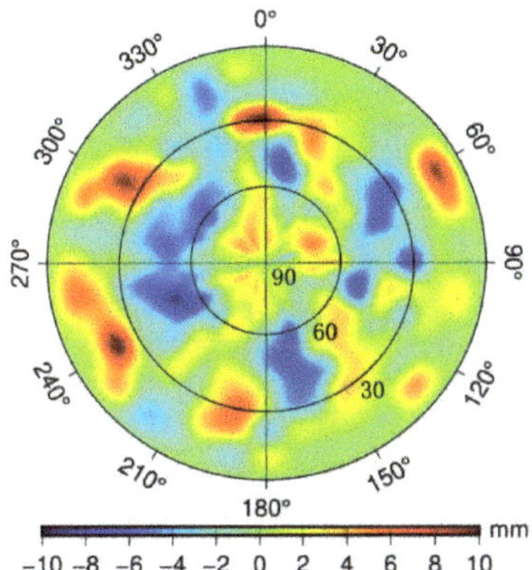

Figure 4. Azimuth-elevation maps of the PCV model of the HY-2C (10° × 10°).

Satellite-borne GPS data are used to estimate the PCV model for the antenna of the HY2_Receiver, and the direct method of estimating PCV is characterized by a more refined model than that for PCV. Suppose each grid point is unknown, we calculate the value of the grid with the orbital parameters directly. The PCV map of HY-2C uses a scale of −10 to 10 mm, but with extreme values of −9.09 mm and 10.37 mm. In Figure 4, we can find that when the elevation angle is small (from 0° to 30°), most of the absolute values of the PCV are small, but there are also large, speckled values with some empty values; when the elevation angle is between 80° and 90°, there will also be a phenomenon of empty values, although they are not obvious. So, the resulting model is more refined, and the map illustrates a spot-like distribution.

The above three schemes are used to determine satellite orbits. Table 6 lists the results of the three types of orbits compared with PSO in the R, T and N directions within 7 days. 3DRMS is used to evaluate the different orbit results.

Table 6. Residuals between RD orbit and PSO in different orbit determination schemes within 7 days.

Scheme	Direction	Min (m)	Max (m)	Mean (m)	STD (m)	RMS (m)	3DRMS (m)
	R	−0.104	0.045	0.001	0.013	0.013	
Scheme 1	T	−0.132	0.091	−0.002	0.027	0.028	0.034
	N	−0.085	0.045	0.003	0.016	0.016	
	R	−0.039	0.051	0.001	0.011	0.011	
Scheme 2	T	−0.087	0.124	−0.002	0.021	0.021	0.027
	N	−0.044	0.045	0.002	0.012	0.012	
	R	−0.046	0.035	0.000	0.010	0.010	
Scheme 3	T	−0.075	0.078	−0.002	0.020	0.020	0.025
	N	−0.038	0.033	−0.001	0.011	0.011	

The accuracy of PSO in the R direction of this orbit is better than 0.015 m [40]. The report from CNES indicates that the ambiguity fixation efficiency of GPS observations for the PSO of HY-2C is as high as 99%, which reflects the excellent performance of China's domestic satellite-borne GPS receivers.

When comparing the RD orbit with PSO, it is necessary to consider unifying the RD orbit and PSO to the time system. Since the time system of PSO for HY-2C provided by CNES is International Atomic Time (TAI) while the RD orbit relates to GPS time (GPST), there is a 19 s deviation between TAI and GPST, i.e., $TAI - GPST = 19\ s$. It is necessary to preprocess the PSO in advance and convert the time system of PSO to that of GPST to facilitate comparisons.

For Scheme 2, decreases of 0.002 m in the R direction, 0.007 m in the T direction, and 0.004 m in the N direction are observed compared with the residual results for Scheme 1. When taking into account PCO corrections, the error in orbit estimation is reduced, and the orbit accuracy is effectively improved. The accuracy of the orbit increase in the T direction is significantly higher than that in the other two directions. A comparison of Scheme 2 and Scheme 3 indicates that the RMS values for Scheme 3 are 0.001 m smaller in the R direction, 0.001 m in the T direction, and 0.001 m in the N direction than those for Scheme 2; additionally, the orbit accuracy is improved after the PCV model is introduced, but the model range is limited. This result is mainly because the PCO parameters estimated at the same time as the dynamic parameters effectively limit the systematic errors in the residuals of the observations, and PCV is relatively small for the corrections of observations.

Compared to Scheme 1, the accuracy of orbit obtained for Scheme 2 and Scheme 3 is improved, and 3DRMS decreases from 0.034 m to 0.027 m and then to 0.025 m, which is attributed to the improvement in the tangential orbit accuracy.

In summary, based on the RD orbit and PSO comparisons, the RMS of the orbital residual in the R direction is 1.0 cm, 2.0 cm in the T direction, and 1.1 cm in the N direction, and the 3DRMS result is 2.5 cm. The HY-2C RD orbit determination method yields the highest accuracy in the R direction and the worst accuracy in the T direction.

The result of comparing the RD orbit with PSO is plotted as a dashed line, as shown in Figure 5.

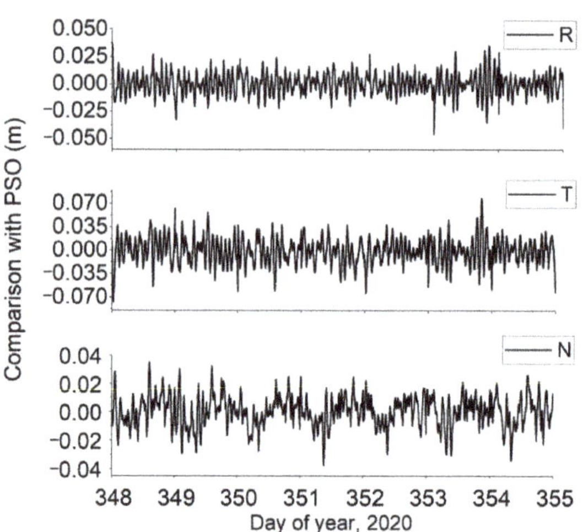

Figure 5. Line chart of orbit residuals from the PSO comparison within 7 days.

Figure 5 shows that the residuals are in the ranges of ±0.04 m in the R direction, ±0.09 m in the T direction, and ±0.05 m in the N direction. The residuals display obvious fluctuations at 0 h and 24 h every day because when HY-2C uses the RD method for orbit determination, the constraints at both ends of the orbit determination arc are relatively weak, resulting in obvious boundary effects. The obviously extreme values show up because the pseudostochastic pulse parameters fail to capture the influence of errors, such as those associated with atmospheric drag and solar radiation pressure. Additionally, the RMS of the residuals for the RD orbit and PSO comparison is plotted as a histogram, as shown in Figure 6.

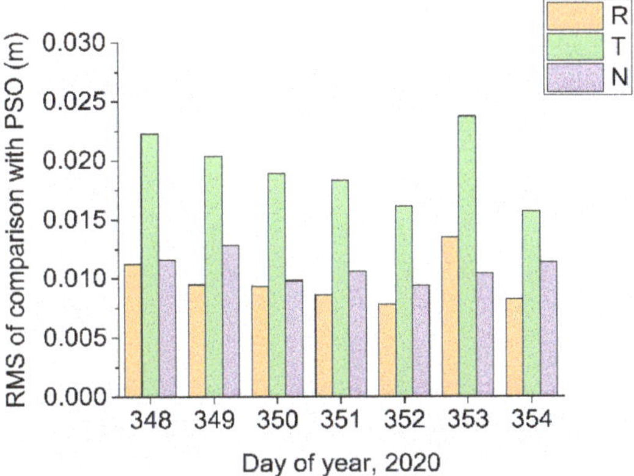

Figure 6. Histogram of residuals from the PSO comparison.

From the results of the RD orbit comparison with PSO, as shown in Figure 6, the RMS value of the difference in the R direction is worse than that in the N direction on DOY 353 in 2020. The RMS values in the R direction on the other 6 days are less than those in the N direction, and the residuals exhibit the largest RMS value in the T direction.

3.2.2. Analysis of Carrier-Phase Residuals

For RD orbit determination, the carrier-phase residuals include modelled errors and nonmodelled errors. Actual PCV values are subtracted from modelled PCV values, and the difference between the two is directly reflected in the linearization after the phase residual is calculated. When the observation model and dynamical model used in LEO POD are extremely consistent with the actual conditions, the carrier-phase residuals are the observation's noise level. The RMS value of the carrier-phase residuals using an ion-free combination can be used as one of the indicators in internal compliance accuracy evaluation [51,56].

The carrier-phase residuals are calculated over 7 days, starting from DOY 348 in 2020. The carrier phase residuals of all GPS satellites are plotted in Figure 7, and summaries are given in Table 7.

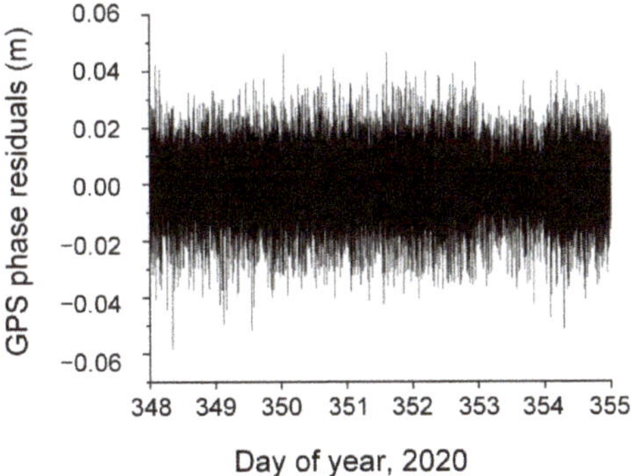

Figure 7. Phase residuals.

Table 7. Phase residual summary statistics.

DOY	Min (m)	Max (m)	Mean (m)	RMS (m)
348	−0.051	0.040	0	0.008
349	−0.042	0.037	0	0.008
350	−0.041	0.043	0	0.008
351	−0.043	0.047	0	0.008
352	−0.040	0.046	0	0.008
353	−0.052	0.037	0	0.008
354	−0.058	0.042	0	0.008

As shown in Figure 7, the residuals of the carrier phase within 7 days fluctuate between −0.06 and 0.05 m, and the residuals are mostly distributed within ±0.02 m, and the residuals fluctuate little, with stable changes. Table 7 suggests that the average RMS value of the 7-day carrier-phase residuals is about 0.008 m. The carrier-phase residual analysis indicates that the HY-2C RD orbit determination strategy is reliable; additionally, data preprocessing can effectively weaken the influence of the cycle slip of the phase data and the RD orbit accuracy obtained is high.

3.2.3. Overlap Validation Accuracy

If both the observation model and the dynamical model can correctly reflect satellite orbit characteristics, the satellite orbits of the two arcs of the orbit determination should exhibit high consistency. Due to the errors in the actual orbit determination based on observations and the model used, the results of comparisons can be used to validate the accuracy of orbits, although the systematic error of orbit determination cannot be detected through the comparison of overlapping orbits [57]. Two separate orbit determination processes are used to calculate the satellite orbits corresponding to two different time periods. The first part of the arc is 0~18 h, the second part of the arc is 12~24 h, and the overlapping time of the two arcs is 6 h. Taking DOY 348 in 2020 as an example, Figure 8 shows the residuals of the orbit comparison during the overlapping period.

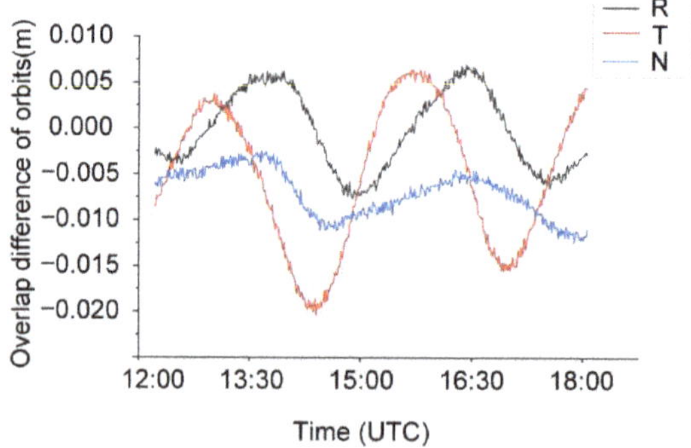

Figure 8. Residuals of overlapping orbits on DOY 348 in 2020.

As shown in Figure 8, the residual fluctuates between −0.0075 m and 0.0075 m in the R direction, between −0.02 and 0.0075 m in the T direction, and between −0.0125 and 0.0025 in the N direction. The results of comparing the overlapping orbits on DOY 348 in 2020 indicate that the orbits display large fluctuations in the T direction and small fluctuations in the N direction. Figure 9 shows the RMS value of the comparison.

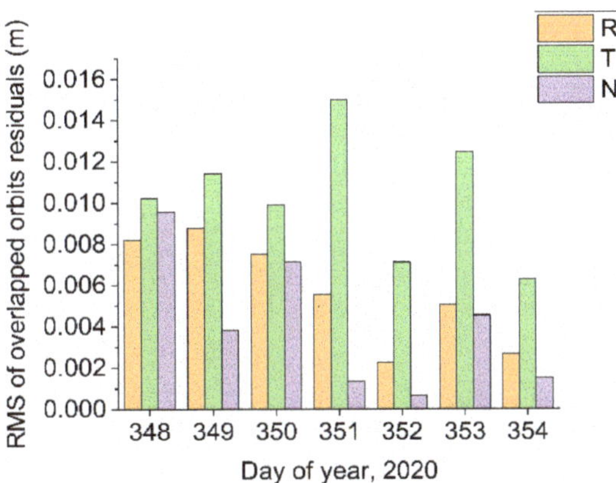

Figure 9. RMS of residuals of overlapping orbits.

As shown in Figure 9, the RMS values of overlapping orbits in the T direction are larger than those in the other two directions, indicating that the overlapping orbit comparison results are the worst in the T direction. The RMS value of the overlapping orbits in the R direction is larger than that in the N direction, except on DOY 348 in 2020. The RMS values in the R direction on the other 6 days are all greater than the RMS values in the N direction. Detailed information on the comparison is given in Table 8.

Table 8. Summary statistics for the residuals of overlapping orbits within 7 days.

Direction	Min (m)	Max (m)	Mean (m)	STD (m)	RMS (m)
R	−0.014	0.042	0	0.006	0.006
T	−0.045	0.031	0.001	0.011	0.011
N	−0.024	0.011	−0.002	0.005	0.006

As shown in Table 8, the maximum difference between the overlapping orbits is 0.042 m in the R direction, 0.031 m in the T direction, and 0.011 m in the N direction; additionally, the RMS value of the 7-day overlapping orbit difference is 0.006 m in the R direction, 0.011 m in the T direction, and 0.006 m in the N direction. The RMS value in the three directions is approximately 0.01 m.

3.2.4. SLR Range Validation

SLR range validation uses LEO satellite coordinates from RD orbits and SLR station coordinates to calculate the distance between stations and LEO satellites, and compares the result with the observations of SLR stations in the corresponding epoch to evaluate the accuracy of the RD method for orbit determination [58]. When conducting SLR range validation, it is necessary to consider the influence of tidal corrections (including ocean tides, solid-earth tides, and polar tides) and plate motions at stations [4]. The correction of observations mainly includes tropospheric delay correction, general relativity correction, centroid compensation correction and station eccentricity correction [59].

SLR observations are used to independently check the RD orbit accuracy of HY-2C. According to [60], the coordinates of SLR stations are based on the Station Location Reference Frame (SLRF) 2014, and relevant data can be obtained at https://ilrs.gsfc.nasa.gov/network/site_information/index.html (accessed on 2 June 2021). The elevation cutoff angle is set to 10° [61] and the tropospheric delay error can be modelled [59].

The report provided by the CNES states that during the period from October 2020 to April 2021, the core stations participating in the HY-2C SLR observation mission included those numbered 7090, 7105, 7810, 7839, 7840, 7941, 7119, and 7501. Among them, few high-elevation-angle observations are available [40]. From DOY 348-DOY 354 in 2020, two of the eight core stations had no data (Station 7501 and Station 7839). In addition to these six SLR core stations, nine SLR stations provided data for validation. Figure 10 shows the flight trajectory of HY-2C and the distribution of stations.

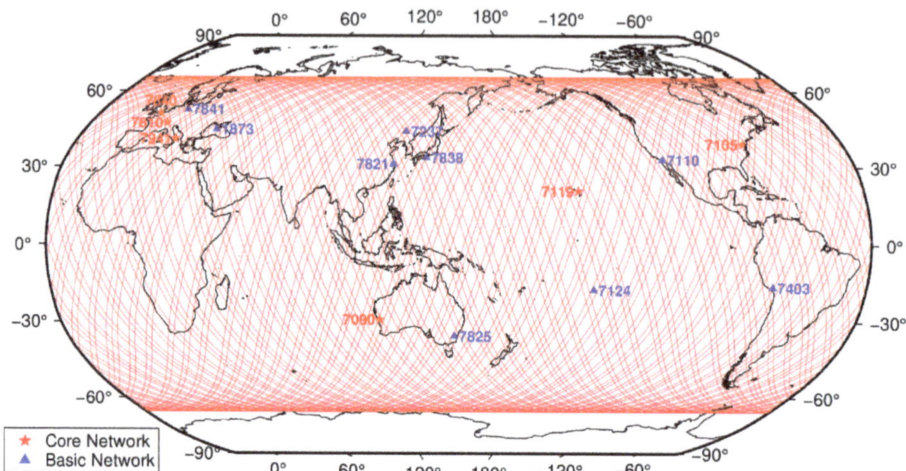

Figure 10. Satellite trajectory and SLR station distribution map.

As shown in Figure 10, the red stars indicate SLR core stations that participated in the laser ranging of HY-2C, and the blue triangles indicate other stations used in the laser ranging of HY-2C.

The range residuals between the SLR NP data and the range differences between HY-2C RD orbit and the ground station coordinates were calculated. The SLR range validation results were plotted as a histogram, as shown in Figure 11. Table 9 lists the statistical results for SLR range validation for HY-2C.

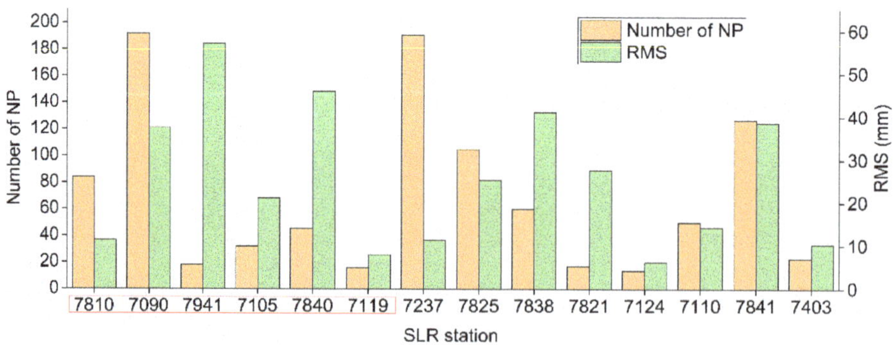

Figure 11. RMS of SLR validation residuals.

Table 9. Summary statistics for the residuals of SLR range validation.

Number of Stations	Number of NP	Min (m)	Max (m)	Mean (m)	STD (m)	RMS (m)
15	1009	−0.077	0.084	0.005	0.029	0.030

As shown in Figure 11, the red boxes denote SLR core stations involved in the HY-2C laser-ranging observation mission. Among the six core stations, the RMS values of the stations except stations 7941 and 7840 are all less than 0.04 m. For the other eight basic stations, except stations 7838 and 7841, the RMS values are all less than 0.03 m, which indicates that the HY-2C orbit-determination results are stable.

From DOY 348 to DOY 354 in 2020, 15 SLR stations participated in the laser-ranging mission of HY-2C and observed 1009 normal points (NPs). The RMS value of residual validation for each SLR station was calculated, and the elevation cutoff angle threshold was set to 10°. The data for station 1873 were removed because of the small amount of station data (only four groups) available on DOY 348. Some observations were removed because of outliers. The number of excluded points was 35, and the total data removal rate was 3.5%.

As shown in Table 9, the RMS value of SLR range validation was 0.030 m. The experimental results showed that the performance of the HY-2C satellite-borne GPS receiver is stable, and the system produces high-quality data. Additionally, the accuracy of RD orbits is higher than 0.03 m.

3.2.5. Validation of Orbits with DORIS Data

In addition to using the above methods to verify the accuracy of RD orbit determination for HY-2C, the satellite orbit can also be validated by using onboard DORIS data.

During DORIS orbit validation, the 3.0 data format is converted to the 2.2 data format. Because of the large amount of phase data in the 3.0 format, only the phase observations at 10 s intervals are used [62].

A new validation method is proposed by using space-borne DORIS data. Because the DGXX receiver uses a specific channel to receive low-altitude data, the elevation angle is generally very low, resulting in inaccurate tropospheric delay calculations. Therefore, the elevation cutoff angle threshold must be set to 10°.

Through experimental tests, it is found that a discontinuity in the observations (that is, observation time intervals greater than 10 s) will cause abnormal values. In the designed DORIS validation algorithm, when the observations from a beacon station include a discontinuity in a certain epoch, the data from the four epochs before and after the discontinuity are eliminated. This approach leads to a reduction in the amount of data used, making the final validation result unreliable. When the results are calculated, it is necessary to exclude stations with insufficient data.

The 7-day satellite-borne DORIS observations used in the validation include data from at most 52 beacon stations and at least 49 stations. Based on the results for each station, the beacon stations with less than 300 observations are eliminated. For convenience, the data volume statistics are based on the total number of 7-day data points from each beacon station. The data from each beacon station are converted to an average range rate. Figure 12 plots the distribution of the DORIS beacon stations used in the validation and the residuals of the validation.

In Figure 12, the colour of the circles indicates the accuracy of the RD orbit of HY-2C gained from the DORIS validation. The RMS value of the verified residual of each beacon station is within 0.01 m/s, but the RMS value of some beacon stations, such as ARFB, HOFC, and YEMB, is greater than 0.008 m/s. The influence of the tropospheric delay is still large when the elevation angle is 15°; if the elevation cutoff angle is set to 15°, too much data will be rejected. However, to assess the reliability of the validation result, it is necessary to keep a sufficient quantity of data. The final elevation angle threshold is set to 10°. Table 10 lists the residual statistical results for the beacon stations in DORIS validation for HY-2C.

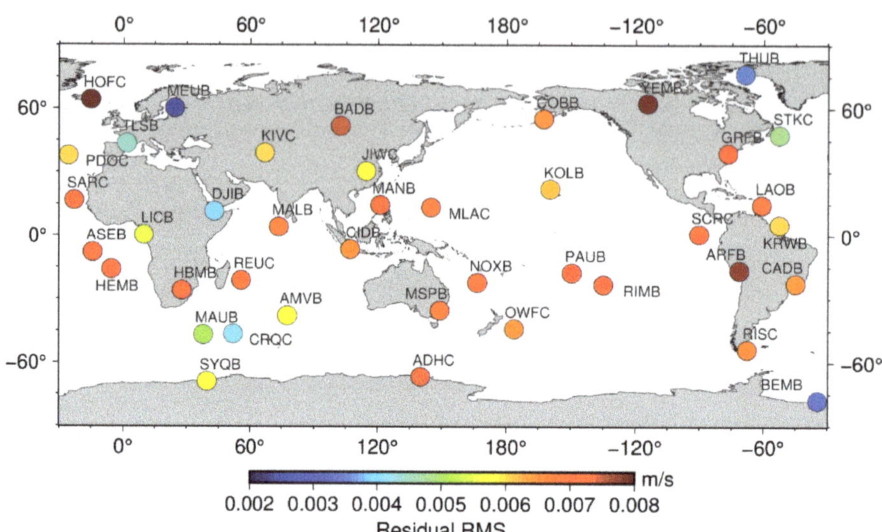

Figure 12. DORIS beacon station distribution map and residuals of DORIS validation.

Table 10. Summary statistics for the residuals of the beacon stations for DORIS validation.

Station Code	Number of Data	Min (m/s)	Max (m/s)	Mean (m/s)	RMS (m/s)
ADHC	1788	−0.0258	0.0283	0	0.0073
AMVB	745	−0.0296	0.0311	0	0.0057
ARFB	504	−0.0272	0.0455	0	0.0077
ASEB	432	−0.0250	0.0212	0	0.0072
BADB	448	−0.0932	0.0220	0	0.0074
BEMB	374	−0.0213	0.0194	0	0.0033
CADB	657	−0.0289	0.0232	0	0.0064
CIDB	436	−0.1410	0.2200	0	0.0065
COBB	2213	−0.0381	0.0340	0	0.0065
CRQC	331	−0.0181	0.0311	0	0.0041
DJIB	567	−0.0247	0.0227	0	0.0040
GRFB	868	−0.0212	0.0224	0	0.0070
HBMB	346	−0.0278	0.0439	0	0.0070
HEMB	692	−0.0248	0.0210	0	0.0070
HOFC	1329	−0.0230	0.1199	0	0.0079
JIWC	805	−0.0291	0.0224	0	0.0058
KIVC	1106	−0.0235	0.0220	0	0.0060
KOLB	844	−0.0222	0.0228	0	0.0061
KRWB	444	−0.0238	0.0299	0	0.0060
LAOB	347	−0.0235	0.0195	0	0.0069
LICB	366	−0.0186	0.0197	0	0.0055
MALB	733	−0.0872	0.0247	0	0.0067
MANB	282	−0.0209	0.0181	0	0.0070
MAUB	511	−0.0277	0.0257	0	0.0050
MEUB	740	−0.0237	0.0217	0	0.0024
MLAC	883	−0.0230	0.0229	0	0.0068
MSPB	769	−0.0210	0.0225	0	0.0067
NOXB	831	−0.0239	0.0239	0	0.0069
OWFC	630	−0.0205	0.0237	0	0.0064
PAUB	635	−0.0201	0.0244	0	0.0071
PDOC	588	−0.0202	0.0204	0	0.0060

Table 10. *Cont.*

Station Code	Number of Data	Min (m/s)	Max (m/s)	Mean (m/s)	RMS (m/s)
REUC	864	−0.0287	0.0278	0	0.0069
RIMB	746	−0.0234	0.0209	0	0.0073
RISC	2166	−0.0343	0.0228	0	0.0065
SARC	568	−0.0251	0.0162	0	0.0073
SCRC	637	−0.0217	0.0233	0	0.0070
STKC	331	−0.0272	0.0240	0	0.0048
SYQB	1370	−0.0325	0.0256	0	0.0057
THUB	329	−0.0154	0.0143	0	0.0035
TLSB	328	−0.0674	0.0659	0	0.0044
YEMB	955	−0.0233	0.0290	0	0.0078

In the checking process, based on the setting of the elevation cutoff angle threshold and the use of observations with many discontinuities, the data for beacon stations such as BETB, DIOB, MAIB, and MIAB are eliminated, with only two groups remaining. The data for SJUC are all eliminated, and the data for beacon stations such as SOFC, WEUC, YASB are also excluded due to the small numbers of observations (less than 300); therefore, the statistical data for the above beacon stations are not included in Figure 12. The 7-day residuals of validation with DORIS data are summarized in Table 11.

Table 11. Summary statistics of the residuals of DORIS validation.

Orbits	Number of Stations	Number of Data	Min (m/s)	Max (m/s)	Mean (m/s)	RMS (m/s)
RD orbits	41	30,538	−0.1110	0.1135	0	0.0063
PSO	41	30,538	−0.1276	0.1258	0	0.0070

By combining Table 10 and the first line of Table 11, it is found that for more than 80% of the beacon stations included in DORIS validation, the RMS values are less than 0.007 m/s, and the RMS value for the overall DORIS validation residual is 0.0063 m/s. The RD orbits of HY-2C are checked with onboard DORIS data, and the HY-2C RD orbit has an accuracy of 0.0063 m/s.

In second line of Table 11, satellite-borne DORIS data are selected to calculate range rate to validate the PSO of HY-2C over 7 days, starting from DOY 348 in 2020. This experiment can prove the reliability of the external validation by using DORIS data on board. It can be seen that the residual RMS value of DORIS validation of PSO is 0.0070 m/s, the result is 0.0007 m/s larger than DORIS validation of RD orbits. These results certify that the accuracy of PSO released by CNES is high [40]. This experiment shows that the proposed method of external validation of RD orbits using DORIS data is reliable and efficient.

4. Conclusions

Based on GPS observations of HY2_Receiver, the quality of the observations is analysed, and the data of the L_1 band are more susceptible to changes in elevation than are those of the L_2 band. The data utilization results suggest that the HY2_Receiver achieves good performance and can be used in POD missions involving LEO satellites. PCO and PCV models are established, and the satellite orbits obtained by using the PCO and PCV models display better accuracy than those obtained without considering the PCO and PCV models. Compared with PSO, the R-direction residual RMS value is 0.01 m, and the 3D RMS value is 0.025 m. The RMS of the carrier-phase residuals is around 0.008 m. The overlap method is applied to assess the accuracy of orbit determination, and the detailed overlap analysis suggests that the RMS value of the residual of overlapping orbits is 0.006 m in the R direction, 0.011 m in the T direction, and 0.006 m in the N direction. The internal-coincidence accuracy-verification results indicate that the HY-2C satellite orbit determination strategy is reliable; notably, the error term is eliminated during the POD process, and the orbit determination result is relatively stable. SLR range validation is analysed in detail, and the RMS value is better than 0.03 m. The results of SLR range validation suggest that the accuracy of the precise orbit of HY-2C reaches the centimetre level. A newly proposed method is used to check orbits based on space-borne DORIS data directly, and that the accuracy of the range rate of the HY-2C RD orbit reaches 0.0063 m/s.

According to the experimental results, the RD orbit of HY-2C reaches centimetre-level accuracy, and the space-borne GPS receiver HY2_Receiver independently developed in China can be used for LEO POD missions. With the improvement and optimization of the model, the RD method for orbit determination will be increasingly applied to LEO satellites, and the orbit accuracy of satellites will be increased in the future.

Author Contributions: Conceptualization, writing—original draft preparation, writing—review and editing and methodology, H.G. and J.G.; software, G.W.; validation, Z.Y., G.W. and L.Q.; data curation, M.L. and H.P.; formal analysis, B.J.; funding acquisition, J.G. All authors have read and agreed to the published version of the manuscript.

Funding: This research was funded by National Natural Science Foundation of China (Grant Nos. 41774001 and 41874091), Autonomous and Controllable Special Project for the Surveying and Mapping of China (Grant No. 816-517), and SDUST Research Fund (Grant No. 2014TDJH101).

Institutional Review Board Statement: Not applicable.

Informed Consent Statement: Not applicable.

Data Availability Statement: Our sincere thanks go to the National Satellite Ocean Application Center for providing space-borne GPS data for HY-2C; CODE for providing GPS satellite orbits, clocks and Earth rotation parameters; the CNES for providing precise orbits for HY-2C; the ILRS for providing SLR data; and the IDS for providing DORIS data.

Acknowledgments: We thank the Astronomical Institute of the University of Bern (AIUB) for providing Bernese GNSS Software. This research is supported by National Natural Science Foundation of China (grant numbers 41774001) and SDUST Research Fund (grant number 2014TDJH101).

Conflicts of Interest: The authors declare no conflict interest.

References

1. Lin, M.S.; Wang, X.H.; Peng, H.L.; Zhao, Q.L.; Li, M. Precise orbit determination technology based on dual-frequency GPS solution for HY-2 satellite. *Eng. Sci.* **2014**, *16*, 97–101. [CrossRef]
2. Jiang, X.W.; Lin, M.S.; Zhang, Y.G. Progress and prospect of Chinese ocean satellites. *J. Remote Sens.* **2016**, *20*, 1185–1198. [CrossRef]
3. Bury, G.; Sośnica, K.; Zajdel, R. Multi-GNSS orbit determination using satellite laser ranging. *J. Geod.* **2019**, *93*, 2447–2463. [CrossRef]
4. Wang, Y.C.; Guo, J.Y.; Zhou, M.S.; Jin, X.; Zhao, C.M.; Chang, X.T. Geometric solution method of SLR station coordinate based on multi-LEO satellites. *Chin. J. Geophys.* **2020**, *63*, 4333–4344. [CrossRef]
5. Strugarek, D.; Sośnica, K.; Arnold, D.; Jäggi, A.; Zajdel, R.; Bury, G.; Drożdżewski, M. Determination of Global Geodetic Parameters Using Satellite Laser Ranging Measurements to Sentinel-3 Satellites. *Remote Sens.* **2019**, *11*, 2282. [CrossRef]
6. Zhao, X.L.; Zhou, S.S.; Ci, Y.; Hu, X.G.; Cao, J.F.; Chang, Z.Q.; Tang, C.P.; Guo, D.N.; Guo, K.; Liao, M. High-precision orbit determination for a LEO nanosatellite using BDS-3. *GPS Solut.* **2020**, *24*, 102. [CrossRef]
7. Zhang, Q.J.; Zhang, J.; Zhang, H.; Rui, W.; Hong, J. The study of HY-2A satellite engineering development and in-orbit movement. *Eng. Sci.* **2013**, *15*, 12–18. [CrossRef]
8. Bertiger, W.I.; Bar-Sever, Y.E.; Christensen, E.J.; Davis, E.S.; Guinn, J.R.; Haines, B.J.; Ibanez-Meier, R.W.; Jee, J.R.; Lichten, S.M.; Melbourne, W.G.; et al. GPS precise tracking of TOPEX/POSEIDON: Results and implications. *J. Geophys. Res. Ocenas* **1994**, *99*, 24449–24464. [CrossRef]
9. Mao, X.Y.; Arnold, D.; Girardin, V.; Villiger, A.; Jäggi, A. Dynamic GPS-based LEO orbit determination with 1 cm precision using the Bernese GNSS Software. *Adv. Space Res.* **2020**, *67*, 788–805. [CrossRef]
10. Kang, Z.; Tapley, B.; Bettadpur, S.; Ries, J.; Nagel, P. Precise orbit determination for GRACE using accelerometer data. *Adv. Space Res.* **2006**, *38*, 2131–2136. [CrossRef]
11. Švehla, D.; Rothacher, M. Kinematic positioning of LEO and GPS satellites and IGS stations on the ground. *Adv. Space Res.* **2005**, *36*, 376–381. [CrossRef]
12. Wu, S.C.; Yunck, T.P.; Thornton, C.L. Reduced-dynamic technique for precise orbit determination of low earth satellites. *J. Guid. Control. Dyn.* **1991**, *14*, 24–30. [CrossRef]

13. Xia, Y.W.; Liu, X.; Guo, J.Y.; Yang, Z.M.; Qi, L.H.; Ji, B.; Chang, X. On GPS data quality of GRACE-FO and GRACE satellites: Effects of phase center variation and satellite attitude on precise orbit determination. *Acta Geod. Geophys.* **2021**, *56*, 93–111. [CrossRef]

14. Hwang, C.; Tseng, T.-P.; Lin, T.; Švehla, D.; Schreiner, B. Precise orbit determination for the FORMOSAT-3/COSMIC satellite mission using GPS. *J. Geod.* **2009**, *83*, 477–489. [CrossRef]

15. Kang, Z.; Nagel, P.; Pastor, R. Precise orbit determination for GRACE. *Adv. Space Res.* **2003**, *31*, 1875–1881. [CrossRef]

16. Haines, B.; Bar-Sever, Y.; Bertiger, W.; Desai, S.; Willis, P. One-Centimeter Orbit Determination for Jason-1: New GPS-Based Strategies. *Mar. Geod.* **2004**, *27*, 299–318. [CrossRef]

17. Bock, H.; Jäggi, A.; Švehla, D.; Beutler, G.; Hugentobler, U.; Visser, P. Precise orbit determination for the GOCE satellite using GPS. *Adv. Space Res.* **2007**, *39*, 1638–1647. [CrossRef]

18. Guo, J.Y.; Qin, J.; Kong, Q.L.; Li, G.W. On simulation of precise orbit determination of HY-2 with centimeter precision based on satellite-borne GPS technique. *Appl. Geophys.* **2012**, *9*, 95–107. [CrossRef]

19. Yunck, T.P.; Bertiger, W.I.; Wu, S.C.; Bar-Sever, Y.E.; Christensen, E.J.; Haines, B.J.; Lichten, S.M.; Muellerschoen, R.J.; Vigue, Y.; Willis, P. First assessment of GPS-based reduced dynamic orbit determination on TOPEX/Poseidon. *Geophys. Res. Lett.* **2013**, *21*, 541–544. [CrossRef]

20. Zhou, X.Y.; Chen, H.; Fan, W.; Zhou, X.; Chen, Q.; Jiang, W. Assessment of single-difference and track-to-track ambiguity resolution in LEO precise orbit determination. *GPS Solut.* **2021**, *25*, 62. [CrossRef]

21. Jäggi, A.; Hugentobler, U.; Beutler, G. Pseudo-Stochastic Orbit Modeling Techniques for Low-Earth Orbiters. *J. Geod.* **2006**, *80*, 47–60. [CrossRef]

22. Guo, J.; Zhao, Q.L.; Li, M.; Hu, Z.G. Centimeter Level Orbit Determination for HY2A Using GPS Data. *Geomat. Form. Sci. Wuhan Univ.* **2013**, *38*, 52–55. [CrossRef]

23. Gong, X.W.; Sang, J.Z.; Wang, F.H.; Li, X.X. A More Reliable Orbit Initialization Method for LEO Precise Orbit Determination Using GNSS. *Remote Sens.* **2020**, *12*, 3646. [CrossRef]

24. Montenbruck, O.; Kroes, R. In-flight performance analysis of the CHAMP BlackJack GPS Receiver. *GPS Solut.* **2003**, *7*, 74–86. [CrossRef]

25. Hwang, C.; Lin, T.-J.; Tseng, T.-P.; Chao, B.F. Modeling Orbit Dynamics of FORMOSAT-3/COSMIC Satellites for Recovery of Temporal Gravity Variations. *IEEE Trans. Geosci. Remote Sens.* **2008**, *46*, 3412–3423. [CrossRef]

26. Cai, Y.R.; Bai, W.H.; Wang, X.Y.; Sun, Y.; Du, Q.F.; Zhao, D.Y.; Meng, X.G.; Liu, C.L.; Xia, J.M.; Wang, D.W.; et al. In-orbit performance of GNOS on-board FY3-C and the enhancements for FY3-D satellite. *Adv. Space Res.* **2017**, *60*, 2812–2821. [CrossRef]

27. Hwang, C.; Tseng, T.-P.; Lin, T.-J.; Švehla, D.; Hugentobler, U.; Chao, B.F. Quality assessment of FORMOSAT-3/COSMIC and GRACE GPS observables: Analysis of multipath, ionospheric delay and phase residual in orbit determination. *GPS Solut.* **2009**, *14*, 121–131. [CrossRef]

28. Moreaux, G.; Lemoine, F.G.; Argus, D.; Santamaría-Gómez, A.; Willis, P.; Soudarin, L.; Gravelle, M.; Ferrage, P. Horizontal and vertical velocities derived from the IDS contribution to ITRF2014, and comparisons with geophysical models. *Geophys. J. Int.* **2016**, *207*, 209–227. [CrossRef]

29. Lemoine, J.-M.; Capdeville, H.; Soudarin, L. Precise orbit determination and station position estimation using DORIS RINEX data. *Adv. Space Res.* **2016**, *58*, 2677–2690. [CrossRef]

30. Dorrer, M.; Laborde, B.; Deschamps, P. DORIS (Doppler orbitography and radiopositioning integrated from space): System assessment results with DORIS on SPOT 2. *Acta Astronaut.* **1991**, *25*, 497–504. [CrossRef]

31. Jayles, C.; Vincent, P.; Rozo, F.; Balandreaud, F. DORIS-DIODE: Jason-1 has a Navigator on Board. *Mar. Geod.* **2004**, *27*, 753–771. [CrossRef]

32. Mercier, F.; Cerri, L.; Berthias, J.-P. Jason-2 DORIS phase measurement processing. *Adv. Space Res.* **2010**, *45*, 1441–1454. [CrossRef]

33. Zelensky, N.P.; Berthias, J.-P.; Lemoine, F.G. DORIS time bias estimated using Jason-1, TOPEX/Poseidon and ENVISAT orbits. *J. Geod.* **2006**, *80*, 497–506. [CrossRef]

34. Doornbos, E.; Willis, P. Analysis of DORIS range-rate residuals for TOPEX/Poseidon, Jason, Envisat and SPOT. *Acta Astronaut.* **2007**, *60*, 611–621. [CrossRef]

35. Li, X.X.; Zhang, K.K.; Meng, X.G.; Zhang, W.; Zhang, Q.; Zhang, X.; Li, X. Precise Orbit Determination for the FY-3C Satellite Using Onboard BDS and GPS Observations from 2013, 2015, and 2017. *Engineering* **2019**, *6*, 904–912. [CrossRef]

36. CNES. HY-2C input data for Precise Orbit Determination. 2021. Available online: https://ids-doris.org/analysis-documents.html (accessed on 24 June 2021).

37. Dach, R.; Brockmann, E.; Schaer, S.; Beutler, G.; Meindl, M.; Prange, L.; Bock, H.; Jäggi, A.; Ostini, L. GNSS processing at CODE: Status report. *J. Geod.* **2009**, *83*, 353–365. [CrossRef]

38. Allahvirdi-Zadeh, A.; Wang, K.; El-Mowafy, A. POD of small LEO satellites based on precise real-time MADOCA and SBAS-aided PPP corrections. *GPS Solut.* **2021**, *25*, 1–14. [CrossRef]

39. Dach, R.; Andritsch, F.; Arnold, D.; Vertone, S.; Thaller, D. *Bernese GNSS Software Version 5.2. User Manual*; Astronomical Institute, University of Bern, Bern Open Publishing: Bern, Switzerland, 2015.

40. Flavien, M.; John, M.; Sabine, H. First POD Results on HY-2C. 2021. Available online: https://ids-doris.org/images/documents/report/AWG202104/IDSAWG202104-Mercier-PODHY2C.pdf (accessed on 24 June 2021).

41. Pearlman, M.; Degnan, J.; Bosworth, J. The International Laser Ranging Service. *Adv. Space Res.* **2002**, *30*, 135–143. [CrossRef]

42. Auriol, A.; Tourain, C. DORIS system: The new age. *Adv. Space Res.* **2010**, *46*, 1484–1496. [CrossRef]

43. Visser, P.; Ijssel, J.V.D. Aiming at a 1-cm Orbit for Low Earth Orbiters: Reduced-Dynamic and Kinematic Precise Orbit Determination. *Space Sci. Rev.* **2003**, *108*, 27–36. [CrossRef]
44. Rodriguez-Solano, C.; Hugentobler, U.; Steigenberger, P. Adjustable box-wing model for solar radiation pressure impacting GPS satellites. *Adv. Space Res.* **2012**, *49*, 1113–1128. [CrossRef]
45. Zhang, B.B.; Wang, Z.; Zhou, L.; Feng, J.; Qiu, Y.; Li, F. Precise Orbit Solution for Swarm Using Space-Borne GPS Data and Optimized Pseudo-Stochastic Pulses. *Sensors* **2017**, *17*, 635. [CrossRef]
46. Saastamoinen, J. Contributions to the theory of atmospheric refraction. *Bull. Géod.* **1972**, *105*, 279–298. [CrossRef]
47. Niell, A.E. Global mapping functions for the atmosphere delay at radio wavelengths. *J. Geophys. Res. Atmos.* **1996**, *101*, 3227–3246. [CrossRef]
48. Vaclavovic, P.; Dousa, J. G-Nut/Anubis: Open-Source Tool for Multi-GNSS Data Monitoring with a Multipath Detection for New Signals, Frequencies and Constellations. In *Gravity, Geoid and Earth Observation*; Springer: Cham, Switzerland, 2015; Volume 143, pp. 775–782.
49. Qi, L.H.; Guo, J.Y.; Xia, Y.W.; Yang, Z.M. Effect of Higher-Order Ionospheric Delay on Precise Orbit Determination of GRACE-FO Based on Satellite-Borne GPS Technique. *IEEE Access* **2019**, *9*, 29841–29849. [CrossRef]
50. Reigber, C.; Schwintzer, P.; Neumayer, K.-H.; Barthelmes, F.; König, R.; Förste, C.; Balmino, G.; Biancale, R.; Lemoine, J.-M.; Loyer, S.; et al. The CHAMP-only earth gravity field model EIGEN-2. *Adv. Space Res.* **2003**, *31*, 1883–1888. [CrossRef]
51. Liu, M.; Yuan, Y.; Ou, J.; Chai, Y. Research on Attitude Models and Antenna Phase Center Correction for Jason-3 Satellite Orbit Determination. *Sensors* **2019**, *19*, 2408. [CrossRef]
52. Yuan, J.J.; Zhou, S.S.; Hu, X.G.; Yang, L.; Cao, J.F.; Li, K.; Liao, M. Impact of Attitude Model, Phase Wind-Up and Phase Center Variation on Precise Orbit and Clock Offset Determination of GRACE-FO and CentiSpace-1. *Remote. Sens.* **2021**, *13*, 2636. [CrossRef]
53. Lu, C.X.; Zhang, Q.; Zhang, K.K.; Zhu, Y.T.; Zhang, W. Improving LEO precise orbit determination with BDS PCV calibration. *GPS Solut.* **2019**, *23*, 1–13. [CrossRef]
54. Jäggi, A.; Dach, R.; Montenbruck, O.; Hugentobler, U.; Bock, H.; Beutler, G. Phase center modeling for LEO GPS receiver antennas and its impact on precise orbit determination. *J. Geod.* **2009**, *83*, 1145–1162. [CrossRef]
55. Shao, K.; Gu, D.F.; Chang, X.; Yi, B.; Wang, Z.M. Impact of GPS receiver antenna GRAPHIC residual variations on single-frequency orbit determination of LEO satellites. *Adv. Space Res.* **2019**, *64*, 1166–1176. [CrossRef]
56. Zhao, Q.L.; Liu, J.N.; Ge, M.R.; Shi, C. Precision Orbit Determination of CHAMP Satellite with cm-level Accuracy. *Geomat. Inf. Sci. Wuhan Univ.* **2006**, *31*, 879–882.
57. Kang, Z.; Bettadpur, S.; Nagel, P.; Save, H.; Poole, S.; Pie, N. GRACE-FO precise orbit determination and gravity recovery. *J. Geod.* **2020**, *94*, 1–17. [CrossRef]
58. Švehla, D.; Rothacher, M. Kinematic and reduced-dynamic precise orbit determination of low earth orbiters. *Adv. Geosci.* **2003**, *1*, 47–56. [CrossRef]
59. Petit, G.; Luzum, B. *IERS Conventions IERS Technical Note*; Verlag des Bundesamtsfür Kartographie and Geodäsie: Frankfurt am Main, Germany, 2010.
60. Guo, J.Y.; Wang, Y.C.; Shen, Y.; Liu, X.; Sun, Y.; Kong, Q.L. Estimation of SLR station coordinates by means of SLR measurements to kinematic orbit of LEO satellites. *Earth Planets Space* **2018**, *70*, 201. [CrossRef]
61. Mendes, V.B. High-accuracy zenith delay prediction at optical wavelengths. *Geophys. Res. Lett.* **2004**, *31*, 189–207. [CrossRef]
62. Kong, Q.L.; Guo, J.Y.; Gao, F.; Han, L.T. Performance Evaluation of Three Atmospheric Density Models on HY-2A Precise Orbit Determination Using DORIS Range-Rate Data. *J. Test. Eval.* **2018**, *47*, 2150–2166. [CrossRef]

remote sensing

Technical Note

A Method of Whole-Network Adjustment for Clock Offset Based on Satellite-Ground and Inter-Satellite Link Observations

Dongxia Wang [1,2], Rui Guo [2,*], Li Liu [2], Hong Yuan [1], Xiaojie Li [2], Junyang Pan [3] and Chengpan Tang [3]

[1] Aerospace Information Research Institute, Chinese Academy of Sciences, Beijing 100094, China
[2] Beijing Satellite Navigation Center, Beijing 100094, China
[3] Shanghai Astronomical Observatory, Chinese Academy of Sciences, Shanghai 200030, China
*   Correspondence: shimbarsalon@163.com

Abstract: The inter-satellite link is an important technology to improve the accuracy of clock offset measurement and prediction for BeiDou Navigation Satellite System (BDS). At present, BDS measures clock offsets of invisible satellite mainly through the "one-hop" reduction mode based on the satellite-ground clock offset of the node visible satellite and the inter-satellite clock offset between the two satellites. However, there exists a systematic deviation caused by the node satellite reduction, and there is still a large room for improvement in clock offset measurement and prediction. Therefore, this paper firstly proposes a method of whole-network adjustment for clock offset based on the satellite-ground and inter-satellite two-way data. The least square method is used to realize the whole-network adjustment of clock offset based on the observations of two sources, and to obtain optimal estimates of different clock offset reduction. Secondly, the evaluation method combining internal and external symbols are proposed by the fitting residual, prediction error and clock offset closure error. Finally, experimental verification is completed based on BDS measured data. In comparison with the "one-hop" reduction method, the fitting residual and prediction error of the whole-network adjustment method reduces about 45.06% and 52.15%, respectively. In addition, inter-satellite station closure error and three-satellite closure error are reduced from 0.69 ns and 0.23 ns to about 0 ns. It can be seen that the accuracy of BDS time synchronization is significantly improved.

Keywords: BDS-3; satellite-ground and inter-satellite link; multi-source data; satellite clock offset; whole-network adjustment

Citation: Wang, D.; Guo, R.; Liu, L.; Yuan, H.; Li, X.; Pan, J.; Tang, C. A Method of Whole-Network Adjustment for Clock Offset Based on Satellite-Ground and Inter-Satellite Link Observations. *Remote Sens.* **2022**, *14*, 5073. https://doi.org/10.3390/rs14205073

Academic Editor: Yunbin Yuan

Received: 11 August 2022
Accepted: 6 October 2022
Published: 11 October 2022

Publisher's Note: MDPI stays neutral with regard to jurisdictional claims in published maps and institutional affiliations.

1. Introduction

The inter-satellite link (ISL) is not only an important feature and technological innovation of BDS, but also an important means to optimize the distribution of regional station, improve the accuracy of spatial signals and increase the frequency of message injection [1–3]. BDS-3 has provided the global services officially on 31 July 2020, and all of the satellites are equipped with ISL [4]. Based on the Time Division Multiple Access (TDMA) technique and the Ka-band technology of BDS ISL strategy, any visible satellites can achieve real-time inter-satellite measurement and communication by the multipoint-to-multipoint time-division two-way links [5,6].

In contrast with GPS and GALILEO systems that update ephemeris and clock offset parameters mainly through the global uniform distribution of stations [7,8], BDS adopts the combined time synchronization strategy which determines satellite clock offset parameters by the satellite-ground link (SGL) for visible satellite and inter-satellite link (ISL) for invisible satellite [9,10]. This strategy was first published in 2018, and usually called as the "one-hop" reduction method [11]. According to this method, when the satellite is visible by the Master Control Center (MCC), the clock offset is measured through the SGL L-band satellite-ground two-way time synchronization; when the satellite is invisible by MCC, the clock offset is calculated by one node satellite (which is visible to both MCC and the invisible node satellite) of the SGL L-band two-way time synchronization and these two satellites of the ISL Ka-band two-way time synchronization, thus clock offsets of invisible satellite can be calculated through the "one-hop" reduction mode by one node visible satellite [11,12]. Compared with only SGL, the tracking coverage is extended by 40% by using ISL [11], the prediction error can be reduced from 3 ns to 1 ns [13], and the orbit result is also greatly improved [14–16].

Some achievements have been published to estimate the affections of the systematic error. Yan [17] and Wang [18] estimate the satellite antenna phase center offsets (PCOs), phase variations (PVs) and the differential code bias (DCB) of BeiDou-3 satellites, and has the results that PCOs and PVs is different with satellites, and DCB value is related to the receiver type. Moreover, Liu [19] proposes a network adjustment model by the correction of inter-satellite clock offsets to detect and analyze the closed residuals, and has the results that the random noise of the inter-satellite clock corrections is reduced by 30% to 50%. These achievements have evaluated the performance of current system and given the application conclusions that the signal in space accuracy has reached up to 0.5 m [2]. However, systematic errors of the measured SGL clock offsets and the calculated ISL clock offsets using different node satellite exist, which affect the accuracy of the time synchronization [12,20].

This paper further studies the whole-network adjustment method by multi-source clock offset of the satellite-ground and inter-satellite observation. Taking SGL and ISL data of all orbiting satellites as a whole, this new method is used to realize the satellite-ground and inter-satellite time synchronization in the entire constellation. The propose is to optimize the time synchronization strategy of BDS, minimize the impact of different devices, and improve the accuracy of BDS broadcast clock offset.

2. Materials and Methods

2.1. Time Synchronization Principle

2.1.1. L-Band Satellite-Ground Two-Way Time Synchronization

BDS achieves satellite-ground time synchronization using L-band two-way time comparison. The satellite and ground station, respectively, generate and broadcast pseudocode ranging signals based on the control of local clocks. Specifically, the ground station obtains the downlink pseudorange at the time corresponding to the ground 1 pulse per second (pps), and the satellite obtains the uplink pseudorange at the time corresponding to the satellite 1 pps. Meanwhile, the satellite sends the observed value of uplink pseudorange to the ground station, and then the ground station calculates the difference between the locally measured downlink pseudorange and the received uplink pseudorange to obtain the clock offset of satellite relative to the ground station. It lays a foundation for completing the satellite-ground two-way time comparison. The principle of L-band satellite-ground two-way time comparison is shown in Figure 1.

The satellite-ground two-way time comparison method is used to obtain the observed satellite-ground clock offset, which can be expressed by a second-order polynomial:

$$\Delta T_s - \Delta T_g = \frac{1}{2c}(\rho_u - \rho_d) - \frac{1}{2c}(R_u - R_d) - \tau_{rel} + \varepsilon \tag{1}$$

where ΔT_s, ΔT_g, respectively, denote the satellite clock offset and the ground station clock offset; ρ_u, ρ_d represent the uplink and downlink pseudorange, respectively; R_u, R_d indicate the theoretical values of uplink and downlink geometric distances, respectively; τ_{rel} represents the relativistic effect delay in the signal propagation path; ε indicates the synthetic measurement noise of the satellite-ground two-way link; and c represents the velocity of light.

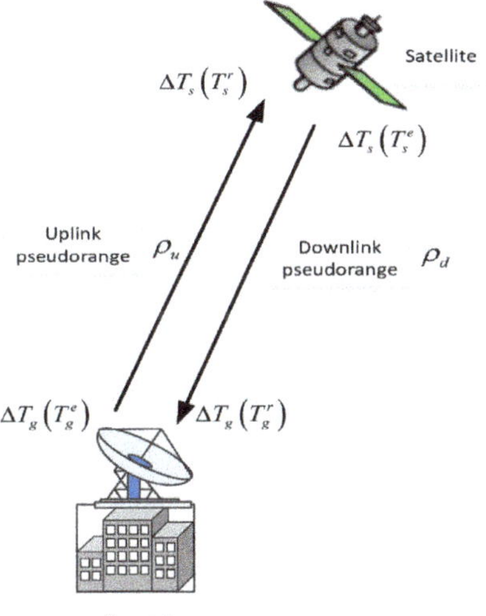

Figure 1. Schematic diagram of satellite-ground time synchronization.

The satellite-ground two-way time comparison method is used to obtain the observed value of satellite-ground clock offset. The satellite-ground clock offset of satellite i can be expressed by a second-order polynomial [21]:

$$\Delta T_i(t) = a_0^i + a_1^i (t - t_0) + a_2^i (t - t_0)^2 \tag{2}$$

where $\Delta T_i(t)$ represents the discrete point of satellite-ground clock offset of satellite i; t denotes the observation time of satellite-ground clock offset; t_0 refers to the reference time of clock offset parameter; a_0^i, a_1^i, a_2^i represent the clock offset, clock rate, and clock drift, respectively.

2.1.2. Ka-Band Inter-Satellite Two-Way Time Synchronization

It is similar to the satellite-ground two-way time comparison method. If satellite i and satellite j send time signals to each other at the clock T_i and T_j, the signal sent by satellite i is received by satellite j after the delay from satellite i to satellite j. On this basis, the pseudorange from satellite i to satellite j can be measured. Similarly, the signal sent by satellite j is received by satellite i after the delay from satellite j to satellite i, so that the pseudorange from satellite j to satellite i can be measured. Next, satellite i and satellite j exchange their respective observation results, and finally calculate the relative clock offset between them. The detailed principle of Ka-band inter-satellite two-way time comparison is shown in Figure 2.

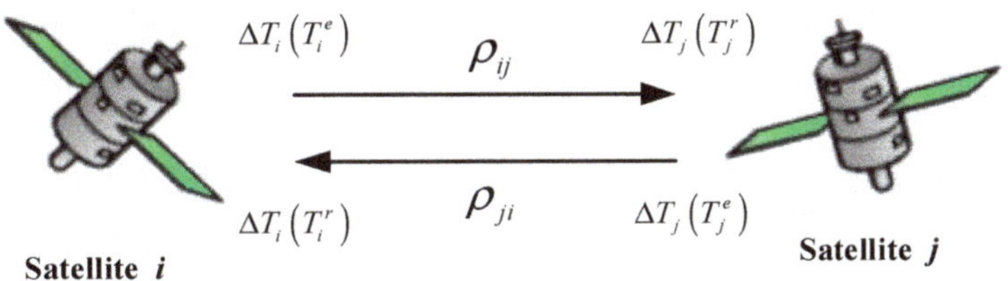

Figure 2. Schematic diagram of inter-satellite time synchronization.

Similar with the satellite-ground two-way time synchronization, the inter-satellite time synchronization of navigation satellites is based on pseudorange measurement of Ka-band two-way time synchronization, and the inter-satellite clock offset can be obtained as:

$$\Delta T_j - \Delta T_i = \frac{1}{2c}\left(\rho_{ij} - \rho_{ji}\right) - \frac{1}{2c}\left(R_{ij} - R_{ji}\right) + \delta \tag{3}$$

where ΔT_i, ΔT_j represent the clock offsets of satellite i and satellite j, respectively; ρ_{ij} refers to the pseudorange of satellite j receiving from satellite i; ρ_{ji} refers to the pseudorange of satellite i receiving from satellite j; R_{ij}, R_{ji} respectively, represent the theoretical values of two-way geometric distances; δ indicates the synthetic measurement noise of the inter-satellite two-way link; and c represents the velocity of light.

The inter-satellite two-way time comparison method is used to obtain the observed inter-satellite clock offset, which can be expressed by a second-order polynomial [21]:

$$\Delta T_{ij}(t) + v_{ij} = a_0^j + a_1^j(t - t_0) + a_2^j(t - t_0)^2 - \left(a_0^i + a_1^i(t - t_0) + a_2^i(t - t_0)^2\right) \tag{4}$$

where $\Delta T_{ij}(t)$ represents the discrete point of inter-satellite clock offset of satellite j relative to satellite i; t denotes the observation time of inter-satellite clock offset; t_0 refers to the reference time of clock offset parameter; v_{ij} indicates the Ka-band inter-satellite link delay calibrated based on the L-band satellite-ground link; a_0^i, a_1^i, a_2^i represent the clock offset, clock rate and clock drift, respectively.

2.2. Whole-Network Adjustment Method

The Whole-network adjustment method uses multi-source clock offset data to obtain the clock offset parameters of satellite-ground and inter-satellite of all satellites. Assuming there is m satellites and n inter-satellite links, the m satellite-ground clock offsets and n inter-satellite clock offsets can be performed by whole-network adjustment method, and the equation can be abbreviated as:

$$\boldsymbol{L} = \boldsymbol{Cx} - \boldsymbol{v} \tag{5}$$

where $\boldsymbol{L} = \begin{pmatrix} \Delta T_i(t) \\ \Delta T_{ij}(t) \end{pmatrix}_{(m+n)*1}$ represents the discrete point of m satellite-ground and n inter-satellite clock offset; $C = \begin{bmatrix} C_i \\ C_{ij} \end{bmatrix}_{(m+n)*3m}$ refers to the coefficient matrix of observation time; $x = \begin{pmatrix} x^1 \\ \vdots \\ x^i \\ \vdots \\ x^m \end{pmatrix}_{3m*1}$ indicates the column vector of clock offset parameters; $x^i = \begin{pmatrix} a_0^i \\ a_1^i \\ a_2^i \end{pmatrix}$ denotes the clock offset parameter of each satellite; $v = \begin{pmatrix} 0 \\ v_{ij} \end{pmatrix}_{(m+n)*1}$ refers to the link delay error [22]. The coefficient matrix is expressed as follows:

$$
C_i = \begin{bmatrix} 1 & t-t_0 & (t-t_0)^2 & 0 & 0 & 0 & \cdots & 0 & 0 & 0 \\ 0 & 0 & 0 & 1 & t-t_0 & (t-t_0)^2 & \cdots & 0 & 0 & 0 \\ \cdots & \cdots & \cdots & \cdots & \cdots & \cdots & & \cdots & \cdots & \cdots \\ 0 & 0 & 0 & 0 & 0 & 0 & \cdots & 1 & t-t_0 & (t-t_0)^2 \end{bmatrix}_{m \times 3m}
$$

$$
C_{ij} = \begin{bmatrix} 0 & \cdots & -1 & -(t-t_0) & -(t-t_0)^2 & 0 & \cdots & 1 & (t-t_0) & (t-t_0)^2 & 0 & \cdots \\ 0 & \cdots & -1 & -(t-t_0) & -(t-t_0)^2 & 0 & \cdots & 1 & (t-t_0) & (t-t_0)^2 & 0 & \cdots \\ \cdots & \cdots & \cdots & \cdots & \cdots & \cdots & \cdots & \cdots & \cdots & \cdots & \cdots & \cdots \\ 0 & \cdots & -1 & -(t-t_0) & -(t-t_0)^2 & 0 & \cdots & 1 & (t-t_0) & (t-t_0)^2 & 0 & \cdots \end{bmatrix}_{n \times 3m}
$$

The principle of least squares is used to obtain the normal equation corresponding to the above equation:

$$
C^{\mathrm{T}}P(L+v) = C^{\mathrm{T}}PCx \tag{6}
$$

where P refers to the weight matrix of the observed column vector, and it can be assigned as the identity matrix when all satellites and atomic clocks are stable according to the classical least square method [23]. The normal equation is solved to obtain the clock offset parameter x of satellite-ground and inter-satellite multi-source data of all satellites.

2.3. Accuracy Evaluation Method

The accuracy evaluation of satellite clock offset is effective to verify the reliability of satellite clock offset data processing results. The optimal clock offset accuracy evaluation uses the theoretical real clock offset to evaluate the calculated clock offset accuracy. However, it is very difficult to obtain the theoretically real clock offset. The clock offset accuracy can be evaluated through different methods such as the inner coincidence accuracy comparison method and the outer coincidence accuracy comparison method. As for the internal coincidence accuracy comparison method, the accuracy is judged based on internal errors that generally refer to the fitting residual or prediction error of clock offset. According to the external coincidence accuracy comparison method, the accuracy is judged based on external information or other link information, which generally refers to the closure error of clock offset, mainly including the inter-satellite station closure error and three-satellite closure error.

2.3.1. Fitting Residual

The discrete points of clock offset are fitted to obtain the fitted clock offset, and the interval between discrete points is 1 s. The differences between the fitted clock offset and the actual clock offset is calculated to obtain the fitting residual of clock offset. The root mean square (RMS) is usually used to represent the fitting residual of clock offset, which can be obtained as follows:

$$
\sigma = \sqrt{\frac{\sum\limits_{i=1}^{n} \left(\Delta T_i^C - \Delta T_i^O\right)^2}{n-1}} \tag{7}
$$

where n represents the number of observation data, ΔT_i^C represents the calculated value of the ith fitted clock offset, and ΔT_i^O denotes the observed value of the ith actual clock offset. Generally speaking, the smaller RMS of the fitting residual indicates the better stability of clock offset, and the higher accuracy of time synchronization. it is noted that the actual clock offset is the post precision satellite clock offset [24].

2.3.2. Prediction Error

The time scale is set to improve the accuracy of satellite broadcast clock offset parameters, so as to ultimately serve users well. The clock offset parameter is calculated by the whole-network adjustment method and used for clock offset prediction. The difference between prediction result and actual clock offset is calculated to obtain the prediction error of clock offset.

$$\Delta T_{error}(t) = \Delta T_{pred}(t) - \Delta T_{actual}(t)$$
$$= a_0 + a_1(t - t_0) + a_2(t - t_0)^2 - \Delta T_{actual}(t) \tag{8}$$

where $\Delta T_{actual}(t)$ refers to the actual clock offset, $\Delta T_{pred}(t)$ denotes the predicted clock offset, and a_0, a_1, a_2 indicate the time difference, frequency difference and frequency drift, respectively, which are the clock offset parameters calculated through the whole-network adjustment.

The satellite clock offset prediction is divided into short-term prediction and long-term prediction. Specifically, first-order polynomial fitting is performed by 2-h clock offset data to predict 1-h result, and statistics the RMSs of 1-h prediction error. In contrast, second-order polynomial fitting is performed by 48-h clock offset data to predict 24-h result, and statistics the RMSs of 24-h prediction error. This paper adopts the short-term prediction method for evaluation.

2.3.3. Closure Error

In order to further verify the effectiveness of the whole-network adjustment method, time synchronization performance is tested according to the closure errors of clock offset. The commonly used closure errors of clock offset include the inter-satellite station closure error and three-satellite closure error. The inter-satellite station closure error is shown in Figure 3.

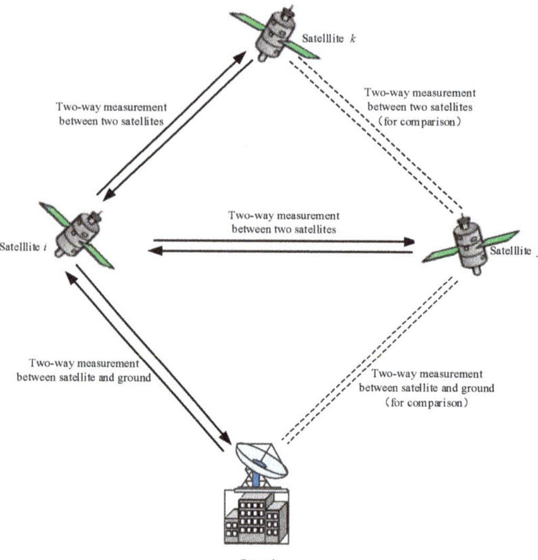

Figure 3. Schematic diagram of inter-satellite station closure error.

Supposing the satellite i is the node satellite, the clock offset of satellite j can be observed indirectly. If there exist the two-way clock offset ΔT_i between satellite i and ground station, and ΔT_{ij} between satellite i and satellite j in the same time period, the clock offset of satellite j can be obtained as:

$$\Delta T_j' = \Delta T_{ij} + \Delta T_i \tag{9}$$

The difference between $\Delta T_j'$ and satellite-ground two-way time synchronization observed value ΔT_j of satellite j is calculated to obtain the residual of the indirect clock offset. Since it is difficult to find the observed values of ΔT_i, ΔT_{ij} and ΔT_j in the same time period, one day of data are acquired. ΔT_i and ΔT_{ij} are fitted to obtain the indirect fitting parameter of $\Delta T_j'$ and then calculate the fitting residual of ΔT_j within the direct observation period.

3. Results

3.1. Data Situation

In order to obtain comprehensive and sufficient calculation results, this paper calculates the whole-network adjustment results of multi-source data according to the L-band satellite-ground two-way time synchronization and Ka-band inter-satellite two-way time synchronization of BDS on 1 March 2022. In addition, this paper makes a comparative analysis of "one-hop" reduction results for the same period. It is worth noting that GEO satellites mainly deliver SBAS, RDSS and other services, so they are without involvement in inter-satellite calculations. Therefore, this paper selects the data of 24 MEO satellites and 3 IGSO satellites for calculation. The data of all the 27 satellites are used for the inter-satellite calculation. The satellite-ground clock offset curve is shown in Figure 4, and ISL establishment during this period is shown in Figure 5.

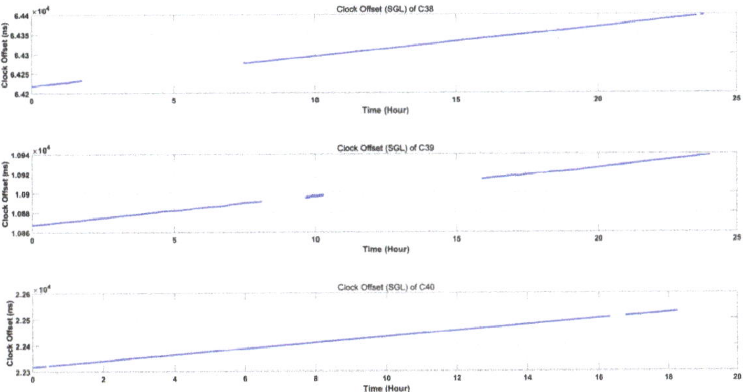

Figure 4. Curve of SGL clock offset.

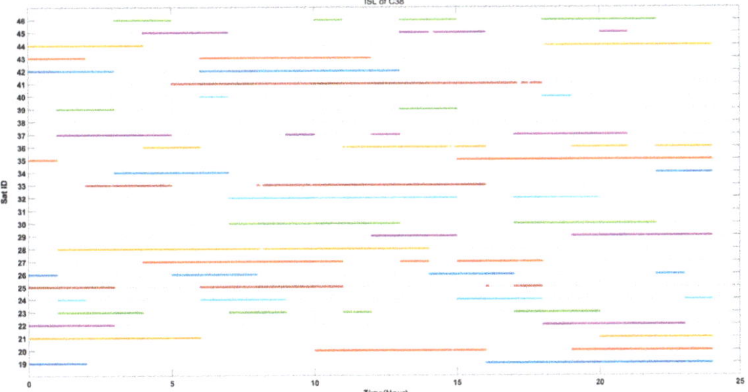

Figure 5. ISL between C38 and other satellites.

As shown in Figure 4, the satellite-ground clock offset curve changes steadily, and we can see from Figure 5 that satellite C38 and other 26 satellites all have established links. Based on such data, we verify the validity of the whole-network adjustment method.

3.2. Fitting Residual

The fitting residuals are calculated based on the clock offset time series. Taking satellites C35, C36 and C37 as example, the fitting residuals using "one-hop" reduction method and the whole-network adjustment method are shown in Figure 6.

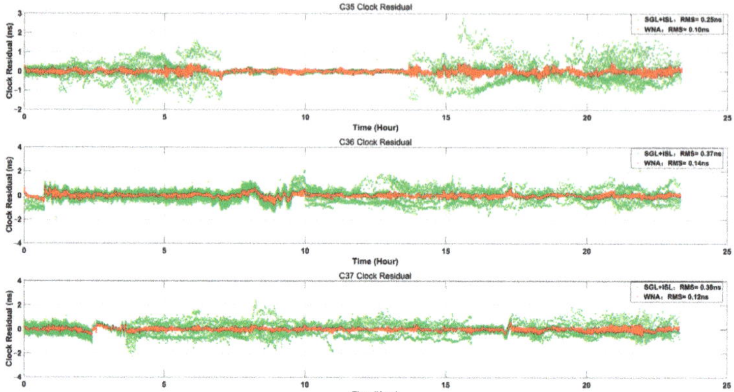

Figure 6. Clock residuals.

As shown in Figure 6, the fitting residuals of "one-hop" reduction method are fluctuating between ±3.5 ns, and the RMSs of fitting residuals of satellites C35, C36 and C37 are 0.25 ns, 0.37 ns, and 0.38 ns, respectively. The fitting residuals of the whole-network adjustment method are fluctuating between ±1 ns, and the RMSs of fitting residuals of satellites C35, C36 and C37 are 0.10 ns, 0.14 ns, and 0.13 ns, respectively.

The RMSs of fitting residuals of all satellites are further calculated, which reflect the internal coincidence accuracy of fitting results. The RMS statistical results of clock offset fitting residuals of 27 satellites are detailed in Table 1.

Table 1. RMS statistical results of fitting residuals (unit: ns).

Satellite	SGL+ISL Method	WNA Method	Improvement Rate
C38	0.21	0.13	19.05%
C39	0.20	0.12	19.05%
C40	0.19	0.12	16.67%
C25	0.34	0.13	50.00%
C26	0.26	0.13	30.95%
C27	0.34	0.13	50.00%
C28	0.34	0.12	52.38%
C29	0.35	0.13	52.38%
C30	0.38	0.15	54.76%
C19	0.32	0.12	47.62%
C20	0.41	0.11	71.43%
C21	0.28	0.10	42.86%
C22	0.31	0.14	40.48%
C23	0.39	0.15	57.14%
C24	0.28	0.12	38.10%
C32	0.36	0.13	54.76%
C33	0.42	0.19	54.76%
C34	0.31	0.12	45.24%

Table 1. *Cont.*

Satellite	SGL+ISL Method	WNA Method	Improvement Rate
C35	0.25	0.10	35.71%
C36	0.37	0.14	54.76%
C37	0.38	0.12	61.90%
C41	0.32	0.10	52.38%
C42	0.22	0.10	28.57%
C43	0.32	0.13	45.24%
C44	0.41	0.17	57.14%
C45	0.29	0.13	38.10%
C46	0.29	0.10	45.24%
Mean	0.32	0.13	45.06%

As shown in Table 1, the RMSs of clock offset fitting residuals of all satellites subject to "one-hop" reduction method between 0.19 ns and 0.42 ns, with a mean of 0.32 ns. The RMSs of clock offset fitting residuals of all satellites subject to whole-network adjustment range between 0.10 ns and 0.19 ns, with a mean of 0.13 ns. Compared with the "one-hop" reduction, whole-network adjustment improves the clock offset fitting residual accuracy by nearly 45.06%.

It is noted that the improvement rate differs widely, the reason is that the value of all satellites subject to "one-hop" reduction method differs widely, and the value subject to the whole-network adjustment method is relatively small and close. This result further proves that the whole network adjustment method eliminates the equipment errors of different satellites.

3.3. Prediction Error

In order to further analyze the effectiveness of whole-network adjustment, the 2-h clock offset data are used for slip prediction in the next one hour. The slip prediction result is compared with the actual observation value to obtain the prediction error. In order to better analyze the effectiveness of different methods, we also use only the satellite-ground data for comparison. Taking satellites C28, C29 and C30 as example, the clock offset prediction errors of only satellite-ground method, "one-hop" reduction method and the whole-network adjustment method are shown in Figure 7.

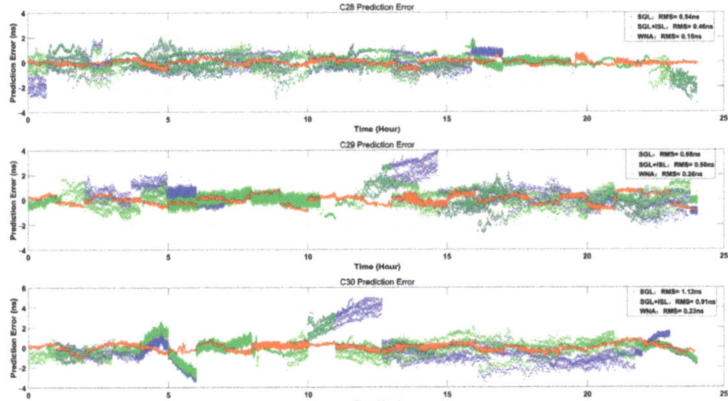

Figure 7. Prediction error.

As shown in Figure 7, the prediction error of only satellite-ground is fluctuating between ±6 ns, and the RMSs of clock offset prediction errors of satellites C28, C29 and C30 are 0.54 ns, 0.68 ns and 1.12 ns, respectively. The clock offset prediction error of "one-hop" reduction method are fluctuating between ±4 ns, and the RMSs of clock offset prediction errors of satellites C28, C29 and C30 are 0.46 ns, 0.50 ns, and 0.91 ns, respectively. The clock offset prediction error of the whole-network adjustment method are fluctuating between ±1 ns, and the RMSs of clock offset prediction errors of satellites C28, C29 and C30 are 0.15 ns, 0.26 ns, 0.23 ns, respectively. The RMSs of clock offset prediction errors of all satellites are further calculated and the statistical results are shown in Table 2.

Table 2. RMS statistical results of prediction error (unit: ns).

Satellite	Only SGL Method	SGL+ISL Method	WNA Method	Improvement Rate 1 (WNA Relative to Only SGL)	Improvement Rate 2 (WNA Relative to SGL+ISL)
C38	0.60	0.49	0.29	51.67%	40.82%
C39	0.40	0.40	0.28	30.00%	30.00%
C40	0.48	0.43	0.19	60.42%	55.92%
C25	0.67	0.51	0.27	59.70%	47.06%
C26	0.48	0.37	0.27	43.75%	27.03%
C27	0.77	0.57	0.24	68.83%	57.89%
C28	0.54	0.46	0.15	72.22%	67.39%
C29	0.68	0.50	0.26	61.76%	48%
C30	1.12	0.91	0.23	79.46%	74.73%
C19	0.80	0.68	0.23	71.25%	66.18%
C20	0.60	0.48	0.22	63.33%	54.17%
C21	0.59	0.51	0.26	55.93%	49.02%
C22	0.68	0.48	0.29	57.35%	39.58%
C23	0.87	0.63	0.33	62.07%	47.62%
C24	0.50	0.49	0.23	54.00%	53.10%
C32	0.76	0.65	0.28	63.16%	56.86%
C33	0.84	0.96	0.34	59.52%	64.64%
C34	1.25	0.49	0.20	84.00%	59.18%
C35	0.33	0.32	0.19	42.42%	40.63%
C36	0.66	0.73	0.26	60.61%	64.38%
C37	0.91	0.51	0.22	75.82%	56.86%
C41	0.49	0.42	0.20	59.18%	52.38%
C42	0.45	0.33	0.21	53.33%	36.36%
C43	0.57	0.59	0.21	63.16%	64.41%
C44	1.66	0.98	0.43	74.10%	56.12%
C45	0.69	0.45	0.22	68.12%	51.11%
C46	0.97	0.32	0.17	82.47%	46.88%
Mean	0.72	0.54	0.25	62.13%	52.15%

As shown in Table 2, the RMSs of predicted only satellite-ground method range between 0.33 ns and 1.66 ns, with a mean of 0.72 ns. The RMSs of predicted clock offset of "one-hop" reduction method range between 0.32 ns and 0.98 ns, with a mean of 0.54 ns. The RMSs of predicted clock offset of the whole-network adjustment method range between 0.15 ns and 0.43 ns, with a mean of 0.25 ns. Compared with the only satellite-ground method and "one-hop" reduction method, the prediction accuracy of the whole-network adjustment method improves about 62.13% and 52.15%, respectively.

3.4. Closure Error

Analyzing the data on 1 March 2022, it can be seen that there are 226 inter-satellite station closure errors and 1710 three-satellite closure errors concerning 27 satellites. Some inter-satellite station closure errors and three-satellite closure errors are detailed in Table 3, and statistical results of all closure errors are itemized in Table 4.

Table 3. RMS statistical results of some inter-satellite station and three-satellite closure errors (unit: ns).

Some Inter-Satellite Station Closure Errors									
Station code	Satellite 1	Satellite 2	Closure error of SGL+ISL method	Closure error of WNA method	Station code	Satellite 1	Satellite 2	Closure error of SGL+ISL method	Closure error of WNA method
1	C38	C39	0.89	2.55×10^{-12}	1	C38	C22	1.56	4.85×10^{-11}
1	C38	C40	1.03	2.78×10^{-12}	1	C38	C23	0.36	6.25×10^{-11}
1	C38	C25	1.07	5.81×10^{-11}	1	C38	C24	1.08	6.79×10^{-11}
1	C38	C26	1.62	6.33×10^{-11}	1	C38	C32	0.65	9.94×10^{-11}
1	C38	C27	0.90	7.28×10^{-12}	1	C38	C33	2.06	5.06×10^{-11}
1	C38	C29	1.02	6.98×10^{-12}	1	C38	C35	0.82	3.21×10^{-11}
1	C38	C30	0.44	2.64×10^{-11}	1	C38	C37	0.81	4.99×10^{-11}
1	C38	C19	0.72	2.42×10^{-11}	1	C38	C41	0.26	5.21×10^{-12}
1	C38	C20	0.80	4.41×10^{-11}	1	C38	C42	0.28	8.31×10^{-12}
Some Three-Satellite Closure Errors									
Satellite 1	Satellite 2	Satellite 3	Closure error of SGL+ISL method	Closure error of WNA method	Satellite 1	Satellite 2	Satellite 3	Closure error of SGL+ISL method	Closure error of WNA method
C38	C39	C25	0.21	4.91×10^{-11}	C38	C40	C28	0.07	1.64×10^{-11}
C38	C39	C21	0.10	1.40×10^{-11}	C38	C40	C30	0.23	1.16×10^{-11}
C38	C39	C32	0.22	4.69×10^{-11}	C38	C40	C19	0.09	2.37×10^{-11}
C38	C39	C36	0.48	4.75×10^{-11}	C38	C40	C24	0.05	3.80×10^{-11}
C38	C39	C41	0.09	5.37×10^{-12}	C38	C40	C34	0.10	5.52×10^{-11}
C38	C39	C42	0.09	1.24×10^{-11}	C38	C40	C41	0.07	4.70×10^{-12}
C38	C39	C44	0.14	1.00×10^{-11}	C38	C40	C43	0.11	1.22×10^{-11}
C38	C39	C46	0.10	4.81×10^{-11}	C38	C40	C46	0.10	4.40×10^{-11}
C38	C40	C25	0.26	4.94×10^{-11}	–	–	–	–	–

Table 4. RMS statistical results of closure errors (unit: ns).

Statistical Results	Closure Error of SGL+ISL Method	Closure Error of WNA Method
Inter-satellite station	0.69	1.34×10^{-10}
Three-satellite	0.23	5.54×10^{-11}

It can be seen from Tables 3 and 4, among 226 inter-satellite station closure errors and 1710 three-satellite closure errors, the closure error of inter-satellite station and three-satellite of "one-hop" reduction method are 0.69 ns and 0.23 ns, respectively. In addition, the closure error of the whole-network adjustment method is 1.34×10^{-10} and 5.54×10^{-11}, respectively, which both approximate 0. The results show that the closure error of inter-satellite station is subject to SGL reduction, while the closure error of three-satellite is subject to ISL measurement only. Therefore, three-satellite closure error is superior to the inter-satellite closure error, indirectly indicating that the SGL measurement error is a limiting factor that improves in the system time synchronization accuracy.

4. Conclusions

Based on the L-band satellite-ground clock offset data and the Ka-band inter-satellite clock offset data, this paper provides a method of whole-network adjustment for clock offset. In addition, the fitting residual, prediction error and closure error are used to evaluate and verify the effectiveness of this new method. This new method makes the RMS of the clock offset fitting residual reach up to 0.13 ns, which is lower than that of "one-hop" reduction by about 45.06%. As well, the RMS of clock offset prediction error hits about 0.25 ns, which is lower than those of only satellite-ground and "one-hop" reduction by about 62.13% and 52.15%, respectively. In addition, the closure error of inter-satellite station and three-satellite are approximate 0, which is much lower than 0.69 ns and 0.23 ns for "one-hop" reduction. This paper enriches the satellite navigation time synchronization system, and can be applied to clock offset measurement, systematic difference calculation and so forth. Furthermore, it provides a theoretical basis for effectively resolving systematic deviations in satellite-ground links and inter-satellite links. Most importantly, it will ultimately contribute to an improvement in the overall performance of BDS in navigation, positioning and timing.

Author Contributions: Conceptualization, D.W. and J.P.; methodology, L.L.; software, R.G. and X.L.; validation, D.W. and R.G.; formal analysis, R.G. and H.Y.; resources, D.W. and C.T.; data curation, D.W. and R.G.; writing—original draft preparation, R.G. and D.W.; writing—review and editing, D.W.; funding acquisition, R.G. All authors have read and agreed to the published version of the manuscript.

Funding: This research was funded by the National Natural Science Foundation of China, grant number 41874043.

Data Availability Statement: The data sources are supported by Beijing Satellite Navigation Center.

Acknowledgments: Thanks for the data sources support of Beijing Satellite Navigation Center, and thanks for the fund support of the National Natural Science Foundation of China (No. 41874043).

Conflicts of Interest: The authors declare no conflict of interest.

References

1. Tan, S.S. Innovative development and forecast of BeiDou system. *Acta Geod. Cartogr. Sin.* **2017**, *46*, 1284–1289.
2. China Satellite Navigation Office. BeiDou Navigation Satellite System Open Service Performance Standard (Version 3.0). 2021. Available online: http://www.beidou.gov.cn/xt/gfxz/202105/P020210526216231136238.pdf (accessed on 10 August 2022).
3. Yang, Y.X.; Gao, W.G.; Guo, S.R.; Mao, Y.; Yang, Y.F. Introduction to BeiDou-3 navigation satellite system. *Navigation* **2019**, *66*, 7–18. [CrossRef]
4. Yang, Y.X.; Mao, Y.; Sun, B.J. Basic performance and future developments of BeiDou global navigation satellite system. *Satell. Navig.* **2020**, *1*, 1–8. [CrossRef]
5. Ruan, R.G.; Jia, X.L.; Feng, L.P.; Zhu, J.; Hu, Y.Z.; Li, J.; Wei, Z. Orbit Determination and Time Synchronization for BDS-3 Satellites with Raw Inter-Satellite Link Ranging Observations. *Satell. Navig.* **2020**, *1*, 8. [CrossRef]
6. Wang, D.X.; Xin, J.; Xue, F.; Guo, R.; Chen, J.P. Prospect and development of GNSS autonomous navigation based on inter-satellite link. *J. Astronaut.* **2016**, *37*, 1279–1288.
7. Ananda, M.P.; Bernstein, H.; Cunningham, K.E.; Feess, W.A.; Stroud, E.G. Global positioning system (GPS) autonomous navigation. In Proceedings of the Position Location and Navigation Symposium, the 1990's-A Decade of Excellence in the Navigation Sciences, Las Vegas, NV, USA, 20 March 1990; pp. 497–508.
8. Avila-Rodriguez, J.A.; Wallner, S.; Hein, G.W.; Eissfeller, B. A vision on new frequencies, signals and concepts for future GNSS systems concepts for future GNSS systems. In Proceedings of the 20th International Technical Meeting of the Satellite Division of the Institute of Navigation, Fort Worth, TX, USA, 25–28 September 2007.
9. Tang, G.F.; Yang, W.F.; Su, R.R.; Xia, A.M. Time Synchronization Method base on Combined Satellite-Ground and Inter-satellite Observation. *Geomat. Inf. Sci. Wuhan Univ.* **2018**, *43*, 183–187.
10. Sun, L.Y.; Wang, Y.K.; Huang, W.D.; Yang, J.; Zhou, Y.F.; Yang, D.N. Inter-satellite Communication and Ranging Link Assignment for Navigation Satellite Systems. *GPS Solut.* **2018**, *22*, 38. [CrossRef]
11. Pan, J.Y.; Hu, X.G.; Zhou, S.S.; Tang, C.P.; Guo, R.; Zhu, L.F.; Tang, G.F.; Hu, G.M. Time synchronization of new-generation BDS satellites using inter-satellite link measurements. *Adv. Space Res.* **2018**, *61*, 145–153. [CrossRef]
12. Yang, Y.F.; Yang, Y.X.; Hu, X.G.; Tang, C.P.; Guo, R.; Zhou, S.S.; Xu, J.Y.; Pan, J.Y.; Su, M.D. BeiDou-3 broadcast clock estimation by integration of observations of regional tracking stations and inter-satellite links. *GPS Solut.* **2021**, *25*, 57. [CrossRef]
13. Chen, J.P.; Hu, X.G.; Tang, C.P.; Zhou, S.S.; Guo, R.; Pan, J.Y.; Ran, L.; Zhu, L.F. Orbit determination and time synchronization for new-generation beidou satellites: Preliminary results. *Sci. Sin. Phys. Mech. Astron.* **2016**, *46*, 119502.
14. Yang, Y.X.; Xu, Y.Y.; Li, J.L.; Yang, C. Progress and performance evaluation of BeiDou global navigation satellite system: Data analysis based on BDS-3 demonstration system. *Sci. China Earth Sci.* **2018**, *61*, 614–624. [CrossRef]
15. Cai, H.L.; Meng, Y.N.; Geng, T.; Xie, X. Initial Results of Precise Orbit Determination Using Satellite-Ground and Inter-Satellite Link Observations for BDS-3 Satellites. *Geomat. Inf. Sci. Wuhan Univ.* **2020**, *45*, 1493–1500.
16. Yang, Y.F.; Yang, Y.X.; Hu, X.G.; Chen, J.P.; Guo, R.; Tang, C.P.; Zhou, S.S.; Zhao, L.Q.; Xu, J.Y. Inter-Satellite Link Enhanced Orbit Determination for BeiDou-3. *J. Navig.* **2020**, *73*, 115–130. [CrossRef]
17. Yan, X.Y.; Huang, G.W.; Zhang, Q.; Wang, L.; Qin, Z.; Xie, S. Estimation of the Antenna Phase Center Correction Model for the BeiDou-3 MEO Satellites. *Remote Sens.* **2019**, *11*, 2850. [CrossRef]
18. Wang, Q.S.; Jin, S.G.; Yuan, L.L.; Hu, Y.J. Estimation and Analysis of BDS-3 Differential Code Biases from MGEX Observations. *Remote Sens.* **2019**, *12*, 68. [CrossRef]

19. Liu, C.; Gao, W.G.; Pan, J.Y.; Tang, C.P.; Hu, X.G.; Wang, W.; Chen, Y.; Lu, J.; Su, C.C. Inter-satellite clock offsets adjustment based on closed-loop residual detection of BDS inter-satellite link. *Acta Geod. Cartogr. Sin.* **2020**, *49*, 1149–1157.
20. Guo, Y.; Gao, S.; Bai, Y.; Pan, Z.; Liu, Y.; Lu, X.; Zhang, S. A New Space-to-Ground Microwave-Based Two-Way Time Synchronization Method for Next-Generation Space Atomic Clocks. *Remote Sens.* **2022**, *14*, 528. [CrossRef]
21. Huang, G.; Zhang, Q.; Xu, G. Real-time clock offset prediction with an improved model. *GPS Solut.* **2014**, *18*, 95–104. [CrossRef]
22. Pan, J.Y.; Hu, X.G.; Tang, C.P. System error calibration for time division multiple access inter-satellite payload of new-generation Beidou satellites. *Chin. Sci. Bull.* **2017**, *62*, 2671–2679. [CrossRef]
23. Bates, J.M.; Granger, C.W. The Combination of forecasts. *Oper. Res. Soc.* **1969**, *20*, 451–468. [CrossRef]
24. Liu, L.; Zhu, L.F.; Han, C.H.; Liu, X.P.; Li, C. The model of radio two-way time comparison between satellite and station and experimental analysis. *Chin. Astron. Astrophys.* **2009**, *33*, 431–439. [CrossRef]

Technical Note

Evaluation of CYGNSS Observations for Snow Properties, a Case Study in Tibetan Plateau, China

Wenxiao Ma [1,2,†], Lingyong Huang [3,4,5,†], Xuerui Wu [6,*], Shuanggen Jin [1,7], Weihua Bai [2,3,4,5] and Xuanran Li [6]

1 Shanghai Astronomical Observatory, Chinese Academy of Sciences, Shanghai 200030, China
2 School of Astronomy and Space Science, University of Chinese Academy of Sciences, Beijing 100049, China
3 National Space Science Center, Chinese Academy of Sciences (NSSC/CAS), Beijing 100190, China
4 Beijing Key Laboratory of Space Environment Exploration, Chinese Academy of Sciences, Beijing 100190, China
5 Key Laboratory of Science and Technology on Space Environment Situational Awareness, Chinese Academy of Sciences (CAS), Beijing 100190, China
6 School of Resources, Environment and Architectural Engineering, Chifeng University, Chifeng 024000, China
7 School of Surveying and Land Information Engineering, Henan Polytechnic University, Jiaozuo 454000, China
* Correspondence: xrwu@shao.ac.cn; Tel.: +86-021-34775291
† These authors contributed equally to this work.

Citation: Ma, W.; Huang, L.; Wu, X.; Jin, S.; Bai, W.; Li, X. Evaluation of CYGNSS Observations for Snow Properties, a Case Study in Tibetan Plateau, China. *Remote Sens.* **2022**, *14*, 3772. https://doi.org/10.3390/rs14153772

Academic Editor: Chung-yen Kuo

Received: 21 June 2022
Accepted: 1 August 2022
Published: 5 August 2022

Abstract: Snow plays an important role in the water cycle and global climate change, and the accurate monitoring of changes in snow depth is an important task. However, monitoring snow properties is still challenging and unclear, particularly in the Tibetan Plateau, which has rough land and uneven terrain. The traditional monitoring methods have some limitations in monitoring snow depth changes, and the Global Navigation Satellite System-Reflectometry (GNSS-R) provides a new opportunity for snow monitoring. This paper employed data from the Cyclone Global Navigation Satellite System (CYGNSS) to discover the effect of snow properties. Firstly, the observations of CYGNSS were used to find the sensitive to snow properties, and the relationships between signal to noise ratio (SNR), leading edge slope (LES), surface reflectivity (SR), and snow depth were studied and analyzed, respectively. It is found that the correlation between the first two parameters and snow depth is poor, while SR can indicate the changes in snow depth, and is proposed as an indicator of SR change, namely, surface reflectivity–difference ratio factor (SR–DR factor). Furthermore, the long-time series data in the Tibetan Plateau (2018–2019) are used to analyze its effects on the time series of the SR–DR factor, while the influences of the soil freeze/thaw (F/T) process and soil moisture are excluded during the analysis. The results indicate that the SR–DR factor can be a good indicator and discriminator for snow depth. Our work shows that space-borne GNSS-R has the potential for the monitoring of snow properties.

Keywords: CYGNSS; GNSS-R; snow depth; the Tibetan Plateau; soil moisture; soil freeze/thaw process

1. Introduction

One of the fundamental components of the water cycle and global energy is the change in snow, which is of great importance for regional climate change, water use, and natural disaster monitoring [1]. The Tibetan Plateau is a sensitive area, and an ecologically fragile zone at risk of climate change, due to its harsh climate and environment, fragile ecology, and frequent snow disasters. It is the region with the highest average elevation in the world. The Tibetan Plateaus is rich in glaciers, snow, and underground ice resources [2], including the headwaters of many rivers in China, and snowmelt water is an important supplementary source of rivers. In the last few years, due to the climate change, the melting of snow and ice on the Tibetan Plateau has accelerated, and the accumulated total amount of snow is also decreasing. The snow on the Tibetan Plateau has an influence not only on the climate of the Chinese mainland, but also on the water circulation and climate systems of the East Asian region. Therefore, the research on snowpack in this area is of great hydrological and climatic importance [3,4]. The Tibetan Plateau is one of the main snow distribution areas in China, and is also an important pastoral area in China. The development of local agriculture and animal husbandry is also closely related to the change in snow cover. As a result, the study of snow cover on the Tibetan Plateau is of great significance for the sustainable development of the regional ecological environment, and agriculture and animal husbandry [5–7].

The Tibetan Plateau is a rough land, and has uneven terrain with an average altitude of more than 4000 m. The ground observation stations are rare with the uneven spatial distribution, and the observation time is discontinuous, which cannot meet the needs of research on large-scale snow distribution characteristics [8].

Satellite observations are an effective way to monitor snow cover. Space-borne optical passive sensors provide useful information for monitoring snow cover, but cannot work in high altitude and cloudy areas, due to the limitation of cloud cover [9,10]. Passive microwave remote sensing data are one of the main means of monitoring snow changes, as this method can penetrate through the porous soil layer, cloud, fog, and so on. It can penetrate a certain depth of surface to obtain information on the surface physical parameters. Therefore, microwave remote sensing is the best technology to obtain regional snow depth and snow water equivalent monitoring [11–13].

Global Navigation Satellite System-Reflectometry (GNSS-R), using the reflected signals from navigation satellites, was used for remote sensing in recent years [14–16]. This method mainly works in the L-band, with strong penetrability, so it is essentially a microwave remote sensing technology. It has the advantages of low cost, low power consumption, and high spatial and temporal resolution, due to the continuous use of navigation satellite signals as a signal source, which does not require the development of a special transmitter. It is a useful supplement to the traditional polar orbit satellites [17].

In 2004 and 2014, the United Kingdom Disaster Monitoring Constellation (UK-DMC) and Techdemosat-1 (TDS-1) test satellites were launched for GNSS-R remote sensing research [18]. In December 2016, the Cyclone Global Navigation Satellite System (CYGNSS) was launched by the National Aeronautics and Space Administration (NASA). The main scientific goal of the system is to measure the ocean surface wind field information during tropical storms and hurricanes [19]. It also provides an important opportunity for the study of land surface parameters, such as soil moisture, vegetation, flood inundation, and wetland monitoring [20–23]. To the best of our knowledge, there is no reported research on snow cover using CYGNSS data. In this study, the CYGNSS data in the Tibetan Plateau from 1 January 2018 to 31 December 2019 are selected for snow cover analysis. This provides a new method for expanding the research field of satellite-borne CYGNSS, and provides a unique remote sensing method for the study of snow characteristics on the Tibetan Plateau. With the advantages of CYGNSS, using it in the remote sensing of the surface snow depth will improve the temporal resolution of snow depth data, and reduce the effects of background brightness temperature and radio frequency interference (RFI).

The data description and processing method are presented in Section 2. The analysis using CYGNSS surface reflectivity (SR) and the surface reflectivity–difference ratio (SR–DR factor) are carried out in Section 3. Finally, the conclusions are given in Section 4.

2. Method

2.1. Data Description

The Tibetan Plateau is the highest plateau in the world, located between 26°N to 39°47′N latitude and 73°19′E to 104°47′E longitude. The topography is complex, and it has large altitude differences in various regions (Figure 1). The average annual temperature decreases from 20 °C in the southeast to below −6 °C in the northwest. In this study, we employed five public datasets to investigate the potential of CYGNSS for the study of snow cover on the Tibetan Plateau: (1) CYGNSS Level-1 (L1) data [24], (2) the Terra and Aqua combined Moderate Resolution Imaging Spectroradiometer (MODIS) Land Cover Climate

Modeling Grid (CMG) (MCD12C1) version 6 data (0.05 °C) [25], (3) long-time series dataset of snow depth in China (1979–2019) [26], (4) Soil Moisture Active and Passive (SMAP) L3 radiometer global daily 36 km EASE-grid soil moisture, version 7 [27], (5) European Centre for Medium-Range Weather Forecasts Reanalysis v5 (ERA5) soil temperature data [28].

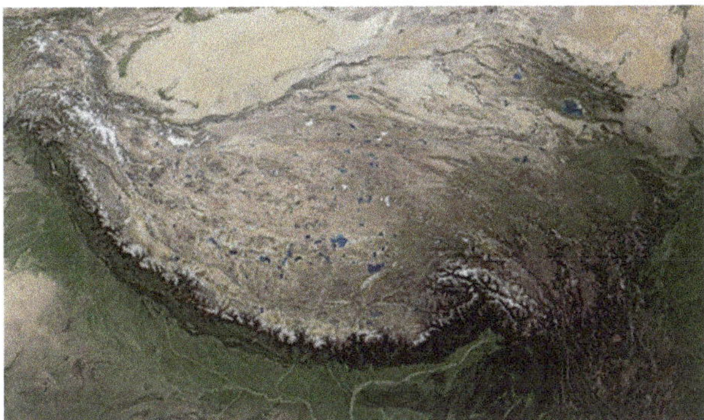

Figure 1. The satellite image of the Tibetan Plateau, Landsat/Copernicus.

2.1.1. CYGNSS Data

CYGNSS was successfully launched in December 2016. Eight small satellites carrying GNSS reflection signal receivers make up its constellation. Each CYGNSS satellite is equipped with four delay-Doppler map instruments (DDMI), and each DDMI includes two left-hand circularly polarized (LHCP) antenna and one right-hand circularly polarized (RHCP) antenna. The LHCP antenna receives GNSS signals from specular and scattering points on the Earth's surface. The RHCP antenna receives GNSS signals directly, to calculate its position and velocity. Each CYGNSS satellite can measure and observe reflections in four directions simultaneously, which means that the CYGNSS system can obtain data from 32 reflection points [29]. The delay-Doppler mapping (DDM) obtained by the receiver is a function of the surface medium, antenna gain, distance, statistical properties, and scattering geometry. The CYGNSS data used in this study were from level 1, version 2.1.

At present, most studies consider that the receiver primarily collects the coherent scattered energy in the first Fresnel region. When only the coherent energy is considered, we can calculate the DDM based on the Friis transmission formula and the Fresnel reflection coefficient of an equivalent smooth surface [30].

$$Pcoh = \frac{P_T G_T \lambda^2 G_R}{(4\pi)^2 (R_R + R_T)^2} \Gamma(s, \theta i, \varepsilon) \tag{1}$$

P_T: The transmitted power of GNSS satellite;
G_T: The GNSS satellite antenna gain at specular point direction;
G_R: The receiver antenna gain;
$P_T G_T$: The GNSS equivalent isotopically radiated power (EIRP);
λ: The wavelength;
R_R and R_T: The distance between the receiver and the specular point and the distance between the transmitter and the specular point, respectively.

According to Equation (1), the reflectivity can be expressed as:

$$\Gamma(s, \theta i, \varepsilon) = \frac{(4\pi)^2 Pcoh(R_R + R_T)^2}{\lambda^2 G_R G_T P_T} \tag{2}$$

where s is the surface rms height, which can characterize the surface roughness, θ_i is the incidence angle and the dielectric constant of the geophysical parameters. Surface reflectivity (SR) is usually a small positive value, which is converted to decibel for visualization and data analysis according to Equation (3). Table 1 shows the main parameters for the L1 data used in this study.

$$dB = 10log_{10}(\Gamma(s, \theta i, \varepsilon)) \tag{3}$$

Table 1. The Cyclone Global Navigation Satellite System (CYGNSS) level 1 data variables.

Name	Comment
ddm_timestamp_utc	DDM sample time
sp_lat	Specular point latitude, in degrees north
sp_lon	Specular point longitude, in degrees east
sp_inc_angle	The specular point incidence angle, in degrees
sp_rx_gain	The receive antenna gain in the direction of the specular point, in dBi
gps_eirp	The effective isotropic radiated power (EIRP) of the L1 C/A code signal within $\pm$ 1 MHz of the L1 carrier radiated by space vehicle, sv_num, in the direction of the specular point, in Watts
rx_to_sp_range	The distance between the CYGNSS spacecraft and the specular point, in meters
tx_to_sp_range	The distance between the GPS spacecraft and the specular point, in meters
power_analog	17×11 array of DDM bin analog power, Watts

It is worth noting that the treatment of the coherent and incoherent scattering is an open issue when dealing with the CYGNSS data [31]. When using CYGNSS data for soil moisture retrieval, the coherent energy comes primarily from specular reflections of water inland within the GNSS-R footprint, which leads to the increase in DDM peak value. The peak energy of the coherent part is many times greater than that in the non-coherent portion. The satellite-based GNSS-R receiver mainly receives coherent scattered signals from the first Fresnel zone around the specular reflection point [32]. In fact, the signals received by GNSS-R receivers, in most cases, contain signals from both coherent and incoherent scattered fields, due to variations in surface roughness. Most of the currently available studies on satellite-based GNSS-R remote sensing of surface soil moisture simplify the scattering mechanism of the L-band on the actual surface, assuming that only coherent scattering occurs on the terrestrial surface, and incoherent scattering is ignored [31]. We removed data with SNRdB less than 2 dB and CYGNSS antenna gain of less than 0 dB. The elevation angle of the specular points is less than 65°.

2.1.2. IGBP Land Cover Classification

The land surface scattering process is complex. In addition to the influence of surface roughness, the surface scattering signals received directly by GNSS-R receivers are also impacted by other factors such as vegetation layer and topography. There are various types of land cover on the Tibetan plateau, and different vegetation covers have different effects on the received signals. In order to reduce their influence on the received signal power, the consequent analysis is performed in relatively uniform land cover categories. In this way, we can ensure that the features in each zone are comparatively unified, so that the differences in the CYGNSS observations caused by various types of surface coverage can be considered.

The MODIS Land Cover Climate Modeling Grid Product (MCD12C1) provided are the sub-pixel proportions of each land cover class in each 0.05° pixel and the aggregated quality assessment information for the IGBP scheme [25]. Here, we have employed the IGBP of 2020 to obtain the land cover information of the Tibetan Plateau region (Figure 2a). The Tibetan Plateau region is dominated by four categories: high-vegetation-covered area; moderate-vegetation-covered area, low-vegetation-covered area, and barren or desert area

(Figure 2b). In this paper, we conducted the analysis of snowpack characteristics within the same land cover type, as far as was possible.

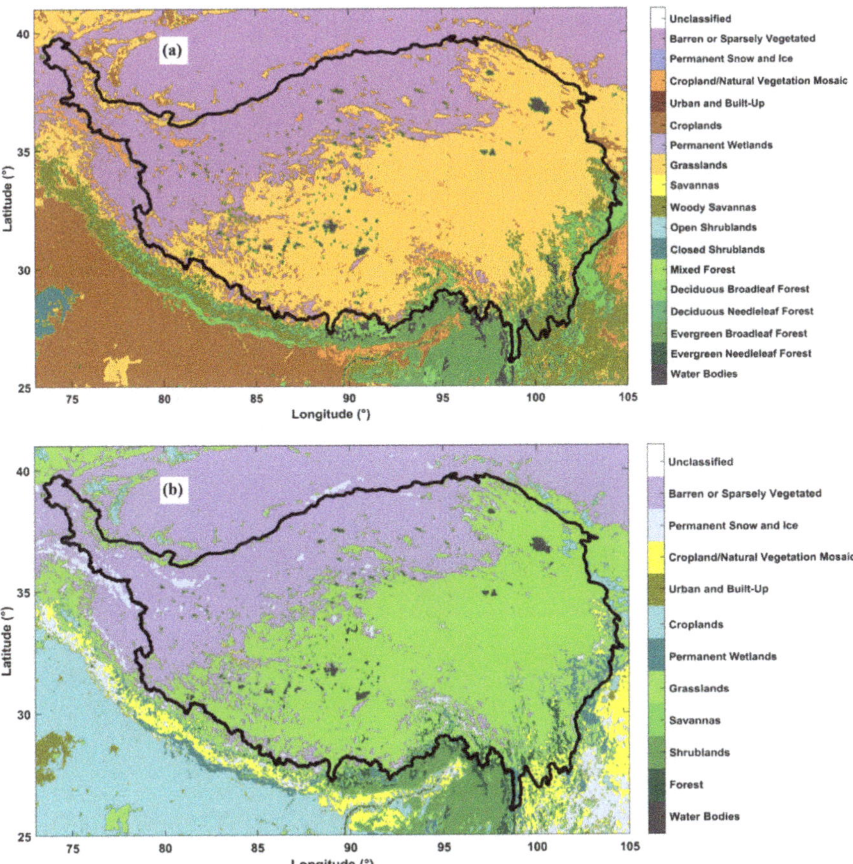

Figure 2. (a) The IGBP land cover type on the Tibetan Plateau; (b) the land cover types on the Tibetan Plateau are reclassified into four types.

2.1.3. Long Time Series Dataset of Snow Depth in China

The long-term series of daily snow depth dataset in China (1979–2019), released by the National Tibetan Plateau Data Center (NTPDC), provides daily snow depth distribution data for China from 1 January 1979 to 31 December 2019. The raw data used to invert this snow depth dataset are from the scanning multichannel microwave radiometer (SMMR) (1979–1987), the special sensor microwave imager (SSM/I) (1987–2007), and the special sensor microwave imager/sounder (SSMI/S) (2008–2019) daily passive microwave bright temperature data (EASE-grid), processed by the National Snow and Ice Data Center (NSIDC). It has a spatial resolution of 25 km, which can be used for climate analysis, hydrological modeling, and water management on a large scale and in long time series [33–35]. During the processing of the data, the brightness temperatures of different sensors are first cross-calibrated, and the observations affected by the snow depth observation errors of ground stations, the surface water, and the liquid water content in the snow layer are removed before the inversion of the snow depth.

Using this dataset, we calculated the snow cover change in the Tibetan Plateau between January 2018 and December 2019, and the corresponding snow depth changes for four different land cover types. The snow accumulation trend under different land cover types is similar. From October every year, the Tibetan Plateau gradually enters the snowfall period, reaching the peak of snowpack in January and February of the following year. The snow gradually melts after April as the temperature rises. Figure 3 shows the snow depth map of the Tibetan Plateau for one month in summer and winter using this dataset.

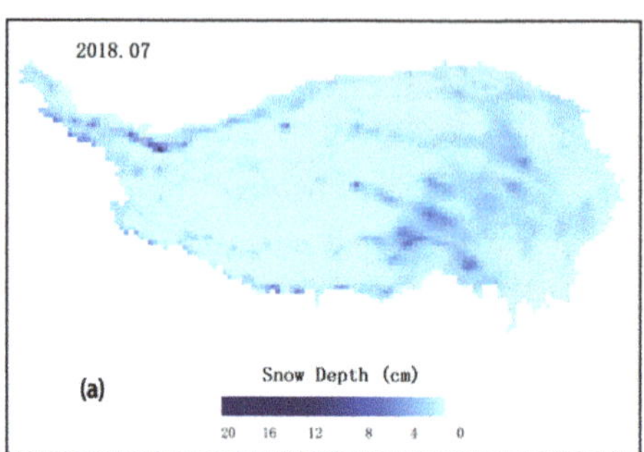

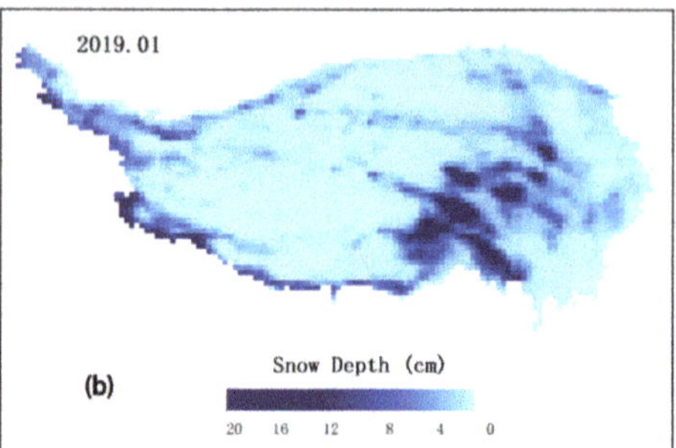

Figure 3. (**a**) The snow depth map of Tibetan Plateau in summer (July 2018); (**b**) the snow depth map of Tibetan Plateau in winter (January 2019).

By comparing the values with the daily reflectivity data of CYGNSS in the same region, the correlation between snow depth and reflectivity can be inferred, and the potential of CYGNSS in snow feature research can be verified.

2.1.4. SMAP L3 Radiometer Global Daily 36 km EASE-Grid Soil Moisture, Version 7

Previous studies show that GNSS signals reflected from the surface are sensitive to changes in soil moisture, and the soil moisture also changes the SR [22]. This soil moisture product provides a composite of daily estimates of global land surface conditions retrieved by the Soil Moisture Active Passive (SMAP) passive microwave radiometer. The data are on a regular latitude/longitude grid, with predictable disparities in space and time. We used the soil moisture data to plot a time series of mean snow depth with SMAP mean soil moisture with different land cover types on the Tibetan Plateau. For more detail, please see the following. By analyzing the variations in soil moisture over the same period, the effect of SR changes due to soil moisture can be excluded; thus, we believe that the variations of SR are due to the changes of snow coverage.

2.1.5. ERA5 Soil Temperature Data

The dataset is the soil temperature at level 1 (in the middle of layer 1). The European Centre for Medium-Range Weather Forecasts (ECMWF) Integrated Forecasting System (IFS) has a four-layer indication for the soil, where layer 1 ranges from 0 cm to 7 cm. Soil temperature is measured in the middle of each layer. When a freeze–thaw transition occurs in soil, the water in the soil changes from solid ice to liquid water. The dielectric constants of water and ice differ significantly, so the reflectivity changes accordingly. By analyzing the changes in snow depth and SR during the period when the surface temperature does not change from positive to negative (or from negative to positive), we assessed whether the snow depth can be monitored by CYGNSS.

2.2. Calculation of the Surface Reflectivity

The incidence angle is an important factor affecting the reflectivity of GNSS constellation; the incidence angle of CYGNSS ranges from 0 to 70°. To improve data quality, the incidence angle greater than 60° is eliminated and the data in the direction of the lowest point are used for normalization. Based on the data of CYGNSS L1 from 1 January 2018 to 31 December 2019 in the Tibetan plateau, we calculated and converted all the values to the dB scale, according to Equations (1)–(3). In this paper, we assume that the energy of CYGNSS data is coherent, and incoherent scattering is not considered in the analysis. The effective sampling range of CYGNSS is between approximately 38°N to 38°S, while the latitude range of the Tibetan Plateau is about 26°N to 40°N, so some areas in the north lack data [29]. Figure 4 shows the surface reflectivity of CYGNSS specular points in the Tibetan Plateau on 1 March 2018. Comparing with the satellite image in Figure 1, we can see that the SR is higher in the area with water and snow. In order to exclude the influence of water bodies on the results, these data should be removed from the analysis.

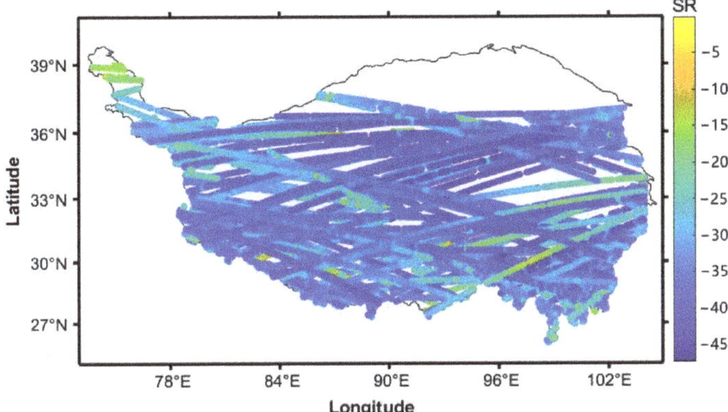

Figure 4. CYGNSS specular point surface reflectivity in the Tibetan Plateau on 1 March 2018.

In order to be consistent with the time series analysis, the raw SR of the specular points was transformed into gridded data. Considering the study area and the spatial resolution of snow data (Section 2.1.3), a grid was generated with a resolution of 0.25 along the geodetic latitude and longitude. The data values for each grid cell were defined as the average value of the specular points falling into that grid.

To test the reaction of SR to the changes in surface parameters on the Tibetan Plateau, the SR at the CYGNSS specular reflection points for January 2019 and July 2019 are given in Figure 5a,b. As seen in the figure, the reflectivity increases in the whole region in July compared to January. To exclude differences in CYGNSS observations due to different ground cover types, the variation in SR in response to surface snow depth is analyzed under the same ground cover types.

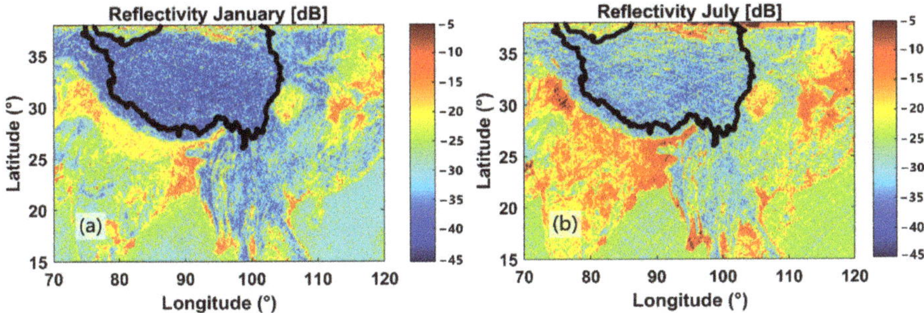

Figure 5. CYGNSS-derived reflectivity at specular points on Tibetan Plateau in January (**a**) and July (**b**) of 2019.

3. Results and Analysis

3.1. Comparison of Surface Reflectivity and Parameters on the Tibetan Plateau

The CYGNSS daily SR and snow depth were analyzed and compared under different land cover types. The snow depth in the Tibetan Plateau region was calculated using the long-term series of daily snow depth dataset in China (1979–2019). Figure 6 shows the time series of CYGNSS SR and snow depth, and we assess the ability of CYGNSS data to monitor changes in snow cover depth by comparing them.

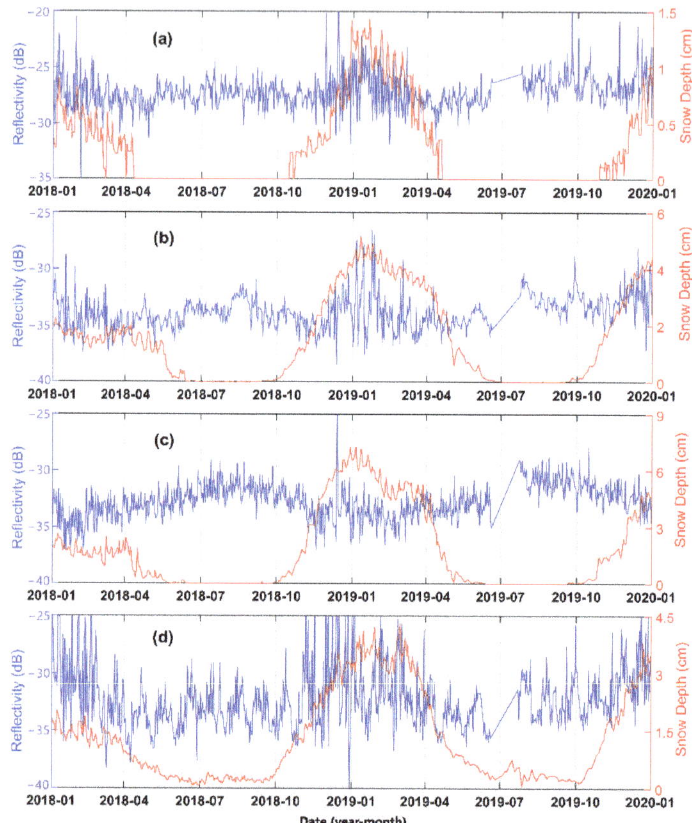

Figure 6. The time series of SR (blue left axis) versus snow depth (red right axis) compared under different land cover type on Tibetan Plateau. (**a**) High-vegetation-covered area; (**b**) moderate-vegetation-covered area; (**c**) low-vegetation-covered area; (**d**) barren or desert area.

The SR varies with the surface dielectric constant, so there are differences in SR for different ground objects. The snow on the Tibetan Plateau is mainly produced from October to April each year, during which the Tibetan region enters a period of high snow frequency, and SR increases as the snow depth increases. A portion of the SMAP data was lost between June 2019 and July 2019, due to data sampling issues with the satellite. Figure 6 shows the variation of CYGNSS reflectivity (blue left axis), which is consistent with the oscillation of snow depth (red right axis) during the months with snow accumulation. In addition, the part of the anomalous oscillation is mainly caused by the attenuation and volume scattering of vegetation. The change trend of SR and snow depth is closest in the high-vegetation-covered area. In the low-vegetation-covered area and barren/desert area, the CYGNSS time series fluctuates more by the noise, but shows an overall increasing trend.

The analysis results show that the CYGNSS SR has a certain correlation with the variation in snow depth, which can be used for snow monitoring on the Tibetan Plateau.

Soil moisture is an important factor affecting SR. To exclude its influence on the results, we plotted the time series of soil moisture using the SMAP surface soil moisture data and examined it with the variation of snow depth over time, in order to investigate the correlation between them. It can be seen from Figure 7 that the interannual variation of soil moisture is closely related to the variation in snow cover, and the soil moisture content fluctuates greatly when the snow cover melts. The low-vegetation-covered area has the largest change in soil moisture. In addition, when the snow cover on Tibetan Plateau is obvious (November 2018–February 2019), the value of soil moisture under different land cover types fluctuates around 0.1 cm^3/cm^3, and does not change significantly. Therefore, we learn that it is not the change in soil moisture that causes the oscillation of SR.

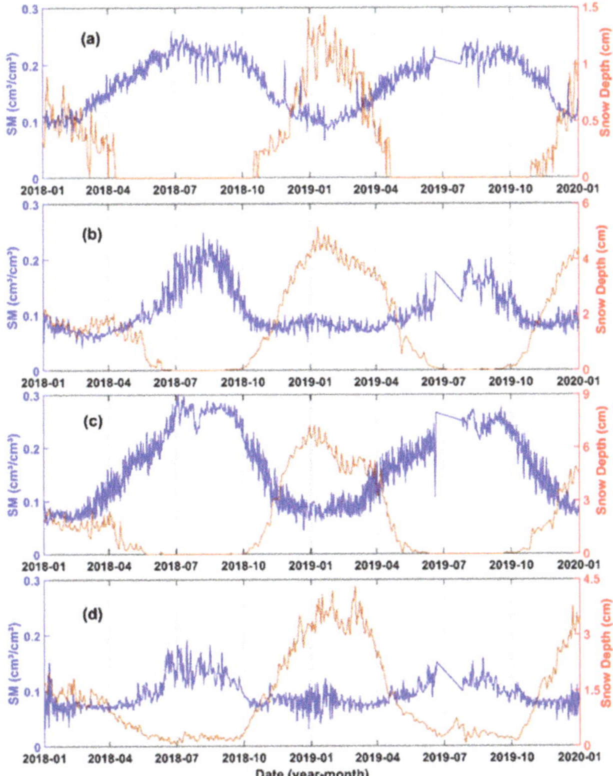

Figure 7. The time series of soil moisture (blue left axis) versus snow depth (red right axis) compared under different land cover type on Tibetan Plateau. (**a**) High-vegetation-covered area; (**b**) moderate-vegetation-covered area; (**c**) low-vegetation-covered area; (**d**) barren or desert area.

To further verify the effect of snow depth on the SR, we also used the ERA5 soil temperature data to obtain the monthly average soil temperature over this four month period (Figure 8). We found that there is no inversion from negative to positive values of surface temperature in most areas of the Tibetan Plateau during this period. The northern region shows the relatively large temperature variation. However, as this region is beyond the detection zone of CYGNSS, we exclude the effect of surface temperature variations.

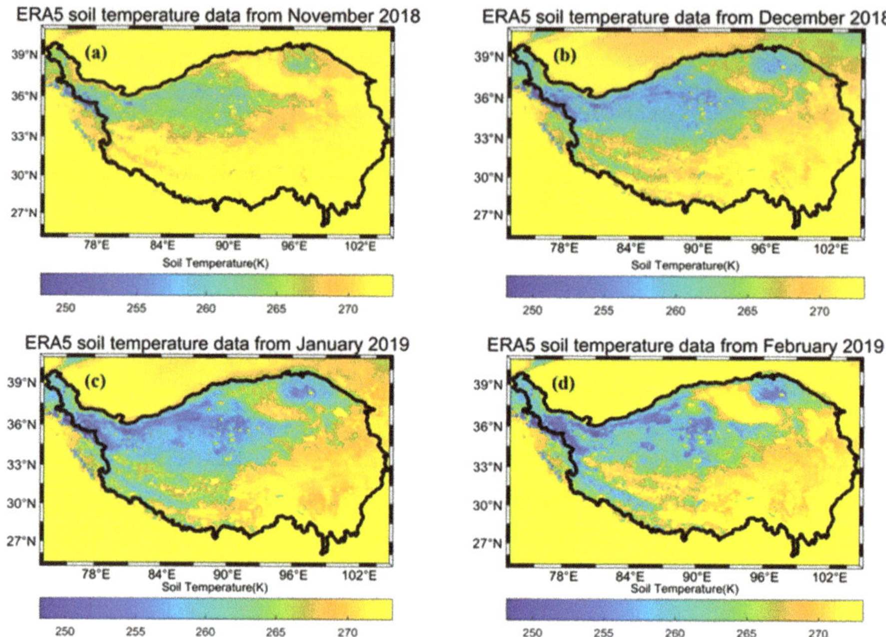

Figure 8. The November 2018 to February 2019 monthly average soil temperature on Tibetan Plateau. (**a**) November 2018, (**b**) December 2018, (**c**) January 2019, (**d**) February 2019.

Table 2 shows the mean values of snow depth, SR, and soil temperature for the four different land cover types over a four month period. It can be seen that the trends of snow depth and SR over time are approximately the same when the soil temperature is negative without more significant changes. For other commonly used evaluation metrics, such as leading edge slope (LES) and signal to noise ratio (SNR), we provide a brief discussion in Appendix A.

Table 2. The mean values of snow depth, SR, and soil temperature for the four different land cover types over a four month period.

Time	Snow Depth (cm)	Surface Reflectivity (dB)	Soil Temperature (°C)
(a) Mixed–Forest			
2018.11	0.30817	−27.36039	−0.43541
2018.12	0.75975	−26.51822	−0.72392
2019.01	1.15806	−25.96435	−1.34392
2019.02	0.99447	−27.32706	−0.82510
(b) Open–Shrubland (Desert)			
2018.11	1.98258	−35.25758	−3.73655
2018.12	3.66645	−33.51377	−7.68061
2019.01	4.54840	−32.83397	−9.16413
2019.02	4.04246	−34.33123	−7.49004

Table 2. *Cont.*

Time	Snow Depth (cm)	Surface Reflectivity (dB)	Soil Temperature (°C)
(c) Grassland			
2018.11	3.38779	−34.03826	−2.11510
2018.12	6.08088	−33.24549	−4.36104
2019.01	6.35053	−33.78475	−5.42212
2019.02	5.02427	−34.37498	−4.49595
(d) Barren/Desert			
2018.11	1.86681	−31.47885	−1.29014
2018.12	2.97565	−30.19818	−7.56332
2019.01	3.60599	−31.14914	−8.23865
2019.02	3.38220	−30.62580	−4.55624

3.2. Surface Reflectivity Difference Ration Factor

In Section 3.1, we evaluated the effects of snow depth on surface reflectivity, and find that it is more sensitive than the other GNSS-R observables, i.e., SNR and LES (Appendix A). Therefore, we provide an indicator in this section to illustrate its effect on the detection of snow properties. Here, we define this indicator as surface reflectivity–difference ratio factor (SR–DR factor) [36]. The factor SR–DR is defined as follows:

$$F(t) = \frac{\Gamma(t) - \Gamma min}{\Gamma max - \Gamma min} \qquad (4)$$

where $\Gamma(t)$ is the SR at time t, and Γ_{max} and Γ_{min} are the maximum and minimum SR in the time range, respectively. It should be noted that within the form of this equation, the snow water equivalent (SWE) can be achieved when the SWE is in quite good relationship with snow depth.

The CYGNSS data for 2019 was studied as a case study. Figure 9 shows the SR–DR factor for the Tibetan Plateau in January, February, March, and July 2019, which has the same resolution of 25 km as SR. Analyzing the factor for the Tibetan Plateau in 2019, we found that the SR–DR factor increases in summer compared to winter for the whole region. It also changes with snow depth in winter when there is snow cover on the Tibetan Plateau.

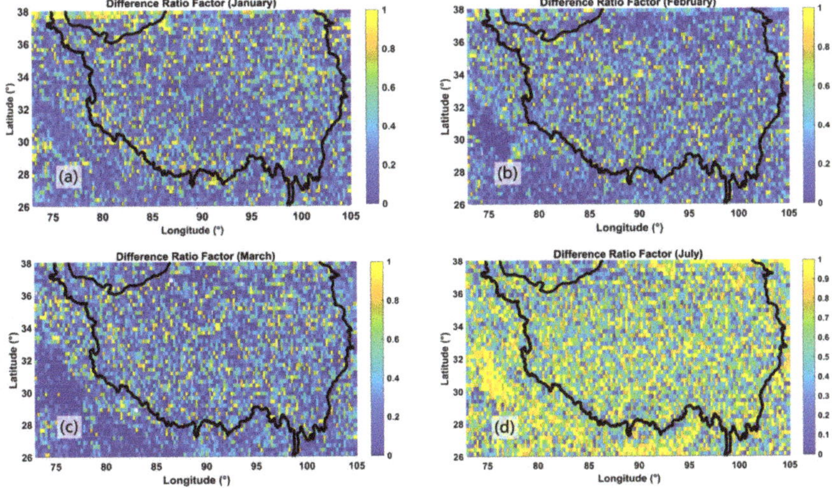

Figure 9. The SR–DR factor on Tibetan Plateau in January (**a**), February (**b**), March (**c**), and July (**d**) of 2019.

Figure 10 shows that the SR–DR factor for January 2019 is mainly concentrated between 0 and 0.4. In February and March 2019, SR–DR factor shows an increase, without a large change in soil moisture during this period. In July, it shows a higher value, mostly greater than 0.4. The snow depth is also lower at this time, due to the arrival of summer.

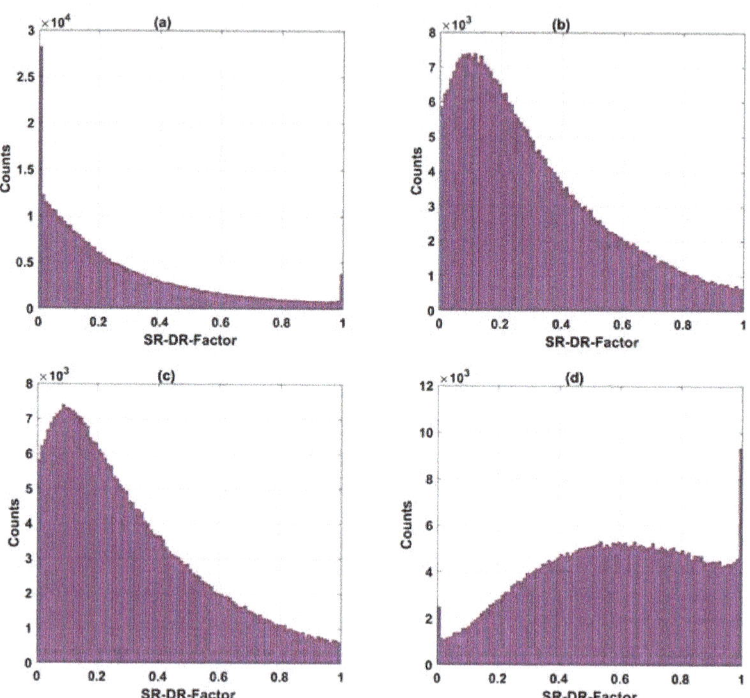

Figure 10. Histogram of the SR–DR factor distribution on the Tibetan Plateau in January (**a**), February (**b**), March (**c**), and July (**d**) of 2019.

We selected a $1° \times 1°$ size area with different land cover types for the analysis of the relationship between the SR–DR factor and snow depth as follows: (a) barren or desert area (79°–80°E, 33°–34°N); (b) low-vegetation-covered area (81.5°–82.5°E, 31.5°–32.5°N); and (c) moderate-vegetation-covered area (95°–96°E, 29°–30°N). We removed the high-vegetation-covered area because this part of the Tibetan Plateau is small and mostly concentrated in mountainous areas with complex topography. The topographic relief and vegetation roughness may have a great impact on the effective surface reflectivity, which will make the results inaccurate. The time with snow cover on the Tibetan Plateau is concentrated from November to March of the following year, so we choose CYGNSS data from November 2018 to March 2019, and the snow depth at the coordinates of the center point of the area is used to express the change in snow accumulation. Figure 11 presents the time series of SR–DR factor versus snow depth in the research area. From the simulation, we can see that the oscillation of the snow depth in the barren area is, in general, consistent with the daily CYGNSS SR–DR factor. The differences become progressively larger as the vegetation cover level increases.

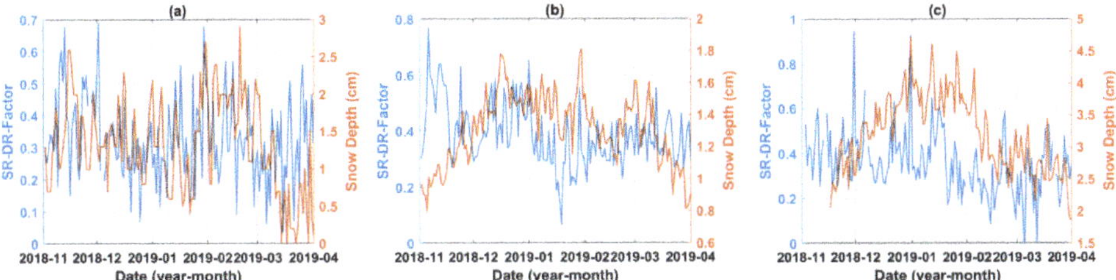

Figure 11. The time series of SR–DR factor versus snow depth. (**a**) Barren or desert area; (**b**) low-vegetation-covered area; (**c**) moderate-vegetation-covered area.

4. Conclusions

Since the launch of GNSS-R satellites, many scholars have used them to study the parameters of wetland dynamics, sea surface wind speed, and soil freeze/thaw. This paper takes CYGNSS as an example to study its potential application in snow monitoring on the Tibetan Plateau. The CYGNSS data collected from January 2018 to December 2019 were processed and analyzed to calculate surface reflectivity and compare it with snow depth. Soil moisture from SMAP and temperature data from ERA5 were used to study the changes in both when the snow depth varied significantly.

With the help of typical GNSS-R observables (LES, SNR, and SR), we found that SR can be a good potential parameter to indicate the snow depth. Therefore, based on the SR, the finally developed indicator, which is the SR–DR factor, shows an apparent good relationship with snow depth. This factor can represent the snow water equivalent to some extent through the ratio form in the equation.

Compared with conventional remote sensing satellites, the sampled data from CYGNSS have high spatial and temporal resolution. This demonstrates the potential of satellite-based GNSS-R to monitor snow depth at a high spatial resolution.

In the analysis of GNSS-R data, it is generally believed that the signal received by the receiver is mainly composed of coherent scattering along the mirror direction, but, in fact, the incoherent scattering affects the results, and their relative contribution depends on the receiver height, surface roughness, and vegetation coverage. In this paper, only the case of coherent scattering is considered.

The CYGNSS was originally intended for sea surface wind field research, and many of its applications on land are exploratory studies. This paper qualitatively proves that satellite-based GNSS-R can be used for snow monitoring, but still lacks a quantitative evaluation to determine the specific value of snow depth as retrieved from SR. It can be further analyzed with other estimators in future research. In addition, the coverage of CYGNSS is mainly concentrated in the middle and low latitudes, so there is a lack of effective data in some areas of the northern Tibetan Plateau. In practical application, it may be necessary to combine the traditional remote sensing satellite data to achieve high-precision snow monitoring.

Author Contributions: W.M. analyzed and interpreted the data and materials, and wrote the original manuscript. L.H. provided the data analysis and interpretation. X.W. conducted the conceptualization suggestions, review, and funding. S.J. performed the suggestion and inspection. W.B. and X.L. conducted the suggestions and funding. All authors have read and agreed to the published version of the manuscript.

Funding: This research was funded by the National Natural Science Foundation of China (No. 42061057&42074042&72004017), Chifeng University, the Laboratory of National Land Space Planning and Disaster Emergency Management of Inner Mongolia (CFXYZD202006), and Innovative Teams of Studying Environmental Evolution and Disaster Emergency Management of Chifeng University in China under grant number cfxykycxtd202006.

Data Availability Statement: The datasets analyzed during the current study are available from NASA, https://podaac.jpl.nasa.gov/dataset/CYGNSS_L1_CDR_V1.0 (accessed on 12 March 2021). The datasets analyzed during the current study are available in the Land Processes Distributed Active Archive Center, https://lpdaac.usgs.gov/products/mcd12c1v006/ (accessed on 13 January 2022). The datasets analyzed during the current study are available in the National Tibetan Plateau Data Center repository, http://data.tpdc.ac.cn/zh-hans/data/df40346a-0202-4ed2-bb07-b65dfcda9368/ (accessed on 10 August 2020). The datasets analyzed during the current study are available in the National Snow and Ice Data Center (NSIDC), https://nsidc.org/data/SPL3SMP/versions/7 (accessed on 22 December 2021). O'Neill, P.E.; Chan, S.; Njoku, E.G.; Jackson, T.; Bindlish, R.; Chaubell, J. 2020. SMAP L3 radiometer global daily 36 km EASE-grid soil moisture, version 7. [Indicate subset used]. Boulder, CO, USA. NASA National Snow and Ice Data Center Distributed Active Archive Center. https://doi.org/10.5067/HH4SZ2PXSP6A (accessed on 23 December 2021). The datasets analyzed during the current study are available in the ECMWF Reanalysis v5—Land, https://www.ecmwf.int/en/forecasts/dataset/ecmwf-reanalysis-v5-land (accessed on 6 January 2021) [25].

Acknowledgments: The authors acknowledge Zhounan Dong in the Shanghai Astronomical Observatory, Chinese Academy of Sciences, and Andrés Calabia in Nanjing University of Information Science and Technology for their help during the data processing.

Conflicts of Interest: The authors declare no conflict of interest.

Appendix A

In addition to the qualitative analysis, we also investigated the relationship between the LES of the delayed power waveform commonly used in satellite-based GNSS-R surveys and the snow depth. The LES-SD scatterplot drawn from November 2018 to February 2019 without limiting the satellite observation angle shows there is some relationship between snow properties and LES or SNR. However, the results are not as obvious as SR; therefore, for we employ SR for our analysis.

Appendix A.1. LES and Snow Properties

LES is one of the most important CYGNSS observables. Figure A1 presents the relationship between LES and snow depth. Figure A1a–c are the corresponding relationships for high-, moderate-, and low-vegetation-covered area, while subfigure d is for the barren or desert area. At present, from the relationship shown in Figure A1, we can see that LES has a poor relationship with the snow depth. Since LES is one of parameters that reflects the surface roughness condition, it cannot present the snow depth information in the CYGNSS pixel, which has a spatial resolution of 7×2.5 km. However, it isa potential good indicator of snow depth for complex mountain terrain.

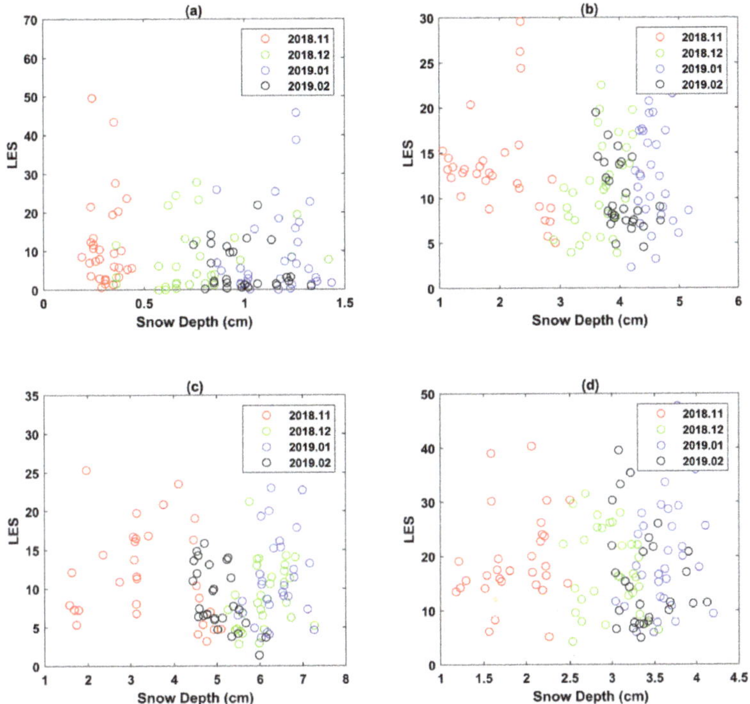

Figure A1. The LES of CYGNSS observations in relation to snow depth in winter. (**a**) High-vegetation-covered area; (**b**) moderate-vegetation-covered area; (**c**) low-vegetation-covered area; (**d**) barren or desert area.

Appendix A.2. SNR with the Snow Properties

SNR is also one of the most important output parameters of the CYGNSS observables. During our analysis, we also compared the relationship between snow depth and SNR (Figure A2). In order to reduce the influence of different land geophysical parameters, we also plotted the relationship for four different surface types, i.e., high-vegetation-covered area (a); moderate-vegetation-covered area (b); low-vegetation-covered area; (d) barren or desert area (c). We can see that the SNR also cannot reveal the relationship between the two parameters. More detailed analysis to retrieve this information for the detection of snow properties will occur on in our future research.

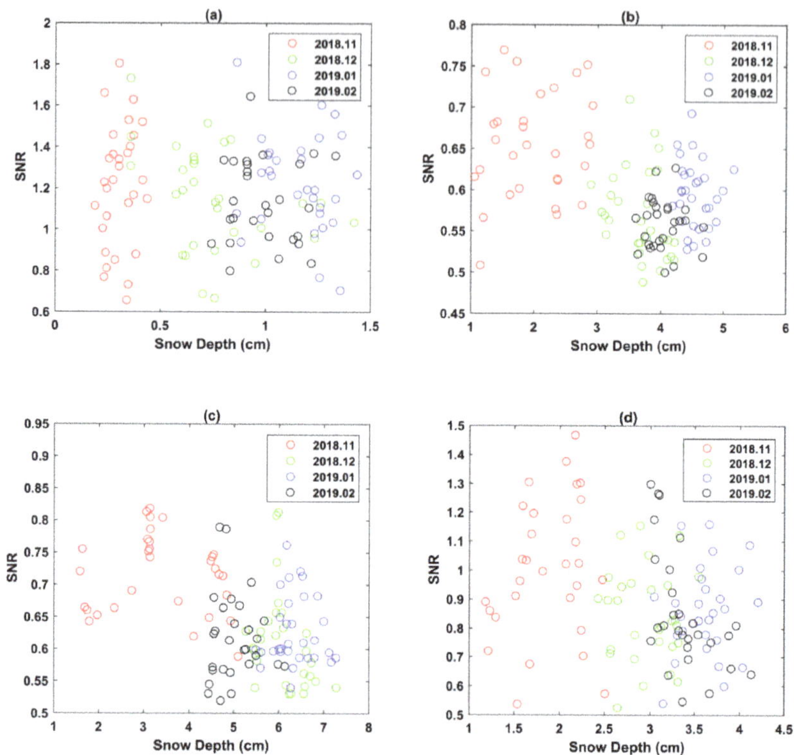

Figure A2. The SNR of CYGNSS observations in relation to snow depth in winter. (**a**) High-vegetation-covered area; (**b**) moderate-vegetation-covered area; (**c**) low-vegetation-covered area; (**d**) barren or desert area.

References

1. Brown, R.D. Northern Hemisphere Snow Cover Variability and Change, 1915–1997. *J. Clim.* **2000**, *13*, 2339–2355. [CrossRef]
2. Qiu, J. China: The third pole. *Nature* **2008**, *454*, 393–396. [CrossRef]
3. Cheng, G.D.; Wu, T.H. Responses of permafrost to climate change and their environmental significance, Qinghai-Tibet Plateau. *J. Geophys. Res. Atmos.* **2007**, *112*, F02S03. [CrossRef]
4. Senan, R.; Orsolini, Y.J.; Weisheimer, A. Impact of springtime Himalayan–Tibetan Plateau snowpack on the onset of the Indian summer monsoon in coupled seasonal forecasts. *Clim. Dyn.* **2016**, *47*, 2709–2725. [CrossRef]
5. Bormann, K.J.; Brown, R.D.; Derksen, C. Estimating snow-cover trends from space. *Nat. Clim. Change* **2018**, *8*, 924–928. [CrossRef]
6. Cui, Y.; Chuan, X.; Lemmetyinen, J.; Shi, J.; Jiang, L.; Peng, B. Estimating snow water equivalent with backscattering at X and Ku band based on absorption loss. *Remote Sens.* **2016**, *6*, 505. [CrossRef]
7. Zhou, J.; Kinzelbach, W.; Cheng, G. Monitoring and modeling the influence of snow pack and organic soil on a permafrost active layer, Qinghai–Tibetan Plateau of China. *Cold Reg. Sci. Technol.* **2013**, *90*, 38–52. [CrossRef]
8. Struzik, P. Japan Aerospace Exploration Agency GCOM-W1 satellite snow depth product: Outcome of the first winter. *J. Appl. Remote Sens.* **2014**, *8*, 4480–4494. [CrossRef]
9. Hall, D.K.; Riggs, G.A.; Salomonson, V.V. MODIS snow cover products. *Remote Sens. Environ.* **2002**, *83*, 181–194. [CrossRef]
10. Gafurov, A.; Bárdossy, A. Cloud removal methodology from MODIS snow cover product. *Hydrol. Earth Syst. Sci.* **2009**, *13*, 1361–1373. [CrossRef]
11. Wulder, M.A.; Nelson, T.A.; Derksen, C. Snow cover variability across central Canada (1978–2002) derived from satellite passive microwave data. *Clim. Change* **2007**, *82*, 113–130. [CrossRef]
12. Kim, S.; van Zyl, J.; McDonald, K.; Njoku, E. Monitoring surface soil moisture and freeze-thaw state with the high-resolution radar of the Soil Moisture Active/Passive (SMAP) mission. In Proceedings of the IEEE Radar Conference, Washington, DC, USA, 10–14 May 2010; pp. 735–739.

13. Derksen, C.; Xu, X.; Dunbar, R.S.; Colliander, A.; Kim, Y.; Kimball, J.S.; Black, T.A.; Euskirchen, E.; Langlois, A.; Loranty, M.M. Retrieving landscape freeze/thaw state from Soil Moisture Active Passive (SMAP) radar and radiometer measurements. *Remote Sens. Environ.* **2017**, *194*, 48–62. [CrossRef]
14. Martin-Neira, M. A passive reflectometry and interferometry system (PARIS): Application to ocean altimetry. *ESA J.* **1993**, *17*, 331–355.
15. Small, E.E.; Larson, K.M.; Braun, J.J. Sensing vegetation growth with reflected GPS signals. *Geophys. Res. Lett.* **2010**, *37*, L12401. [CrossRef]
16. Boyd, D. Inversion Study of Simulated and Physical Soil Moisture Profiles using Multifrequency Soop-Sources. In Proceedings of the IGARSS 2019—2019 IEEE International Geoscience and Remote Sensing Symposium, Yokohama, Japan, 28 July–2 August 2019; pp. 5259–5262.
17. Masters, D.; Axelrad, P.; Katzberg, S. Initial results of land-reflected GPS bistatic radar measurements in SMEX02. *Remote Sens. Environ.* **2004**, *92*, 507–520. [CrossRef]
18. Comite, D.; Cenci, L.; Colliander, A.; Pierdicca, N. Monitoring Freeze-Thaw State by Means of GNSS Reflectometry: An Analysis of TechDemoSat-1 Data. *IEEE J. Sel. Top. Appl. Earth Obs. Remote Sens.* **2020**, *13*, 2996–3005. [CrossRef]
19. Steve, P. NASA Intensifies Hurricane Studies with CYGNSS. *Earth Obs.* **2013**, *25*, 13–21.
20. Ruf, C.S.; Chew, C.; Lang, T.; Morris, M.G.; Nave, K.; Ridley, A.; Balasubramaniam, R. A New Paradigm in Earth Environmental Monitoring with the CYGNSS Small Satellite Constellation. *Sci. Rep.* **2018**, *8*, 8782. [CrossRef]
21. Chew, C.; Reager, J.T.; Small, E. CYGNSS data map flood inundation during the 2017 Atlantic hurricane season. *Sci. Rep.* **2018**, *8*, 9336. [CrossRef]
22. Kim, H.; Lakshmi, V. Use of Cyclone Global Navigation Satellite System (CyGNSS) Observations for Estimation of Soil Moisture. *Geophys. Res. Lett.* **2018**, *45*, 8272–8282. [CrossRef]
23. Wu, X.R.; Dong, Z.D.; Jin, S.G. First Measurement of Soil Freeze/Thaw Cycles in the Tibetan Plateau Using CYGNSS GNSS-R Data. *Remote Sens.* **2020**, *12*, 2361. [CrossRef]
24. CYGNSS. CYGNSS Level 1 Climate Data Record Version 1.0. Ver. 1.0. PO. DAAC, CA, USA; 2020. Available online: https://podaac.jpl.nasa.gov/dataset/CYGNSS_L1_CDR_V1.0 (accessed on 12 March 2021). [CrossRef]
25. Friedl, M.; Sulla-Menashe, D. MCD12C1 MODIS/Terra+Aqua Land Cover Type Yearly L3 Global 0.05Deg CMG V006. 2015, Distributed by NASA EOSDIS Land Processes DAAC. Available online: https://lpdaac.usgs.gov/products/mcd12c1v006/ (accessed on 13 January 2022). [CrossRef]
26. Che, T. Long-Term Series of Daily Snow Depth Dataset in China (1979–2020). National Tibetan Plateau Data Center, 2015. Available online: https://data.tpdc.ac.cn/en/data/df40346a-0202-4ed2-bb07-b65dfcda9368/ (accessed on 10 August 2020). [CrossRef]
27. O'Neill, E.P.; Chan, S.; Njoku, E.G.; Jackson, T.; Bindlish, R.; Chaubell, J. SMAP L3 Ra-Diometer Global Daily 36 km EASE-Grid Soil Moisture, Version 7. [Soil Moisture]. NASA National Snow and Ice Data Center Distributed Active Archive Center: Boulder, CO, USA, 2020. Available online: https://nsidc.org/data/spl3smp/versions/7 (accessed on 22 December 2021). [CrossRef]
28. Muñoz Sabater, J.; ERA5-Land Hourly Data from 1981 to Present. Copernicus Climate Change Service (C3S) Climate Data Store (CDS). 2019. Available online: https://cds.climate.copernicus.eu/cdsapp#!/dataset/10.24381/cds.e2161bac (accessed on 6 January 2021). [CrossRef]
29. Ruf, C.; Chang, P.S.; Clarizia, M.P.; Gleason, S.; Jelenak, Z. *CYGNSS Handbook*; Michigan Pub: Ann Arbor, MI, USA, 2016; p. 154.
30. Al-Khaldi, M.M. Time-Series Retrieval of Soil Moisture Using CYGNSS. *IEEE Trans. Geosci. Remote Sens.* **2019**, *57*, 4322–4331. [CrossRef]
31. Munoz-Martin, J.F.; Onrubia, R.; Pascual, D.; Park, H.; Camps, A.; Rüdiger, C.; Walker, J.; Monerris, A. Untangling the Incoherent and Coherent Scattering Components in GNSS-R and Novel Applications. *Remote Sens.* **2020**, *12*, 1208. [CrossRef]
32. Garrison, J.L.; Katzberg, S.J.; Howell, C.T. Detection of ocean reflected GPS signals: Theory and experiment. In Proceedings of the IEEE SOUTHEASTCON'97.'Engineering the New Century', Blacksburg, VA, USA, 12–14 April 1997.
33. Che, T.; Li, X.; Jin, R.; Armstrong, R.; Zhang, T.J. Snow depth derived from passive microwave remote-sensing data in China. *Ann. Glaciol.* **2008**, *49*, 145–154. [CrossRef]
34. Dai, L.Y.; Che, T.; Ding, Y.J. Inter-calibrating SMMR, SSM/I and SSMI/S data to improve the consistency of snow-depth products in China. *Remote Sens.* **2015**, *7*, 7212–7230. [CrossRef]
35. Dai, L.Y.; Che, T.; Ding, Y.J.; Hao, X.H. Evaluation of snow cover and snow depth on the Qinghai–Tibetan Plateau derived from passive microwave remote sensing. *Cryosphere* **2017**, *11*, 1933–1948. [CrossRef]
36. Carreno-Luengo, H.; Ruf, C.S. Retrieving Freeze/Thaw Surface State From CYGNSS Measurements. *IEEE Trans. Geosci. Remote Sens.* **2022**, *60*, 1–13. [CrossRef]

Technical Note

On-Orbit Calibration of the KBR Antenna Phase Center of GRACE-Type Gravity Satellites

Zhiyong Huang [1,2,*], **Shanshan Li** [1], **Lingyong Huang** [2] **and Diao Fan** [1]

[1] Institute of Geospatial Information, Information Engineering University, Zhengzhou 450001, China; zzy_lily@sina.com (S.L.); fandiao2311@mails.jlu.edu.cn (D.F.)

[2] State Key Laboratory of Geo-Information Engineering, Xi'an 710054, China; hlylj87@126.com

* Correspondence: geo_hzy@126.com

Abstract: The coordinates of the KBR (K-band ranging system) antenna phase center of GRACE-type gravity satellites in the satellite Science Reference Frame should be precisely known, and the determination accuracy should reach 0.3 mrad in the Y (pitch) and Z (yaw) directions. Due to the precision limitation of ground measurement and the change of space environment during orbit, the KBR antenna phase center changes. In order to obtain more accurate KBR antenna phase center coordinates, it is necessary to maneuver the satellite to achieve the on-orbit calibration of the KBR antenna phase center. Based on the in-orbit calibration data of KBR of GRACE-FO satellites, a new method is proposed to estimate the antenna phase center of KBR using the inter-satellite range acceleration as the observation value. The antenna phase center of KBR is solved by the robust estimation method, and the obtained calibration results are better than 72 μm in the Y and Z directions and better than 1.3 mm in the X direction, which is 50% better than the least squares estimation algorithm. The accuracy of KBR calibration results obtained by using the data of positive maneuvers or mirror (negative) maneuvers, respectively, does not meet 0.3 mrad. It is shown that mirror maneuvers are required for KBR calibration of a GRACE-type gravity satellite to obtain antenna phase center estimation results that meet the accuracy requirements. The calibration algorithm proposed in this paper can provide reference for KBR antenna phase center calibration of Chinese GRACE-type gravity satellites.

Keywords: GRACE; GRACE follow-on; gravity satellite; KBR antenna phase center; calibration; M-estimation

Citation: Huang, Z.; Li, S.; Huang, L.; Fan, D. On-Orbit Calibration of the KBR Antenna Phase Center of GRACE-Type Gravity Satellites. *Remote Sens.* **2022**, *14*, 3395. https://doi.org/10.3390/rs14143395

Academic Editor: Xiaogong Hu

Received: 12 June 2022
Accepted: 11 July 2022
Published: 14 July 2022

Publisher's Note: MDPI stays neutral with regard to jurisdictional claims in published maps and institutional affiliations.

1. Introduction

GRACE consists of two near-circular polar orbit satellites orbiting at an altitude of approximately 500 km in coplanar orbits, carrying a KBR (K-band ranging) instrument that measures inter-satellite range on the micron scale. However, in the recovery of the gravitational field, the required observation values are the inter-satellite range and range rate between the center of mass (CoM) of the satellite, so it is necessary to convert the range measurements between the KBR antenna phase center (APC) into the range observation values between the CoM of the satellite. Accurate satellite attitude and KBR APC coordinates are needed to ensure that the accuracy of range measurements is not lost during the conversion process.

The KBR APC vector of satellites relative to the CoM of satellites (CoM-to-APC [1]) is generally accurately measured before it enters orbit. However, the accuracy of ground measurement is limited and affected by the measurement accuracy of the whole satellite structure. After the satellite enters orbit, both the satellite body and the APC are affected by stress release, platform vibration, and space temperature change, leading to changes in the CoM-to-APC vector [2]. It is necessary to calibrate the CoM and KBR APC of the satellite periodically after the satellite is in orbit to obtain an accurate CoM-to-APC vector. JPL (Jet

Propulsion Laboratory), the official agency of GRACE and GRACE-FO satellites, typically writes initial and calibrated CoM-to-APC vectors into Level-1B VKB1B products.

The CoM of the satellite can always coincide with the proof mass of the accelerometer by periodic CoM calibration and CoM-Trim (the Mass Trim Assembly Mechanism (MTM) on all three axes of the GRACE-type satellite is adjusted accordingly in order to realize the compensation of CoM offset; this is called a CoM-Trim event). Therefore, it is more advantageous to set the origin of the Science Reference Frame (SRF) at the proof mass of the accelerometer. VKB1B products in the SRF can be accurately determined by periodic KBR phase center calibration, which can be used for KBR Level-1A to Level-1B data processing. In order to ensure that the KBR APC error does not affect the accuracy of inter-satellite range and range rate, the CoM-to-APC vector determination accuracy should reach 0.3 mrad in the Y (pitch) and Z (yaw) directions [3].

Although the determination accuracy of the mounting matrix of the star camera (QSA1B, rotation from star camera frames into the SRF), the estimation accuracy of the CoM (VCM1B, vector offset file for CoM solution from calibration maneuvers or tracking model in the SRF) of the satellite, and the realization accuracy of the inter-satellite pointing (QCP1B, rotation from the combined SCA "pilot" frame to the KBR pointing frame) will all have an impact on the KBR range measurement, this paper does not discuss the influence of the above three factors, and assumes that they all meet the accuracy requirements.

For the problem of KBR calibration, two maneuvering modes are proposed [3], linear drift maneuvering and periodic oscillation maneuvering, for on-orbit calibration of the KBR APC. Inter-satellite range and inter-satellite range rate are used in the calculation, and polynomial fitting is adopted to eliminate the low-frequency residual errors of precision orbit determination (POD). However, phase deviation parameters need to be introduced when the inter-satellite range is used to estimate the KBR APC. In [4], a nonlinear Kalman filter is used to estimate the KBR APC, and the simulated observation value does not consider the influence of orbit low-frequency error. The authors of [5,6] proposed the use of a predictive Kalman algorithm to estimate the star camera quaternion of the satellite, and the use of an extended Kalman filter algorithm to achieve the on-orbit calibration of the KBR APC. However, when the frequency of star camera observation data is high, the improvement of attitude estimation accuracy by the Kalman algorithm is not obvious, and the influence of orbit low-frequency error is not considered in this method.

To counteract the effects of residual low frequency errors and time-varying gravity fields, this paper proposes using the inter-satellite range acceleration as the observation value to obtain a high precision KBR APC estimation result. At the same time, the inter-satellite range and the inter-satellite range rate are taken as the observation values, and polynomial fitting is carried out. The KBR APC calibration is calculated with three observation values at the same time, and the result with higher estimation accuracy is selected as the final KBR APC calibration result. In view of the parameter estimation algorithm, this paper proposes using the M-estimation (robust estimation) method [7,8]. Through an IGGG-3 weighting scheme [9,10], the influence brought by the gross error of observation value and the gross error of design matrix can be overcome.

2. KBR Calibration Maneuver Scheme of a GRACE-Type Gravity Satellite

Due to its complexity, an analytical solution for the determination of the KBR APC can only be given under quite a number of restrictions. Therefore, experimental methods are generally used to determine the APC. In the ground stage, the calibration of the KBR APC is achieved by the method of one phase center swinging with certain regularity relative to another phase center [3].

This concept is based on rotating the satellite back and forth along a certain axis: the KBR antenna generates maneuvering signals with the satellite and the KBR APC reference point is found at the location of the observed minimum change in generating phase response [3]. However, only maneuvers in the direction perpendicular to the inter-satellite pointing (pitch axis (Y), yaw axis (Z)) can generate the sensitive inter-satellite ranging signal, and the maneuvers in the rotation axis (X) will not generate the ranging signal. In order to generate strong signals, the offset angle is generally set in advance, and then the satellite is shaken with a certain amplitude and period, because the ranging signal variation caused by the attitude maneuver of the satellite increases with the increase of the offset angle [3].

The KBR calibration of GRACE and GRACE-FO satellites adopted the calibration maneuver scheme proposed by Wang [3]. Using the satellite sequence of events (SOE) files recorded by JPL, the dates of KBR calibration and calibration events of GRACE-FO and GRACE are listed in Tables 1 and 2.

Table 1. KBR calibration events of GRACE-FO.

Date	Satellite	Type of Maneuvers
24 July 2020	GRACE-C	Test KBR calibration maneuver (wiggle test)
26 August 2022	GRACE-D	Test KBR calibration maneuver (wiggle test)
17 September 2020	GRACE-C	+pitch, −pitch, +yaw, −yaw
28 September 2020	GRACE-D	+pitch, −pitch, +yaw, −yaw

Table 2. KBR calibration events of GRACE.

Date	Satellite	Type of Maneuvers
16 March 2002	GRACE-AB	No Level-1B data
8 April 2002 to 9 April 2002	GRACE-AB	Test KBR calibration maneuver
5 June 2002	GRACE-B	No Level-1B data
10 February 2003 to 20 February 2003	GRACE-B	−pitch, +pitch, −yaw, +yaw
12 March 2003 to 21 March 2003	GRACE-A	−pitch, +pitch, −yaw, +yaw
5 April 2004	GRACE-B	−pitch, +pitch, −yaw, +yaw

Taking the KBR calibration maneuver of GRACE-C on 17 September 2020 and GRACE-D on 28 September 2020 as an example, four sub-maneuvers were carried out for each satellite, respectively: + pitch, −pitch, +yaw, and −yaw, where "+" means that the satellite offset angle is +2 degrees and "−" means that the satellite offset angle is −2 degrees. After setting the offset angle of 2 degrees, a sinusoidal maneuver signal with a period of 250 s and an amplitude of 1 degree is applied to the satellite, which generally lasts for 15 cycles with a total of 3750 s. The maneuver execution time required to achieve offset angle and return to normal inter-satellite pointing is about 30 s, and the total maneuver time is 3780 s. Figure 1 shows the inter-satellite pointing angle of GRACE-C during a single +pitch maneuver.

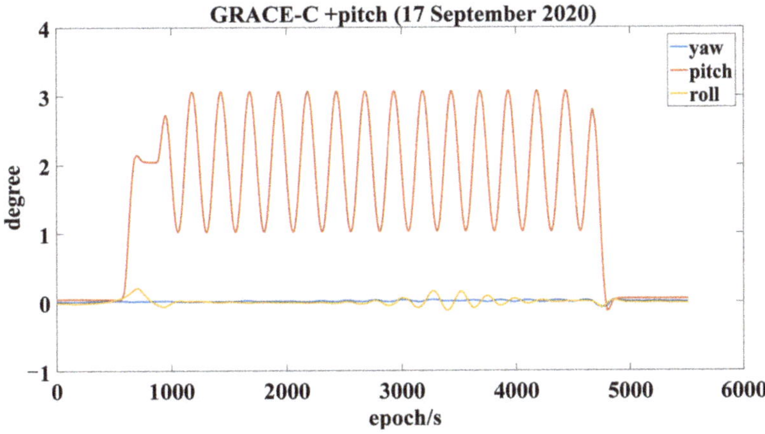

Figure 1. Inter-satellite pointing during single +pitch maneuver.

On-orbit calibration maneuvers of KBR are performed by a magnetorquer (MTQ) and ACTs (attitude control thrusters) of an attitude and orbit control system (AOCS). The subsystem of attitude and orbit control automatically calculates the required rotation torque, that is, the combination of the rotation torque of the attitude control thrusters and MTQ is used to drive the satellite to complete the designed attitude maneuver.

The inter-satellite pointing angle of GRACE and GRACE-FO during KBR calibration was calculated [11]. The intersatellite pointing angle comprises α_A and α_B in Figure 6, which represent the satellite KBR Frame (KF, K-Band Frame) and the line between the centers of mass of the satellites (LOSF, Line-of-Sight Frame). Figure 2 shows the inter-satellite pointing angles of GRACE-C and GRACE-D.

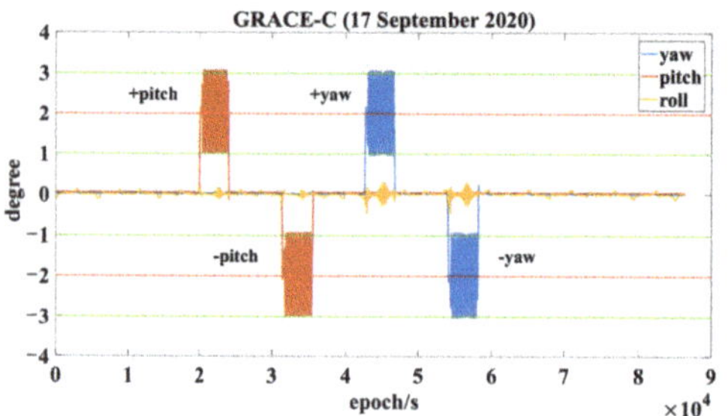
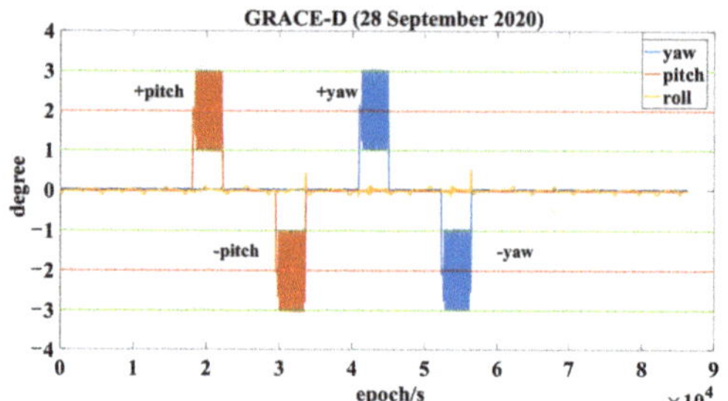

Figure 2. GRACE-FO KBR calibration maneuvers (September 2020).

In this paper, the inter-satellite pointing angles of the GRACE satellites during the KBR calibration period (except the time when Level-1B observation data were missing) were calculated, respectively. As can be seen from Figure 3, the maneuver strategy determined by GRACE in 2002 was not clear, and there is about 1 degree deviation in the pitch and roll directions; the author speculated that the maneuver strategy was explored and improved (implemented by AOCS) after this unqualified sub-maneuver.

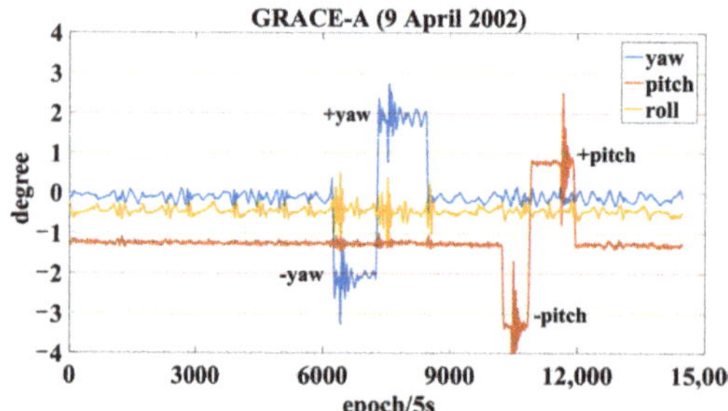
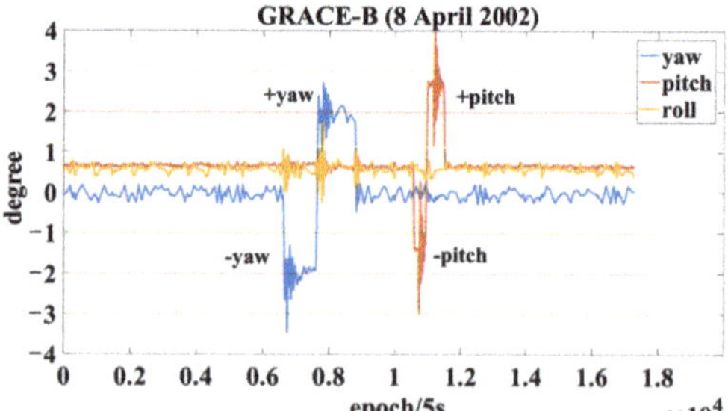

Figure 3. GRACE KBR calibration maneuvers (April 2002).

As can be seen from Figures 4 and 5, it adopted a maneuver strategy consistent with that of the in-orbit GRACE-FO satellite in Figure 2. However, its maneuver amplitude in the pitch direction reached about 1.5 degrees and its maneuver amplitude in the yaw direction reached about 1.7 degrees, while the amplitude of GRACE-FO in the KBR calibration maneuver was 1 degree.

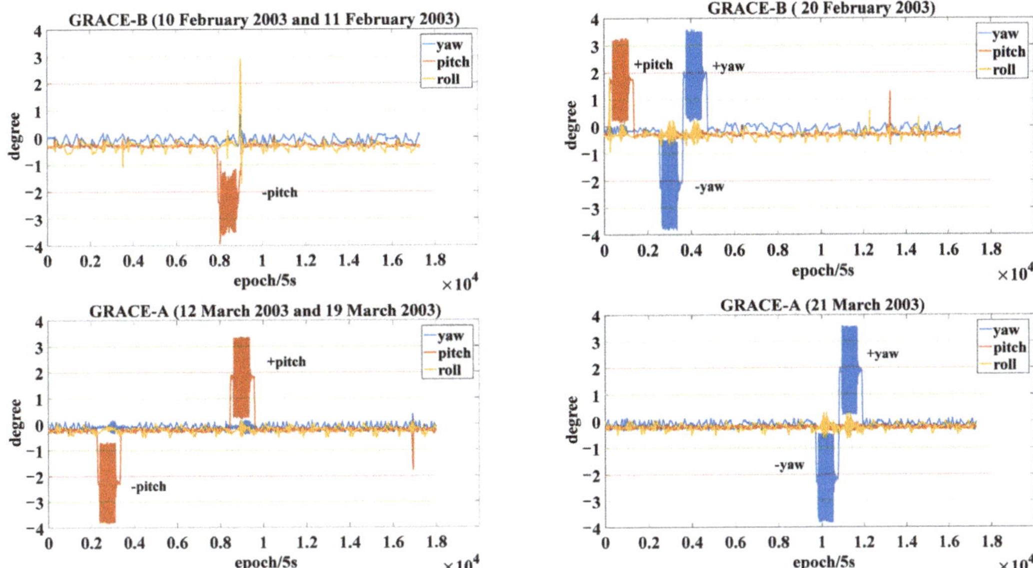

Figure 4. GRACE KBR calibration maneuvers in February 2003 and March 2003.

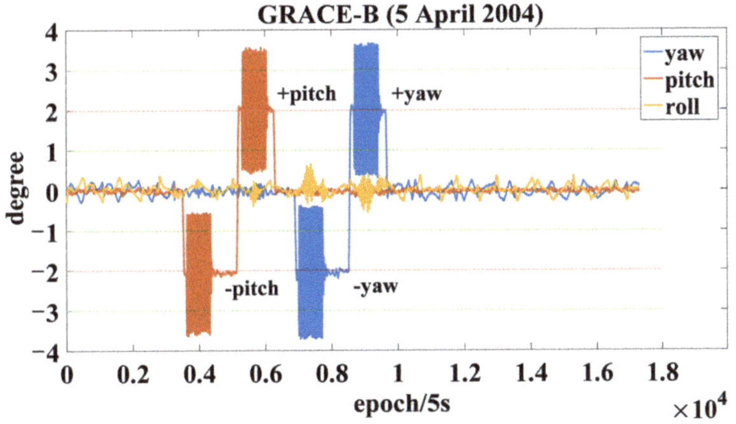

Figure 5. GRACE KBR calibration maneuvers of GRACE-B (April 2004).

The potential multipath effect of the KBR ranging signal can be reduced by using a positive and negative maneuver as mirror maneuvers of each other. When the mirror maneuver is not included, the KBR calibration of GRACE-FO contains the following four sub-maneuver signals.

Sub-maneuver 1: The following signal θ^1_{pitch} around the pitch axis (Y) is applied to GRACE-C, where superscript 1 represents GRACE-C, superscript 2 represents GRACE-D, θ_0 represents constant deviation angle, A represents the angular amplitude of the periodic signal, and f represents the frequency of the periodic signal. The maneuver signal is expressed as

$$\theta^1_{pitch} = \theta_0 + A\sin(2\pi f t) \tag{1}$$

During the pitch maneuver of GRACE-C, the GRACE-C deviation from the nominal roll and yaw angle should be kept as small as possible, and GRACE-D must maintain its nominal attitude. In this case, the KBR ranging measurements are highly sensitive to the CoM-to-APC vector of the Z and X direction of GRACE-C. Therefore, a single GRACE-C pitch maneuver can calculate the CoM-to-APC vector in the Z and X directions of GRACE-C. However, due to the higher signal-to-noise ratio in the Z direction than in the X direction, the accuracy of the estimated Z direction APC vector is better than that in the X direction. When the KBR range value is converted to the range of the CoM of the satellite, the KBR APC estimation error in the X direction of the two satellites is absorbed by the constant phase deviation of the KBR ranging measurements, which does not affect the gravity field reversion accuracy.

Sub-maneuver 2: The following signal θ^1_{yaw} about yaw axis (Z) is applied to GRACE-C, and other parameters have the same meaning as sub-maneuver 1.

$$\theta^1_{yaw} = \theta_0 + A\sin(2\pi f t) \tag{2}$$

Sub-maneuver 3: The following signal θ^2_{pitch} about pitch axis (Y) is applied to GRACE-D, and other parameters have the same meaning as sub-maneuver 1.

$$\theta^2_{pitch} = \theta_0 + A\sin(2\pi f t) \tag{3}$$

Sub-maneuver 4: The following signal θ^2_{yaw} about yaw axis (Z) is applied to GRACE-D, and other parameters have the same meaning as sub-maneuver 1.

$$\theta^2_{yaw} = \theta_0 + A\sin(2\pi f t) \tag{4}$$

3. Estimation Algorithm of the KBR APC

The position vectors of the satellite in the inertial system are, respectively, r_A and r_B, the rotation matrix transformed from the SRF to the inertial system is R_A and R_B, and the vector between the CoM of the satellite [12] is

$$u = r_B - r_A \tag{5}$$

The distance between the CoM of the satellites is

$$\rho_{com} = \| u \| \tag{6}$$

However, KBR observes the range between the KBR APC, as shown in Figure 6. Through vector conversion, the KBR observation range can be expressed as

$$\rho_{KBR} = \| (r_B + R_B c_B) - (r_A + R_A c_A) \| \tag{7}$$

where $c_i (i = A, B)$ is the corrected KBR APC vector (CoM-to-APC) in the SRF. More succinctly, it can be expressed as

$$\rho_{KBR} = \| u + v \| \tag{8}$$

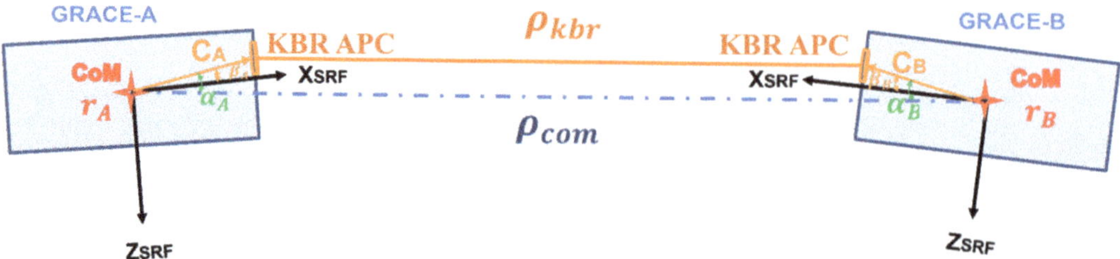

Figure 6. Observation geometry of the GRACE KBR measurement.

Among them,

$$v = R_B c_B - R_A c_A \tag{9}$$

Approximately, the KBR range measurements can be expressed as the distance between the CoM of GRACE-A and GRACE-B minus the APC correction (that is, the projected distance of the APC vectors of the two satellites in the LOSF (the line between r_A and r_B)):

$$\rho_{KBR} = \rho_{COM} - \rho_{AOC} \tag{10}$$

and ρ_{AOC} can be expressed as

$$\rho_{AOC} = \langle R_A c_A, e \rangle + \langle R_B c_B, e \rangle = -e^T R_A \cdot c_A + e^T R_B \cdot c_B \tag{11}$$

where e is the unit vector of the satellite CoM LOSF (Line-of-Sight Frame) obtained by

$$e = \frac{u}{\parallel u \parallel} \tag{12}$$

Then, the observation equation of KBR calibration is obtained by

$$\rho_{KBR} - \rho_{COM} = A \cdot x = \begin{bmatrix} e^T R_A & -e^T R_B & 1 \end{bmatrix} \cdot \begin{bmatrix} x_{pc_A} \\ x_{pc_B} \\ b \end{bmatrix} \tag{13}$$

where A is the design matrix, $\begin{bmatrix} x_{pc_A} & x_{pc_B} & b \end{bmatrix}^T$ is the estimated value of the KBR APC and the phase deviation term to be obtained, and $\rho_{KBR} - \rho_{COM}$ contains maneuver signals applied during KBR calibration, but it contains a large low-frequency error. Firstly, the error of orbit 1 cycle per revolution (cpr) contained in ρ_{KBR} and ρ_{COM} cannot be completely offset; secondly, the inter-satellite ranging signal in ρ_{KBR} measures the current gravity field signal, while the gravity signal contained in ρ_{COM} (the difference of the reduced dynamic orbit) is greatly affected by the prior gravity field. Therefore, the residual effect of the low frequency error will still exist after the minus: $\rho_{KBR} - \rho_{COM}$. Polynomial fitting is considered to remove its influence first, and then KBR APC parameters are estimated to avoid excessive parameters caused by estimating polynomial parameters and KBR APC parameters at the same time.

When the difference of the satellite velocity is $\dot{u} = \dot{r}_B - \dot{r}_A$, the velocity between the CoM of the satellite can be expressed as

$$\dot{\rho}_{COM} = \langle e, \dot{u} \rangle \tag{14}$$

When the difference of the satellite accelerations is $\ddot{u} = \ddot{r}_B - \ddot{r}_A$, then the acceleration between the CoM of the satellite can be expressed as

$$\ddot{\rho}_{COM} = \langle e, \ddot{u} \rangle + \langle \dot{e}, \dot{u} \rangle \tag{15}$$

Then, the inter-satellite range rate measured by KBR can be expressed as

$$\dot{\rho}_{KBR} = \dot{\rho}_{COM} - \dot{\rho}_{AOC} \tag{16}$$

The observation equation of KBR calibration based on inter-satellite range rate is

$$\dot{\rho}_{KBR} - \dot{\rho}_{COM} = A \cdot x = D \begin{bmatrix} -e^T R_A & e^T R_B \end{bmatrix} \cdot \begin{bmatrix} x_{pc_A} \\ x_{pc_B} \end{bmatrix} \tag{17}$$

where D represents the partial derivative matrix of the polynomial; the third-order polynomial is used in this paper.

The inter-satellite range acceleration measured by KBR can be expressed as

$$\ddot{\rho}_{KBR} = \ddot{\rho}_{COM} - \ddot{\rho}_{AOC} \tag{18}$$

The KBR calibration observation equation based on inter-satellite range acceleration is

$$\ddot{\rho}_{KBR} - \ddot{\rho}_{COM} = A \cdot x = DD\begin{bmatrix} -e^T R_A & e^T R_B \end{bmatrix} \cdot \begin{bmatrix} x_{pc_A} \\ x_{pc_B} \end{bmatrix} \tag{19}$$

Observation Equations (13), (17), and (19) are constructed according to the above equation, and the estimated KBR APC values of the two satellites can be calculated according to the three observation values based on the LS estimation and M-estimation.

In the LS estimation and the initial weighting of the M-estimation, the weight matrix of the observed value is taken as the identity matrix in this paper. When constructing the observed values of inter-satellite range, range rate, and range acceleration, only light time correction term is carried out.

4. KBR Calibration Results of GRACE-Type Gravity Satellites

The RL04 version Level-1B GNI1B, SCA1B, and KBR1B data during the KBR calibration period of the GRACE-FO satellite were downloaded from ftp://isdcftp.gfz-potsdam.de on 5 April 2022 [13], and their meanings are shown in Table 3. At the same time, the RL02 version GNV1B, SCA1B, and KBR1B data of GRACE satellites during the KBR calibration stage were downloaded [14]. The data sampling rate was 5 s, and the data meaning is similar to that in Table 3.

Table 3. GRACE-FO data product for KBR calibration.

Data	Description
GNI1B	Reduced-dynamic orbit data in inertial frame, 1 Hz
SCA1B	Compressed/combined star camera data, 1 Hz
KBR1B	Biased inter-satellite ranges and their first two time derivatives, range rate, and range acceleration, 5 Hz

4.1. Low Frequency Error Processing

In a single maneuver, 700 epochs were selected in total for calculation. In order to avoid the influence of large maneuvers, 125 s before and after maneuvers were removed. Firstly, the inter-satellite pointing angle, angular velocity, and angular acceleration of +pitch maneuver and +yaw maneuver of GRACE-C were calculated. The corresponding KBR ranging value (KBR), range rate (KBRR), and range acceleration (KBRA) were calculated and taken as the observed value (O). Then, the inter-satellite baseline, baseline rate, and baseline acceleration determined by POD (precision orbit determination) were taken as the calculated values (C), and the differences between the KBR measurements (O) and the inter-satellite baseline (C) was obtained (O−C): represented as "KBR-POD", "KBRR-POD velocity", and "KBRA-POD acceleration", respectively.

As can be seen from Figure 7, O−C has a good consistency with the applied pitch and yaw maneuver signals. However, the "KBRA-POD acceleration" contains significant high-frequency noise, which may affect the estimation accuracy of KBR APC. "KBR-POD" and "KBRR-POD velocity" still contain low-frequency periodic signals, which need to be fitted. In this paper, the 15-order and 13-order polynomial fitting is carried out, respectively, and the filtered results (after removing the best fitting polynomials) are obtained in Figure 8, where Figure 8a–d are the results of polynomial fitting on the single +pitch, +yaw maneuver data of GRACE-C and Figure 8e,f are the filtering results of observation values of "KBR-POD" and "KBRR-POD velocity" obtained after polynomial fitting on all the KBR calibration data of eight sub-maneuvers. As can be seen from Figure 8a–f, the low frequency errors were effectively removed.

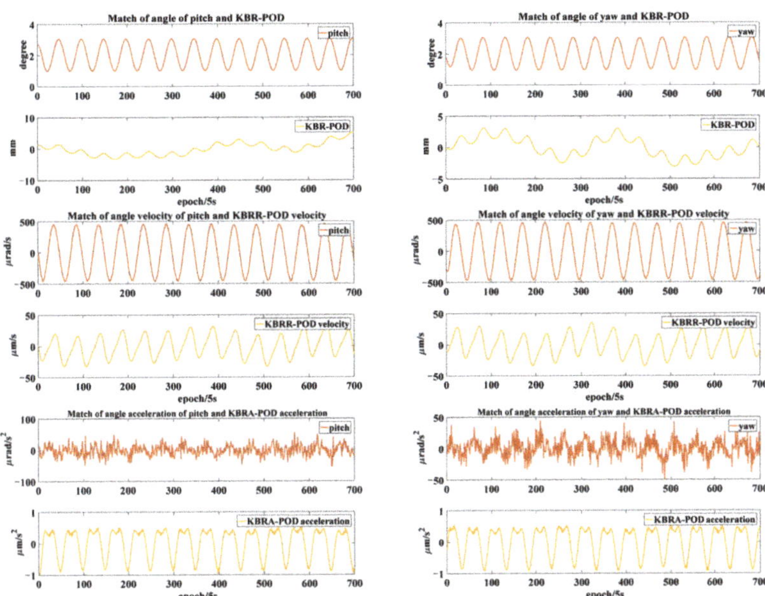

Figure 7. Match of attitude variation and three types of observations of KBR-POD.

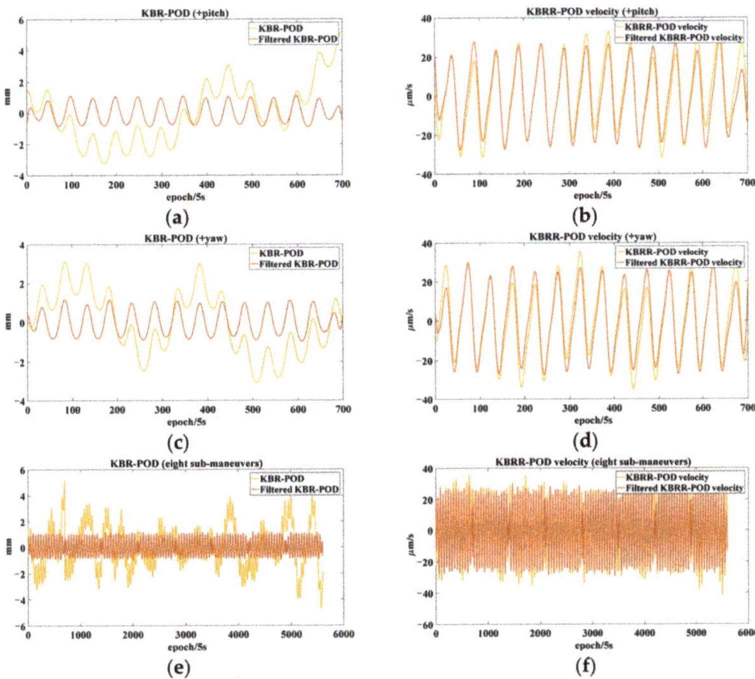

Figure 8. Polynomial fitting to remove low frequency errors: (**a**) polynomial fitting of KBR-POD (+pitch); (**b**) polynomial fitting of KBRR-POD velocity (+pitch); (**c**) polynomial fitting of KBR-POD (+yaw); (**d**) polynomial fitting of KBRR-POD velocity (+yaw); (**e**) polynomial fitting of KBR-POD; and (**f**) polynomial fitting of KBRR-POD velocity.

For the three kinds of observation values, the first and last 50 s observation values were removed, and then Fourier transform was carried out. The polynomial fitting effect was analyzed in terms of frequency, as shown in Figure 9a–d, corresponding to Figure 8a–d. The maneuvering frequency signal of 40 mHz and twice the maneuvering frequency signal of 80 mHz were retained, and the low-frequency noise was effectively suppressed. While, as shown in Figure 8d, higher amplitudes can be seen for the filtered results for nearly all frequencies above 1 mHz, this may be due to errors from the polynomial fitting. As shown in Figure 9e,f, analyzing the "KBRR-POD acceleration" observation value from the frequency domain, the polynomial fitting is not required and the low-frequency noise interference is small.

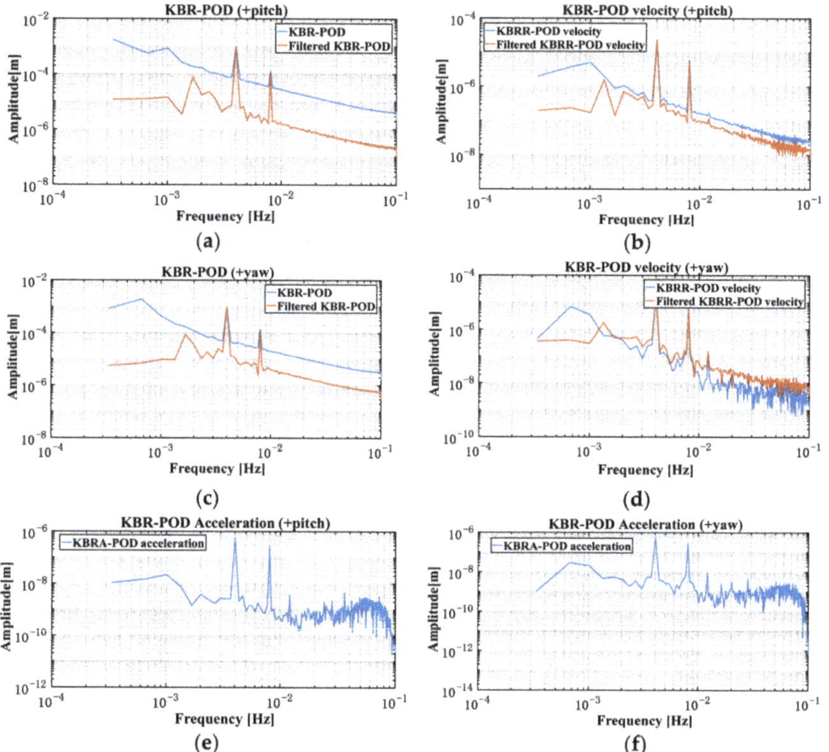

Figure 9. Polynomial fitting to remove low frequency errors (FFT): (**a**) polynomial fitting of KBR-POD (+pitch); (**b**) polynomial fitting of KBRR-POD velocity (+pitch); (**c**) polynomial fitting of KBR-POD (+yaw); (**d**) polynomial fitting of KBRR-POD velocity (+yaw); (**e**) KBRA-POD acceleration (+pitch); and (**f**) KBRA-POD acceleration (+yaw).

4.2. KBR APC Estimation Based on LS Estimation and M-Estimation

The filtered observation values of "KBR-POD" and "KBRR-POD velocity" and the unfiltered observation values of "KBRA-POD acceleration" were used to estimate the vector (CoM-to-APC) of KBR APC in the SRF, corresponding to the angle β_A and β_B in Figure 6. The least squares fitting curves and fitting residual results by the three estimation methods were obtained. It can be seen from Figure 10 that all the three methods can achieve good fitting effects. In this calculation, we stripped out the observations of 50 epochs at the beginning and the end, and combined the data of four sub-maneuvers and four mirror sub-maneuvers.

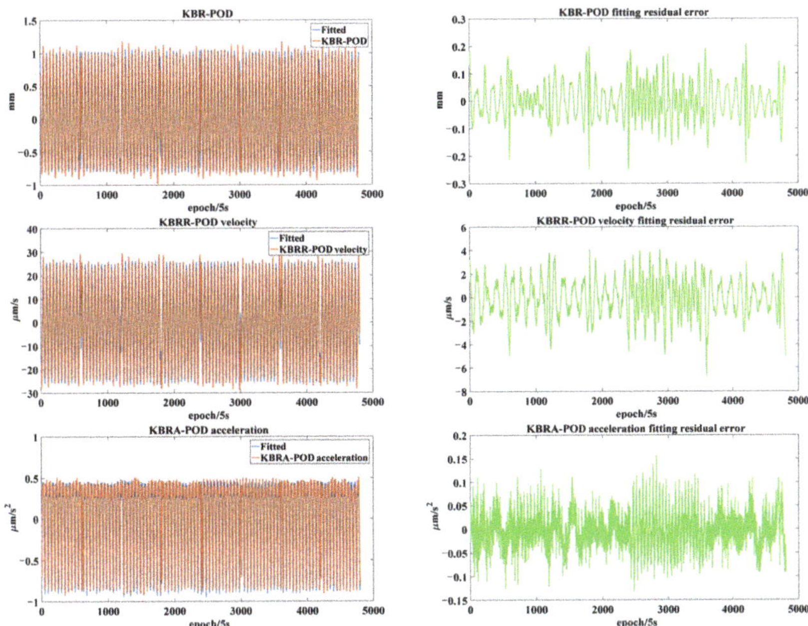

Figure 10. The fitting curves and the fitting residuals obtained by the three estimation methods.

The GRACE-C and GRACE-D APC vectors estimated by the three observation values obtained from the eight sub-maneuvers were denoted as X_{APC}, Y_{APC}, and Z_{APC}, respectively, and the formal errors were calculated according to the least squares accuracy estimation theory. As can be seen from Table 4, Y_{APC} and Z_{APC} estimated by "KBR-POD" had high accuracy. However, due to the need to solve the phase deviation term (Bias) of the ranging value, the accuracy of the estimated X_{APC} and phase deviation term was low. The reason is that the collinearity of X_{APC} and the phase deviation term leads to the ill-conditioned normal equation, which indirectly leads to the unreliable estimated Y_{APC} and Z_{APC}, and the results are deviated by the other two estimation methods. The X_{APC} estimated by "KBRA-POD acceleration" had a high accuracy, and the calibration result obtained by "KBRA-POD acceleration" was selected as the final KBR calibration result in this paper. Among them, the LS estimation accuracy of Y_{APC} and Z_{APC} was about 138 μm, about 0.096 mrad, which meets the index of 0.3 mrad. The LS estimation accuracy of X_{APC} was 2.4 mm.

Table 4. KBR calibration results and formal errors based on LS estimation.

	KBR-POD		KBRR-POD Velocity		KBRA-POD Acceleration	
	Calibrated Value	**Formal Error**	**Calibrated Value**	**Formal Error**	**Calibrated Value**	**Formal Error**
	Use KBR calibration data based on 4 sub-maneuvers and 4 mirror sub-maneuvers					
C-X_{APC}	1457.1 mm	3.2 mm	1465.0 mm	3.0 mm	1439.3 mm	2.4 mm
C-Y_{APC}	136.5 μm	54.5 μm	−382.0 μm	150.1 μm	−504.1 μm	133.3 μm
C-Z_{APC}	−708.2 μm	54.5 μm	172.5 μm	152.5 μm	152.0 μm	135.4 μm
Bias	2947.9 mm	4.6 mm	–	–	–	–
D-X_{APC}	1491.8 mm	3.3 mm	1499.4 mm	3.0 mm	1474.6 mm	2.5 mm
D-Y_{APC}	378.4 μm	54.7 μm	325.0 μm	154.6 μm	179.6 μm	137.4 μm
D-Z_{APC}	774.9 μm	54.6 μm	1280.9 μm	154.7 μm	1256.7 μm	137.4 μm
Bias	2947.9 mm	4.6 mm	–	–	–	–

In order to verify the effectiveness of the mirror maneuver, the paper only uses the positive maneuver and mirror maneuver for KBR APC calibration data, respectively, to obtain the calibration results. The results are shown in Table 5. The Y_{APC} and Z_{APC} values are obviously large, and the estimation accuracy (formal error) does not meet the accuracy requirements of 0.3 mrad. The results show that the KBR calibration results are easily affected by multipath error when the mirror maneuver is not adopted, which leads to the inaccurate KBR APC estimation results. According to the positive and negative sign characteristics of Y_{APC} and Z_{APC} calibrated values in the calibration results, they are just opposite, which is the reason why correct results can be obtained by combining positive and negative maneuvers.

Table 5. KBR calibration results and formal errors based on LS-estimation and different maneuver data.

	KBR-POD		KBRR-POD Velocity		KBRA-POD Acceleration	
	Calibrated Value	Formal Error	Calibrated Value	Formal Error	Calibrated Value	Formal Error
	Use KBR calibration data based on 4 sub-maneuvers					
C-X_{APC}	1374.6 mm	37.5 mm	1411.3 mm	17.2 mm	1399.6 mm	7.5 mm
C-Y_{APC}	−3.6 mm	1.3 mm	−2.3 mm	0.6 mm	−2.2 mm	0.3 mm
C-Z_{APC}	3.1 mm	1.3 mm	2.3 mm	0.6 mm	2.0 mm	0.3 mm
Bias	2780.0 mm	54.4 mm	–	–	–	–
D-X_{APC}	1406.4 mm	39.3 mm	1433.0 mm	18.0 mm	1440.1 mm	7.9 mm
D-Y_{APC}	−3.1 mm	1.4 mm	−2.2 mm	0.7 mm	−1.5 mm	0.3 mm
D-Z_{APC}	4.1 mm	1.4 mm	3.6 mm	0.7 mm	2.6 mm	0.3 mm
Bias	2780.0 mm	54.4 mm	–	–	–	–
	Use KBR calibration data based on 4 mirror sub-maneuvers					
C-X_{APC}	1389.0 mm	32.4 mm	1401.4 mm	16.4 mm	1399.1 mm	7.2 mm
C-Y_{APC}	1.7 mm	1.1 mm	1.8 mm	0.6 mm	1.2 mm	0.3 mm
C-Z_{APC}	−2.3 mm	1.1 mm	−2.3 mm	0.6 mm	−1.7 mm	0.3 mm
Bias	2853.5 mm	46.7 mm	–	–	–	–
D-X_{APC}	1465.4 mm	33.6 mm	1464.6 mm	17.0 mm	1437.1 mm	7.5 mm
D-Y_{APC}	1.0 mm	1.2 mm	1.7 mm	0.6 mm	1.9 mm	0.3 mm
D-Z_{APC}	0.3 mm	1.2 mm	39.7 μm	0.6 mm	−0.2 mm	0.3 mm
Bias	2853.5 mm	46.7 mm	–	–	–	–

Since the observed values may still contain gross errors, in order to further improve the estimation accuracy, this paper uses the method of M-estimation to estimate the KBR APC, combining the data of four positive maneuvers and four mirror maneuvers. The estimation results are shown in Table 6. The KBR APC accuracy estimated by the three observation values was greatly improved. Among them, the estimation accuracy of the proposed inter-satellite range acceleration estimation in the Y and Z directions was improved from 138 μm to 72 μm, and the X direction was improved from 2.5 mm to 1.3 mm, with an increase of nearly 50%.

Table 6. KBR calibration results and formal errors based on M-estimation.

	KBR-POD		KBRR-POD Velocity		KBRA-POD Acceleration	
	Calibrated Value	Formal Error	Calibrated Value	Formal Error	Calibrated Value	Formal Error
	Use KBR calibration data based on 4 sub-maneuvers and 4 mirror sub-maneuvers					
C-X_{APC}	1449.3 mm	1.5 mm	1458.9 mm	1.7 mm	1443.7 mm	1.2 mm
C-Y_{APC}	228.0 μm	28.8 μm	−257.7 μm	82.5 μm	−370.6 μm	64.7 μm
C-Z_{APC}	−464.2 μm	24.7 μm	480.7 μm	85.7 μm	145.0 μm	65.4 μm
Bias	2939.2 mm	2.2 mm	–	–	–	–
D-XAPC	1490.8 mm	1.6 mm	1495.9 mm	1.6 mm	1481.7 mm	1.3 mm
D-YAPC	460.3 μm	27.6 μm	-95.6 μm	81.4 μm	183.0 μm	65.9 μm
D-ZAPC	803.7 μm	28.8 μm	1215.1 μm	86.4 μm	1393.1 μm	71.6 μm
Bias	2939.2 mm	2.2 mm	–	–	–	–

VKB1B results of GRACE-FO recommended by this paper obtained from M-estimates are shown in Table 7, while the SOE file gives GRACE-FO VKB1B in May 2019, indicating a significant change in the Z_{APC} of GRACE-D. The authors recommend that this change be taken into account during KBR data preprocessing.

Table 7. Results of GRACE-FO KBR APC.

VKB1B (September 2020)	X_{APC}	Y_{APC}	Z_{APC}
GRACE-C	1443.7 mm	-370.6 μm	145.0 μm
GRACE-D	1481.7 mm	183.0 μm	1393.1 μm
VKB1B (SOE)	X_{APC}	Y_{APC}	Z_{APC}
GRACE-C	1444.4 mm	-170.0 μm	448.0 μm
GRACE-D	1444.5 mm	54.0 μm	230.0 μm

At the same time, the VKB1B of GRACE recommended by this paper (in 2004, GRACE-A did not perform the KBR calibration maneuver, so the APC of GRACE-A could not be estimated) was calculated by using the method of M-estimation, and VKB1B results calculated by JPL recorded in SOE files are shown in Table 8. The results of the Y and Z directions estimated in this paper are close to those of JPL's solution, which shows that the KBR calibration algorithm proposed in this paper is feasible.

Table 8. Results of GRACE KBR APC.

VKB1B (March 2003)	X_{APC}	Y_{APC}	Z_{APC}
GRACE-A	1468.5 mm	-664.3 μm	2263.7 μm
GRACE-B	1477.1 mm	516.6 μm	2922.2 μm
VKB1B (April 2004)	X_{APC}	Y_{APC}	Z_{APC}
GRACE-A	–	–	–
GRACE-B	1422.2 mm	369.1 μm	3952.4 μm
VKB1B (SOE)	X_{APC}	Y_{APC}	Z_{APC}
GRACE-A	1445.1 mm	-423.3 μm	2278.7 μm
GRACE-B	1444.4 mm	576.1 μm	3304.1 μm

5. Discussion

The above maneuver strategy can be used to estimate the KBR APC of GRACE-type satellites. Although any other maneuvers can also be used for KBR calibration (as shown in Figure 3 or the linear drift maneuver method), the sinusoidal maneuver strategy is a more reliable KBR calibration maneuver scheme. The positive maneuver and negative maneuver (mirror maneuver) mode is very important for KBR calibration and can effectively eliminate the multipath influence of the KBR ranging signal. The results of KBR calibration with combined positive and negative maneuvers are reasonable and accurate.

In estimating the KBR APC, we can choose the strategy of using CoM-to-APC in the fixed X direction as the known value and only solving the Y and Z direction APC. In fact, when the APC vectors of three directions are solved simultaneously, the accuracy of the X direction is the worst. Therefore, the APC of the X direction is generally not updated, but the measured value of the ground is used. In addition, the APC vectors of the Y and Z direction obtained by the two strategies (fixed X direction or not) are consistent. Therefore, the APC vectors of the three directions were solved at the same time in this paper, but the APC vector of the X direction was not recommended.

Although the application of "KBRA-POD acceleration" as the observation value can effectively eliminate the influence of low-frequency error, it amplifies the influence of high-frequency noise, and there is no effect method to eliminate the high-frequency noise contained in the observation value and the design matrix at the same time. In fact, as shown in Figure 7, better accuracy can be obtained by using "KBRA-POD acceleration" as an observation value to estimate the CoM-to-APC if the maneuvering strategy implemented by the attitude and orbit control system results in smoother angular acceleration. By calculating the CoM-to-APC of the three observation values at the same time, the results can be better verified and the estimated results with better integrity can be obtained. In the estimation of GRACE satellite APC, due to the ill-posed problem of the normal equation, the solution results deviated from the results of the other two algorithms. Finally, the solution results obtained by "KBRR-POD velocity" observation values with higher accuracy were adopted. "KBR-POD" is not recommended for subsequent calculation.

Compared with the least squares estimation method, the improved accuracy of the M-estimation method is obvious. The reason is that there may still be gross error or high frequency error in the observed value and design matrix, and the weight of these observed value can be reduced by using M-estimation. Because the weight matrix of the observed value is taken as the identity matrix in this paper, other weighted least squares strategies can also resist gross errors and improve accuracy, which is worth further study.

6. Conclusions

This paper summarized the KBR calibration algorithm of GRACE-type gravity satellites, proposed estimating the APC of KBR using inter-satellite range, range rate, and range acceleration simultaneously, and verified the KBR calibration results based on three kinds of Level-1B observation data of GRACE and GRACE-FO. The results show that the APC of KBR obtained by LS estimation using "KBRA-POD acceleration" was 138 μm in the Y and Z directions, and 2.5 mm in the X direction. When M-estimation was used, the accuracy was 72 μm in the Y and Z directions and 1.3 mm in the X direction. In this paper, KBR calibration calculation was performed on the data of positive and mirror (negative) maneuvers. It was shown that the mirror maneuver is required for the KBR calibration of GRACE-type gravity satellites in order to obtain the APC estimation results that meet the accuracy requirements.

In this paper, positive and mirror (negative) maneuvers are recommended for KBR calibration. During KBR calibration, range rate and range acceleration can be used as observation values to estimate the KBR APC, and the optimal solution results are selected as the final VKB1B recommended value. M-estimation is recommended for parameter estimation.

Author Contributions: Conceptualization, Z.H.; data curation, Z.H.; formal analysis, Z.H.; funding acquisition, S.L.; investigation, S.L. and D.F.; methodology, Z.H.; project administration, S.L. and L.H.; resources, S.L.; software, Z.H.; supervision, S.L.; validation, Z.H.; visualization, D.F.; writing—original draft, Z.H.; writing—review and editing, S.L., D.F.; and L.H. All authors have read and agreed to the published version of the manuscript.

Funding: This study was funded by the National Natural Science Foundation of China (No. 42174007).

Data Availability Statement: The data used to support the findings of this study are available from the corresponding author upon request.

Acknowledgments: We are very grateful to the GFZ and JPL for providing the GRACE and GRACE-FO Level-1B data.

Conflicts of Interest: The authors declare no conflict of interest.

References

1. Horwath, M.; Lemoine, J.-M.; Biancale, R.; Bourgogne, S. Improved GRACE science results after adjustment of geometric biases in the Level-1B K-band ranging data. *J. Geod.* **2010**, *85*, 23–38. [CrossRef]
2. Kornfeld, R.P.; Arnold, B.W.; Gross, M.A.; Dahya, N.T.; Klipstein, W.M.; Gath, P.F.; Bettadpur, S. GRACE-FO: The Gravity Recovery and Climate Experiment Follow-On Mission. *J. Spacecr. Rocket.* **2019**, *56*, 931–951. [CrossRef]
3. Wang, F. Study on Center of Mass Calibration and K-Band Ranging System Calibration of the GRACE Mission. Ph.D. Thesis, The University of Texas at Austin, Austin, TX, USA, 2003.
4. Zhang, H.; Zhao, Y.; Sun, K.; Liang, L.; Gu, X. Study on on-orbit calibration of phase center of intersatellite microwave ranging system. *Shanghai Aerosp.* **2010**, *27*, 55. (In Chinese) [CrossRef]
5. Xin, N.; Qiu, L.; Zhang, L.; Ding, Y. An On-orbit Calibration algorithm of KBR phase center. *Chin. J. Space Sci. Technol.* **2014**, *34*, 50–56. (In Chinese)

6. Xin, N.; Qiu, L.; Zhang, L.; Ding, Y. Algorithm for KBR System Phase Center In-orbit Calibration of Gravity Measurement Satellite. *Spacecr. Eng.* **2014**, *23*, 24–30. (In Chinese)
7. Huber, P.J. Robust Estimation of a Location Parameter. In *Breakthroughs in Statistics*; Springer: New York, NY, USA, 1992; pp. 492–518. [CrossRef]
8. Zhou, J. Classical Error Theory and Robust Estimation. *Acta Geod. Cartogr. Sin.* **1989**, *18*, 115–120. (In Chinese)
9. Sui, L.; Song, L.; Chai, H. *Error Theory and Foundation of Surveying Adjustment*; Surveying and Mapping Publishing House: Beijing, China, 2003. (In Chinese)
10. Yang, Y. Parameter M estimation and posteriori accuracy review. *Bull. Surv. Mapp.* **1991**, *6*. (In Chinese)
11. Bandikova, T.; Flury, J.; Ko, U.-D. Characteristics and accuracies of the GRACE inter-satellite pointing. *Adv. Space Res.* **2012**, *50*, 123–135. [CrossRef]
12. Ellmer, M. Contributions to GRACE Gravity Field Recovery: Improvements in Dynamic Orbit Integration Stochastic Modelling of the Antenna Offset Correction, and Co-Estimation of Satellite Orientations. Ph.D. Thesis, Graz University of Technology, Graz, Austria, 2018. [CrossRef]
13. Wen, H.Y.; Kruizinga, G.; Paik, M.; Landerer, F.; Bertiger, W.; Sakumura, C.; Mccullough, C. *Gravity Recovery and Climate Experiment Follow-On (GRACE-FO) Level-1 Data Product User Handbook*; Technical Report JPL, D-56935; NASA Jet Propulsion Laboratory: La Cañada Flintridge, CA, USA; California Institute of Technology: Pasadena, CA, USA, 2019.
14. Case, K.; Kruizinga, G.; Wu, S. *Grace Level 1b Data Product User Handbook*; Technical Report JPL, D-22027; NASA Jet Propulsion Laboratory: La Cañada Flintridge, CA, USA; California Institute of Technology: Pasadena, CA, USA, 2010.

Technical Note

Research on the Rotational Correction of Distributed Autonomous Orbit Determination in the Satellite Navigation Constellation

Wei Zhou [1], Hongliang Cai [1], Ziqiang Li [2,*], Chengpan Tang [3], Xiaogong Hu [3] and Wanke Liu [2]

[1] Beijing Institute of Tracking and Telecommunications Technology (BITTT), 26 Beiqing Road, Beijing 100094, China; zhouwei_0611@163.com (W.Z.); caibanyu@126.com (H.C.)

[2] School of Geodesy and Geomatics, Wuhan University, Wuhan 430079, China; wkliu@sgg.whu.edu.cn

[3] Shanghai Astronomical Observatory, Chinese Academy of Sciences, Shanghai 200030, China; tcp@shao.ac.cn (C.T.); hxg@shao.ac.cn (X.H.)

* Correspondence: ziqiangli@whu.edu.cn; Tel.: +86-15827546913

Citation: Zhou, W.; Cai, H.; Li, Z.; Tang, C.; Hu, X.; Liu, W. Research on the Rotational Correction of Distributed Autonomous Orbit Determination in the Satellite Navigation Constellation. *Remote Sens.* 2022, 14, 3309. https://doi.org/10.3390/rs14143309

Academic Editor: Yunbin Yuan

Received: 16 June 2022
Accepted: 6 July 2022
Published: 9 July 2022

Publisher's Note: MDPI stays neutral with regard to jurisdictional claims in published maps and institutional affiliations.

Abstract: The autonomous orbit determination of the navigation constellation uses only bidirectional ranging data of the inter-satellite link for data processing. The lack of space-time benchmark information related to the Earth inevitably causes overall rotational uncertainty in the constellation, leading to a decrease in orbit accuracy and affecting user positioning accuracy. This study (1) introduces a method for rotation correction in distributed autonomous orbit determination based on inter-satellite bidirectional ranging; (2) conducts constellation autonomous orbit determination and time synchronization processing experiments based on inter-satellite ranging data for the 24 medium Earth orbit (MEO) satellites in the Beidou-3 global satellite navigation system (BDS-3); and (3) makes comparative analyses on the accuracy of autonomous orbit determination based on three rotation correction cases, including a no-rotation-correction case, independent satellite constraints case, and global satellite constraints case. The experimental results are described as follows. For the no-rotation-correction case, the prediction error of the orbital inclination angle (iot, i) for the entire constellation on the 30th day was 2.11×10^{-7}/rad, the prediction error of the right ascension of the ascending point (Omega, Ω) was 2.25×10^{-7}/rad, and the average root mean square (RMS) of the user range error (URE) for the entire constellation orbit was 1.41 m. In the autonomous orbit determination experiment with independent constraints on satellites, the prediction error of i for the entire constellation on the 30th day was 5.43×10^{-7}/rad, the prediction error of Ω was 2.03×10^{-7}/rad, and the average RMS of the orbital URE for the entire constellation was 1.09 m. In the autonomous orbit determination experiment with global satellite constraints, the prediction error of i for the entire constellation on the 30th day was 5.31×10^{-7}/rad, the prediction error of Ω was 1.95×10^{-7}/rad, and the RMS of the orbital URE for the entire constellation was 0.94 m. According to the analysis of the above experimental results, compared with the autonomous orbit determination under the no-rotation-correction case, the adoption of an algorithm for independent satellite constraints to correct the overall constellation rotation weakens the constellation rotation influence; however, it may destroy the overall constellation configuration, which affects the stability of autonomous orbit determination. Finally, the algorithm based on global satellite constraints both impairs the influence of constellation rotation and maintains the overall constellation configuration.

Keywords: Beidou satellite navigation system; distributed autonomous orbit determination; inter-satellite link; overall constellation rotation; algorithm for the correction of independent satellite constraints; algorithm for the correction of global satellite constraints

1. Introduction

In the 1980s, M.P. Ananda et al. proposed the concept of autonomous orbit determination of the navigation constellation [1]. This technology ensures the long-term, autonomous, and stable operation and service capabilities of a navigation system, based on the distance measurement and communication functions of the inter-satellite link through the autonomous operation of the constellation and on-orbit ephemeris updates during long-term absence of ground system support. Subsequently, the initial theory, design, and data research work on autonomous orbit determination was conducted in June 1990 [2]. In July 2020, the Beidou-3 global satellite navigation system (BDS-3) began to operate and provide global service. The Ka-band inter-satellite link payload carried by the BDS-3 satellites can realize inter-satellite pseudorange measurement and inter-satellite communication, which enables research on the autonomous orbit determination of the constellation [3].

Autonomous navigation is first introduced for GPS Block IIR satellites and its design index requirement is that the user range error (URE) is less than 6 m within 180 days [4]. The autonomous orbit determination of navigation constellation uses only bidirectional ranging data of the inter-satellite link and lacks space-time benchmark information related to the Earth. Therefore, this technology cannot eliminate or suppress accumulation of the overall rotation errors of the constellation, making it difficult to operate autonomously for a long time [5–10]. Therefore, the design index requirement of GPS Block IIF satellites is changed to a URE of less than 2 m in a 60-day autonomous navigation [11]. Ananda et al. proved the unobservability of the overall rotation of the satellite constellation with inter-satellite measurements only and proposed an algorithm to constrain the right ascension of the ascending point by analyzing the long-predicted reference ephemeris provided by the Operational Control Segment (OCS) [12]. Abusali et al. pointed out that the orbit determination errors in the tangential and normal directions caused by the overall constellation rotation up to kilometer level, and the influence of errors in the predicted Earth Orientation Parameter (EOP), reach to meter level [13]. Gill et al. applied simulation data to study autonomous orbit determination [14]. With the development of BDS, several scholars researched the autonomous orbit determination technology of the navigation constellation by simulation, showing that the uncertainty of the right ascension of the ascending point Ω at the orbital plane will lead to overall rotation errors in orbit determination results [15–22]. In addition, other scholars also successively carried out related work, including algorithm research and simulation analysis [23–29]. The aforementioned studies aided in carrying out the research in this paper. This paper focuses on the problem of the overall constellation rotation correction in autonomous orbit determination, three correction cases are designed, and experiment results of different cases are analyzed and compared using inter-satellite observations.

2. Autonomous Orbit Determination Model

2.1. Observation Model for Autonomous Orbit Determination

Based on the concurrent spatial time division duplexing technology of the phased array antenna, bidirectional one-way distance measurement between the BDS-3 satellites was completed within 3.0 s [30]. At different times within these 3.0 s, the observation equations of the two satellites SAT_A and SAT_B can be expressed as.

$$\rho_{AB} = |R_B(t_1) - R_A(t_1 - \Delta t_1)| - \delta t_A + \delta t_B + c \cdot \tau_A^{send} + c \cdot \tau_B^{rec} + \Delta\rho_{AB} + \varepsilon_{AB} \tag{1}$$

$$\rho_{BA} = |R_A(t_2) - R_B(t_2 - \Delta t_2)| + \delta t_A - \delta t_B + c \cdot \tau_B^{send} + c \cdot \tau_A^{rec} + \Delta\rho_{BA} + \varepsilon_{BA} \tag{2}$$

where R_A and R_B are the positions of the two satellites SAT_A and SAT_B, respectively; c is the speed of light; δt_A and δt_B are the clock offsets of the two satellites SAT_A and SAT_B, respectively; Δt_1 and Δt_2 are the propagation times of ρ_{AB} and ρ_{BA}, respectively; τ^{send} and τ^{rec} are the emission delay and reception delay of the satellite, respectively; $\Delta\rho_{AB}$ and $\Delta\rho_{BA}$ are the error terms that can be modeled, including the relativistic effect and so on; and ε_{AB} and ε_{BA} are the noises of the respective observed values.

In addition, it is necessary to further reduce the inter-satellite bidirectional observations at different times to the same time to participate in the satellite orbit determination. First, utilizing the predicted orbit and the predicted satellite clock offset parameters, the observed values measured at t_1 and t_2 are reduced to the nearest full 3 s at time t_0. Considering that the bidirectional observed values are completed in a very short time, the prediction errors for the orbit and the clock offset can be ignored.

$$\rho_{AB}(t_0) = \rho_{AB} + |R_B(t_0) - R_A(t_0)| - |R_B(t_1) - R_A(t_1 - \Delta t_1)| + \\ c \cdot (\delta t_B(t_0) - \delta t_A(t_0)) - c \cdot (\delta t_A(t_1) - \delta t_B(t_1 - \Delta t_1)) \tag{3}$$

$$\rho_{BA}(t_0) = \rho_{BA} + |R_A(t_0) - R_B(t_0)| - |R_A(t_2) - R_B(t_2 - \Delta t_2)| + \\ c \cdot (\delta t_A(t_0) - \delta t_B(t_0)) - c \cdot (\delta t_A(t_2) - \delta t_B(t_2 - \Delta t_2)) \tag{4}$$

Then, adding the bidirectional pseudorange at time t_0, the observation equation that only contains the satellite orbit parameters can be obtained by sorting

$$\frac{\rho_{AB}(t_0) + \rho_{BA}(t_0)}{2} = |R_B(t_0) - R_A(t_0)| + \frac{c \cdot \tau_A^+}{2} + \frac{c \cdot \tau_B^+}{2} + \varepsilon^+ \tag{5}$$

The above equation is the orbit determination observation equation in autonomous orbit determination, where τ^+ is the sum of the emission delay and reception delay of the satellite, which is referred to as the time delay sum parameter in autonomous orbit determination.

In this contribution, we adopt an extended Kalman filter (EKF) to estimate the satellite's orbit parameters to ensure the real-time performance, in each of which we simultaneously estimate all satellite orbit parameters in a distribute processing mode.

2.2. Rotation Correction Model

In autonomous orbit determination based on only inter-satellite bidirectional ranging, the state parameters of all satellites in the constellation must be estimated, which means a rank deficiency for the solution of the parameters and the lack of the necessary starting benchmark. Thus, it is impossible to determine the absolute positions of all the satellites. In autonomous orbit determination, the observed and real inter-satellite distances of the constellation are the same, meaning the constellation estimation errors cannot be observed theoretically. The unobserved constellation estimation errors of autonomous orbit determination by inter-satellite ranging measurements mainly refer to the unobservability of constellation rotation; on one hand, inter-satellite ranging cannot correct some of the orbital elements and, on the other hand, the unobservability is brought about by using the prediction of EOP [27].

In the current distributed autonomous orbit determination, the two orbital plane orientation parameters of the right ascension for the ascending point Ω and the orbital inclination angle i are adopted as the constraint conditions to limit the overall constellation rotation satellite specifically, which is the algorithm for the correction of independent satellite constraints. Since the effect (angle) of the constellation rotation error on each satellite is consistent, the independent constraint algorithm eliminates this consistency; therefore, the result is suboptimal. After estimating and obtaining the constellation rotation error of the orbit solved by filtering relative to the reference orbit, the rotation correction algorithm for the global satellite constraints can then directly correct the orbit parameters obtained in autonomous orbit determination to realize the suppression of the overall rotation error of the constellation, which guarantees the integrity of the constellation.

2.2.1. Theoretical Analysis of the Influence of the Overall Constellation Rotation

At a certain time, the position and velocity of a certain satellite in the constellation are

$$X = \begin{pmatrix} x & y & z & \dot{x} & \dot{y} & \dot{z} \end{pmatrix}^T \tag{6}$$

If there is an overall rotation error with a rotation quantity of $\theta\left(\theta_\alpha, \theta_\beta, \theta_\gamma\right)$ in the constellation for autonomous orbit determination, then according to the criteria for coordinate transformation, the estimated value X' of the satellite position state is

$$
X' = \begin{pmatrix} x' \\ y' \\ z' \\ \dot{x}' \\ \dot{y}' \\ \dot{z}' \end{pmatrix} = \begin{bmatrix} R(-\theta) & 0 \\ 0 & R(-\theta) \end{bmatrix} \begin{pmatrix} x \\ y \\ z \\ \dot{x} \\ \dot{y} \\ \dot{z} \end{pmatrix} = R^* X \tag{7}
$$

where R represents the rotation matrix of the constellation; then,

$$
R(-\theta) = R_\gamma(-\theta_\gamma) R_\beta(-\theta_\beta) R_\alpha(-\theta_\alpha)
$$

$$
R_\alpha(-\theta_\alpha) = \begin{bmatrix} 1 & 0 & 0 \\ 0 & \cos\theta_\alpha & -\sin\theta_\alpha \\ 0 & \sin\theta_\alpha & \cos\theta_\alpha \end{bmatrix}
$$

$$
R_\beta(-\theta_\beta) = \begin{bmatrix} \cos\theta_\beta & 0 & \sin\theta_\beta \\ 0 & 1 & 0 \\ -\sin\theta_\beta & 0 & \cos\theta_\beta \end{bmatrix}
$$

$$
R_\gamma(-\theta_\gamma) = \begin{bmatrix} \cos\theta_\gamma & -\sin\theta_\gamma & 0 \\ \sin\theta_\gamma & \cos\theta_\gamma & 0 \\ 0 & 0 & 1 \end{bmatrix}
$$

where R_α, R_β, and R_γ represent the rotation matrices around the X-axis, Y-axis, and Z-axis, respectively (same below). When θ is a small quantity, we have

$$
R(-\theta) \approx \begin{bmatrix} 1 & -\theta_\gamma & \theta_\beta \\ \theta_\gamma & 1 & -\theta_\alpha \\ -\theta_\beta & \theta_\alpha & 1 \end{bmatrix} \tag{8}
$$

Therefore,

$$
R^* \approx \begin{bmatrix} 1 & -\theta_\gamma & \theta_\beta & 0 & 0 & 0 \\ \theta_\gamma & 1 & -\theta_\alpha & 0 & 0 & 0 \\ -\theta_\beta & \theta_\alpha & 1 & 0 & 0 & 0 \\ 0 & 0 & 0 & 1 & -\theta_\gamma & \theta_\beta \\ 0 & 0 & 0 & \theta_\gamma & 1 & -\theta_\alpha \\ 0 & 0 & 0 & -\theta_\beta & \theta_\alpha & 1 \end{bmatrix} \tag{9}
$$

Noting that

$$
X' = R^* X \approx \begin{pmatrix} x + z\theta_\beta - y\theta_\gamma \\ y - z\theta_\alpha + x\theta_\gamma \\ z + y\theta_\alpha - x\theta_\beta \\ \dot{x} + \dot{z}\theta_\beta - \dot{y}\theta_\gamma \\ \dot{y} - \dot{z}\theta_\alpha + \dot{x}\theta_\gamma \\ \dot{z} + \dot{y}\theta_\alpha - \dot{x}\theta_\beta \end{pmatrix} = \begin{pmatrix} x \\ y \\ z \\ \dot{x} \\ \dot{y} \\ \dot{z} \end{pmatrix} + \begin{bmatrix} 0 & z & -y \\ -z & 0 & x \\ y & -x & 0 \\ 0 & \dot{z} & -\dot{y} \\ -\dot{z} & 0 & \dot{x} \\ \dot{y} & -\dot{x} & 0 \end{bmatrix} \begin{pmatrix} \theta_\alpha \\ \theta_\beta \\ \theta_\gamma \end{pmatrix} \tag{10}
$$

$$
H = \begin{bmatrix} 0 & z & -y \\ -z & 0 & x \\ y & -x & 0 \\ 0 & \dot{z} & -\dot{y} \\ -\dot{z} & 0 & \dot{x} \\ \dot{y} & -\dot{x} & 0 \end{bmatrix} \tag{11}
$$

$$X' \approx X + H\theta \tag{12}$$

That is, the satellite position and velocity state errors caused by the constellation rotation error are

$$\delta X = X' - X \approx H\theta = \begin{pmatrix} z\theta_\beta - y\theta_\gamma \\ x\theta_\gamma - z\theta_\alpha \\ y\theta_\alpha - x\theta_\beta \\ \dot{z}\theta_\beta - \dot{y}\theta_\gamma \\ \dot{x}\theta_\gamma - \dot{z}\theta_\alpha \\ \dot{y}\theta_\alpha - \dot{x}\theta_\beta \end{pmatrix} \tag{13}$$

2.2.2. Algorithm for Independent Satellite Constraints

The constraint conditions for the algorithm for independent satellite constraints are shown as follows:

$$\begin{cases} i = \tilde{i} \\ \Omega = \widetilde{\Omega} \end{cases} \tag{14}$$

where $\tilde{i}$ and $\widetilde{\Omega}$ are the orbital inclination angle and the right ascension of the ascending point of the reference orbit, respectively, and i and Ω are the corresponding parameters to be estimated. Considering i and Ω as functions of state X to be estimated (Cartesian coordinates and velocity in the inertial system) and linearly expanding at the approximate value X_0, we have

$$\begin{cases} i_0 + \frac{\partial i}{\partial X}\Big|_{X_0} (X - X_0) = \tilde{i} \\ \Omega_0 + \frac{\partial \Omega}{\partial X}\Big|_{X_0} (X - X_0) = \widetilde{\Omega} \end{cases} \tag{15}$$

where i_0 and Ω_0 are obtained by calculation based on X_0, and

$$\begin{bmatrix} \frac{\partial i}{\partial X} \\ \frac{\partial \Omega}{\partial X} \end{bmatrix} = \begin{bmatrix} \frac{\partial i}{\partial \vec{r}} & \frac{\partial i}{\partial \dot{\vec{r}}} \\ \frac{\partial \Omega}{\partial \vec{r}} & \frac{\partial \Omega}{\partial \dot{\vec{r}}} \end{bmatrix} = \begin{bmatrix} \frac{\partial i}{\partial x} & \frac{\partial i}{\partial y} & \frac{\partial i}{\partial z} & \frac{\partial i}{\partial \dot{x}} & \frac{\partial i}{\partial \dot{y}} & \frac{\partial i}{\partial \dot{z}} \\ \frac{\partial \Omega}{\partial x} & \frac{\partial \Omega}{\partial y} & \frac{\partial \Omega}{\partial z} & \frac{\partial \Omega}{\partial \dot{x}} & \frac{\partial \Omega}{\partial \dot{y}} & \frac{\partial \Omega}{\partial \dot{z}} \end{bmatrix} \tag{16}$$

When the one-step prediction value from the filtering results of the previous epoch to the start time of the current epoch is selected as the linearized expansion point X_0, Equation (16) can be written as a general expression form of the constraint equation

$$C\delta X = M \tag{17}$$

where

$$C = \begin{bmatrix} \frac{\partial i}{\partial X} \\ \frac{\partial \Omega}{\partial X} \end{bmatrix}_{X_0}, \quad M = \begin{pmatrix} \tilde{i} - i_0 \\ \widetilde{\Omega} - \Omega_0 \end{pmatrix}$$

Therefore, the complete orbit filtering model of the algorithm for independent constraints on satellites in distributed autonomous orbit determination is

$$\begin{cases} \delta X_k = \Phi_{k,k-1}\delta X_{k-1} + W_{k-1} \\ Z_k = H_k\delta X_k + Y_k + V_k \\ C_k\delta X_k = M_k \end{cases} \tag{18}$$

2.2.3. Algorithm for Global Satellite Constraints

The estimated value of the inclination angle i and the right ascension of the ascending point Ω in the orbit of the orbit elements obtained in the autonomous orbit determination of Satellite m is $\sigma'_m(i'_m, \Omega'_m)$, and the estimated value of the inclination angle i and the right ascension of the ascending point Ω among the real orbital elements of the satellite is $\sigma_m(i_m, \Omega_m)$. Due to the unobservable satellites of constellation rotation, when an overall rotation of a rotation quantity $\theta\left(\theta_\alpha, \theta_\beta, \theta_\gamma\right)$ occurs, the following relationship will exist between σ'_m and σ_m:

$$\sigma = \tilde{\sigma} + H\theta \tag{19}$$

In the above equation, the partial derivative matrix of H Kepler orbital roots to the constellation rotation error is given without additional derivation:

$$H = \frac{\partial\sigma}{\partial\theta} = \begin{pmatrix} \frac{\partial i}{\partial\theta_\alpha} & \frac{\partial i}{\partial\theta_\beta} & \frac{\partial i}{\partial\theta_\gamma} \\ \frac{\partial \Omega}{\partial\theta_\alpha} & \frac{\partial \Omega}{\partial\theta_\beta} & \frac{\partial \Omega}{\partial\theta_\gamma} \end{pmatrix} = \begin{pmatrix} \cos\Omega_m & \sin\Omega_m & 0 \\ -\sin\Omega_m \cot i_m & \cos\Omega_m \cot i_m & 1 \end{pmatrix} \tag{20}$$

To estimate the rotation quantity $\theta\left(\theta_\alpha, \theta_\beta, \theta_\gamma\right)$, the errors in the inclination angle i and in the right ascension of the ascending point Ω of at least two satellite orbital elements are needed. In fact, in estimating a more optimal constellation rotation quantity θ, it is smaller the more satellites there are, and these satellites should be distributed as evenly as possible on each orbital plane. If the errors in the inclination angle i and in the right ascension of the ascending point Ω of the left and right satellites in the constellation are known, and Equation (19)of each satellite is listed, we have

$$U = X - H\theta \tag{21}$$

In the equation, $U = \left(U_1, U_2, \ldots, U_m, \ldots, U_n\right)^T$, $H = \left(H_1, H_2, \ldots, H_m, \ldots, H_n\right)^T$, and $U_m = \sigma'_m - \sigma_m - H_m\theta$. The optimal estimated value of θ can be determined under the least squares criterion.

$$\hat{\theta} = \left(H^T P H\right)^{-1} H^T P X \tag{22}$$

After estimating and obtaining the constellation rotation error of the orbit solved by filtering relative to the reference orbit, the results obtained in autonomous orbit determination can be directly corrected.

3. Analysis of the Inter-Satellite Measurement Situation

3.1. Analysis of the Link Establishment Situation for Inter-Satellite Links

According to the constellation configuration, there are three kinds of visible relationships among BDS-3 MEO satellites, namely, continuously visible, non-continuously visible, and invisible, which are related to the continuity of inter-satellite measurement, the number of established links, the orbital positions of satellites, and time slot route planning, etc. Z.L., J.X. et al. analyzed the inter-satellite observation conditions [30]. Figure 1 shows the 30-day time series of the number of satellites with established links for the PRN36 satellite. The fluctuation is considerable, with a minimum of 0 and a maximum of up to 20. Table 1 shows the statistics for the number of established links. In general, the minimum number of established links is 0, the maximum is 19.96, and the average number of established links is 14.45.

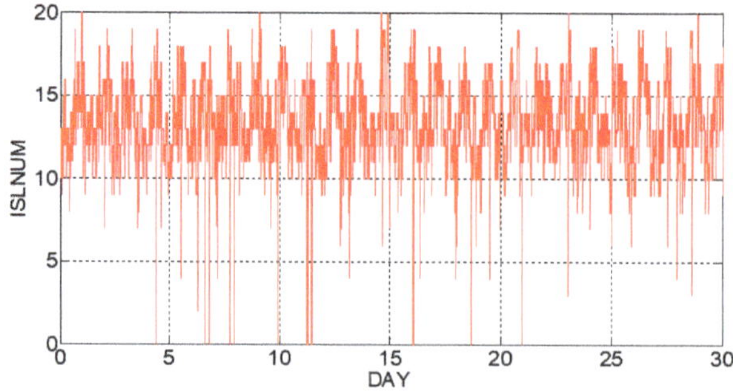

Figure 1. The 30-day link-tracking number of PRN36 for BDS-3.

Table 1. The statistical situation in the number of established links for each BDS-3 satellite.

PRN	Maximum Number of Established Links	Minimum Number of Established Links	Mean
25	20	0	14.55
26	20	0	14.55
27	20	0	14.41
28	20	0	14.25
29	20	0	14.20
30	20	0	13.68
19	20	0	14.56
20	20	0	14.48
21	20	0	14.41
22	20	0	14.73
23	20	0	14.97
24	20	0	14.95
32	20	0	15.06
33	19	0	13.98
34	20	0	13.92
35	20	0	14.21
36	20	0	14.45
37	20	0	14.58
38	20	0	14.66
39	20	0	14.80
40	20	0	14.01
41	20	0	14.18
42	20	0	14.91
43	20	0	14.09
Mean	19.96	0	14.45

3.2. Constellation Configuration Analysis for Autonomous Orbit Determination

When the number of established links between satellites to be assessed is greater than or equal to three, the corresponding position dilution of precision (PDOP) value can be calculated to reflect the quality of the geometric configuration of inter-satellite link establishment. Figure 2 presents the changing situation for the PDOP values of the PRN36 satellite during a consecutive 30-day period, and Table 2 shows the PDOP statistics of all 24 MEO satellites in BDS-3. The average minimum PDOP of all satellites is approximately 0.74, the average maximum PDOP is approximately 4.41, and the mean PDOP value is approximately 0.99. It can be concluded that the overall structure of the geometric figure for inter-satellite link establishment is improved [26].

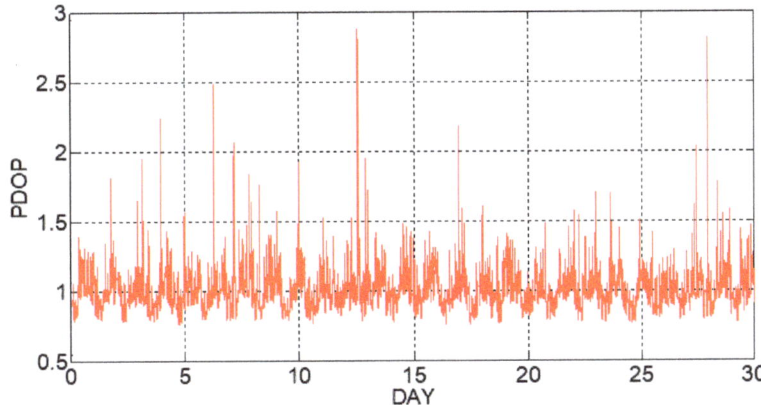

Figure 2. The changing situation of PDOP values for PRN36 satellites as the positioning objects of BDS-3.

Table 2. The statistical situation of PDOP values for the BDS-3 satellites.

PRN	Minimum	Maximum	Mean
25	0.75	3.33	0.94
26	0.73	4.59	0.95
27	0.73	3.93	0.96
28	0.74	3.54	0.97
29	0.73	4.50	0.96
30	0.73	4.04	0.98
19	0.75	2.89	0.95
20	0.73	5.57	0.95
21	0.75	5.31	0.96
22	0.74	4.95	0.96
23	0.74	4.23	0.94
24	0.74	4.63	0.94
32	0.73	3.07	0.94
33	0.75	3.39	0.98
34	0.74	5.64	0.98
35	0.73	4.71	0.96
36	0.75	2.91	0.96
37	0.75	5.96	0.95
38	0.73	4.42	0.94
39	0.74	5.70	0.94
40	0.74	4.04	1.06
41	0.75	4.21	1.77
42	0.75	3.93	0.94
43	0.75	4.62	0.97
Mean	0.74	4.41	0.99

4. Orbit Determination Results

4.1. Processing Cases and Strategies for Autonomous Orbit Determination

Distributed autonomous orbit determination processing was adopted in this paper to assess the accuracy of the autonomous orbit determination of BDS-3 under different rotation correction methods.

In the general processing of precise orbit determination, parameters to be estimated include initial satellite position and velocity, solar radiation pressure parameters, clock offset parameters (clock offset and clock speed), as well as other parameters induced by measurements. Targeted optimization was carried out on the autonomous orbit determination algorithm in this paper. Table 3 shows the specific models and relevant strategies adopted in orbit determination. Among them, the Empirical CODE orbit Model (ECOM) was adopted for modeling solar radiation pressure perturbation, and a longer arc segment of the satellite-Earth-satellite joint orbit determination result was used to estimate the solar radiation pressure parameters more precisely. In this way, solar radiation pressure parameters were no longer estimated, thereby ensuring fewer estimated parameters and a smaller computational load.

4.2. Data Processing Strategies for Autonomous Orbit Determination

In data processing for autonomous orbit determination in this paper, necessary correction of observation was first carried out. Then, time reduction processing was carried out according to the predicted orbit of the current epoch. Afterwards, gross errors in observations were detected and eliminated and high-quality observations were acquired. Finally, the autonomous orbit determination was conducted and the data processing flow is shown in Figure 3.

Table 3. Processing strategies of autonomous orbit determination (AOD).

Parameter	Model
Observed values	Observation data of inter-satellite links
Observation interval	1 min
Satellite transceiver delay	Not estimated, calibrated numerical values are adopted
Gravity field model	Earth Gravitational Model 2008 (EGM2008) model to the 8th order
Tidal correction	Only solid tides are considered
Solar radiation pressure model	ECOM model, parameters not estimated
EOP	International Earth Rotation and Reference Systems Service (IERS) prediction of EOP (Bulletin A)
Gravitational force of N body	Considers the gravitational forces of the sun and moon
Parameters of the initial orbit	Broadcast ephemeris orbit
Parameters of the initial clock offset	Broadcast ephemeris clock offset
Estimator	Extended Kalman filter (EKF)
Parameters to be estimated	Only the position and velocity parameters of each satellite are estimated

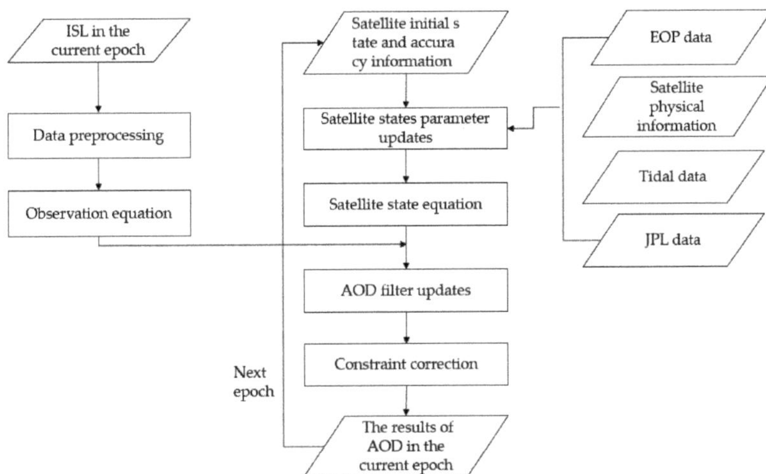

Figure 3. AOD data processing flowchart.

4.3. Analysis of Experimental Results

To verify the effect of the aforementioned rotation correction methods on the accuracy of autonomous orbit determination solutions, the following three kinds of a priori constraint information were adopted in data filtering processing.

Case 1: No rotation correction.

Case 2: Adoption of an algorithm for independent constraints to correct the rotation.

Case 3: Adoption of an algorithm for overall constraints onboard the satellite to correct the rotation.

A priori constraint information was generated for the three aforementioned cases, respectively, and autonomous orbit determination simulation processing was carried out with inter-satellite ranging data from 16 October 2020 to 14 November 2020 of 24 MEO satellites of BDS-3.

Using the satellite-ground and inter-satellite joint orbit determination results as reference, the rotation errors acquired by Case 1, Case 2, and Case 3 were analyzed. The three Euler angles (α, β, γ) of the constellation rotation parameters were calculated, and the results are presented in Figure 4. Figure 5 shows the prediction errors of UT1-UTC in EOP. Compared with the no-rotation-correction case, both the algorithm for independent constraints and the algorithm for global constraints could constrain the constellation rotation on the X and Y axes better, while no significant improvement was shown on the Z-axis. Moreover, the errors were basically consistent with the prediction errors of UT1-UTC in EOP given in Figure 6, which occurred because the errors on the Z-axis in the overall rotation of the constellation were mainly caused by the prediction errors of UT1-UTC, and the errors could not be eliminated without external reference input.

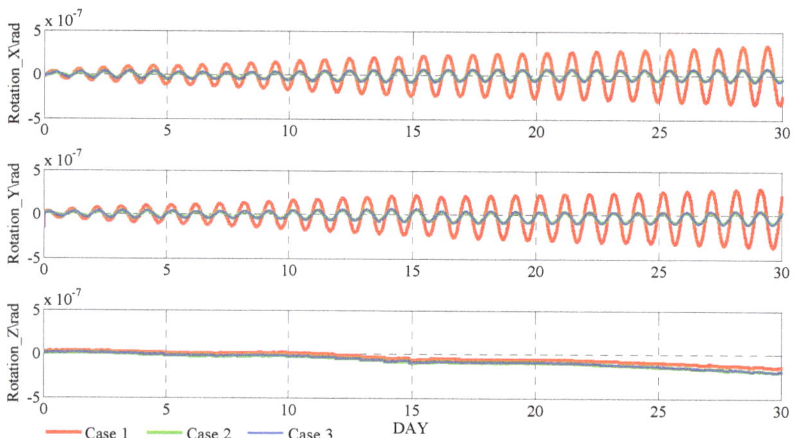

Figure 4. Euler angles of the constellation rotation in each direction under different autonomous orbit determination cases.

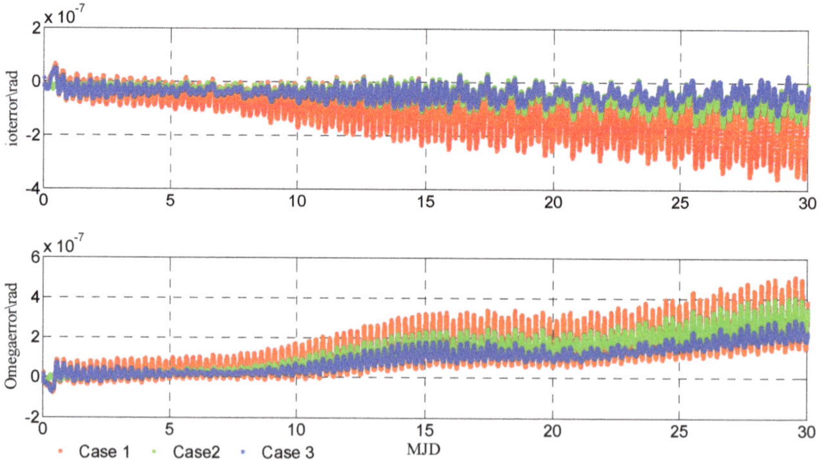

Figure 5. The i angle and Ω angle errors of PRN25 under different cases for autonomous orbit determination.

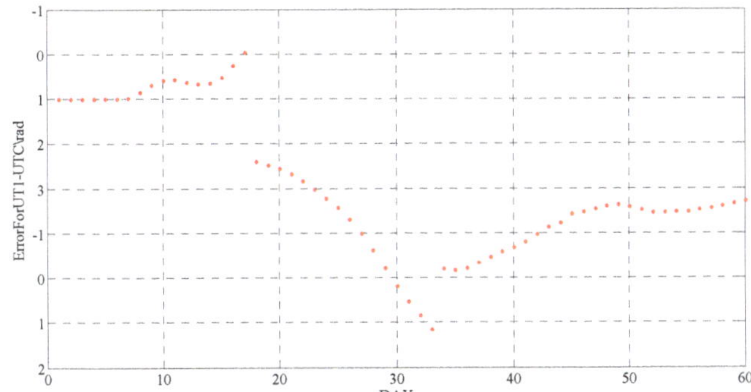

Figure 6. 30-day prediction errors of UT1-UTC in EOP.

Furthermore, the orbit inclination angle *i* (iot) and right ascension of the ascending point Ω (Omega) under the three cases were calculated. Using the post-processed precise orbits of satellite-ground and inter-satellite joint orbit determination as references, the angle errors relative to the precise ephemeris error were obtained. Due to the space limitations of the article, only the time series of the PRN25 and PRN37 satellites are given here, as shown in Figures 5 and 7. Table 4 shows the prediction errors in the *i* and Ω for each satellite on the 30th day under the different cases. It can be seen from Figures 7 and 8, when the rotation was not corrected, the *i* and Ω of PRN25 and PRN37 became divergent gradually, and the trends of the two satellites were basically consistent. After adopting the algorithm for independent satellite constraints, the *i* errors of the two satellites could be better constrained, and there was also a correction effect on Ω. However, the margin of error for each satellite after correction was inconsistent, which affected the integrity of the constellation as a rigid body. The algorithm for global constraints on satellites corrected the *i* and Ω well, and the errors after correction were basically consistent, which did not affect the integrity of the constellation as a rigid body. Without any correction, the average RMS of the *i* prediction errors for the entire constellation on the 30th day was 2.11×10^{-7}/rad, and the average RMS of the Ω prediction errors was 2.25×10^{-7}/rad. After adopting the algorithm for independent satellite constraints, the *i* prediction error was 5.43×10^{-8}/rad, and the Ω prediction error was 2.03×10^{-7}/rad. When the algorithm for global satellite constraints was applied, the average *i* prediction error was 5.31×10^{-8}/rad, and the average Ω prediction error was 1.95×10^{-7}/rad. The two constraint algorithms could weaken the *i* and Ω prediction errors of the constellation to some extent, and the algorithm for global satellite constraints achieved the best performance.

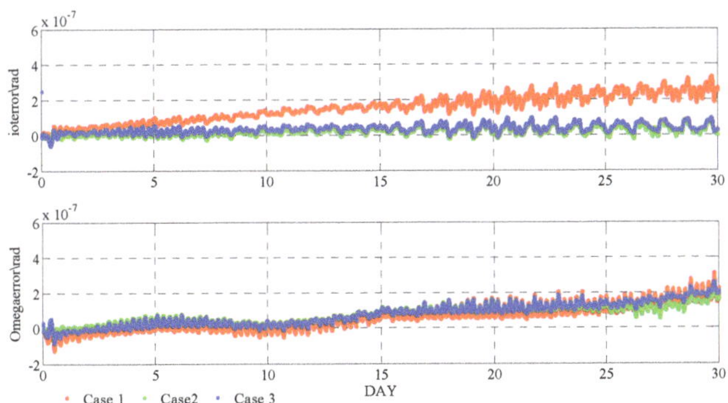

Figure 7. The *i* angle and Ω angle errors of PRN37 under different cases for autonomous orbit determination.

Table 4. The i and Ω prediction errors on the 30th day of autonomous orbit determination under different cases.

PRN	i/rad			Ω/rad		
	Case 1	Case 2	Case 3	Case 1	Case 2	Case 3
25	1.99×10^{-7}	2.86×10^{-8}	4.65×10^{-8}	3.07×10^{-7}	1.94×10^{-7}	2.23×10^{-7}
26	2.19×10^{-7}	3.70×10^{-8}	5.33×10^{-8}	3.06×10^{-7}	2.17×10^{-7}	2.29×10^{-7}
27	2.65×10^{-7}	5.72×10^{-8}	6.06×10^{-8}	1.75×10^{-7}	1.59×10^{-7}	1.92×10^{-7}
28	2.65×10^{-7}	5.83×10^{-8}	6.14×10^{-8}	1.80×10^{-7}	2.03×10^{-7}	1.92×10^{-7}
29	2.53×10^{-7}	4.16×10^{-8}	5.43×10^{-8}	1.97×10^{-7}	1.62×10^{-7}	2.03×10^{-7}
30	2.57×10^{-7}	4.91×10^{-8}	5.77×10^{-8}	1.84×10^{-7}	1.26×10^{-7}	1.93×10^{-7}
19	1.59×10^{-7}	2.98×10^{-8}	4.83×10^{-8}	1.41×10^{-7}	1.63×10^{-7}	1.30×10^{-7}
20	1.58×10^{-7}	4.32×10^{-8}	4.46×10^{-8}	1.49×10^{-7}	1.78×10^{-7}	1.52×10^{-7}
21	1.59×10^{-7}	5.99×10^{-8}	4.49×10^{-8}	1.50×10^{-7}	1.28×10^{-7}	1.57×10^{-7}
22	1.52×10^{-7}	4.16×10^{-8}	4.62×10^{-8}	1.51×10^{-7}	1.35×10^{-7}	1.30×10^{-7}
23	2.04×10^{-7}	4.09×10^{-8}	5.07×10^{-8}	2.96×10^{-7}	2.11×10^{-7}	2.17×10^{-7}
24	2.12×10^{-7}	4.39×10^{-8}	5.18×10^{-8}	3.21×10^{-7}	2.30×10^{-7}	2.32×10^{-7}
32	1.74×10^{-7}	6.73×10^{-8}	5.98×10^{-8}	1.65×10^{-7}	1.58×10^{-7}	1.48×10^{-7}
33	1.55×10^{-7}	5.43×10^{-8}	4.46×10^{-8}	1.41×10^{-7}	1.58×10^{-7}	1.58×10^{-7}
34	2.59×10^{-7}	3.52×10^{-8}	5.92×10^{-8}	1.89×10^{-7}	2.55×10^{-7}	1.99×10^{-7}
35	2.61×10^{-7}	5.24×10^{-8}	5.82×10^{-8}	1.97×10^{-7}	2.23×10^{-7}	2.05×10^{-7}
36	2.02×10^{-7}	5.66×10^{-8}	5.13×10^{-8}	3.16×10^{-7}	2.00×10^{-7}	2.29×10^{-7}
37	2.22×10^{-7}	5.07×10^{-8}	5.70×10^{-8}	3.02×10^{-7}	1.92×10^{-7}	2.22×10^{-7}
41	1.50×10^{-7}	5.84×10^{-8}	4.47×10^{-8}	1.53×10^{-7}	1.43×10^{-7}	1.50×10^{-7}
42	1.58×10^{-7}	5.48×10^{-8}	4.31×10^{-8}	1.50×10^{-7}	1.58×10^{-7}	1.56×10^{-7}
43	2.58×10^{-7}	4.47×10^{-8}	5.91×10^{-8}	1.94×10^{-7}	2.48×10^{-7}	2.01×10^{-7}
44	2.64×10^{-7}	5.95×10^{-8}	6.02×10^{-8}	1.87×10^{-7}	2.23×10^{-7}	1.99×10^{-7}
45	2.08×10^{-7}	9.51×10^{-8}	5.37×10^{-8}	2.98×10^{-7}	3.24×10^{-7}	2.25×10^{-7}
46	2.17×10^{-7}	8.82×10^{-8}	5.66×10^{-8}	3.19×10^{-7}	3.12×10^{-7}	2.31×10^{-7}
RMS	2.11×10^{-7}	5.43×10^{-8}	5.31×10^{-8}	2.25×10^{-7}	2.03×10^{-7}	1.95×10^{-7}
STD	1.58×10^{-7}	1.55×10^{-8}	6.04×10^{-9}	6.78×10^{-7}	5.15×10^{-8}	2.91×10^{-8}

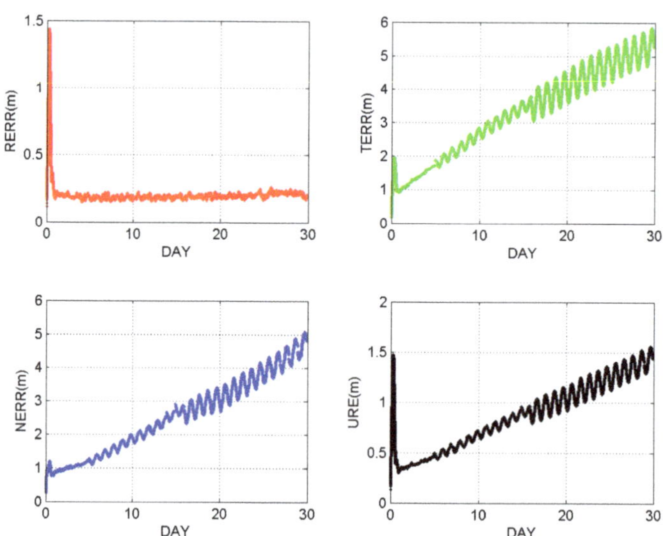

Figure 8. The autonomous navigation constellation average orbit error with no-rotation-correction case.

The differences between the autonomous orbit determination results and the reference orbits were analyzed, and the radial errors (RERR), tangential errors (TERR), normal-direction errors (NERR), and comprehensive errors (user range errors, URE) of the constellation are given in Figures 8–10. The 30-day accuracy statistics for autonomous orbit determination by the three cases are shown in Table 5.

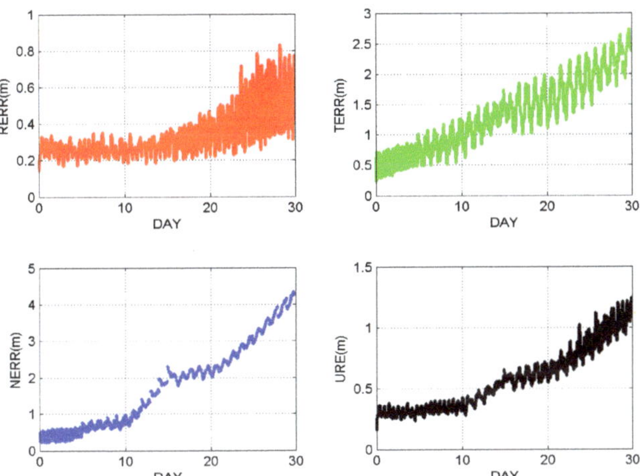

Figure 9. The autonomous navigation constellation average orbit error with independent constraints to correct the rotation case.

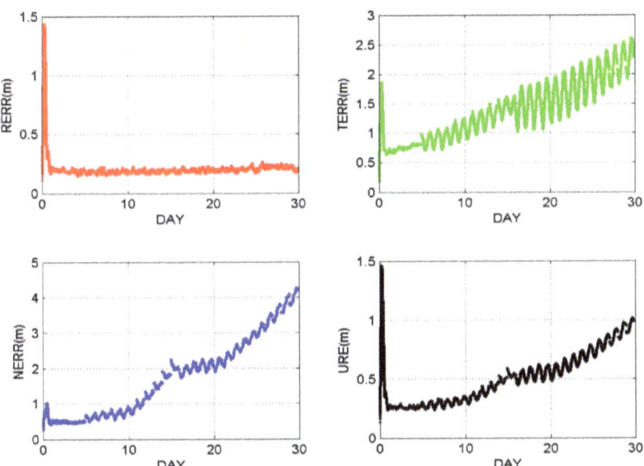

Figure 10. The autonomous navigation constellation average orbit error with overall constraints onboard the satellite to correct the rotation case.

Table 5. The autonomous navigation constellation average orbit error with different cases.

Case	RERR (m)	TERR (m)	NERR (m)	URE (m)
Case 1	0.17	4.70	4.55	1.41
Case 2	0.54	2.21	4.07	1.09
Case 3	0.17	2.14	4.02	0.94

For the 24 BDS-3 MEO satellites, the orbit-only URE is expressed as:

$$URE = \sqrt{0.96 * \text{RERR}^2 + 0.04 * \left(\text{TERR}^2 + \text{NERR}^2\right)} \tag{23}$$

As can be seen, without rotation correction, the average RMS of the RERRs for the entire constellation in 30 days was approximately 0.17 m, the TERR was 4.70 m, the NERR was 4.55m, and the orbit accuracy was approximately 1.41 m. After adopting the algorithm for independent satellite constraints to carry out rotation correction, the average RMS of RERR was 0.54 m, the RMS of TERR was 2.21 m, the RMS of NERR was 4.07 m, and the orbit accuracy of was 1.09 m. By adopting the algorithm for global satellite constraints, the average RMS of RERR was 0.17 m, and the NERR and TERR were 2.14 m and 4.02 m, respectively, and the accuracy of orbit determination for the entire constellation was 0.72 m. Among them, the 30-day accuracy in autonomous orbit determination of without rotation correction was the worst, and the accuracy under the algorithm for global constraints on satellites was the best. The orbit accuracy of the three cases of autonomous orbit determination deteriorated with time gradually, because the errors created by UT1-UTC prediction in autonomous orbit determination cannot be eliminated relying on only the observed values of inter-satellite links but without external reference constraints. The orbit error trend was basically consistent with the prediction errors of UT1-UTC in Figure 4.

Compared with Case 1, the algorithm for independent satellite constraints significantly decreased TERR and NERR in the orbit determination. However, the RERR showed an increasing trend. The reason is that the rotation parameters of each satellite in the independent constraint algorithm are calculated separately, which caused variations in the geometry configuration of the inter-satellite ranging network [11]. When the Ω variations of two satellites were the same, the inter-satellite distance was not to be changed. Conversely, a change in the inter-satellite distance was unavoidable. Table 4 shows the Ω error of the predicted orbit for each satellite. It can be seen that the Ω of each satellite was different and it would cause errors in the inter-satellite distance, which would indirectly lead to orbital RERR. Compared with the algorithm for independent satellite constraints, the algorithm for global satellite constraints achieved a smaller improvement in the tangential and normal directions, but it constrained the RERR well. This result illustrates that the algorithm for independent satellite constraints did not destroy the configuration of autonomous orbit determination. Therefore, it would not lead to an increasing orbit RERR.

The following conclusions were summarized through this study: (1) Under the premise of no external constraints, the accuracy of autonomous orbit determination deteriorates with time. (2) Both algorithms for rotation correction can weaken the influence of the overall constellation rotation. (3) The algorithm for independent satellite constraints will affect the integrity of the constellation configuration as a rigid body to a certain extent, which may lead to an increasing RERR. However, the algorithm for global satellite constraints will not destroy the constellation configuration. (4) Among the three cases, the 30-day accuracy in autonomous orbit determination using the algorithm for global satellite constraints was the best.

5. Conclusions

In this paper, the models for constellation rotation errors during autonomous orbit determination were analyzed theoretically. Aiming at the drawbacks of the algorithm for independent satellite constraints in distributed autonomous orbit determination, an algorithm for global satellite constraints was proposed. Based on the inter-satellite observation data for the 24 MEOs in the BDS-3, BDS-3 autonomous orbit determination experiments were designed under three schemes for correcting the rotation and the orbit determination results were assessed. The results are described as follows:

(1) The BDS-3 inter-satellite ranging data show satisfactory continuity, and the average number of established links for a single satellite was approximately 14.4. The geometric configuration of inter-satellite link establishment was good, with an average PDOP of approximately 0.99.

(2) Among three rotation-correction cases, the algorithm for independent satellite constraints and the global satellite constraints constrained the constellation rotation on the X and the Y axes better. However, this method did not improve the Z-axis rotation errors caused by the prediction errors in UT1-UTC.

(3) Under the three rotation-correction schemes, distributed autonomous orbit determination processing can be carried out. Compared with precise ephemeris, it shows that without external constraints, the accuracy of autonomous orbit determination deteriorates with time. With no correction algorithm applied, the average RMS of the 30-day orbit URE of autonomous orbit determination was 1.41 m. Using the algorithm for independent satellite constraints, the average RMS of the 30-day orbit URE was 1.09 m. Using the algorithm for global satellite constraints, the average RMS of the 30-day orbital URE was 0.94 m, and this scheme achieves stable and reliable autonomous orbit determination results.

(4) Both the algorithm for independent satellite constraints and the algorithm for global satellite constraints can weaken the influence of the overall constellation rotation. The former will affect the integrity of the constellation configuration as a rigid body to a certain extent, while the later solves this problem perfectly. Among the three schemes, the 30-day accuracy of autonomous orbit determination using the algorithm for global constraints on satellites was the best.

The work conducted and the conclusions obtained in this paper provide some reference for improving the accuracy of autonomous orbit determination. However, in a case where the ground reference benchmarks are missing, the constellation rotation errors of autonomous orbit determination caused by UT1-UTC predictions using the inter-satellite link ranging data cannot be eliminated. To follow up, further research and performance analysis on autonomous orbit determination technology for the navigation constellation based on anchoring support should be carried out.

Author Contributions: W.Z., Z.L. and W.L. conceived and designed the experiments; W.Z., H.C. and Z.L. performed the experiments and analyzed the data; W.Z., H.C. and Z.L. wrote the paper; C.T., X.H. and W.L. reviewed the paper. All authors have read and agreed to the published version of the manuscript.

Funding: This research received no external funding.

Data Availability Statement: The data presented in this study are available on request from Beijing Institute of Tracking and Telecommunications Technology (BITTT). The data are not publicly available due to the confidentiality of the data.

Conflicts of Interest: The authors declare no conflict of interest.

References

1. Ananda, M.P.; Berstein, H.; Bruce, R.W. Autonomous Navigation of the Global Positioning System Satellites. In Proceedings of the AIAA Guidance and Control Conference, Seattle, WA, USA, 20–22 August 1984; pp. 1–11.
2. Abusali, P.A.M.; Tapley, B.D.; Schutz, B.E. Autonomous Navigation of Global Positioning System Satellites Using Cross-Link Measurements. *J. Guid. Control. Dyn.* **2015**, *21*, 321–327. [CrossRef]
3. Tan, S. Development and Thought of Compass Navigation Satellite System. *J. Astronaut.* **2008**, *29*, 391–396.
4. Ananda, M.P.; Bernstein, H.; Cunningham, K.E.; Feess, W.A.; Stroud, E.G. Global Positioning System (GPS) autonomous navigation. In Proceedings of the IEEE Position Location & Navigation Symposium, Las Vegas, NV, USA, 20 March 1990.
5. Rajan, J.A.; Orr, M.; Wang, P. On-Orbit Validation of GPS IIR Autonomous Navigation. In Proceedings of the ION 59th Annual Meeting/CIGTF 22nd Guidance Test Symposium, Albuquerque, NM, USA, 23–25 June 2003.
6. Chen, Z.; Shuai, P.; Qu, G. Development of Satellite Navigation Systems (Part 1). *Aerosp. China* **2007**, *9*, 24–29.
7. Chen, Z.; Shuai, P.; Qu, G. Development of Satellite Navigation Systems (Part 2). *Aerosp. China* **2007**, *10*, 22–25.
8. Chen, Z.; Shuai, P.; Qu, G. Analysis of technical characteristics and development trend of modern satellite navigation systems. *Sci. Chins (Ser. E) Technol. Sci.* **2009**, *39*, 686–695.
9. Menn, M.D.; Bernstein, H. Ephemeris Observability Issues in the Global Positioning System Autonomous Navigation (AUTONAV). In Proceedings of the Position Location and Navigation Symposium, New York, NY, USA, 11–15 April 1994; pp. 677–680.
10. Rajan J, A. Highlights of GPS II-R Autonomous Navigation. In Proceedings of the ION 58th Annual Meeting/CIGTF 21st Guidance Test Symposium, Albuquerque, NM, USA, 24–26 June 2002; pp. 354–363.

11. Fisher, S.C.; Ghassemi, K. GPS IIF-the next generation. *Proc. IEEE* **1999**, *87*, 24–47. [CrossRef]
12. Gill, E. Precise Orbit Determination of the GNSS-2 Space Segment from Ground-Based and Satellite-to-Satellite Tracking. In Proceedings of the 2nd European Symposium on Global Navigation Satellite Systems, Toulouse, France, 20–23 October 1998.
13. Wolf, R. Satellite Orbit And Ephemeris Determination Using Inter Satellite Links. Ph.D. Thesis, Bundeswehr University Munich, Munich, Germany, 2000.
14. Eissfeller, B.; Zink, T.; Wolf, R.; Hammesfahr, J.; Hornbostel, A.; Hahn, J.H.; Tavella, P. Autonomous Satellite State Determination by Use of Two-Directional Links. *Int. J. Satell. Commun.* **2000**, *18*, 325–346. [CrossRef]
15. Fernández, F.A. Inter-satellite ranging and inter-satellite communication links for enhancing GNSS satellite broadcast navigation data. *Adv. Space Res.* **2011**, *47*, 786–801. [CrossRef]
16. Liu, L.; Liu, Y. Study on the problem of rank deficiency in autonomous orbit determination with relative measurements between satellites. *J. Spacecr. TTC Technol.* **2000**, *29*, 13–16.
17. Liu, Y.; Liu, L. On the Problem of Orbit Determination with Satellite-Satellite Tracking. *J. Spacecr. TTC Technol.* **2000**, *19*, 7–14.
18. Liu, Y.; Liu, L. On the Problem of Joint Orbit Determination with Satellite-Satellite Tracking and Ground Tracking. *Publ. Purple Mt. Obs.* **2000**, *19*, 117–120.
19. Zhang, Y. Study on the Methods of Autonomous Orbit Determination for the Constellation Based on Inter-satellite Observation. Ph.D. Thesis, National University of Defense Technology, Changsha, China, 2005.
20. Chen, J.; Jiao, W.; Ma, J.; Song, X. Autonav of Navigation Satellite Constellation Based on Crosslink Range and Orientation Parameters Constraining. *Geomat. Inf. Sci. Wuhan Univ.* **2005**, *30*, 439–443.
21. Chen, J.; You, Z.; Jiao, W. Research on Autonav of Navigation Satellite Constellation Based on Crosslink Range and Inter-satellites Orientation Observation. *J. Astronaut.* **2005**, *26*, 43–46.
22. Cai, Z.; Han, C.; Chen, J. Constellation Rotation Error Analysis and Control in Long-term Autonomous Orbit Determination for Navigation Satellites. *J. Astronaut.* **2008**, *29*, 522–528.
23. Liu, W. Research and Simulation on Autonomous Orbit Determination and Combined Orbit Determination of Navigation Satellites. Ph.D. Thesis, Wuhan University, Wuhan, China, 2008.
24. Wang, F.; Liu, W.; Lin, X. Distributed Autonomous Orbit Determination of Global Navigation Constellation Via Inter-Satellite Pseudo-Ranging Measurements. In Proceedings of the Electronic Collection of the 2nd China Satellite Navigation Conference, Shanghai, China, 18 May 2011.
25. Zheng, J.; Lin, Y.; Chen, Z. GPS Crosslink Technology and Autonomous Navigation Algorithm. *Spacecr. Eng.* **2009**, *18*, 28–35.
26. Chen, Y.; Hu, X.; Zhou, S.; Song, X.; Huang, Y.; Mao, Y.; Huang, C.; Chang, Z.; Wu, S. A new autonomous orbit determination algorithm based on inter-satellite ranging measurements. *Sci. Sin. Phys. Mech. Astron.* **2015**, *45*, 079511.
27. Du, Y. Study on the Whole Rotation and Suppression Method of Autonomous Orbit Constellation of Distributed Navigation Satellite. Ph.D. Thesis, Wuhan University, Wuhan, China, 2015.
28. Gong, X.; Li, Z.; Liu, W.; Wang, F. GPS satellite-Earth joint orbit determination based on pseudorange observation values measured by ground stations. In Proceedings of the CPGPS2010, Shanghai, China, 18–20 August 2010; p. 8.
29. Gong, X. Research on Centralized Autonomous Realtime Orbit Determination and Time Synchronization of BDS. Ph.D. Thesis, Wuhan University, Wuhan, China, 2013.
30. Li, Z.; Xin, J.; Guo, R.; Li, X.; Tang, C.; Tian, Y. Feasibility Analysis of Autonomous Orbit Determination of BDS Satellites with Inter-Satellite Links. *Geomat. Inf. Sci. Wuhan Univ.* **2022**, *47*, 55–60.

Technical Note

Middle- and Long-Term UT1-UTC Prediction Based on Constrained Polynomial Curve Fitting, Weighted Least Squares and Autoregressive Combination Model

Yuguo Yang [1], Tianhe Xu [1,*], Zhangzhen Sun [2], Wenfeng Nie [1] and Zhenlong Fang [1]

[1] Institute of Space Science, Shandong University, Weihai 264209, China; 201920776@mail.sdu.edu.cn (Y.Y.); wenfengnie@sdu.edu.cn (W.N.); zhlfang@mail.sdu.edu.cn (Z.F.)
[2] Rongcheng Micro & AI Technology Industry Research Institute, Weihai 264300, China; sunzz@sdu.edu.cn
* Correspondence: thxu@sdu.edu.cn; Tel.: +86-631-562-2731

Abstract: Universal time (UT1-UTC) is a key component of Earth orientation parameters (EOP), which is important for the study of monitoring the changes in the Earth's rotation rate, climatic variation, and the characteristics of the Earth. Many existing UT1-UTC prediction models are based on the combination of least squares (LS) and stochastic models such as the Autoregressive (AR) model. However, due to the complex periodic characteristics in the UT1-UTC series, LS fitting produces large residuals and edge distortion, affecting extrapolation accuracy and thus prediction accuracy. In this study, we propose a combined prediction model based on polynomial curve fitting (PCF), weighted least squares (WLS), and AR, namely, the PCF+WLS+AR model. The PCF algorithm is used to obtain accurate extrapolation values, and then the residuals of PCF are predicted by the WLS+AR model. To obtain more accurate extrapolation results, annual and interval constraints are introduced in this work to determine the optimal degree of PCF. Finally, the multiple sets prediction experiments based on the International Earth Rotation and Reference Systems Service (IERS) EOP 14C04 series are carried out. The comparison results indicate that the constrained PCF+WLS+AR model can efficiently and precisely predict the UT1-UTC in the mid and long term. Compared to Bulletin A, the proposed model can improve accuracy by up to 33.2% in mid- and long-term UT1-UTC prediction.

Keywords: UT1-UTC; prediction model; polynomial curve fitting; annual constraint; interval constraint; WLS+AR

Citation: Yang, Y.; Xu, T.; Sun, Z.; Nie, W.; Fang, Z. Middle- and Long-Term UT1-UTC Prediction Based on Constrained Polynomial Curve Fitting, Weighted Least Squares and Autoregressive Combination Model. *Remote Sens.* **2022**, *14*, 3252. https://doi.org/10.3390/rs14143252

Academic Editor: Xiaogong Hu

Received: 10 June 2022
Accepted: 4 July 2022
Published: 6 July 2022

Publisher's Note: MDPI stays neutral with regard to jurisdictional claims in published maps and institutional affiliations.

1. Introduction

Earth orientation parameters (EOPs), including precession–nutation, universal time (UT1-UTC), and length of day (LOD), as well as polar motion (PM x,y), are essential to realizing the transformation between the celestial and terrestrial reference frames, which has important applications in astro-geodynamics, deep space exploration, and high-precision space navigation and positioning [1,2]. Based on space geodetic techniques, e.g., Very Long Baseline Interferometry (VLBI) [3], Satellite Laser Ranging (SLR) [4], Global Navigation Satellite System (GNSS) [5], and Doppler Orbitography and Radio positioning Integrated by Satellite (DORIS) [6], the International Earth Rotation and Reference Systems Service (IERS) comprehensively calculates EOPs and publishes them regularly [7,8]. However, due to the complexity of observation and data processing, it is difficult to obtain EOP in real time, with the delay being in the range of several hours to a few days. Therefore, high-precision EOP prediction is particularly important for real-time applications. Among the five EOPs, the UT1-UTC, which is affected by irregular amplitude and phase variations of annual, semi-annual, and shorter-period oscillations, is the most difficult to predict, especially during extreme events of El Niño/Southern Oscillation [9–11].

Since the UT1-UTC series is discontinuous, leap seconds and solid Earth zonal tides [12] need to be removed before predicting to obtain a continuous time series, i.e.,

UT1R-TAI [13]. Various stochastic methods and techniques have been applied to UT1-UTC prediction. Kosek et al. [14] first used the autocovariance (AC) method to improve polar motion and UT1-UTC predictions. Schuh et al. [2] introduced the artificial neural network (ANN) into the UT1-UTC prediction, significantly improving the medium- and long-term prediction accuracy of UT1-UTC. Kosek et al. [10] compared the accuracy of UT1-UTC predictions based on the least squares (LS) extrapolation combined with autocovariance (AC), autoregressive (AR), autoregressive moving average (ARMA), and neural network (NN) methods, respectively. Moreover, the prediction method combining the LS extrapolation with multivariate autoregressive (MAR) and the accuracy of UT1-UTC prediction have been discussed by Niedzielski and Kosek [11]. The Earth orientation parameter prediction comparison campaign (EOP PCC) was performed under the auspices of the IERS in 2005–2009 to evaluate the accuracy and reliability of different prediction methods [1]. The results showed that the three techniques, i.e., Kalman filter [15], wavelet decomposition+autocovariance (DWT+AC) [13,16], and adaptive transformation from atmospheric angular momentum (AAM) to LODR (i.e., the LOD after removing the solid Earth zonal tides) [1], had the best accuracy for UT1-UTC prediction. Thereafter, Xu et al. [17] verified that the combined LS+AR+Kalman model could effectively improve the accuracy of UT1-UTC prediction. Dill et al. [18] used 6-day effective angular momentum (EAM) forecasted values for the UT1-UTC prediction based on the LS+AR model. Obviously, most of the above prediction methods are based on the combination of least squares extrapolation and stochastic models using only information of the UT1-UTC series or considering EAM. However, due to the irregular variations of amplitude and phase in the UT1-UTC series periodic oscillations [10,19], there is a large residual when using the LS fitting model with annual and semi-annual period terms, which affects the accuracy of the extrapolation [20]. In addition, as shown in Section 3.2 of the present study, the distortions at the end of the LS fitting residual series, namely, the edge effects [21], also reduce the extrapolation accuracy of the trend term, which in turn affects the UT1-UTC prediction.

Different from the LS fitting model with time-varying periods, we use the polynomial curve fitting (PCF) method to determine the linear trend terms and extrapolations in this research. The PCF model can better describe the trend of discrete data, with the advantages of high efficient and easy implementation [22]. It is also an effective method for dealing with the edge effect by increasing constraints [23,24]. Afterwards, the PCF residuals of UT1-UTC, containing period terms, is predicted by the combined weighted least squares (WLS) and AR models [25], and the values predicted by UT1-UTC are the sum of the PCF extrapolations and the residual predictions. Therefore, the UT1-UTC prediction model based on the combination of PCF extrapolation, WLS and AR model, denoted as PCF+WLS+AR, is proposed.

In this study, we first introduce the PCF algorithm to fit and extrapolate the C04 series. The annual and interval constraint methods to determine the optimal polynomial fitting order are then analyzed. Afterwards, the WLS+AR model is used to predict the PCF residuals. Finally, multiple sets of experiments are carried out based on the EOP 14C04 series to compare the prediction accuracy with the LS+AR model as well as the Bulletin A.

2. Methodology

2.1. Polynomial Curve Fitting

Consider a data set $(x_i, y_i)(i = 1, 2, \cdots, m)$, the approximate expression is a polynomial $P_n(x) = \sum_{k=0}^{n} a_k x^k, (n < m)$, where n is the maximum degree of the polynomial, a_k is the fitting coefficient, and $P_n(x)$ is the n degrees polynomial fitting function. When the square of the deviations is minimized, that is

$$R^2 = \sum_{i=1}^{m} [y_i - P_n(x_i)]^2 = \min \tag{1}$$

it is called a polynomial curve fitting [22].

The condition for R^2 to be a minimum is that

$$\frac{\partial\left(R^2\right)}{\partial a_j} = 0 \quad (j = 1, \cdots, n) \tag{2}$$

so

$$\sum_{i=1}^{m}\left[y_i - \sum_{j=1}^{n} a_j x_i^j\right] x_i^k = 0 \quad (k = 0, 1, \cdots, n) \tag{3}$$

These lead to the equations

$$\begin{bmatrix} m & \sum_{i=1}^{m} x_i & \cdots & \sum_{i=1}^{m} x_i^n \\ \sum_{i=1}^{m} x_i & \sum_{i=1}^{m} x_i^2 & \cdots & \sum_{i=1}^{m} x_i^{n+1} \\ \vdots & \vdots & \ddots & \vdots \\ \sum_{i=1}^{m} x_i^n & \sum_{i=1}^{m} x_i^{n+1} & \cdots & \sum_{i=1}^{m} x_i^{2n} \end{bmatrix} \begin{pmatrix} a_0 \\ a_1 \\ \vdots \\ a_n \end{pmatrix} = \begin{pmatrix} \sum_{i=1}^{m} y_i \\ \sum_{i=1}^{m} x_i y_i \\ \vdots \\ \sum_{i=1}^{m} x_i^n y_i \end{pmatrix} \tag{4}$$

The coefficient matrix can be obtained according to Equation (4), and the function of PCF is given by $P_n(x) = \sum_{k=0}^{n} a_k x^k$ [26].

2.2. Weighted Least Squares

The weighted least squares (WLS) model is established based on the LS model by adding a suitable weight matrix. Compared to the LS method, the WLS model can better reflect the strong influence of recent data on the prediction values [25,27]. For the PCF residuals prediction of UT1-UTC, the WLS model is written as follows:

$$f(t) = \alpha_0 + \alpha_1 t + \sum_{i=1}^{k} \left(B_i \cos\left(\frac{2\pi t}{R_i}\right) + C_i \sin\left(\frac{2\pi t}{R_i}\right)\right) + \omega \tag{5}$$

where α_0 is the constant term, α_1 is the linear term, B_i and C_i are the coefficients for periodic terms, R_i is the corresponding periodic, t is the time of UTC, k is the number of periodic terms, ω is the random error.

According to the error theory and surveying adjustment, Equation (5) can be written in the form of an error equation,

$$V = A\widehat{X} - L \tag{6}$$

where $\widehat{X} = [\alpha_0, \alpha_1, B_1, C_1, \cdots, B_k, C_k]^T$ is the estimated parameter vector, L is the original observation matrix, n is the length of the series, and the coefficient matrix A can be expressed as

$$A = \begin{bmatrix} 1 & t_1 & t_1^2 & \sin\left(\frac{2\pi t_1}{R_1}\right) & \cos\left(\frac{2\pi t_1}{R_1}\right) & \cdots & \sin\left(\frac{2\pi t_1}{R_k}\right) & \cos\left(\frac{2\pi t_1}{R_k}\right) \\ 1 & t_2 & t_2^2 & \sin\left(\frac{2\pi t_2}{R_1}\right) & \cos\left(\frac{2\pi t_2}{R_1}\right) & \cdots & \sin\left(\frac{2\pi t_2}{R_k}\right) & \cos\left(\frac{2\pi t_2}{R_k}\right) \\ \vdots & \vdots & \vdots & \vdots & \vdots & \vdots & \vdots & \vdots \\ 1 & t_n & t_n^2 & \sin\left(\frac{2\pi t_n}{R_1}\right) & \cos\left(\frac{2\pi t_n}{R_1}\right) & \cdots & \sin\left(\frac{2\pi t_n}{R_k}\right) & \cos\left(\frac{2\pi t_n}{R_k}\right) \end{bmatrix} \tag{7}$$

Based on the least squares criterion, the parameters $\widehat{X}$ can be denoted as

$$\widehat{X} = (A^T P A)^{-1} A^T P L \tag{8}$$

where P is the weight matrix, which is a diagonal matrix. For the setting of the weight matrix, it is necessary to highlight the influence of recent data and does not affect the overall trend when data fitting. Based on this, it can be designed as

$$
P = \begin{bmatrix} \frac{\sigma_0^2}{\sigma_1^2}w(1) & 0 & 0 & 0 \\ 0 & \frac{\sigma_0^2}{\sigma_2^2}w(2) & 0 & 0 \\ 0 & 0 & \ddots & 0 \\ 0 & 0 & 0 & \frac{\sigma_0^2}{\sigma_n^2}w(n) \end{bmatrix}^e
$$

(9)

where $w(i), i = 1, 2, \cdots, n$ is the reciprocal of the distance from the point to the forecast starting point, σ_0^2 is the unit weight variance, σ_i^2 is the variance of the ith observation (available from the IERS 14C04). e is the power, whose value is determined according to the principle of minimum prediction error. Through experimental comparison and analysis in this study, the value of e is taken as 10.

2.3. AR Model

AR model is the description of the relationship between a random series $z_t (t = 1, 2, \cdots, N)$ before time t and the white noise of current time t. Its expression can be written as

$$
z_t = \sum_{i=1}^{p} \varphi_i z_{t-i} + \omega_t
$$

(10)

where $\varphi_1, \varphi_2, \cdots, \varphi_p$ are the autoregressive coefficients, ω_t is the white noise with zero means, p is the AR model order. The above equation denoted by $AR(p)$ is well-known as the AR model of the order. In this study, the final prediction error criterion is adopted to determine the order.

2.4. Error Analysis

To evaluate the prediction accuracy, the Mean Absolute Error (MAE) is used, which can be expressed as

$$
E_i = P_i - O_i
$$

(11)

$$
MAE_j = \frac{1}{n} \sum_{i=1}^{n} (|E_i|)
$$

(12)

where P_i is the predicted value of $i-$th prediction, O_i is the corresponding observation value, E_i is the real error (Assumed the observation value as true value), n is the total prediction number, MAE_j is the MAE at span j.

3. The Constrained PCF+WLS+AR Prediction Model

3.1. Data Description

The latest version of the EOP 14C04 was released by IERS on 1 February 2017 [28], and is available at http://hpiers.obspm.fr/eoppc/eop/eopc04/ (accessed on 3 May 2021). The new C04 is consistent with ITRF2014, and the EOPs have been updated since 1984. Figure 1 shows a comparison of UT1-UTC and UT1-TAI adopted by IERS from 1984 to now. Since existing studies have shown that the optimal base sequence length for UT1-UTC prediction is between 10 and 18 years [11,17,29,30], 12 years was chosen empirically as the base sequence length in this study.

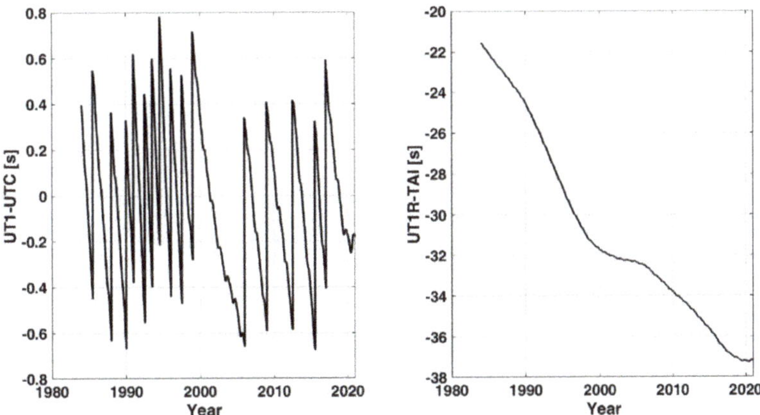

Figure 1. The UT1−UTC and UT1R−TAI series from 1984 to now (data from IERS website).

3.2. Application of PCF in UT1R-TAI Modeling

The UT1R-TAI is a non-stationary series due to complex time variation. Based on the PCF algorithm, we extract trend terms from the original series to improve the applicability between the WLS+AR prediction model and the residual series. Figure 2 shows the fitted series and their residuals based on the LS and PCF algorithms for UTIR-TAI from 2000 to now, respectively. Compared with the LS fitting, the PCF has smaller fitted residuals, which are more consistent with the overall trend of UT1R-TAI series. After spectral analysis using fast Fourier transform (FFT), there are significant annual and semi-annual periodic terms in the PCF residuals shown in Figure 3.

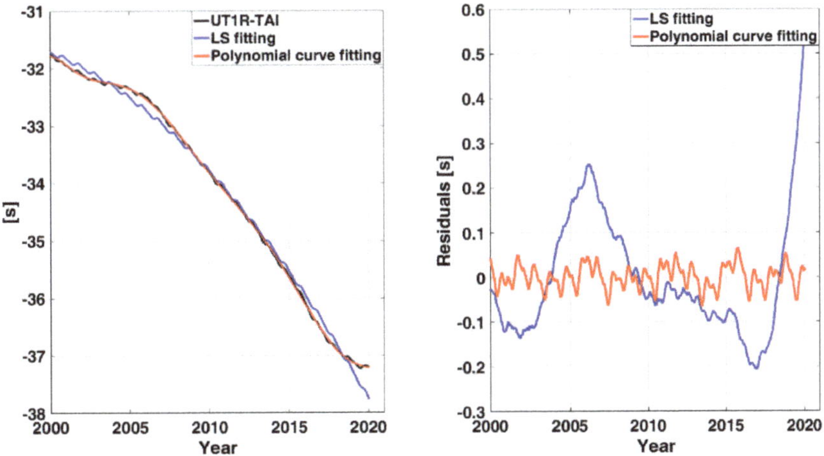

Figure 2. The original UT1R−TAI series and its two fitting series with different residuals since 2000.

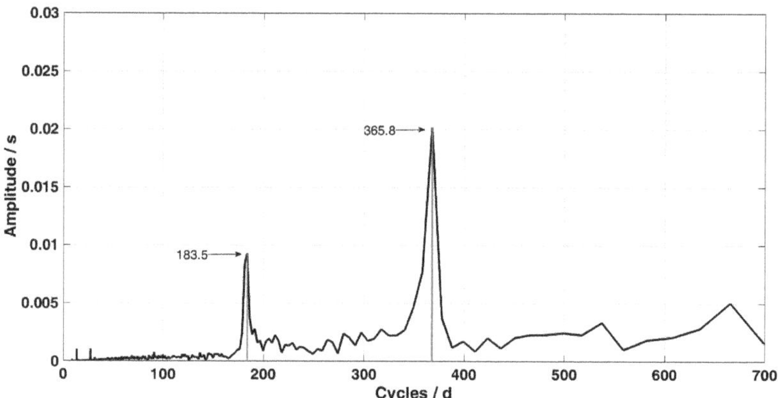

Figure 3. Frequency spectrum of the PCF residuals based on UT1R-TAI using fast Fourier transform (FFT).

For discrete data, the standard deviation (STD) of the fitted residuals is generally used to determine the optimal degree of PCF. The equation can be formed as follows:

$$S^2 = \frac{1}{n-1}\sum_{i=1}^{n}\left(f_{pcf}(i) - O_{ut}(i)\right)^2 \tag{13}$$

where S is the standard deviation, f_{pcf} is the fitted value of UT1R-TAI, and O_{ut} is the original value of UT1R-TAI.

Figure 4 shows the standard deviation of PCF at different degrees for UT1R-TAI from 2000 to now. It is important to note that the standard deviation of the PCF residuals is basically constant after the degree reaching a threshold. As is well known, the curve will exhibit high-frequency oscillations with increasing degree, leading to overfitting. Therefore, the optimal PCF degree for UT1R-TAI since 2000 can be chosen as 8.

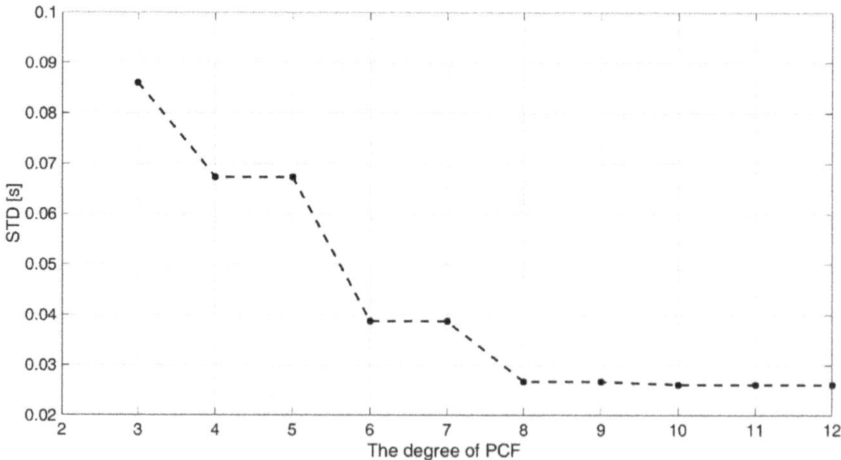

Figure 4. The STD of PCF at different degrees for UT1R-TAI from 2000 to now.

Considering that the UT1-UTC is susceptible to strong El Niño, which occurred three times after 2000 (i.e., 2006–2007, 2010–2011, 2014–2016), we separately determine the optimal PCF degree based on multiple sets of 12-year UT1-UTC series in this study.

As can be seen in Figure 5, the optimal PCF degree for the 12-year UT1-UTC series is smaller, with values between 4 and 8, than that for the 20-year series (shown in Figure 4). Another significant piece of information in Figure 5 is that the optimal degree for different base series is inconsistent, a fact which should be paid more attention in PCF extrapolation. The purpose of using PCF in this study is to improve the extrapolation accuracy of trend term in UT1R-TAI series, and multiple sets of experiments were carried out to further verify the accuracy of the optimal degree for PCF extrapolation.

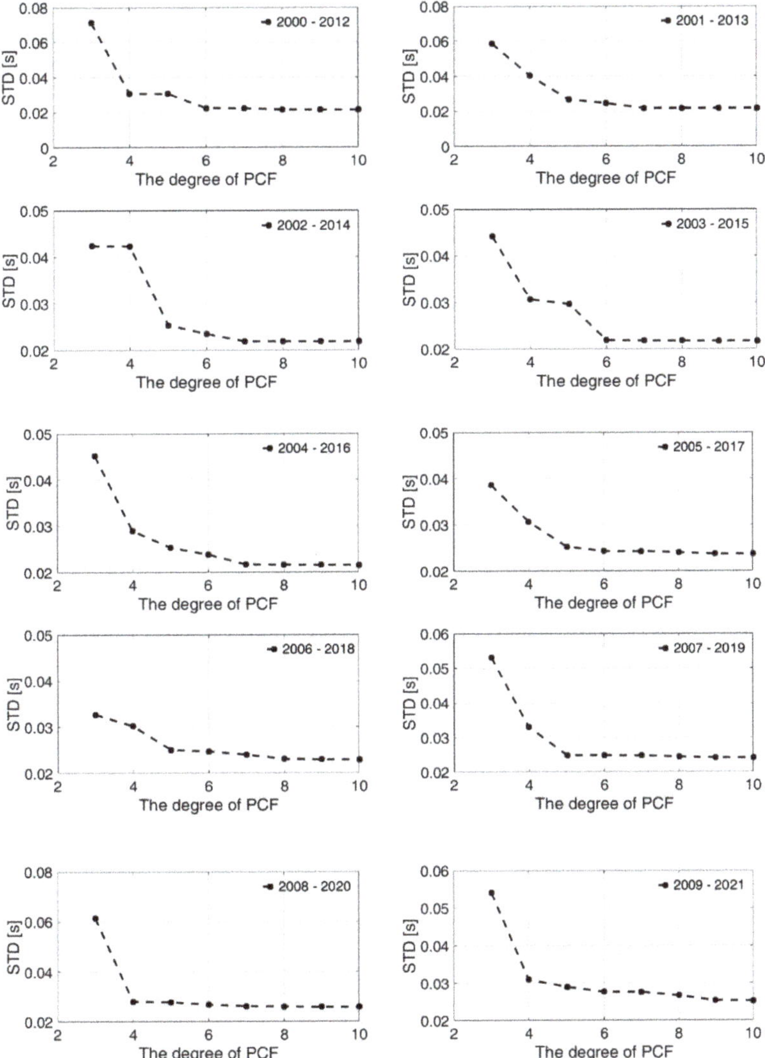

Figure 5. The standard deviation of PCF at different degrees for UT1R−TAI of 12 years.

Figure 6 shows the PCF extrapolation results of different degrees over six time periods. It can be seen that there is a big difference between the extrapolation results using a few fitting degrees with similar accuracy. The 1-year extrapolated values from 1 January 2014 using 4-, 6- and 8-degree PCF, from 1 January 2015 using 4-degree PCF, from 1 January 2016 using 4-, 5- and 7-degree PCF, from 1 January 2017 using 5- and 8-degree PCF, from 1

January 2018 using 5- and 6-degree PCF, and from 1 January 2019 using 4-degree PCF are in good agreement with the original observations. Compared with the results in Figure 5, the optimal fitting and extrapolation PCF degree based on the 12-year UT1R-TAI series are different, but 4- or 5-degree PCF have better accuracy in the two applications. In view of the difference of 12-year UT1-UTC series used in each prediction, the optimal degree for PCF extrapolation should be first determined. In this study, we first propose the annual constraint method to determine the optimal degree of PCF.

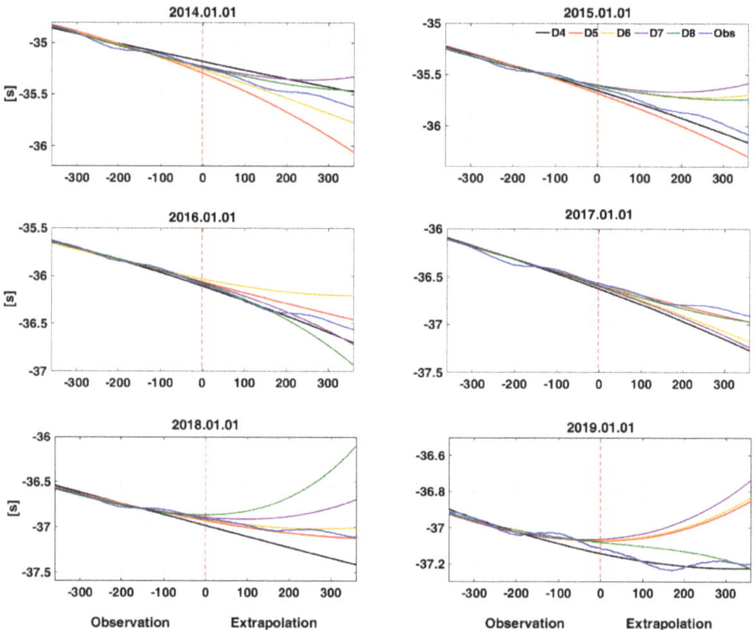

Figure 6. Comparison of polynomial curve fitting at different degrees for extrapolation. D4, D5, D6, D7 and D8 represent the fitted and extrapolated results of the original series with 4-, 5-, 6-, 7- and 8-degree PCF, respectively.

3.2.1. Determination of Degree of PCF by Annual Constraint Method

As shown in Figure 2, the long-term trends of the UT1R-TAI series approximate a straight line. Although there are still annual and semi-annual periodic terms in the PCF residuals, the residual values are relatively small compared to the observations, which is not considered in the determination of PCF degree. Therefore, we use the observations at the start time of the prediction and the previous span of time to constrain the prediction at the next span of time, as shown in Figure 7. Since the prediction span is 1 year in this study, this constraint method is defined as an annual constraint.

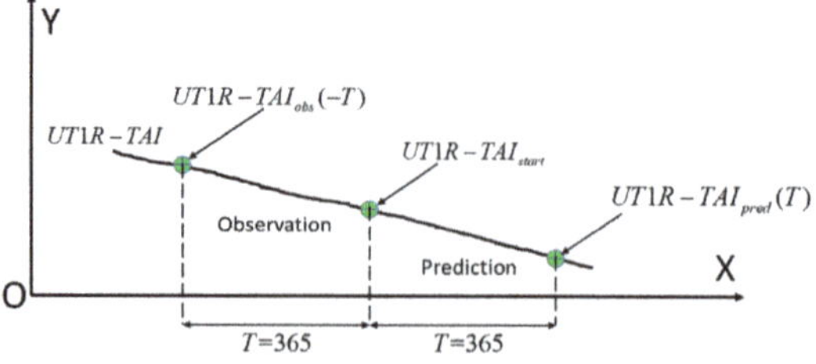

Figure 7. The distribution of points in predictive constraint method.

Using linear symmetry features [31], the constrained value can be obtained from

$$[UT1R - TAI_{pred}(T)] \approx [UT1R - TAI'_c(T)] = 2 * [UT1R - TAI_{start}] - [UT1R - TAI_{obs}(-T)] \tag{14}$$

where $[UT1R - TAI_{pred}(T)]$ is the prediction value at a span of 365 days, $[UT1R - TAI'_c(T)]$ is the constrained value, which is called the annual constraint. $[UT1R - TAI_{start}]$ is the prediction start time, $[UT1R - TAI_{obs}(-T)]$ is the observation from 365 days before the start time.

To verify the availability and accuracy of the annual constraint, the PCF+WLS+AR prediction model is established with a different degrees in this study, and the effects of different PCF degrees on the prediction results were compared with the 365-day UT1R-TAI prediction. According to the previous study (Figure 6), 4 or 5 is the optimal degree for PCF extrapolation in most cases. Therefore, the optimal PCF degree was selected as either 4 or 5 based on the annual constraint method, and the PCF+WLS+AR model was used to predict the UTIR-TAI in this study. Figure 8 shows a comparison of UT1R-TAI prediction in different time periods based on the PCF+WLS+AR model.

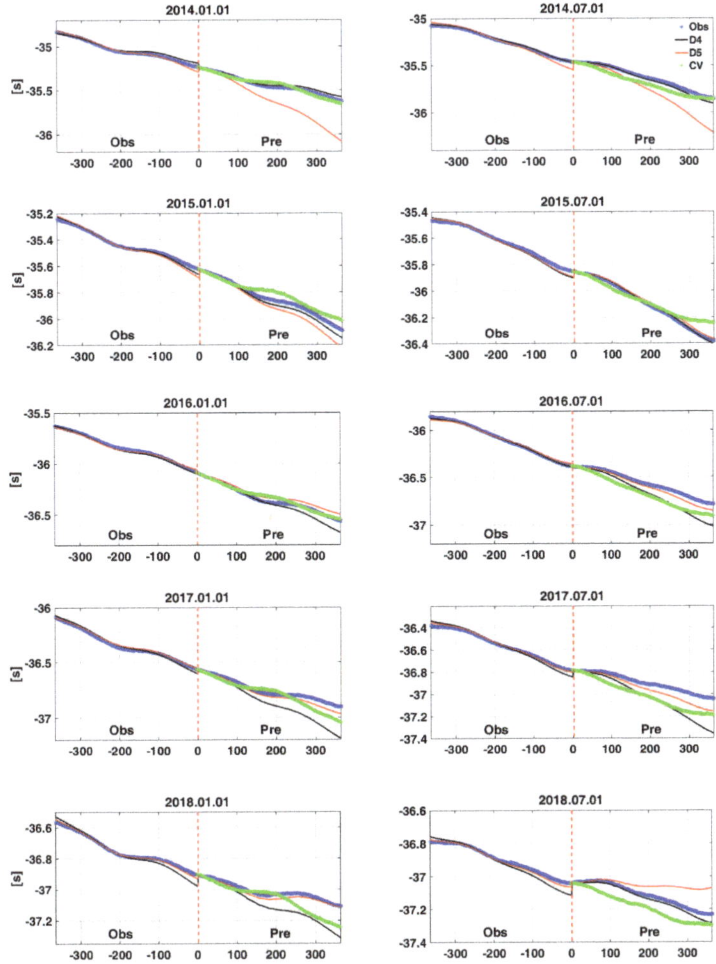

Figure 8. The comparison of UT1R−TAI prediction in different time periods based on the PCF+WLS+AR model.

In Figure 8, the blue line represents the UT1R-TAI series, the green line is the constraint value (CV) series calculated based on the annual constraint method in this study, the thin black and red lines correspond to the prediction series based on the PCF+WLS+AR model on 4 and 5 degrees, respectively. Comparing the predictions and constraint values for the 365-day span, we selected the optimal PCF degree according to the closest to constraint values principle. In addition, we also compared the optimal degree prediction with the original series at different time periods and determined an identification as either T or F. As can be seen in Table 1, the predictions of degree selected based on the annual constraint are optimal for most of the time periods, except for 1 January 2018. This may be due to the fact that UT1-UTC is affected by the internal and external factors of the Earth, deviating sharply from the original trend during 2018–2019. Therefore, we need to further optimize the selection strategy for PCF degree.

Table 1. The chosen degree of PCF depends on the predictive constraint method.

Time	Optimal Degree of PCF by Annual Constraint	True/False
1 January 2014	4	T
1 July 2014	4	T
1 January 2015	4	T
1 July 2015	5	T
1 January 2016	5	T
1 July 2016	5	T
1 January 2017	5	T
1 July 2017	5	T
1 January 2018	4	F
1 July 2018	4	T

3.2.2. Optimization of PCF Degree Using the Interval Constraint Method

As shown in the previous analysis, UT1-UTC may exhibit violent oscillations in the series, with a serious influence of internal and external factors at a certain stage, resulting in large errors in the annual constraint method. Therefore, in this study, the interval constraint method is introduced to further optimize the determination of PCF degree by calculating the annual variation.

Figure 9 shows the annual increment of the UT1R-TAI from 2000 to 2014. It can be seen that the annual variation of UT1R-TAI from 2000 to 2006 is relatively small, about -0.05 s~-0.2 s. However, the annual increment of UT1R-TAI suddenly decreased to -0.3 s~-0.4 s in 2007, and remained at this magnitude for several years thereafter. Considering the strong El Niño that occurred in 2006, this drastic variation may have been caused by this event. The real cause is beyond the scope of this study, and will not be studied in detail. In addition, it is not difficult to find that the mean annual variation of UT1R-TAI is about -0.3 s, with a fluctuation of 0.15 s. Therefore, -0.3 0.15 s can be chosen as the constraint interval for annual increment. The annual constraint method determines the PCF degree N (4 or 5) based on the difference between the annual prediction value and the CV, while the interval constraint method optimizes the PCF degree on the basis of the difference between the annual prediction value and the prediction starting value, i.e., the predicted annual increment. N is the optimal degree if the predicted annual increment is within the constraint interval. In other cases, the optimal PCF degree is the one closed to the prediction annual increment of -0.3 s, when the prediction annual increments based on the PCF+WLS+AR model with degree 4 and 5 are both outside the interval. Additionally, when the results of the annual constraint and interval constraint methods are inconsistent, the optimal degree based on the interval constraint is preferred.

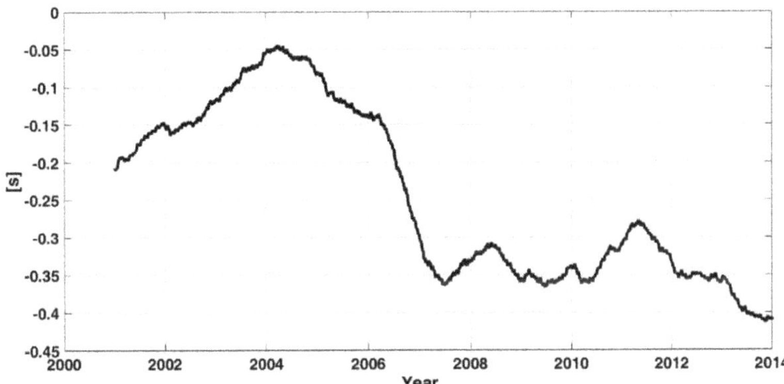

Figure 9. The annual increment of UT1R-TAI from 2000 to 2014.

3.3. The Processing of the Constrained PCF+WLS+AR Model

In this study, we proposed the model for UT1-UTC prediction as shown in Figure 10. The prediction process is demonstrated by the following steps.

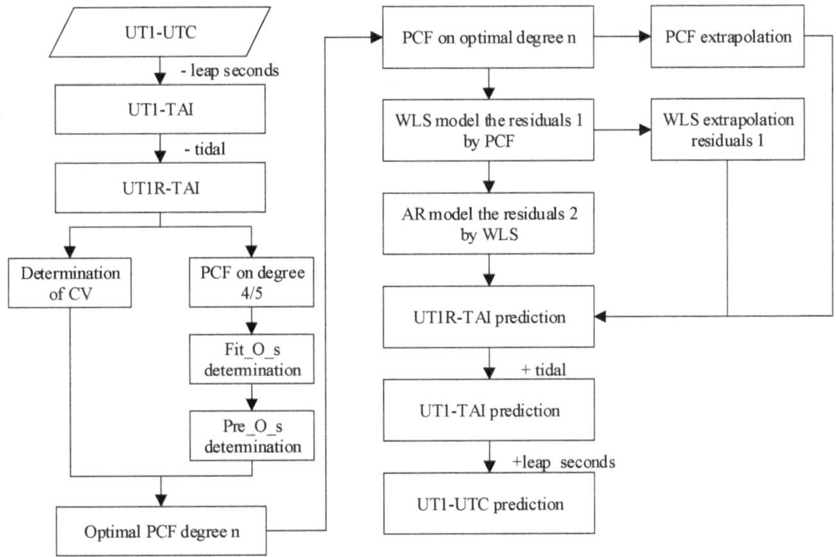

Figure 10. The process of the PCF+WLS+AR prediction model.

(1) Determination of the PCF model.
1. Based on the annual constraint method, the CV is calculated and compared with the annual prediction value to determine the initial optimal PCF degree;
2. Optimize with interval constraints to obtain the final optimal PCF degree;
3. Use the optimal degree for polynomial curve fitting and extrapolation.
(2) Construction of WLS model.
4. Model the residuals of PCF and obtain the WLS extrapolation.
(3) Residual prediction based on AR model.
5. Model the residuals of WLS.
(4) UT1-UTC prediction using the combined PCF+WLS+AR model.

6. Add the leap seconds and solid Earth zonal tides.

It is important to note that Pre_O_s is the difference between the PCF+WLS+AR model on degree 4/5 and the UT1R-TAI value at the starting point and Fit_O_s is the difference between the PCF model and the UT1R-TAI value at the starting point of prediction.

4. Results and Discussion

In this study, the accuracy of the constrained PCF+WLS+AR prediction model was validated based on multiple sets of prediction experiments using the EOP 14C04 series between 2014 and 2019. The experimental scheme was as follows.

- Case 1: UT1-UTC prediction based on the LS+AR model;
- Case 2: UT1-UTC prediction based on the constrained PCF+WLS+AR model (considering annual constraint);
- Case 3: UT1-UTC prediction based on the constrained PCF+WLS+AR model (considering annual constraint and interval constraint);
- Case 4: Bulletin A results achieved by IERS.

The four cases above represent the LS+AR model, the two PCF+WLS+AR models proposed in this paper, and Bulletin A provided by IERS, respectively. The prediction span was 360 days with a weekly sliding window, and the accurate statistical period was from 3 January 2014 to 6 August 2019. Figure 11 shows the MAE of UT1-UTC prediction in four cases. Compared to the other models, the MAE of the proposed constrained PCF+WLS+AR models exhibited fewer errors in mid- and long-term prediction (90~360 days). Since the PCF+WLS+AR model better takes into account the overall trend of the base series, it obtains a more accurate long-term trend and long-period term than the LS model during extrapolation, thus improving the mid- and long-term UT1-UTC prediction accuracy. Combined with the UT1-UTC prediction accuracy statistics presented in Table 2, the improvement value relative to Bulletin A was positive after 90 days in the future, and the improvement gradually increased with the progression of the prediction span, reaching a maximum of 33.2%. However, for short-term prediction, Bulletin A showed a smaller MAE in the four cases. As shown in Table 2, it can be seen that Bulletin A clearly outperformed the other models in terms of short-term prediction, especially the first 20 days in the future. This advantage is primarily due to the fact that Bulletin A takes into account the effects of Atmospheric Angular Momentum (AAM) and Oceanic Angular Momentum (OAM). Kalarus et al. [1] showed that the main contribution of AAM and OAM to UT1-UTC is its high-frequency component. Therefore, the additional inputs, i.e., AAM and OAM, can improve the modeling of the high-frequency component of UT1-UTC, thus improving the short-term prediction accuracy of UT1-UTC. Compared with the LS+AR model, except that the first 60 days of future have basically similar accuracy, the PCF+WLS+AR model demonstrates a more obvious improvement in terms of mid- and long-term prediction.

Table 2. Comparison of prediction accuracy for 4 cases with the unit [ms]. The improvement represents the prediction accuracy of Case 3 relative to Case 4.

Span	Case 1	Case 2	Case 3	Case 4	Improvement
1	0.036	0.037	0.036	0.056	35.5%
10	1.049	1.124	1.083	0.452	−139.4%
20	2.460	2.749	2.596	1.601	−62.1%
30	3.544	4.151	3.802	2.910	−30.7%
60	7.396	8.722	7.616	7.284	−4.6%
90	12.390	13.191	11.332	11.271	−0.5%
120	18.338	17.867	15.077	15.626	3.5%
180	33.148	29.346	23.954	25.962	7.7%
240	51.349	41.785	33.250	39.485	15.8%
300	71.272	52.021	40.979	56.666	27.7%
360	92.197	65.819	50.934	76.217	33.2%

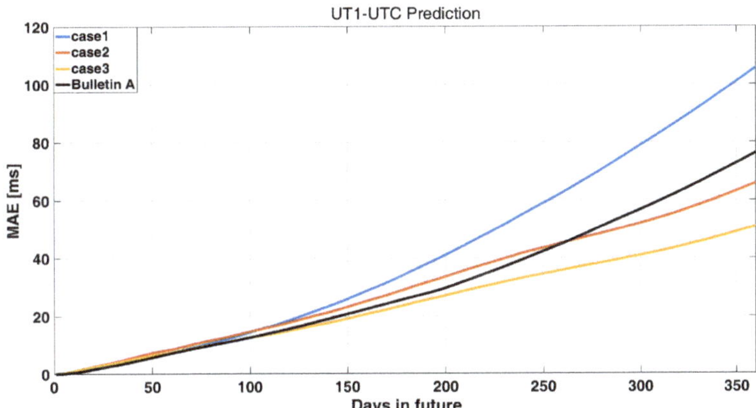

Figure 11. Mean absolute errors (MAE) of the UT1-UTC prediction using 4 cases with the unit [ms]. Case 1, Case 2, and Case 3 denote the LS+AR model (blue line), the PCF+WLS+AR model with annual constraints (orange line), and the PCF+WLS+AR model with annual and interval constraints (yellow line), respectively. Bulletin A is indicated by the black line.

Figure 12 presents the Absolute Errors (AE) of 360-day prediction between 2014 and 2019 for these four cases in order to better understand the prediction performance and its causes. It can be seen that the PCF+WLS+AR model considering annual constraint and interval constraint demonstrates a significant improvement in mid- and long-term UT1-UTC prediction accuracy compared to the LS+AR model and Bulletin A. In the results of Bulletin A, the long-term prediction from the beginning of 2015 has high errors, which may have been caused by the superstrong El Niño warm event occurring between 2015 and 2016. Obviously, our proposed method was effective for prediction with high accuracy during this period. This is due to the fact that we accurately fit the base series and continuously optimize the polynomial degree by means of annual and interval constraints to obtain the optimal extrapolation values. However, with the time-varying of PCF degree, there is also a sudden change in adjacent starting prediction errors.

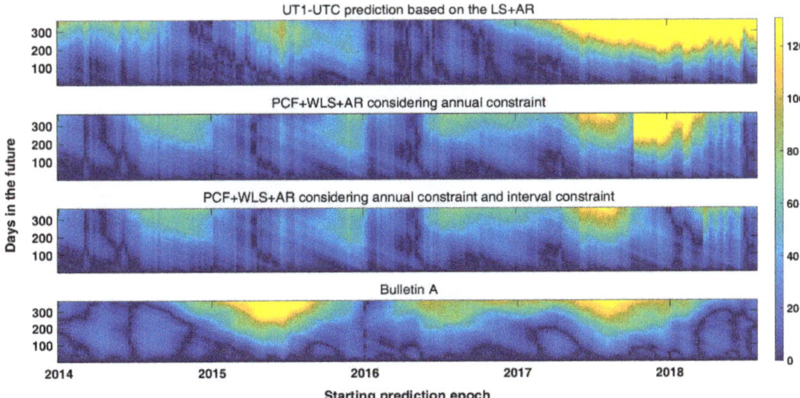

Figure 12. Absolute errors (in mas) of the UT1-UTC prediction using LS+AR model, the PCF+WLS+AR model with annual constraints, the PCF+WLS+AR model with annual and interval constraints and Bulletin A. The four subplots represent the prediction accuracy of Case 1, Case 2, Case 3, and Case 4 from top to bottom.

5. Conclusions

UT1-UTC is a quantitative parameter describing the Earth's rotation rate, which is the most difficult to predict in the five EOPs due to the complex and variable periodic term. Among existing prediction models, LS is the most widely used for fitting trend terms and periodic terms. In this study, based on the characteristics of UT1-UTC and its inherent periodic and trend terms, a constrained PCF+WLS+AR prediction model is proposed. The proposed model fully takes into account the trend and long-period terms of the UT1R-TAI series, as well as its annual variation and the strong influence of recent data. This is an important enhancement to high-precision modeling and PCF extrapolation. To obtain the optimal PCF degree, we introduce two constraint methods, namely annual constraint and interval constraint. Finally, multiple sets of prediction experiments based on EOP 14C04 series were carried out to verify the accuracy of this proposed model. The comparison with prediction accuracy of other methods indicates that the constrained PCF+WLS+AR model can efficiently and precisely predict the UT1-UTC in the middle and long term. Compared to Bulletin A, released by IERS, the proposed model demonstrates improved accuracy of UT1-UTC prediction by up to 33.2% in the middle and long term (90–360 days). However, Bulletin A demonstrates much smaller errors in short-term prediction, especially in the ultra-short term.

Future work may focus on the strong relation between AAM, OAM, and UT1-UTC. A prediction model such as the constrained PCF+WLS+MAR, based on multiple inputs of EOP, AAM and OAM, will be established to improve short-term UT1-UTC prediction accuracy. At the same time, a better constraint method for PCF needs to be further explored.

Author Contributions: Y.Y., T.X. and Z.S. conceived the experiments. Y.Y. and Z.S. processed the data and drew the pictures. W.N. and Z.F. investigated the results. Y.Y. wrote the whole manuscript. T.X. and W.N. executed the review and editing. All authors have read and agreed to the published version of the manuscript.

Funding: This study is under the support of the National Natural Science Foundation of China (Grant No. 41874032), State Key Laboratory of Geodesy and Earth's Dynamics, Innovation Academy for Precision Measurement Science and Technology, Chinese Academy of Sciences (SKLGED2021-3-4).

Data Availability Statement: All data for this paper referred to in the data description.

Acknowledgments: The authors are grateful to IERS for the EOP 14C04 solution. It is worth stating that all the prediction models in this experiment were implemented based on our self-developed software compiled in the MATLAB platform. Interested readers may contact the authors by email.

Conflicts of Interest: The authors declare no conflict of interest.

References

1. Kalarus, M.; Schuh, H.; Kosek, W.; Akyilmaz, O.; Bizouard, C.; Gambis, D.; Gross, R.; Jovanović, B.; Kumakshev, S.; Kutterer, H.; et al. Achievements of the Earth orientation parameters prediction comparison campaign. *J. Geod.* **2010**, *84*, 587–596. [CrossRef]
2. Schuh, H.; Ulrich, M.; Egger, D.; Müller, J.; Schwegmann, W. Prediction of Earth orientation parameters by artificial neural networks. *J. Geod.* **2002**, *76*, 247–258. [CrossRef]
3. Schuh, H.; Behrend, D. VLBI: A fascinating technique for geodesy and astrometry. *J. Geodyn.* **2012**, *61*, 68–80. [CrossRef]
4. Pearlman, M.R.; Degnan, J.J.; Bosworth, J.M. The international laser ranging service. *Adv. Space Res.* **2002**, *30*, 135–143. [CrossRef]
5. Hein, G.W. Status, perspectives and trends of satellite navigation. *Satell. Navig.* **2020**, *1*, 22. [CrossRef]
6. Willis, P.; Fagard, H.; Ferrage, P.; Lemoine, F.G.; Noll, C.E.; Noomen, R.; Otten, M.; Ries, J.C.; Rothacher, M.; Soudarin, L. The international DORIS service (IDS): Toward maturity. *Adv. Space Res.* **2010**, *45*, 1408–1420. [CrossRef]
7. Gambis, D. Monitoring Earth orientation using space-geodetic techniques: State-of-the-art and prospective. *J. Geod.* **2004**, *78*, 295–303. [CrossRef]
8. Nilsson, T.; Heinkelmann, R.; Karbon, M.; Raposo-Pulido, V.; Soja, B.; Schuh, H. Earth orientation parameters estimated from VLBI during the CONT11 campaign. *J. Geod.* **2014**, *88*, 491–502. [CrossRef]
9. Gross, R.S.; Marcus, S.L.; Eubanks, T.M.; Dickey, J.O.; Keppenne, C.L. Detection of an ENSO signal in seasonal length-of-day variations. *Geophys. Res. Lett.* **1996**, *23*, 3373–3376. [CrossRef]
10. Kosek, W.; Kalarus, M.; Johnson, T.; Wooden, W.; McCarthy, D.; Popinski, W. A comparison of UT1-UTC forecasts by different prediction techniques. *Proc. Journeys Syst. Ref. Spatiotemporal* **2004**, 140–141.

11. Niedzielski, T.; Kosek, W. Prediction of UT1–UTC, LOD and AAM χ_3 by combination of least-squares and multivariate stochastic methods. *J. Geod.* **2007**, *82*, 83–92. [CrossRef]
12. McCarthy, D.D.; Petit, G. *IERS Conventions (2003)*; International Earth Rotation And Reference Systems Service (IERS): Cologne, Germany, 2004.
13. Kosek, W.; Kalarus, M.; Johnson, T.; Wooden, W.; McCarthy, D.; Popinski, W. A comparison of LOD and UT1-UTC forecasts by different combined prediction techniques. *Artif Satell* **2005**, *40*, 119–125.
14. Kosek, W.; McCarthy, D.; Luzum, B. Possible improvement of Earth orientation forecast using autocovariance prediction procedures. *J. Geod.* **1998**, *72*, 189–199. [CrossRef]
15. Gross, R.S.; Eubanks, T.M.; Steppe, J.A.; Freedman, A.P.; Dickey, J.O.; Runge, T.F. A Kalman-filter-based approach to combining independent Earth-orientation series. *J. Geod.* **1998**, *72*, 215–235. [CrossRef]
16. Kosek, W. *Future Improvements in EOP Prediction*; Springer: Berlin/Heidelberg, Germany, 2012; pp. 513–520.
17. Xu, X.Q.; Zhou, Y.H.; Liao, X.H. Short-term earth orientation parameters predictions by combination of the least-squares, AR model and Kalman filter. *J. Geodyn.* **2012**, *62*, 83–86. [CrossRef]
18. Dill, R.; Dobslaw, H.; Thomas, M. Improved 90-day Earth orientation predictions from angular momentum forecasts of atmosphere, ocean, and terrestrial hydrosphere. *J. Geod.* **2018**, *93*, 287–295. [CrossRef]
19. Chen, J.; Wilson, C. Hydrological excitations of polar motion, 1993–2002. *Geophys. J. Int.* **2005**, *160*, 833–839. [CrossRef]
20. Shen, Y.; Guo, J.Y.; Liu, X.; Kong, Q.L.; Guo, L.X.; Li, W. Long-term prediction of polar motion using a combined SSA and ARMA model. *J. Geod.* **2018**, *92*, 333–343. [CrossRef]
21. Zheng, D.; Chao, B.; Zhou, Y.; Yu, N. Improvement of edge effect of the wavelet time–frequency spectrum: Application to the length-of-day series. *J. Geod.* **2000**, *74*, 249–254. [CrossRef]
22. Burden, R.L.; Faires, J.D.; Burden, A.M. *Numerical Analysis*; Cengage Learning: Boston, MA, USA, 2015.
23. Keren, D.; Gotsman, C. Fitting curves and surfaces with constrained implicit polynomials. *IEEE Trans. Pattern Anal. Mach. Intell.* **1999**, *21*, 31–41. [CrossRef]
24. Peck, J. Polynomial curve fitting with constraint. *SIAM Rev.* **1962**, *4*, 135–141. [CrossRef]
25. Sun, Z.; Xu, T. Prediction of earth rotation parameters based on improved weighted least squares and autoregressive model. *Geod. Geodyn.* **2012**, *3*, 57–64.
26. Weisstein, E.W. Least Squares Fitting. 2002. Available online: https://mathworld.wolfram.com/ (accessed on 10 July 2020).
27. Cohen, A.; Migliorati, G. Optimal weighted least-squares methods. *SMAI J. Comput. Math.* **2017**, *3*, 181–203. [CrossRef]
28. Bizouard, C.; Lambert, S.; Gattano, C.; Becker, O.; Richard, J.-Y. The IERS EOP 14C04 solution for Earth orientation parameters consistent with ITRF 2014. *J. Geod.* **2019**, *93*, 621–633. [CrossRef]
29. Jia, S.; Xu, T.H.; Sun, Z.Z.; Li, J.J. Middle and long-term prediction of UT1-UTC based on combination of Gray Model and Autoregressive Integrated Moving Average. *Adv. Space Res.* **2017**, *59*, 888–894. [CrossRef]
30. Modiri, S.; Belda, S.; Hoseini, M.; Heinkelmann, R.; Ferrándiz, J.M.; Schuh, H. A new hybrid method to improve the ultra-short-term prediction of LOD. *J. Geod.* **2020**, *94*, 23. [CrossRef]
31. Bigun, J. *Optimal Orientation Detection of Linear Symmetry*; Linköping University Electronic Press: Linköping, Sweden, 1987.

MDPI

St. Alban-Anlage 66

4052 Basel

Switzerland

www.mdpi.com

Remote Sensing Editorial Office

E-mail: remotesensing@mdpi.com

www.mdpi.com/journal/remotesensing

www.ingramcontent.com/pod-product-compliance
Lightning Source LLC
LaVergne TN
LVHW071618170726
843515LV00009B/2422